ODOR AND VOC CONTROL HANDBOOK

Other Environmental Engineering Books from McGraw-Hill

Aldrich • POLLUTION PREVENTION ECONOMICS: FINANCIAL IMPACTS ON BUSINESS AND INDUSTRY
American Water Works Association • WATER QUALITY AND TREATMENT
Baker, Herson • BIOREMEDIATION
Brunner • HAZARDOUS WASTE INCINERATION, SECOND EDITION
Callahan, Green • HAZARDOUS SOLVENT SOURCE REDUCTION
Cascio, Woodside, Mitchell • ISO 14000 GUIDE: THE NEW INTERNATIONAL ENVIRONMENTAL MANAGEMENT STANDARD
Chopey • ENVIRONMENTAL ENGINEERING IN THE PROCESS PLANT
Cookson • BIOREMEDIATION ENGINEERING: DESIGN AND APPLICATION
Corbitt • STANDARD HANDBOOK OF ENVIRONMENTAL ENGINEERING
Curran • ENVIRONMENTAL LIFE-CYCLE ASSESSMENT
Fiksel • DESIGN FOR ENVIRONMENT: CREATING ECO-EFFICIENT PRODUCTS AND PROCESSES
Freeman • HAZARDOUS WASTE MINIMIZATION
Freeman • STANDARD HANDBOOK OF HAZARDOUS WASTE TREATMENT AND DISPOSAL
Freeman • INDUSTRIAL POLLUTION PREVENTION HANDBOOK
Harris, Harvey • HAZARDOUS CHEMICALS AND THE RIGHT TO KNOW
Harrison • ENVIRONMENTAL, HEALTH AND SAFETY AUDITING HANDBOOK, SECOND EDITION
Hays, Gobbell, Ganick • INDOOR AIR QUALITY: SOLUTIONS AND STRATEGIES
Jain et al. • ENVIRONMENTAL ASSESSMENT
Kaletsky • OSHA INSPECTIONS: PREPARATION AND RESPONSE
Kolluru • ENVIRONMENTAL STRATEGIES HANDBOOK
Kolluru et al. • RISK ASSESSMENT AND MANAGEMENT HANDBOOK: FOR ENVIRONMENTAL, HEALTH AND SAFETY PROFESSIONALS
Kreith • HANDBOOK OF SOLID WASTE MANAGEMENT
Levin, Gealt • BIOTREATMENT OF INDUSTRIAL AND HAZARDOUS WASTE
Lund • THE MCGRAW-HILL RECYCLING HANDBOOK
Majumdar • REGULATORY REQUIREMENTS FOR HAZARDOUS MATERIALS
Marriott • ENVIRONMENTAL IMPACT ASSESSMENT: A PRACTICAL GUIDE
Rossiter • WASTE MINIMIZATION THROUGH PROCESS DESIGN
Smallwood • SOLVENT RECOVERY HANDBOOK

ODOR AND VOC CONTROL HANDBOOK

Harold J. Rafson Editor in Chief

McGraw-Hill

New York San Francisco Washington, D.C. Auckland Bogotá
Caracas Lisbon London Madrid Mexico City Milan
Montreal New Delhi San Juan Singapore
Sydney Tokyo Toronto

Library of Congress Cataloging-in-Publication Data

Odor and VOC control handbook / Harold J. Rafson, editor in chief.
p. cm.
Includes index.
ISBN 0-07-052523-4 (alk. paper)
1. Odor control—Handbooks, manuals, etc. 2. Volatile organic compounds—Environmental aspects—Handbooks, manuals, etc. 3. Air quality management—Handbooks, manuals, etc. I. Rafson, Harold J.
TD886.036 1998
628.5′3—dc21 98-4221
CIP

McGraw-Hill
A Division of The ***McGraw-Hill*** *Companies*

1 2 3 4 5 6 7 8 9 0 DOC/DOC 9 0 3 2 1 0 9 8

ISBN 0-07-052523-4

The sponsoring editor for this book was Robert Esposito, the editing supervisor was David E. Fogarty, and the production supervisor was Tina Cameron. It was set in Times Roman by Estelita F. Green of McGraw-Hill's Professional Book Group composition unit.

Printed and bound by R. R. Donnelley & Sons Company.

This book is printed on recycled, acid-free paper containing a minimum of 50% recycled, de-inked fiber.

CONTENTS

Preface ix
Acknowledgments xi

Chapter 1. From Conservation to Sustainability—The Evolution of Environmental Protection *Anthony Earl* 1.1

Chapter 2. Fundamentals: Chemistry and Characteristics of Odors and VOCs *Thomas R. Card* 2.1

Chapter 3. Health Effects and Exposure Assessments of VOCs *Lorraine M. Conroy and Linda Forst* 3.1

Chapter 4. Analytical Methods 4.1

4.1. Chemical Methods *Frank H. Jarke* 4.1
4.2. Sensory Test Methods *Samuel S. Cha* 4.16
4.3. Monitoring for Volatile Organic Compounds *Richard Trzupek* 4.33
4.4. Statisticals *Dennis R. Weier* 4.54

Chapter 5. Modeling 5.1

5.1. Emission Modeling—Meteorology, Topography, and Dispersion *Kalliat T. Valsaraj* 5.1
5.2. Emission Models *Richard L. Corsi and David A. Olson* 5.11

Chapter 6. Air Pollution Prevention 6.1

6.1. Industrial Pollution Prevention *Marcus J. Healey, Mark S. Walsh, Douglas Kretkowski, and Jeffrey Lewis* 6.1
6.2. Wastewater *Gayle Van Durme* 6.36

Chapter 7. Foul Air Collection and Conveyance *Phillip Wolstenholme* 7.1

Chapter 8. Emission Control Technology 8.1

8.1. The Addition of Chemicals to Liquid to Control Odors *Edward A. Bonanni* 8.2
8.2. Masking and Odor Neutralization *Timothy W. Planker* 8.18
8.3. Condensation *Richard E. Waldrop* 8.25
8.4. Thermal Oxidation *Gene H. Anguil* 8.31
8.5. Adsorption 8.66
8.5.1. Carbon *Charles Parmele and Teresa Kovalcson* 8.66
8.5.2. Fog *Robert N. Rafson* 8.92
8.6. Chemical Scrubbing 8.102
8.6.1. Packed Columns *Charles M. McGinley* 8.102
8.6.2. Mist Scrubbing *Harold J. Rafson* 8.111
8.6.3. Other Designs *Robert P. G. Bowker* 8.131
8.7. Dry Chemical Scrubbers *Richard L. Davie* 8.139
8.8. Biological Systems 8.150
8.8.1. Biofilters *George J. Skladany, Marc A. Deshusses, Joseph S. Devinny, A. Paul Togna, and Todd S. Webster* 8.192
8.8.2. Aeration Basins *Robert P.G. Bowker* 8.203
8.9. Hybrid Systems *Avinash N. Patkar* 8.216
8.10. Particulates 8.216
8.10.1. Physical Collection *Robert R. Banks* 8.216
8.10.2. Electrostatic Precipitators *Chen-Lu Yang and David C. Meier* 8.225

Chapter 9. Evaluating Emission Control Systems *Harold J. Rafson and Joseph R. Popeck* 9.1

Chapter 10. Regulations 10.1

10.1. U.S. Federal and State Laws *G. Vinson Hellwig* 10.2
10.2. Odor and VOC Regulations in Canada *Michael Rix* 10.11

Chapter 11. Air Pollution Control Management *Joseph P. Botos* 11.1

Chapter 12. Stepwise Examples of Compliance *Suresh R. Relwani and Mellissa A. Hickey* 12.1

Chapter 13. Social and Business Issues 13.1

13.1. Public Relations Aspects *Alan Leahigh* 13.1
13.2. Comparative Risk *Ken Jones, Sarah T. Fox, Sue Thomas, and Michele Witten* 13.9
13.3. Pollution Liability and Liability Insurance *Harlow G. Ballard* 13.40
13.4. Environment, Ecology, and Ethics *Jim Kenney* 13.48

Acronyms A.1

Glossary G.1

Contributor Biographies C.1

Index follows Contributor Biographies

PREFACE

This Handbook contains the work of about 45 contributors, still it has a personal viewpoint. I take responsibility (or credit) for the structure and subject matter, the selection of the contributors, and editorial efforts aiming for a clear, coherent manuscript.

That Handbook concept goes beyond the typical range of an engineering text, to deal with the broader question of solving odor and VOC control issues and to select, for the technology portions, writers who bring to the project a great deal of practical experience. Here are a few highlights of the Handbook:

Chapter 1. Written by a former governor of Wisconsin and head of the Department of Natural Resources, and now a partner of an environmental law firm. He provides a wonderfully varied and clear historical overview of the subject.

Chapter 2. If you give serious thought to the nature of the chemical contaminants which you are trying to control, you will likely find the surest path to a workable solution. This chapter aims to get the reader to think about fundamentals which govern control technologies. It was written by a noted consultant who has firm technical credentials.

Chapter 3. This chapter aims to add perspective to a discussion of the health effects of pollutants. It was written by very capable industrial hygienists.

Chapter 4. It is necessary to quantify the contamination problem and the effectiveness of control technologies. This is dependent on the analytical technologies and the methods of handling the data. It is not the object of this chapter to make the engineer an analyst, but rather to make her or him familiar with the analytical technologies available. This subject is divided into chemical analysis and sensory analysis (because there are unique aspects concerning odor control), using continuous monitoring and statistical analysis of the data. The section on statistics is meant to open your thinking on handling data. Each subject is written by an expert in the field.

Chapter 5. With any emission, its transit to neighbors affects the impact on the community. For VOCs, it is desirable to estimate emission rates. These are complex subjects which have had a great deal of study, and these sections, written by professors working on these subjects, are an introduction to the subject matter.

Chapter 6. This chapter deals with pollution prevention, and more with the industrial process than with add-on controls. Rather than discuss all industrial processes, several examples have been chosen. Since wastewater treatment is common to municipal and many industrial processes, it is discussed as well. The authors are specialists in their fields.

Chapter 7. This chapter deals with collection of emissions. An experienced engineer involved in such designs has written a very useful chapter.

Chapter 8. This chapter deals with the technology for pollution control. It aims to provide a broad discussion of options. This is the heart of the engineering portion of the Handbook. There are so many comments to make that I have written a separate introduction to Chap. 8. This section covers the most important of the wide range of technologies. In the interest of making the book useful, experts, many of them equipment suppliers, were selected as contributors. The reader is urged to bear this in mind and to complete his or her own investigations with added information from a wide selection of suppliers in the control technology field.

Chapter 9. This is a summary and overview chapter. A lot of information has been provided to this point, and a review of the selection or process is made.

Chapter 10. Air pollution control is driven, to a great degree, by the regulations. In this chapter specialists review U.S. and Canadian laws, and reference is made to sources of information on regulations in many countries.

Chapter 11. This chapter deals with environmental management. An engineer who has worked out what should be done may not be able to get it done without management support. The broad picture of what is involved with a corporate all-inclusive plan that is proactive and includes monitoring is important. It is written by an executive who is responsible for such a successful program with a large multinational corporation.

Chapter 12. This chapter gives a few examples that have used the know-how described in prior chapters. One example primarily deals with VOC regulations. The other example focuses on a complete project to arrive at odor control. It is written by a consultant who is actively working on such projects.

Chapter 13. This chapter considers subjects which are important and often overlooked in technology books. The first section is public relations, which is very important to keep emission problems with neighbors in realistic terms. The second section reviews where the regulations seem to be headed. It represents broadening of the involvement in decisions by scientists, citizens, and local authorities. The third section covers insurance, a useful tool in dealing with pollution issues. And finally, there is a consideration of ethics. The decision to pollute has ethical implications that should be thought about. This is an essay, and it expresses a viewpoint on the ethical issues.

It has been a complex task to assemble this Handbook which treats such a broad range of subject matter, yet I felt it was necessary to provide a broad perspective to the engineer. This Handbook may be revised in the future, as regulations and technologies change. I look forward to hearing from readers. Write to me at the publisher (McGraw-Hill), and all your comments will be read carefully. Ultimately developments depend on those people working in the field, and as I continue to be active, if I can be helpful through making contacts, or providing information, I will be pleased to encourage progress in this exciting technological field.

Harold J. Rafson

ACKNOWLEDGMENTS

Putting together a book of this size and scope is like jumping into very deep water, particularly for someone who has never taken on a project like this before.

I want to thank the contributors, who brought to the project their special knowledge and talents, their good humor (most of the time), and their willingness to make changes in scope and content as the project required. Some were old friends and project partners, others are new friends whom I had never met or worked with before this project. It was a delight to work with them all—even though the encounters were mostly virtual.

I want to thank the publisher and editors at McGraw-Hill and their staff, specifically, Sybil Parker, Bob Esposito, and Zoe Foundotos. They each helped me as a first-time author to get through the details, and I learned a great deal.

I wish to thank my wife, whose unflagging assistance and advice pulled me through. I also want to thank innumerable others who recommended contributors, reviewed chapters, or aided in some other way. It has been a great experience, and hopefully, the resulting book will be helpful.

CHAPTER 1

FROM CONSERVATION TO SUSTAINABILITY—THE EVOLUTION OF ENVIRONMENTAL PROTECTION

Anthony Earl
Quarles & Brady
Madison, Wisconsin

From the time of Teddy Roosevelt, Gifford Pinchot, and John Muir until the late 1950s, concern with the natural environment and the out-of-doors came under the rubric of *conservation.* Conservation meant protection of "the Great Outdoors." Thus, it involved promotion of national parks, national forests, hunting, fishing, and other outdoor recreational opportunities. It did not involve government regulation as we know it today. Indeed, its impact on economic activities was quite limited. National forests were managed to sustain the lumber and paper industries, public lands in the plains were available for grazing cattle, and mining claims could be staked at minimal cost on public lands throughout the west. Nevertheless, it cannot be argued that the conservation movement did not afford protection to much of the environment. Without public ownership of vast tracks of lands throughout the country, these lands and the species that inhabited them would surely have been adversely affected by the growth and development which occurred throughout this century.

Although the conservation movement enjoyed broad public support, the events of the economic boom years that immediately followed World War II persuaded many people that conservation was not enough. People began to think in terms of protecting the environment as well as having some of it set aside for public use.

A number of extraordinary events helped to catalyze this change in public thinking. Illustrative were news stories reporting that the Cuyahoga River in Cleveland had caught fire. This would have sounded like something from Ripley's *Believe It or Not,* but photographs and news stories that were published all around the country proved that the unthinkable had happened—a river had caught on fire. Other news stories detailed the conclusion of scientists that Lake Erie was dying. The once famous fishery on the lake was all but gone; beaches were closed, and Lake Erie was overloaded with human and industrial pollution not only from enterprises along its shore, but also from those upstream. About this same time, Rachel Carson's book *Silent Spring* became a best-seller. Carson outlined in persuasive detail how relatively small, modest, incremental human activities (e.g., the spraying of pesticides) could have unanticipated cumulative results whose implications (including the gradual disappearance of songbirds) dramatically outweighed any modest benefits that might have resulted from the use of those pesticides.

These and other events of the late 1950s and early 1960s led to the emergence of the environmental movement. For many people, the transition from conservation to environmental protection was not easy. Traditional supporters of the conservation movement including sports organizations, rod and gun clubs, and others were skeptical of this move. The environmental movement involved not just "users" of the outdoors, but a vast array of people who simply felt that our resources deserved better protection than they were getting. The movement had a special appeal for young people which was recognized by U.S. Senator Gaylord Nelson (D., Wisconsin) who called for the celebration of an "Earth Day" and urged students and others to spend the day studying about and considering the environment and what they could do to protect it. That celebration has become an annual event and is observed throughout the country.

The early popularity and political potency of the environmental movement are reflected in the amazing amount of significant environmental legislation which was considered and passed by Congress and signed by President Nixon in the early 1960s. This legislation had broad bipartisan support. What is impressive on reflection is that this bipartisan support did not result from the fact that the legislation which was passed was modest or watered down; indeed, it was significant and far-reaching. Among the most important actions taken by Congress during this time were

- The establishment of the U.S. Environmental Protection Agency (EPA) with broad and far-reaching powers to establish minimal environmental standards which were applicable to all 50 states
- Creation of the Council on Environmental Quality with its power to require Environmental Impact Statements for actions which would have a significant impact on the environment
- Passage of the Clean Water Act which called for ensuring the waters of the nation would be fishable and swimmable within a given time
- Passage of the Clean Air Act which established strict standards on air emissions so as to protect the public health
- Passage of the Resource Conservation Recovery Act which established new and tougher standards for the handling and disposal of solid waste.

These actions by Congress are illustrative of only some of the most dramatic actions that were taken by Congress during this period. During this time, the states were not to be outdone, and environmental activism at the state level led to the creation of environmental protection agencies in virtually every state. In some states, environmental protection responsibility was given to those agencies responsible for traditional conservation activities (parks, forests, fish, and wildlife). In other states, environmental protection agencies were freestanding. In either instance, states took unto themselves broad powers relative to air, water, and solid waste. An infrastructure of laws, regulations, and, most important, regulators quickly grew up all around the country to ensure the success of the broad environmental measures which had been passed into law.

The citizen activism which led to these political actions was not limited to the U.S. Capitol or the state Houses. Citizen activists were busy and vocal at the national, state, and local levels and were not satisfied merely with the passage of laws or the creation of agencies to oversee the enforcement of those laws. Activists who did not believe enforcement of those laws was forceful enough had no hesitancy about bringing lawsuits against not only those whose actions were responsible for polluting the environment, but also those who were responsible for enforcing the laws to protect the environment.

The generic environmental laws referred to above relied heavily on "command and control" to protect the environment. "Command and control" is highly proscriptive. It describes in detail what can or cannot be done and relies heavily on the application of the law and technology to improve and thus clean up processes which are seen as defiling the environment.

The command-and-control philosophy of the 1960s coupled with popular support and citizen activism led to a number of important successes.

Industries and municipalities located on waterways built new wastewater treatment plants and sewerage treatment plants which substantially reduced the amount of suspended solids, biological oxygen demand (BOD), and pathogens discharged into the nation's waters. As the amount of pollutants which had been discharged was sharply reduced, many rivers, streams, and lakes (including Lake Erie) showed signs of remarkable recovery. Some public bathing beaches were reopened; recreational fishing and boating resumed; and waterfronts once again became attractions rather than attractive nuisances.

On the air side, belching smokestacks which were once symbols of prosperity were cleaned up with scrubbers and precipitators, and the plague of soot in urban areas was largely eliminated. At the same time, automobile manufacturers were required to install catalytic converters to clean up automotive emissions, and within a few years air quality, which had been declining steadily, began to improve. The number of people exposed to unhealthy air no longer expanded, but began to constrict.

On the solid waste front, the ubiquitous local "dump" slowly, but surely, became a thing of the past. Dumps were almost always inappropriately located; had no management; attracted rodents and other pests; generated foul odors; and were generally unhealthy, unattractive, and the cause of overall environmental degradation. In their place came landfills which are appropriately located, scientifically engineered, and carefully managed. In time, enthusiasm for recycling sprung up all over the country so that today many items which had previously been thrown away and ended up in landfills are recycled and put to beneficial use.

Although it is clear these early efforts were very successful, it has become equally obvious that problems affecting the environment are considerably more complex and more costly to solve than had been initially anticipated. For example, early efforts aimed at cleaning up surface water paid relatively little attention to nonpoint source pollution or to the even more difficult problem of protection of the groundwater. Our air protection efforts which focused largely on particulate matter were inadequate as regards heavy metals, toxics, and dioxins and did take into account the problems associated with the long-range transport of air pollutants. On the solid waste front, substituting sanitary landfills for dumps was the easy part. The difficult part of cleaning up the thousands of polluted dump sites all over the country will require a staggering effort far more expensive and difficult than the move from the dumps to sanitary landfills.

Any balanced review of our experience under command-and-control laws must acknowledge important successes that have occurred since their imposition. Furthermore, these efforts were able to employ relatively simple technologies (a wastewater treatment plant, an air scrubber, a liner for a sanitary landfill), and they involved relatively modest expenditures. In the early going, the economic impacts of command and control were relatively modest. However, as time went on and the complexity of pollution abatement became more apparent, those costs rose dramatically.

With all the successes, there is a negative side which must be acknowledged as well. The early environmental laws with their reliance on the law (command and control) got at the "easy part" of environmental pollution. By the easy part, I mean removing suspended solids and BOD from watercourses; removing particulate matter from the air; and closing town dumps. The more difficult environmental problems do not easily lend themselves to command-and-control solutions. There is no readily available technology for dealing with nonpoint source pollution. There is no easy and inexpensive way to deal with contamination in bottom sediments; there is no easy answer for bioaccumulative toxics in the waterstream or bottom sediments. These and other far more complex and difficult environmental problems are not dealt with easily, if at all, by command-and-control models.

Another problem with the command-and-control regime is that it is largely reactive. It focuses on cleaning up an emission stream rather than minimizing or eliminating the pollution before it gets into the discharge, whether it be air or water. As time has gone on, those with experience in protection of the environment have come to place greater and greater emphasis on the desirability of minimizing, if not altogether eliminating, pollution from the processes that create it. As I will indicate later, market forces are far more successful in providing incentives to minimize or eliminate pollution than is an exclusive reliance on command and control.

Another negative result of the early heavy reliance on command and control was the creation of an adversarial system between the regulated community and regulators. In the early days of command and control, most of the expertise was in the hands of regulators at the federal and state levels. Most of those in the regulated community were without such expertise and, accordingly, relied on lawyers rather than scientists or engineers to deal with the new requirements they were being forced to employ. Accordingly, rather than both sides attempting to find a common-ground solution to protection of the environment, regulators laid down "the letter of the law" as they read it, and those being regulated went to the courthouse to resist. Gradually, as the regulated community began to recognize the need to equip itself with its own experts (either in-house or as consultants), this situation began to change. This action by the regulated community was illustrative of a philosophical shift on its part. The regulated community began to explicitly adopt a compliance philosophy. Thus, the dialogue between the regulators and the regulated moved from "You must do this." "We won't, and we will see you in court." to "This is what must be done." "We agree; but let's talk about the means to be employed to get there and the time within which it must be done."

Before I make too much of this new attitude on the part of the regulated and the regulators, let me acknowledge that much of the residual from the early adversarial relationships of the 1960s and 1970s remains.

The move to a compliance philosophy on the part of the regulated community was made considerably more attractive by the introduction of market incentives as a means to protect the environment. The regulated community understands the market and given the opportunity, will use it to its advantage.

The most notable use to date of market forces to protect the environment is found in the Acid Rain Title of the Clean Air Act Amendments (CAAA) of 1990. Under the terms of that title, an overall national cap on sulfur dioxide (SO_2) emissions was established. That cap was then set to ratchet down until SO_2 emissions were only half what they were at the time of the passage of the CAAA. Every large emission source (principally, electric utilities) was assigned an allocation of the total emissions allowed. The goal was a 50 percent reduction in overall emissions. The emitters covered by the act were given the widest possible latitude to attain compliance. For example, they could switch to cleaner fuels; they could replace dirty plants; they could move to more renewable energy sources; and perhaps most importantly, they could benefit from the activities of other emitters who had overcomplied by buying some of the excess emission credits from those overcompliers. This latter technique, called *tradable emission credits,* provided considerable incentive for utilities to overcomply. Not only did they avoid sanctions for noncompliance, but they had a salable commodity—an emission credit—which could be sold or traded to those utilities which were having a more difficult time bringing down their emissions.

The Acid Rain Title of the Clean Air Act Amendments of 1990 has been by far the most successful portion of that act. According to EPA statistics, over 60 percent of all the air pollution reductions which have been achieved under all the titles of the Clean Air Act Amendments have resulted from the Acid Rain Title alone, and what is more remarkable, these emission reductions have resulted from only 5 percent of the total expenditures for compliance under the act as a whole. It is important to note that the use of market forces does not mean the abandonment of strict regulation. In the case of the Acid Rain Title described above, substantial reductions in emissions were required by a certain date. The market comes into play when those subject to the law are allowed to use whatever techniques make the greatest economic sense in their efforts to achieve environmental compliance.

The use of market incentives is bound to become even more attractive as the emphasis on environmental protection becomes more global. Market incentives, for example, make it attractive for industries under obligation to reduce emissions to invest in rain forest protection and the cleanup of environmental problem areas in the developing world and in third world countries. To encourage this kind of activity makes sense for a number of reasons:

- It reflects the global nature of many of our pollution problems (climate change, for example).
- It provides for the export of our technology and know-how to developing in third world nations.
- In forest preservation, we not only provide for carbon sequestration, but also provide important protection for biodiversity.
- Modest expenditures yield substantial environmental benefits.

Let me make it clear that I believe these activities ought to be only a part of the overall compliance strategy of domestic enterprises. These activities should be seen not as being in lieu of doing the right thing here at home, but as additional efforts to ensure that we will get the greatest amount of environmental protection for every dollar spent, and in that sense these efforts deserve to be encouraged.

Perhaps the most important benefit of the use of market incentives is that they will lead to a broader acceptance of the concept of sustainable development. By *sustainable development,* I mean moving business directions toward a healthy environment and a healthy economy simultaneously. This concept recognizes that environmental protection and a healthy economy are not polar opposites, but are in many ways dependent upon each other. Indeed, most people of goodwill in the business community, in the environmental movement, and in government recognize that there can be no long-term economic growth unless such growth is environmentally sustainable. Increasingly, the public wants products, services, and processes that make a positive contribution to sustainability.

The evolution from conservation through environmental protection to sustainable development has not always been easy, but it has been inexorable. The difficulty has arisen because as we have moved through this transition, the economic implications have become considerably greater. The conservation movement had no impact on privately owned property and had relatively modest implications for the business community. The environmental movement which began in the early 1960s had explicit implications for privately owned property; and although the economic costs were modest in the early stages of command and control, costs of compliance rose dramatically in recent years to the point where they have become one of the stated reasons for resistance to environmental regulations. Sustainable development deals with both private and public property and reaches far beyond traditional political boundaries. Further, it has the capacity to harness economic interests on behalf of the environment, both domestically and internationally. As such, it is a compelling and exciting concept.

If sustainable development is to succeed, the business community must explicitly accept the fact that its long-term economic interests require a sustainable development. Further, business must employ scientists, engineers, and technicians who are committed to this view. The environmental community must be willing to acknowledge the shortcomings of command and control and the benefits of harnessing economic incentives to advance overall environmental interests. Government for its part must modify existing statutes, regulations, and codes to encourage the use of economic incentives and promote the concept of sustainability.

I am persuaded that is the direction in which we are headed and that it will lead us to a cleaner, healthier, more prosperous world.

CHAPTER 2

FUNDAMENTALS: CHEMISTRY AND CHARACTERISTICS OF ODORS AND VOCs

Thomas R. Card
Environmental Management Consulting
Fall City, Washington

INTRODUCTION 2.1
- **Compounds of Concern** 2.1
- **Why Chemical Properties Are Important** 2.1

CHEMICAL PROPERTIES OF ODORS AND VOCs 2.2
- **Basic Chemical Properties** 2.2
- **Sources of Chemical Property Data** 2.3
- **Prediction of Chemical Properties** 2.3
- **Chemical Property Discussion** 2.3
- **Ionization** 2.7

VOLATILIZATION/ADSORPTION 2.10
- **Mass-Transfer Concepts** 2.10

SORPTION 2.12
APPENDIX 2.12
REFERENCES 2.36

INTRODUCTION

This chapter introduces the chemical and physical properties of the compounds of concern for this handbook. These chemical properties dictate the appropriate management and control strategies for any specific compound or groups of compounds. For example, when scrubbing gases, oxidizable compounds can be selectively absorbed over nonoxidizable compounds by the addition of an oxidizing agent.

Compounds of Concern

The compounds of concern here are classified by chemical and physical properties, not regulatory properties. Table 2.1 lists the compounds of concern by their chemical properties.

Why Chemical Properties Are Important

The primary function of most environmental control measures is to separate unwanted compounds from compounds that are either desired or of no concern. The method of this separation can be physical, if the compounds are in different phases or have different

TABLE 2.1 Chemical Properties of the Common Compounds of Concern

	Highly	Some	Not much
Chemical reactivity (oxidizable)	Hydrogen sulfide	Ammonia Most hydrocarbons	Carbon disulfide
Biological reactivity	Hydrogen sulfide Ammonia Most hydrocarbons Alcohols Ketones		Chlorinated hydrocarbons
Aqueous solubility	Alcohols		Chlorinated hydrocarbons
Sorbability	Hydrogen sulfide Very high-molecular-weight hydrocarbons	High-molecular-weight hydrocarbons	Low-molecular-weight chlorinated hydrocarbons

particle sizes. If the compounds are dissolved in either gases or liquids, then the separation must be chemically based. The difference in chemical properties of the target compounds from the other compounds in solution determines the available methods to effect this separation.

CHEMICAL PROPERTIES OF ODORS AND VOCS

Basic Chemical Properties

Gases and Liquids. Most gas-phase compounds do not influence the individual properties of other gas-phase compounds. However, in the liquid phase most compounds are highly interactive. Therefore gas-phase properties are mostly independent of the particular mixture of compounds, while liquid-phase properties are highly dependent on what other compounds are in solution or are in separate liquid phases.

Polar Compounds. Organic compounds for the most part form covalent chemical bonds. This type of bonding produces compounds that have a uniform electric charge throughout the compound. Compounds that only have carbon and hydrogen (methane, octane, benzene) are usually fully covalent. Some organic compounds have functional groups that tend to concentrate the electric charge (alcohols and acids). This charge difference increases the energy needed to separate the molecules from the liquid to the gas phase. Therefore polar compounds are less volatile, in general, than nonpolar compounds. In addition, inorganic solvents (water) are usually very polar; therefore polar organic compounds normally have much higher aqueous solubility than nonpolar compounds. Alcohols dissolve much better in water than oils do.

Ionization. Both organic and inorganic compounds (acids and bases) can ionize in aqueous solutions. Ionization is an extremely rapid reaction that reaches equilibrium essentially instantaneously for all the compounds of concern in this handbook. This equilibrium can be predicted based on ionization constants specific to a compound (at a certain temperature) and aqueous pH. Ionization greatly increases the solubility, because only the nonionized species can volatilize.

Reactivity. Most organic compounds are readily oxidizable at elevated temperatures. The addition of functional groups such as halogens and of sulfur reduces the reactivity substantially. Most of these compounds will readily oxidize biologically; however, the presence of halogens will reduce biological oxidizability.

Strong oxidants, such as ozone, hydrogen peroxide, and chlorine, will not oxidize organic compounds at low temperatures (in gas phase or aqueous solution) unless there is a catalyst present or an ultraviolet light source or the organic compound has a reactive functional group (acid, alcohol).

Hydrogen sulfide is readily oxidizable by most chemical oxidants and by biological systems. Hydrogen sulfide will react with dissolved oxygen, but substantially more slowly than with the stronger oxidants.

Ammonia is readily biodegradable, but will not chemically oxidize in solution. It will readily oxidize in the gas phase at elevated temperatures.

Sources of Chemical Property Data

Chemical property data are available from many sources. A fairly complete data set is provided in the Appendix of this chapter from the U.S. Environmental Protection Agency (1994). This data set includes molecular weight, density, vapor pressure at 25°C, Henry's law coefficients, gas- and liquid-phase diffusion coefficients, boiling points, Antoine coefficients, octanol/water partition coefficients, biological oxidation rates, and hydrolysis rates. These data are acceptable for most regulatory models, but they are not the best data possible. The most accurate data are generally available from peer-reviewed journal articles that can be located by using the American Chemical Society's Chemical Abstracts Service.

Prediction of Chemical Properties

All the chemical properties discussed in this chapter can be predicted using various types of quantitative structural activity relationships (QSARs). These methods rely on estimating the influence of the position and number of elements that are bonded to the carbon structure (Hansch and Leo, 1995).

Chemical Property Discussion

In the following there will be a brief discussion of these chemical and physical properties and how they affect the control of air pollution.

Vapor Pressure. Vapor pressure is the gas-phase concentration that is in equilibrium with a pure liquid phase at a particular temperature. In general, compounds that are nonpolar have higher vapor pressures because they require lower energy levels for the molecules to escape the liquid phase. It is very important to note that vapor pressure is only an indication of the gas/liquid equilibrium of pure components; it does not give a complete indication of the gas/liquid equilibrium if the compound is dissolved in a second compound (water). Table 2.2 provides some examples of vapor pressures of common compounds of concern.

Vapor pressure is highly dependent on temperature. The relationship between temperature and vapor pressure is shown by the Antoine equation

TABLE 2.2 Molecular Weight and Vapor Pressure (at 25°C) of Common Compounds of Concern

Compound	Molecular weight	Vapor pressure (mmHg)
2,4 D	221	0.0828
Acetaldehyde	45.06	870
Acetone	58	266
Benzene	78.1	95.2
Dichlorobenzene	147	1.2
Ethanol	46.1	50
Formaldehyde	30	3500
Hydrazine	32	14.4
Hydrogen sulfide	34.1	15,200
Methanol	32	114
Methylene chloride	85	438
MEK	72.1	100
Octane	114.3	17
Phenol	94.1	0.341
Styrene	104.2	7.3
Perchloroethene	165.83	19
Toluene	92.4	30
Vinyl chloride	62.5	2660

$$\ln P_0 = A - \frac{B}{T + C} \tag{2.1}$$

where P_0 = vapor pressure (mmHg)
T = temperature, °C
A, B, C = constants provided in the Appendix (Table 2.6)

Figure 2.1 illustrates the dependence of vapor pressure on temperature. This diagram shows that the boiling point for compounds is the temperature where the vapor pressure equals the atmospheric pressure.

Prediction of Vapor Pressure. Vapor pressure can be predicted from the boiling point for compounds by using the following equation:

$$\ln P_v = 19\left(1 - \frac{T_{\text{bp}}}{T}\right) + 8.5 \ln \frac{T_{\text{bp}}}{T} \tag{2.2}$$

This equation loses accuracy for compounds that are polar or have hydrogen bonding. There are more sophisticated, QSAR-type methods for the prediction of vapor pressure for those compounds.

Air Pollution Significance. A knowledge of the volatility of a compound greatly affects the options for odor and volatile organic compound (VOC) control. If the gaseous contaminant is condensable at 90°F, then condensation using water which is generally available at temperatures of 80°F or less is applicable. This can be accomplished by direct or indirect contact condensers, and the choice is affected by the solubility of the contaminant.

The range of application of condensation can be extended by the use of chilled water, heat-transfer compounds, or refrigerants. This technology usually finds application where it is worthwhile to recover a volatilized solvent.

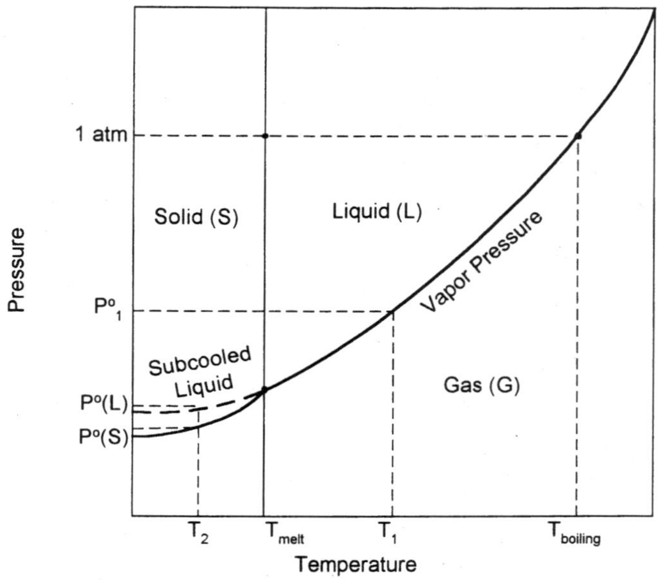

FIGURE 2.1 Phase diagram for organic liquids showing temperature dependence.

When condensation alone is inadequate to achieve acceptable levels of exhaust contaminant, then it is common to use multiple systems in series, such as condensation followed by adsorption, chemical scrubbing, or thermal oxidation.

The volatility of a compound is critical in determining the effectiveness of many treatment systems. As an example, hexane is highly volatile, and adsorption is ineffective since hexane volatilizes from the adsorbent. In such cases, thermal oxidation may be the control technology of last resort.

Solubility in Water. *Water solubility* is defined as the concentration in the aqueous phase that is in equilibrium with the pure component phase. Compounds that ionize or are polar tend to be more soluble in water than compounds that are not polar or do not ionize. Organic compounds that are entirely carbon and hydrogen tend not to be very soluble in water. The solubility of two liquid phases is not generally as an important property for environmental concerns as the gas-phase/liquid-phase solubility described in the next section.

Gas/Liquid Partitioning (Henry's Law Coefficient). The portion of a compound that partitions between separate physical phases is a function of the chemical activity in each phase. For the case of a compound partitioning between an aqueous and a gas phase, the partition coefficient is called *Henry's law coefficient.* Henry's law coefficient is equal to the vapor pressure divided by the aqueous solubility. Henry's law coefficient can appear in many forms with many different units. It is very important that the correct form with the correct units be used for any calculations. Some of the most common forms are

1. Mole fraction gas phase over mole fraction liquid phase (common units are atmospheres per mole fraction).

2. Mole fraction over mole concentration (common units are atmospheres per mole per cubic meter).
3. Mass concentration gas phase over mass concentration liquid phase (grams per cubic meter over grams per cubic meter).

Compounds that have high Henry's law coefficients are the compounds that have high vapor pressures and low solubilities. The normal classification for compounds is that if they have a Henry's law coefficient of above 0.01 (milligrams per liter over milligrams per liter), they are considered volatile. If Henry's law coefficient is between 0.0001 and 0.01, the compounds are considered semivolatile. If Henry's law coefficient is below 0.0001, the compound is generally considered nonvolatile from the aqueous phase.

There is a wide discrepancy of Henry's law coefficients reported in the literature for many compounds, especially those with low volatilities. It is important to carefully choose the data that you use in regards to Henry's law coefficients. The values presented in the Appendix are official Environmental Protection Agency (EPA) values, but they are necessarily correct. The EPA tends to rely too heavily on property prediction and not enough on peer-reviewed laboratory data. Table 2.3 provides some examples of Henry's law coefficients for common compounds.

TABLE 2.3 Relative Solubility in Water (Henry's Law Coefficients) for Some Common Compounds

Compound	Gas-phase concentration [ppm/(mg/L)]
Octane	33,900
Oxygen	22,600
Hydrogen sulfide	255
Benzene	71
PCBs	2.96
Ammonia	0.78
Ethanol	0.66

Predicting Henry's Law Coefficients. Henry's law coefficients can be predicted from vapor pressure and solubility data. Henry's law coefficient is equal to the vapor pressure of the gas divided by the aqueous saturation solubility. However, both properties must be either for a pure substance or water-saturated and are rarely reported as such.

Henry's law coefficients can also be predicted by using quantitative structural activity relationships. Hine and Mookerjee (1975) and Nirmalakhandan and Speece (1988) offer fairly easily implemented techniques to predict Henry's law as a function of the chemical bond structure of the compound.

Air Pollution Significance. The ability of a compound to dissolve in water is the critical factor in determining whether the compound is suitable for control by liquid scrubbing. The relative solubility of a compound will determine

1. If the compound can be controlled by water scrubbing alone
2. If the compound can be controlled by an aqueous solution that is chemically enhanced to improve solubility (see the ionization paragraphs below)
3. If the compound will adsorb and react with a liquid-phase oxidant

Ionization

Ionization is very important because it allows many compounds (hydrogen sulfide, ammonia) to have very concentrated solutions, far beyond their pure-water solubilities. Acids and bases (and their salts) will ionize in aqueous solutions. As an example, one of the more common odorous compounds ionizes as

$$H_2S \Leftrightarrow H^+ + HS^- \Leftrightarrow 2H^+ + S^= \tag{2.3}$$

This ionization reaction is essentially instantaneous. The equilibrium concentration of each of the species, as a function of pH, can be calculated from the following equations. The total amount of hydrogen sulfide species in solution is represented by

$$C_t = [H_2S] + [HS^-] + [S^=] \tag{2.4}$$

where C_t = total concentration of sulfide species in moles per liter and the bracketed terms ($[xxS^{xx}]$) represent individual species concentration in moles per liter.

The portion of the total that is un-ionized α_0 is

$$\alpha_0 = \frac{[H_2S]}{C_t} = \frac{1}{1 + K_1/[H^+] + K_1K_2/[H^+]^2} \tag{2.5}$$

The portion of the total that is un-ionized α_1 is

$$\alpha_1 = \frac{[HS^-]}{C_t} = \frac{1}{[H^+]/K_1 + 1 + K_2/[H^+]} \tag{2.6}$$

The portion that is completely ionized α_2 is

$$\alpha_2 = \frac{[S^=]}{C_t} = \frac{1}{[H^+]^2/K_1K_2 + [H^+]/K_2 + 1} \tag{2.7}$$

The hydrogen ion concentration is calculated from the pH by

$$[H^+] = 10^{-pH} \tag{2.8}$$

The first ionization constant represents

$$K_1 = \frac{[H^+][HS^-]}{[H_2S]} \tag{2.9}$$

The second ionization constant is represented by

$$K_2 = \frac{[H^+][S^=]}{[HS^-]} \tag{2.10}$$

where all bracketed quantities are concentrations in moles per liter. Values of ionization constants (at 20°C) for compounds of interest are shown in Table 2.4.

Figure 2.2 shows how the concentration of the ionized species of hydrogen sulfide is a function of pH. Note that at a neutral pH (pH = 7), about one-half of the hydrogen sulfide is ionized. To determine the equilibrium gas-phase concentration of an ionized species, an ionization Henry's law coefficient can be determined by multiplying the un-ionized fraction coefficient α_0 by the normal Henry's law coefficient H.

Air Pollution Significance. If a compound will ionize in solution, then the performance and economics of liquid scrubbing systems can generally be enhanced. For example, the removal of ammonia and hydrogen sulfide in gas streams is very depen-

TABLE 2.4 Ionization Constants at 20°C for Common Compounds

Compound	Chemical formula	Ionization constant(s)
Hydrogen sulfide	H_2S	$K_1 = 1.0 \times 10^{-7}$ $K_2 = 1.3 \times 10^{-13}$
Hypochlorous acid	HClO	$K_1 = 3.2 \times 10^{-8}$
Ammonium hydroxide	NH_4OH	$K_1 = 1.8 \times 10^{-5}$

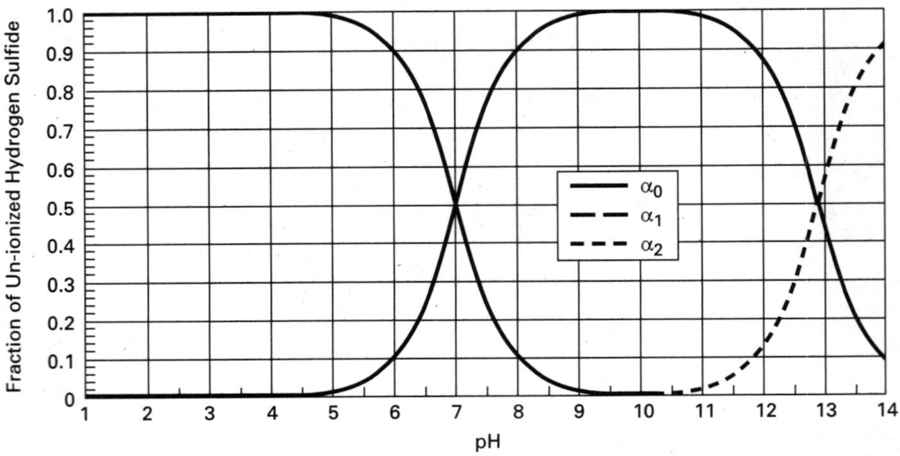

FIGURE 2.2 Ionization species for hydrogen sulfide as a function of pH. [See Eqs (2.4) to (2.7) for α_0, α_1, and α_2.]

dent on the fact that these gases will ionize in solution. The addition of either acid (for ammonia removal) or caustics (for hydrogen sulfide removal) greatly increases the ability of liquid scrubbers to remove these compounds.

Reactivity. Some of the compounds of concern are highly stable, and others are very reactive. The reactivity of a compound can be used as a tool for control if the products of the reaction are easier to control than the reactants.

Oxidation and Reduction Potential. All the elements (with the arguable exception of the noble gases) are subject to either oxidation or reduction. Oxidation increases the valence of the compound, and reduction reduces the valence of the compound. As an example, carbon can exist as carbon dioxide (fully oxidized), carbon monoxide (partially oxidized), elemental carbon (neutral), or methane (completely reduced).

Examples of processes that utilized chemical oxidation reduction include oxidizing hydrogen sulfide to sulfur or sulfate using chemical oxidants such as air, oxygen, hydrogen peroxide, sodium hypochlorite, chlorine, or potassium permanganate. The speed of the reaction is a function of the oxidation reduction potential of the reaction. The sulfide/chorine reaction has a very high potential and is very fast. The sulfide/oxygen reaction has lower potential and is much slower. Reaction can be speeded up by using inorganic catalysts (iron for peroxide, nickel for chlorine) or enzymes (organic catalysts). Biological organisms use enzymes to oxidize many compounds.

Biological Oxidation. Biological organisms are designed to oxidize (or reduce) compounds at ambient temperatures as part of their energy source. The biological reactions are quite complex. Local environmental conditions (oxygen levels, nutrient levels, species distribution, temperature, presence of food, and presence of toxics) have a dominating effect on the success of the reaction. Many times the contaminant compounds are just sorbed, but appear to be removed.

In general, most organic, sulfur, and nitrogen compounds can be oxidized (or reduced) biologically. The compounds most resistant to oxidation are high-molecular-weight compounds and halogenated organic compounds. In addition, some noncarbohydrate (oils, chlorinated compounds) compounds may not be degradable unless they are present in low concentrations or present with abundant carbohydrate compounds.

Many different mathematical models are used to represent the kinetics of biological reactions. One of the most widely used is the Monod model

$$R_b = \frac{-kX_a C_1 V}{K_s + C_1} \tag{2.11}$$

where R_b = mass rate of biodegradation (mg/s)
k = maximum specific substrate utilization (mg/mgVSS • s) (VSS is volatile suspended solids)
X_a = active cell concentration (mgVSS/m^3)
C_1 = substrate concentration (mg/m^3)
V = reactor volume (m^3)
K_s = Monod half-velocity constant (mg/m^3)

There are many cases where the substrate concentration is very low ($K_s \gg C_1$). In these cases this equation simplifies to

$$R_b = -k_1 X_a C_1 V \tag{2.12}$$

where $k_1 = k/K_s$. Values of k_1 and K_s are presented in the Appendix. However, keep in mind that actual biodegradation rates are a function of the exact circumstances of the reaction and may be several orders of magnitude different from the Appendix values.

Biorates can be predicted using many techniques. All the techniques have caveats, the most important being that if the environment for biodegradation is not adequate, no degradation will occur. The EPA (1994) uses a very simplistic prediction method based on the octanol water coefficient:

$$K_1 = 0.135 K_{\text{ow}}{}^{0.38} \tag{2.13}$$

where K_1 = the first-order biorate constant (liters per hour per gram organism) and K_{ow} is the octanol/water partition coefficient. Others (Hine and Mookerjee, 1975) have developed far more sophisticated methods. However, actual biodegradation rates are still a very strong function of environmental conditions (oxygen level, substrate, organism type, nutrients) for most compounds.

Hydrolysis. Hydrolysis occurs where a chemical reacts with water. This is generally not very important in odor and VOC control because not many compounds are water-reactive.

Air Pollution Significance. Reactivity is very important in the selection of air pollution control equipment. Compounds that are not biodegradable cannot be removed in biological systems. When there are mixtures of gases, and some gases are reactive and others are not, then the gases can be separated. A good example is the removal of hydrogen sulfide at wastewater treatment facilities. The ventilation air at wastewater treatment facilities often contains large concentrations of carbon dioxide. The normal concentration of atmospheric carbon dioxide is 300 to 400 parts per million (ppm). The ventilation air at wastewater treatment facilities ranges from 500 to 5000 ppm because of the large amount of biological respiration. Liquid scrubbers that use caustic alone will sorb both hydrogen sulfide and carbon dioxide in large quantities. If an oxidant is used, the sorption of carbon dioxide can be reduced to essentially zero, saving on chemical costs. The oxidant makes the scrubber selective for hydrogen sulfide because it is reactive and carbon dioxide is not.

VOLATILIZATION/ADSORPTION

Transfer to and from the gas phase is one of the most important phenomena in VOC and odor control technology. The volatilization (or adsorption) of a compound from an aqueous phase into gas phase is defined by

$$R_v = -K_t\left(C_l - \frac{C_g}{H_c}\right)A \tag{2.14}$$

where R_v = mass-transfer rate (mg/s)
K_t = overall mass-transfer rate (m/s)
C_l = liquid-phase concentration (mg/m^3)
C_g = gas-phase concentration (mg/m^3)
H_c = Henry's law coefficient (mg/m^3 per mg/m^3)
A = surface area (m^2)

Mass-Transfer Concepts

Mass transfer across a gas-liquid interface is a function of chemical properties of the transported substance (diffusion rates and partition coefficients) and the physical system (surface areas and boundary-layer thicknesses). For a given system, with a constant surface area, the mass-transfer rate will be controlled by the thickness of the laminar layers. Highly turbulent systems have thinner layers (and usually more surface area), therefore the mass transfer is higher. As turbulence increases, the importance of diffusion decreases and mass transfer becomes a function of bulk turbulence levels.

There are two main mass-transfer theories for gas/liquid systems. Two-film theory is most applicable to systems with lower levels of turbulence. Figure 2.3 illustrates the concept of two-film theory. The basis of this theory is that the mass-transfer coefficients k_l and k_g are equal to the diffusion rate divided by the laminar film thickness (δ_g and δ_l).

The second theory—the penetration (or surface renewal) theory—assumes that the bulk fluid turbulence is the primary system driver. This theory assumes that mass-transfer coefficients are equal to the square root of the diffusion coefficient times the renewal rate. The renewal rate, like the film thickness, is essentially impossible to measure directly, and therefore the values are determined indirectly from experimentation. The two-film theory does not

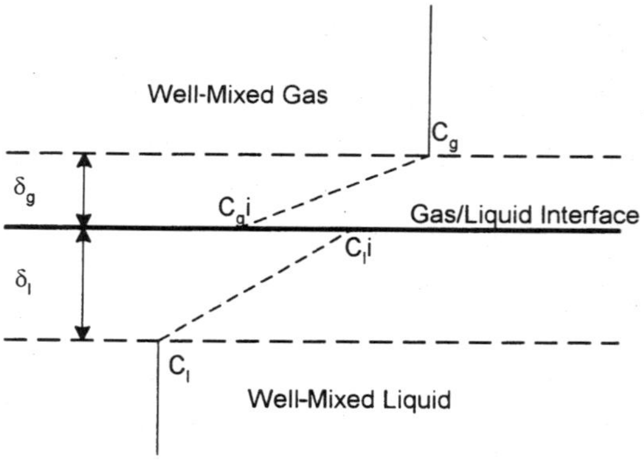

FIGURE 2.3 Illustration of the two-film theory of mass transfer.

account for accumulation in the laminar layers whereas the penetration theory does. In general, the two-film theory is more appropriate for lower-turbulence, steady-state systems. Penetration theory is in general more appropriate for high-rate, transitory systems.

These theories are important and allow the mass-transfer rate of one compound to be related to that of another compound. This allows the use of oxygen transfer data, which are generally available, to predict the mass transfer of other compounds, such as benzene. The ratio of mass-transfer coefficients is the ratio of the diffusion coefficients raised to a power. This exponent would be 1 if the system were controlled completely by two-film theory and 0.5 if the system were controlled by penetration theory. For most environmental systems, the exponent varies between 0.55 and 0.70 with the lower exponents valid for the more turbulent systems.

Diffusion Coefficients. The chemical properties that have the most control over microscale mass transfer are the liquid- and gas-phase diffusion coefficients. These values are provided in the Appendix. They have units of area over time.

Mass-Transfer Coefficients. The overall mass-transfer coefficient is calculated by

$$K_t = \frac{1}{1/K_L + 1/(H_c K_g)} \tag{2.15}$$

where K_L is the liquid-phase mass-transfer coefficient (m/s) and K_G is the gas-phase mass-transfer coefficient. Both K_L and K_G are normally estimated from empirical equations that are correlated to diffusion coefficients and turbulence levels. Compounds that are liquid-phase-controlled are highly correlated to liquid conditions, and compounds that are gas-phase-controlled highly correlated to gas conditions. For example, the dissolution of oxygen into or out of water is a liquid-phase-controlled process because Henry's law coefficient of oxygen is very large, making the second term in the denominator very small. Therefore gas conditions (wind speed) have very little to do with oxygen transfer. On the other hand, water vapor (evaporation) is completely gas-phase-controlled (Henry's law coefficient is zero), and gas-phase conditions (wind speed) completely control volatilization.

SORPTION

Many compounds will sorb onto solid surfaces in either the liquid or the gas phase. In the liquid phase, a derived chemical property called the *octanol/water partition coefficient* expresses the affinity of a particular compound to partition from the aqueous phase onto a solid surface.

Compounds tend to sorb better when they have low water solubilities and higher molecular weights. This rule has some major exceptions. Two of the compounds that sorb onto activated carbon from the gas phase very poorly are methane and methylene chloride.

One problem with gas-phase sorption for many air pollution control situations is that water vapor sorbs very well onto activated carbon as well. Therefore humid streams will show less capacity for sorption of contaminants than dry streams. A quick way to estimate this effect is to compare the gas-phase isotherm for a compound with the liquid-phase isotherm. In a gas stream that has a relative humidity of 100 percent, the sorption can be estimated by partitioning the compound into the liquid phase and then calculating the sorption based on the liquid-phase isotherm.

APPENDIX

Tables 2.5 and 2.6 present chemical property data* for a fairly comprehensive list of properties for some organic compounds. The following list gives abbreviations used in these tables.

MWT	molecular weight in g/gmol
DENS.	density in g/cm^3
VP@25C	vapor pressure in mmHg at 25°C
HL@25C	Henry's law coefficient in atm/(mol/m^3)
Dl	aqueous diffusion coefficient in cm^2/s
Dv	diffusion coefficient in air in cm^2/s
Bpt	boiling point in °C
A,B,C	Antoine coefficients
LN(OW)	base-10 logarithm of octanol/water partition coefficient
K_{max}	maximum biorate constant in g/s per gram biomass
k_1	K_{max}/K_s where K_s is the half saturation rate in g/m^3
RT HYD	hydrolysis rate constant. per second

*From EPA (1994).

TABLE 2.5 Compound Properties for Organic Compounds, I

Compound name		MWT	DENS.	VP@25C	HL@25C	Dl	Dv	Bpt
2,4 D	C	221	1.41	0.0828648	0.0000227	0.0000073	0.0231	240.1
2,4,5 T	C	255.48	1.41	5.250e-09	6.348e-12	0.0000067	0.0192	276
50% PEG	H	115	1	0	0.0003	0.0000088	0.0645	0
Acenaphthene	A	154.21	1.07	0.005	0.00771	0.0000077	0.0421	278
Acenaphthylene	A	152.21	1.02	0.022952	0.000114	0.0000075	0.04386	265
Acetal	H	118.17	0.8254	41.76573	0.000353	0.0000077	0.0677	102.7
Acetaldehyde	O	45.06	0.788	870	0.0000877	0.0000141	0.124	21
Acetaldol	H	81.11	1	0.1450557	3.350e-08	0.0000108	0.10487	185
Acetamide	N	73.1	1.15	86.06205	1.800e-09	0.0000125	0.1148	80.1
Acetic acid	O	60.05	1.05	15.4	0.0000056	0.000012	0.113	118
Acetone	O	58	0.79	266	0.000025	0.0000114	0.124	56.2
Acetonitrile	N	41.03	0.78	90	0.0000199	0.0000166	0.128	81.6
Acetophenone	O	120.16	1.03	0.297	0.0000092	0.0000087	0.06	202.3
Acetyl chloride	C	78.5	1.11	288	0.000199	0.0000115	0.099	52
Acetylaminofluorene,2-	A	223.27	1.02	0.0012	0.0000013	0.000006	0.02654	300
Acetylmethylphthalate 4	A	252.22	1.02	0	0.011	0.0000056	0.0226	0
Acetyl-2-thiourea,1-	N	118.15	1.15	0	0.1933	0.0000094	0.0585	0
Acifluorfen	A	361	1	0.0022605	0.00151	0.0000044	0.0145	306.0001
Acrolein	O	56.1	0.84	244	0.0000822	0.0000122	0.105	53
Acrylamide	N	71.09	1.12	0.012	2.610e-10	0.0000106	0.097	125
Acrylic acid	O	72.07	0.97	5.2	0.0000002	0.0000106	0.098	141.6
Acrylonitrile	N	53.1	0.81	114	0.000098	0.0000134	0.122	77.4
Adenine	N	135.13	1.15	0.1191207	8.850e-11	0.0000087	0.0487	220
Adiponitrile	H	108.14	0.951	0.0018275	1.310e-08	0.0000089	0.0718	295
Aldicarb	P	190.29	1.18	0.0001	3.168e-08	0.0000072	0.0305	287
Aldrin	P	364.93	1.18	0.000006	0.000496	0.0000049	0.0132	398
Allyl alcohol	O	58.1	0.85	23.3	0.000018	0.0000114	0.114	97
Allyl chloride	C	76.53	0.94	368	0.00927	0.0000108	0.11653	45
Allyl ether, diallyl ether	H	98.14	0.805	58.94663	0.0022624	0.0000085	0.0882	94
Alpha methyl styrene	A	118.18	1.02	0.076	0.00591	0.0000114	0.264	166
Ametryn	H	227.35	1	0.0758461	1.880e-11	0.0000059	0.0261	228.4
Aminobiphenyl,4-	N	169.23	1.15	0.0004959	0.0000003	0.0000076	0.0361	302
Aminophenol(-o)	N	109.12	1.15	0.511	0.0000037	0.0000086	0.0774	174
Aminophenol(-p)	N	109.12	1.15	0.893	0.0000197	0.0000024	0.0774	164
Aminopyridine,4-	N	94.12	1.15	0.002	0.1933	0.0000108	0.0802	273
Amyl acetate(-n)	O	130.18	0.88	5.42	0.000464	0.0000012	0.064	148.4
Aniline	N	93.1	1.02	1	0.0000018	0.0000083	0.07	184
Anisidine,o-	A	123.15	1.096	0.0551848	0.0000017	0.0000089	0.0565	225
Anthracene	A	178.23	1.25	0.0000013	0.0675	0.0000077	0.0324	0
Anthraquinone	O	210.24	1.43	3.000e-08	3.200e-09	0.0000076	0.0245	380
Aziridineethylene imine	H	43.1	1	160	0.000454	0.0000158	0.2646	56
Benzal chloride	C	127	1.26	0.07	0.00741	0.0000095	0.051	207
Benzaldehyde	O	106.13	0.97	1	0.0000423	0.0000091	0.073	178.9
Benzene	A	78.1	0.87	95.2	0.0055502	0.0000098	0.088	80.1
Benzidine	A	184.23	1.02	0.0000004	1.360e-11	0.000015	0.034	400
Benzofuran 2,3	A	118.14	1.072	2.009953	0.000237	0.000009	0.0603	174
Benzoic acid	O	122.13	1.27	0.00704	1.820e-08	0.000008	0.0536	122.12

TABLE 2.5 Compound Properties for Organic Compounds, I (***Continued***)

Compound name		MWT	DENS.	VP@25C	HL@25C	Dl	Dv	Bpt
Benzonitrile	N	103.07	1.15	1	0.0000136	0.0000102	0.0706	0
Benzophenone	O	182.23	0.97	0.0001763	0.00911	0.0000066	0.0353	0
Benzotrichloride	C	195.47	1.38	0.2	0.000981	0.0000078	0.0275	221
Benzoyl chloride	C	113.57	1.41	214	0.000188	0.0000109	0.0567	0
Benzo(a)anthracene	A	228.3	1.11	0.0000002	1.380e-09	0.000009	0.051	435
Benzo(a)pyrene	A	252.3	1.11	0.000568	1.380e-09	0.000009	0.043	312
Benzo(b)fluoranthene	A	252.32	1.02	1562.767	0.0000201	0.0000056	0.0226	4
Benzo(k)fluoranthene	A	252.32	1.02	9.590e-11	0.011	0.0000056	0.0226	480
Benzyl alcohol	O	108.15	0.97	0.15	0.0000006	0.000009	0.0712	0
Benzyl chloride	C	126.6	1.1	1.21	0.000319	0.0000078	0.075	179.4
Biphenyl	B	154.2	1.18	1	0.000408	0.0000082	0.0404	254
Bisphenol(a)	O	228.28	0.97	0.0759058	0.00228	0.0000057	0.0264	232
Bis (1,1,2,2-tetrachloropropyl) ether	C	377.7	1.41	875	43.5	0.0000053	0.0116	21.3
Bis(2-chloroethyl)ether	C	143	1.22	1.4	0.000013	0.0000075	0.0692	178
Bis(2-ethylhexyl) phthalate	H	390.56	1	0.0000002	0.0000003	0.0000037	0.0351	384
Bis(chloromethyl)ether	C	115	1.32	30	0.0000903	0.0000094	0.0573	104
Bromacil	H	261.11	1	0.0877172	137.4758	0.0000054	0.021917	239
Bromobenzene	C	157.02	1.4952	4.1344	0.00486	0.0000093	0.0355	0
Bromochloromethane	C	129.39	1.41	146.3781	25.9	0.00001	0.0474	0
Bromodichloromethane	C	163.8	1.97	59.2	0.2050041	0.0000106	0.0298	90
Bromoform	C	252.77	2.89	5.6	0.000532	0.0000103	0.0149	149.5
Bromomethane	C	94.95	1.41	1590	0.0068591	0.0000121	0.0728	4.6
Bromotoluene 4	C	171	1.3959	1.0868	0.00241	0.0000085	0.0326	0
Bromoxynil	A	276.92	1	0.0017175	2.140e-08	0.0000052	0.0203	281
Butadiene-(1,3)	H	54.09	0.76	2100	0.0713	0.0000108	0.249	−4.4
Butane	H	58.12	0.76	1522.124	0.2910089	0.0000112	0.1896	−0.5
Butanedinitrile	N	80.09	1.15	6	0.0000049	0.0000118	0.1008	265
Butanol iso	O	74	0.79	10	0.0000022	0.0000093	0.086	0
Butanol(s)	O	74.14	0.97	10	0.0000127	0.0000112	0.1207	0
Butanol-1	O	74.1	0.81	6.5	0.0000089	0.0000093	0.08	0
Butene	H	56.1	0.6255	2218.377	0.0411	0.0000102	0.2166	−6.47
Butyl acetate(-n)	O	116	0.88	15	0.000164	0.0000081	0.0675	0
Butyl acrylate	O	128.2	0.8986	5.8	0.000611	0.0000077	0.0584	145
Butyl benzene	A	134.22	1.02	1	0.0883	0.0000081	0.0519	0
Butyl carbitol	H	162.23	0.96	0.00468	0.0811	0.000007	0.0414	0
Butyl cellosolve	H	118.2	0.9	1.61	0.0000003	0.0000081	0.0651	170
Butylamine	N	73.14	0.7327	72	0.0000165	0.0000096	0.1385	77.9
Butylene glycol-(1,3)	O	90.14	1.004	0.06	0.0000036	0.0000102	0.0902	207.5
Butylisobutyrate	H	102.13	0.891	68.94406	0.00719	0.0000088	0.0799	90
Butyraldehyde	O	72.11	0.97	104.2284	0.000258	0.0000114	0.1255	74.8
Butyraldehyde iso	O	72.12	0.794	170	0.000147	0.0000101	0.1365	63
Butyric acid	O	88.1	0.97	0.84	0.0000017	0.0000101	0.0945	163.5
c10 linear	H	168	0.75	2.009953	0.0512	0.0000059	0.0443	174
Caprolactam	N	113.16	1.02	0.0072722	3.600e-09	0.000009	0.06545	270
Captan	A	300.57	1	0.00006	0.0000468	0.0000049	0.0183	312.111
Carbaryl sevin	N	201.22	1.232	0.00004	0.0000324	0.0000071	0.0278	274

TABLE 2.5 Compound Properties for Organic Compounds, I (***Continued***)

Compound name		MWT	DENS.	VP@25C	HL@25C	Dl	Dv	Bpt
Carbendazim	H	191.18	1	0.3060232	0.0000025	0.0000065	0.0327	214
Carbon disulfide	S	76.1	1.26	366	0.019152	0.00001	0.104	46.3
Carbon tetrachloride	C	153.8	1.59	113	0.0302007	0.0000088	0.078	76.8
Carbonyl sulfide	O	60	1	3062.283	0.0000991	0.000013	0.1617	−13.11
Catechol	A	110	1.02	0.0153211	4.020e-11	0.0000092	0.068	245.5
Chloracetophenone,2-	A	154.59	1.32	0.044	0.0000009	0.0000087	0.0383	244
Chloral	O	147.4	1.51	50	0.00097	0.0000097	0.0385	97.6
Chloramben	C	171.58	1.41	3.418682	0.0000341	0.0000085	0.0323	147
Chlorambucil	C	304.23	1.41	0	10.6	0.000006	0.0153	0
Chlordane	P	409.8	1.11	0.00001	0.0000367	0.0000044	0.0118	391
Chlornaphazine	C	268.2	1.41	0	10.6	0.0000065	0.018	0
Chloro 2 butene,1 trans	H	90.55	0.9295	87.02964	0.00187	0.0000097	0.0926	84
Chloroacetaldehyde	C	78.5	1.11	60	0.000026	0.0000115	0.099	85
Chloroacetic acid	C	94.5	1.41	0.13832	6.480e-08	0.0000121	0.0733	189
Chloroallyl alcohol 2	C	92.53	1.41	1.5	0.0000183	0.0000123	0.0755	146
Chloroaniline(2)	C	127.6	1.41	1	0.0168	0.0000101	0.0483	0
Chloroaniline(3)	C	127.57	1.213	1	0.0168	0.0000093	0.0515	209
Chloroaniline,p-	C	127.57	1.41	0.015	10.6	0.0000101	0.0483	232
Chloroazobenzene	C	216.7	1.41	0.0757537	0.0108	0.0000074	0.0237	0
Chlorobenzene	C	112.6	1.11	11.8	0.003762	0.0000087	0.073	132
Chlorobenzilate	C	325.2	1.41	0.0000022	9.414e-08	0.0000058	0.0141	400
Chlorobenzophenone (para)	C	216.67	1.41	0.0000026	3.400e-09	0.0000074	0.0237	330
Chlorobenzotrifluoride, p	H	120.53	1	103.0833	0.0652549	0.0000086	0.0605	79.6001
Chlorobenzyl alcohol -(m)	C	142.59	1.41	0.0050423	0.0000029	0.0000095	0.0415	237
Chlorobenzylate	O	325.2	0.97	2.664e-09	5.080e-10	0.0000046	0.0168	415
Chlorobutadiene,1	H	88.54	0.96	160.0874	0.010102	0.00001	0.0943	68
Chloroethane (ethyl chloride)	C	64.51	1.41	1200	0.0120961	0.0000115	0.271	12.4
Chloroethylene	C	63.5	1.41	2946.029	0.00544	0.0000154	0.1299	0
Chlorohydrin, a 3 chloro 1,2 propanediol	H	110.54	1.345	0.024749	5.350e-09	0.0000107	0.06	213
Chloromethyl methyl ether	C	80.51	1.41	223.079	0.0000864	1.34e-5	0.092	59.1
Chloronaphthalene,2-	C	162.62	1.41	0.017	0.0182	0.0000088	0.0347	256
Chloronitrobenzene(-o)	C	157.56	1.52	0.0760185	0.00788	0.0000094	0.0351	242
Chloronitrobenzene, p	A	157.57	1.534	0.1022261	0.0000914	0.0000094	0.0349	236
Chlorophenol-2	C	128.6	1.26	1.4	0.0000083	0.0000095	0.0501	175.6
Chlorophenol-3	C	128.6	1.24	0.5	0.0000033	0.0000094	0.0505	214
Chlorophenol-4	C	128.6	1.31	0.18	0.0000011	0.0000097	0.0493	217
Chloroprene	C	88.5	0.958	273	0.0009293	0.00001	0.104	0
Chloropropane-1	C	78.54	0.89	350	0.013	0.0000103	0.1153	0
Chloropropane-2	C	78.54	0.87	523	0.017	0.0000101	0.1164	0
Chloropropene 3	C	75.6	1.41	361	0.359	0.0000139	0.1008	0
Chloropropionitrile,3-	C	89.53	1.41	2.5	0.0000051	0.0000125	0.0791	0
Chloropropylene-2	C	76.53	1.41	0	0.3589956	0.0000138	0.099	0
Chlorotoluene-4	C	126.6	1.07	2.8	0.00466	0.0000087	0.055	162
Chrysene	A	228.2	1.11	5.760e-10	1.180e-09	0.0000062	0.0248	488

TABLE 2.5 Compound Properties for Organic Compounds, I (***Continued***)

Compound name		MWT	DENS.	VP@25C	HL@25C	Dl	Dv	Bpt
Cresol	O	108	1.03	0.3	0.0000016	0.0000093	0.0694	195
Cresol(-m)	O	108.1	1.03	0.08	0.0000007	0.00001	0.074	202
Cresol(-o)	O	108.1	1.05	0.24	0.0000016	0.0000083	0.074	190.8
Cresol(-p)	O	108.1	1.03	0.11	0.0000007	0.00001	0.074	203
Crotonaldehyde	O	70.09	0.85	30	0.0000154	0.0000102	0.0903	99
Cumene hydroperoxide	H	152.2	1.03	0.733844	0.0000312	0.0000076	0.0436	158
Cumene (isopropylbenzene)	A	120.2	0.86	4.6	0.0131	0.0000071	0.065	153
Cyanogen	N	52.04	0.95	3980	0.00496	0.0000137	0.20355	−21.2
Cyclohexane	H	84.2	0.78	100	0.0137	0.0000091	0.0839	81
Cyclohexanol	O	100.2	0.95	1.22	0.0000045	0.0000083	0.214	161
Cyclohexanone	O	98.2	0.95	4.8	0.0000041	0.0000086	0.0784	157
Cyclohexylamine	N	99.17	1.15	8.664001	0.0000423	0.0000104	0.0745	0
Cyclopentadiene	H	66.1	0.82	4	19.3	0.0000109	0.1525	40
Cymene,para	H	134.22	0.86	1.75365	0.0183	0.0000073	0.056	177
Dazomet	H	162.27	1	0.2642225	0.0000021	0.0000072	0.0406	217
DDE,p,p′-	C	318.03	1.41	0.0000065	0.000068	0.0000059	0.0144	348.111
DDT	P	354.49	1.18	0.0000002	0.114	0.000005	0.0137	260
Diazinon	H	304.36	1	0.0000118	2.150e-08	0.0000049	0.018	369.2
Diazomethane	N	42.04	1.15	3811.993	0.000013	0.0000175	0.260532	−23
Dibenzofurans	A	222	1.02	0.006698	0.00399	0.000006	0.0267	287
Dibenzopyrene1,2,7,8	A	302.4	1.02	0	0.011	0.000005	0.018	0
Dibromochloromethane	C	208.29	2.451	15	0.000783	0.0000105	0.0196	120.2
Dibromoethane-1,2	C	187.88	1.41	11.78	0.0109	0.0000081	0.0287	0
Dibromomethane	C	173.85	1.41	48	0.000998	0.0000084	0.0318	0
Dibromo-3-chloropropane,1,2	C	236.36	1.41	0.8	0.0000236	0.000007	0.0212	196
Dibromo-4-hydroxybenzonitrile, 3,5	A	290.93	1	0.0167509	0.0000005	0.000005	0.0191	244.0
Dibutylphthalate	O	278.3	1.043	0.0001	0.0000003	0.0000079	0.0438	340
Dichloro 2-propanol 1,3	C	129	1.35	0.27	0.00046	0.0000098	0.0484	174
Dichloro propanol 2,3	C	129	1.35	7	0.0000234	0.0000098	0.0484	182
Dichloroaniline 2,3	A	162	1	0.0033803	0.0000005	0.0000072	0.0407	272
Dichloroaniline(2,3)	A	162.03	1	0.0114387	0.0000018	0.0000072	0.0407	252
Dichlorobenzene(1,2) (-o)	C	147	1.31	1.5	0.00194	0.0000079	0.069	179
Dichlorobenzene(1,4) (-p)	C	147	1.29	1.2	0.00317	0.0000079	0.069	173.4
Dichlorobenzidine,3,3″-	C	253.13	1.41	0.0013037	0.0000027	0.0000067	0.0194	287.111
Dichlorobenzonitrile,2,6-	P	172.01	1.18	0.0005	0.0000063	0.0000076	0.0349	270
Dichlorobenzophenone p,p	C	251.11	1.41	0.0000029	0.0000047	6.8e-6	0.0196	353
Dichloroethane(1,1) ethylidenedichloride	C	98.96	1.17	591	0.0056	0.0000105	0.0742	0
Dichloroethane(1,2)	C	99	1.26	80	0.0011769	0.0000099	0.104	0
Dichloroethene 1,2 trans	C	96.94	1.41	331	0.0644767	0.0000119	0.0707	47.7
Dichloroethene(1,1) vinylidene chloride	C	96.97	1.213	630	0.0259005	0.0000104	0.09	31.9
Dichloroethyl ether	C	143.02	1.41	8.230494	0.0000205	0.0000095	0.0413	142
Dichloroethylene(1,2) cis	C	96.95	1.28	200	0.0154982	0.0000113	0.0736	60.7
Dichloromono-fluoromethane	C	102.92	1.41	1360	921	0.0000115	0.0650	9

TABLE 2.5 Compound Properties for Organic Compounds, I (*Continued*)

Compound name		MWT	DENS.	VP@25C	HL@25C	Dl	Dv	Bpt
Dichlorophenol	A	163.01	1	3.600194	0.0000501	0.0000071	0.0404	145
Dichlorophenol 2,5	A	163.01	1	0.1120569	0.0000016	0.0000071	0.0404	211
Dichlorophenol(2,4)	C	163.01	1.41	0.12	0.0000048	0.0000088	0.0346	210
Dichlorophenol(2,6)	C	163	1.41	0.034	0.0000048	0.0000088	0.0347	220
Dichlorophenoxyacetic acid(2,4)	C	221	1.41	290	0.0621	0.0000065	0.0588	47.2
Dichloropropane 1,2	C	112.99	1.156	40	0.002862	0.0000087	0.0782	95.5
Dichloropropene(1,3)	C	111	1.2	43	0.0035499	0.00001	0.0626	112
Dichloropropylene,1,2- (cis)	C	110.97	1.41	375.841	0.00898	0.000011	0.0586	92.5
Dichloropropylene,1,2- (trans)	C	110.97	1.41	113.8602	0.011	0.000011	0.0586	77
Dichloro-2-butene 1,2	C	124.9	1.41	0	10.6	0.0000103	0.0498	0
Dichloro-2-butene(1,4)	C	125	1.19	2.87	0.000259	0.0000081	0.0725	158
Dichloro-2-butene, 1,4	H	125	1.188	18.3067	0.00165	0.0000093	0.0534	123
Dichlorvos	A	220.98	1.415	0.012	0.0000003	0.0000073	0.02315	227.111
Dieldrin	P	380.93	1.18	0.0000002	0.0000584	0.0000047	0.0125	0
Diethanolamine	N	105.14	1.0881	0.0005463	1.800e-12	0.0000098	0.0704	268.8
Diethyl amine	N	73.14	1.15	224.959	0.00731	0.0000125	0.1147	0
Diethyl ether	O	74.14	0.97	534.1944	0.000265	0.0000086	0.0782	245
Diethyl sulfate	S	154.19	1.177	0.0342	0.0000061	0.0000081	0.0409	208
Diethyl (N,N) aniline	N	149.23	0.93	0.00283	5.740e-08	0.0000059	0.0513	0
Diethylbenzene p	A	134.22	1.02	0.9956	0.00671	0.0000081	0.0519	0
Diethylene glycol diethyl ether	H	162.23	0.909	1.214403	0.0000021	0.0000068	0.0424	185
Diethylene glycol dimethyl ether	H	162.23	0.937	3.443058	0.0000015	0.0000069	0.0418	162
Diethylene glycol monobutyl ether	H	162.23	0.967	7.556084	2.160e-08	0.000007	0.0412	231
Diethylene glycol monoethyl ether	H	134.18	0.999	0.0501	4.860e-08	0.000008	0.0524	202
Diethylene glycol monoethyl ether acetate	H	192.23	1	0.113302	0.0000006	0.0000065	0.0325	189
Diethylene glycol monomethyl ether	H	120.15	1.01	8.298628	4.140e-08	0.0000086	0.0605	194
Diethylhydrazine N,N	H	88.2	1	0	0.0003	0.0000103	0.0932	0
Diethylthiophosphate benzo m ethyl ether	H	253.31	1	0.0000095	2.120e-08	0.0000055	0.0227	392.001
Diisobutylene	H	112.22	0.708	44.69303	0.1177248	0.0000073	0.0778	101
Diisodecyl phthalate	H	446.7	1	0.0000001	0.000408	0.0000039	0.0111	451.01
Diisopropyl benzene (para)	A	162.28	1.02	1	0.107	0.0000072	0.0402	0
Diisopropyl ketone	H	114.19	0.806	12.99374	0.000568	0.0000078	0.0717	124
Diisopropylamine	H	101.19	0.722	73.17715	0.000307	0.0000078	0.0887	84
Dimethoxy-(3,3′)-benzidine	O	244.32	0.97	0.0758996	0.00244	0.0000055	0.0242	311
Dimethyl amine	N	45.09	1.15	1520	0.0000052	0.0000167	0.2342	6.9
Dimethyl benzylamine N,N	N	135.23	1.15	0.07587	0.00135	0.0000087	0.0487	0
Dimethyl benz(a)ant 7,12	A	212.28	1	0	0.0003	0.0000061	0.0285	0

TABLE 2.5 Compound Properties for Organic Compounds, I (***Continued***)

Compound name		MWT	DENS.	VP@25C	HL@25C	Dl	Dv	Bpt
Dimethyl carbamoyl chloride	C	107.54	1.168	1.757571	0.0000004	0.00001	0.0662	167
Dimethyl disulfide	S	94.2	1.046	29.488	0.0000015	0.0000101	0.0834	109
Dimethyl formamide	N	73.09	0.9445	4	0.0000002	0.0000103	0.0939	153
Dimethyl hydrazine(1,1)	H	60.1	0.791	157	0.0000016	0.0000109	0.106	63.9
Dimethyl phthalate	A	194.2	1.19	0.0002	0.000001	0.0000063	0.0568	283.8
Dimethyl sulfate	S	126.14	1.26	0.1	0.000004	0.0000096	0.0514	188
Dimethyl sulfide	S	62.12	1.26	420	0.00545	0.0000146	0.140	36.2
Dimethyl trisulfide	H	126	1	2.200398	3.034249	0.0000083	0.057	172
Dimethylacetamide	H	45.082	0.688	1483.699	0.0000102	0.0000123	0.286963	7
Dimethylaminoazo-benzene,4	N	212	1.15	0.0006401	7.320e-08	0.0000066	0.0268	298.111
Dimethylaniline N,N	N	128	0.956	0.708	0.0000139	0.0000141	0.152982	193
Dimethylbenzidine 3,3	A	212.28	1.02	0.0000014	1.352e-09	0.0000062	0.0283	384
Dimethylbenz(a) anthracene(7,12)	A	256.33	1.02	1.620e-09	2.730e-10	0.000005	0.0461	477
Dimethylethylamine	N	73.19	1.11	20	0.000385	0.0000122	0.1163	0
Dimethylphenol(2,4)	O	122.16	1.04	0.0573	0.000921	0.0000087	0.0584	211.5
Dimethylphenol(3,4)	O	122.17	0.97	0	0.0072	0.0000083	0.0602	0
Dimethylsulfone	H	94.33	1	0.0923178	2.270e-08	0.0000099	0.0849	238
Dimethylsulfoxide	H	58.08	1	0	0.0003	0.0000132	0.1696	0
Dimethylsulfoxide	H	78.13	1	1.008534	0.0000005	0.0000111	0.111	189
Dinitrobenzene m	N	168.1	1.56	0.05	0.000022	0.0000076	0.279	0
Dinitrophenol 2,4	N	184	1.68	0.014	0.0000051	0.0000091	0.0273	250
Dinitrotoluene 2,6	N	182.14	1.15	6	0.0000091	0.0000073	0.0327	0
Dinitrotoluene(2,4)	N	182.1	1.31	0.0051	0.0000072	0.0000071	0.203	0
Dinitro-o-cresol(4,6)	N	198	1.15	0.018	0.0000014	0.0000069	0.0293	254.7
Dioxane(1,4)	O	88.2	1.03	37	0.0000055	0.0000102	0.229	0
Diphenylhydrazine(1,2)	B	184.23	1.19	0.000052	0.0000002	0.0000074	0.0317	220
Diphenylmethane	B	168.23	1.19	0.0003899	0.0000363	0.0000078	0.0358	0
Dipropylamine	N	101.22	1.15	189.9606	0.000253	0.0000103	0.0724	0
Dipropylbutral	H	156	1	0.2041487	0.0000867	0.0000073	0.0428	209.8
Dipropylene glycol	H	146	1.023	0.0388421	1.590e-08	0.0000077	0.0463	206
Di-n-octyl phthalate	H	390.56	0.76	4.849e-09	0.137	0.0000036	0.0151	490
Dodecane	H	170	0.75	0.2747982	0.0664	0.0000059	0.0436	216.2
Endosulfan	P	406.95	1.18	0.00001	0.0000191	0.0000046	0.0115	390
Endrin	P	380.93	1.18	0.0000002	0.0000004	0.0000047	0.0125	445
Epichlorohydrin	C	92.5	1.18	17	0.0000335	0.0000098	0.086	117
Epoxybutane 1,2	C	72.11	0.826	207.912	0.000461	0.0000103	0.1343	61
Ethane	H	30	0.76	2970	0.0494	0.0000167	0.5139	0
Ethanol	O	46.1	0.79	50	0.0000303	0.000013	0.123	0
Ethanolamine(mono-)	N	61.09	1.02	0.4	0.0000003	0.0000114	0.107	172
Ethoxyethanol-2	O	90	0.9	5.4	0.0000064	0.0000096	0.0947	135
Ethyl acrylate	O	100	0.92	40	0.000254	0.0000086	0.077	100
Ethyl carbamate	C	89.09	1.41	0.6218955	1.170e-08	0.0000126	0.0796	183
Ethyl ether	O	74.12	0.71	520	0.00068	0.0000093	0.074	34.5
Ethyl morpholine, ethyl diethylene oxime	H	115.18	1	0.4095041	0.0000002	0.0000088	0.0644	208

TABLE 2.5 Compound Properties for Organic Compounds, I (***Continued***)

Compound name		MWT	DENS.	VP@25C	HL@25C	Dl	Dv	Bpt
Ethyl toluene, 4	H	120.2	0.861	3.44306	0.0128	0.0000078	0.0649	162
Ethyl vinyl ether	H	72.11	0.754	573.6588	0.00213	0.0000098	0.1396	33
Ethylacetate	O	88.1	0.9	100	0.000128	0.0000097	0.0732	77
Ethylamine	N	45	1.15	1057.349	0.0000256	0.0000168	0.2349	0
Ethylbenzene	A	106.2	0.87	10	0.0078806	0.0000078	0.075	136.2
Ethylene diamine	N	60.11	1.15	10.18362	0.0000085	0.0000141	0.152523	126
Ethylene dibromide	C	187.88	2.7	14	0.00065	0.0000119	0.0217	131.6
Ethylene glycol	O	62.08	1.11	0.126	1.800e-09	0.0000122	0.108	0
Ethylene glycol dimethyl ether	H	90.12	0.867	83.73	0.000035	0.0000094	0.0961	85
Ethylene glycol monobutyl ether	H	118.18	0.903	0.339572	0.0000005	0.0000082	0.065	171
Ethylene glycol monobutyl ether acetate	H	176	1	3.416916	0.0000049	0.0000068	0.0365	208
Ethylene glycol monoethyl ether acetate	H	148.17	1	1.746256	0.0000018	0.0000076	0.0458	143
Ethylene glycol monoethyl ether cellosol	H	90.12	0.93	2.746598	0.0000011	0.0000098	0.0932	135
Ethylene glycol monomethyl ether	H	76.1	1	4.928497	0.0000008	0.0000112	0.114805	124.5
Ethylene glycol monomethyl ether acetate	H	134.14	1	5.066	0.0000022	0.000008	0.0529	124
Ethylene glycol monophenyl ether	H	138.17	1.102	5.042345	6.840e-08	0.0000084	0.0482	237
Ethylene glycol monopropyl ether	H	104.16	1	3.072762	0.0000009	0.0000093	0.0739	133
Ethylene oxide	O	44	0.87	1250	0.0002381	0.0000145	0.104	0
Ethylene thiourea	N	102.17	1.15	0.3716955	1.520e-10	0.0000102	0.0715	198.222
Ethylhexylacrylate,2-	H	184.28	0.88	1.271859	0.00294	0.0000061	0.0364	184
Ethylphenol,3-	P	122.17	1.18	1	0.0000001	0.0000094	0.0553	0
Ethyl(2) hexanol	O	130.22	0.8344	0.36	0.0000617	0.0000073	0.0592	184
Fluoranthene	A	202	1.02	0.0177	0.067	0.0000064	0.0302	250
Formaldehyde	O	30	0.97	3500	0.0000003	0.0000198	0.178	−14
Formamide	O	45	0.97	342.0861	0.0000012	0.0000151	0.2509	0
Formic acid	O	46.03	1.22	42	0.0000007	0.0000014	0.079	100.7
Freon 11, fluorotrichloromethane	C	137.37	1.494	795.2637	0.0527	0.00001	0.0426	23.7
Freon 12 dichlorodifluoromethane	C	120.92	1.41	5000	0.401	0.0000105	0.052	−29.8
Freon 12, dichlorodifluoromethane	C	120.91	1	6768.665	0.7809416	0.0000085	0.0603	−41.001
Fumaric acid	O	116.07	0.97	0.0760844	0.0000017	0.0000086	0.0646	0
Furan	O	68.08	0.94	596	0.00534	0.0000122	0.104	31.4
Furfural	O	96.09	1.16	2	0.0000811	0.0000104	0.0872	161.7
Generic organic material	H	100	1	25	0.000018	0.0000096	0.0782	125
Glutaric acid	O	132.13	0.97	0.0002911	2.060e-08	0.000008	0.0542	340
Glycerin (glycerol)	O	92.09	1.25	0.00016	1.300e-08	0.0000115	0.0798	216

TABLE 2.5 Compound Properties for Organic Compounds, I (*Continued*)

Compound name		MWT	DENS.	VP@25C	HL@25C	Dl	Dv	Bpt
Glyoxal	O	58.04	1.27	221.1501	0.000011	0.0000153	0.1544	51
Glyphosate	H	169.07	1	0.1363822	7.800e-09	0.000007	0.0385	186.001
Guthion	H	317.34	1	0.0002913	1.680e-09	0.0000048	0.0171	340.001
Gylcidol	O	74.08	1.11	0.925	0.0000009	0.0000122	0.114272	162
Heptachlor	C	373.35	1.57	0.0003	0.0023	0.0000057	0.0112	339
Heptane iso	H	100.21	0.76	66	4.35	0.0000071	0.187	0
Heptane(-n)	H	100.02	0.68	46	2.01998	0.0000076	0.0926	0
Hexachlorobenzene	C	284.8	2.04	0.0048	0.0017001	0.0000059	0.0542	322
Hexachlorobutadiene	C	260.8	1.67	0.15	0.0103002	0.0000062	0.0561	215
Hexachlorocyclohexane (gamma isomer)	C	290.83	1.87	0.0000109	0.0000078	0.0000073	0.0142	390
Hexachlorocyclopenta-diene	C	272.77	1.7	0.081	10.6	0.0000072	0.0161	239
Hexachloroethane	C	237	2.09	0.65	0.0083501	0.0000068	0.0025	186
Hexafluoroacetone	O	166.02	1	57353.46	162.7182	0.0000071	0.0394	−101
Hexamethylene 1,6 diisocyanate	N	168.27	1.04	0.0382011	0.0000003	0.0000072	0.038	255
Hexamethylphosphor-amide	N	179.2	1.03	0.1190317	1.280e-13	6.9e-6	0.0352	233
Hexane(-n)	H	86.2	0.66	150	0.7680137	0.0000078	0.2	69
Hexanoic acid	O	116.06	0.9265	0.0763529	0.0000011	0.0000084	0.0659	205.7
Hexanol-1	O	102.18	0.82	0.812	0.0000182	0.0000075	0.059	158
Hexen-2-one 5	O	98.16	0.847	10.85701	0.00008	0.0000088	0.0862	128
Hydrazine	N	32.06	1.0081	14.4	0.0000007	0.000019	0.4164	113.5
Hydrogen sulfide	C	34.1	1.41	15200	0.023	0.0000161	0.176	−60.2
Hydroquinone	O	110.11	1	0.0006957	1.440e-09	0.000009	0.06853	286
Hydroxyacetic acid	O	76.05	0.97	189.8651	0.00019	0.000011	0.11638	216.001
Indeno(1,2,3-cd)-pyrene	N	276.34	1.15	1.000e-09	5.070e-13	0.0000057	0.019	536
Isobutylene	H	56.11	0.626	2250.243	0.0366	0.0000102	0.216458	−6.9
Isodecanol	H	152.2	1	0.0264044	0.0000053	0.0000074	0.0442	212
Isophorone	H	138.21	0.92	0.439	0.0000066	0.0000068	0.0623	215
Lindane hexachlorocy-clohexane	C	290.85	1.41	0.0309093	0.0021	0.0000062	0.0162	259
Maleic açid	O	116.07	1.59	0.075846	1.470e-08	0.0000115	0.0523	260.001
Maleic anhydride	O	98.1	0.93	0.0013406	0.0000002	0.0000111	0.095	200
Methacrylic acid	O	86.1	1.0153	0.1	0.0000113	0.0000105	0.0958	163
Methanol	O	32	0.79	114	0.0000052	0.0000164	0.15	0
Methomyl	H	162.2	1	0.00005	0.0000008	0.0000072	0.0407	144.0001
Methoxychlor	O	345.65	0.97	0.0001552	0.000259	0.0000045	0.0156	350
Methyl 1-pentene 2	H	84	0.76	270.7512	10.5	0.000009	0.113	0
Methyl acetate	O	74.1	0.934	235	0.000102	0.00001	0.104	56.9
Methyl acrylate	O	86.1	0.97	88	0.0000001	0.0000102	0.0976	77
Methyl amine	N	31.06	1.15	1520	0.00538	0.000021	0.417	0
Methyl aziridine 2	N	57.1	1.15	0	0.1933	0.0000145	0.164502	0
Methyl benzyl alcohol 4	O	122.17	1.015	0.0864688	0.0000139	0.0000086	0.059	215
Methyl chloride	C	50.5	0.95	3830	0.0088201	0.0000065	0.126	−24
Methyl cholanthrene 3	A	268.36	1.02	0.000759	0.000134	0.0000054	0.0209	0
Methyl cyclohexane	H	92.2	0.76	43	0.979	0.0000085	0.0986	0
Methyl etherdimethyl ether	O	46.08	0.97	3980	0.00318	0.0000149	0.242	−23.6

TABLE 2.5 Compound Properties for Organic Compounds, I (*Continued*)

Compound name		MWT	DENS.	VP@25C	HL@25C	Dl	Dv	Bpt
Methyl ethyl ketone, 2 butanone	O	72.1	0.82	100	0.00013	0.0000098	0.0808	79.6
Methyl formate	O	60.05	0.97	500	0.13	0.0000127	0.1635	0
Methyl hydrazine	N	46.09	0.866	49.6	0.0000004	0.0000139	0.253053	87
Methyl iodide	M	141.94	1.65	400	0.00253	0.0000104	0.039	42.4
Methyl isobutyl ketone	O	100.2	0.8	15.7	0.0003901	0.0000078	0.075	115.8
Methyl isocyanate	N	57.06	1.15	461.743	0.0105	0.0000145	0.1647	39.1
Methyl isopropyl ketone	H	86.15	0.805	189.8966	0.000458	0.0000092	0.1057	94
Methyl mercaptan	S	48.1	0.999	1728.82	0.00363	0.0000148	0.2244	1
Methyl methacrylate	O	100.1	0.95	39	0.0001409	0.0000086	0.077	101
Methyl morpholine	H	101.15	0.92	17.7398	0.0000033	0.000009	0.0798	117
Methyl napthalene(1-)	A	142.19	1.02	0.0539	0.00071	0.0000078	0.048	0
Methyl napthalene(2-)	A	142.19	1.02	0.06772	0.000058	0.0000078	0.048	0
Methyl parathion	P	263.23	1.18	0.0000098	6.789e-08	0.0000059	0.02	0
Methylene chloride, dichloromethane	C	85	1.34	438	0.003	0.0000117	0.101	39.8
Methylene diphenyl diisocyanate	N	236	1.15	0.0006439	4.860e-08	0.0000062	0.0233	327.111
Methylene diphenyl-amine (MDA)	H	198	1	0.000351	2.810e-08	0.0000064	0.0313	337
Methylenedianiline 4,4	N	172	1.15	0.0004037	0.0000005	0.0000075	0.0353	305.111
Methylene-bis (2-chloroaniline),4,4′-	N	267.15	1.15	0.0000694	0.0000003	0.0000058	0.0199	331
Methylstyrene (-4)	A	118.19	1.02	1.6188	0.00591	0.0000088	0.0616	0
Methyl-tertiary-butyl ether	O	83.1	0.97	185.949	0.0005551	0.0000105	0.1024	64
Monomethyl formanide	O	59.07	1.011	0.7626558	9.670e-10	0.0000132	0.1647	184
Morpholine	N	87.12	1	10	0.0000573	0.0000096	0.091	129
Nabam	H	256.35	1	0.0000004	2.500e-16	0.0000054	0.0224	416.0001
Naphthalene	A	128.2	1.14	0.23	0.0004831	0.0000075	0.059	218
Naphthol,alpha-	O	144.18	0.97	0.002204	2.540e-08	0.0000076	0.0482	0
Naphthol,beta-	O	144.17	0.97	10	1.390e-08	0.0000076	0.0482	285
Naphthylamine,alpha-	N	143.19	1.15	0.0005366	5.000e-08	0.0000084	0.0451	300.8
Naphthylamine,beta-	N	143.19	1.15	0.0003781	3.520e-08	0.0000084	0.0451	306.1
Neopentyl glycol	O	104	0.97	0.0341692	1.600e-08	0.0000092	0.0751	208
Nitro m xylene, 2	A	151.17	1.117	0.1869559	0.000427	0.000008	0.0425	224
Nitroaniline p	N	138.14	1.15	0.0015	0.000227	0.0000086	0.0473	0
Nitrobenzene	N	123.1	1.2	0.3	0.0000239	0.0000086	0.076	0
Nitrobiphenyl,4-	B	199.21	1.19	0.0002913	0.0000073	0.000007	0.0286	340
Nitroglycerin	N	227.09	1.6	0.0036	6.000e-19	0.0000078	0.0211	260
Nitromethane	N	61.05	1.15	27.8	0.0235	0.000014	0.1491	0
Nitrophenol,2-	N	139.11	1.15	0.0013423	0.0000001	0.0000085	0.0469	216
Nitrophenol,4-	N	139.11	1.4	0.0001	1.314e-09	0.0000096	0.043	279
Nitropropane 2	N	89.09	0.9876	101.517	0.000119	0.0000101	0.0923	120.3
Nitrosodimethylamine n	N	74.08	1.15	8	0.0000008	0.0000124	0.1126	152
Nitrosomorpholine	N	116.14	1.15	0.0990759	8.540e-08	0.00001	0.059	225
Nitroso-n-methylurea n	N	103.1	1.15	0.8041194	2.590e-08	0.0000102	0.0706	178
Nitrotoluene (-p)	N	137.13	1.15	1	0.000408	0.0000086	0.0478	0
Nitrotoluene, m	A	137.13	1.063	0.1384812	0.0000684	0.0000082	0.0495	230
Nitrotoluene, o	A	137.14	1.163	0.1778758	0.0000879	0.0000087	0.0475	225

TABLE 2.5 Compound Properties for Organic Compounds, I (***Continued***)

Compound name		MWT	DENS.	VP@25C	HL@25C	Dl	Dv	Bpt
Nitrotoluene, o	A	137.13	1.163	0.1778758	0.0000879	0.0000087	0.0476	225
Nonanol, n	H	144.26	0.827	0.021735	0.0000045	0.0000069	0.0518	215
Octane	H	114.3	0.7	17	3.87	0.0000071	0.0763	125.7
Octanol 1	O	130.3	0.8259	0.124	0.0000434	0.0000073	0.0593	0
Octanol 2	o	130.3	0.8207	0.15808	0.0000118	0.0000073	0.0595	0
Octanol 3	O	130.3	0.8216	0.09424	0.0000071	0.0000073	0.0595	0
Octanol 4	O	130.3	0.8192	0.12464	0.0000094	0.0000073	0.0595	0
Oil	H	170	0.76	0	3.5	0.0000059	0.0433	0
Oxalic acid	O	90.04	1.65	0.0006077	3.600e-09	0.0000137	0.0736	155
Paraldehyde	O	132.3	0.99	25.3	0.0000367	0.000008	0.0536	125
Parathion	P	291.3	1.26	0.0000378	0.0000006	0.0000058	0.017	375
PCBs (aroclors)	B	292	1.19	0.0041	0.000864	0.000008	0.0175	306.111
Pentachlorobenzene	C	250.34	1.61	0.0046	0.0073	0.0000063	0.057	277
Pentachloroethane	C	202.3	1.67	4.4	0.021	0.0000073	0.066	162
Pentachloronitrobenzene	C	295.36	1.41	0.0000991	0.000385	0.0000061	0.0159	357
Pentachlorophenol	C	266.4	1.98	0.005	0.0000882	0.0000061	0.056	310
Pentadiene 1,2	H	67.1	0.76	367.458	0.0119	0.0000103	0.154	0
Pentaerythritol	H	136.15	0.659	0.0098698	3.700e-10	0.0000062	0.0621	227
Phenol	O	94.1	1.07	0.341	0.0000013	0.0000091	0.082	182
Phenylcyclohexanone 4	O	175	0.97	1.4668	0.00875	0.0000067	0.0373	0
Phenylene diamine(-m)	N	108.14	1.14	0.028	1.080e-08	0.0000099	0.0663	284
Phenylene diamine(-o)	N	108.14	1.15	0.008	1.080e-08	0.0000099	0.066	258
Phenylene diamine(-p)	N	108.14	1.15	0.0046	1.260e-09	0.0000099	0.06615	267
Phenylphenol P	O	170.2	0.97	10	0.0032	0.0000068	0.0387	0
Phosgene (decomposes)	O	98.92	1	1390	0.0140401	0.0000011	0.108	8.1
Phosphine	S	34	1	2000	0.2270003	0.0000182	0.381	−87.78
Phthalic acid	O	166.14	1.59	121	0.0132	0.0000068	0.064	72.6
Phthalic anhydride	O	148.1	1.33	0.0015	0.0000008	0.0000086	0.071	284
Phthalimide	N	147.1	1.15	118	0.0114	0.0000083	0.0435	69.5
Pinene(alpha-)	H	136.2	0.86	5	0.0262	0.0000073	0.0549	156
Piperazine	H	86.14	1	3.72	0.0000422	0.0000104	0.091	146
Propane	H	44.09	0.76	760	0.022	0.0000132	0.285	0
Propane sultone,1,3-	H	122.14	1.392	0.0011886	9.010e-09	0.0000103	0.0515	300
Propanoic acid	O	74.08	0.97	8.553405	0.0000018	0.0000112	0.120812	141.1
Propanol	O	60.1	0.8037	20.824	0.0000019	0.0000114	0.176	97.2
Propanol iso	O	60.09	0.79	42.8	0.000015	0.0000104	0.098	82.4
Propene	H	42.08	0.61	8168.507	0.0427	0.0000119	0.334106	−47
Propiolactone b	O	72.1	0.97	3.4	0.0000001	0.0000114	0.125	162.3
Propionaldehyde	O	58.08	0.81	300	0.0000598	0.0000114	0.102	49.5
Propionic acid	O	74.09	0.97	10	0.0000487	0.0000112	0.1208	0
Proporur (Baygon)	N	209.24	1.15	0.0171961	5.770e-08	0.0000067	0.0273	255.111
Propyl acetate iso	O	102.13	0.87	73.954	0.000317	0.0000087	0.0807	88.2
Propyl amine iso	N	59.08	0.694	460	0.000358	0.0000105	0.192	33
Propyl ether iso	O	102.18	0.97	150	0.00224	0.0000093	0.0769	68.4
Propylene	H	42.12	0.76	7600	2.11	0.0000136	0.305	0
Propylene chlorohydrin	H	94.54	1.103	3.072759	0.0000024	0.0000105	0.0811	133
Propylene glycol	O	76.11	1.04	0.3	0.0000015	0.0000102	0.093	188
Propylene oxide	O	58.1	0.83	525	0.0003559	0.00001	0.104	33.9

TABLE 2.5 Compound Properties for Organic Compounds, I (*Continued*)

Compound name		MWT	DENS.	VP@25C	HL@25C	Dl	Dv	Bpt
Propylenimine 1,22 methyl aziridine	N	57.09	1.15	153.257	0.0000094	0.0000145	0.1645	66
Propyl(-n) acetate	O	102.13	0.89	35	0.000294	0.0000088	0.0799	101.6
Propyl(-n) benzene	A	120.19	0.86	2.5	0.00659	0.0000078	0.065	0
Propyn-1-ol 2(proparlgyl)	O	56.06	0.97	11.6	0.0000086	0.0000133	0.181	113.6
Pyridine	O	79.1	0.98	20	0.0000236	0.0000076	0.091	115.5
Quinoline	A	129.16	1.02	0.095	0.0000003	0.0000083	0.0546	237.7
Quinone	N	108.09	1.318	172.1972	0.000001	0.0000107	0.0624	63.111
Resorcinol	O	110.12	0.97	0.00026	1.880e-08	0.0000087	0.078	280
Silvex	P	269.51	1.18	0	4420	0.0000058	0.0194	0
Sodium formate	M	68.01	1.65	0.0007599	1.700e-09	0.0000163	0.11052	0
Styrene	A	104.2	0.9	7.3	0.0026048	0.000008	0.071	145
Styrene oxide	O	120.15	1.054	0.798015	0.0000893	0.0000089	0.0594	194
Succinic acid	O	118.09	0.97	0.0007615	1.740e-09	0.0000085	0.0631	0
Tamaron (methamidiphos)	H	141.12	1	127.3185	0.0000045	0.0000078	0.0489	69.0001
Terephthalic acid	A	166.13	1	0.0039109	1.040e-08	0.0000071	0.0394	296.5
Terpineol, alpha	H	154.24	1	2.071076	0.000516	0.0000074	0.0434	140
Tetrachlorobenzene (1,2,3,4)	C	215.9	1.41	0.019	0.0027	0.0000074	0.0239	254
Tetrachlorobenzene (1,2,3,5)	C	215.9	1.41	0.03	0.00426	0.0000074	0.0239	246
Tetrachlorobenzene (1,2,4,5)	C	215.9	1.86	0.03	0.00426	0.0000088	0.0211	246
Tetrachlorodibenzo-p-dioxin(2,3;7,8)	C	321.96	1.41	0.0000012	0.0000398	0.0000058	0.0143	421
Tetrachloroethane(1,1,2,2)	C	168	1.59	6.5	0.0002495	7.9e-6	0.071	146.2
Tetrachloroethene	C	165.83	1.624	19	0.0177003	0.0000082	0.072	121.4
Tetrachlorophenol(2,3,4,6)	C	231.9	1.41	0.89	0.0000045	0.0000071	0.0217	164
Tetrachlorophenol(2,3,5,6)	C	231.9	1.41	0.01	110.9999	0.0000071	0.0217	0
Tetraethyl lead	M	323.45	1.653	0.35	0.0809	0.0000064	0.0132	200
Tetraethyldithiopyrophos-phate	S	322.34	1.26	0.0000004	7.200e-09	0.0000055	0.015	335
Tetraethylene pentamine	H	189.31	1	0.0521293	1.500e-13	6.5e-6	0.0331	229
Tetrafluoromethane	C	104.46	1	382429.8	60.40877	0.0000093	0.0736	−176.0
Tetrahydrofuran	O	72.12	0.88	72.1	0.000049	0.0000105	0.098	67
Tetralin	H	132.22	1	0.40318	0.00188	0.0000081	0.0534	0
Tetranitromethane	N	196.03	1.15	8	0.1933	0.000007	0.0297	126
Thiourea	N	76.12	1.41	145	0.00016	0.0000138	0.107	182
Thiourea,1-(o-chlorophenyl)-	N	186.66	1.15	0.000002	0.1933	0.0000072	0.0317	323
Toluene	A	92.4	0.87	30	0.0064201	0.0000086	0.087	110.7
Toluene diamine(2,4)	N	122	1.11	1.589889	1.260e-09	0.0000091	0.0569	284
Toluene diisocyanate(2,4)	N	174.16	1.2	0.08	0.0000002	0.0000062	0.061	251
Toluenediamine(2,6)	N	122.17	1.15	0	0.1933	0.0000092	0.0559	0
Toluenediamine(3,4)	N	122.17	1.15	0.0052068	3.280e-09	0.0000092	0.0559	265
Toluenesulfonyl chloride	A	190.65	1	0.0042605	0.0000022	0.0000065	0.0328	295
Toluic acid (para-)	O	136.16	0.97	0.00031	5.600e-09	0.0000078	0.052	275
Toluic aldehyde	O	120.14	1.03	0.16	0.000253	0.0000087	0.06	204
Toluidine m	A	107.16	0.999	0.1877656	0.0000016	0.0000092	0.0711	203
Toluidine P	A	107.17	1.046	1	0.0000191	0.0000094	0.069758	200.6

TABLE 2.5 Compound Properties for Organic Compounds, I (***Continued***)

Compound name		MWT	DENS.	VP@25C	HL@25C	Dl	Dv	Bpt
Toluidine (-0)	A	107.17	0.989	0.242	0.0000024	0.0000091	0.0714	200.4
Toxaphene	H	414	1.11	0.3	0.00489	0.0000043	0.0116	384
Tributyl phosphoro-trithioate SSS	O	314.54	0.97	0.0007587	0.000157	0.0000047	0.0175	470
Tributyl tin acetate	H	230.4	1	0.0263411	0.00696	0.0000058	0.0257	262
Tributylphosphate	O	266.32	0.97	127	0.007543	0.0000052	0.0216	289
Trichlorobenzene 1,2,3	C	181.46	1.41	0.2169527	0.00787	0.0000082	0.03	221
Trichlorobenzene 1,2,4	C	181.5	1.41	0.18	0.0019201	0.0000082	0.03	213
Trichlorobenzene 1,3,5	C	181.5	1.41	0.23	0.0209	0.0000082	0.03	208.5
Trichloroethane 1,1,1 methyl chloroform	C	133.4	1.33	123	0.0174002	0.0000088	0.078	75
Trichloroethane 1,1,2	C	133.4	1.3	25	0.0008239	0.0000088	0.078	74
Trichloroethylene	C	131.4	1.4	75	0.0102002	0.0000091	0.079	87
Trichlorofluoromethane	C	137.37	1.49	796	0.0583	0.0000097	0.087	23.8
Trichlorophenol 2,4,5	P	197.45	1.18	0.0399045	0.0000087	0.000007	0.0291	249
Trichlorophenol 2,4,6	O	197.46	1	0.0073	0.0000177	0.0000064	0.0314	244.5
Trichloropropane 1,1,1	C	147.43	1.41	3.1	0.029	0.0000079	0.071	107
Trichloropropane(1,1,2)	C	147.43	1.41	6.64	0.029	0.0000093	0.0396	140
Trichloropropane(1,2,2)	C	147.38	1.41	1.37	0.029	0.0000093	0.0397	124
Trichloropropane(1,2,3)	C	147.4	1.41	3	0.028	0.0000079	0.071	156
Trichloro-1,2,2-trifluoro-ethane,1,1,2-	C	187.38	1.41	270	0.3915875	0.0000081	0.0288	47.7
Triethanolamine	N	149.19	1.15	0.01	1.450e-10	0.0000082	0.0427	335.4
Triethylamine	N	101.22	0.7326	400	0.000125	0.0000079	0.0881	89.6
Triethylene glycol dimethyl ether	H	178.17	0.986	0.277505	4.680e-08	0.0000067	0.0361	216
Trifluoroethane(1,1,1)	C	84	1.41	9240	84	0.000013	0.0866	−47.3
Trifluralin	N	335.29	1.15	0.0014133	0.00016	0.000005	0.0149	314
Triisobutylene	H	170.32	1	0.4095041	0.0917	0.000007	0.0381	208
Triisopropylamine	H	185.34	1	0.216942	0.00343	0.0000066	0.0341	221.001
Trimethyl benzene, 123	A	120.2	0.89	1.920766	0.0112	0.000008	0.064	175
Trimethylbenzene (1,3,5)	A	120.2	1.02	2.4	0.147	0.0000087	0.0602	0
Trimethylpentane 2,2,4	H	114.22	0.76	40.6	3.338133	0.0000075	0.0733	99.2
Trinitrotoluene(2,4,6)	N	227.1	1.15	0.046	0.0000137	0.0000064	0.0245	0
Tripropylene glycol	H	160.26	1	0.0047115	9.530e-08	0.0000072	0.0413	238
Urea	N	60.06	1.34	6.69	0.000264	0.0000137	0.122	133
Urethane	N	89.09	1.15	10	0.0000586	0.0000106	0.0866	−40.3
Vinyl acetate	O	86.09	0.93	115	0.0005078	0.0000092	0.085	73
Vinyl acetylene	H	52.08	1	14664.62	0.0962	0.0000141	0.1992	−66
Vinyl bromide	C	106.96	1.517	969.1209	0.00674	0.0000118	0.0598	18
Vinyl chloride	C	62.5	0.91	2660	0.0264963	0.0000123	0.106	−13.9
Vinylidene chloride see 1,1dichloroethen	C	96.97	1.213	630	0.0259005	0.0000104	0.09	31.9
Warfarin	P	308.33	1.18	0.00001	4420	0.0000054	0.0163	356
Xylene	A	106.2	1.02	8.5	0.00525	0.0000093	0.0714	140
Xylene(-m)	A	106.16	0.86	8	0.0074341	0.0000078	0.07	139
Xylene(-o)	A	106.2	0.88	7	0.004878	0.00001	0.087	144.4
Xylene(-p)	A	106.16	0.86	9.5	0.0074402	0.0000084	0.0769	138.4
Xylenol(3,4)	O	122.17	0.97	0.019	3.940e-08	0.0000083	0.0602	225
Xylidinedimethylaniline	A	121.18	0.98	0.0819225	0.0000028	0.0000084	0.0606	218

TABLE 2.6 Compound Properties for Organic Compounds, II

Compound name	A	B	C	LN(OW)	K_{max}	k_1	RT HYD
2,4 D	6.908157	1746.881	193.6385	3.228066	10.76	2.275134	0
2,4,5 T	16.13832	7280.503	273.16	2.854958	10.76	1.64144	0
50% PEG	0	0	0	−4.232786	15.3	0.00332	0
Acenaphthene	7.728	2534.234	245.576	3.92	31.1	4.16815	0
Acenaphthylene	7.72819	2534.234	245.576	4.07	31.1	2.7	0
Acetal	6.849664	1279.732	219.7466	2.287809	15.3	0.999325	0
Acetaldehyde	8.005	1600.017	291.809	0.43	82.42	0.1966685	0
Acetaldol	8.206501	2072.293	204.1096	0.1735975	15.3	0.1571452	0
Acetamide	7.156695	1300.449	224.0406	−0.775268	9.7	0.0685	0
Acetic acid	7.387	1533.313	222.309	−0.31	14	0.98	0
Acetone	7.117	1210.595	229.664	−0.24	1.3	1.15	0
Acetonitrile	7.119	1314.4	230	−0.34	9.7	0.1002612	0
Acetophenone	7.385889	1891.5	217.885	1.58	17.56	0.5379447	0
Acetyl chloride	6.948	1115.954	223.554	−0.308331	10.76	0.1030783	0
Acetylaminofluorene,2-	7.25633	2110.188	182.2596	3.482134	31.1	2.841536	0
Acetylmethylphthalate 4	0	0	0	2.224537	31.1	0.9455041	0
Acetyl-2-thiourea,1-	0	0	0	−0.643182	9.7	0.0768	0
Acifluorfen	6.934575	1974.716	181.1196	5.0533	15.3	11.2357	0
Acrolein	7.212867	1297.327	246.6905	−0.09	7.8	0.36	0
Acrylamide	11.29315	3939.877	273.16	0.8008422	9.7	0.2720544	0
Acrylic acid	5.652	648.629	154.683	0.31	17.56	0.1770657	0
Acrylonitrile	7.11	1335.674	238.207	−0.92	18	0.75	0
Adenine	7.221354	1812.02	197.4596	−0.16	9.7	0.1173636	0
Adiponitrile	7.213714	2072.084	183.2096	1.675912	15.3	0.5850384	0
Aldicarb	6.927094	1908.791	184.7296	2.174818	15.3	0.9052528	0
Aldrin	9.357671	4347.02	273.16	5.669268	15.3	19.26059	0
Allyl alcohol	9.148151	2319.925	273.16	0.17	17.56	0.1566514	0
Allyl chloride	7.576274	1493.914	273.16	0.95454	10.76	0.3112157	0
Allyl ether, diallyl ether	6.845911	1250.578	221.3996	2.51897	15.3	1.223344	0
Alpha methyl styrene	6.92366	1486.88	202.4	3.463532	31.1	2.795661	0
Ametryn	7.225121	1843.155	195.8636	1.204255	15.3	0.3872162	0
Aminobiphenyl,4-	7.500307	2235.337	181.8796	3.605957	9.7	3.166695	0
Aminophenol(-o)	9.228947	2838.639	273.16	0.5815316	9.7	0.2245521	0
Aminophenol(-p)	−3.357	699.157	−331.343	0.5815316	9.7	0.2245521	0
Aminopyridine,4-	9.589092	3663.803	273.16	0.28	9.7	0.1724783	0
Amyl acetate(-n)	8.06791	2186.68	273.16	1.708488	17.56	0.6019543	0
Aniline	6.9502	1467.242	177.115	0.9	7.1	21	0
Anisidine,o-	7.464861	1932.244	196.5096	1.605756	15.3	0.5502056	0
Anthracene	8.91	3761	273.16	4.45	31.1	2.2	0
Anthraquinone	7.289244	2411.734	167.0596	3.729464	17.56	3.528086	0
Aziridine ethylene imine	9.389235	2142.318	273.16	0.1757197	15.3	0.1574373	0
Benzal chloride	9.49223	3174.547	273.16	2.152145	10.76	0.8874708	0
Benzaldehyde	8.461938	2523.011	273.16	1.48	17.56	0.4928757	0
Benzene	6.905	1211.033	220.79	2.15	19	1.4	0
Benzidine	7.5424	2625.8	163.256	1.81	31.1	0.6578671	0
Benzofuran 2,3	6.88042	1520.652	206.1996	2.625319	15.3	1.342646	0
Benzoic acid	9.033	3333.3	273	1.86	17.56	0.6872873	0
Benzonitrile	6.74631	1436.72	181	2.046477	9.7	0.8090959	0
Benzophenone	7.34966	2331.4	185	4.525765	17.56	7.08167	0
Benzotrichloride	8.326414	2691.007	273.16	2.92	10.76	1.737565	0
Benzoyl chloride	7.9245	2372.1	273.16	1.959114	10.76	0.7495522	0

TABLE 2.6 Compound Properties for Organic Compounds, II (***Continued***)

Compound name	A	B	C	LN(OW)	K_{max}	k_1	RT HYD
Benzo(a)anthracene	6.9824	2426.6	156.6	5.61	31.1	18.28721	0
Benzo(a)pyrene	9.245506	3724.363	273.16	5.98	31.1	25.27833	0
Benzo(b)fluoranthene	6.809279	952.6317	238.4996	6.843004	31.1	53.78881	0
Benzo(k)fluoranthene	11.10133	6191.376	273.16	6.84	31.1	53.64763	0
Benzyl alcohol	7.19817	1632.593	172.79	1.675986	17.56	0.5850764	0
Benzyl chloride	7.53029	1923.019	234.3478	2.3	17.75	1.010042	0
Biphenyl	7.6317	2167.862	207.61	4.266275	19	5.643243	0
Bisphenol(a)	8.64308	2910.876	273.16	4.651325	17.56	7.904033	0
Bis(1,1,2,2-tetrachloro-propyl) ether	7.812044	1452.055	273.16	4.322158	10.76	5.926039	0
Bis(2-chloroethyl)ether	8.210002	2404.325	273.16	1.58	10.76	0.5379447	0
Bis(2-ethylhexyl)phthalate	10.83704	5228.522	273.16	5.3	0.77	0.35	0
Bis(chloromethyl)ether	8.178545	1998.099	273.16	−0.38	10.76	0.0968	0
Bromacil	6.907701	1743.059	193.8496	8.145402	15.3	168.115	0
Bromobenzene	6.86064	1438.817	205.441	3.229374	10.76	2.277738	0
Bromochloromethane	6.49606	942.267	192.587	0.9726127	10.76	0.3161761	0
Bromodichloromethane	7.9655	1846.561	273.16	1.88	10.76	0.6994205	0
Bromoform	7.988101	2158.654	273.16	2.3	10.76	1.010042	0
Bromomethane	7.566313	1301.449	273.16	1.1	10.76	0.3534569	0
Bromotoluene 4	7.00762	1612.35	206.36	3.536277	10.76	2.979392	0
Bromoxynil	7.531281	2171.211	185.8696	1.989251	15.3	0.7695807	0
Butadiene-(1,3)	7.216854	1144.753	269.0367	1.871113	15.3	0.6940029	0
Butane	3.18243	0	0	1.859488	15.3	0.6869795	0
Butanedinitrile	8.860728	3218.161	273.16	0.35	9.7	0.1833727	0
Butanol iso	7.32705	1248.48	172.92	0.75	7.8	0.11	0
Butanol(s)	7.4768	1362.39	178.77	0.75	7.8	0.11	0
Butanol-1	7.4768	1362.39	178.77	0.83	7.8	0.11	0
Butene	6.805603	918.4548	240.4889	1.865313	15.3	0.6904899	0
Butyl acetate(-n)	7.127	1430.418	210.745	1.79263	17.56	0.6479443	0
Butyl acrylate	8.141759	2199.925	273.16	2.105592	17.56	0.8520479	0
Butyl benzene	6.98317	1577.965	201.378	4.028965	31.1	4.585114	0
Butyl carbitol	7.74114	2056.904	195.655	2.202231	15.3	0.9272288	0
Butyl cellosolve	6.956	1399.903	172.154	1.554456	15.3	0.5260546	0
Butylamine	8.649454	2025.139	273.16	0.88	9.7	0.2915654	0
Butylene glycol-(1,3)	9.583491	3221.718	273.16	−0.04555	17.56	0.1297253	0
Butylisobutyrate	6.84419	1237.194	222.1596	2.53453	15.3	1.240113	0
Butyraldehyde	6.358544	913.59	185.48	1.125761	17.56	0.3615145	0
Butyraldehyde iso	7.983729	1715.402	273.16	1.126484	17.56	0.3617431	0
Butyric acid	8.064204	2263.387	273.16	0.7236805	17.56	0.2542928	0
c10 linear	6.88042	1520.652	206.1996	4.463722	15.3	6.707476	0
Caprolactam	7.243459	1997.954	187.9596	0.8215773	9.7	0.2770353	0
Captan	10.25674	4316.929	273.16	2.92452	15.3	1.74445	0
Carbaryl sevin	11.59653	4768.904	273.16	2.36	9.7	1.064485	0
Carbendazim	6.897335	1657.229	198.5996	3.384036	15.3	2.607808	0
Carbon disulfide	6.942	1169.11	241.59	2.16	15.3	0.8935917	0
Carbon tetrachloride	6.9339	1242.43	230	2.72	1.5	1.5	0
Carbonyl sulfide	7.117189	968.5864	241.7505	−0.275866	15.3	0.1060484	0
Catechol	7.514789	2030.245	192.6146	0.94	31.1	0.3072813	0
Chloracetophenone,2-	8.6451	2981.1	273.16	2.594686	15.3	1.307137	0
Chloral	7.734462	1799.546	273.16	1.490009	15.3	0.4972112	0

TABLE 2.6 Compound Properties for Organic Compounds, II (***Continued***)

Compound name	A	B	C	LN(OW)	K_{max}	k_1	RT HYD
Chloramben	7.427135	1629.082	211.3296	1.593383	10.76	0.5442809	0
Chlorambucil	0	0	0	3.274369	10.76	2.3692	0
Chlordane	9.300818	4263.922	273.16	2.78	15.3	1.537236	0
Chlornaphazine	0	0	0	3.950025	10.76	4.279104	0
Chloro 2 butene,1 trans	6.841615	1217.14	223.2996	1.349343	15.3	0.43963	0
Chloroacetaldehyde	8.360256	1962.524	273.16	−0.219135	10.76	0.1114454	0
Chloroacetic acid	7.55016	1723.365	179.98	−0.621223	10.76	0.0783	0
Chloroallyl alcohol 2	9.545546	2793.597	273.16	0.3672189	10.76	0.1861564	0
Chloroaniline(2)	7.56265	1998.6	220	0	0.27	0.86	0
Chloroaniline(3)	7.55939	2073.75	215	0	0.27	0.86	0
Chloroaniline,p-	9.386585	3286.465	273.16	1.83	0.27	0.86	0
Chloroazobenzene	−1.120589	0	0	5.026951	10.76	10.97962	0
Chlorobenzene	6.978	1431.05	217.55	2.5	0.39	10	0
Chlorobenzilate	9.669569	4569.931	273.16	6.169221	10.76	29.82998	0
Chlorobenzophenone (para)	11.15672	4991.71	273.16	4.209192	10.76	5.368304	0
Chlorobenzotrifluoride, p	6.839732	1202.451	224.1356	2.560925	15.3	1.269087	0
Chlorobenzyl alcohol -(m)	8.235456	2309.119	194.2296	2.436677	10.76	1.138353	0
Chlorobenzylate	8.326468	3133.646	160.4096	5.37252	17.56	14.8561	0
Chlorobutadiene,1	6.8348	1163.8	226.3396	2.454152	15.3	1.155892	0
Chloroethane (ethyl chloride)	6.986	1030.01	238.61	1.43	10.76	0.472	0
Chloroethylene	6.89117	905.01	239.48	0.6267082	10.76	0.2336062	0
Chlorohydrin, a 3 chloro 1,2 propanediol	8.222207	2199.559	198.7896	−0.956336	15.3	0.0584	0
Chloromethyl methyl ether	6.831054	1134.219	228.0306	−0.21	10.76	0.1123397	0
Chloronaphthalene,2-	8.657456	3056.778	273.16	4.12	10.76	4.965281	0
Chloronitrobenzene(-o)	8.376652	2831.246	273.16	2.603606	10.76	1.317378	0
Chloronitrobenzene, p	6.906466	1732.741	194.4196	2.603606	15.3	1.317378	0
Chlorophenol-2	6.877	1471.61	193.17	2.15	15	0.8858069	0
Chlorophenol-3	7.900331	2445.317	273.16	1.403233	10.76	0.4608566	0
Chlorophenol-4	8.510922	2759.663	273.16	2.480466	6.5	1.182815	0
Chloroprene	6.161	783.45	179.7	0.57	10.76	0.2222978	0
Chloropropane-1	6.92648	1110.19	227.94	0.9487147	10.76	0.3096333	0
Chloropropane-2	7.771	1582	288	1.048029	10.76	0.3377439	0
Chloropropene 3	5.29716	418.375	128.168	0.9545268	10.76	0.311212	0
Chloropropionitrile,3-	7.32973	1732.55	211.79	0.422814	10.76	0.1954358	0
Chloropropylene-2	0	0	0	1.985027	10.76	0.7667412	0
Chlorotoluene-4	8.177249	2304.805	273.16	3.536277	10.76	2.979392	0
Chrysene	10.68596	5940.981	273.16	5.61	31.1	1.4	0
Cresol	8.850432	2794.746	273.16	0	23	17	0
Cresol(-m)	7.508	1856.36	199.07	0.97	23.21	17	0
Cresol(-o)	7.426974	1744.32	194.444	1.98	22.78	17	0
Cresol(-p)	7.035	1511.08	161.85	1.94	23.21	17	0
Crotonaldehyde	8.536501	2104.827	273.16	1.092311	17.56	0.3510868	0
Cumene hydroperoxide	8.191144	1950.172	209.2396	2.308645	15.3	1.017711	0
Cumene (isopropylbenzene)	6.963	1460.793	207.78	3.497709	31.1	2.880526	0
Cyanogen	3.59986	0	0	0.8077466	9.7	0.2737029	0
Cyclohexane	6.841	1201.53	222.65	2.529003	15.3	1.234131	0
Cyclohexanol	6.255	912.87	109.13	1.576838	17.56	0.5364586	0
Cyclohexanone	7.84918	2137.192	273.16	0.81	11.49	0.2742431	0

TABLE 2.6 Compound Properties for Organic Compounds, II (***Continued***)

Compound name	A	B	C	LN(OW)	K_{max}	k_1	RT HYD
Cyclohexylamine	6.68954	1229.42	188.8	1.997921	9.7	0.7754406	0
Cyclopentadiene	6.823	1071	271.6	2.028058	15.3	0.7961605	0
Cymene,para	6.881701	1530.862	205.6296	4.158295	15.3	5.134476	0
Dazomet	6.898588	1667.51	198.0296	3.518476	15.3	2.933347	0
DDE,p,p′-	10.32563	4625.263	273.16	5.69	10.76	19.61316	0
DDT	15.19374	6564.769	273.16	6.19	15.3	30.37728	0
Diazinon	7.284863	2370.844	169.1116	3.821204	15.3	3.822968	0
Diazomethane	6.800292	864.7323	243.6296	−0.124468	9.7	0.1210698	0
Dibenzofurans	6.927094	1908.791	184.7296	5.695732	31.1	19.71178	0
Dibenzopyrene 1,2,7,8	0	0	0	8.199224	31.1	176.2214	0
Dibromochloromethane	7.288803	1733.934	273.16	2.09	10.76	0.035	0
Dibromoethane-1,2	6.72148	1280.82	201.75	1.137044	10.76	0.3651012	0
Dibromomethane	1.68123	0	0	0.6958743	10.76	0.2481805	0
Dibromo-3-chloropropane, 1,2	7.886061	2348.271	273.16	0.2066454	10.76	0.1617557	0
Dibromo-4-hydroxybenzonitrile, 3,5	7.514083	2024.306	192.8996	2.294361	15.3	1.005071	0
Dibutylphthalate	9.386544	3955.114	273.16	5.2	0.4	1	0
Dichloro 2-propanol 1,3	9.783355	3086.549	273.16	−0.480189	10.76	0.0886872	0
Dichloro propanol 2,3	6.746815	1759.657	273.16	−0.269502	10.76	0.1066405	0
Dichloroaniline 2,3	7.48674	2116.834	187.5796	3.024556	15.3	1.904024	0
Dichloroaniline(2,3)	7.47752	2038.121	191.3796	3.024556	15.3	1.904024	0
Dichlorobenzene(1,2) (-o)	6.882553	1537.672	205.2496	3.38	2.5	0.58	0
Dichlorobenzene(1,4) (-p)	7.199	1690.291	218.09	3.39	6.4	2.3	0
Dichlorobenzidine,3,3′-	7.493614	2176.46	184.7085	3.51	10.76	2.911671	0
Dichlorobenzonitrile,2,6-	10.40392	4086.263	273.16	3.75325	15.3	3.602285	0
Dichlorobenzophenone p,p	10.5374	4794.257	273.16	6.046987	10.76	26.80424	0
Dichloroethane(1,1) ethylidenedichloride	6.992756	1176.864	228.838	1.79	10.76	2.3	0
Dichloroethane(1,2)	7.068385	1292.54	225	1.45	2.1	0.98	0
Dichloroethene 1,2 trans	6.9651	1141.9	231.9	1.48	10.76	0.4928757	0
Dichloroethene(1,1) vinylidene chloride	6.9722	1099.4	237.2	0	10.76	0.9	0
Dichloroethyl ether	7.6924	1990.8	235.347	0.685801	10.76	0.2460026	0
Dichloroethylene(1,2) cis	7.0223	1205.4	230.6	2.202489	10.76	0.9274379	0
Dichloromonofluoromethane	7.590301	1328.834	273.16	1.83253	10.76	0.6709645	0
Dichlorophenol	7.466021	1635.588	211.7096	3.241122	15.3	2.301272	0
Dichlorophenol 2,5	7.498359	1893.996	199.1696	3.241122	15.3	2.301272	0
Dichlorophenol(2,4)	7.497876	1890.059	199.3596	2.75	25	3.3	0
Dichlorophenol(2,6)	6.899838	1677.797	197.4596	3.241122	10.76	2.301272	0
Dichlorophenoxyacetic acid(2,4)	8.500344	1800.279	273.16	2.792008	10.76	1.553473	0
Dichloropropane 1,2	6.98	1380.1	222.8	2.28	17	1.4	0
Dichloropropene(1,3)	6.80731	1327.64	230.1337	1.98	10.76	0.7633762	0
Dichloropropylene,1,2- (cis)	6.845289	1245.558	221.6846	2.735832	10.76	1.478961	0
Dichloropropylene,1,2- (trans)	6.838623	1193.779	224.6296	2.568767	10.76	1.277825	0
Dichloro-2-butene 1,2	0	0	0	0.8726078	10.76	0.2896856	0
Dichloro-2-butene(1,4)	8.312499	2341.934	273.16	0.8726078	10.76	0.2896856	0

TABLE 2.6 Compound Properties for Organic Compounds, II (*Continued*)

Compound name	A	B	C	LN(OW)	K_{max}	k_1	RT HYD
Dichloro-2-butene, 1,4	6.858446	1347.971	215.8896	0.8726076	15.3	0.2896855	0
Dichlorvos	9.964257	3543.651	273.16	0.9334952	31.1	0.3055373	0
Dieldrin	−6.744684	0	0	6.906134	15.3	56.84358	0
Diethanolamine	8.252672	2454.928	188.1876	−0.66325	9.7	0.0755	0
Diethyl amine	5.8016	583.3	144.1	1.242518	9.7	0.4003997	0
Diethyl ether	6.92032	1064.07	228.8	1.639246	17.56	0.5665672	0
Diethyl sulfate	7.838081	2173.876	230.36	−0.29	15.3	0.1048994	0
Diethyl (N,N) aniline	7.466	1993.57	218.5	3.562567	9.7	3.048722	0
Diethylbenzene p	6.9982	1588.31	201.97	4.158206	31.1	5.134077	0
Diethylene glycol diethyl ether	7.86919	2094.43	231.887	2.934824	15.3	1.760248	0
Diethylene glycol dimethyl ether	7.083	1556.26	210.37	1.781736	15.3	0.641797	0
Diethylene glycol monobutyl ether	8.232163	2281.689	195.3696	2.202231	15.3	0.9272288	0
Diethylene glycol monoethyl ether	8.216066	2149.489	200.8796	1.33416	15.3	0.4338283	0
Diethylene glycol monoethyl ether acetate	8.208758	2090.436	203.3496	1.267111	15.3	0.4091089	0
Diethylene glycol monomethyl ether	8.211577	2113.133	202.3996	0.9001058	15.3	0.2967401	0
Diethylhydrazine N,N	0	0	0	0.6255218	15.3	0.2333639	0
Diethylthiophosphatebenzo m ethyl pether	6.96702	2275.218	164.7794	4.326276	15.3	5.947427	0
Diisobutylene	6.84893	1274.031	220.0696	3.671378	15.3	3.353253	0
Diisodecyl phthalate	6.987966	2483.235	153.5677	9.650949	15.3	627.6503	0
Diisopropyl benzene (para)	6.9933	1663.88	194.41	4.928876	31.1	10.07671	0
Diisopropyl ketone	7.177119	1459.448	215.6996	2.171245	15.3	0.9024271	0
Diisopropylamine	7.158473	1314.511	223.2996	2.210661	15.3	0.9340937	0
Dimethoxy-(3,3′-benzidine	7.051436	2436.322	273.16	3.150456	17.56	2.125762	0
Dimethyl amine	7.08212	960.242	221.67	0.05483	9.7	0.1416346	0
Dimethyl benzylamine N,N	−1.119923	0	0	2.526956	9.7	1.231922	0
Dimethyl benz(a)ant 7,12	0	0	0	7.456762	15.3	92.02836	0
Dimethyl carbamoyl chloride	7.197148	1616.599	207.5296	1.044858	10.76	0.3368081	0
Dimethyl disulfide	6.8257	1303.5	218.4	3.410681	15.3	2.669322	0
Dimethyl formamide	6.928	1400.87	196.43	−0.041408	9.7	0.1301963	0
Dimethyl hydrazine(1,1)	7.408	1305.91	225.53	0.595338	15.3	0.2272812	0
Dimethyl phthalate	4.52232	700.31	51.42	1.87	2.2	3.1	0
Dimethyl sulfate	7.4455	1843.343	217.055	0.318303	15.3	0.1783568	0
Dimethyl sulfide	7.1509	1195.58	242	1.70649	15.3	0.6009024	0
Dimethyl trisulfide	6.879565	1513.849	206.5796	5.829383	15.3	22.15714	0
Dimethylacetamide	7.124729	1039.441	237.9296	0.05483	15.3	0.1416346	0
Dimethylaminoazobenzene,4-	7.498566	2219.947	182.6185	3.521143	9.7	2.940198	0
Dimethylaniline N,N	7.72649	2110.009	242.88	−0.88	9.7	0.0624	0
Dimethylbenzidine 3,3	7.535748	2561.72	166.2996	4.169629	31.1	5.185646	0

TABLE 2.6 Compound Properties for Organic Compounds, II (***Continued***)

Compound name	A	B	C	LN(OW)	K_{max}	k_1	RT HYD
Dimethylbenz(a) anthracene(7,12)	10.57967	5775.389	273.16	7.456761	31.1	92.02824	0
Dimethylethylamine	7.08212	960.242	221.67	1.126405	9.7	0.3617182	0
Dimethylphenol(2,4)	8.926483	2930.103	273.16	2.42	10.7	4.7	0
Dimethylphenol(3,4)	0	0	0	0	5.5	1.05	0
Dimethylsulfone	6.90729	1739.619	194.0396	−0.264333	15.3	0.107124	0
Dimethylsulfoxide	0	0	0	0.0361875	15.3	0.139343	0
Dimethylsulfoxide	6.886806	1571.757	203.3496	0.3634962	15.3	0.185551	0
Dinitrobenzene m	4.337	229.2	−137	1.52229	9.7	0.5114555	0
Dinitrophenol 2,4	7.516903	2048.072	191.7596	1.54	8	0.62	0
Dinitrotoluene 2,6	4.372	380	−43.6	2.05	9.7	0.8115939	0
Dinitrotoluene(2,4)	7.981089	3074.44	280.23	2.01	9.7	0.7836799	0
Dinitro-o-cresol(4,6)	8.884898	3169.326	273.16	2.85	9.7	1.634334	0
Dioxane(1,4)	7.350545	1517.53	238.065	1.22037	17.56	0.392715	0
Diphenylhydrazine(1,2)	13.8359	5402.621	273.16	3.03	19	1.913115	0
Diphenylmethane	6.291	1261	105	4.798269	19	8.988508	0
Dipropylamine	2.278649	0	0	2.110636	9.7	0.8558166	0
Dipropylbutral	7.216752	1774.272	199.3976	3.731519	15.3	3.534435	0
Dipropylene glycol	8.218305	2167.685	200.1196	0.1654824	15.3	0.1560334	0
Di-n-octyl phthalate	7.001507	2621.431	146.1596	9.2	15.3	423.0154	0
Dodecane	6.898254	1664.768	198.1816	5.331716	15.3	14.33503	0
Endosulfan	9.318406	4269.167	273.16	3.55	15.3	3.015383	0
Endrin	9.681486	4883.984	273.16	5.6	15.3	18.1279	0
Epichlorohydrin	8.22943	2086.816	273.16	0.03	10.76	0.1385906	0
Epoxybutane 1,2	6.83185	1140.529	227.6696	1.441351	10.76	0.4764868	0
Ethane	6.82915	663.72	256.68	0.9913252	15.3	0.3213955	0
Ethanol	8.321	1718.21	237.52	−0.32	8.8	0.9	0
Ethanolamine(mono-)	7.456	1577.67	173.37	−0.773026	9.7	0.0686	0
Ethoxyethanol-2	7.874	1843.5	234.2	0.6863381	17.56	0.2461183	0
Ethyl acrylate	7.964455	1897.011	273.16	1.5014	17.56	0.5021918	0
Ethyl carbamate	7.364662	1737.454	204.4896	−0.447976	10.76	0.0912	0
Ethyl ether	6.92	1064.07	228.8	1.639246	17.56	0.5665672	0
Ethyl morpholine, ethyl diethylene oxime	6.894822	1636.683	199.7396	1.924971	15.3	0.7274908	0
Ethyl toluene, 4	6.875279	1479.869	208.4796	3.772945	15.3	3.664899	0
Ethyl vinyl ether	6.820354	1047.859	232.9896	1.645062	15.3	0.5694574	0
Ethylacetate	7.101	1244.95	217.88	0.9244851	17.58	0.303138	0
Ethylamine	7.05413	987.31	220	0.17986	9.7	0.1580087	0
Ethylbenzene	6.975	1424.255	213.21	3.15	6.8	2.1	0
Ethylene diamine	7.337209	1521.051	215.3196	−0.269471	9.7	0.1066435	0
Ethylene dibromide	7.34485	1675.301	244.82	1.6	10.76	0.5474415	0
Ethylene glycol	8.0908	2088.9	203.5	−0.914444	17.56	0.0606	0
Ethylene glycol dimethyl ether	7.232956	1377.29	232.43	1.134076	15.3	0.3641545	0
Ethylene glycol monobutyl ether	8.19856	2008.895	206.7696	1.554456	15.3	0.5260546	0
Ethylene glycol monobutyl ether acetate	7.2159	1659.2	191.339	1.48746	15.3	0.4961033	0
Ethylene glycol mono-ethyl ether acetate	7.377117	1641.85	209.2748	0.619354	15.3	0.2321078	0

TABLE 2.6 Compound Properties for Organic Compounds, II (*Continued*)

Compound name	A	B	C	LN(OW)	K_{max}	k_1	RT HYD
Ethylene glycol monoethyl ether cellosol	8.177948	1846.634	213.6096	0.2502244	15.3	0.1680427	0
Ethylene glycol monomethyl ether	8.171905	1799.523	215.6046	0.2522626	15.3	0.1683427	0
Ethylene glycol monomethyl ether acetate	7.52829	1641.851	209.2747	0.185323	15.3	0.1587658	0
Ethylene glycol monophenyl ether	8.235456	2309.119	194.2296	1.889739	15.3	0.705406	0
Ethylene glycol monopropyl ether	8.176798	1837.652	213.9896	1.120402	15.3	0.3598234	0
Ethylene oxide	7.128	1054.54	237.76	−0.3	4.2	0.91	0
Ethylene thiourea	7.211493	1731.503	201.5974	−0.641652	9.7	0.077	0
Ethylhexylacrylate,2-	6.884682	1554.707	204.2996	3.773185	15.3	3.665669	0
Ethylphenol,3-	7.468	1856	187	2.4	15.3	1.102401	0
Ethyl(2) hexanol	9.114945	2850.004	273.16	2.643529	17.56	1.364211	0
Fluoranthene	6.373	1756	118	5.33	31.1	1.5	0
Formaldehyde	7.195	970.6	244.1	1.94	5	0.25	0
Formamide	2.602059	19.62512	264.1977	−0.553912	17.56	0.0831	0
Formic acid	7.581	1699.2	260.7	−0.924103	17.56	0.0601	0
Freon 11, fluorotrichloromethane	6.816687	1017.234	234.7566	2.576381	15.3	1.286367	0
Freon 12 dichlorodifluoromethane	3.698946	0	0	2.16	10.76	0.8935916	0
Freon 12, dichlorodifluoromethane	6.795418	806.5787	247.0498	2.402047	15.3	1.104378	0
Fumaric acid	−1.118697	0	0	−0.445571	17.56	0.0914	0
Furan	6.975	1060.87	227.74	1.853527	17.56	0.6834056	0
Furfural	6.575	1198.7	162.8	1.578185	17.56	0.5370913	0
Generic organic material	6.859312	1354.71	215.5096	0	19	1.1	0
Glutaric acid	6.947832	2093.118	174.6596	0.0219418	17.56	0.1376169	0
Glycerin (glycerol)	6.165	1036	28	−1.432496	17.56	0.0385	0
Glyoxal	8.130124	1472.782	229.5696	−1.11614	15.3	0.0508	0
Glyphosate	8.207067	2076.833	203.9194	−0.494987	15.3	0.0875	0
Guthion	6.94768	2093.121	174.6594	2.708063	15.3	1.443459	0
Gylcidol	9.224102	2760.353	273.16	−0.381763	17.56	0.0966636	0
Heptachlor	8.961409	3722.308	273.16	5.05	10.76	11.20331	0
Heptane iso	6.8391	1335.41	237.2	3.162377	15.3	2.148051	0
Heptane(-n)	6.89677	1264.9	216.54	3.161653	15.3	2.146692	0
Hexachlorobenzene	9.55388	3248.572	203.07	5.47	10.76	16.17883	0
Hexachlorobutadiene	7.485238	1956.415	215.333	3.74	10.76	3.560762	0
Hexachlorocyclohexane (gamma isomer)	6.966292	2268.189	165.1596	0.936952	10.76	0.3064629	0
Hexachlorocyclopentadiene	8.41529	2834.547	273.16	3.99	10.76	4.431426	0
Hexachloroethane	7.2284	1347.978	132.91	3.456234	10.76	2.777865	0
Hexafluoroacetone	7.105094	665.0909	258.4496	4.214555	15.3	5.393552	0
Hexamethylene 1,6 diisocyanate	6.914242	1798.169	190.8096	1.27331	9.7	0.4113342	0
Hexamethylphosphoramide	6.90523	1722.428	194.9896	−0.000432	9.7	0.1349489	0
Hexane(-n)	6.876	1171.17	224.41	2.727623	15.3	1.468376	0

TABLE 2.6 Compound Properties for Organic Compounds, II (*Continued*)

Compound name	A	B	C	LN(OW)	K_{max}	k_1	RT HYD
Hexanoic acid	9.477544	3158.919	273.16	1.591822	17.56	0.5435379	0
Hexanol-1	7.86	1761.26	196.66	1.774724	17.56	0.6378716	0
Hexen-2-one 5	7.178988	1474.01	214.9396	1.344229	17.56	0.4376675	0
Hydrazine	8.683764	2243.776	273.16	−1.37	9.7	0.0407	0
Hydrogen sulfide	6.99392	768.13	249.09	1.173106	10.76	0.3768052	0
Hydroquinone	9.778832	3857.106	273.16	0.59	17.56	0.2262222	0
Hydroxyacetic acid	3.82111	459.9652	273.16	−1.097389	17.56	0.0516	0
Indeno(1,2,3-cd)-pyrene	9.813007	5609.269	273.16	7.66	9.7	109.9392	0
Isobutylene	6.805457	917.0536	240.5706	1.796808	15.3	0.6503171	0
Isodecanol	8.22165	2195.003	198.9796	2.920139	15.3	1.737775	0
Isophorone	7.962584	2480.726	273.16	1.7	15.3	0.5975001	0
Lindane hexachlorocyclohexane	6.915866	1811.968	190.0496	0.936952	10.76	0.3064629	0
Maleic acid	7.956923	2706.393	273.16	−0.445571	17.56	0.0914	0
Maleic anhydride	12.6834	4638.2	273.16	0	4.08	1	0
Methacrylic acid	11.26555	3657.094	273.16	0.6610121	17.56	0.2407243	0
Methanol	7.897	1474.08	229.13	−0.7	18	0.2	0
Methomyl	20.8751	7506.508	273.16	−0.301311	15.3	0.1037134	0
Methoxychlor	6.951463	2128.043	172.7596	6.571831	17.56	42.42731	0
Methyl 1-pentene 2	6.83529	1121.3	229.687	2.664926	15.3	1.389991	0
Methyl acetate	7.065	1157.63	219.73	−0.09	19.87	0.1247767	0
Methyl acrylate	6.838646	1193.779	224.6296	0.9331476	17.56	4.3	0
Methyl amine	7.3369	1011.5	233.3	−0.530324	9.7	0.0848	0
Methyl aziridine 2	0	0	0	0.6105353	9.7	0.2303237	0
Methyl benzyl alcohol 4	9.069858	3021.253	273.16	2.385993	17.56	1.088972	0
Methyl chloride	7.093	948.58	249.34	1.92	10.76	0.7243333	0
Methyl cholanthrene 3	−3.119754	0	0	7.430058	31.1	89.90299	0
Methyl cyclohexane	6.823	1270.76	221.42	2.963777	15.3	1.805412	0
Methyl ether dimethyl ether	3.59986	0	0	0.6286275	17.56	0.2339989	0
Methyl ethyl ketone, 2 butanone	7.11243	1305.006	229.2666	0.28	2	0.2	0
Methyl formate	3.027	3.02	−11.9	0.0940167	17.56	0.1465752	0
Methyl hydrazine	6.5762	1007.5	181.4	−0.478592	9.7	0.0888	0
Methyl iodide	7.657383	1507.3	273.16	1.69	0	0.5922948	0
Methyl isobutyl ketone	6.672	1168.4	191.9	1.38	0.74	0.4515826	0
Methyl isocyanate	6.822807	1067.989	231.8306	2.84454	9.7	1.626545	0
Methyl isopropyl ketone	5.483397	955.5712	273.16	0.9050759	15.3	0.2980333	0
Methyl mercaptan	6.808206	942.8276	239.0696	1.173106	15.3	0.3768052	0
Methyl methacrylate	6.5168	1052.176	188.37	1.295875	17.56	4.3	0
Methyl morpholine	7.173846	1433.993	217.0296	1.357002	15.3	0.4425861	0
Methyl napthalene(1-)	7.03592	1826.948	195	4.132105	24.03	5.018151	0
Methyl napthalene(2-)	7.0685	1840.264	198.4	4.132105	31.1	5.018151	0
Methyl parathion	−5.008742	0	0	2.04	15.3	0.8045235	0
Methylene chloride, dichloromethane	6.968411	1074.291	222.995	1.25	22	0.4	0
Methylene diphenyl diisocyanate	6.942754	2048.169	177.1085	3.308513	9.7	2.441051	0

TABLE 2.6 Compound Properties for Organic Compounds, II (***Continued***)

Compound name	A	B	C	LN(OW)	K_{max}	k_1	RT HYD
Methylene diphenylamine (MDA)	6.946537	2082.649	175.2296	3.379612	15.3	2.597734	0
Methylenedianiline 4,4	7.501696	2247.655	181.2885	2.633374	9.7	1.352142	0
Methylene-bis (2-chloroaniline),4,4′-	7.513116	2350.365	176.3696	4.900831	9.7	9.832447	0
Methylstyrene (-4)	7.0112	1535.1	200.7	3.463532	31.1	2.795661	0
Methyl-tertiary-butyl ether	6.85249	1103.737	222.72	1.90147	17.56	0.712684	0
Monomethyl formanide	7.204987	1679.083	204.2996	−0.678949	15.3	0.0745	0
Morpholine	7.71813	1745.8	235	−1.08	9.7	0.0524	0
Nabam	7.303247	2548.42	160.2196	−1.646165	15.3	0.0319	0
Naphthalene	7.3729	1968.36	222.61	3.37	42.47	1	0
Naphthol,alpha-	7.28421	2077.56	184	3.076284	17.56	1.992183	0
Naphthol,beta-	7.347	2135	183	2.84	17.56	1.620096	0
Naphthylamine,alpha-	7.49977	2230.587	182.1076	2.22	9.7	0.9417578	0
Naphthylamine,beta-	7.502136	2251.572	181.1006	2.25	9.7	0.9668059	0
Neopentyl glycol	8.219421	2176.788	199.7396	0.3583827	17.56	0.1847226	0
Nitro m xylene, 2	6.901502	1691.519	196.6996	3.067034	15.3	1.976124	0
Nitroaniline p	9.5595	4039.73	273.16	1.182549	9.7	0.3799313	0
Nitrobenzene	7.115	1746.6	201.8	1.84	11	2.3	0
Nitrobiphenyl,4-	6.947679	2093.118	174.6596	3.945698	19	4.262934	0
Nitroglycerin	9.636311	3601.771	273.16	0.6459801	9.7	0.2375789	0
Nitromethane	7.28166	1446.94	227.6	0.0639644	9.7	0.1427711	0
Nitrophenol,2-	11.86137	4392.938	273.16	1.79	9.7	0.6464548	0
Nitrophenol,4-	10.95784	4459.823	273.16	1.91	9.7	0.7180231	0
Nitropropane 2	7.271697	1530.544	228.62	1.283951	9.7	0.4151816	0
Nitrosodimethylamine n	7.5239	1974.062	273.16	−0.47	9.7	0.0894	0
Nitrosomorpholine	7.183389	1813.598	196.5096	−0.44	9.7	0.0918	0
Nitroso-n-methylurea n	7.362276	1718.378	205.4396	0.185811	9.7	0.1588336	0
Nitrotoluene (-p)	6.9948	1720.39	184.9	2.45499	9.7	1.156741	0
Nitrotoluene, m	6.903989	1712.12	195.5596	2.45499	15.3	1.156741	0
Nitrotoluene, o	6.901917	1694.951	196.5096	2.45499	15.3	1.156741	0
Nitrotoluene, o	6.901917	1694.951	196.5096	2.45499	15.3	1.156741	0
Nonanol, n	8.223319	2208.672	198.4096	3.076832	15.3	1.993139	0
Octane	6.918	1351.99	209.15	3.595688	15.3	3.138369	0
Octanol 1	12.0701	4506.8	319.9	2.642804	17.56	1.363345	0
Octanol 2	6.3888	1060.4	122.5	2.643529	17.56	1.364211	0
Octanol 3	5.2215	560.3	64.7	2.643529	17.56	1.364211	0
Octanol 4	5.7396	760.5	89.5	2.643529	17.56	1.364211	0
Oil	0	0	0	5.331716	15.3	14.33503	0
Oxalic acid	16.86465	5987.326	273.16	−1.280351	17.56	0.044	0
Paraldehyde	7.286656	1754.238	273.16	2.303315	17.56	1.012976	0
Parathion	9.10235	4032.563	273.16	3.055472	15.3	1.956233	0
PCBs (aroclors)	−6.785018	−5599.125	273.16	7.308553	19	80.83541	0
Pentachlorobenzene	9.054626	3396.595	273.16	5.96656	10.76	24.98281	0
Pentachloroethane	6.74	1378	197	2.718297	10.76	1.456443	0
Pentachloronitrobenzene	6.954095	2152.518	171.4296	5.645749	10.76	18.86828	0
Pentachlorophenol	8.119069	3054.752	273.16	5.01	130	3.4	0
Pentadiene 1,2	6.9182	1104.991	228.85	2.068878	15.3	0.8251115	0

TABLE 2.6 Compound Properties for Organic Compounds, II (***Continued***)

Compound name	A	B	C	LN(OW)	K_{max}	k_1	RT HYD
Pentaerythritol	8.229961	2263.417	196.1296	−1.547323	15.3	0.0348	0
Phenol	7.133	1516.79	174.95	1.46	97	13	0
Phenylcyclohexanone 4	−0.708	106.4	−146.6	3.692797	17.56	3.41669	0
Phenylene diamine(-m)	7.984773	2843.732	273.16	0.8427868	9.7	0.2822246	0
Phenylene diamine(-o)	9.250531	3383.349	273.16	0.8427868	9.7	0.2822246	0
Phenylene diamine(-p)	9.309744	3472.661	273.16	0.8427868	9.7	0.2822246	0
Phenylphenol P	8.6575	3022.8	216.1	3.822525	17.56	3.827388	0
Phosgene (decomposes)	6.842	941.25	230	1.179	15.3	0.3787535	0
Phosphine	3.301009	0	0	2.344077	15.3	1.049757	0
Phthalic acid	7.879505	1728.354	273.16	0.8225744	17.56	0.2772771	0
Phthalic anhydride	8.022	2868.5	273.16	−0.62	17.56	0.0784	0.000555
Phthalimide	8.300783	1857.213	273.16	0.8427868	9.7	0.2822246	0
Pinene(alpha-)	6.8525	1446.4	208	3.937396	15.3	4.232081	0
Piperazine	8.57357	2386.184	273.16	0.4268957	15.3	0.196135	0
Propane	6.80338	804	247.04	1.425412	15.3	0.4698876	0
Propane sultone,1,3-	7.256331	2110.189	182.2596	−2.82	15.3	0.0114	0
Propanoic acid	6.866276	1409.059	212.4506	0.2895945	17.56	0.1739324	0
Propanol	7.84767	1499.21	204.64	0.4913113	17.56	0.2075073	0
Propanol iso	8.117	1580.92	219.61	−0.16	15	0.075	0
Propene	6.794065	787.2845	248.1896	1.43124	15.3	0.4722896	0
Propiolactone b	7.982566	2221.617	273.16	0.4909006	17.56	0.2074327	0
Propionaldehyde	7.2088	1235.771	237.47	0.691664	17.56	0.2472678	0
Propionic acid	6.403	950.2	130.3	0.2895945	17.56	0.1739324	0
Proporur (Baygon)	7.236959	1942.436	190.7885	2.322595	9.7	1.030209	0
Propyl acetate iso	6.843416	1231.175	222.5016	1.359286	17.56	0.4434716	0
Propyl amine iso	11.00768	2488.127	273.16	0.894336	9.7	0.2952457	0
Propyl ether iso	6.834991	1165.131	226.2636	2.653113	17.56	1.375699	0
Propylene	6.77811	770.85	245.51	1.43124	15.3	0.4722896	0
Propylene chlorohydrin	8.176797	1837.652	213.9896	−0.003456	15.3	0.1345924	0
Propylene glycol	8.208243	2085.9	203.5396	−0.479632	17.56	0.0887	0
Propylene oxide	7.0671	1133.267	236.1054	0.235614	17.56	0.1659081	0
Propylenimine 1,2 2 methyl aziridine	7.1502	1249.718	226.7196	0.6105354	9.7	0.2303237	0
Propyl(-n) acetate	7.016	1282.28	208.6	1.358564	17.56	0.4431913	0
Propyl(-n) benzene	6.95142	1491.297	207.14	3.594934	31.1	3.1363	0
Propyn-1-ol 2(proparlgyl)	8.993223	2364.043	273.16	0.2602102	17.56	0.1695174	0
Pyridine	7.041	1373.8	214.98	0.65	35.03	0.238416	0
Quinoline	7.3838	2300.405	273.16	2.03	31.1	0.7975147	0
Quinone	7.148885	1239.346	227.2685	1.016473	9.7	0.3285457	0
Resorcinol	6.924308	1884.547	186.0596	0.8	17.56	0.271854	0
Silvex	0	0	0	4.072398	15.3	4.76272	0
Sodium formate	−3.119232	0	0	−0.439411	0	0.0919	0
Styrene	6.945357	1437.432	208.38	3.16	31.1	0.11	0
Styrene oxide	6.888923	1588.822	202.3996	2.644778	17.56	1.365701	0
Succinic acid	−3.118326	0	0	−0.412145	17.56	0.0941278	0
Tamaron (methamidiphos)	7.309743	1307.183	226.1496	0.5316042	15.3	0.2149536	0
Terephthalic acid	6.930848	1941.731	182.9246	0.8225744	15.3	0.2772771	0
Terpineol, alpha	8.180823	1869.102	212.6596	3.182377	15.3	2.185972	0
Tetraorobenzene (1,2,3,4)	8.872679	3158.681	273.16	5.206064	10.76	12.84254	0

TABLE 2.6 Compound Properties for Organic Compounds, II (***Continued***)

Compound name	A	B	C	LN(OW)	K_{max}	k_1	RT HYD
Tetrachlorobenzene(1,2,3,5)	8.821956	3084.413	273.16	5.206064	10.76	12.84254	0
Tetrachlorobenzene(1,2,4,5)	8.821956	3084.413	273.16	4.51	10.76	6.984653	0.42
Tetrachlorodibenzo-p-dioxin(2,3,7,8)	6.977444	2377.268	159.2696	6.64	10.76	45.03498	0
Tetrachloroethane(1,1,2,2)	6.893793	1354.506	192.43	2.56	6.2	0.68	0
Tetrachloroethene	6.976	1386.92	217.53	2.6	6.2	0.68	0
Tetrachlorophenol(2,3,4,6)	9.168764	2748.848	273.16	4.762268	10.76	8.709778	0.35
Tetrachlorophenol(2,3,5,6)	−1.999987	0	0	4.762268	10.76	8.709778	0
Tetraethyl lead	8.56581	2689.922	273.16	2.392857	0	1.095533	0
Tetraethyldithiopyrophosphate	11.84911	5454.172	273.16	1.656811	15.3	0.5753421	0
Tetraethylene pentamine	7.386298	1913.731	195.7494	0.1216321	15.3	0.15016	0
Tetrafluoromethane	6.882466	386.9427	272.6996	2.89165	15.3	1.694992	0
Tetrahydrofuran	6.995	1202.29	226.25	1.440628	17.56	0.4761852	0.604
Tetralin	7.07055	1741.3	208.26	3.763707	15.3	3.635396	0
Tetranitromethane	8.350471	2183.276	273.16	0.4822038	9.7	0.2058602	0
Thiourea	4.247091	621.8834	273.16	0.8612069	9.7	0.2868102	0.46
Thiourea,1-(o-chlorophenyl)-	11.46513	5117.637	273.16	2.166816	9.7	0.898937	0
Toluene	6.954	1344.8	219.48	2.69	73.48	2.4	0
Toluene diamine(2,4)	7.492203	2164.173	185.2996	1.454882	9.7	0.4821615	0.07
Toluene diisocyanate(2,4)	6.912613	1784.378	191.5696	6.803719	9.7	51.97128	0
Toluenediamine(2,6)	0	0	0	1.454882	9.7	0.4821615	0
Toluenediamine(3,4)	7.483529	2089.257	188.9096	1.454882	9.7	0.4821615	0
Toluenesulfonyl chloride	6.930257	1936.527	183.2096	2.188507	15.3	0.9161608	0
Toluic acid (para-)	10.50106	4177.125	273.16	2.105137	17.56	0.8517082	0.81
Toluic aldehyde	9.005018	2922.234	273.16	2.285832	17.56	0.9975978	0
Toluidine m	7.454377	1846.317	200.6896	2.115259	15.3	0.8592852	0
Toluidine P	7.77229	2317.386	273.16	1.39	31.1	0.4555512	0
Toluidine (-0)	7.19724	1682.94	191.138	2.115272	31.1	0.8592948	0
Toxaphene	5.707643	1857.691	273.16	3.3	15.3	2.422935	0
Tributyl phosphorotrithioate SSS	6.901398	2987.955	273.16	4.803134	17.56	9.026857	0
Tributyl tin acetate	6.91708	1822.323	189.4796	5.463342	15.3	16.08485	0
Tributylphosphate	6.0038	1755.628	273.16	5.430991	17.56	15.63593	0
Trichlorobenzene 1,2,3	6.900254	1681.226	197.2696	4.445513	10.76	6.601452	0
Trichlorobenzene 1,2,4	7.7056	2242.67	252.836	3.98	10.76	4.392822	0.55
Trichlorobenzene 1,3,5	8.598745	2754.108	273.16	4.445513	10.76	6.601452	0
Trichloroethane 1,1,1 methyl chloroform	6.827401	1147.14	218.5387	2.49	3.5	0.74	0.39
Trichloroethane 1,1,2	7.1921	1480.319	229.0943	2.17	3.5	0.74	0
Trichloroethylene	6.518	1018.6	192.7	2.29	3.9	0.88	0.12
Trichlorofluoromethane	6.884	1043.004	236.88	2.53	10.76	1.235208	0
Trichlorophenol 2,4,5	8.577468	2974.575	273.16	4.001704	15.3	4.477043	0
Trichlorophenol 2,4,6	9.696309	3528.119	273.16	3.69	15.3	0.26	0
Trichloropropane 1,1,1	11.56902	3302.916	273.16	0.2066454	10.76	0.1617557	0.07
Trichloropropane(1,1,2)	8.218203	2205.203	273.16	1.654597	10.76	0.5742285	0.009
Trichloropropane(1,2,2)	11.14517	3282.281	273.16	1.426414	10.76	0.4702994	0
Trichloropropane(1,2,3)	6.903	788.2	243.23	0.2066454	10.76	0.1617557	0

TABLE 2.6 Compound Properties for Organic Compounds, II (***Continued***)

Compound name	A	B	C	LN(OW)	K_{max}	k_1	RT HYD
Trichloro-1,2,2-trifluoroethane,1,1,2-	7.637469	1526.226	273.16	3.456217	10.76	2.777823	0
Triethanolamine	8.287391	2762.499	175.5336	−1.75	9.7	0.0291	0
Triethylamine	6.958867	1271.96	223.262	2.359257	9.7	1.063793	0
Triethylene glycol dimethyl ether	6.898169	1664.082	198.2196	2.429506	15.3	1.131232	0
Trifluoroethane(1,1,1)	3.965647	0	0	2.703412	10.76	1.437597	0
Trifluralin	6.937691	2002.526	179.5996	5.355146	9.7	14.63195	0
Triisobutylene	6.894822	1636.683	199.7396	5.303749	15.3	13.98851	0
Triisopropylamine	6.900255	1681.23	197.2694	4.965421	15.3	10.40413	0
Trimethyl benzene, 123	6.880847	1524.055	206.0096	3.999725	15.3	4.469296	0
Trimethylbenzene (1,3,5)	7.07436	1569.22	209.58	3.999725	31.1	4.469296	0
Trimethylpentane 2,2,4	6.797857	1249.485	219.7695	3.567022	15.3	3.06063	0
Trinitrotoluene(2,4,6)	−1.337234	0	0	0	4.4	0.45	0
Tripropylene glycol	8.236003	2313.693	194.0396	2.126133	15.3	0.8675004	0
Urea	8.555151	2304.696	273.16	−0.469833	9.7	0.0894	0.15
Urethane	7.421	1758.21	205	0.4310978	9.7	0.1968575	0.51
Vinyl acetate	7.21	1296.13	226.66	0.930263	17.56	0.3046745	0
Vinyl acetylene	6.790945	726.4801	251.7996	1.652999	15.3	0.5734261	0
Vinyl bromide	6.814487	998.5032	235.8396	1.066461	10.76	0.343235	0
Vinyl chloride	6.9907	969.0518	250.5856	0.06	10.76	0.1422767	0.88
Vinylidene chloride (see 1,1dichloroethen)	6.972	1099.4	237.2	0	10.76	0.9039771	0
Warfarin	9.979672	4466.33	273.16	5.387373	15.3	15.05042	0
Xylene	7.940135	2090.317	273.16	3.15	40.8	1.8	0.091
Xylene(-m)	7.009	1462.266	215.11	3.2	31.1	2.219942	0.07
Xylene(-o)	6.998	1474.679	213.69	2.95	40.79	1.8	0.12
Xylene(-p)	6.99	1453.43	215.31	3.15	31.1	2.124915	0
Xylenol(3,4)	7.07919	1621.45	159.26	0.1	17.56	0.1473445	0
Xylidine dimethylaniline	7.461541	1904.87	197.8396	2.727331	15.3	1.468	0

REFERENCES

Environmental Protection Agency, *Air Emissions Models for Waste and Wastewater,* EPA-453/R-94-080a, 1994.

Hansch C., and A. Leo, *Exploring QSAR: Fundamentals and Applications in Chemistry and Biology,* American Chemical Society, 1995.

Hine, J., and P. K. Mookerjee, "The Intrinsic Hydrophilic Character of Organic Compounds Correlation in Terms of Structural Contributions," *Journal of Organic Chemistry,* 40:292–298, 1975.

Nirmalakhandan, N. N., and R. E. Speece, "QSAR Model for Predicting Henry's Law Constant," *Environmental Science and Technology,* 22:1349–1357, 1988.

Okey, R. W., and H. D. Stensel, "A QSAR-Based Biodegradability Model—A QSBR," *Water Research,* 30(9), 1996.

CHAPTER 3
HEALTH EFFECTS AND EXPOSURE ASSESSMENTS OF VOCs

Linda Forst
Division of Environmental and Occupational Health Sciences and Department of Emergency Medicine, University of Illinois at Chicago

Lorraine M. Conroy
Division of Environmental and Occupational Health Sciences, University of Illinois at Chicago

VOLATILE ORGANIC COMPOUNDS 3.1
- Sources of Exposure 3.1
- VOCs in the Body 3 1
- Chemical Groups 3.2
- Principles of Toxicology 3.3

EXPOSURE ASSESSMENT 3.4
- Basic Characterization 3.4
- Qualitative Risk Assessment and Prioritization 3.5
- Environmental Monitoring 3.6
- Interpretation and Decision Making 3.8
- Recommendations for Control 3.10

HEALTH EFFECTS 3.13
- Organ Systems and Toxicant Susceptibility 3.13
- Toxicity of Specific VOCs 3.17
- Indoor Air Pollution 3.24

RESOURCES 3.25

REFERENCES 3.27

VOLATILE ORGANIC COMPOUNDS

Sources of Exposure

Exposure to volatile organic compounds (VOCs) occurs in the workplace. They are also the most common accidental release in the outdoor environment. One study examined the blood of individuals living close to hazardous waste sites and found no difference in VOC levels between the exposed population and controls (Hamar, 1996). There are ongoing studies to look at citizens living near toxic waste sites in order to determine blood VOC levels and to follow them over time to detect potential adverse health effects (Gist, 1995).

VOCs in the Body

By definition, it is the volatility of these organic compounds which renders them most susceptible to inhalation by exposed workers. This most commonly occurs in manufacturing or other industries which rely on the VOCs to perform a number of functions, including degreasing, lubricating, fueling, stripping or thinning paint, and cleaning. Recently, the role of VOCs in indoor air pollution and attendant symptoms is being examined. Ingestion is rare in occupational settings, although it has occurred because of mislabeled containers and in suicide attempts. The question of ingestion is more frequently entertained in environmental settings where, e.g., the water supply may be contaminated by local industry or by a toxic waste site.

Because of their lipid solubility, VOCs are readily absorbed by the lung, gastrointestinal tract, and, in liquid form, the skin. They may bioaccumulate in lipid tissues in the body. In general, most of the internal dose is eliminated quickly, although a small fraction is eliminated more slowly. Cigarette smoking is a major source of VOCs found in the body and is the major confounder in studies of workers and potentially exposed citizens (Ashley, 1996).

Chemical Groups

VOCs are organic hydrocarbons that can volatilize to form an organic vapor in air. Hydrocarbons are molecules that consist of hydrogen and carbon. Hydrocarbons with 1 to 4 carbon atoms exist in the gas phase at ordinary temperature and pressure. Hydrocarbons with 5 to 12 carbon atoms exist in the liquid or solid phase at ordinary temperature and pressure. Hydrocarbons with more than 12 carbon atoms do not volatilize sufficiently to reach atmospheric concentrations in the gas phase (Waldbott, 1973).

Hydrocarbons can be characterized into groups based on the carbon bonds formed in the molecule. Alkanes are hydrocarbons in which all the carbon atoms form single bonds. Alkanes conform to the general formula C_nH_{2n+2}, where n is the number of carbon atoms. Examples of alkanes are methane, ethane, propane, butane, and pentane (Mortimer, 1971).

Cycloalkanes are ring structures with only single carbon bonds. Examples of cyclohexanes are cyclopropane and cyclohexane. Petroleum is the principal source of alkanes and cycloalkanes (Mortimer, 1971).

Alkenes or olefins are aliphatic hydrocarbons with a double-bonded carbon-carbon in the molecule. Examples of alkenes are ethene or ethylene, propene, and butene. Alkanes are considered saturated compounds because all the valance electrons of each carbon atom are single-bonded. Alkenes are considered unsaturated because of the double bond. Alkenes have the general formula C_nH_{2n}.

Alkynes are molecules with carbon-carbon triple bonds. The general formula is C_nH_{2n-2}. Examples of alkynes include alkyne or acetylene, butyne, and pentyne. Aromatic hydrocarbons are ring structures based on that of benzene (C_6H_6). The six carbon atoms are arranged in a ring with one hydrogen bond for each carbon. Substitution of one or more hydrogen atoms with methyl or ethyl groups results in other aromatic hydrocarbons. Examples include toluene ($C_6H_5CH_3$); *o*-, *m*-, and *p*-xylene [$C_6H_4(CH_3)_2$]; and ethylbenzene ($C_6H_5CH_2CH_3$). A phenyl radical is formed when one hydrogen atom is removed from a benzene molecule. Several rings can be fused together to form other molecules. One example is naphthalene ($C_{10}H_8$) where two rings are fused (Mortimer, 1971).

Chlorinated or brominated hydrocarbons are those in which a hydrogen atom has been replaced with a chlorine or bromine atom. Alcohols are those molecules in which a hydrogen atom is replaced by a hydroxyl group. The elimination of one water molecule from an alcohol molecule results in formation of an olefin compound. Elimination of one water molecule from two molecules of alcohol results in formation of an ether molecule (Mortimer, 1971).

Carbonyl compounds are those in which a carbon atom forms a double bond with an oxygen atom. If the carbonyl group is bonded to one alkyl group and one hydrogen molecule or to two hydrogen atoms, an aldehyde is formed. Examples include formaldehyde

(COH_2) and acetaldehyde ($COHCH_3$). If the carbonyl group is bonded to two alkyl groups, a ketone is formed (Mortimer, 1971). Examples include acetone [$CO(CH_3)_2$] and methyl ethyl ketone. Oxidation of the aldehyde group results in formation of a carboxyl group. Compounds containing the carboxyl group are weak acids.

Amines are compounds derived from ammonia with one or more of the hydrogen atoms replaced by alkyl groups. Examples include methylamine, dimethylamine, and trimethylamine. Amides are compounds produced when ammonia reacts with carboxylic acids. One example is acetamide ($CONH_2CH_3$).

Principles of Toxicology

To understand adverse health effects of volatile organic compounds, it is essential to understand basic principles and definitions of toxicology.

Pharmacokinetics refers to the quantitative time course of toxins in the body and depends on *absorption, distribution, biotransformation,* and *excretion.*

Absorption of toxic chemicals by exposed individuals can occur via three different *routes of entry.* These include *inhalation* of gases, vapors, and particulates into the respiratory tract; *ingestion* into the gastrointestinal tract; and *dermal* (*skin*) *absorption* of a liquid. Logically, the more volatile the compound, the more likely it is to be inhaled. VOCs, by definition, are most likely to be inhaled in the occupational setting. Skin absorption of the liquid form is also common. Ingestion of these compounds occurs in environmental situations, including water contamination and accidental or intentional ingestion. The more fat-soluble a compound, the more likely it is to be absorbed into the body by all three routes.

Chemical toxins can exert their effects at the point of entry with direct, *local irritation*; others are absorbed into the bloodstream and act at distal sites by (1) direct toxicity to a susceptible *target organ*; (2) by *sensitization* or *allergy,* generally after multiple episodes of exposure, causing a reaction manifested by any or all of the following: skin rash, bronchospasm (an asthmalike picture), hypotension, and sudden death; or (3) by an *idiosyncratic reaction* which has to do with susceptibility based on the individual's genetic makeup; this type of reaction may initiate a cascade of biochemical and microanatomical changes that manifest in a toxic syndrome.

The *distribution* of a chemical toxin in the body is based on the charge/fat solubility, the size of the molecule or ion, and its interaction with naturally occurring enzymes, protein carriers, and other chemicals in the body. A toxin's distribution greatly influences its toxicity. For example, some chemicals are absorbed by organs they do not damage (such as lead in bone), thereby reducing their effective circulating dose in the body. Others concentrate in their target organ, where they exert their adverse effects.

Many chemicals undergo *biotransformation* in the body by existing enzyme systems. This process can render them more or less toxic, and can form charged or uncharged molecular structures which determine how a given chemical will be handled by the body. Fat-soluble molecules may *bioaccumulate,* or concentrate, in fatty tissues of the body where quantity can grow over time; a stressor, such as rapid weight loss, can cause a sudden increase in circulating levels of toxicants in the body.

The more rapidly a chemical is *eliminated* from the body, most frequently in the kidney or gastrointestinal tract, the less time it has to exert its toxic effect. The elimination *half-life* refers to the amount of time required for excretion of one-half of the internal exposure concentration. Fat-soluble substances, which tend to be absorbed in fatty tissue, are less available for *excretion.* They may be excreted slowly over time, and only if transformed to a water-soluble product.

Acute toxicity refers to short-term adverse health effects; *chronic toxicity* refers to chronicity of exposure which results in a delayed toxic manifestation. The occurrence of the latter is presumably due to accumulation of dose over time or to the eventual override of the *host's* (exposed individual's) natural defense mechanisms.

The quantity and duration of exposure determine the *dose* assumed by the host. In general, there is a *dose-response relationship* between toxin exposure and symptom or disease manifestation; i.e., the greater the dose, the more likely it is that an adverse effect will be seen (and the more severe the effect, as the dose increases). For many chemicals, there appears to be a *threshold,* frequently expressed in milligrams (or quantity) of toxin per kilogram of the host's body weight, above which toxic effects are first manifested. This threshold value is best described for pharmaceuticals which have been carefully tested in animal models as well as humans; for industrial or environmental chemicals, this threshold is derived from animal studies and from case reports of intoxicated humans where exposure levels have been known. There may be no threshold below which carcinogenicity will not occur.

EXPOSURE ASSESSMENT

There are several reasons for performing an occupational exposure assessment. The American Industrial Hygiene Association (AIHA) (1991) gives "to assess potential health risks faced by all workers, to differentiate between acceptable and unacceptable exposures, and to control unacceptable exposures" as the first such reason. Other reasons include establishing and documenting an historical record of exposure levels and ensuring and demonstrating compliance with legal standards.

The industrial hygienist is concerned with solving occupational health problems by a three-step approach:

1. *Recognition* of the problem or potential hazard
2. *Evaluation* of the potential hazard
3. *Control* of the hazard

These three steps involve answering three questions:

1. What are the potential health hazards?
2. Are the potential health hazards really hazardous?
3. How can the hazard be eliminated or minimized?

The steps involved in an occupational exposure assessment (AIHA, 1991) are as follows: (1) basic characterization of the workplace, workforce, and environmental agents; (2) qualitative risk assessment and prioritization; (3) environmental monitoring; (4) interpretation and decision making; (5) recommendations and reporting; and (6) reevaluation.

Basic Characterization

The basic characterization step involves understanding the process leading to potential exposures; developing an inventory of chemical, physical, and biological agents; characterizing the workforce including job descriptions, tasks, and number of workers; and characterizing the environmental agents by examining the available health effects data and current regulations and exposure limits.

Potential health hazards can be *biological* (bacteria, viruses, rickettsia, fungi, parasites, enzymes, and products of genetic engineering); *chemical* [aerosols (dusts, mists, fumes), gases, and vapors in air, liquids, and solids which contact the skin]; or *physical* (heat, cold, noise, vibration, acute and cumulative trauma, ionizing radiation, and nonionizing radiation).

Potential health hazards are usually identified during a walk-through or preliminary survey. The survey involves (1) becoming familiar with processes; (2) maintaining an inventory of physical, chemical, and biological agents encountered; (3) reviewing the different job activities of a work area; (4) recording the number of employees at each job or station; and noting (5) safety supervision and controls; (6) first aid and medical services; (7) the different types of operations conducted; (8) raw materials; (9) processing aids; (10) products; (11) by-products; (12) intermediate chemicals; (13) methods of solid, liquid, and gaseous waste disposal; (14) plant layout showing location of operations or process equipment; (15) sources of radiant heat; (16) abnormal temperatures and humidity; (17) excessive noise; (18) improper or inadequate illumination; (19) ultraviolet (uv) radiation; (20) microwaves; (21) lasers; and (22) X and gamma rays.

The preliminary survey should also note the types of control measures and their effectiveness. Control measures include local exhaust ventilation, general ventilation, protective respiratory devices, protective clothing, and shielding from heat, uv light, or other forms of radiant energy. General guides to effectiveness include the presence of dust on floors and ledges, holes in ductwork, fans not operating or operating in wrong direction, and manner in which personal protective equipment is treated by workers (worn, maintained, cleaned, stored).

The survey is best accomplished by following the industrial process through the plant from raw materials to finished product. The industrial hygienist should be accompanied by someone familiar enough with the process to answer questions and explain the process or steps that may not be evident. This person could be the production superintendent, chief process engineer, plant chemist, department supervisor, or safety director. The nurse or physician may be able to point out departments or areas which have had problems.

There may be contributory or intermittent operations not in evidence at the time of the survey. These may include preliminary treatment of raw materials, disposal of by-products or waste products, periodic maintenance and rebuilding, and warehouse operations. It is important to investigate these procedures or operations as well as proposed new operations or pilot plant work.

The preliminary survey also serves to identify locations where, or workers whose, exposures should be evaluated to determine if additional controls are needed. All workers are assigned to a similar exposure group based on the basic characterization. The similar exposure group may be developed using a job description approach, a task-based approach, a chemical-based approach, or some combination of these. A job description approach would group all workers with the same job description (e.g., chemical operator or printer) into one similar exposure group. A task-based approach would group all workers performing the same tasks into one similar exposure group (e.g., sample collection). A chemical-based approach treats all workers exposed to a single chemical as a similar exposure group (e.g., benzene may be a priority chemical). Similar exposure groups may be best classified by using a combination of approaches; e.g., all chemical operators who work with benzene may be classified together.

Qualitative Risk Assessment and Prioritization

The purpose of the qualitative risk assessment is to prioritize hazards for quantitative evaluation using environmental sampling. The qualitative assessment involves estimating the exposure and evaluating potential health effects. Exposure scenarios where there is a high exposure rating and a high health effects rating will have higher priority for further evaluation and control than those conditions with a low exposure rating and/or a low health effects rating.

A distinction must be made between *toxicity* and *hazard*. The toxicity of a substance describes the nature, degree, and extent of undesirable effects. It is a basic biological property of a material and reflects its inherent capacity to produce injury. Hazard describes the likelihood of this toxicity's being manifest, the probability or likelihood of injury resulting from actual use of the substance in the quantity and manner proposed.

In evaluating the hazard of a particular substance or agent, it is necessary to know not only its toxicity but also the dose. It is possible to have a high degree of hazard associated with a low order of toxicity and a low degree of hazard associated with a high order of toxicity depending on the dose. An example of a high-toxicity but low-hazard condition is the use of a mercury-in-glass thermometer. Mercury is a highly toxic metal, but use of a thermometer involves a small amount of mercury sealed in glass. When used properly, the hazard is negligible. An example of a low-toxicity but high-hazard situation is the use of nitrogen for blanketing chemical storage tanks. Nitrogen is a nontoxic gas, constituting 80 percent of the atmosphere; however, in concentrations high enough to displace oxygen, asphyxiation becomes possible.

The exposure rating can be based on observations and professional judgment regarding the quantity of material used, the manner in which the agent is used, the number of people affected, etc. Other methods for estimating exposure include use of previous monitoring data, published data for similar processes, and modeling. Mass-balance modeling on a particular chemical can be used. Often, to simplify the modeling, assumptions of steady-state or worst-case conditions are made. An example of this type of modeling is shown in Fig. 3.1.

Environmental Monitoring

The next step is to determine the intensity of exposure. Dose can be divided into two parts: *exposure* of individuals to the substance or agent and *absorption* into tissue. Exposure is easier to measure than dose. As a result, evaluation is usually based on measurement of exposure. Methods of determining exposure level vary with type of hazard.

The evaluation of a potential hazard can be divided into two steps: measurement of the substance or agent and determination of whether the measured level constitutes a hazard.

To determine airborne exposure to chemical agents, including volatile organic compounds, air samples are collected and analyzed. Air sampling will give a measure of *concentration* of a particular contaminant in air, say, milligrams per cubic meter or parts per million. With air sampling, several questions must be answered:

1. *Where to sample?*
   a. *Personal sampling* involves attaching monitoring equipment near the worker's breathing zone. The inlet to the sampler is usually clipped to the worker's lapel or collar.
   b. *Area sampling* is used when direct measurements of worker exposure are not necessary. Area sampling can be used to gather information about the general air quality in a given space, to evaluate control methods, and as an indirect assessment of worker exposure.
2. *Whom to sample?* Personnel exposure is best determined by monitoring a representative sample from each of the similar exposure groups identified and prioritized in the earlier steps, i.e., representative workers for each job task. It may be necessary to sample workers at a particular operation thought to cause problems. Some standards require routine monitoring of workers exposed to a certain chemical.
3. *How long to sample? How many samples to take?* A *single sample* can be collected over the entire work shift. Such a sample gives an 8-h time-weighted average (8-h

a. Carbon disulfide is used in a laboratory to desorb organic compounds from charcoal sampling tubes prior to gas chromatography. An accidental spill of 0.5 L of CS_2 occurs in the laboratory. The laboratory is 7 m wide by 10 m long by 3 m high. A worst-case estimate of the resulting concentration can be made if we assume there is no ventilation and all the CS_2 evaporates.

$$C = \frac{\text{spill volume L}}{\text{room volume m}^3} \times \frac{1000\text{ mL}}{\text{L}} \times \frac{\text{sp gr g/mL}}{\text{g/mL H}_2\text{O}} \times \rho_{\text{H}_2\text{O}}\text{ g/mL} \times \frac{\text{gmol}}{\text{MW g}} \times \frac{\text{molar volume L}}{\text{gmol}} \times \frac{\text{m}^3}{1000\text{ L}} \times 10^6$$

$$= \frac{0.5\text{ L}}{210\text{ m}^3} \times \frac{1000\text{ mL}}{\text{L}} \times \frac{1.26\text{ g/mL}}{\text{g/mL H}_2\text{O}} \times 1\text{ g/mL H}_2\text{O} \times \frac{\text{gmol}}{76.1\text{ g}} \times \frac{24.45\text{ L}}{\text{gmol}} \times \frac{\text{m}^3}{1000\text{ L}} \times 10^6 = 964\text{ ppm}$$

b. Trichloroethylene is used as a vapor degreasing solvent. About 20 L of solvent is used in one 8-h shift. The shop is 10 m by 20 m by 5 m, and the effective air exchange rate is approximately 0.5. The steady-state concentration can be used as an upper estimate of 8-h TWA concentration. Steady-state concentration can be estimated using a completely mixed space model.

$$C_{ss}\text{ (ppm)} = \frac{G\text{ (m}^3\text{/min)}}{kQ\text{ (m}^3\text{/min)}} \times 10^6$$

where C_{ss} = steady-state concentration
G = solvent vapor generation rate
Q = ventilation rate
k = mixing factor (to account for incomplete mixing, typically 0.1 to 0.3)

$$G = \frac{20\text{ L}}{8\text{ h}} \times \frac{1\text{ h}}{60\text{ min}} \times \frac{1000\text{ mL}}{1\text{ L}} \times \frac{1.46\text{ g/mL}}{\text{g/mL H}_2\text{O}} \times 1\text{ g/mL H}_2\text{O} \times \frac{\text{gmol}}{131.4\text{ g}} \times \frac{0.02445\text{ m}^3}{\text{gmol}} = 0.0113\text{ m}^3\text{/min}$$

$$\frac{kQ}{V} = \frac{0.5}{\text{h}} \rightarrow kQ = 0.5 \times 10\text{ m} \times 20\text{ m} \times 5\text{ m} = 500\text{ m}^3\text{/min}$$

$$C_{ss} = \frac{0.0113\text{ m}^3\text{/min}}{500\text{ m}^3\text{/min}} \times 10^6 = 23\text{ ppm}$$

FIGURE 3.1 Modeling calculations.

TWA) concentration but gives no information on temporal or spatial variation. Taking *consecutive samples* over the entire work shift gives the 8-h TWA and some indication of temporal variation. If the location of the worker during each sampling period is also recorded, some indication of spatial variation may also be possible, although area sampling at various locations will give a better indication of spatial variation. Consecutive sampling takes more effort and is more costly. Sampling can also take place for shorter times, not necessarily covering the entire work shift. These samples can be useful in determining *short-term exposures.*

If the sampling is done to establish compliance with *short-term exposure limits,* the sample should be taken to represent the worst case. Sampling to determine compliance with 8-h TWA standards must take place over the entire work shift.

4. *When to sample?* It has been found that in many situations, airborne concentrations may be different for each shift, and sampling should be done during each to deter-

TABLE 3.1 Summary of Contaminant Collection Devices

Air contaminant			Collector	
Gases	High H_2O solubility	Acid	Absorber	NaOH
		Base	Absorber	H_2SO_4
	Low H_2O solubility		Absorber	Specific reagent
			Badge	
	Inert		Container	Flexible
				Rigid
Vapor	Nonpolar		Adsorber	Charcoal
				Porous polymer
				Badge
	Polar		Adsorber	Silica gel
				Porous polymer
				Badge
Particulate	Total		Filter	Membrane
				Fiber
	Size fraction		Cyclone and filter	Membrane
				Fiber
			Cascade impactor	
			Impinger	
			Filter and microscopy	Membrane
				Nucleopore

mine hazard potential. Also where large seasonal temperature differences exist, sampling should be done in both winter and summer. Normally there is more general ventilation in summer, a factor which tends to dilute the concentration of the contaminant.

5. *How to sample?* Air sampling involves the use of a sampling train. At a minimum, the air sampling train must have a collector, airflow controller, and air mover. The sampling train may also include flow-measuring devices, timers, water traps, etc. The pieces of the train are usually connected using Tygon or Teflon tubing. Selection of an appropriate collector depends on the contaminant being evaluated. Table 3.1 gives some general types of collectors. Specific sampling methods are given by National Institute of Occupational Safety and Health (NIOSH) (1994) and Occupational Safety and Health Administration (OSHA) (1996b).

Interpretation and Decision Making

Occupational Safety and Health Administration

Permissible Exposure Limits. In the United States, OSHA promulgates and enforces exposure standards. The standards for airborne exposure are called *permissible exposure limits* (*PELs*). PELs are set for *8 h/day exposure, ceiling exposure,* and in some cases *exceedance of the 8-h TWA for a specified time.* The ceiling exposure is a level not to be exceeded regardless of the 8-h TWA exposure.

OSHA initially adopted the 1968 American Conference of Governmental Industrial Hygienists (ACGIH) threshold limit values (TLVs) as chemical exposure standards. New standards or changes to the initial standards must be made through a formal rule-making procedure with input from industry, government (NIOSH), public interest groups, and individuals (P.L. 91-596, 1970). The procedure is very lengthy, and for that reason very few new standards for chemical exposure have been promulgated (approximately 30 in the 25 years since OSHA was established).

OSHA standards are given in *29 CFR 1910, General Industry Standards. Subpart Z* covers Toxic and Hazardous Substances. *Standard 1910.1000* is the original standards (1968 TLVs). The PELs are given in Tables Z-1, Z-2, and Z-3 of the standard. Table Z-3 deals with mineral dusts. VOC standards are given in Tables Z-1 and Z-2. New standards or changes to existing standards are given in 1910.1001 to 1910.1045.

If workers are exposed to more than one contaminant with similar health effects, then the effects are considered additive and the equivalent exposure for the mixture E_m is calculated as the sum of the ratios of the measured exposure concentration to the allowable concentration. If $E_m < 1.0$, then the exposure to the mixture does not exceed the standard or guideline. If $E_m \geq 1.0$, then the standard or guideline is exceeded. Examples of these types of calculations are shown in Fig. 3.2.

a. Three consecutive samples are collected during the evaluation of a printer's toluene exposure. The results of the sampling are as follows:

Sample	Sampling time (h)	Concentration (ppm)
1	2	35
2	2	39
3	4	23

The 8-h TWA concentration can be calculated as follows:

$$C_{\text{TWA}} = \frac{2\text{ h} \times 35\text{ ppm} + 2\text{ h} \times 39\text{ ppm} + 4\text{ h} \times 23\text{ ppm}}{2\text{ h} + 2\text{ h} + 4\text{ h}} = 30\text{ ppm}$$

The measured concentration can be compared with the OSHA standard (PEL) and the ACGIH guideline (TLV). The PEL is 200 ppm as an 8-h TWA, and the TLV is 50 ppm as an 8-h TWA; therefore, the measured concentration is below the PEL and TLV.

b. A worker is exposed to three compounds during the course of the (8-h) work shift. The measured 8-h TWA concentrations and the applicable standards and guidelines are shown below. To determine if the standards and guidelines were exceeded, the effective concentration of the mixture E_m must be calculated.

VOC	Measured concentration (ppm)	PEL (ppm)	TLV (ppm)
Acetone	250	1000	750
MEK	10	200	200
MBK	5	100	5

Using the PELs gives

$$E_m = \frac{250\text{ ppm}}{1000\text{ ppm}} + \frac{10\text{ ppm}}{200\text{ ppm}} + \frac{4\text{ ppm}}{100\text{ ppm}} = 0.34 \leq 1.0$$

Using the TLVs gives

$$E_m = \frac{250\text{ ppm}}{750\text{ ppm}} + \frac{10\text{ ppm}}{200\text{ ppm}} + \frac{4\text{ ppm}}{5\text{ ppm}} = 1.18 > 1.0$$

The PEL for the mixture is not exceeded; however, the TLV for the mixture is exceeded.

FIGURE 3.2 Example calculations using OSHA PELs and ACGIH TLVs.

Substance-Specific Standards. There are substance-specific standards for a number of chemicals. VOCs with substance-specific standards include benzene, formaldehyde, and methylene chloride. In addition to setting PELs, these substance-specific standards set *action levels (ALs)*. Exposures above the AL require employers to perform other activities in order to be in compliance with the standards. These other requirements can include periodic exposure monitoring; routine physical examinations with specific tests, e.g., chest X rays or pulmonary function tests; biological monitoring; medical removal provisions; specific control measures; recordkeeping, labeling, training, and other forms of hazard communication.

Hazard Communication Standard. In addition to substance-specific standards, *29 CFR 1910.1200* is the *hazard communication standard* which requires employers to provide information and training to employees regarding hazardous chemicals to which they may be exposed at work. The standard requires chemical manufacturers, distributors, and importers to perform a hazard determination on each chemical. All employers are required to (1) have a written hazard communication program outlining how they will achieve compliance with the standard and listing all hazardous chemicals used; (2) label all containers of hazardous chemicals; (3) maintain a Material Safety Data Sheet for each chemical and make these available to all employees; and (4) provide training and information to employees. The training and information must include the requirements of the standard, the signs and symptoms of exposure, proper work practices and use of personal protective equipment, procedure to follow in an emergency, and exposure limits.

American Conference of Governmental Industrial Hygienists Threshold Limit Values. The ACGIH, a professional society, establishes exposure guidelines called *threshold limit values*. TLVs are not legally binding and are set by the TLV committee. The committee membership is limited to IHs from government or academic institutions. The committee reviews information on contaminants and reaches a consensus. Since the TLVs are not legally enforceable and do not have the same lengthy rule-making procedure as OSHA standards, changes and new guidelines are more common. Many TLVs are lower than OSHA standards for the same substances, although not all.

There are three types of exposure guidelines given by ACGIH: *TLV-TWA, TLV-STEL,* and *TLV-C.* The TLV-TWA is a time-weighted average assuming 8 h/day, 40 h/week exposure "to which nearly all workers may be repeatedly exposed day after day with adverse effect" (ACGIH, 1996). The TLV-STEL is a 15-minute short-term exposure limit "to which workers can be exposed continuously for a short period of time without suffering from (1) irritation, (2) chronic or irreversible tissue damage, or (3) narcosis of sufficient degree to increase the likelihood of accidental injury, impair self-rescue, or materially reduce work efficiency, and provided the TLV-TWA is not exceeded" (ACGIH, 1996). Worker exposure to TLV-STEL should not occur more than 4 times per day, and there should be at least a 60-minute interval between TLV-STEL exposures. TLV-C is the instantaneous concentration that should never be exceeded regardless of TLV-TWA. Table 3.2 lists the OSHA PELs and ACGIH TLVs for selected VOCs.

Recommendations for Control

Control at the Sources

Substitution. Toxic materials, equipment, or processes can sometimes be replaced to reduce exposure potential.

TABLE 3.2 OSHA Standards and ACGIH Guidelines for Selected VOCs

		OSHA PEL (OSHA, 1996a)					ACGIH TLV (ACGIH, 1996)		
VOC	Standard	TWA (ppm)	STEL (ppm)	Ceiling (ppm)	Exceedance (ppm)	AL (ppm)	TWA	STEL/C	Odor (ppm)
Benzene	1910.1028	1	5	—	—	0.5	10, A2		2
							NOIC		
							0.5, A1	2.5, A1	
Toluene	1910.1000 (Z-2)	200	—	300	500 ppm 10 min	—	50, A4 (skin)	—	2
Xylene	1910.1000 (Z-1)	100	—	—	—	—	100, A4	150, A4	0.5
Phenol	1910.1000 (Z-1)	5 (skin)	—	—	—	—	5, A4 (skin)	—	0.3
Chlorobenzene	1910.1000 (Z-1)	75	—	—	—	—	10, A3	—	0.2
n-Hexane	1910.1000 (Z-1)	500	—	—	—	—	50	—	—
Formaldehyde	1910.1048	0.75	2	—	—	0.5	—	C 0.3, A2	1
Acetone	1910.1000 (Z-1)	1000	—	—	—	—	750	1000	2
							NOIC		
							500, A4	750, A4	
MEK	1910.1000 (Z-1)	200	—	—	—	—	200	300	5
MBK	1910.1000 (Z-1)	100	—	—	—	—	5 (skin)	—	—
Methylene chloride	1910.1052	25	125	—	—	12.5	50, A3	—	200
Perchloroethylene	1910.1000 (Z-2)	100	—	200	300, 5 min in any 3 h	—	25, A3	100, A3	5
Trichloroethylene	1910.1000 (Z-2)	100		200	300, 5 min in any 2 h	—	50, A5	100, A5	20
1,1,1-Trichloroethane	1910.1000 (Z-1)	350	—	—	—	—	350, A4	450, A4	20
1,1,2-Trichloroethane	1910.1000 (Z-1)	10 (skin)	—	—	—	—	10, A4 (skin)	—	—
Vinyl chloride	1910.1017	1	5	—	—	0.5	5, A1	—	—

TWA—8-h time-weighted average

STEL—short-term exposure limit

Skin—indicates the potential importance of skin absorption to the total dose.

C—ceiling; the instantaneous concentration should never exceed this level regardless of the TWA concentration.

A1—confirmed human carcinogen; A2—suspected human carcinogen; A3—animal carcinogen; A4—not classifiable as human carcinogen; A5—not suspected as human carcinogen.

NOIC—substance is listed in the Notice of Intended Changes. Adopted values are given in the first line, intended changes in second line.

Odor—approximate odor threshold

MATERIALS. Less toxic, less flammable, less dusty, less volatile material may be substituted for more hazardous materials. Examples include shotblasting instead of sandblasting; freon instead of methyl bromide or ammonia as a refrigerant; fiberglass or cellulose instead of asbestos; water and detergents instead of organic solvents; pelletized or briquetted forms of materials that are less dusty than powder forms.

PROCESS. The process can be changed in a way that reduces exposure. Examples include using a conveyor instead of a person to move materials; paint dipping instead of spraying; airless or electrostatic spraying instead of air spraying; wet techniques for dry techniques; chemical cleaning for sand blasting; vapor degreasing in controlled tanks instead of hand cleaning in open air; wet methods instead of dry methods; automation; and machine application of lead oxide to battery grids.

EQUIPMENT. Equipment may be changed to safer types. Examples include changing from gasoline-powered forklifts to liquid propane or, even better, battery-powered; mechanical gauges for mercury-containing manometers and thermometers; welded pipe for flanged sections to prevent leaks; and mechanical pump seals for gasket pump seals to prevent leaks.

Care must be taken not to replace one toxic condition for one that is also toxic or dangerous.

Isolation. This method of control involves physically isolating the material, process, or operation. Isolation can be at the source or worker, or between the source and worker. Examples include enclosing noisy equipment; automated sandblasting rooms; insulating hot water lines; automated plating operations; chemical process operations in closed systems; pneumatic conveying of dusty material in enclosed ducts; shielding high-temperature sources; shielding radiation sources; enclosures at sampling ports; glove boxes used for working with highly toxic chemicals, radioisotopes, or biological agents.

Problem: Access is often necessary for repair or maintenance. Isolation often replaces many low-level exposures with a few high exposures.

Local Exhaust Ventilation. Local exhaust ventilation (LEV) is the capture or control of contaminants at the point of generation. This is the preferred method for controlling contaminant concentrations. LEV consists of hoods or enclosures, ductwork to transport the contaminated air, an air cleaner, in some cases, and a fan. Hoods are of three general types: booths or enclosures which enclose, totally or partially, the operation or source of contaminant (spray paint booth, laboratory fume hood); receiving hoods which use some phenomena of the process to aid in capture (grinding wheel hood, canopy hood); and exterior hoods which are hoods located near but not around the process. These hoods must "reach" out to capture contaminants. Exterior hoods are used most commonly because they do not interfere with process or production. They are also the least effective or require the largest volume of airflow for control.

Capture efficiency is not usually 100 percent, and many factors affect performance: distance of hood from source (especially important with exterior hoods); worker activity—reaching into zone of influence of hood (e.g., laboratory hoods—working near front decreases efficiency, reaching in and out also decreases efficiency); drafts that significantly affect exterior hood performance but also can affect booths or partial enclosures and receiving hoods; temperature of source; physical state of contaminant—particles or gases or vapors; and direction and velocity of contaminant release.

Generally, LEV systems should enclose the source as much as possible, and hoods should be located so contaminant is moved away from the breathing zone of workers.

Control between Source and Worker

Isolation or Enclosures. A material, process, or operation can be isolated physically to eliminate or reduce hazardous exposures to workers. Examples include controlled-access areas, some workers are exposed but others are not; shields between worker and source, e.g., welding curtains; air-conditioned crane cabs; and well-ventilated or air-conditioned control rooms.

Dilution Ventilation. Dilution ventilation is the addition of clean or cooled or heated air to an area in an effort to improve air quality. Toxic contaminants are diluted by mixing with clean air. Dilution ventilation is best for uniform releases of small quantities of low-toxicity gases and vapors into areas that are separated from workers to allow for dilution before exposure. It is usually used for many small sources (fugitive emissions). Identifiable point sources are controlled better and less expensively with local exhaust ventilation.

Control at the Worker. Personal protective equipment should be used when other control methods are not feasible or when temporary and unusual conditions exist. Personal protective equipment can be divided into two general categories: those devices which are designed to prevent a catastrophic event (hard hats, fall protection, safety shoes) and those used to reduce a continuous exposure to an acceptable level (respiratory protection, hearing protection).

Administrative Controls. Administrative controls are work practices or management techniques used to limit exposure. These are used to supplement other types of controls. These are examples of administrative controls: Training should be conducted initially and at regular intervals, and anytime there is a change in process, work practices, new materials, new health and safety equipment. Workers need to be trained in safe work practices; emergency response; hazards associated with duties, tasks, processes, and products; available control measures (how they work, how they should be maintained, how to recognize if they are not effective, not working); environmental monitoring—continuous monitoring equipment (can be equipped with warning signals if concentration is high), dosimeters, periodic monitoring programs; employee monitoring—routine physical examinations, biological monitoring; scheduling—rotation of workers during shift so no one is exposed to "dirty" jobs all the time (dirty jobs are sometimes scheduled for off shifts when the number of workers exposed is less); preventive maintenance—both production equipment and monitoring equipment should be maintained; good housekeeping practices; restrictions on entry into certain areas; restrictions on eating, drinking, smoking in work areas; hygiene facilities—showers, change rooms, lunch rooms.

HEALTH EFFECTS

Organ Systems and Toxicant Susceptibility

Central Nervous System. Comprised of the brain, spinal cord, and surrounding membranes, the central nervous system (CNS) has a special defense mechanism called the blood-brain barrier. The blood-brain barrier has anatomic and physiologic controls which prevent absorption of toxicants into the central nervous system. These include (1) tight junctions of capillary (small blood vessel) cells which prevent diffusion of toxins across capillary walls, (2) a high density of nerve cell components which acts to further seal the capillary walls, and (3) a relatively low amount of protein in the CNS, reducing oncotic pressure in the brain. Newborns have undeveloped blood-brain barriers and are therefore more susceptible to CNS toxins.

Decreased oxygen flow to the brain can be induced by primary oxygen loss (e.g., replacement of oxygen by carbon monoxide), by poor blood flow (damage to the heart pump or to blood vessels, preventing flow to the brain), or by cellular damage. Toxicants may also have directly adverse effects on specific structures of the brain.

Exposure to toxicants can damage structures which control sensory function (e.g., touch/pain sensation, hearing, vision), motor function (movement), cognitive and intellectual function, and behavior. While brain imaging studies, such as computerized tomography (CT) scan or magnetic resonance imaging (MRI), can evaluate gross structural abnormalities, functional tests are generally required to evaluate the more subtle effects of chemical toxicants. Neurobehavioral (or neuropsychological) examination entails administering a battery of tests which evaluate memory, intellect, executive function, coordination, and psychological status. Various toxins cause identifiable patterns of abnormalities; however, there are no abnormalities which are specific to individual toxins. Other factors, such as CNS infection or head trauma, may show abnormalities which are similar to those of chemical toxicants.

The area of the brain or spinal cord that is affected by a chemical toxicant is what determines the clinical manifestations of damage. Because nervous tissue does not easily repair itself or regenerate, damage is more frequently irreversible than is the case for other organ systems.

Most lipid- (fat-) soluble agents have relatively easy access to the CNS; these include VOCs. The earliest manifestations of acute encephalopathy (brain damage) include mood alterations (disinhibition), sleepiness, incoordination, abnormal gait (ataxia), headache, and seizure. The more severe the exposure, the more compromised the level of consciousness. If the exposure is excessive and acute, the proximate relationship between exposure and illness is obvious. Chronic, lower-level exposure to the VOCs or other CNS toxicants can lead to a slower, more subtle progression of the same symptoms. Significant mood alterations and even psychiatric diagnoses may ensue. Cognitive deficits can eventually turn into profound dementia.

Most chemical solvents and *all the VOCs are known to cause both acute and chronic CNS disease.*

Peripheral Nervous System. The peripheral nervous system (PNS) may be damaged by *n-hexane, MBK,* heavy metals, ethanol, and a variety of naturally occurring diseases. *MEK* can potentiate the effect of *MBK* in causing peripheral neuropathy. Dying back of peripheral nerves is caused by these toxins. The feeling of numbness and tingling in a "stocking-glove" distribution are typical (deficits of sensation). Over time, ankle reflexes and muscle wasting occur, due to loss of innervation of the muscles (motor deficits). Loss of facial nerve function has been described with *trichloroethylene* exposure. Peripheral neuropathy is very gradually reversible, provided toxin exposure is stopped. Electromyogram (measuring muscle function, the health of which relies on nerves that serve the muscle) and nerve conduction velocities (how quickly nerves conduct electricity) are the best tests of peripheral nerve function.

Respiratory System. The respiratory system consists of lung tissue which is fed air (oxygen) by branching airways, called the bronchial tree. As air moves into the lung tissue, oxygen diffuses into a fine capillary (blood vessel) system as carbon dioxide diffuses out. This delicate system also allows the diffusion of other gases and solvents across the lung into the vasculature of the body. The fat-soluble VOCs easily move into the airways and across the lung tissue into the bloodstream. Many of them have a local, irritating effect on the airways as well.

Obstruction of airflow through the bronchial tree may be fixed, as is the case with *chronic obstructive pulmonary disease.* While cigarettes are the prototypical respiratory tract irritants, chronic exposure to other respiratory irritants may also cause irreversible airway disease. *Bronchial asthma* is, by contrast, reversible inflammation of the airways, leading to transient episodes of bronchospasm and reduced airflow. The disorder may be allergically based, and has been described for a number of chemical toxicants, including toluene diisocyanate, platinum, trimellitic anhydride, and phthalic anhydride; *formaldehyde* has been implicated in causation of asthma. Asthma may also be caused by chemical irritants (as opposed to allergens). Until recently, this irritant effect was presumed to be transient, and was given the name *reactive airways disease.* This latter condition is indistinguishable from asthma on a clinical basis. Complete resolution and disappearance of bronchospasm after a few months of intermittent illness allows this diagnosis to be made in retrospect.

In addition to reversible and irreversible airflow obstruction, the lung tissue itself may be damaged. Acute damage can lead to leakage of fluid from the capillaries into the lung tissue and can occur hours to days following a severe, acute exposure to a number of locally damaging chemicals. For the sake of completeness, it should be mentioned that exposure to certain fibers and particulates (e.g., asbestos, silica sand) can lead to scarring of the lung tissue and restriction of expansion and recoil of the lung during normal breathing.

Longstanding airway obstruction may be evident on chest X ray. Pulmonary function testing, which measures flows, volume, and diffusing capacity of gases through the lung tissue, is the best indicator of the respiratory system's functioning.

Most VOCs can cause irritation of the respiratory tract, leading to (1) throat irritation and cough in mild exposure, (2) irritation of the airways, leading to spasm and wheezing which may be transient or permanent, or (3) damage to the lung itself, leading to leakage of fluid into lung tissue and impairing diffusion of oxygen across the tissue.

Liver. The liver, which is part of the gastrointestinal system, is the major organ responsible for biotransformation (alteration) of exogenous chemicals, drugs, and dietary substances for purposes of detoxification, availability for required bodily functions, or formation of water-soluble metabolites which can be eliminated by the body. A series of enzyme systems exist within the liver to facilitate these processes. Cascades of chemical reactions can be initiated by certain chemicals, and there is tremendous interaction between exogenous substances leading to synergistic or antagonistic processes. In some cases, the liver may create metabolic products which are more toxic than their parent compounds.

Most toxins affect the liver in predictable ways and are related to dose. There are, however, idiosyncratic reactions which occur sporadically and unpredictably. Some agents, such as carbon tetrachloride, injure the liver cells either acutely or chronically. With acute toxicity, liver inflammation can lead to tissue necrosis and subsequent organ failure. If the liver repairs itself, it may function normally, or it can become scarred and shrunken. In situations where there has been chronic, low-level inflammation, normal liver tissue can be replaced with fat (steatosis), and the liver can also become scarred (cirrhosis). Manifestations of chronic toxicity may not become apparent for years.

Acute liver toxicity can manifest itself in symptoms of abdominal pain in the right upper abdomen, with signs of jaundice and liver enlargement. The injured liver spills several enzymes which can be measured in the blood (*biomarkers* of liver toxicity). There are a number of tests of liver function as well. These include measurable synthetic or metabolic products of the liver and clearance tests in which dyes are taken up and processed by the liver. Anatomic or histologic (microscopic tissue) observations may be made by CT scan, ultrasound, and examination of liver biopsy specimens under the microscope. Some hepatotoxins have been implicated in liver cancer causation (see Table 3.3).

VOCs causing liver disease include carbon tetrachloride (worse with ingestion, over inhalation), tetrachloroethane, tetrachloroethylene, trichloroethylene, 1,1,1-trichloroethane, 1,1,2-trichloroethane, chloroform, vinyl chloride, and toluene.

TABLE 3.3 Target Organs for Volatile Organic Compounds

Compounds	Target organ	Carcinogenicity
	Simple aromatics	
Benzene	CNS, blood, bone marrow, mucous membrane, skin	Yes
Toluene	CNS, mucous membrane, liver, kidney, skin	No
Xylene	CNS, eyes, GI tract, blood, liver, kidneys, skin	No
Phenols and cresols	Liver, kidneys, skin, CNS, respiratory system, eyes	No
Chlorobenzene	Respiratory system, eyes, skin, CNS, liver	Possible
	Aliphatics	
Methane	Simple asphyxiant	No
n-Hexane	PNS, CNS, respiratory system, skin	No
	Halogenated aliphatics	
Trihalomethanes	CNS, kidney, liver	Probable
(Chloroform, dichlorobromomethane, dibromochloromethane, tribromomethane)		
Trichloroethylene	Respiratory system, heart, liver, kidneys, CNS, skin	Possible
Tetrachloroethylene (perc)	Liver, kidneys, eyes, upper respiratory tract, CNS	
Methylene chloride	Skin, cardiovascular, eyes, CNS	Possible
Vinyl chloride	Liver, CNS, blood, respiratory system, lymph system	Angiosarcoma
Alcohols, aldehydes, ethers, carboxylic acids (increased toxicity with increased chain length)		
Formaldehyde	Eye, nose, throat, lung (asthma)	Possible
Methyl *tert*-butyl ether	Headache, eye, nose	
	Ketones	
Acetone	Respiratory system, skin	No
Methyl ethyl ketone (MEK)	Respiratory system, skin, CNS	No
Methyl butyl ketone (MBK)	Eye, throat, CNS, depression, PNS	No

CNS—central nervous system (brain, spinal cord)
PNS—peripheral nervous system

Kidney. The kidneys provide a major route for the excretion of toxins. Like the liver, they contain enzyme systems which metabolize and detoxify systemic toxins. Twenty-five percent of the body's blood supply traverses them. They have significant concentrating ability, allowing exogenous chemicals to accumulate at very high concentrations. Yet they are relatively resistant to chemical injury, as evidenced by low incidence of occupational kidney disease as compared to other organ systems of the body.

Most chemical toxins interfere with the kidneys' excretory function. Tests for products of urination (e.g., protein) and blood in the urine can aid in the diagnosis of acute renal failure. *Biomarkers* of kidney dysfunction can be found in the blood. Anatomic and physiologic studies include X ray, CT scan, ultrasound, radiologic dye excretion studies, and kidney biopsy.

VOCs which are well known to cause renal disease include carbon tetrachloride (renal failure seen in 1 to 10 days after exposure with complete shutdown of urination and the need for dialysis), 1,1,1-trichloroethane, tetrachloroethylene, and trichloroethylene. Inhalation of toluene and other petroleum distillates may lead to fulminant renal failure, but may also cause transient abnormalities of renal function.

Hematopoietic System. Red and white blood cells and platelets are produced by stem (template) cells located in the bone marrow. Stem cells are primitive, or poorly differentiated cells which undergo division, reproduction, and maturation of progeny cells which are released into the bloodstream to carry oxygen (red cells), fight infections and conduct surveillance for foreign substances (white cells), and clot (platelets or thrombocytes).

Benzene is the only VOC which has been shown to be toxic to bone marrow. Exposure may result in depression or cessation of blood cell production (aplastic anemia), production of an overabundance of abnormal white blood cells (leukemia), or scarring of the bone marrow with abnormal cell forms produced.

Skin and Mucous Membranes. The principal abnormalities of skin seen in exposure to the VOCs include defatting (drying) of the skin and contact dermatitis, which is generally a direct irritant reaction, rather than an allergic (immunologically mediated) one. The same is true for the mucous membranes (eyes, nose, mouth, and throat) which are subject to irritation by volatile toxins. The eye is susceptible to painful corneal abrasion, either from splashes or from highly concentrated, irritating vapors.

All the VOCs can be irritating to the skin, particularly when contacted in liquid form. *Formaldehyde* has been shown to cause an allergy-based contact dermatitis. Because of their fat solubility, and after damaging protective layers of skin, the VOCs are readily absorbed through the skin into the bloodstream. Table 3.3 gives the target organ systems for selected VOCs.

Toxicity of Specific VOCs

Table 3.4 describes toxic effects of specific VOCs at measured or estimated concentrations that have been seen in (1) acute, accidental exposures, (2) experimental, controlled exposures, and (3) occupational settings. What follows is a narrative description of the overall effects of these chemicals.

Aromatics

Benzene. Benzene is absorbed by all three routes in the body. It accumulates in fatty (adipose) tissue because of its lipid solubility. Benzene in the blood can be rapidly excreted by lung exhalation. Biotransformation occurs in the liver, with phenol and dihydroxybenzenes formed. The metabolites are conjugated with sulfates and excreted in the urine. Peak urine concentration is at 4 to 8 h after exposure.

Benzene causes chromosomal damage and gene toxicity. Rare, acute intoxication causes imbalance, vomiting, impaired consciousness, lung inflammation, rapid heart rate, and collapse. Inhalation of lesser amounts of the solvent causes headache, flushing, chest tightness, shortness of breath, visual disturbances, and convulsions. Skin absorption is unusual.

Chronic exposure has been shown to cause abnormalities in blood and bone marrow function. Bone marrow depression (decreased production of all types of blood cells) and development of leukemia have been demonstrated. There is no apparent threshold (minimal exposure level) for bone marrow toxicity. The other major effect of chronic exposure is CNS toxicity, with behavioral and psychomotor deficits. It may also cause imbalance, vertigo, and hearing deficits through its effect on the ear.

TABLE 3.4 Toxic Effects of Inhaling Volatile Organic Compounds

VOC	Inhalation concentration (ppm)	Exposure time	Health effects	Exposure source*
Benzene	210	6–14 yr	Bone marrow toxicity	3
	250–500	Continuous	Vertigo, drowsiness, headache, nausea	3
	3,000	Continuous	Euphoria, nausea, staggering gait, coma	
	20,000	Brief, intermittent	Eye, respiratory irritation	2
Toluene	10–40	Continuous	No symptoms	2
	80–300	Years	No abnormal liver tests	3
	100	6 h	Eye/nose irritation, headache, dizziness, “high” feeling, normal neurobehavioral tests	2
	117 (average)	Years	No neurologic abnormalities	3
	200	8 h	Mild fatigue, weakness, confusion, skin tingling	2
	600	8 h	Euphoria, headache, dizziness, nausea, pupil dilation	2
	800	8 h	Above symptoms but more pronounced and persisting for days, muscle fatigue, nervousness, insomnia	2
	Glue sniffer	3 yr	Reversible lung, kidney, liver function tests	3
Xylene	100	8 h	No symptoms	2
	200	8 h	Nose, throat, eye irritation	2

Xylene	230	8 h	Not objectionable	2
	60–350	Intermittent	Giddiness, anorexia, vomiting	3
	460	15 min	Tearing, lightheadedness	2
Phenol	48 (with 8 ppm formaldehyde exposure)	Brief, intermittent	Nose, throat, eye irritation	3
	4	Continuous	No respiratory irritation	3
Chlorobenzene	200	Intermittent	Eye, nose irritation	2
n-Hexane	54–200	1 yr	Polyneuropathy	3
	<500		No symptoms	2
	400–600	Years	Polyneuropathy	3
	420–1280	>5 yr	Changes in color vision, retinal pigmentation	3
	>1,000	Continuous	Symptoms of narcosis	2
	1,500	Continuous	Eye, throat irritation, nausea	2
	2,000	10 min	No effects	2
	5,000	10 min	Dizziness, confusion	2

*Exposure source: (1) accidents, (2) experimental chambers, and (3) estimated from occupational settings.

TABLE 3.4 Toxic Effects of Inhaling Volatile Organic Compounds (*Continued*)

VOC	Inhalation concentration (ppm)	Exposure time	Health effects	Exposure source*
Formaldehyde	0.1–0.3	Brief	Mild mucous membrane irritation	2
	1–3	10 min	No reactive airways	2
	1–3.2	Seconds	Reactive airways	2
	2–3	Continuous	Tingling in mucous membranes	2
	4–5	Continuous	Mucous membrane irritation	2
	4–5	30 min	Intolerable irritation	2
	10	Minutes	Intolerable	2
	10–20	Minutes	Cough, difficulty breathing	2
	50–100	Minutes	Serious lung injury	2
Acetone	500	Minutes	Eye, nose, throat irritation	2
	500	Minutes	Aware of odor, no effects	2
	1,000	Minutes	Eye irritation	2
	>12,000	Minutes	Dizziness, weakness, loss of consciousness	1
MEK	<200 (combined with alcohol, toluene, acetone)	Months–years	Polyneuropathy	3

MBK	1,000	Minutes	Eye, nose irritation	2
	147–516 (peaks to 763)	6–12 mo	Polyneuropathy	3
Methylene chloride	75–100	Years	Subjective neurologic symptoms, normal EKG, EMG/NCV, neuropsych tests	3
	250	7.5 h	Normal neuropsych tests	2
	300–800	40 min	Abnormal neuropsych tests	2
Methylene chloride	500–1,000	1–2 h	Lightheadedness	2
	500–3,600	Years	Toxic encephalopathy	3
	50,000	Seconds	Death	3
	"High"	Acute	Fatigue, weakness, nausea, pulmonary edema	1
	"High" unmeasured	1–3 h	Respiratory irritation, decreased hemoglobin, coma, death	1
Perchloroethylene	60–450	2–20 yr	CNS depression, EEG abnormalities	3
	100	7 h	Eye, throat irritation, somnolence, slurred speech	2
	<300	Years	No liver enzyme abnormalities	3
	600	10 min	Dizziness, incoordination, tingling around mouth	2

*Exposure source: (1) accidents, (2) experimental chambers, and (3) estimated from occupational settings.

TABLE 3.4 Toxic Effects of Inhaling Volatile Organic Compounds (*Continued*)

VOC	Inhalation concentration (ppm)	Exposure time	Health effects	Exposure source*
Perchloroethylene	275 and 1,100	180 min and 30 min	Loss of consciousness, abnormal liver function tests	3
	2,000	5 min	Mild CNS depression	2
	5,000	6 min	Vertigo, confusion, nausea	2
1,1,1-Trichloroethane	>200	6 mo–6 yr	No impairment	3
	350	Acute	Psychomotor performance abnormalities	3
	900–1,000	20 min	Lightheadedness, incoordination, eye irritation	2
	2,000	5 min	Disturbed equilibrium	1
	10,000	30 min	Marked incoordination	1
	20,000	60 min	Coma, death	1
1,1,2-Tricholorethane			None reported in relation to exposure levels	
Vinyl chloride	300		Abnormal liver function tests	2
	500	7.5 h	No clinical or neurologic abnormalities	2
	2,000	5 min	Dizziness, lightheadedness, nausea, dulling of vision, auditory cues	2

*Exposure source: (1) accidents, (2) experimental chambers, and (3) estimated from occupational settings.

Source: Proctor et al. (1988).

Toluene. Toluene causes depression of the central nervous system, evidenced by fatigue, weakness, confusion, euphoria, headache, dizziness, and nausea. The symptom of euphoria is what has led to its abuse by glue sniffers. Long-term inhalation can cause severe degeneration of brain tissue; mild central nervous system abnormalities have been seen in workers chronically exposed to toluene, as compared to unexposed controls. Splashing on the skin leads to defatting and irritation. There is a case report of corneal abrasion following splashing of toluene into the eyes. Transient kidney and liver damage has been seen in glue sniffers. At this time, there is no evidence of carcinogenicity of toluene.

Xylene. Xylene is most commonly absorbed by inhalation or skin absorption. It is irritating to the eyes and mucous membranes; if inhaled for prolonged periods, it is locally irritating to the lungs. At high concentrations, it can cause headache, vertigo, ataxia ("drunkenness"), and narcosis. While chronic effects of xylene have been described, contamination with other VOCs was likely; blood abnormalities (e.g., leukemia) have not been seen with xylene exposure.

Phenol. Acute toxicity usually occurs through skin splashes with this commonly used disinfectant; it has relatively low volatility. Corrosive burns can occur on the skin, and corneal abrasion when splashed on the eye. Phenol's lipid solubility allows its absorption by all three routes. Symptoms of acute toxicity include cardiovascular collapse (hypotension), muscle weakness, and respiratory failure. Systemic absorption can cause liver and kidney toxicity. Chronic exposure was shown to cause weight loss, poor appetite, vomiting, diarrhea, headache, and vertigo.

Chlorobenzenes. This group of compounds causes eye, nose, and throat irritation. They are severe CNS depressants, causing narcosis, coma, and death in acute exposure. Headache, dizziness, and stupor are seen in long-term, lower-level exposure, with marked damage to liver and kidney tissue. While dichlorobenzene isomers are less narcotic, *p*-dichlorobenzene has been shown to cause cancer in animal models.

Aliphatics

n-Hexane. *n*-Hexane is readily absorbed through inhalation or dermal absorption, and acts as an irritant to the eyes and respiratory tract. If ingested, it is extremely toxic to the gastrointestinal tract, causing direct irritation, with consequent vomiting and risk of aspiration into the lungs. It is probably the most toxic of the alkanes, causing severe peripheral neuropathy, manifesting in numbness and tingling of the extremities and muscle weakness.

Alcohols, Aldehydes, and Ethers

Formaldehyde. Formaldehyde gas is produced from liquid formalin and by heating or hydrolysis of the solid polymer, paraformaldehyde. Exposure is considerable in the machining and fabrication of particleboard and laminates which are made of formaldehyde resins. Eye, skin, and respiratory tract irritation can occur from vapor exposure. It is a known skin sensitizer (causing allergic contact dermatitis) and is implicated in causation of asthma. Recently, animal carcinogenicity has been discovered.

Ketones

Acetone. Acetone is extremely volatile and causes slight eye and nose irritation in high concentrations. It has not been shown to be toxic to the nervous system. Acetone is rapidly absorbed through inhalation and has a half-life elimination of 3 h. Headache, drowsiness, loss of appetite, nausea, vomiting, limb weakness, dizziness, and mucous membrane irritation have been described, although these effects completely resolve within a few days.

Methyl Ethyl Ketone (MEK). MEK is similar to acetone, although it has a higher odor threshold and is more irritating to the eyes, mucous membranes, and skin. At high concentrations it can cause CNS toxicity. Cases of peripheral neuropathy have been described, although exposure was to a mixture of solvents, including MEK. It likely potentiates the peripheral neuropathy caused by MBK.

Methyl n-Butyl Ketone (MBK). While MBK is similar to the other ketones, it is less volatile and less acutely toxic. Very high exposures can cause mucous membrane and eye irritation. Chronic exposure can lead to peripheral neuropathy, similar to that seen in *n*-hexane exposure, with numbness and tingling of the extremities as the first symptoms. Motor neuropathy (foot drop) may ensue. Peripheral neuropathy may be potentiated by concomitant MEK exposure and can be reversed with exposure prevention.

Chlorinated Hydrocarbons

Methylene Chloride. Methylene chloride is irritating to the eye, skin, and respiratory tract. It is also a mild CNS depressant. In high doses, it can cause simple asphyxiation by displacement of oxygen in the air; it is also metabolized to carbon monoxide in the body, preventing oxygenation of body tissues. It has been shown to be carcinogenic in animal models.

Perchloroethylene (Tetrachloroethylene). Used as the major agent in the dry cleaning industry, "perc" has been shown to cause severe CNS depression and liver damage. It is carcinogenic in laboratory animals and has been shown to cause peripheral neuropathy in animal models. Its milder effects include mucous membrane irritation and flushing of the face.

1,1,1-Trichloroethane. This compound can cause CNS depression as well as abnormalities of heart rhythm leading to sudden death. It is a severe CNS depressant, causing lightheadedness, incoordination, and headache. It has also been shown to cause liver and kidney damage in animals and humans, although the latter effect is not as pronounced as that caused by its isomer, 1,1,2-trichloroethane.

1,1,2-Tricholoroethane. This molecule may be inhaled or absorbed in liquid form through the skin. It may cause CNS depression. It has also been shown to cause liver and kidney damage and to be carcinogenic in animal models.

Vinyl Chloride. Vinyl chloride is best known for its association with the development of angiosarcoma (a type of cancer) of the liver. It causes degeneration of the bones of the tips of the fingers and toes, destruction of the blood vessels of the fingers (leading to poor circulation and cold sensitivity), low platelet count (platelets are blood cells that assist clotting), and liver function abnormalities. Acute exposure can lead to CNS depression. In addition to angiosarcoma, vinyl chloride is associated with the development of a variety of tumors in laboratory animals; in epidemiologic studies of humans, increases in cancers of the respiratory tract, the central nervous system, and the lymphatic and hematopoietic systems have been seen.

Indoor Air Pollution

Since the 1980s there has been increasing interest in the problems of indoor air pollution. Enclosed in modern buildings with sealed windows, well-controlled heating and cooling systems, offices decorated with synthetic materials, and with operation of a wide array of office machines, workers have complained of an array of symptoms, including eye, nose, and throat irritation; headache; visual disturbances; nausea; and breathing abnormalities. Chamber and field studies of indoor environments implicate low-level VOCs in causation. While measurement of VOCs in low ranges is technically difficult, many authors have recommended guidelines for indoor VOC concentrations of 250 to 300 $\mu g/m^3$ (Samet and Spengler, 1992).

RESOURCES

Books

Brief, R. S.: *Basic Industrial Hygiene—A Training Manual.* New York: Exxon Corporation, 1975.

Burgess, W. A.: *Recognition of Health Hazards in Industry, A Review of Materials and Processes,* 2d ed. New York: Wiley, 1995.

Clayton, G. D., and F. E. Clayton, eds.: *Patty's Industrial Hygiene and Toxicology,* 4th ed. New York: Wiley, 1991.

Gosselin, R. E., R. P. Smith, H. C. Hodge, and J. E. Braddock: *Clinical Toxicology of Commercial Products,* 5th ed. Baltimore, MD: Williams and Wilkins, 1984.

Lists consumer products by trade and generic names, chemical formulations, and ingredients; covers first aid and emergency treatment of poisonings; provides names, addresses, and telephone numbers of manufacturers.

Handbook of Organic Industrial Solvents, 5th ed. Chicago: Alliance of American Insurers, 1980.

The Industrial Environment—Its Evaluation and Control. DHEW/NIOSH. Washington: Government Printing Office, 1973.

Klaassen, C. D., M. O. Amdur, and J. Doull, eds.: *Cassarett and Doull's Toxicology: Basic Science of Poisons.* New York: McGraw-Hill, 1996.

Definitive and detailed text on how toxicants are handled in the body. Requires some understanding of human physiology, biochemistry, and pharmacologic principles.

Levy, B. S., and D. H. Wegman: *Occupational Health, Recognizing and Preventing Work-Related Disease.* Boston: Little, Brown, 1983.

NIOSH Manual of Analytical Methods, 4th ed. P.M. Eller, ed. DHHS (NIOSH) publication 94-113. Washington: Government Printing Office, 1994.

NIOSH Pocket Guide to Chemical Hazards. DHHS (NIOSH) publication 94-116. Washington: Government Printing Office, 1994.

Many industrial chemicals are listed in tabular form with odor thresholds, threshold limit values, toxicologic effects, and recommended respiratory protection.

OSHA General Industry Standards (29 CFR 1910). Washington: Government Printing Office, 1996.

OSHA Industrial Hygiene Technical Manual. OSHA Instruction CPL 2-2.20B. Washington: U.S. Department of Labor, 1990.

OSHA Technical Manual (Instruction TED 1.15). Salt Lake City, UT: U.S. Department of Labor, OSHA Office of Science and Technology Assessment, May 24, 1996.

Proctor, N. H., J. P. Hughes, and M. L. Fischman: *Chemical Hazards of the Workplace.* Philadelphia, PA: JB Lippincott Company, 1988.

Organized in alphabetical order, contains one- to three-page summaries of adverse health effects and salient studies in vitro, in laboratory animals and in humans, of occupationally used chemicals.

Threshold Limit Values for Chemical and Physical Agents. Cincinnati, OH: American Conference of Governmental Industrial Hygienists, 1996.

Computerized Database Suppliers

Canadian Centre of Occupational Health and Safety, 250 Main Street East, Hamilton, Ontario, L8N 1H6, Canada (800-263-8340).

Micromedex Inc., 600 Grant Street, Denver, CO 80203 (800-525-9083).

National Library of Medicine through MEDLARS (800-638-8480).

Information on acute poisonings and chemical hazard response and access to a number of databases.

SilverPlatter Information, Inc., 100 River Ridge Drive, Norwood, MA 02062 (800-343-0064).

Computerized Databases

MEDLINE—biomedical journals for health professionals.

NIOSHTIC—occupational safety and health journals.

POISINDEX—description of adverse health effects of drugs and toxicants.

TOXLINE—human and animal toxicity studies, pollution effects, analytical methods.

TOXNET—factual information and references on chemicals.

Hotlines

Chemical Manufacturers Association (800-CMA-8200).

EPA pesticide information network (800-858-7378).

NIOSH (800-35-NIOSH).

Regional and state poison control centers (look on inside cover of local telephone book)

Organizations and Agencies

Agency for Toxic Substances and Disease Registry, ATSDR-Chamblee, 1600 Clifton Rd, NE, Atlanta, GA 30333 (http://atsdr1.atsdr.cdc.gov:8080/atsdrhome.html).

Agricultural Health and Safety Center, University of California at Davis (http://www-oem.ucdavis.edu).

American Conference of Governmental Industrial Hygienists (ACGIH), 6500 Glenway Avenue, Cincinnati, OH 45211 (http://www.acgih.org).

Association of Occupational and Environmental Clinics, 1030 15th St, NW, Suite 410, Washington, DC 20005 (202-347-4976).

Canadian Centre for Occupational Health and Safety (http://www.ccohs.ca).

Center for Environmental Health and Safety, Southern Illinois University (http://www.cehs.siu.edu).

Centers for Disease Control and Prevention (CDC) (http://www.cdc.gov).

Department of Defense (DOD), Safety and Health (http://www.acq.osd.mil/ens/sh).

Department of Energy (DOE) (http://www.doe.gov).

Duke Occupational and Environmental Medicine (http://occ-env-med.mc.duke.edu/oem/).

Environmental Protection Agency (EPA), Public Information Center, PM-211B, 401 M Street SW, Washington, DC 20460 (http://www.epa.gov).

FAIRS: Florida Agricultural Information Retrieval System (FAIRS) (http://hammock.ifas.ufl.edu).

Government Printing Office (http://www.access.gpo.gov).

International Labour Organization (ILO) (also available through WHO) (http://www.unicc.org/ilo/).

National Institute of Environmental Health Sciences (http://www.niehs.nih.gov).

National Institute for Occupational Safety and Health (NIOSH), 4676 Columbia Pkwy, Cincinnati, OH, 45226-1998 (http://www.cdc.gov/niosh/homepage.html).

National Institutes of Health (NIH) (http://www.nih.gov).

National Library of Medicine (NLM) (also available through NIH) (http://www.nlm.nih.gov).

National Technical Information System (NTIS) (http://www.fedworld.gov).

Occupational Safety and Health Administration, U.S. Department of Labor, 200 Constitution Ave, Washington, DC 20210 (Ten regional offices exist in Boston, New York, Philadelphia, Georgia, Chicago, Dallas, Kansas City, Denver, San Francisco, and Seattle.) (http://www.osha.gov).

Swedish National Institute of Occupational Health (NIOH) (http://www.niwl.se).

U.S. Army, Logistics Management Institute, Army Industrial Hygiene (http://reml.lmi.org/Armyih).

U.S. Bureau of Labor Statistics (also available through DOL) (http://stats.bls.gov/blshome.html).

U.S. Department of Labor (DOL) (http://www.dol.gov).

World Health Organization (WHO) (http://www.who.ch).

REFERENCES

ACGIH: *Threshold Limit Values for Chemical and Physical Agents.* Cincinnati, OH: American Conference of Governmental Industrial Hygienists, 1996.

AIHA: *A Strategy for Occupational Exposure Assessment.* N.C. Hawkins, S.K. Norwood, and J.C. Rock, eds. Fairfax, VA: American Industrial Hygiene Association, 1991.

Ashley, D. L., M. A. Bonin, F. L. Cardinali, J. M. McCraw, and J. V. Wooten: Measurement of Volatile Organic Compounds in Human Blood. *Env Health Perspect 104* (Suppl 5):871–7, 1996.

Gist, G. L., and J. R. Burg: Methodology for Selecting Substances for the National Exposure Registry. *J Expo Anal Environ Epid 5*(2):197–208, 1995.

Hamar, G. B., M. A. McGeekin, B. L. Phifer, and D. L. Ashley: Volatile Organic Compound Testing of a Population Living near a Hazardous Waste Site. *J Expo Anal Environ Epid 6*(2):247–55, 1996.

Mortimer, C. E.: *Chemistry, A Conceptual Approach,* 2d ed. New York: D. Van Nostrand Co., 1971.

NIOSH Manual of Analytical Methods, 4th ed. P.M. Eller, ed. DHHS (NIOSH) publication 94-113. Washington: Government Printing Office, 1994.

OSHA General Industry Standards (*29 CFR 1910*). Washington: Government Printing Office, 1996a.

OSHA Technical Manual (Instruction TED 1.15). Salt Lake City, UT: U.S. Department of Labor, OSHA Office of Science and Technology Assessment, May 24, 1996b.

Proctor N. H., J. P. Hughes, and M. L. Fischman: *Chemical Hazards of the Workplace.* Philadelphia, PA: J.B. Lippincott, 1988.

Public Law 91-596, Occupational Safety and Health Act of 1970.

Samet, J. M., and J. D. Spengler: Indoor Air Pollution. In *Environmental and Occupational Medicine.* W. Rom, ed. Boston: Little, Brown, 1992.

Waldbott, G. L.: *Health Effects of Environmental Pollutants.* St. Louis, MO: G.V. Mosby Co., 1973.

CHAPTER 4
ANALYTICAL METHODS

SECTION 4.1

CHEMICAL METHODS

Frank H. Jarke
Mostardi-Platt Associates, Inc.
Elmhurst, Illinois

SAMPLING TECHNIQUES 4.2
  Capture Techniques 4.2
  Concentration Techniques 4.4
LABORATORY ANALYTICAL TECHNIQUES 4.7
  Gas Chromatography 4.7
  Gas Chromatography—Mass Spectrometry 4.8
  High-Performance Liquid Chromatography 4.9
  GC Sniffing 4.9
  Titrations 4.9
FIELD TECHNIQUES 4.10
  Organic Vapor Analyzers 4.10
  Monitors 4.10
  Absorption Tubes 4.10
  Portable GCs and GC-MSs 4.11
FINDING A COMPETENT LABORATORY 4.11
METHODS 4.11

All odors are chemicals; however, not all chemicals are odorous. This fact forms the basis for determining the best analytical method to use in the investigation of either an odor analysis problem or a chemical analysis problem. Chapter 2 dealt with the characteristics of odors and VOCs, and we learned that chemicals become airborne according to their vapor pressure at the temperature of the source. We also learned that some chemicals, when inhaled through the nose, produce a sensation defined by humans as odor when these chemicals strike the olfactory bulb. Whether you are trying to determine the chemical or chemicals responsible for producing an odor or you are trying to determine the amount of chemical being released from a source, it is imperative that you understand the methods available to you and how best to employ them to identify the chemical.

There are two major separate and distinct activities involved in identifying an unknown chemical, whether odorous or nonodorous. The first is to determine how you will sample the source. The second deals with how you will analyze the chemical species that have been sampled. In the former case, you must have some knowledge of what it is that you are trying to detect, the approximate concentration and the source. For example, if you are trying to determine the chemical components of lemon oil, you might use one of several techniques available for capturing these chemical compounds and returning them to the laboratory for analysis. However, if you are trying to determine the odorous chemical components of lemon essence, you will be limited to only those sampling methods that allow you to capture the chemicals in a form representative of what a human nose senses when exposed to lemon essence. Similarly, if you are interested in the composition of the chemicals being emitted from a coal-fired boiler, you might sample the stack emissions near the outlet of the stack and could, after analyzing the sample, use modeling to determine the exposure level to a population at a certain distance from the source. On the other hand, you might want to verify the results of your model by actually collecting ambient samples at various distances from the source to verify the concentrations at those locations.

There are also two distinct methodologies for the analytical determination of airborne chemicals. One set of analytical methodologies is used in the field. These methods allow one to capture analytical information at the source of the chemical emission. The second is laboratory methods. Laboratory methods require that the sample be captured in the field and brought to the laboratory for analysis. The type of sampling methodology employed is dependent on the subsequent analytical method that is going to be used to analyze the sample and the concentration required to obtain results.

In this chapter we review both the sampling techniques and the analytical techniques that allow the engineer to make an intelligent decision regarding a wide range of chemical analyses. We cover first the sampling techniques available, then the laboratory analytical methods, and finally the field analytical methods.

SAMPLING TECHNIQUES

In this section the variety of sampling techniques are discussed. These techniques fall into two categories: capture techniques and concentration techniques. Both have advantages and disadvantages, as discussed.

Capture Techniques

Capture techniques are just that—they capture the sample as it exists at the point at which the sample is taken. These techniques employ inert flexible bags, steel canisters, and glass bombs. The major advantage of capture techniques is that a large sample can be captured and subsequently analyzed multiple times. This makes the analytical process less risky in that analytical and operator errors can be minimized. The major disadvantage is that interactions between the chemicals in the sample cannot be prevented between sampling and analysis, nor can interactions between the bag and the sample. Therefore, capture techniques work best when used to sample chemically stable sources that have high vapor pressures at ambient temperature. A second disadvantage is that no concentration of the sample occurs. Therefore, the chemicals of interest in the source must be at a high enough concentration for the analytical technique to detect.

Inert Flexible Bags. Inert flexible bags are usually made out of an inert material such as Tedlar. They come in a variety of sizes from 1 L up to 100 L. They are usually square and have a valve or quick-connect sealed into one side. The major difficulty with inert flexible bags is the technique for getting the sample in the bag without contaminating it. Flexible bags are generally prepared for use by inflating with an inert gas such as compressed zero-grade air and evacuating with a vacuum pump several times. This process will generally reduce the background concentration of chemicals in the bag to below most analytical method detection limits. If the bag has previously been used on a "dirty" source, however, it should be discarded or reserved for use only on other dirty sources. Once the bag has been cleaned, it is ready for sampling.

The best way to use flexible bags in sampling is to fill them using a *lung*. A lung is an outer rigid leak-free container that is big enough to hold the flexible bag fully expanded. The deflated flexible bag is attached to an appropriate connection that passes through the lid of the lung. The lid is put in place with the bag inside the container. A second connection on the lid is attached to a vacuum pump. The connection on the lid of the lung to the flexible bag is then attached to the source. The vacuum pump is used to evacuate the outer container. By reducing the pressure in the outer container, the flexible bag will fill with contamination-free sample. Use of a vacuum pump, even with an inert diaphragm, to fill the flexible bag is not recommended for most sources.

A second method of filling the flexible bag can be used for positive-pressure sources. In this case, low-pressure sources can be attached directly to the bag at an appropriate shutoff valve and filled directly from the source. The pressure of the source cannot be much above atmospheric pressure, or else the bag may inflate so rapidly that it explodes.

While flexible bags are generally inert, they do not ensure that the sample will arrive at the laboratory unaltered. Flexible bags leak! They are easily damaged, especially in field applications. The tiniest pinhole can allow the sample to escape or become contaminated with ambient air during the transit time to the laboratory. Also, the material that the flexible bag is made of allows gases to permeate through the material. A perfect example is the new type of helium-filled balloons now available at party outlets. These balloons stay quite nice-looking for a few days, but slowly lose the helium contents and deflate. The same thing happens to gases in a flexible bag. The lighter gases such as methane will permeate the walls and not only reduce the sample size, but also alter the chemical composition of the sample.

Flexible bags are also not suitable for sources that are much above ambient temperature. The gas composition of a source sampled into a flexible bag at 200°F will be much different when analyzed in the laboratory at 70 to 80°F. The vapor pressure of the gases at the lower temperature will be lower, and consequently some amount will plate out on the walls. Also, not all the gases in the sample will have the same vapor pressure, and therefore the chemical composition of the gas will be altered.

In summary, inert flexible bags can be very effectively used in sampling near ambient temperature sources for gases with chemical concentrations above 1 ppm. They are also effective in sampling of odorous sources with high odor thresholds for subsequent analysis by panels of judges for odor threshold determinations or intensity measurements. Also, they can be an inexpensive first step in determining the chemical composition of unknown sources.

Steel Canisters. Steel canisters come in all sizes from 100 mL to the more familiar cylinders used for commercial gases holding hundreds of liters of compressed gas. The shape is generally cylindrical, resulting in a very strong sampling container. The wall thickness can be tailored to the pressure of the source being sampled.

Steel canisters have the advantage of being much more rugged than flexible bags. They also are much easier to use. Not only does the cleaning step involve repeated pressurization with zero-grade air and subsequent evacuation, but also heat can be applied to remove any stubborn low volatile chemicals. The heating of a steel canister needs to be done with extreme caution and monitored with a pressure gauge to ensure that the canister does not explode. As with the flexible bag, steel canisters work well only with stable ambient sources.

Steel canisters come in two varieties. One is used to sample moderately high-pressure sources. Unlike the flexible bag, the sample container will not explode under moderately high pressures. Steel canisters are also available with thicker walls for use on very high-pressure sources.

A second type of steel canister is the passivated canister or Summa canister. These steel canisters have had the interior wall treated with acid to remove any active sites and reduce the chance of the sample's decomposing on the wall of the container. These canisters have found application in sampling ambient atmospheres. They can be fitted with a control device that limits how fast the container fills and thus allows the duration of sampling to be controlled. The major disadvantage with steel canisters used for atmospheric sampling is that to remove the sample, the sample has to be pressurized to at least 2 atm, which dilutes the sample by one-half and reduces the amount of detectable compounds in the sample by one-half.

Glass Bombs. Glass bombs also come in a variety of sizes, but are generally less than 1 L. Glass bombs are usually cylindrical. They are quite fragile and can be difficult to work with in the field. They cannot tolerate pressures as high as those of the steel canister, but higher than those of the flexible bag. Glass bombs generally have inert valves at either end of the cylinder and a rubber septum in the side of the cylinder. They have the advantage of not having to be evacuated for positive-pressure sources. Simply attach one end to the source, open both valves, and allow gas to flow through the glass bomb for a few minutes. Close both valves, and the sample is collected. For ambient use, the glass bomb does have to be evacuated first. For sampling of ambient sources, only one valve of the evacuated cylinder is opened, and only long enough to equilibrate with the source. The disadvantage of the glass bomb is that samples are generally withdrawn for analysis with a gastight syringe. For glass bombs pressurized above atmospheric pressure, this is not a problem; but for samples at atmospheric pressure, the gas composition can be altered slightly as multiple samples are withdrawn.

Concentration Techniques

Concentration techniques allow for the collection of gaseous molecules by passing air through these media. The advantage of concentration techniques over the previously reviewed capture techniques is that the concentration of the molecules can be increased without affecting their relative concentrations. This can greatly aid in the detection of very low concentrations of gaseous constituents. For example, the detection limit for most molecules by gas chromatography (GC) using an injection volume of 1 mL is about 1 ppm or 1 mg/m^3. Using concentration techniques, the gaseous components of 1 L of air can be concentrated using a concentration technique. This entire concentrated sample can then be introduced into the GC, thus improving the detection limit to 1 part per billion (ppb) or 1 $\mu g/m^3$. In this section the various media that can be used for concentrating samples and the advantages and disadvantages of each will be reviewed.

Porous Polymers. Porous polymers are a class of materials that, as the name describes, are very porous. This means that there is a great amount of space between the molecules in these materials—so much space that other molecules easily become trapped in these spaces, also called *interstitial spaces.* This property makes porous polymers ideal for adsorbing molecules and later releasing them without altering the chemistry. There are a wide variety of materials that can be classified as porous polymers including Tenax, XAD resin, and most GC column packing materials.

These materials are usually packed into glass or metal tubing using glass or quartz wool to seal the ends of the tubing to contain the material. To collect a sample of air, simply attach a vacuum pump to one end of the tubing and applying the vacuum, pass air from the source to be sampled through the media in the tube. Several properties of these media need to be kept in mind.

First, each type of medium has an affinity for certain molecules. For example, Tenax works well with compounds that have at least six carbon atoms or greater. The reason is that the larger molecules become entrapped in the interstitial spaces more easily than the molecules below six carbon atoms. Although some amount of the lower-molecular-weight molecules will become entrapped during the sampling process, others will escape out the other end and be lost. Therefore, their concentration in the original source will not be accurately quantitated. This ability for some molecules to "break through" can be quantitated. Let us assume that it has been decided to use one of the porous polymers and trap design of 100-mL volume. The breakthrough volume for this arrangement can be determined by exposing the trap containing the media to a source of known concentrations of, e.g., an analogous series of alkanes—methane through dodecane. Traps are exposed to varying volumes of this known gas from 100 mL to 1 L. Plotting the concentration of each constituent would show at what volume these molecules break through. It would also demonstrate the detection limit that could be achieved for the gases that did not break through. After passing 100 mL of gas through the porous polymer, all the gases would be quantitatively sampled, but the detection limit would be about the same as that in taking a sample in a flexible bag. After passing 1 L of gas through the porous polymer, the methane would be detected at one-tenth the concentration that was present in the original sample. The hexane through dodecane would be detected and quantified with a detection limit 10 times greater than that with a flexible bag. After 10 L of gas had been passed through the porous polymer, the hexane might be only 90 percent retained, but the detection limit would now be 100 times lower than that for a sample collected in a flexible bag.

Through such experiments one can easily determine the optimum conditions of sample volume and breakthrough volume that allow the compounds of interest to be collected quantitatively. Also, two sample tubes can be connected in series. The second tube has the same breakthrough volume as the first tube, but since no molecules get through the first tube until their breakthrough volume is exceeded, the second tube provides a backup or effectively doubles the breakthrough volumes of all molecules being sampled. Analysis of both the first sample tube and then the second can provide quantitative results for the lighter molecules that may break through the first tube. To check the effectiveness of the second tube, all of the molecules that are collected in the second tube should be significantly less than the amount found in the first tube. If the concentrations are equal, that molecule was not collected quantitatively.

The elution or removal of the molecules trapped on a porous polymer is quite important if the detection limit is to be enhanced significantly over that of the capture techniques. If molecules that have been concentrated on a porous polymer are all removed or eluted at the same time and analyzed at the same time, then the concentration factor is maintained. If, however, the elution technique and subsequent analysis result in a dilution of the sample, then the concentration effect is lost and the detection limit is adversely affected. The preferred elution technique for a porous polymer that ensures the maximum concentration and therefore the minimum detection limit is thermal desorption. Simply heat the trap very rapidly and desorb the molecules into the detector. Techniques involving solvent elution and subsequent introduction into the detector are not effective in maintaining the effect that concentration produced.

The major drawback to collection on porous polymers and subsequent elution by thermal desorption is that each sample represents an attempt by the analyst to get everything to work correctly. Also, if the tube is broken in transit, which is a problem with glass sampling tubes, then the sample is lost.

In spite of the drawbacks of porous polymers, they can be very effective in collecting samples and identifying trace constituents that may be important in the odor of a material as well as in determining the trace toxic constituents that may be present.

Charcoal. Charcoal is very similar to porous polymers, except for one major difference. Thermal elution techniques work very poorly with charcoal. Molecules tend to decompose when thermally eluted. Usually, molecules that are adsorbed on charcoal are eluted with a good solvent such as carbon disulfide. Heat is also used to enhance the solubility of molecules in the solvent, and an incubation period is used. Therefore the use of charcoal as a collection medium does not usually enhance the detection limit over that of flexible bags. The advantage of charcoal as a collection medium is that it is usually available in small glass cartridges that are much smaller than flexible bags and therefore easier to ship to the laboratory.

Charcoal has not found much use in the detection of odors as a collection medium, but more in the control of offending odors or in the use of some designs of olfactometers. Its use is primarily in source testing where it can be employed on site and easily shipped back to the laboratory for analysis. Charcoal also has larger breakthrough volumes for most molecules than porous polymers, which slightly offsets the loss of detection limit.

Polyurethane Foam (PUF). Polyurethane foam is an alternative material for collecting molecules, and several methods have been developed for molecules such as pesticides and polychlorinated biphenyls (PCBs). PUF works like porous polymers do in that the spaces in the foam trap large molecules. Although PUF cannot be thermally desorbed, it can be eluted with a number of solvents, and the adsorbed molecules quantitatively detected. It is a very special class of porous polymers and should not be used except for the collection of molecules described in methods specific to this medium.

Cryogenic Techniques. Cryogenic techniques rely on the freezing point of molecules. A U-shaped tube is immersed in a material at below ambient temperature. This can be ice, dry ice, liquid nitrogen, or liquid helium. Any material can be used if its temperature is lower than the freezing point of the molecule of interest. The advantage of this technique is that certain molecules can be preferentially collected, or all the molecules that freeze above the temperature of the cryogenic medium can be collected. There is one major problem with cryogenic techniques. If the source has a lot of water, then with most cryogenic techniques, the water will freeze and plug the U-tube, thus halting the collection process. Cryogenic techniques are used in several methods, and this is discussed in the section on Methods.

Impingers. Impingers are collection devices that contain a liquid. The sample is drawn through a tube that is immersed in the liquid. Molecules of interest can be collected in the liquid for subsequent analysis. Acids and bases can be used as the liquid to collect acid and base gases, respectively. Impingers can also be used in series with different liquids in each impinger to collect different gases. Impingers can be used to remove interfering gases from a source before presenting the gas to the detector. The major drawback to impingers is that a huge excess of liquid has to be used to ensure the quantitative collection of the target gas. This results from the poor contact between the bubbles of gas exiting the immersion tube and the liquid. Glass frits are frequently used on the end of the immersion tube to produce the finest bubbles, so that the contact between the bubbles and the liquid can be maximized. Cooling the impingers can enhance the solubility of the gas in the liquid by reducing the vapor pressure of the gas. The methods employing impingers are discussed in greater detail in the section on Methods.

Treated Media. Treated media are glass fiber or cellulose filter media, porous polymers, or other media that have been soaked in a solution which has an affinity for a particular molecule or group of molecules. These highly specialized sampling media are used to measure a single compound or class of compounds. As the gas is drawn into the tube with a vacuum source, it reacts with the solution on the substrate. Some tubes are even available that contain a solution that produces a color change so that the amount of target gas can be detected by the color change produced on the medium as the sample is collected. If a known quantity of gas is pulled through the tube, for example, 100 mL, then the tube can be calibrated; and the tube is actually marked in concentration so that the concentration of the target gas can be read directly from the collection tube. Other sampling tube configurations require the material that is preferentially collected be eluted with a solvent for subsequent analysis in the laboratory.

LABORATORY ANALYTICAL TECHNIQUES

There are several laboratory analytical techniques available for the investigation of odors and VOCs, gas chromatography (GC), gas chromatography–mass spectrometry (GC-MS), ion-specific electrodes, and acid-base titrations. All these techniques are very useful and produce results that are quite satisfactory in these investigations.

As stated before, all odors are chemicals. The majority of odorous chemicals are organic, although several important ones are not considered organic, such as ammonia and hydrogen sulfide. VOCs are all organic, as the name implies. GC and GC-MS provide the best solution for analysis of organic compounds, as explained below. Ammonia and hydrogen sulfide can also be analyzed by GC and GC-MS. However, several alternative methods such as ion-specific electrodes and acid-base titrations can also be used.

Gas Chromatography

GC is the workhorse of the analytical laboratory. GC is an elegant technique that was invented near the end of World War II. It allows one to separate a complex mixture into its individual components and to identify the compounds. This has a tremendous advantage over other techniques, because a mixture containing hundreds of compounds can be investigated and the components identified. No other analytical technique allows for this type of diversity.

The GC technique is quite simple. The instrument consists of an injector in which the sample is introduced. This is nothing more than a miniature block heater with a hole down the middle. At one end is a device that allows introduction of the sample, usually a rubber septum. At the other end is a connection to the column. The purpose of the injector is to ensure that as much of the sample as possible is vaporized when introduced to the interior of the block heater. The vaporized sample is known as the *mobile phase.*

The column can be made of any number of materials including copper tubing, steel tubing, glass tubing, or more recently quartz optical fibers of various internal diameters. Inside the column are two components: a stationary phase and a support. In the case of large-diameter tubing of glass, steel, or copper, the support is usually an inert material such as pumice or firebrick that has been ground to a fine mesh. Coated on the support is the stationary phase that is usually a material such as soap. With quartz optical fibers, also called open tubular columns, the stationary phase is simply coated on the walls of the fiber. The internal diameter of these columns is quite small, which allows the molecules to have better contact with the walls than for the large "packed" columns.

One end of the column is attached to the injector oven; the other end is attached to the detector oven. The detector oven is a block heater containing the detector. The detector can be a thermal conductivity detector (TCD), flame ionization detector (FID), an electron capture detector (ECD), a photoionization detector (PID), an electrolytic detector, or a mass spectrometer. Each detector has certain properties that allow for the detection of certain molecules. The selection of the detector and the column is quite important in the determination of the target compounds. For example, TCD works best on permanent gases such as methane and ethane. The FID is the best all-around detector and is used for most general work. The ECD detects chlorinated compounds.

The operation of the GC is quite simple. The sample is introduced into the injector. The temperature of the injector oven vaporizes the compounds that make up the sample into the gaseous state. A carrier gas such as helium that is introduced into the sidewall of the injector oven sweeps the gaseous molecules into the column. As the compounds are swept through the column, each compound is preferentially absorbed and desorbed from the stationary phase depending on its affinity for the stationary phase. That is, some of the molecules prefer to be in the gas phase, and others prefer to be absorbed in the stationary phase. This results in the partitioning of the compounds, or a separation. The compounds that prefer the gas phase reach the end of the column first; the compounds that prefer to be absorbed reach the end of the column later. As the compounds emerge from the end of the column, they are detected. In the case of the TCD, the compounds alter the conductivity of a hot filament. In the case of the other detectors, the compounds are ionized in a flame or by a radioactive source and are detected by an electrometer. The signal from the detector is fed to a recorder that plots response versus time. The result is the familiar gas chromatogram that consists of a baseline tracing in which each compound is represented by a peak. The area of each peak can be used to determine the concentration of the compound in the original sample.

Each peak that emerges from a GC column can be identified by comparing the time that it takes to traverse the column, called the retention time, to a known set of retention times for various compounds. For a given combination of column stationary phase, injector, and detector oven temperatures, known compounds will have specific retention times. The retention time becomes the marker that identifies that compound uniquely and allows the GC technique to qualitatively identify unknown compounds in mixtures. This is known as a secondary analytical technique, in that the identification is by inference. While the peak from the sample has the same retention time as the known compound under the same conditions, there is no way to verify this compound's identity. This is the major drawback of the GC technique. In the next section a technique is described that removes the uncertainty associated with not being able to positively identify the unknown.

Gas Chromatography—Mass Spectrometry

GC-MS is a hybrid of a separation technique, GC, and a specialized detector, MS. The coupling of these two techniques has greatly enhanced the analytical capabilities for complex mixtures.

Mass spectrometry is a unique tool in that it produces a fingerprint of any molecule introduced for analysis. A mass spectrometer operates under high vacuum. Molecules exiting the end of a GC column are first introduced into an evacuated chamber. Since the GC column is at approximately 30 lb/in^2, this first chamber is primarily for the purpose of removing most of the molecules and reducing the pressure to about 1 torr. Molecules in this first chamber are then drawn into a second chamber, at much lower pressure, through a skimmer or small hole. Upon entering the second chamber, the molecules that are traveling as a beam enter the ion region of the mass spectrometer. Electrons from a hot filament bombard the molecules with 70 eV energy, which breaks the molecule into a number of positively charged fragments. These fragments are repelled by a negatively charged plate through a series of lenses that focus the charged fragments into the mass analyzer, which consists of four charged rods. Two opposing rods are negative, and the other two are positive. The charge on these rods alternates very rapidly. The rods also have an increasingly larger direct current applied. As the charge fragments move into this region, they are alternately attracted to one rod or the other and quickly begin a spiral motion through the length of the rods. As the direct current is ramped up, the lighter fragments exit the rod chamber first, while the large fragments are destroyed on the rods. Later the heavier fragments emerge. The fragments are detected by an electrometer, and this signal is fed to a recorder. The resulting mass spectrogram is a plot of the ion intensity versus the voltage applied to the rods. This fingerprint is unique for each compound. This fingerprint is then compared to a computerized library of more than 50,000 compounds and can be identified with very high precision.

The ability of the GC to separate a complex mixture into its component parts and the MS to uniquely identify the individual components makes the GC-MS the most versatile of all analytical techniques.

High-Performance Liquid Chromatography

High-performance liquid chromatography (HPLC) is a technique that is used with some methods discussed at the end of this section. HPLC is essentially the same as gas chromatography except that the carrier is a liquid instead of a gas. Samples to be tested are injected in a manner similar to GC; however, the sample is injected into a stream of liquid. This liquid can be a single liquid that does not vary during the analytical process (isocratic elution operation), or one whose concentration can be altered during the analytical process to effect the elution of the compounds in the sample (gradient elution operation). The molecules are separated by the combination of the effects of the column packing and the carrier solution. This method is usually used with molecules that have been collected in solution and that are too fragile for the temperature conditions of the GC or too large to pass through a GC column.

GC Sniffing

GC sniffing is a special application of gas chromatography that is very useful in the identification of odor-causing compounds in a complex mixture. Since odor is a subjective property only definable by humans, the mere separation of a complex mixture into its component parts by GC is not enough to determine which components might be odorous. The human nose needs to be applied.

In GC sniffing, the detector end of a GC column is split with one-half of the gas exiting the column going to the detector, and the other half exiting through a tube in such a way that the operator can smell each peak as it exits the column, as indicated by the detector, and can annotate the chromatograph with an odor descriptor.

This technique is only qualitative, since the compounds that are detected as odorous in the GC sniffing technique may be above the odor threshold whereas in the original environment, these same compounds may be below the odor threshold.

Titrations

In some collection methods discussed in detail at the end of this section, the compounds of interest are sampled into impinger solutions. These solutions are then titrated to determine the amount of the compound of interest.

FIELD TECHNIQUES

Field techniques allow for the measurement of VOCs and odors directly in the field or at the source. There are a wide variety of techniques from simple devices to very complex devices. In this section we cover all these techniques.

Organic Vapor Analyzers

Organic vapor analyzers (OVAs) are simply gas chromatographs without the column. Since they do not allow the separation of complex mixtures into component parts, they only measure total organic vapors. Several types are available and generally differ only in the type of detector used. They are quite simple to use. A gas of known concentration is used to standardize the unit. For example, a gas such as isobutane at 100 ppm can be used. The OVA is zeroed with ambient air, and then the standard gas is introduced. The unit now reads total organic vapor concentrations as isobutane. A built-in pump in the OVA samples the unknown gas, and the gas passes over an ionizer such as a FID, radioactive source (rare), or a photoionization source such as a uv lamp. The ionized gas is then detected by the electrometer, and the signal is read out on a meter.

These units are ideal for survey work in which the investigator is trying to isolate areas that are producing high emissions of organic vapor. While the readings do not give any indication of the organics that are causing the response, the reading can be very useful.

The detection limit of these devices is usually about 500 ppb. This sensitivity, while not very useful in odor-related situations, can be quite useful in industrial hygiene and survey work. The device also has the ability to be very portable, and the sampling port can be placed very close to the source.

Monitors

A large number of monitors are available. Most are specific to a certain chemical, e.g., ammonia, hydrogen sulfide, and total organics. They are generally not portable but are rack-mounted and intended to be used over a long time to monitor a specific source. These devices are usually brought to a location, and the source gas is introduced through a heated line or some other device. These devices work very well with specific sources of VOCs or odors, but do not work at all well with nonpoint sources or in ambient applications. These are discussed in greater detail in the section on continuous emission monitors (Sec. 4.3).

Absorption Tubes

Absorption tubes are actually similar to the collection devices discussed earlier. The collection medium is impregnated or coated with a chemical that reacts preferentially with a specific chemical. An indicator is also added so that a color change is produced.

These devices are quite simple and produce semiquantitative results. A tube is selected that responds to the range of interest, and air is drawn through the tube with a hand pump. If the amount of air drawn through is precisely what is recommended for the particular tube, then the calibrations on the side of the tube give an accurate reading of the concentration in the air sampled. These devices are used most frequently in the industrial hygiene area.

Portable GCs and GC-MSs

There are a wide variety of commercial GCs and GC-MSs that have been designed very compactly expressly for use in field applications. They operate exactly as the laboratory instruments, but have much simpler electronics and fewer bells and whistles. These devices, however, require that a discreet sample be introduced. This is usually accomplished with a gastight syringe or a valving system. Portable GCs and GC-MSs simply take the laboratory out to the field and allow the investigator to obtain laboratory-quality data without having to ship samples to the laboratory and wait several days for the results.

FINDING A COMPETENT LABORATORY

In the previous sections, we discussed the various techniques available to collect and analyze samples for VOCs and odors. Now that you have an understanding of the methods, you are in a better position to talk intelligently with a laboratory about your needs. But how do you find a competent laboratory to perform the sampling and testing described? The majority of commercial laboratories specialize in the analysis of water, wastewater, waste, and soil. They are very familiar with these media, but most do not know how to handle air samples, which is the subject of this book. There are several laboratories that specialize in air testing. The simplest way is to talk to laboratories in your area. The laboratory community is generally aware of laboratories to which customers have been referred for testing that the laboratories are not capable of performing. They want to keep their current customers and any future customers happy, and the best way is to refer them to competent laboratories for test methods that they do not perform.

Also, there are listings of approved laboratories compiled by the Environmental Protection Agency (EPA). Further, regional EPA personnel or state environmental management personnel may have knowledge of qualified local laboratories. Buyers' guides are another source of advertisements and recommendations of laboratories by test method or field of testing.

METHODS

In this section, the various methods published that rely on the techniques discussed in this section are briefly listed for the reader's reference. A brief description of each method is given in order to help the reader decide if this method would help achieve a particular goal. Each method is designed for a specific class of compounds. While space does not allow for the listing of the compounds covered by each method, readers should acquaint themselves with these methods and choose the method that is specific for the compound or compounds of interest. An attempt has been made to include some of the limitations where appropriate.

"Compendium of Methods for the Determination of Toxic Organic Compounds in Ambient Air." U.S. Environmental Protection Agency. EPA-600/4-89-017 (Supplements: 600/4-87-006, 600/4-87-013)

The methods described in this section are used for collecting and identifying ambient VOCs. The major limitation of these methods is that they generally must be used for ambient air, or for sources that are not at elevated temperatures. The problem was discussed earlier, but bears repeating. Molecules at elevated temperatures will condense when the temperature is drastically reduced. If the method relies on the molecules remaining in the vapor state, which many of these methods do, then these methods are not acceptable.

Method TO-01—Determination of Volatile Organic Compounds in Ambient Air Using Tenax Adsorption and Gas Chromatography–Mass Spectrometry. Method TO-01 uses a cartridge of Tenax to absorb molecules in the boiling range of about 80 to 200°C. The cartridges are then desorbed directly into a GC or GC-MS for analysis. The temperature of the source must be near ambient, or else the molecules will not be absorbed. This method works very well for molecules above six carbon atoms and is especially useful for aromatics and chlorinated compounds.

Method TO-02—Determination of Volatile Organic Compounds in Ambient Air by Carbon Molecular Sieve Adsorption and Gas Chromatography–Mass Spectrometry. Carbon molecular sieve is an alternative to Tenax. As with method TO-01, it uses a cartridge containing carbon molecular sieve, which is effective in collecting lower-boiling-point compounds than Tenax. The boiling range of compounds is from −15 to 120°C. This method is particularly effective with compounds that are highly volatile and with nonpolar organics.

Method TO-05—Determination of Aldehydes and Ketones in Ambient Air Using High-Performance Liquid Chromatography. Method TO-05 uses impingers filled with a solution of DNPH (dinitrophenylhydrazine) to collect the sample. This method is more tolerant of sources above ambient temperature, but the temperature should still be of concern. It is specific for aldehydes and ketones. These compounds react with the DNPH to form diphenylnitrohydrazone derivative that can then be analyzed in the laboratory using HPLC. This method is an alternative to EPA method 0011, discussed later in this section. The main difference between the two methods is that the DNPH solution used in method 0011 is saturated while the solution in method TO-05 is a much more dilute solution. The recovery process is also different, with method TO-05 forming a visible emulsion at stoichiometries above about 5:1 sample to DNPH. The recovery process will not quantitatively recover this emulsion using the recovery method described in method TO-05. It is recommended that if the source being sampled has high concentrations of aldehydes, method 0011 be used.

Method TO-06—Determination of Phosgene in Ambient Air Using High-Performance Liquid Chromatography. This method is specific for phosgene. A midget impinger containing an aniline/toluene mixture is used to collect the sample. The sample is subsequently analyzed by HPLC. Caution should be exercised in any sampling around phosgene that has the odor of newly mowed grass. This compound is deadly, and the odor threshold is above the OSHA-recommended limit for this compound.

Method TO-07—Determination of N-Nitrosodimethylamine in Ambient Air Using Gas Chromatography. This method utilizes a cartridge containing Theromosorb/N. This is also a porous polymer much like others that have been discussed. In this particular case, the sample is eluted not with temperature, but with various solvents. Elution of absorbent cartridges with solvents reduces the concentration effect that was obtained during the sampling process. For example, if 100 L of gas has been sampled and the concentration of *N*-nitrosodimethylamine is 1 ppb, then using 10 mL of eluent to remove the sample raises the detection limit by a factor of 10. Thermal elution or cryofocusing is always preferred, but if elution with solvents is the only alternative, then the amount of air that must be sampled must be great enough to ensure that during the elution process, enough of the analyte of interest remains to be detected.

Method TO-08—Determination of Phenol and Methyl Phenols (Cresols) in Ambient Air Using High-Performance Liquid Chromatography. This method can be used for the determination of phenols and cresols in air at approximately 1 to 5 ppb. This method uses midget impingers containing sodium hydroxide followed by analysis by HPLC.

Method TO-11—Determination of Formaldehyde in Ambient Air Using Adsorbent Cartridge Followed by High-Performance Liquid Chromatography. This method is specific for formaldehyde and other aldehydes and ketones. This method uses a DNPH-coated silica gel support. Air is passed over the sample in a cartridge, and the same stable compounds as formed in method TO-05 are formed. The sample is eluted with acetonitile followed by analysis using HPLC. As previously mentioned, the effects of concentrating the sample on the solid support are reduced by the amount of eluent used in recovering the sample.

Method TO-14—Determination of Volatile Organic Compounds in Ambient Air Using Summa Polished Canister Sampling and Gas Chromatography. Method TO-14 uses an empty steel container, discussed previously, to collect the sample. The sample container is then pressurized to approximately 5 lb/in^2, and the sample is withdrawn and cryofocused on the GC column. Cryofocusing simply means that the GC column is at or near liquid nitrogen temperature (77 K). This sampling and analysis method is applicable to a wide range of VOCs. It is not applicable for sources that are at elevated temperatures, because the molecules can condense at the lower temperature and fail to be removed during the analysis process. This could distort the relative concentrations of contaminants in the source. The advantage this method has over cartridges that are eluted by thermal desorption is that more than one sample can be extracted from the steel container. So there is less chance of a sample's being lost during the analysis process.

"Test Methods for Evaluating Solid Waste," 3d Ed. Report No. SW-846. U.S. Environmental Protection Agency, Office of Solid Waste and Emergency Response. Washington: 1986

These methods are designed primarily for determining the destruction and removal efficiency of control technologies in sources. They were designed primarily with incinerators in mind and so are less susceptible to the temperature of the source. They also are capable of sampling much larger volumes of air than the ambient methods.

EPA Method 0010—Modified Method 5 Sampling Train. This method utilizes a complicated apparatus consisting of a probe, a filter, and several impingers to collect the compounds of interest. The probe is inserted into the source. The filter collects any particulate matter above about 0.45 μm. Gaseous compounds pass through the filter and are collected in the impingers. This method is particularly effective for semivolatile compounds.

EPA Method 0011—Sampling for Formaldehyde Emissions from Stationary Sources. This method is used for determining formaldehyde and other aldehydes and ketones emissions from sources. Air from the source is pulled through impingers containing saturated DNPH. The dinitrophenylhydrozone derivatives are formed and analyzed by HPLC.

EPA Method 0030—Volatile Organic Sampling Train (VOST). This method uses an apparatus that samples a source and collects the sample on a VOST cartridge containing Tenax. This method is similar to the TO-01 method except for the isokinetic sampling apparatus. Compounds in the boiling range of 30 to 100°C are effectively collected by this method. The analysis consists of GC or GC-MS.

EPA Method 5040—Analysis of Sorbent Cartridges from Volatile Organic Sampling Train. This method is the analytical method for samples collected using the VOST described above. This method utilizes thermal desorption to elute compounds from the VOST cartridge directly onto the GC column. This method should be reviewed to provide the reader with information about the analytical process and the extensive quality assurance and control measures that should accompany the analysis of samples.

EPA Method 5041—Protocol for the Analysis of Sorbent Cartridges from Volatile Organic Sampling Train: Wide-Bore Capillary Column Technique. This is another analytical method used to analyze sorbent cartridges. This method is specific to technology known as wide-bore capillary column chromatography. Method 5040 refers to the use of packed column chromatography while this method utilizes capillary column methods.

U.S. Environmental Protection Agency. *Code of Federal Regulations. Title 40, Part 61, Appendix B. Washington: Office of the Federal Register, July 1, 1987*

These methods were adopted for sampling existing sources. They are adapted from methods already discussed and are included here for completeness.

EPA Method 106—Determination of Vinyl Chloride from Stationary Sources. Method 106 is specific to vinyl chloride. A sample is collected in a Tedlar bag and returned to the laboratory for analysis using GC.

"Standards of Performance for New Stationary Sources." Compilation. U.S. Environmental Protection Agency, EPA-340/1-77-015

These methods were designed for sampling new sources.

EPA Method 15—Determination of Hydrogen Sulfide, Carbonyl Sulfide, and Carbon Disulfide from Stationary Sources. This is an on-site method that can be used for the compounds listed in the title. A sample is drawn from a source using a probe and a heated transfer line. The sample is carried to the inlet of a GC where it is sampled and analyzed using a flame photometric detector.

EPA Method 16—Determination of Hydrogen Sulfide, Dimethyl Mercaptan, Dimethyl Sulfide, and Dimethyl Disulfide from Stationary Sources. This method is similar to method 15 except for the compounds being analyzed.

EPA Method 18—Measurement of Gaseous Organic Compound Emissions by Gas Chromatography. This method utilizes a GC at the source. A sample is withdrawn using a probe and heated line to carry the sample to the GC. The sample is then analyzed by whatever chromatographic conditions are appropriate for the compounds of interest. This method is quite complicated. The compounds of interest in the gas stream must be known in advance. This is usually done by a preliminary investigation to determine what compounds may be present. Then the chromatographic conditions are chosen to optimize the analytical process to ensure that the compounds of interest are all detected quantitatively and resolved during the analytical process.

NIOSH Manual of Analytical Methods, Part 1: *NIOSH Monitoring Methods,* Vol. 1. U.S. Department of Health, Education, and Welfare, National Institute for Occupational Safety and Health. Cincinnati, OH: Revised 1985

These methods are primarily designed for use in workplace sampling. The methods attempt to obtain information on worker exposure during an 8-h work period. Hundreds of methods have been developed and can be quite useful and are very specific for the compounds of interest. As with the toxic organic methods, these methods are primarily designed for use in ambient air and are not recommended for elevated-temperature sources. They are very effective with ambient sources, however. The reader is referred to the manual of methods, and no examples are provided.

Stationary Source Tests Methods, Volume 3: *Methods for Determining Emissions of Toxic Air Contaminants from Stationary Sources.* State of California Air Resources Board, Monitoring and Laboratory Division. Sacramento: 1989

The California Air Resources Board has developed methods of its own. These methods may differ from some of the methods already discussed; however, the reader is referred to this resource as an alternative to what has already been provided.

SECTION 4.2

SENSORY TEST METHODS

Samuel S. Cha
TRC Environmental Corporation
Windsor, Connecticut

BACKGROUND 4.16
- **Threshold** 4.16
- **Intensity and Pervasiveness** 4.18
- **Character** 4.20
- **Hedonic Tone** 4.21

HUMAN RESPONSE TO ODORS AND ODOR PERCEPTION 4.21
- **Sensitization, Desensitization, and Tolerance of Odors** 4.21
- **Odor Mixtures** 4.21
- **Other Factors Affecting Odor Perception** 4.21
- **Odor and Health Effects** 4.22

COMMUNITY SURVEYS 4.22

ODOR EMISSION SAMPLING 4.24
- **Source Sampling** 4.24
- **Flow Rate Measurement** 4.24

SENSORY ODOR MEASUREMENT 4.26
- **Odor Panel Selection and Screening** 4.27
- **Sample Evaluation—Detection Threshold** 4.28
- **Sample Evaluation—Intensity** 4.29
- **Sample Evaluation—Character** 4.30
- **Parameters Affecting Odor Measurement** 4.31

REFERENCES 4.31

BACKGROUND

An odor is a sensation produced by chemical stimulation of chemoreceptors in the olfactory epithelium in the nose. The chemicals that stimulate the olfactory sense are called *odorants,* although people frequently refer to odors when odorants are meant.

Olfactory studies have brought to light interesting information regarding odor perception. There are two sensory channels responsible for human detection of inhaled chemical substances in the environment. These are odor perception and the *common chemical sense* (CCS). Odor perception is a function of cranial nerve I, whereas CCS is a function of cranial nerve V. The CCS is described as pungency and associated sensations such as irritation, prickliness, burning, tingling, and stinging. Unlike the olfactory structure, the CCS lacks morphological receptor structures, and it is not restricted to the nose or the oral cavity (Cometto-Muñiz and Cain, 1991).

Several dimensions of human responses to the odor sensation can be scientifically characterized: threshold, intensity, pervasiveness, character, and hedonic tone.

Threshold

Threshold, or *detectability,* refers to the theoretical minimum concentration of odorant stimulus that is necessary for perception in some specified percentage of the population, usually 50 percent mean. A threshold value is not a fixed physiological fact or a physical constant, but is rather a statistical value representing the best estimate from a group of individual scores. Two types of thresholds can be evaluated: *detection* and *recognition.*

The detection threshold is the lowest concentration of odorant that will elicit a sensory response in the olfactory receptors of a specified percentage, usually 50 percent, of the population of a group being tested. Refer to Fig. 4.2.1. The detection threshold is identified as an awareness of an odor not necessarily involving its recognition.

The recognition threshold is defined as the minimum concentration recognized as having a characteristic odor quality by a specified percentage of the population, usually 50 percent. It differs from the detection threshold as it is the point at which a specific odor character description can be attributed to the sensory response.

If only a single odorant compound is involved, the odor dilution-to-threshold ratio can be expressed by the concentration. However, in measuring environmental odors, which are generally complex mixtures of compounds, the threshold cannot be expressed as a concentration of the mixture. Instead, threshold can only be expressed as dilution-to-threshold (*D/T*) ratios; it is dimensionless. A *D/T* ratio of 100 means that a given volume (i.e., 1 ft^3) of odorous air requires 100 volumes (that is, 100 ft^3) of odor-free air to dilute it to threshold, or a barely detectable odor. This concept is depicted in Fig. 4.2.1 where 50 percent panel response occurs at 235*D/T* for one sample and 344*D/T* for the other sample.

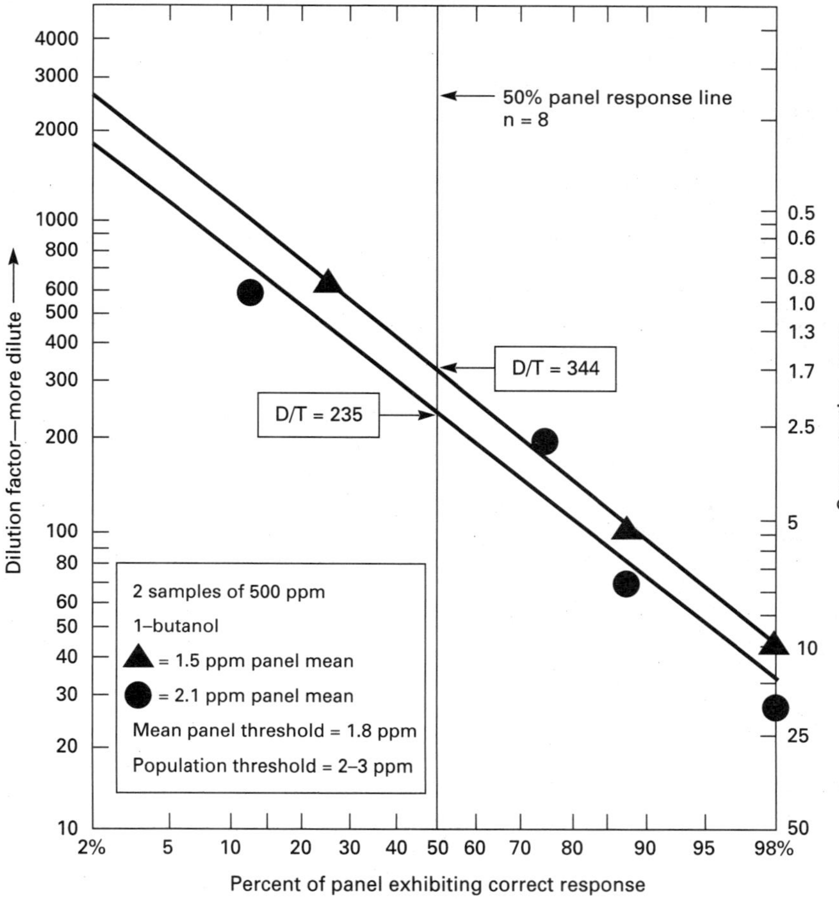

FIGURE 4.2.1 Odor detection dilution-to-threshold ratio.

Over the years, many terms have been used in the literature to express the concentration of odor. These include

Odor unit (OU) = 1 volume of odorous air at odor threshold; often the volume is defined in terms of cubic feet

$$\frac{\text{Odor units}}{\text{ft}^3} = \frac{\text{volume of sample diluted to threshold (ft}^3\text{)}}{\text{original volume of sample (ft}^3\text{)}}$$

$$\text{Odorant quotient: } Z_t{}^{\dagger} = \frac{C_0}{C_t} = \frac{\text{odorant concentration of sample}}{\text{odorant concentration at threshold}}$$

Both odor units/ft^3 and Z_t are dimensionless and are synonyms for dilution-to-threshold ratio, or *D/T*; odor unit is a unit of volume and is not a synonym for *D/T* (ASTM, 1993; Turk, 1973).

Another term, *odor emission rate,* is often used to describe the severity of the potential downwind impact, i.e., the problem a typical odor source could create at downwind locations. The odor emission rate, expressed in unit volume per minute, is the product of the dilution-to-threshold ratio (*D/T,* dimensionless) of the odorous air and the airflow rate (ft^3/min or its equivalent).

Listings of threshold values for pure chemicals are common in many environmental resource references. It is important to understand the limitations of reported threshold values, which can vary as much as several orders of magnitude, suggesting that thresholds are limited in usefulness. However, review of threshold value methodology finds that when threshold values are subjected to basic methodological scrutiny, the range of experimentally acceptable values is considerably reduced. There are two documents of odor threshold reference guides available recently. One, published by the American Industrial Hygiene Association (AIHA) (1989), is based on a literature review for 183 chemicals with occupational health standards. The other document was published by the Environmental Protection Agency (EPA) (1992). It compiled odor thresholds for the 189 chemicals listed in the Clean Air Act Amendments of 1990. Other threshold compilations available are by Amoore (1983), Fazzalari (1978), Ruth (1986), and van Gemert and Nettenbreijer (1977).

Intensity and Pervasiveness

Odor *intensity* refers to the perceived strength of the odor sensation. Intensity increases as a function of concentration. The relationship of the perceived strength (intensity) and concentration is expressed by Stevens (1961) as a psychophysical power function as follows:

$$S = kI^n$$

where S = perceived intensity of odor sensation
I = physical intensity of stimulus (odorant concentration)
n = constant
k = constant

When the above equation is expressed in logarithms,

$$\log S = \log k + n \log I$$

†The letter Z is for H. Zwaardemaker, a Dutch scientist, who was the earliest investigator to use dilution ratios for odor measurement (ASTM, 1993).

then

$$n = \text{slope} \qquad \text{and} \qquad k = y \text{ intercept}$$

An intensity function on logarithmic coordinates of the standard odorant 1-butanol is shown in Fig. 4.2.2. The slope of the function, also called the *dose-response* function,

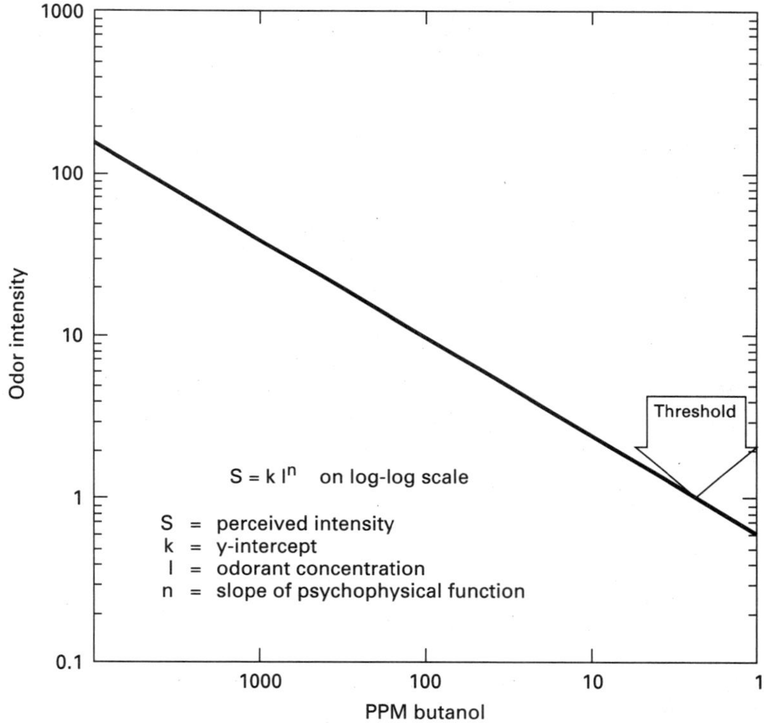

FIGURE 4.2.2 Intensity-concentration curve (1-butanol).

varies with type of odorant. Odor pollution control is concerned with the dose-response function, or the degree of dilution necessary to decrease the intensity. This is often called *pervasiveness* and can be described by the slope. A low slope value indicates that the odor requires greater relative dilution for it to dissipate; a high slope value indicates that the odor can be reduced by dilution more quickly. Compounds with low slope values are hydrogen sulfide, butyl acetate, and amines. Compounds with high slope values are ammonia and aldehydes. This explains why hydrogen sulfide, butyl acetate, and amines can be detected far away from the odor origin, while ammonia and aldehydes can be detected relatively near to the odor origin.

Reference Scales. The reference scale for measuring odor intensity has been standardized as the American Society for Testing and Materials (ASTM) Standard Practice E 544 (ASTM, 1993). The method proposes a standard reference odorant, 1-butanol, set in a series of known concentrations. The advantages of this method are that (1) it allows comparison of subjective odor intensities between laboratories, (2) it allows for odor control regulations to be expressed in terms of perceived intensity rather than odor thresholds, and (3) it allows cross-modality comparisons (i.e., sound and odor). One disadvantage is that some find it difficult to compare odors which have a different odor character from the reference standard.

Category Scales. Category scales were the first to be developed to measure odor intensity. One of the most widely used scales was developed for a 1930 study of odor used as gas alarms. The scale has six simple categories:

0 no odor

1 very faint odor

2 faint odor

3 easily noticeable odor

4 strong odor

5 very strong odor

The limitations of category scaling are that there are a finite number of categories to choose from and they are open to bias through subjective number preferences or aversions.

Magnitude Estimation. In magnitude estimation, the observer creates his or her own scale which is based on a specific reference point. The data from several observers are then normalized to the reference point. Magnitude estimation is a form of ratio scaling and has many advantages over category scaling (i.e., an unlimited range and greater sensitivity); however, it does require a more sophisticated observer and statistical analysis.

Character

The other dimension of odor is its character, or what the odor smells like. ASTM publication DS 61 (Dravnieks, 1985) presents character profiles for 180 chemicals using a 146-descriptor scale. The scale uses such terms as *fishy, nutty, moldy, rancid, sewer,* and *ammonia.* This dimension is useful in air pollution control to describe the source or process responsible for community odors because different odor characters are associated with various processes and industries. Table 4.2.1 provides a short list of the industrial odor character descriptions.

TABLE 4.2.1 A Short List of Industrial Odor Descriptors

Odor character descriptor	Potential sources
Nail polish	Painting, varnishing, coating
Fishy	Fish operation, rendering, tanning
Asphalt	Asphalt plant
Plastic	Plastics plant
Damp earth	Sewerage
Garbage	Landfill, resource recovery facility
Weed killer	Pesticide, chemical manufacturer
Gasoline, sulfur	Refinery
Airplane glue	Chemical manufacturer
Household gas	Gas leak
Rotten egg	Sewerage, refinery
Rotten cabbage	Pulp mill, sewerage sludge
Cat urine	Vegetation

Hedonic Tone

Hedonic tone, also referred to as *acceptability,* is a category judgment of the relative pleasantness or unpleasantness of the odor. Acceptability of the odor is based on the combination of frequency of occurrence, odor character, and odor intensity. In the context of air pollution field work, the hedonic tone is often irrelevant. Even pleasant odors may become objectionable if they persist long enough.

HUMAN RESPONSE TO ODORS AND ODOR PERCEPTION

Olfactory acuity in the population follows a bell-curve normal distribution of sensitivity, ranging from hypersensitive to insensitive and, for a few, all the way to anosmic (unable to smell). Individual odor threshold scores can be distributed around the mean value to several orders of magnitude.

Human responses to odor perception can follow patterns associated with both the olfactory and the CCS functions. These are discussed below to facilitate the understanding of what prompts odor complaints and the difficulties associated with odor identification and measurement.

Sensitization, Desensitization, and Tolerance of Odors

Repeated exposure to an odor can result in either an enhanced reaction described as *sensitization* or a diminishing reaction defined as *tolerance.* If a person or a population becomes sensitized to an odor, the result may be an increase in complaints regarding the odor. On the other hand, tolerance of an odor may contribute to one's unawareness of the continuous exposure to a potentially harmful substance. To illustrate the complexity of odor perception, short-term quickly repetitive or continuous-pungency stimulation results in an increased response to odors. Also, a person may be more sensitive to one odorant than to another. Such differences are often caused by repeated exposure to a particular odor. For example, this would not be uncommon for chemists or manufacturers who have been exposed daily to an odorant over time. As a consequence, they are very sensitive to the odorant. On the other hand, fatigue, or desensitization may also play an important role in odor perception. Repeated exposure to odorants would result in a desensitization to the odorants, such that an observer may no longer detect an odor although it is strongly detectable by another.

Odor Mixtures

An additional problem in odor identification is that generally odors are mixtures of compounds in different concentration levels. In a comparison, the perceived intensity of two chemicals, each at a certain concentration in a mixture, is lower than each chemical alone at the same concentration. This is known as *hypoadditivity.* On the other hand, the CCS responses such as pungency of the two chemicals in a mixture are mainly additive (Cometto-Muñiz and Cain, 1991).

Other Factors Affecting Odor Perception

Human response to odor is also based upon the community, meteorology, and topography interactions. An odor with an acceptable hedonic tone and low intensity may generate complaints when there is a lack of familiarity with the odor character, therefore causing concern with toxicity. Conversely, in some instances high intensities of hedonically unpleasant odors are tolerated because they are considered socially acceptable. Even pleasant odors such as perfumes or roasting coffee are considered objectionable if they are allowed to persist for a long time or are frequently present.

Demographically, awareness of air pollution problems increases with education, income/occupational level, and age. Some of this awareness is also based on economic factors, such as property values.

Odor and Health Effects

Odors may serve as a warning agent of potentially dangerous materials. The warning property of odor has long been recognized. In the Middle Ages, odors were held responsible for disease rather than the result of disease and poor hygiene. Physicians protected themselves by theoretically purifying the air in a sickroom with perfumed water. In this century, the property of odor as a warning agent has been utilized on a worldwide scale through the addition of pungent odorants such as ethyl mercaptan to odorless natural gas. The perceived connection between odors and disease persists. With national attention focused on waste and chemical spills, in the absence of specific information to the contrary, the average person will conclude that a bad smell is likely to be unhealthy.

The relationship between odor and health effects should be clarified. Many odorants are perceptible at concentrations far below harmful concentration levels. For example, hydrogen sulfide (H_2S) has been detected at concentrations as low as 0.01 part per billion (ppb), whereas the acceptable exposure limit recognized by the health risk professionals or the threshold limit value–time-weighted average (TLV-TWA) recommended by the American Conference of Governmental Industrial Hygienists (ACGIH) for this compound is 10 parts per million (ppm)—six orders of magnitude greater. Thus detection of the "rotten egg" character of H_2S is not necessarily an indicator of a potential health effect. Indeed, at concentrations near the TLV, the odor intensity of H_2S is unbearable. Similarly, the TLVs for cresol and ethyl acrylate are four orders of magnitude greater than their threshold values.

The nose is not a suitable screening device to determine the presence or the absence of a health risk. Although detection of an odor indicates that a chemical exposure has occurred, a more detailed sampling investigation should be conducted (Amoore, 1983).

Studies which have reviewed community odor and health problems reveal that a variety of common ailments are related to exposure to chemicals. However, in most cases, the identified chemicals were well below the thresholds for toxicity. This suggests that the detection of unpleasant odors may cause adverse physiological and neurogenic responses such as nausea, stress, and low concentration levels, but one must not discount that these effects are indeed a result of chemical exposure. This indicates that further studies are necessary to define allowable exposures and how they relate to the detection of odors (Shusterman, 1992; Ziem and Davidoff, 1992).

By definition, chemicals that are hazardous to health are considered toxic whether odorous or not. They are therefore controlled under existing laws such as the Toxic Substances Act and the Clean Air Act and related regulations.

COMMUNITY SURVEYS

There are two purposes of the community odor survey. The first is to identify possible sources of odor in the community other than the source(s) of concern and to provide some qualitative data on the odor levels that are observed downwind of the odor source(s) during various meteorological conditions. The second purpose is to develop data in order to validate odor complaints and the dispersion odor model. The validated model could then be used to predict odor levels that may be expected to occur under any given set of meteorological conditions.

The survey process involves walking or driving slowly around the neighborhood in the vicinity of the plant(s) of interest and recording the locations, odor levels, odor characters, and intensities of the observed odors as well as the wind speed, wind direction, ambient temperature, and sky conditions during the survey time period. The pattern of impact of an odor can be helpful in tracking the odor to its source. Due to meteorological stability, odor surveys should be conducted during various times of the day. For example,

Stable conditions	Early morning (6:00 to 8:00 a.m.)
	Evening (after 8:00 p.m.)
Unstable conditions	Between 9:00 a.m. and 7:00 p.m.

Odors are more easily observed due to inversion weather conditions which are most likely to occur during the stable meteorological conditions. This is because mixing by the wind is limited; therefore, odors are not easily diluted and dispersed.

The wind direction is noted by the observation of nearby trees or flags and the use of a compass. The wind direction is always recorded as the direction from which the wind is blowing. The wind speed can be measured using a handheld anemometer, and the temperature determined by a thermometer. The sky conditions include descriptions such as hazy, rainy, partly cloudy, and clear.

The odor levels are measured by using the odor classification scale, shown in Table 4.2.2. The scale is numbered 0 through 5, in which 0 represents no odor and 5 represents a very strong odor.

ASTM recommends a similar method; it suggests that the odor intensity be determined using a reference scale with *n*-butanol as the reference compound (ASTM

TABLE 4.2.2 Odor Classification Scale

Category scale		Odor annoyance
0	No odor	No detectable odors
1	Very faint	An odor that would ordinarily not be noticed by the average person, but could be detected by an experienced inspector or a very sensitive individual
2	Faint	An odor that is so weak that the average person might detect it if attention were called to it, but that would not otherwise attract her or his attention
3	Distinct, easily noticeable	An odor of moderate intensity that would be readily detected and might be regarded with disfavor (a possible nuisance in inhabited areas)
4	Strong	An odor that would force itself upon the attention and one that might make the air very unpleasant (a probable nuisance, if found in inhabited areas)
5	Very strong	An odor of such intensity that the air would be absolutely unfit to breathe (term to be used only in extreme cases)

Standard Practice E-544). This method uses *n*-butanol vapor as a reference odorant at eight concentrations ranging from approximately 15 and 2000 ppm air. These concentrations represent different odor intensities ranging from slightly above threshold (1) to very strong (8). Intensity ratings from 1 to 3 would not be expected to have an odor impact, odors described as intensity ratings 4 through 6 are a possible nuisance, and odors between 6 and 8 are a probable nuisance. An example odor survey record form for use during a community odor survey is presented in Table 4.2.3.

ODOR EMISSION SAMPLING

Source Sampling

Odor emission sampling is no more than typical integrated bag sampling. Samples are collected in Tedlar gas sampling bags using a "bag in the drum" technique. A vacuum is pulled on the drum which forces sample air into the sample bag. This is illustrated in Fig. 4.2.3. The sample bags are preconditioned with the sample air prior to sample collection to equilibrate the inner surface of the bag with the material being sampled. Without preconditioning the bag, odor levels may be significantly less due to adsorptive losses on the inner surface of the bag. Sample storage time, i.e., the time from when the sample is drawn until it is evaluated, must also be kept as short as possible to avoid deterioration of the sample. Ideal sample storage times are less than 4 h. Acceptable storage times for most sample mixtures are between 8 and 24 h, or no more than one overnight.

If the odorous air is very humid and it is suspected that condensation will occur, an additional method of predilution may be employed. The bag can be partially filled with a known volume of odor-free, dry air. The odorous air sample can then be collected into the remaining portion of the sample bag.

Another possible method of sample collection employs the use of a flux chamber, an EPA-recommended surface emission isolation device. The odorous air sample is drawn into the chamber from an opening in the side of the chamber and is collected in a sampling bag in the chamber. The flux chamber is extremely reliable in the collection of area source samples and the determination of the emission rate. The unit is placed on the surface, and air is added to the chamber at a controlled, fixed rate. The odor emission rate is calculated by knowing the flow rate of added air and the exposed surface area of the chamber.

Flow Rate Measurement

Flow rate measurements of the sources being sampled can be made using several different methods. The most often used method for point sources follows EPA Source Testing Reference Methods 1 and 2 (40 CFR Part 60), which include the use of a Pitot tube and an inclined manometer. One can also use other instruments such as the vane anemometer. Whichever device is used, the area of the source must be measured in order to complete the calculations of flow rates in cubic feet per minute (ft^3/min, or cfm).

For flow rate estimates of area sources, two possible methods are suggested. The first is to measure or estimate the area of the source and multiply that value by the constant 0.1 ft/s, which represents the upward emission rate of the odorous emissions. This value must be converted to cubic feet per minute to be consistent with all flow measurements. The second method also involves measurement or estimation of the height of the plume, measurement of the wind speed (in feet per minute), and measurement of the length of the downwind side

TABLE 4.2.3 Odor Survey Record

Odor Survey Data			**Client:**		
Project No.:		Sky condition:			
Date:		Wind direction:			
Conducted by:		Wind speed: mi/h			
Time period:		Temp.: deg. F			
Survey no.	Location	Time	Odor character	Intensity	Potential odor sources

Intensity: Possible values of intensity are 1 to 5.
Wind direction: Direction from which the wind is blowing.

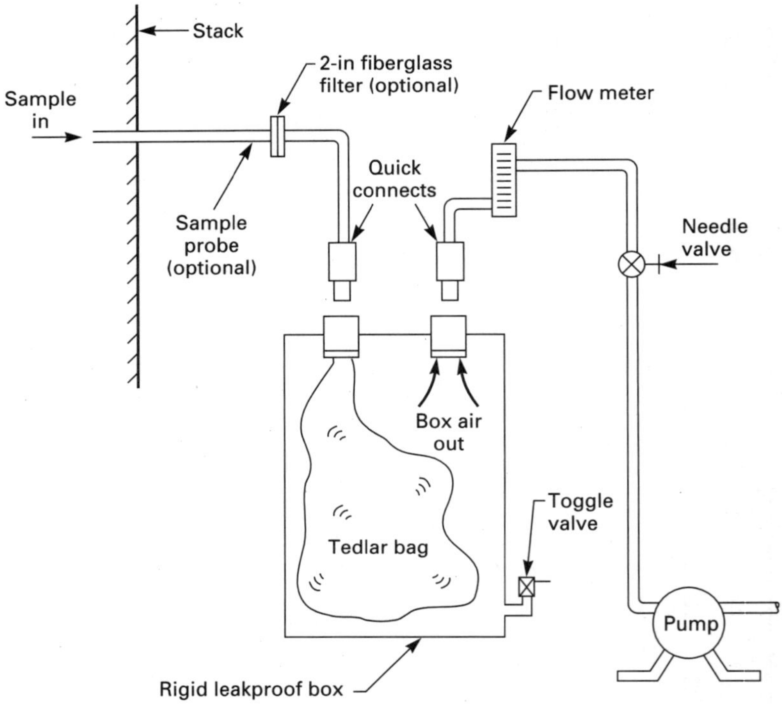

FIGURE 4.2.3 Integrated bag sampling train schematic.

of the source. Area sources should always be sampled at the downwind end. The flow rate, expressed in cubic feet per minute, will be the product of these three values.

SENSORY ODOR MEASUREMENT

The science of odor measurement uses many techniques. A famous early olfactometer (i.e., the instrument used to quantitate odor emissions) was developed by a Dutch researcher, Zwaardemaker, as early as 1895. Odor threshold values are sometimes referred to as *Z* scores in his memory. Recent efforts have been made to coordinate the research efforts in the field of odor measurement. The ASTM Sensory Evaluation Committee E-18 and the Air and Waste Management Association (AWMA) Odor Committee have both reviewed the problems generated by uncoordinated efforts and have made recommendations. Accordingly, the methodology presented in this book incorporates the recommendations of the following references:

- ASTM E 679-91. Standard Practice for the Determination of Odor Taste Thresholds by a Forced Choice Ascending Concentration Series Method of Limits
- ASTM E 544-75. Standard Practices for Referencing Suprathreshold Odor Intensity
- W. H. Prokop, Developing Odor Control Regulations: Guidelines and Considerations, *JAPCA,* 1978, 28: 9–22 (APCA Committee TT-4 position paper)

A comprehensive list of criteria for good threshold measurement techniques is as follows:

1. Panel size of at least six per group
2. Panelist selection based on odor sensitivity
3. Panel calibration
4. Consideration of vapor pressure of the odorant
5. Diluent comparability
6. Presentation mode
7. Analytical measurement of odorant concentration
8. Calibration of flow rate and face velocity (for olfactometers)
9. Consideration of threshold type
10. Ascending presentation series
11. Repeated trials
12. Forced-choice procedure
13. Concentration step increasing by a factor of no more than 2 or 3

These criteria are explained further in the next section.

Odor Panel Selection and Screening

The quantification of odors requires a panel of human judges since an odor is a sensation which occurs in response to a chemical stimulus (odorant). The only requirements for odor panelists are that they have a "normal" sense of smell and are able to follow directions. Individual "normalcy" is determined through screening tests.

There are typically three sources of odor panelists. In some cases, plant personnel are available. It is preferable that such panelists be administrative staff rather than production people, who may have anosmia (i.e., inability to smell) because of their constant exposure to the odors in question. When public relations are a major consideration, panelists may be recruited from the local community. This often helps to convince the neighbors that the company is sincere in its desire to resolve the odor situation. Panelists can also be recruited from temporary agencies.

An odor panel must be comprised of at least six people. Panelists must not be sensitive to odors or unable to smell odors. The panelists must be screened prior to their participation in the evaluation procedure to ensure that they have a normal sense of smell. Screening consists of two steps, as recommended in ASTM STP 440:

1. *Triangle test.* This is a threshold test for the panelist's ability to detect 1-butanol vapors in an aqueous solution contained in odorless Teflon bottles. Initially, a subthreshold concentration of 1-butanol in water is presented to the panelist along with two bottles containing only water. The panelist is required to sniff the headspace of each bottle and choose correctly the bottle containing the 1-butanol solution.
2. *Intensity test.* Panelists are tested for their ability to match odor intensities. The 1-butanol series is arranged in order from weak to strong. First the panelist is asked to become familiar with the series. Then one of the bottles is removed from the series, and the remaining bottles are arranged in a straight line. The panelist is asked to reinsert the bottle which has been removed within two places of its correct position.

The instructions to make the 1-butanol series are found in ASTM E 544-75.

Panelists are given a list of guidelines to follow for each day that they will be evaluating the samples. These guidelines restrict certain foods and the use of perfumes, etc., to ensure that unwanted odors are not brought into the room. This list of guidelines is presented in Table 4.2.4.

TABLE 4.2.4 Odor Panel Guidelines

On the day of an odor evaluation:

1. *Do not* wash your hair with strong, perfumed shampoo.
2. *Do not* wear perfume, scented powder, aftershave, or cologne.
3. *Do not* eat strongly flavored foods, especially Mexican or southern European.
4. *Do not* wear strongly scented clothing (i.e., leather clothing).

During the hour before an evaluation:

1. *Do not* smoke.
2. *Do not* drink coffee or tea or strong soda.
3. Avoid eating anything for at least one-half hour before the evaluation.
4. *Do not* chew gum or mints.

Sample Evaluation—Detection Threshold

The odor samples are evaluated for the threshold of detection by a dynamic dilution technique. The instrument which is used is called a *dynamic dilution forced-choice olfactometer.* The system uses carbon-filtered ambient air to produce a series of dilutions of the emission sample. All the instrument connections and tubing are either stainless steel or Teflon, to prevent any contamination of the sample. Outside air is pulled into the system by a pump and flows after the pump to a carbon filter. The odorous sample is attached to a quick-connect and pulled into the olfactometer by a venturi. The flow of the clean air remains constant, while flowmeters are used to create the sample dilutions. The clean air and the sample are mixed for presentation at the sniff cups. For quality-control purposes, the olfactometer is calibrated on a regular basis.

The dilutions are presented to the odor panelists (i.e., judges) in an ascending series, in threefold concentration steps. The dilution ratios are approximately $^1/_{3000}$, $^1/_{1000}$, $^1/_{300}$, $^1/_{100}$, $^1/_{30}$, $^1/_{10}$. Each dilution level is presented to the panelists by means of three identical sniff cups. The airflow to the sniff cups is maintained at 3 L/min, the rate of breathing. Two of the cups have only carbon-filtered air while the third has the diluted odorous air. Panelists are asked to choose which of the three differed from the other two, i.e., which has the odor. In this forced-choice procedure, panelists must choose a cup whether or not a difference can be discerned. The cup which contains the diluted odor is changed for each dilution level to avoid bias through the presentation techniques. Odorous exhaust from the olfactometer is removed by an exhaust line directed outside the room. Figure 4.2.4 illustrates the dynamic dilution forced-choice triangle olfactometer designed by the late Dr. A. Dravnieks, and manufactured by IIT Research Institute (IITRI).

The odor detectability is reported as the dilution-to-threshold ratio *D/T*. This value represents the ratio of the volume of odor-free air that must be added to the odorous air to reach threshold (i.e., barely perceptible). For example, a *D/T* value of 100 means that 100 volumes of odor-free air must be added to 1 volume of odorous air to reach threshold. The detection threshold represents the dosage of odorant necessary for 50 percent

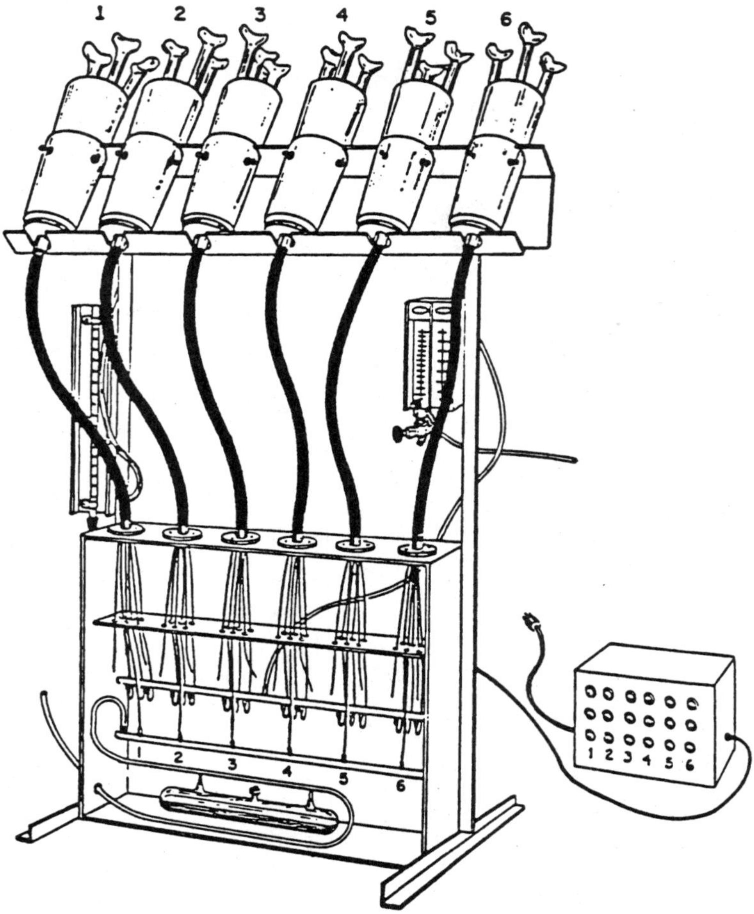

FIGURE 4.2.4 Dynamic dilution forced-choice olfactometer.

of the panel to detect a difference between the odorous stimuli and the odor-free air. On the triangle olfactometer, this is the point at which the panelist correctly selects the cup which contains the odor.

The *D/T* ratio is most often calculated by log dilution level versus probability plots. The correct responses are plotted against the odorant dilutions for a best-fit, least-squares straight-line function.

Sample Evaluation—Intensity

The perceived odor intensity can be measured by two methods. In both methods, suprathreshold levels of 1-butanol (standard reference ASTM E 544-75) are presented in twofold concentration steps. Panelists compare ports on the triangle olfactometer and indicate the comparable level. The intensity is reported by port numbers 1 through 8.

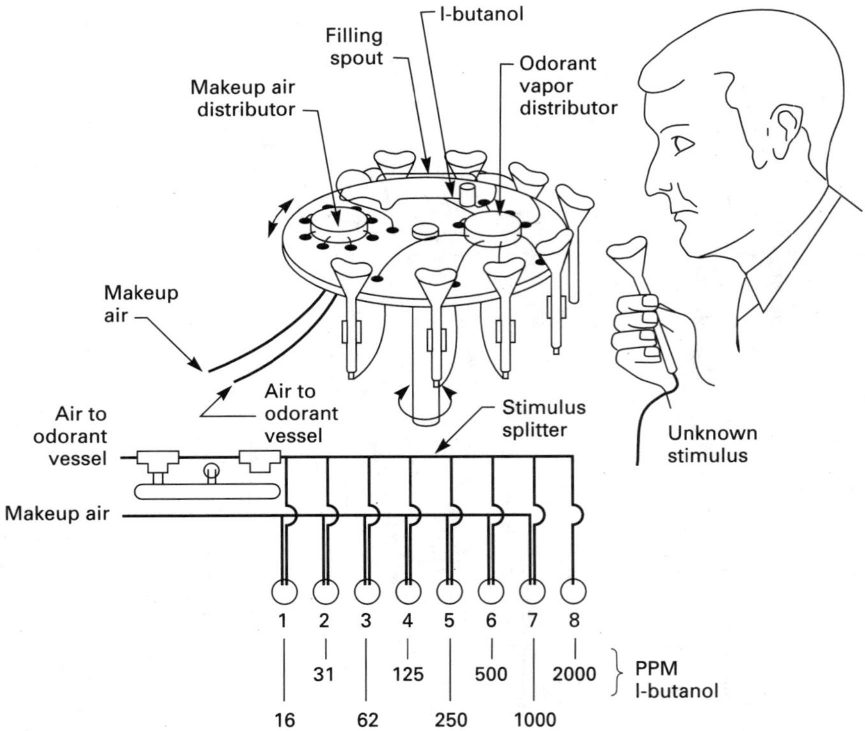

FIGURE 4.2.5 Butanol olfactometer.

Ports 1 through 3 are considered weak odors, ports 4 through 6 are moderate odors, and ports 7 and 8 are strong odors. The reported values are averaged for a mean intensity score. The first method employs a dynamic dilution binary-scale olfactometer arranged in a "lazy-Susan" configuration. The butanol olfactometer is depicted in Fig. 4.2.5, which was also designed by Dr. A. Dravnieks.

The second method is measurement by static scale. Rather than air as the diluent, odor-free distilled water is used. A series of eight solutions are prepared in either glass or Teflon (odor-free) bottles and are maintained at ambient temperature. Between sniffing, the bottles remain covered to ensure equilibration between the solution and the air headspace above it. The bottles are gently shaken by each panelist prior to sniffing to ensure equilibrium.

Sample Evaluation—Character

Odor character may be measured with a comprehensive scale of 146 odor descriptors developed by ASTM Sensory Evaluation Committee E18. The descriptor profiling form is discussed in the section Background. Panelists rate each odorant using all descriptors that apply on a scale from 0 to 5. The final scores are geometric means of percentage frequency with which a particular descriptor was chosen and percentage similarity score. Therefore, the odor descriptor represents both selection frequency and degree of similarity.

Parameters Affecting Odor Measurement

Some parameters are known to significantly affect odor sample measurement results: the dilution system, sample presentation, panelist selection, training, and panel size as well as panelist olfactory acuity.

It has been demonstrated that different odor measurement systems will produce different measured ratios for the same sample. A round-robin test conducted among practitioners indicated an order-of-magnitude difference when evaluating samples by the same means. One reason for this variability is that the flow rates at the dilution olfactometer's sniff ports are known to have significant influence on a panelist's perceptibility. Among the practitioners, there is continuing disagreement over the "correct" flow rate. The IITRI system described before uses 0.5 L/min. The Europeans are using 20 L/min currently. Others are in between. Generally, panels should consist of trained participants. Most dynamic dilution olfactometers require a panel size at least 6 to 10 panelists. The physical condition of the panelist is important. Panelists must be free from conditions that affect the olfactory system. Prolonged use of the sense of smell can cause olfactory fatigue. Repeated smelling of the same odor has the same effect. For prolonged test periods, rest periods are necessary.

REFERENCES

AIHA: *Odor Thresholds for Chemicals with Established Occupational Health Standards,* American Industrial Hygiene Association, 1989.

Amoore, J. E., and E. Hautala: "Odor as an Aid to Chemical Safety: Odor Thresholds Compared with Threshold Limit Values and Volatiles for 214 Industrial Chemicals in Air and Water Dilution," *J. Applied Toxicology,* 3:272–290, 1983.

ASTM: "Standard Practice for Referencing Suprathreshold Odor Intensity," E 544-75, *Annual Book of ASTM Standards,* vol. 15.07, American Society for Testing and Materials, 1993, reapproved 1988.

ASTM: "Standard Practice for Determination of Odor and Taste Thresholds by a Forced-Choice Ascending Concentration Series Method of Limits," E 679-91, *Annual Book of ASTM Standards,* vol. 15.07, American Society for Testing and Materials, 1993.

Cometto-Muñiz, J. E., and W. S. Cain: "Influence of Airborne Contaminants on Olfaction and the Common Chemical Sense," chap. 49 in *Smell and Taste in Health and Disease,* T. V. Getchell, et al., eds., Raven Press, New York, 1991.

Dravnieks, A.: "Atlas of Odor Character Profiles," ASTM Data Series 61, American Society for Testing and Materials, 1985.

Environmental Protection Agency: "Reference Guide to Odor Thresholds for Hazardous Pollutants Listed in the Clean Air Amendments of 1990," EPA 600/R-92/047, 1992.

Fazzalari, F. A.: *Compilation of Odor and Taste Threshold Values Data,* DS 48A, American Society for Testing and Materials, 1978.

Ruth, J. H.: "Odor Thresholds and Irritation Levels of Several Chemical Substances: A Review," *American Industrial Hygiene Association J.,* 47: A-142–A-151, 1986.

Shusterman, D.: "Critical Review: The Health Significance of Environmental Odor Pollution," *Arch. Environmental Health,* 47(1): 76–87, 1992.

Stevens, S. S.: "The Psychophysics of Sensory Function," in *Sensory Communications,* W. A. Rosenblith, ed., MIT Press, Cambridge, Mass., 1961.

Turk, A.: "Expressions of Gaseous Concentration and Dilution Ratios," *Atmospheric Environment,* 7: 967–972, 1973.

van Gemert, L. J., and A. H. Nettenbreijer: "Compilation of Odour Threshold Values in Air and Water," National Institute for Water Supply, and Central Institute for Nutrition and Food Research TNO, Netherlands, 1977.

Ziem, G. E., and L. L. Davidoff, "Illness from Chemical Odors: Is the Health Significance Understood?" *Arch. Environ. Health,* 47(1): 88–91, 1992.

SECTION 4.3

MONITORING FOR VOLATILE ORGANIC COMPOUNDS

Richard Trzupek
Air Solutions, Inc.
Oak Brook, Illinois

TYPES OF MONITORING SYSTEMS 4.33
Direct Measurement Techniques 4.34
Data Acquisition Systems 4.40
CHOOSING A MONITORING TECHNIQUE 4.49
EPA TEST METHODS 4.50
Appendix A Methods 4.50
Method 25A 4.50
Method 25 4.51
Method 18 4.51
Method 23 4.52
SW-846 Methods 4.52
Other Methods 4.52
SUMMARY 4.53

Monitoring emissions of volatile organic compounds (VOCs) is a unique exercise, fraught with potential pitfalls. There are many more choices for measuring VOCs than there are for the other gaseous criteria pollutants that are routinely monitored (NO_x, SO_2, and CO). Moreover, because that group of compounds that we generically call VOCs includes thousands of unique compounds, nearly every monitoring application will call for a singular evaluation—a fresh look—to ensure that the data gathered are accurate and accomplish the intended purpose.

Four primary criteria should be considered in choosing a measurement system for VOCs:

1. The intended purpose of monitoring
2. The chemical characteristics of the VOCs to be monitored
3. The nature of the emissions source
4. The chemical and physical matrix in which the VOCs are contained

The manner in which these criteria influence the choice of measurement system is discussed in greater detail below. First, however, we discuss the various ways in which VOC emissions may be monitored.

TYPES OF MONITORING SYSTEMS

System type	Example	Utility
Continuous emissions monitor	FID analyzer system	Constant direct measurement
Stack test	EPA method 25A	One-time check
Parametric emissions monitor	Combustion temperature monitoring	Constant indirect measurement

In general, VOC emissions can be monitored in one of two ways: direct measurement or indirect measurement. If emissions are monitored directly as part of a one-time source test, then stack test methods are used. If emissions are monitored directly and continuously, the system is referred to as a *continuous emissions monitoring* (CEM) system. If some process parameter is measured, as a surrogate for direct measurement, and the magnitude of VOC emissions can be logically and defensibly inferred from this measurement, the system is referred to as a *parametric emissions monitoring* (PEM) system. The first two techniques are direct measurement methods. The last describes indirect measurement methods.

The choice of direct or indirect technique will largely be influenced by the factors of cost and the purpose for which the measurement is intended. In general, indirect techniques are less costly and less maintenance-intensive than direct measurement techniques. Furthermore, indirect measurement often involves only the recording of a simple process parameter which is already being monitored as part of normal production or maintenance practice. Accordingly, the attendant effort in performing indirect monitoring is typically less.

However, when a regulatory or process-related need demands the utmost in accuracy, indirect measurement methods may not be practical. For example, if one wants to know the concentration of methyl ethyl ketone (MEK) in an emissions stream to within 10 parts per million by volume (ppmv), as part of a permit condition or for quality control purposes, it is unlikely that any substitute for direct measurement of MEK using an appropriate analyzer could be found. On the other hand, if one simply needs to assure a regulatory authority or quality control department that the concentration of MEK was consistently less than 100 ppmv (but the exact concentration was not of interest), then simpler and cheaper substitutes to direct measurement could probably be found.

Direct Measurement Techniques

A variety of analytical tools exist for the direct measurement of VOCs. The technology used is, in general, well established and well understood. However, it cannot be overemphasized that the chemical characteristics of the individual VOCs being measured and the nature of the matrix in which they exist can have an enormous impact on the accuracy of the analytical tool to be used. Even the best analyzers, using the most well-established technologies, can yield misleading results if the nature of the analyte and its attendant matrix is not well understood.

Three primary types of analyzers are employed for direct, continuous measurement of VOC emissions:

1. Flame ionization detector (FID) systems
2. Infrared analyzers (IR) and Fourier transform infrared (FTIR) analyzers
3. Gas chromatography (GC) systems

Because FIDs are used most often in direct measurement systems for VOCs, this type of analyzer is covered in detail. The other system types have more specialized and less common applications. Accordingly, only a general overview of their operation and utility is provided.

Flame Ionization Detector Systems. FID systems are, by far, the most common type of system employed when direct, continuous measurement of VOC emissions is desired. FID technology is well established, the analyzers are relatively simple to operate, ana-

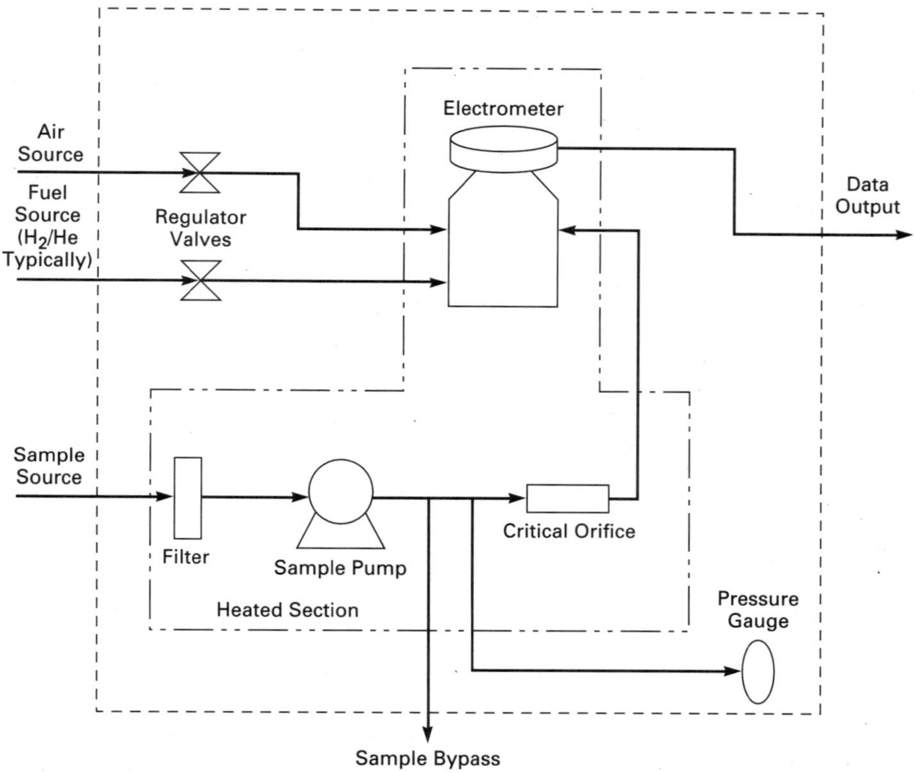

FIGURE 4.3.1 Flame ionization detection analyzer.

lytical results are generally reproducible, and the technology lends itself to the measurement of a wide variety of VOCs at a relatively low cost.

FID Analyzer Operation. In an FID analyzer, VOCs pass through a hydrogen flame, carbon atoms are ionized to carbon cations, and the concentration of carbon cations present is measured by an electrometer. Figure 4.3.1 shows, in schematic form, the most important elements of an FID analyzer. The flame source is fueled by a mixture of hydrogen and air. The hydrogen fuel is supplied from a compressed-gas cylinder and may be pure hydrogen or a hydrogen/helium mix, depending on the specifications of the analyzer being used. In general, it is thought that use of a hydrogen/helium fuel lessens some potential interference effects (described below).

Fuel air may be supplied from a compressed-gas cylinder or, alternately, by an air pump provided as part of the unit. Zero-grade air is sufficient for most applications. If the unit generates air internally, using an air pump, the air must be pretreated to remove any VOCs present in the ambient air. VOC removal from fuel air in such a system is generally accomplished by means of a carbon filter. If a carbon filter is used to clean fuel air, it is vital that the filter be regularly changed in accordance with the FID manufacturer's instructions. Failure to do so will result in positive interference caused by ambient VOCs.

In most applications, a heated air sample is drawn into the FID analyzer by an internal sample pump. A filter is used to remove particulate contaminants which could plug orifices and the flame tip. The pump should be of sufficient capacity to draw the sample over the length of sampling line to be used. For most applications (and assuming roughly ambient static pressure at the source), sample pumps are capable of drawing gas through at least 200 ft of sampling line. The majority of the gaseous sample is discharged through a bypass loop, with only a small portion being sent to the detector section. Because most of the sample drawn is exhausted unburned through the bypass, if an FID system is used to monitor gases with potentially toxic compounds in them, the bypass exhaust should be treated or directed to a safe area.

That portion of sample gas which is analyzed will pass through a critical orifice which maintains a constant sample rate. If the orifice plugs or decays or if the pressure against the orifice changes (compared to the pressure at which the unit was calibrated), VOC readings will not be valid. The detector measures a mass of carbon ions passing through the electrometer, and it is assumed that this mass is contained in a constant volume of air. A change in flow through the critical orifice will change the air volume passing through the detector, and base assumptions will no longer hold. Accordingly, most manufacturers provide a pressure readout with which the sample pressure can be monitored. FID pressure should be checked on a routine basis when an analyzer is in service.

The portion of the sample gas to be analyzed passes into the detector section of the analyzer where VOCs are oxidized in a hydrogen flame in the manner previously described. The concentration of carbon cations present is continually monitored by the electrometer. Some type of analog and/or digital readout then enables the concentration of VOC present to be displayed. Concentrations may also be recorded on a strip chart or some type of data acquisition system.

It is important to note that all the components of the analyzer described above are normally heated to a temperature high enough to prevent the condensation of both water and any organic species present. For most applications, an analyzer temperature of 170 to 190°C is sufficient. Whatever the target, the operator should ensure that the analyzer chosen can achieve and maintain the desired temperature. It should also be apparent that the heating requirements of an FID analyzer lead to a need for a relatively large amount of electricity. At a minimum, an operator should anticipate using 15 A of 120-V power for each analyzer used.

The following are some tips for FID analyzers:

- Hydrogen/helium (mixed) fuel is generally preferred.
- If zero air is supplied by an air pump, maintain a clean charcoal filter.
- Vent the analyzer bypass exhaust safely.
- Monitor the analyzer sample pressure frequently.
- Be sure the analyzer temperature is adequate for the sample matrix.
- Have at least 15 A of 120-V power available per analyzer.

FID System Construction. Figure 4.3.2 describes the general arrangement of a complete VOC analysis system using an FID analyzer. The analyzer and its supporting components (fuel, air, and data acquisition system) are described above.

The sample line used to deliver gas to the analyzer is normally of Teflon construction and is heated to at least 105°C. If the sample line is used in higher latitudes, where it may be exposed to severe cold, extra insulation and/or heating elements may be necessary to maintain the desired temperature. The final sample line temperature chosen should be dictated by the gas matrix. If, for example, the gas to be analyzed is essentially ambient air with low (<1000 ppmv) concentrations of relatively volatile components, the aforementioned temperature of 105°C should be more than sufficient. If, however, the gas contains significant concentrations of higher-boiling-point compounds, a higher sample line temperature may be required.

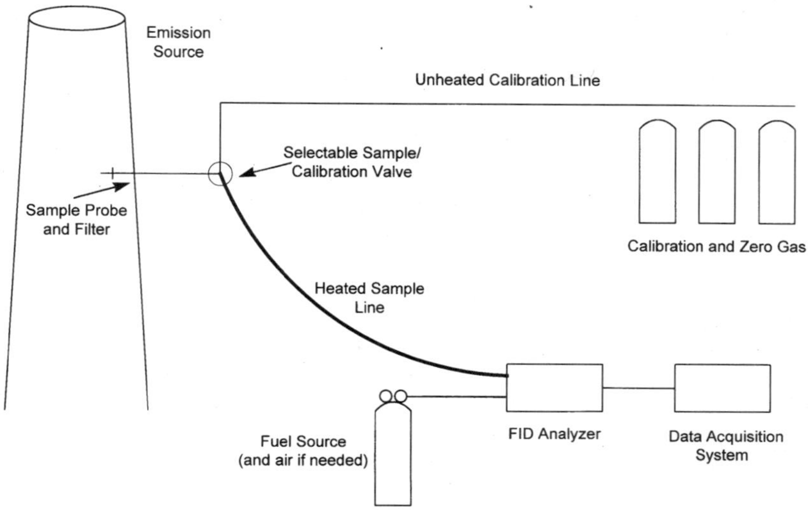

FIGURE 4.3.2 Flame ionization detection analysis system.

The gas sample is typically extracted from the source by means of a stainless-steel sampling probe. An in-stack filter, normally of sintered metal construction, is desirable in order to provide primary removal of particulate matter. To ensure optimum system performance, both the probe and the filter require periodic maintenance. This may be in the form of a "back-purge" with pressurized air to clean the filter and probe. Alternately, removal of either component and its replacement may sometimes be necessary.

Calibration of the system is best accomplished by delivery of calibration gases at a point as near as possible to the sample probe insertion point. This enables the operator to calibrate the entire system (rather than simply the analyzer). When such a calibration arrangement is used, system leaks are accounted for and can be corrected.

The calibration gases used are typically certified concentrations of either propane or methane contained in either zero air or zero nitrogen. Pure zero air or zero nitrogen is normally used as calibration gas. The choice of air or nitrogen is an important one, especially where very low concentrations ($<$10 ppmv as propane) are to be analyzed. The effects of *oxygen synergism* (see below) may be somewhat mitigated in these situations by choosing a calibration gas in which the dilutant oxygen concentration most closely resembles the analyte gas.

The choice of calibration gas (methane, propane, or other) is also important. If the analyzer is calibrated to methane, all concentrations reported by the system will be *as methane* (C1, molecular weight = 16). Alternately, calibrating the analyzer with propane will yield *as propane* concentrations (C3, molecular weight = 44). FID data should always be reported in terms of the species used for calibration. For example, if an analyzer shows 500 ppmv as methane but the analyte in question is butane, then the actual concentration of butane present is 500/(4/1) = 125 ppmv. Since the analyzer essentially counts carbon atoms, the actual concentration must be adjusted for the carbon number ratio between the analyte and the calibration gas. The formula used is as follows:

$$V_f = \frac{V_r}{C_a/C_c}$$

where V_f = final VOC concentration, ppmv of analyte
V_r = raw VOC concentration, as read on analyzer, ppmv of calibration gas
C_a = carbon number of analyte
C_c = carbon number of calibration gas

FID Analyzer Interferents. Figure 4.3.3 graphically describes the way in which carbon atoms are ionized in an FID. The presence of carbon cations, caused by oxidation in the hydrogen flame, produces a deflection in an electromagnetic field which, in turn, is reported as VOC concentration. When certain other atoms are present in a molecule, most notably as part of oxygenated or halogenated hydrocarbons, the electrometer signal may be dampened by the presence of these highly electronegative species.

The exact amount of negative interference caused by oxygenated or halogenated compounds is somewhat specific to the design, construction, and optimization of the analyzer in use. However, it is safe to say that, in general, the higher the ratio between oxygen or halogen content and total molecular weight, the greater the effect will be. Methanol, e.g., has an extremely high oxygen–molecular weight ratio (16:32), and the FID response can be dampened by a factor of 4 or more as a result. In other words, rather than responding with a readout of 400 ppmv (as methane) in the presence of 400 ppmv of methanol, the analyzer may only show an apparent concentration of 100 ppmv.

In most cases, the effect is not this pronounced. In addition, the negative interference of oxygenated species or halogenated species can be accounted for by use of a response factor. If the analyte is known, the response factor can be determined precisely by the use of an appropriate calibration gas.

If the analyte is not known (as, e.g., in the exhaust gas of a thermal oxidizer controlling VOCs) a conservative assumption as to VOC constituents may be made in some situations. Consider a thermal oxidizer used to control an inlet concentration of 1000 ppmv (as propane) VOCs with a regulatory destruction requirement of 90 percent. This means that, to meet regulatory requirements, no more than 100 ppmv (as propane) of VOCs may be detectable in the outlet gas. In practice, a thermal oxidizer will ordinarily reduce the VOC concentration in the outlet gas to < 2 ppmv (as propane). Since it is impractical to determine the exact composition of the VOCs in the exhaust, it is not possible to be sure that the VOC concentration detected is entirely accurate. However, since the

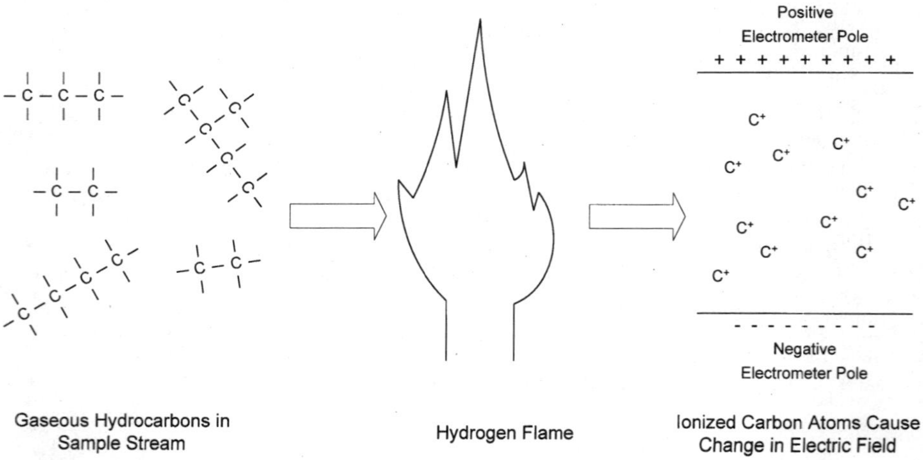

FIGURE 4.3.3 Flame ionization detection—principles of operation.

apparent concentration is over 50 times below the compliance level, the source operator can safely make the conservative assumption that all VOCs in the exhaust are present as a species with an extremely high oxygen–molecular weight ratio. If all emissions were reported adjusted for the response factor of methanol, e.g., a value of <8 ppmv (as methanol) would be documented. This value would still be more than 6 times below the compliance level required and would more than reasonably account for the presence of any oxygenated species. Use of such a response factor adjustment should always be approved by a regulatory authority before it is employed for compliance purposes.

Reduced levels of molecular oxygen in the exhaust gas can also produce negative interference in an FID analyzer. This effect is known as *oxygen synergism.* It can be mitigated somewhat by optimization of the FID by the manufacturer or a trained technician, by the use of hydrogen/helium fuel and by the choice of calibration gases with oxygen contents similar to the analyte gas matrix.

Infrared and FTIR Analyzers. Infrared (IR) and Fourier transform infrared (FTIR) analyses both involve taking advantage of a compound's (or several compounds') ability to adsorb infrared light. In the case of IR, a single wavelength is chosen, generally corresponding to the peak absorbence of the analyte that is to be found in highest concentrations. In the case of FTIR, multiple "scans" of the entire infrared spectrum (or a large portion thereof) are performed in order to obtain information on a number of different compounds simultaneously.

In both IR and FTIR systems, care should be taken to evaluate potential interference effects from the sample matrix. In some cases, negative interference caused by the sample matrix can be severe and may lead to virtual blinding of the analyzer with respect to the target compound(s). This can occur if another compound or compounds absorb infrared light at the target wavelength in sufficient quantity to obscure the analyte.

In recent years, instrument manufacturers have made substantial strides in overcoming the problem of interference in IR and FTIR. It is a problem that is exacerbated because moisture (i.e., water molecules) is a particularly good absorber of infrared light across a wide portion of the infrared band. Removing moisture prior to analysis is, from a VOC monitoring point of view, usually an unacceptable solution to the problem. Techniques which might remove moisture through condensation tend to condense out at least some portion of the target organic compounds. Techniques which lower moisture content through dilution with dry air do not change the moisture–organic compound ratio and thus do not substantially improve the analysis. Techniques which remove moisture through the use of permeable membranes often risk removal of some organic compounds as well.

Some moisture is present in almost all gas matrices in significant concentrations (even ambient air typically contains 1 to 3 percent moisture), it is vital that an IR or FTIR system used for organic analysis either operate in a portion of the infrared spectrum where moisture interference is not substantial or have the ability to accurately adjust for moisture interference effects. Depending on the target compounds, species besides moisture can also have a marked effect on organic analysis. When using an infrared technique, one must thoroughly understand the capabilities and limitations of the system chosen and the nature of the sample gas matrix.

A great advantage of using FTIR techniques is the ability to generate data for a vast number of compounds simultaneously. In the controlled environment of a laboratory, FTIR has been used for many years to qualify and quantify a number of compounds in a mixture at once. The demands of field sampling and monitoring, however, are greater than the capabilities of most laboratory equipment. Fortunately, recent technological developments and regulatory need have allowed a number of instrument manufacturers to build reliable and accurate field-capable FTIR analyzers. Costs for an FTIR system are high relative to other techniques, but where compound-specific monitoring is required and the system is appropriately matched to the source, the value of FTIR can be unmatched.

Single-wavelength IR systems are, by contrast, relatively inexpensive and simple. Such systems are generally employed where a small number of analytes are of interest (preferably one or two) and where the sample gas matrix is very predictable and consistent. These factors enable the operator to accurately calibrate the instrument for the compound(s) in question. The absence of even one of these factors will usually significantly degrade the utility of the data gathered.

A final note in regard to IR and FTIR techniques involves maintenance. Any optical technique, especially when used in a severe environment, can be affected significantly by contamination of the optical surfaces. Even a small amount of dust collected on mirrors, detectors, or light sources can have a substantial impact on data accuracy. Therefore, it is always recommended that the manufacturer's maintenance instructions be faithfully adhered to and if the source or ambient environment is especially dirty, exceeded.

Gas Chromatography. Gas chromatography (GC) techniques are among the most time-tested methods to come out of the organic chemistry laboratory. As is the case with all analytical techniques, the transition to field service imposes much greater demands on the method; but if properly used, on-line GC can be an effective monitoring tool.

A GC technique involves two steps. Separation is accomplished through the use of appropriate columns, separation temperature(s), and carrier gas pressures. Detection involves the use of an appropriate detector. The choice of column, analytical conditions, and detector involves a number of complex issues and should not be made by anyone except an individual thoroughly experienced with the technique. In addition, an understanding of the gas matrix to be analyzed and potential interferents is, as always, critical.

GC techniques are generally not used to measure total VOC concentrations. Rather, they are used to accurately monitor the concentrations of a number of different organic compounds when the compounds present are expected to be consistent in terms of species and range of concentration. When a system is properly designed, there are few interference effects (although these are always possible). Costs are comparable to those for FTIR techniques, but since the technology is more established for field use, GC is sometimes preferred.

GC methods also have the advantage of being able to detect very low concentrations of target compounds relatively easily. It is not unusual to achieve detection limits in the ppbv range with the proper choice of detector and column.

The most popular detector choices include

- *Flame ionization detectors.* As discussed earlier, these detectors are most appropriate for basic hydrocarbon species. In the controlled environment of GC, FIDs can yield very low detection limits.
- *Electron counter detectors (ECDs).* Generally an ECD is used to measure small concentrations of halogenated (or other highly electronegative) compounds.
- *Flame photometric detectors (FPDs).* An FPD is generally used to measure small concentrations of organic sulfur or organic phosphor compounds.

Data Acquisition Systems

Data acquisition systems (DASs) are too often an afterthought in the process of building a VOC analytical system. Care should be taken to ensure that the DAS chosen is capable of recording the data gathered by the analyzer and presenting the data in a fashion which is useful to the operator.

The most basic type of DAS is the traditional strip chart recorder. A strip chart recorder enables one to visually identify emission trends and instantaneous concentrations. It is, however, a poor choice of DAS if one needs to average the data gathered over some period or perform other data manipulations.

Data loggers are a more popular choice of DAS when long-term data storage and manipulation are an issue. Data loggers are self-contained, software-driven systems that record data, perform calculations on the data, and provide some means of retrieving values. Certain data logging systems may also allow the operator to automate certain CEM system functions, such as calibrations, sample line purging, and leak checks.

There are a number of manufacturers of data loggers and a wide variety of system capabilities available. Basic distinctions between different data loggers include system flexibility, ease of access, and range of functionality.

System flexibility refers to the ability of the data logging system to be reconfigured to new input or output parameters. The architecture of certain systems may be so rigid that changes in any parameters desired by the operator may be difficult or impossible. In other systems, both the hardware and the software design may easily accommodate such changes or even system expansion.

The issue of system flexibility is closely related to the issue of system access. Certain data logging systems may be virtually inaccessible to the end user, in respect to making changes to system parameters. Such inaccessibility may be unintentional (and simply the result of the manufacturer's indifference to the importance of user-friendliness) or intentional (in an effort by the manufacturer to protect the integrity of the system). In any case, the potential operator is encouraged to carefully consider the system access and flexibility that will be required and to evaluate potential systems in terms of these issues.

Range of functionality refers to the type and number of functions available within a system. At its most basic, a data logger will function as an analog-to-digital (A/D) converter, and the analyzer's output signal will be both displayed and recorded in some data file. Beyond this basic function, the capabilities of a given data logging system will vary. Functions which may be available include

- *Alarm functions,* which allow the system to alert an operator if acquired data or calibration data exceed some preset range
- *Averaging functions,* which allow the operator to average data over one (or many) fixed time period(s)
- *Data storage functions,* which allow the system to record data to internal memory or an external device (such as a floppy disk) within some preset storage capacity
- *Mathematical functions,* which manipulate the data from one unit (such as parts per million by volume) to another unit (such as pounds per hour)
- *Output functions,* which allow the system to generate a signal that can operate external equipment, such as calibration system solenoids or control device valves

Advances in computer technology and software design have recently led several companies away from traditional, stand-alone data logging systems and into systems which are integrated into personal computers (PCs) or networks. In this type of system, the only hardware the manufacturer will generally supply is a simple A/D converter. The digital signal generated is then sent to a computer in which the manufacturer's software enables the PC to perform all the functions of a traditional data logger. Because the power, memory, and speed of a PC are usually far superior to the hardware provided in a stand-alone data logger, these systems will generally allow the user much greater capability and more flexibility. It is not unusual, e.g., for a PC-driven system to enable the operator to generate an "on-screen" graphical representation of data (not unlike a strip chart recorder) while also recording and storing a prodigious amount of data. While such systems will often require more expertise to modify, the rapid advances in the simplicity of software and ever-growing computer literacy within the workplace make them an attractive choice in many situations.

Indirect Measurement Systems. Indirect measurement systems, also known as parametric emissions monitoring systems, are a popular means of ensuring control system compliance with regulatory requirements without resorting to the expense and (in many cases) maintenance burdens often associated with a CEM system. In most circumstances, the components and data used in PEM systems are already in place, having been provided by the control device manufacturer as part of the system package. In ideal circumstances, converting the existing components to a PEM system will only require formally designating the system as a PEM system with the regulatory authority.

Most PEM systems will be used in conjunction with a control device of some sort. When a control device is not used for VOC control, parametric monitoring is often still possible, but is relatively simple to perform. An example of parametric monitoring in an uncontrolled situation would be monitoring paint usage and VOC percentage in order to calculate VOC emissions on a total-loss basis. Means to perform this type of parametric monitoring are, for the most part, self-apparent and are not considered further in this section.

PEM systems used to monitor control device performance will be closely associated with the type of control device in use. Four types of control devices, and the PEM system commonly used with each, are considered below:

- Thermal oxidizers
- Catalytic oxidizers
- Carbon adsorbers
- Wet scrubbers

When considering a PEM system, one must distinguish between the objective of demonstrating compliance (or, in the case of process monitoring, ensuring that data fall below some predetermined de minimus level) and that of reliably determining VOC emissions rates within some desirable margin of error. Certain PEM techniques are perfectly suited to demonstrating compliance with a control requirement, such as achieving 90 percent control of VOCs entering the device, but would not allow the operator to determine VOC emissions on a pounds per hour basis without an unacceptably wide margin of error. Each type of control device and associated PEM system is considered on both bases.

Thermal Oxidizers. By far, the most common PEM method used in conjunction with a thermal oxidizer is the monitoring of combustion chamber temperature. In a regulatory compliance scenario, the control efficiency of the oxidizer is usually measured using an Environmental Protection Agency (EPA) VOC air test method (see "EPA Test Methods," following) over three 1-h periods. Each test corresponds to some measured VOC loading condition to the control device and to a predetermined combustion chamber temperature. Assuming that the emissions test demonstrates compliance with the permit control efficiency requirement, maintenance of the measured combustion chamber temperature becomes de facto demonstration of compliance with the control efficiency requirement.

In many cases, the difference between the permitted control efficiency requirement and the tested control efficiency is so great that the assumed correlation between combustion chamber temperature and compliance is a safe one, even if oxidizer performance degrades somewhat over time. It is not at all uncommon for a thermal oxidizer to function at VOC destruction efficiencies in excess of 99 percent. Where a permit requires only 90 percent, or even 95 percent, control, the magnitude of difference between actual and required performance will justify a PEM system in most circumstances. However, measurement of oxidizer combustion chamber temperature does not generally yield accurate information in regard to VOC mass emission rate or concentration. Differences in VOC loading rates, atmospheric changes, and control device aging may all significantly affect the actual VOC emission rate, even though that emission rate is still well below the level needed to demonstrate compliance with control efficiency requirements.

Monitoring of combustion chamber temperature is typically provided by the thermal oxidizer manufacturer as part of the control device package. From a system performance point of view, these data are necessary to ensure system integrity and safety. In most cases, combustion chamber temperature is recorded on a strip chart recorder of some type.

When using combustion chamber temperature as a surrogate for demonstrating compliance with a control efficiency requirement, the operator of the oxidizer must still exercise some degree of caution and vigilance. Concerns chiefly relate to the issues of thermocouple placement, thermocouple and recording device calibration, system leakage, and data interpretation.

Regulatory requirements may sometimes state or imply that combustion chamber temperature monitoring take place at the portion of the combustion chamber likely to achieve the lowest temperature (generally the far end of the combustion chamber in relation to the burner). Not all thermal oxidizer manufacturers conform to such a standard, however. Therefore the operator should ensure that thermocouple placement is consistent with permit conditions and local regulations.

Likewise, it is often assumed that both the thermocouple and the temperature-recording device maintain accuracy throughout long years of service. Neither assumption will necessarily hold true and will be less likely to be true as time passes. In many cases, the service agreement with the control device manufacturer will include periodic maintenance checks of these items. The operator should ensure the integrity and accuracy of both on a routine basis.

System leakage is an issue that is most often important in systems equipped with shell-and-tube heat exchangers, chiefly recuperative thermal oxidizers. In such a system, significant leaks may develop in the heat exchanger over time as a result of thermal stress. If this occurs, then untreated, VOC-laden air may bypass the combustion chamber. In that case, even though the combustion chamber temperature may be accurate and may indicate compliance with the destruction efficiency requirements, in fact, if enough untreated air bypasses the heat exchanger, the system may not be meeting target destruction efficiencies. There are several ways to avoid this pitfall including periodic heat-exchanger inspections, informal (i.e., not witnessed by the regulatory agency) checks of VOC concentrations in the exhaust stack of the unit, and gas flow balancing studies.

System leakage is a somewhat different, but equally important, issue in regenerative thermal oxidizer systems. In this type of system, one of the most vulnerable points is the valves which switch inlet and outlet gas flow between media beds (or, in the case of a single-bed system, reverses gas flow). If valves deteriorate over time, the amount of untreated, VOC-laden air that escapes through valve leakage may become significant.

The final issue—data interpretation—is important in most systems and particularly for a system in which the regulatory authority may make an especially rigorous interpretation of compliance data. What is and is not a demonstration of compliance is not always readily apparent.

Figure 4.3.4 is a representation of a portion of a theoretical thermal oxidizer's combustion temperature monitoring plot. In this case it is assumed that compliance was demonstrated by means of an emissions test during which the oxidizer was operated at 1400°F. This temperature becomes the compliance level, indicated by the dashed line. The system operator in this case has chosen to set the normal operating temperature at 1450°F—50°F above the compliance level.

The reasons for this action are clear when one looks at the temperature monitoring plot. Variations in VOC loading to the system can cause momentary drops in system operating temperature as the oxidizer controls adjust to the new condition. Production stoppages and an upset in the VOC emitting process are shown to produce momentary drops in combustion chamber temperature below the target 1450°F. Had the oxidizer set point been 1400°F, a rigorous interpretation of permit conditions might have led to the conclusion that the unit was out of compliance during these periods. If a source is located in a state or district with an especially strict regulatory authority or if the corporate entity chooses to exercise a greater degree of caution on its own, oxidizer set points in excess of the regulatory requirement may be chosen for this reason.

The flame failure event depicted in Fig. 4.3.4 is a different sort of circumstance. The operating temperature in this case drops to well below 1000°F, a point at which it would be almost impossible to assume that any significant portion of VOCs is being destroyed. In many systems, the VOC source is interlocked to the thermal oxidizer so that if the combustion chamber drops below some predetermined set point, the source ceases to operate (and therefore ceases to emit VOCs). If this were the case in this example, the operator would be well served to note that process shutdown corresponded to the flame failure, to ensure protection in the case of future inspection of oxidizer operating records.

Catalytic Oxidizers. Combustion chamber temperature monitoring, as utilized within thermal oxidizers, has no practical utility within a catalytic oxidizer system.

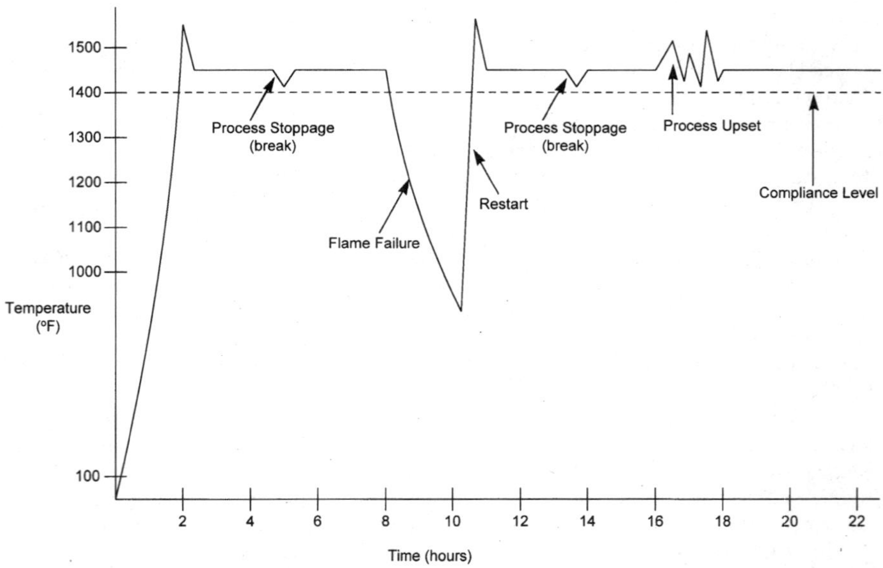

FIGURE 4.3.4 Thermal oxidizer.

Since the operating temperature of the catalyst bed (above the base bed temperature) is dependent on VOC loading, the catalyst temperature will vary with VOC load.

Most manufacturers of catalytic oxidizers provide for monitoring of catalyst temperature at the inlet and outlet of the catalyst bed. The temperature rise across the bed is proportional to the VOC loading to the system. By monitoring the temperature rise, system performance can be ensured, and so this is often the PEM system of choice for a catalytic oxidizer.

Both the way in which temperature rise data are utilized and the manner in which monitoring requirements are written are sometimes misunderstood. It is unfortunately common for a regulatory authority to require some minimum temperature rise across the catalyst as a surrogate means of monitoring compliance with destruction efficiency requirements. Such a requirement is usually inappropriate for a catalytic system.

Because temperature rise will vary with VOC load, requiring a de minimus temperature rise implies that the control device must always be operated above some corresponding minimum VOC load level to ensure compliance with control requirements. In fact, at lower-load conditions, a properly maintained catalytic oxidizer will still likely perform at an adequate control level. Lower VOC loading means not that the device is failing to function within limits, only that the heat content of the inlet stream is insufficient, even though the VOCs in it are completely oxidized, to achieve the target temperature rise.

Figure 4.3.5 illustrates this concept by using a plot of temperature rise over time for a theoretical catalytic oxidizer. Catalyst temperature is seen to vary with different process conditions. In all cases, the oxidizer may be exceeding control device requirements. There is, in fact, likely little difference in destruction efficiencies at any condition. Even at that point of the performance curve labeled "Process Down Time," where a decline in temperature is observed, the oxidizer may be working at peak efficiency. The minimal amount of heating value provided by the gas stream in this case is insufficient to overcome the thermodynamic cooling provided by radiant heat loss at the top of

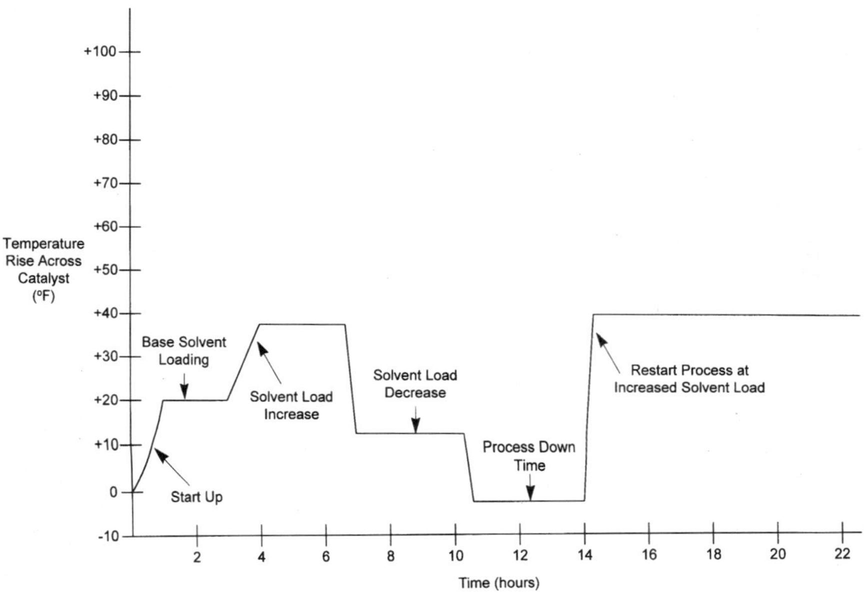

FIGURE 4.3.5 Catalytic oxidizer.

the bed and the cooling capacity of the non-VOC components of the gas stream passing through it.

There are two surrogate means to demonstrate continuing catalytic oxidizer performance that may be used. One is periodic testing of the catalyst in order to ensure adequate catalyst activity. The other is defining the minimum temperature rise to be achieved at a certain corresponding VOC load condition. For example, rather than specifying maintenance of "a temperature rise of +20°F" in order to demonstrate compliance, the specification may call for "a temperature rise of +20°F whenever a VOC loading of 45 pounds of VOC per hour is achieved."

As is the case with thermal oxidizers, PEM systems can be an adequate means of demonstrating compliance with a catalytic oxidizer destruction efficiency requirement, but they are rarely adequate to determine a mass emission rate or VOC outlet concentration. Similarly, caveats in terms of temperature recording system maintenance, heat-exchanger leakage (when a heat exchanger is used), and data interpretation also apply.

Carbon Adsorbers. To all appearances, parametric monitoring of a carbon adsorber system should be a simple exercise. A mass balance should be all that is needed. The amount of VOCs recovered can be compared to the amount of VOCs used (typically determined from purchasing records), and the control efficiency thus easily determined. Because the control device does not modify the chemical constituents of the VOC stream in any way, and because an actual mass balance is used, both the control efficiency of the absorber and the mass emission rate of VOCs present should be easy to determine.

The above assumptions are all true, but in practice care must be taken in how use and recovery data are utilized as part of a PEM plan. The difficulty in monitoring carbon adsorber performance by these parameters chiefly lies in the averaging periods chosen.

Figure 4.3.6 shows a solvent recovery plot for a theoretical carbon adsorption system used to control a VOC emitting process. In this process, a constant amount of

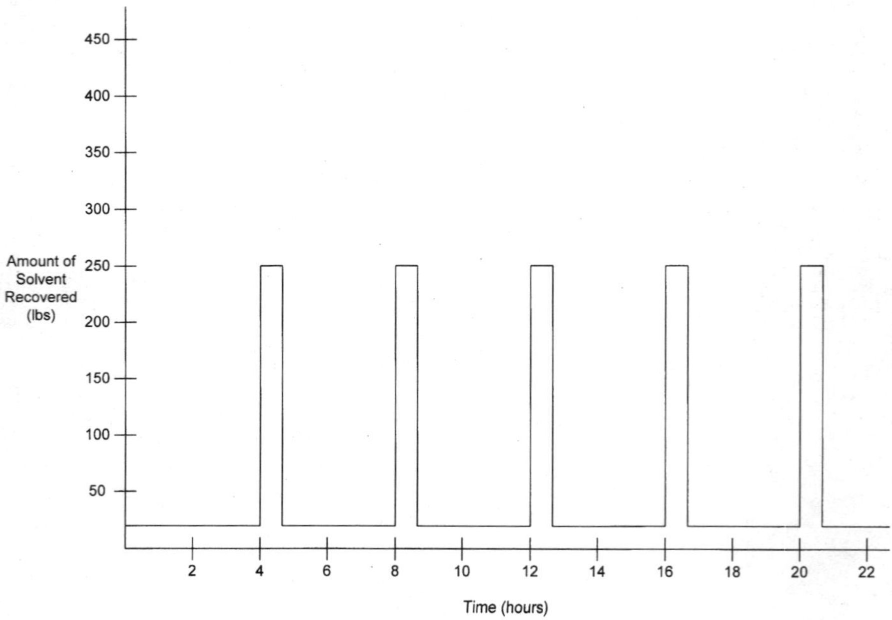

FIGURE 4.3.6 Carbon adsorber—solvent recovery profile.

VOC—70 lb/h—is delivered to the carbon adsorber. The carbon beds are desorbed every 4 h, and as the plot shows, about 250 lb of solvent is recovered per desorption cycle. This corresponds to a recovery rate of 250/4 = 62.5 lb/h. The control efficiency is therefore 62.5/70 × 100 = 89.3 percent.

Many carbon adsorption systems do not, however, allow the operator to easily determine the amount of solvent recovered during each desorption cycle. Moreover, the amount of VOCs used in a process rarely is at so constant a rate or so easily accessible to be accurately determined on an hourly basis. The more typical way in which carbon adsorber performance records are kept is represented in Figure 4.3.7.

Figure 4.3.7 shows data for the same theoretical absorber referred to above and represented in Fig. 4.3.6. However, the data plotted are purchases of solvent (which is assumed to be 100 percent VOC) and the amount of VOC removed from the adsorber recovery tank. It can be seen that neither solvent purchases nor recovery takes place at predictable intervals or in regular amounts. Rather, the amount and timing of solvent purchases may depend on the whims of the purchasing department, the plant budget, or the immediate needs of production. Similarly, the time at which the solvent recovery tank is unloaded (and the amount of recovered solvent measured and recorded) may depend on competing production needs, personnel availability, or the day of the week. In any case, the important point is that both operations are somewhat unpredictable within a certain range.

In this example, it can be seen that the period of time chosen for demonstration of compliance on a mass-balance basis is critical. Even choosing the use of a weekly average to compare purchases and recovery would lead to an erroneous conclusion. In the first 7 days of this plot, the solvent recovery tank is recovered once and 4032 lb of solvent is recovered. In the same week, 9750 lb of solvent is purchased. A comparison of these two data points shows an apparent carbon adsorber efficiency level of 4032/9750 × 100 = 41.4 percent! The second week is confusing on the other end of the spectrum. No solvent purchases are recorded, but 6048 lb is recovered. In this case, the

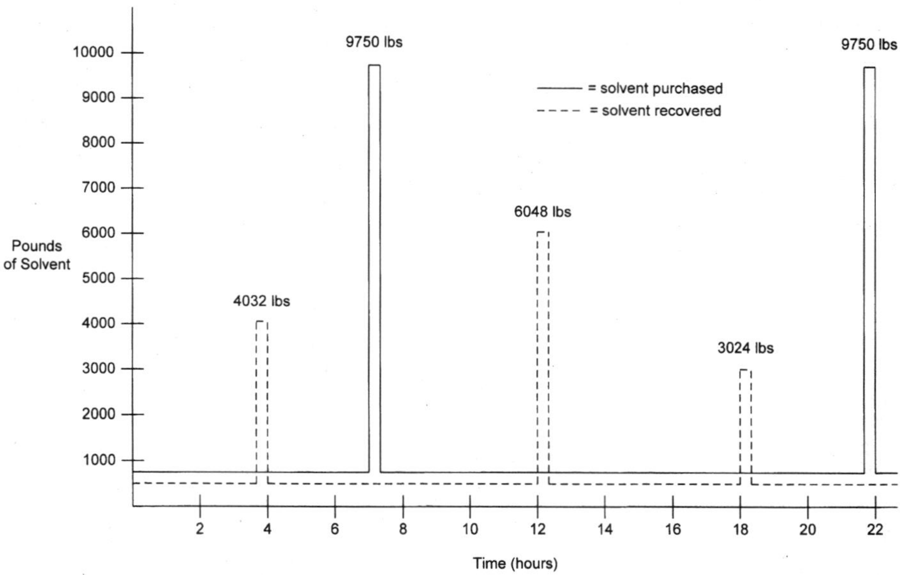

FIGURE 4.3.7 Carbon adsorber—recovery versus purchase analysis.

carbon adsorber efficiency is incalculable, since it requires the operator to divide by zero.

If compliance is to be demonstrated in the example system, it can be logically shown only by choosing an averaging period sufficiently long to smooth out the irregularities in purchasing and recovery activities. A 30-, 60-, or 90-day rolling average may be necessary to accurately demonstrate compliance. Alternately, the source may choose to install a CEM system to monitor VOCs at the inlet and outlet of the adsorber continuously and not rely on use-versus-recovery data at all. Finally, the source may choose to rigorously control purchasing and recovery activities on a fixed schedule in order to meet a shorter averaging-period requirement.

A final note about carbon adsorber mass-balance calculations: The above examples have not considered any VOC which may be solubilized in the condensate stream from the adsorber. Most organic solvents have at least a small solubility in water, and in some cases, the solubility can be substantial. If the amount of steam used (and therefore condensate generated) is especially large, or the solubility in water of the VOCs used is especially great, this portion of the mass-balance equation may be of significance in determining compliance.

Wet Scrubbers. Effective parametric monitoring of packed column wet scrubbers typically involves monitoring the pressure drop across the scrubber. A given pressure drop, at a given VOC loading rate, is assumed to correspond to a minimum control efficiency. The pressure drop–control efficiency relationship is usually established through a one-time emissions test of the system. As is the case with oxidizers, a PEM system can be an effective tool for demonstrating continuous compliance, but is rarely adequate to determine VOC emission rates or concentrations with any acceptable degree of accuracy.

Monitoring of the pressure drop across the scrubber is accomplished by means of a differential-pressure measurement device. A draft gauge, which measures pressure differential as function of the displacement of a fluid of known specific gravity, is one device commonly used. Mechanical differential-pressure gauges are also common.

When mechanical gauges are used, periodic calibration and maintenance are of greater concern. Mechanical gauges have the advantage of sometimes being equipped with data output capability, which enables the operator to continually record the pressure differential. If a draft gauge is used, the operator will have to rely on periodic inspection of the gauge to ensure that the system is operating within an acceptable range. Whichever type of device is used, it is normally provided by the scrubber manufacturer as part of the system installation package. The operator of a recirculating wet scrubber should ensure that the type of monitoring device chosen will most effectively meet her or his need to demonstrate continuing compliance of the control system.

The operator of a wet scrubber system should also be aware that modifications in the blowdown rate of the system will have an effect on VOC emissions. Such a modification, and corresponding change in system efficiency, would not affect the pressure drop across the system. For example, consider a theoretical scrubber that is set up as a "once-through" (100 percent blowdown) system. That is, only fresh water is used in the scrubber (no water recirculates). This system is tested at a VOC loading of 40 lb/h and is determined to have an efficiency of 93 percent at a pressure drop of 3.5 inches of water (inH_2O).

A few months later, to conserve water, maintenance adjusts the system to an 80 percent blowdown rate (20 percent of the scrubber water is recirculated through the system). The pressure drop across the scrubber will remain the same, but 20 percent of the scrubber water used will now be initially laden with some percentage of VOC. Some portion of this additional VOC is likely to volatilize in the scrubber and reduce system efficiency. Accordingly, the assumption that a pressure drop of 3.5 inH_2O across the scrubber corresponds to an efficiency of 93 percent is no longer valid.

CHOOSING A MONITORING TECHNIQUE

Having examined the most common techniques of monitoring VOCs, we can now reconsider the criteria for choosing a monitoring system. The critical factors previously identified are as follows:

1. *The intended purpose of monitoring.* When the purpose of monitoring is to establish an actual mass emission rate or concentration, CEM techniques are usually most appropriate. Within certain conditions, a mass-balance PEM system on a carbon adsorber can also fulfill this function. If compound-specific data are required, an FTIR or gas chromatography CEM system is most often used. If total VOC data only are required, an FID (or less commonly IR) system is most often used.

When the purpose of monitoring is to establish compliance with a minimum control efficiency requirement only, a PEM system will generally be less expensive, easier to use and require less maintenance.

2. *The chemical characteristics of the VOCs to be measured.* The chemical characteristics of the VOCs to be measured are important if a CEM system is used. In all cases, the relative volatility and concentrations of the VOCs should always be a concern. The operator of the system should be sure that sampling, transport, and analysis temperatures are adequate to maintain the target VOCs in the gas phase at all times.

In an FID system, the presence or potential pressure of large concentrations of oxygenated or halogenated hydrocarbons may require modified data interpretation; or if the regulatory authority does not allow such adjustments, use of an FID system may not be desirable.

In a GC system, the characteristics of the VOCs to be measured will affect the choice of column, system parameters, maintenance requirements, and detector. Not all VOC streams will be easily monitored by use of GC because of some of or all of these factors.

In an IR or FTIR system, the absorbence characteristics of the VOCs to be measured are of primary importance. The operator of the system should ensure that the instrument chosen is sufficiently sensitive to distinguish the VOCs present at their expected concentrations and with adequate detection limits.

3. *The nature of the emissions source.* The nature of the emissions source monitored is of primary importance when one is considering a PEM system. Each type of common VOC control device implies the use of a PEM system uniquely suited to it. In each case, the limitations of the PEM system must be well understood, and plans must be made to ensure that the system is adequately maintained.

4. *The chemical and physical matrix in which the VOCs are contained.* The gas matrix in which the VOCs are contained is important, no matter what type of system is used. The effects of interfering species can be significant in any monitoring scheme. Severe conditions, especially high particulate loading, will tax the capabilities of any monitoring system, and increased system maintenance activities should be anticipated accordingly. If particulate loadings are too high, or cannot be sufficiently removed before analysis, the use of certain especially sensitive techniques such as FTIR and GC may not be desirable.

The potential effects of the gas matrix on a detector in a CEM system should be considered. Variable oxygen levels may cause the generation of anomalous data in an FID system which is used to monitor very low concentrations of VOCs. Similarly, competing compounds with absorbence spectra overlapping those of target compounds may cause difficulties in FTIR and IR systems.

EPA TEST METHODS

A number of EPA test methods exist for the measurement of VOCs. Since these are often used to establish the accuracy of a CEM or PEM system, a few words in regard to EPA methods are in order. The complexity and wide range of application of these methods preclude all but a cursory overview in the context of this discussion. However, the operator of a VOC monitoring system should at least be aware of the basic principles involved in each technique.

The most common EPA air test methods can be found in the Code of Federal Regulations, Title 40, Part 60, Appendix A and in the EPA's *Handbook for the Analysis of Solid Waste,* SW-846. The former methods are commonly referred to as *appendix A methods.* The latter are referred to as *SW-846 methods.*

Appendix A Methods

Appendix A test methods are commonly used in the initial performance test of a VOC control system. Most states and regulatory districts have incorporated these methods into their regulations, and most will specify them in a construction or operating permit. There are a host of emissions-testing companies capable of performing some or all of them. The reader should be cautioned, however, that these methods are highly technique-oriented. The validity and accuracy of the results obtained can vary greatly depending on the experience of the emissions tester along with the method and the type of source tested. This is especially true if the nature of the source, the nature of the emissions stream, or potential interferents vary from the norm.

A typical Appendix A emissions test will consist of three 1-h test runs. Normally, the test program will take 6 to 10 h per source to complete, when setup and pretest and posttest calibration and quality assurance activities are considered. In most cases, the regulatory authority will wish to correlate the emissions test to process conditions. Accordingly, the operator of the VOC source should ensure that all relevant process parameters are recorded and are within the range desired.

Method 25A

Method 25A is the EPA's basic FID emissions test method. The equipment used and the testing protocol are essentially equivalent to those specified in the earlier discussion of FID systems as a CEM system. The calibration scheme is somewhat more rigorous using method 25A than is ordinarily required of a CEM system, however.

EPA guidance dictates that method 25A not be used to demonstrate compliance with a control device efficiency requirement when a destructive control device is used (i.e., a thermal or catalytic oxidizer) and the outlet concentration is over 50 ppmv (as methane). In such cases, method 25 is preferred by the agency.

The motivation behind this guidance centers on concern over the response factor effect on FID analyzers. In a destructive control device, it is assumed that oxygenated species can be created which may not be accurately analyzed due to negative interference in an FID system. Since method 25 is theoretically "interference-free" and since the stated detection limit of method 25 is 50 ppmv (as methane), the agency believes that the method is more appropriate for destructive control devices with outlet concentrations at such a level.

In practice, method 25 (see below) is also interference-prone, often more so than method 25A. For this reason and because corrections can be applied to method 25A data, as described earlier, and a conservative error margin applied to FID data, some experts in the field still prefer method 25A to method 25 in all instances.

Method 25

Method 25 occurs in two distinct steps. The collection phase occurs at the source, where VOCs are condensed within a U-tube trap immersed in dry ice. The analytical phase occurs sometime later in the laboratory, where VOCs are revolatilized, oxidized to CO_2, reduced to methane, and finally analyzed by GC or FID as methane.

Figure 4.3.8 shows the basic steps in the method 25 procedure. Like all emissions-testing methods, it is very technique-oriented. The expertise required to perform it, in both the collection and analytical phases, makes it among the most difficult of all EPA methods to execute accurately. Only the most experienced operators and chemists should perform method 25, and even then, the nature of the emissions source and VOC stream should be carefully considered.

In the sampling phase, the most common errors relate to the small sample volume collected. Typically 3 to 4 L of gas is gathered over each 1-h test period. The corresponding amount of VOC collected, therefore, will typically amount to only a few milligrams. Accordingly, any organic contamination within the test train can result in significant positive errors. Similarly, any leakage in the test train (and there is no posttest leak check performed in the method) can result in negative errors.

In the analytical phase, the efficiency of the catalyst used to oxidize VOCs to CO_2 can vary substantially according the chemistry of each VOC. Certain VOCs are only minimally oxidized and, consequently, will not completely show up in the final analysis as a VOC. Finally, the calibration, maintenance, and cleanliness of the analytical system can all have marked effects on the quality of data generated.

Method 18

Method 18 is essentially the EPA's outline of a GC method. It contains quality assurance guidelines and allows the operator a wide range of latitude in the precise choice of technique used. Method 18 samples may be collected on some absorption media, such as charcoal, Tenax resin, or even water. Grab samples in Tedlar bags may also be collected. After collection, the sample will be analyzed by GC using a column, conditions, and detector appropriate to the analytes. Alternately, a method 18 test can consist of online GC in which a gas sample is delivered directly to the analytical system at specified intervals over the 1-h (sometimes longer) test period.

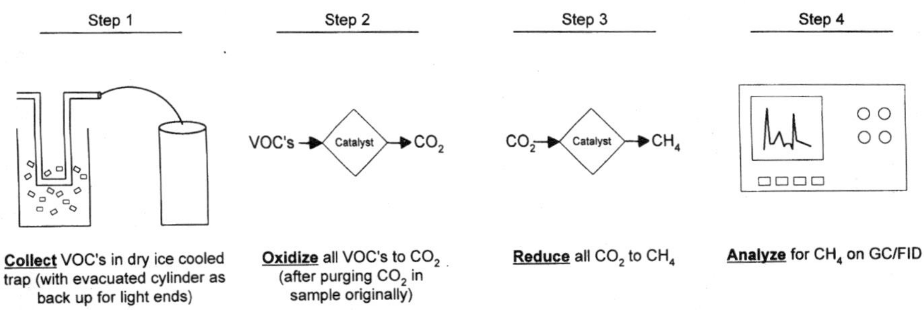

FIGURE 4.3.8 Principle of EPA method 25.

Like any GC method, method 18 is not appropriate for determining total VOC emissions if the VOC mix is especially complex or even partially unknown. It is most often used when one or two target organic constituents are of special interest and can be easily and reliably delivered to a GC.

Method 23

Method 23 is the EPA method specified for the collection and analysis of dioxins and furans. It is a time-consuming, complex, and expensive method to perform. Costs of a method 23 test series on the order of $50,000 to $100,000 per source are not unusual. This cost is chiefly due to the extreme quality assurance measures required in the method and the expense of the exacting GC/mass spectroscopy (GC/MS) analysis performed on the collected samples.

Method 23 samples are collected on a resin. Sampling times are generally specified by the regulatory authority; 3 to 4 h per test run is typical. Sample recovery and analysis are performed in the laboratory.

The technique will generally yield detection limits in the picogram-per-sample range. The GC/MS analysis can also yield data on a number of other organic species present which were collected. If the source operator desires data on other compounds present in the VOC mix, a method 23 sampling event may be an ideal opportunity to gather such data. If, on the other hand, a source operator wishes to limit the scope of analysis and avoid the generation of data on compounds which can be later used in a "witch hunt," a method 23 sampling event can potentially be a concern. The source operator should define the scope of the analytical portion of the project prior to submitting samples to the laboratory.

SW-846 Methods

Commonly used SW-846 VOC test methods include method 0010, for the determination of semivolatile organic compounds, and method 0030, for the determination of volatile organic compounds. Both methods involve collection on a medium and subsequent GC/MS analysis. Method 0010 is very similar to method 23, but is focused on (and validated for) a larger list of analytes. Method 0030, while also utilizing discrete collection on a resin, uses a different sampling train, smaller sample volumes, and a substantially different analytical scheme. The target compounds in method 0030 have higher vapor pressures than those targeted by method 0010.

In both methods, care should be taken to note the validated and acceptable accuracy of analysis. For the SW-846 analytical techniques used in conjunction with method 0010, for example, the acceptable recovery range for surrogate, quality assurance compounds "spiked" onto the sample train can be as small as 10 percent recovery and as high as 150 percent recovery. This range of acceptable recovery percentages is a reflection of the fact that mass spectroscopy, while a valuable qualitative tool, may often be a poor quantitative tool. Accordingly, care should be taken to consider the likely margin of error and expertise of the analytical laboratory when one is evaluating data gathered using either of these two techniques.

Other Methods

The variability in the chemistry of organic compounds and gas matrices will sometimes necessitate the employment of methodologies other than those listed above. National

Institute for Occupational Safety and Health (NIOSH) methods, Occupational Safety and Health Administration (OSHA) methods, EPA ambient air methods ("TO methods"), state-specific methods (particularly methods from the state of California), and custom-designed methods are all employed from time to time. Various trade organizations, environmental agencies in other countries, environmental chemistry texts, and a number of research organizations also sometimes serve as sources of test methods.

Finally, the reader will note that the EPA has proposed two protocols intended to establish standards for VOC CEM systems. These have been drafted by the agency as PS 101 and PS 102. The former is intended to establish standards for FID CEM systems. The latter is intended to establish standards for GC CEM systems. Standard procedures for the operation of FTIR systems are also under development by the agency.

SUMMARY

Monitoring of VOC emissions requires a great deal of research and consideration before a system is selected. No off-the-shelf system will be adequate for every situation, and conversely, a number of systems can often be made to fit any one situation. Careful examination of the intended purpose of monitoring, the characteristics of the VOCs to be monitored, the nature of the emissions source, and the gas matrix to be sampled is vital to selecting a successful monitoring system.

SECTION 4.4

STATISTICS

Dennis R. Weier
Battelle—Pacific Northwest National Laboratory
Richland, Washington

INTRODUCTION 4.54
DATA COLLECTION PLANNING 4.55
TYPES OF DATA 4.55
MEASURES OF LOCATION AND DISPERSION 4.56
GRAPHICAL DISPLAYS 4.58
SAMPLING ISSUES 4.60
SAMPLING DISTRIBUTIONS 4.62
INTERVAL ESTIMATES 4.63
COMPARISON TO A STANDARD 4.65
TOLERANCE LIMITS 4.65
CONTROL CHARTS 4.66
CONFIDENCE INTERVAL FOR A PROPORTION 4.67
HYPOTHESIS TESTS 4.68
RELATIONSHIP OF CONFIDENCE INTERVALS AND HYPOTHESIS TESTS 4.73
DATA PROBLEMS 4.74
NONPARAMETRIC METHODS 4.75
REGRESSION AND CORRELATION 4.77
EXPERIMENT DESIGN 4.79
STATISTICAL SOFTWARE 4.80
COMMON EMISSIONS APPLICATIONS 4.80
Odor Testing 4.80
Emissions Compliance 4.81
REMINDERS 4.81
REFERENCES 4.82

INTRODUCTION

Decisions based on data must be made in the face of uncertainty or variability. Measurements of a process characteristic of interest can have many sources of uncertainty, e.g., the actual variability of the process, inconsistencies in the analytical measurement process, and inherent differences in samples due to the sampling process. While efforts can be made to minimize the influences of such sources of uncertainty on making good decisions, the uncertainty in data can never be completely eliminated.

Such uncertainties complicate decision making since true process behavior is not actually known, but rather simply estimated. Making decisions in the face of uncertainties is the business of the field of statistics. Identifying and estimating the uncertainties, minimizing the associated impact through efficient sampling designs, and making the best decisions possible are the goal.

With the limited space available, this section is intended to be a practical guide on elementary statistical applications. The goal is not to make the reader a statistician or to furnish a self-contained do-it-yourself manual. That is not practical in a single section. Instead, it is hoped that the reader will be able to recognize the types of tools needed for different types of applications so that appropriate additional statistical resources such as references or software can be identified. Common statistical tools are discussed, and then in the final section some specific emissions applications are discussed to illustrate when the various tools could be applied. Readers with competent statistical backgrounds may prefer to simply refer to the final section. Readers who are not familiar with statistics should be willing to read some sections twice, to become familiar with the terminology and to understand the important issues.

DATA COLLECTION PLANNING

Often data are collected with little regard for how they will ultimately be used. Throughout the country huge amounts of environmental data have been accumulated with much serving little use. In recognition of this, the Environmental Protection Agency (EPA) has proposed a planning process for use prior to the collection of environmental data. The importance of using this process, or something similar, cannot be overstated. EPA document QA/G-4, *Guidance for the Data Quality Objectives Process,* provides guidance on the data quality objectives (DQO) approach which is discussed only briefly here. As topics are discussed, issues will be pointed out that should have been resolved in the DQO planning for the task.

The DQO process involves early communication with regulators and other stakeholders. A DQO team consisting of project personnel, subject matter experts, and regulators follows a seven-step process:

1. State the problem.
2. Identify the decision.
3. Identify inputs to the decision.
4. Define the boundaries of the study.
5. Develop a decision rule.
6. Specify tolerable limits on decision errors.
7. Optimize the design for obtaining data.

The process involves considerable discussion in each step. It serves as an excellent communication tool and helps all parties involved address the specific issues about the what, when, where, how much, and, most importantly, the why of data collection prior to the actual collection activities. The latter steps become more statistical in nature and attempt to ensure that reasonable quantities of data are obtained on which to base the decisions described in earlier steps. The development of the various steps can be fairly straightforward or quite involved depending on the complexity of the application considered. The reader who is in a position of making decisions about data collection is advised to seek more information on the DQO process. A good source for such information can be found at http://etd.pnl.gov:2080/DQO.

TYPES OF DATA

Appropriate statistical tools depend on the type of data. The first level of data is *nominal,* or *categorical.* As the name suggests, no inherent ordering of different data "values" is present. For example, say the responses for marital status on a survey are

1. Married
2. Single
3. Divorced
4. Widowed

No ordering should be associated with such data, yet many times when answers from a set of respondents are summarized, a mean response of 2.73 will be reported. Hopefully the nonsensical nature of such a number is obvious. Better summary information would consist of the proportions or percentages of respondents providing each response.

The second level of data is *ordinal.* Typically these data involve rankings or orderings, such as the common

1. Strongly disagree
2. Disagree
3. No opinion
4. Agree
5. Strongly agree

Now a mean response of 2.73 has somewhat more meaning, but the use of such measures may still be misleading. Is the difference between a 2 and a 3 (disagree and no opinion) really the same as the difference between a 4 and a 5 (agree and strongly agree)? That is actually the assumption being made when statistics such as the mean or standard deviation are used. Proportions or percentages may again be more appropriate measures for this type of data as well.

For *interval* data, differences are meaningful. The difference between 15 and 25°C is exactly the same as the difference between 65 and 75°C. However, is 100°C twice as hot as 50°C? How about between 50 and 100°F? The problem here is that ratios of numbers (for example, 100/50 = 2) are not meaningful. However, the typical statistical summary measures are appropriate for this type of data. A mean daily temperature of 72.4°F makes sense.

Ratio data are what we typically encounter in applications such as emissions monitoring. Indeed 100 parts per million (ppm) is twice as much as 50 ppm; 99 lb is 3 times as heavy as 33 lb; and 5 min is one-half as long as 10 min. For this type of data, all statistical measures are appropriate. Note that the common emissions measure of percentage of removal is another example of ratio data; 90 percent removal is twice as efficient as 45 percent removal. Such percentage of removal values should not be confused with estimated proportions which involve count data. For example, when 12 out of 50 samples give results above a standard, the proportion exceeding the standard is 12/50 = 0.24, or 24 percent.

MEASURES OF LOCATION AND DISPERSION

In typical statistical applications, an underlying *population* of numbers represents a phenomenon of interest. Statistical measures that describe the population are called *parameters.* If one could somehow obtain all the values in the population, the parameters could be computed; but this is impractical or, more likely, impossible. Instead a *sample* of values from the population is observed, and corresponding sample *statistics* are computed to estimate the population parameters. Since these sample statistics are subject to many sources of variability, how they behave with respect to the unknown underlying population parameters is of interest. In particular, how close the statistics should lie to the parameters they estimate is the basis of *statistical inference* topics such as confidence intervals and hypothesis testing.

Commonly used measures of location are the mean, median, and other percentiles. The *mean* of a sample, typically denoted by $\bar{x}$, is the arithmetic average of the sample values and estimates the generally unknown, underlying population mean typically denoted by μ. That is, for a set of sample values $x_1, x_2, \ldots, x_n$,

$$\bar{x} = \frac{x_1 + x_2 + \cdots + x_n}{n}$$

The *median* of a population is the value that has one-half of the population values less than itself and the other one-half greater than itself. It is estimated by the *sample median.* For the sample values $x_1, x_2, \ldots, x_n$ the median is the $(n + 1)/2$ value in the ordered sample listing, if n is odd, or is the average of $n/2$ and $(n/2) + 1$ items, if n is even. For example, if $n = 9$, the median is the 5th largest (or smallest) of the ordered values, while for $n = 10$, the median is the average of the 5th and 6th largest values.

Percentiles of populations or samples can be defined in a manner similar to the median. In fact, the median is the 50th percentile. We later refer to the first and third *quartiles,* typically denoted by Q1 and Q3. These are the percentile values which have, respectively, 25 and 75 percent of the values less than themselves.

The most common measures of dispersion or spread of a set of values are the range, variance, and standard deviation. The *sample range* is simply the difference between the maximum and minimum values in the sample. The sample *variance,* denoted by s^2, estimates the population variance, denoted by σ^2, and is computed as

$$s^2 = \frac{(x_1 - \bar{x})^2 + (x_2 - \bar{x})^2 + \cdots + (x_n - x)^2}{\ }$$

The square root of the sample variance is the sample *standard deviation s,* and it estimates the population standard deviation σ.

The mean and standard deviation, the most commonly used statistics, are readily computed for a set of data by most software packages and even hand calculators. If you get a different result for the standard deviation than other sources when using a calculator, the explanation may well be that the calculator uses n in the denominator rather than $n - 1$. The preferred value is $n - 1$ since it is an *unbiased* estimator; that is, on average, it equals σ^2. That is not the case for the estimator that uses n in the denominator.

The roles of the mean and standard deviation are illustrated in the following diagram of the commonly used *normal distribution.* Note that values occur in normal or other continuous distributions according to probabilities represented by an associated probability density function. For a normal distribution this function is represented by the curve in Fig. 4.4.1. The probability of a value's lying in a particular interval is equal to

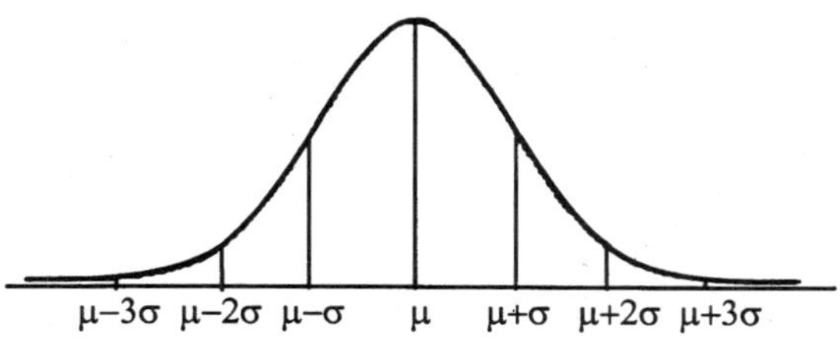

FIGURE 4.4.1 Normal distribution.

the area under the curve above the interval. Thus the entire area under the curve must be 1.0.

If a population of values has a normal distribution, then it is known that approximately 67 percent of the values will lie within 1 standard deviation of the mean (between $\mu - \sigma$ and $\mu + \sigma$), approximately 95 percent will lie within 2 standard deviations, and over 99 percent will lie within 3 standard deviations. Note that since 50 percent of the values lie between the first and third quartiles, Q1 and Q3, respectively, would lie a little closer to μ than do $\mu - \sigma$ and $\mu + \sigma$. Since the normal distribution is symmetric, the median is equal to the mean μ, since 50 percent of the values lie on either side of the mean. Tables are available in standard statistics tests that allow the computation of areas under the curve (probabilities) for any intervals.

Note, in contrast, the relative locations of the quartiles, median, and mean for the lognormal distribution in Fig. 4.4.2. The larger values that skew the distribution to the

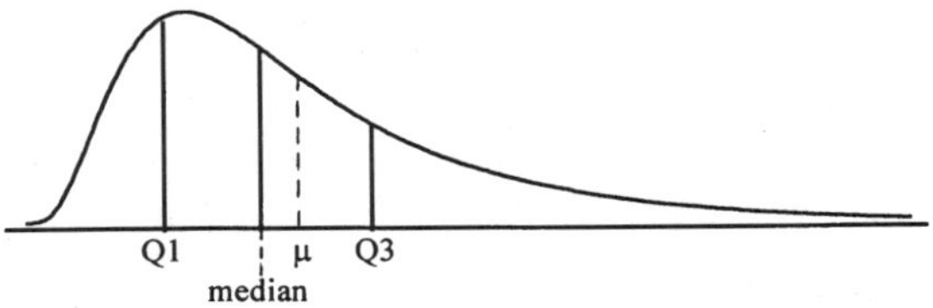

FIGURE 4.4.2 Lognormal distribution.

right have a greater impact on the mean than the median, so the mean will be larger. This is due to the fact that the mean is a function of the magnitude of every value and is thus quite sensitive to very large (or very small) values. The median simply divides the values into two stacks of equal size. The magnitude of extreme values is therefore not as important.

Since the lognormal distribution is skewed (heavy-tailed) to the right, it is commonly used to model environmental data which often show a similar skewness to the right with occasional large values—more than would be expected with a normal distribution. The lognormal distribution is convenient since log-transforming lognormal values gives values that follow a normal distribution. Many of the common statistical methods are based on the assumption of an underlying normal distribution, so log transformations of environmental data are sometimes useful to meet normality assumptions. For a summary of properties of lognormal distributions, see Gilbert (1987).

Another parameter of interest is a *proportion.* Many applications involve estimating the probability of an event or the proportion of time it would be expected to happen. Such a population proportion p is estimated by the sample proportion x/n, where n is the total number of values or cases observed and x is the number of them that satisfy the condition of interest. Recall the earlier comment that percentage of removal is not an example of a proportion problem since it does not involve counting the number of "successes" in so many trials.

GRAPHICAL DISPLAYS

Graphically displaying data is always recommended in conjunction with the computation of summary sample statistics. Understanding of a phenomenon is greatly facilitated by "looking at the data" in this manner. Without graphing data, decisions are often unduly influenced by outlying data values or other unusual features of the data which can be easily detected with graphical tools.

A particularly useful graphical tool for quickly summarizing a set of data is called a *box-and-whisker plot* and is available on many types of software packages. An example is given in Fig. 4.4.3. The interpretation of the box-and-whisker plot is straightforward. The 'x' indicates the sample mean, which for the data represented by this plot is approximately 21.0. The median appears to be about 18.0. The first and third quartiles appear

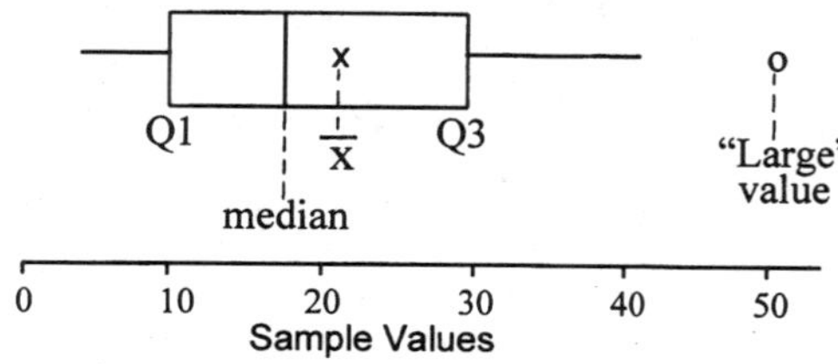

FIGURE 4.4.3 Box-and-whisker plot.

to be 10.0 and 30.0, respectively, so about 50 percent of the sample values lie between these values. The minimum value falls at the end of the left-hand "whisker" and is about 3.0 while the maximum value is about 50.0. This latter value is flagged since it falls "relatively far" from the other data. The second-largest value is at the end of the right-hand whisker and is about 41.0. The length of the whiskers and where particularly large (or small) values are flagged are computed based on the variability indicated in the data by the *interquartile range* (the distance between the quartiles). Standard statistics texts discuss this topic. The contrast in box-and-whisker plots for normally and lognormally distributed data is indicated in Fig. 4.4.4.

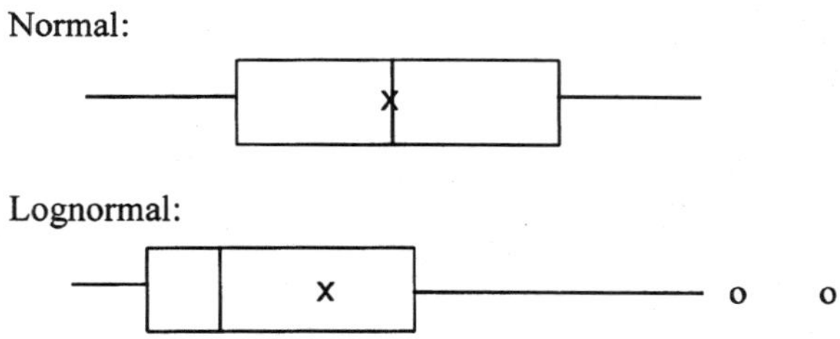

FIGURE 4.4.4 Box-and-whisker plots for normal and lognormal data.

Multiple box-and-whisker plots are extremely useful for comparing multiple sets of data. Consider the example in Fig. 4.4.5 which could represent trichloroethylene (TCE) emissions in samples from nine different stacks. Note that adding a horizontal reference

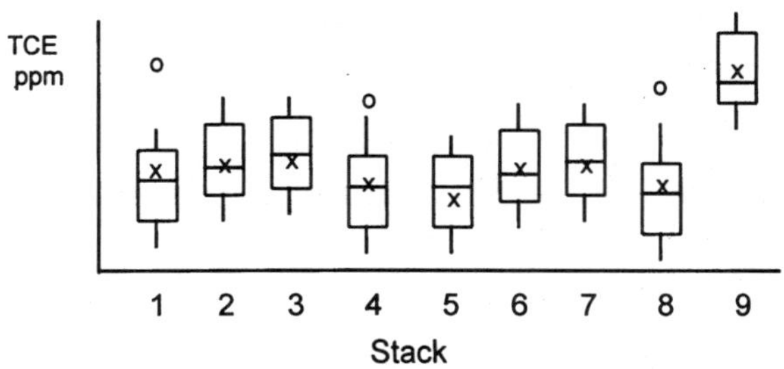

FIGURE 4.4.5 Multiple box-and-whisker plots.

line at a regulatory standard value in Fig. 4.4.5 would be useful for comparisons to the standard. If comparisons to background data are of interest, as might be the case for soil sampling, then the background data could be represented as a box-and-whisker plot on the same diagram. This is an excellent way to visualize potential differences.

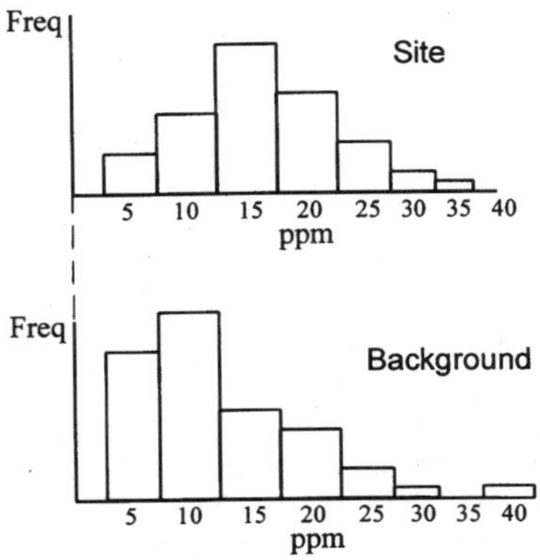

FIGURE 4.4.6 Multiple histograms.

Another useful graphical tool that is quite common is the *histogram. Multiple histograms* can also be used for comparing different sets of data. In Fig. 4.4.6 data from a potentially contaminated site are compared to background data. The height of the bar over each midpoint parts per million (ppm) value represents the number of values observed in the corresponding interval. This diagram would suggest that the site is somewhat more contaminated than the background since more of the values occur in higher ranges.

Another useful graphical tool is a *probability plot.* Since normality of underlying data is often a required feature to apply certain statistical tools, testing the data for this characteristic is of interest. Several types of statistical tests are available for testing for normality, but the graphical probability plot is an alternative approach that is simple to use. Consider the examples in Fig. 4.4.7. For both graphs the sample values are plotted against normal probabilities. The x values lying along the straight line in the top graph indicate that the data reasonably can be assumed to have come from a normal distribution. The data that generated the bottom graph, however, would fail this normal distribution test because the x values fall too far from the line. An obvious disadvantage of this approach to testing for normality is the subjective nature of determining how well the x values match up with the straight line. How close is close enough?

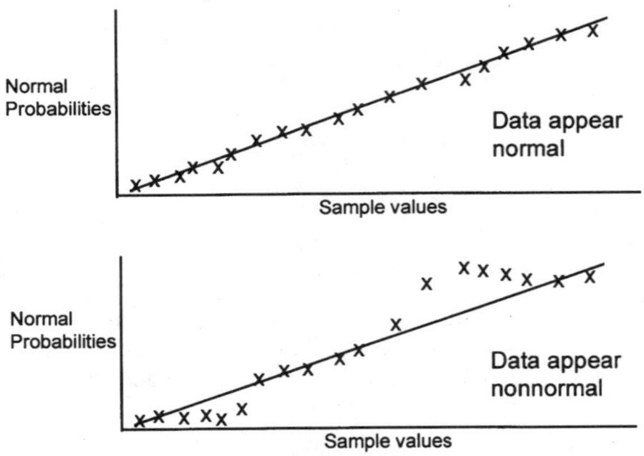

FIGURE 4.4.7 Normal probability plots.

These and other types of graphical tools are available on most software packages. Again, the importance of taking a quick look at the data with these or other graphical tools cannot be overstated.

SAMPLING ISSUES

The underlying statistical methodology that supports the use of common statistical tools typically starts out, "Given a set of independent and identically distributed (IID) random variables x_1, x_2, ..., x_n, then...." The validity of the subsequent statistical statement depends on this IID assumption. which is appropriate when a *simple random sample* of values is selected from an underlying population. A simple random sample is obtained when the selection process makes every possible sample of size n equally likely to be selected. For any other type of sampling approach, the standard statistical tools may not apply without some type of modification. Thus sampling methods are extremely important in making good data-based decisions.

Developing an optimal sampling approach is a primary goal of the DQO process discussed earlier. Before this can be done, as the DQO process illustrates, thorough discussion by involved parties must determine the problem to be solved, the decision to be made, the associated statistical criteria to be used, and some level of error probability that can be tolerated. Given this type of information, an adequate sampling design, including the sample size, can be created. In many cases this may require the support of a statistician.

Simple random samples are typically the most useful and can be obtained in a number of ways. Random selections of items or monitoring times would provide such samples. Many times *systematic samples* are taken by using a standardized monitoring schedule, or among items, picking a random starting point and then taking every *k*th item thereafter. Values resulting from such systematic samples are typically treated as if they were obtained from simple random sampling. This is appropriate as long as the phenomenon being measured does not display some type of repeating pattern that might coincide with a monitoring schedule or the systematic selection of every *k*th item. If such a pattern were present, biased samples that could lead to incorrect decisions could result.

A useful sampling approach that can reduce costs is *composite sampling*. Multiple individual samples are combined and thoroughly mixed before analytical testing is performed. An analytical result is then the average over the set of samples that made up the composite. Several such composite samples generate the set of results to be used for the decision making. Many times this averaging process is very useful in reducing the sampling variability in an application and lowering analysis costs.

Other sampling methods that make use of known features of a phenomenon are available. They include

- *Cluster sampling*. Items occur in groupings, so groupings are randomly selected; then either items within the selected grouping are all used or a subset is randomly selected.
- *Stratified sampling*. The population is divided into strata of like items or areas, and a simple random sample is selected from each stratum.
- *Sequential sampling*. Sampling continues with statistics recomputed after each sample or set of samples until one of specified alternative conditions is met.
- *Grid sampling*. Systematic sampling of an area is achieved by using a grid of sampling locations. This is typically used in conjunction with detecting contamination "hot spots."
- *Adaptive sampling*. Subsequent sample locations are based on results obtained from preliminary samples; the extent of areas of contamination is of interest.

These sampling approaches are not discussed further here, but they and others are thoroughly discussed in statistical texts such as that by Cochran (1977) or Thompson (1992). The various approaches may have advantages over simple random sampling since more information about the phenomenon of interest is being used in sample selection. However, this in turn complicates the subsequent statistical analyses. For this reason such sampling approaches typically require the involvement of a statistician.

All the sampling methods mentioned so far are types of *probability sampling*; i.e., selections are made based on some type of underlying probability model. This model allows the associated applications of statistical tools for making inferences about the underlying population of values. Alternative sampling approaches are typically called *judgmental sampling,* since sample locations or times are selected due to expectations of their "special nature" and the likelihood of gaining more useful information at such locations than what might be obtained at randomly selected times or locations. A good example is the selection of soils believed to be contaminated. The difficulty with such sampling approaches is the lack of an underlying probability model on which to base statistical conclusions. Such judgmental sampling is thus not recommended for statistical decision making.

In applications such as emissions compliance, the sampling activities and associated computations can be relatively complex. If a standard regarding the annual total weight of an emission analyte is to be met, the sampling approach is not at all obvious. How should the operating schedule of the process generating the emission analyte be used? Should the level of operation be used, since emissions vary according to the process? Should samples be composited? These are exactly the types of questions that should be answered by going through the DQO process. What it means to meet the standard must be carefully spelled out and agreed upon in advance. In particular, what constitutes statistical evidence that the standard is being met must be mutually understood by all parties. Only then can reasonable sampling approaches be developed.

SAMPLING DISTRIBUTIONS

When a sample of values is selected from a population and a sample statistic computed, statistical inference regarding the associated population parameter depends on how the possible values of the statistic vary from the true value of the underlying parameter. How these values vary is called the *sampling distribution* of the statistic, and this sampling distribution is the basis for statistical statements regarding the parameter value.

Consider as a first application the sampling distribution of the sample mean $\bar{x}$. Say we take a simple random sample of size n from a population with mean μ and variance σ^2. Depending on the actual sample obtained, the potential values that $\bar{x}$ can take on are hopefully "close" to μ. In fact, straightforward statistical and mathematical properties show that the expected value (mean) of potential $\bar{x}$ values is μ, and the variance of the potential $\bar{x}$ values is σ^2/n. Note that this means that regardless of the sample size, the average value expected for $\bar{x}$ is μ, and the larger n is, the closer $\bar{x}$ would generally be to μ, since $\bar{x}$ has smaller variance for larger n. Note also that these facts do not depend at all on the shape or distribution of the underlying population of values. If we add the fact that the population follows a normal distribution, then we also know that the shape of the distribution of $\bar{x}$ is also that of a normal distribution, in this case with mean μ and variance σ^2/n.

These are the types of features regarding sampling distributions of sample statistics that are used to draw inferences about population parameters. Such inference includes confidence intervals and hypothesis testing, topics which are discussed in subsequent sections. Similar properties of the sampling distributions of the sample variance, or a sample proportion, are used for inference for their corresponding parameters, but they are not discussed here. Standard statistical tests contain such information.

Two more issues regarding the use of $\bar{x}$ to estimate μ are of importance. What good is knowing that the variance of $\bar{x}$ is σ^2/n if we do not know the value of σ^2 (which we probably do not, since to be estimating μ, we must not know its value or that of σ^2)? Also what if the underlying population of values is not normally distributed? Those two questions must be answered to complete a description of the sampling distribution of $\bar{x}$ and to describe how statistical inference regarding μ is performed.

As stated, if the underlying population of values is normal with mean μ and variance σ^2, the sampling distribution of $\bar{x}$ is normal with mean μ and variance σ^2/n. Thus the quantity $(\bar{x} - \mu)/(\sigma/\sqrt{n})$ has a standard normal distribution (mean of 0 and variance of 1). With σ unknown, its estimate s, the sample standard deviation, can be used instead, giving $(\bar{x} - \mu)/(s/\sqrt{n})$, which has *Student's t distribution* with $n - 1$ degrees of freedom, again provided that the underlying population is normal. This fact will be used in subsequent sections.

If the underlying population of values cannot be reasonably assumed to be normal, then we are somewhat stuck. One aid is the *central limit theorem,* which says that if the sample size is sufficiently large, the distribution of the sample mean can be taken to be normal regardless of the shape of the underlying distribution. Thus the distribution of the expression $(\bar{x} - \mu)/(s/\sqrt{n})$ can be reasonably approximated by a t distribution providing the sample size is sufficiently large (the magic number that is "sufficiently large" is typically listed as $n = 30$). The remaining situation of small sample sizes with non-normal populations does not lend itself to this standard approach. One hope would be that a data transformation could be made to make the data appear to be more likely to have come from a normal distribution. In particular, if a population of values has a lognormal distribution, then a log transformation on the values will give a population with a normal distribution. Inference based on the normal distribution could then take place after the log transformation.

One does have to be careful, however, about using the log inverse transformation on inference results to get back to the original data scale. Some estimates such as percentiles will transform correctly; however, others such as the mean will not. Gilbert (1987) provides a good discussion of parameter estimation for lognormal distributions.

Note that normality "tests" were briefly mentioned under normal and lognormal probability plotting. They are discussed in a later section on hypothesis testing.

INTERVAL ESTIMATES

The sample statistics $\bar{x}$ and s provide point estimates of the population parameters μ and σ. Point estimates alone give no indication of how "far off" they might be from the true underlying values. A remedy to this situation involves the use of *interval estimates* based on statistical *confidence intervals.*

The most common confidence interval application is that for the mean. Assume that the t distribution for $(\bar{x} - \mu)/(s/\sqrt{n})$, as described in the previous section, is appropriate (i.e., either n is large or the underlying population distribution is normal). Then the probability that $(\bar{x} - \mu)/(s/\sqrt{n})$ takes on a value between $-t_{\alpha/2}$ and $t_{\alpha/2}$ is $1 - \alpha$. This is illustrated in the diagram of a t distribution in Fig. 4.4.8. Standard statistical texts or

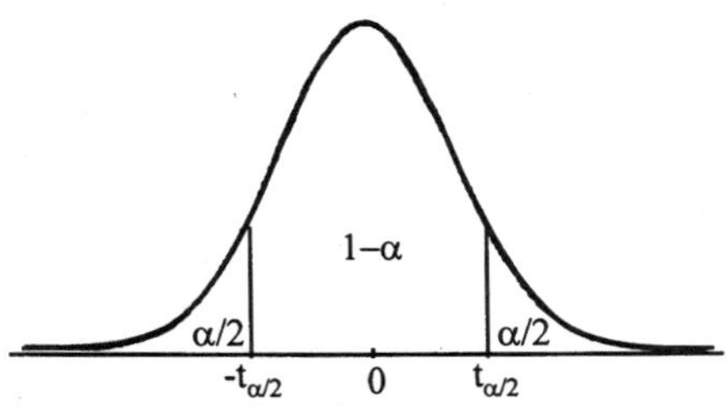

FIGURE 4.4.8 The t distribution.

books of mathematical tables contain tables of these t distribution percentiles for different values of n and α.

Algebraic manipulation of the relationship

$$-t_{\alpha/2} < \frac{\bar{x} - \mu}{} < t_{\alpha/2}$$

which occurs with probability $1 - \alpha$, shows it is equivalent to

$$\bar{x} - t_{\alpha/2} s/\sqrt{n} - \mu < \bar{x} + t_{\alpha/2} s/\sqrt{n}$$

Thus the interval $\bar{x} \pm t_{\alpha/2} s/\sqrt{n}$ is expected to contain μ with probability $1 - \alpha$. Therefore given a random sample $x_1, x_2, \ldots, x_n$, from a population with mean μ, resulting in the point estimates $\bar{x}$ and s, a $1 - \alpha$ confidence interval for μ is $\bar{x} \pm t_{\alpha/2} s/\sqrt{n}$, where $t_{\alpha/2}$ is obtained from a t distribution table.

As an example, say a set of 12 stack emissions gives a sample mean and standard deviation for TCE content of 22.3 ppm and 7.6, respectively, and the underlying distribution of TCE emissions being sampled can reasonably be assumed to be normal. To compute a 95 percent confidence interval, $\alpha = 0.05$, and $t_{0.025}$ with $n - 1 = 11$ degrees of freedom is found in a t distribution table to be 2.201. Thus the 95 percent confidence interval is $22.3 \pm 2.201(7.6)/\sqrt{12}$ which gives (17.47, 27.13). Note that the length of such an interval will decrease as either the sample size increases or the estimate of the standard deviation decreases.

The interpretation of such an interval is typically an interesting point of discussion. Many times the oversimplified statement is made that the probability is 0.95 that μ is between 17.47 and 27.13. That is not correct. It is true that the probability that μ will be between $\bar{x} - t_{0.025} s/\sqrt{12}$ and $\bar{x} + t_{0.025} s/\sqrt{12}$ is 0.95, but the random quantities are $\bar{x}$ and s, not μ. Thus prior to observing the specific values of $\bar{x}$ and s, the probability statement makes sense. Imagine a whole set of possible intervals landing in the vicinity μ depending on the values of $\bar{x}$ and s; 95 percent of them will actually contain μ. We then pick one interval by selecting the sample that generates the specific values of 22.3 and 7.6, and then the probability is "used up." Now the unknown fixed value μ is either in the interval or is not, and we do not know which is true. We would certainly expect that it "should be" since it is true that 95 percent of similarly drawn samples will generate values of $\bar{x}$ and s that lead to intervals that will contain μ. However, the term *probability* should not be used after the sample results are observed. Often the statement is made, "We are 95 percent *confident* that the computed interval contains μ," and this is certainly appropriate, provided the underlying meaning is understood.

If this all seems like a matter of semantics (and maybe it is), imagine holding a die in your hand. The probability you will roll a 3 is ⅙. Now roll the die, and look at the resulting number. Do you think the probability is still ⅙ that it is a 3? No, the probability it is a 3 is either 1.0, if it is indeed a 3, or 0.0, if it is not. Similarly a computed confidence interval either "worked" or it did not; unfortunately we never know which is the truth since μ is, and will remain, unknown.

All this discussion was about the simplest case of showing how a confidence interval for a mean is derived. Similar confidence intervals can be computed for variances, standard deviations, proportions, percentiles, differences in means (the interval containing 0.0 indicates the means may be equal), differences in proportions (same comment for containing 0.0), ratios of variances (the interval containing 1.0 indicates the variances may be equal), regression slopes and intercepts, correlations, and many others. Standard statistical texts give derivations, formulas, and tables for these types of confidence intervals.

Prior to a discussion of hypothesis testing which is closely related to confidence intervals, four topics involving confidence interval applications are briefly discussed:

comparisons to a standard, tolerance intervals, control charts, and confidence intervals for a proportion.

COMPARISON TO A STANDARD

Environmental compliance may involve comparison to a regulatory standard to ensure it is not exceeded. Hopefully the ambiguity in the previous sentence was noticed. *What* is not supposed to exceed the standard? The mean? The median? Every single measured value ever obtained, that is, in effect, the maximum? An upper percentile? This again illustrates the importance of the DQO planning process. Such questions should be thoroughly discussed by all involved parties and answered to their common agreement. Only then can statistical consideration address efficient sampling designs for accumulating the data on which to base decisions.

The author believes that the intent of the EPA in most applications is that a regulatory standard not be exceeded on the average (unfortunately, interpretation of regulations is often left to local regulators, so different interpretations can be found across the country). The averaging approach is definitely used for Resource Conservation and Recovery Act (RCRA) hazardous waste characterization in EPA guidance document SW-846, *Test Methods for Evaluating Solid Waste,* chapter 9. The impact of disposing of a large amount of waste should not be based on the worst-case analytical value, but rather on the average. Similarly, public health risk assessments are based on average contaminant levels, not the worst case. In both of these applications, however, commonly applied methods do not simply use point estimates of the mean, but rather use the more conservative upper confidence bounds for the mean. Given that a sample mean does not exceed a standard, we cannot be sure that the underlying population mean fails to exceed the standard since we have no estimate of the uncertainty in the estimate. However, if an upper confidence bound for a mean does not exceed the associated standard, then we can have "high confidence" that the underlying true population mean must not exceed it.

An upper confidence bound for a mean is simply the upper limit of a one-sided confidence interval. From earlier discussions about the confidence limits for a mean, it can be seen that such a $1 - \alpha$ upper bound can be obtained by using the appropriate t_α value instead of $t_{\alpha/2}$ in the upper limit without a lower limit. This gives an interval that again contains the true mean with probability $1 - \alpha$. In other words, a value is obtained that bounds the true mean with $1 - \alpha$ "confidence." Again under the proper conditions for ensuring the normality of the sample mean, the value $\bar{x} + t_\alpha s/\sqrt{n}$ gives a $1 - \alpha$ *upper confidence bound* for the mean. As discussed in a subsequent section on hypothesis testing, application of upper confidence bounds on the mean is a very good way to demonstrate compliance with a standard. It should certainly be discussed with regulators.

TOLERANCE LIMITS

Tolerance limits can be interpreted as confidence limits on population percentiles. Say we want to estimate the 95th percentile of a population. An approach that does not depend on the underlying distribution (called *nonparametric,* as discussed later) is to use the corresponding sample percentile. For a 95th percentile, that is the sample value which is greater than or equal to at least 95 percent of the sample values yet is less than or equal to at least 5 percent of the sample values. For an underlying normal distribution, an estimate of the 95th percentile could be $\bar{x} + 1.645s$ since $\mu + 1.645\sigma$ is the 95th percentile. These are again only point estimates with no indication of how far off they might be from the true underlying values they are trying to estimate. A tolerance limit quantifies the degree of uncertainty in the estimate, as does an upper confidence limit for a mean.

Two quantities must be specified for a tolerance limit: the tolerance level and the confidence level. Thus a 90/95 tolerance limit is that value for which we are 90 percent confident that it exceeds 95 percent of the population values. For this tolerance limit, 95 percent is the tolerance level (typically called $1 - \lambda$) and 90 percent is the confidence limit ($1 - \alpha$). Upper tolerance limits are of the form $\bar{x} + ks$, where the tolerance factor k is given in tables and is a function of the desired tolerance and confidence levels and the sample size [see Odeh and Owen (1980)]. Note that for small sample sizes the value of k can be very large (considerably larger than t_α) and use of the *upper tolerance limit* (*UTL*) very conservative, probably too much so. The nature of this conservatism, of course, depends on the type of decision being made. In addition, the inference results will be highly dependent on the underlying population distribution, even for large sample sizes. The appropriateness of normal distribution theory for large samples established by the central limit theorem is valid only for the measurement of central tendency, i.e., the mean. If one is estimating the tails of the population distribution, the form of that distribution remains important even for large samples. The use of upper tolerance limits is discussed in the hypothesis-testing section in comparisons to background.

CONTROL CHARTS

Statistical control charts can be used with confidence interval type of criteria to monitor data over time. Figure 4.4.9 contains an example. The centerline and upper control limit (UCL) and lower control limit (LCL) are computed based on accumulated histor-

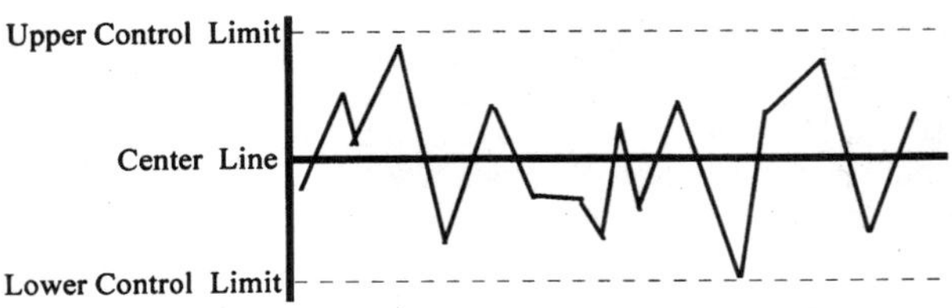

FIGURE 4.4.9 Control chart.

ical data and represent the performance of a process against which ongoing process performance is to be compared. Say, e.g., that results from weekly emission samples are reported on a monthly basis. Say DQO discussions have led to monitoring criteria for the monthly average TCE emissions which will be estimated by the mean of four weekly samples. The overall sample mean and standard deviation could be estimated from historical data from a period during which the process performance was considered adequate. Ongoing performance is to be compared to that performance.

The centerline is simply the sample mean of the historical data. Control limits reflect the variability we expect for the sample mean, that is, $s/\sqrt{n}$, where s is the standard deviation of the historical data. Using this estimate of expected variability, the control limits could be taken to be equivalent to the upper and lower confidence limits for the mean based on the t distributions if the underlying distribution of values can again be assumed to be normal (recall $\bar{x} \pm t_{\alpha/2}s/\sqrt{n}$). Many times this formality is not used, and the t distribution value is simply replaced by 2.0 to obtain a *warning control limit* and 3.0 for an *action control limit.* As an illustration, consider that the sample mean and standard deviation are, respectively, 10.0 and 6.0 ppm. Then $s/\sqrt{n} = 2.0$ ppm, and the control chart in Fig. 4.4.10 could be used. Each month the four weekly sample results are averaged,

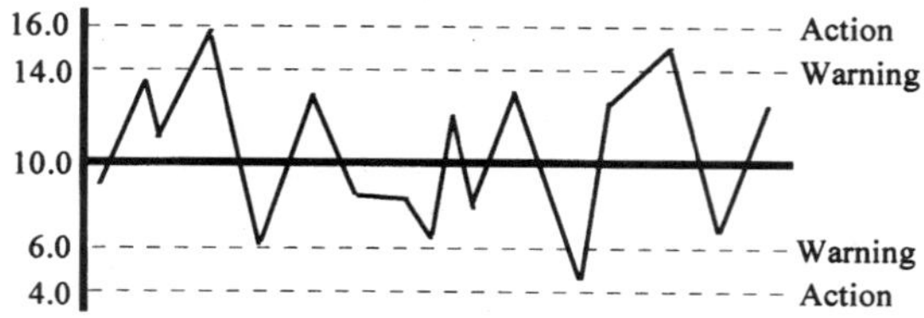

FIGURE 4.4.10 Control chart with action and warning limits.

and a point is added to the control chart. As long as values lie within the warning limits, the process is taken to be performing as expected. Values between the warning and action limits suggest that special attention may be desirable. Values outside the action limits suggest process change, and identification of the reason may be required.

Other criteria may be applied as well, such as a significant number of consecutive points (typically five or six) increasing or decreasing or lying on the same side of the centerline. Each of these conditions could also provide an early indication of process change.

Such control charting methods are common tools in the area of *quality control.* Other features besides process means can also be followed in a control chart. In manufacturing applications, numbers or percentages of defects or defectives are often of interest. One-at-a-time charts that plot each individual sample value are also commonly used, as are range charts that monitor process variability. Control limits and centerlines are computed much as confidence limits are in each of these application types.

Note that appropriate control limits are computed to ensure a reasonable *false-alarm rate,* i.e., to quantify the chance of raising an alarm about a process that is performing in the appropriate manner. Such false alarms will result, but at a predetermined rate, due to the inherent sources of variability in data-based estimation. Standard quality control texts such as that by Montgomery (1991) contain information on these topics.

CONFIDENCE INTERVAL FOR A PROPORTION

A common confidence interval application is the estimation of an underlying population proportion p. Given a sample size n with x of the n sample exhibiting the characteristic of interest, a $1 - \alpha$ confidence interval for the true underlying proportion of the population that exhibits that characteristic is

$$\frac{x}{n} \pm z_{\alpha/2} \sqrt{(x/n)(1 - x/n)/n}$$

where $z_{\alpha/2}$ is the appropriate percentile of a normal distribution, as discussed earlier for the t distributions. Again tables of z values are available in standard statistical texts or books of mathematical tables.

As an example, in a sample of 100 emission samples, 28 exceeded an associated standard. What is an estimate of the proportion of time the standard is exceeded? The point estimate is 28/100 = 0.28; but again, as for any point estimate, this gives no idea of how far off this value could be from the true underlying proportion. Instead, consider the 95 percent confidence interval $0.28 \pm 1.96\sqrt{0.28(0.72)/100}$, which is computed as (0.192, 0.368). Note that 1.96 is the value obtained from a normal distribution table. Recall the previous discussion of the interpretation of confidence intervals. We are 95 percent confident that this interval contains the true underlying proportion.

Note that although normal probabilities are used in the computation, an underlying normal distribution is not required in this application. In fact, this is a *nonparametric* or *distribution-free* method that does not depend on an underlying distribution. Discussion is available in standard statistical texts, but stated simply, the confidence interval is derived from *binomial distribution* probabilities, and the binomial distribution in turn can be approximated by a normal distribution provided that p is not "too close" to 0 or 1. Note also that this is the application that has been used when you read of presidential polls, newspaper surveys, or television program ratings that are said to be "scientifically accurate to 4 percent." The 4 percent represents the error term that is added to and subtracted from the point estimate x/n to get the confidence interval.

HYPOTHESIS TESTS

An alternative to using confidence intervals as a way to summarize data for statistical inference decision making to use *hypothesis testing*. In reality, the two inference methods are equivalent and simply represent decisions in a slightly different way.

Consider, as an example, emissions monitoring in which several samples are taken, and the average TCE content is determined. Say that a standard of interest is 50.0 ppm. Again the DQO process should determine the way that this standard is to be applied. Let us assume that the standard is intended to be a performance measure to be met on average. Occasional values exceeding the standard do not indicate a violation in themselves; rather, the average of the samples is to be compared to the standard.

A hypothesis about the emissions values could be stated in either of two ways depending on the degree of conservatism thought appropriate (once again, the DQO process). Say that everyone clearly believes that the process is meeting the emission standard on average, so rather than overreact whenever marginal data results occur, we raise an alarm only if the statistical evidence is sufficiently strong to reject that belief. That is, we raise an alarm only when the mean of the emission results comes back so much higher than the standard that we can no longer defend our belief that the process emissions remain below the standard on average. Let's call this approach A.

The alternative approach, which is more conservative in terms of protecting the public from potential harm from emissions, is to assume that the process is running with mean emissions above the standard unless the data clearly show that the emissions data are so low that certainly the standard is met on average. Call this approach B.

Consider the following:

Approach	Null hypothesis	Alternative hypothesis
A	H_0: $\mu \leq 50.0$	H_a: $\mu > 50.0$
B	H_0: $\mu \geq 50.0$	H_a: $\mu < 50.0$

In a hypothesis-testing framework, under approach A we assume we are in compliance with the standard (the null hypothesis that $\mu \leq 50.0$ is true) unless we see sufficient evidence in the data to conclude that this cannot be the case, so instead we must be out of compliance (the alternative, that $\mu > 50.0$ is true). Under approach B exactly the opposite is true. We assume under the null hypothesis that we are out of compliance unless the data show otherwise, and the alternative hypothesis is true that we are in compliance. Clearly approach B is more conservative. The data must prove that the standard is being met. In other words, we are guilty until we can prove ourselves innocent.

Under the framework of approach A, consider the table in Fig. 4.4.11 where the null and alternative hypotheses are, respectively, H_0: $\mu \leq 50.0$ and H_a: $\mu > 50.0$. Let the column heading indicate the true, unknown state of our emissions. Thus in the first column we are in compliance with the standard while in the second column we are not. The row headings are then the decisions we make based on the data we observe.

Decision \ Truth	$\mu \leq 50$	$\mu > 50$
Accept Ho ($\mu \leq 50$)	OK	Type II error
Reject Ho ($\mu > 50$)	Type I error (false alarm)	OK (Power)

FIGURE 4.4.11 Hypothesis-testing decision errors.

If we are truly in compliance (first column), we might decide that we are in compliance and make a correct decision (top left box), or we might decide that we are out of compliance (bottom left box). The latter decision constitutes an error in this case, and this type of error, when a true null hypothesis is incorrectly determined to be false, is called a *type I error*.

If, on the other hand, we are truly *not* in compliance (second column), we might decide that we are not in compliance and make a correct decision (bottom right box), or we might decide that we are in compliance (top right box). Now this latter decision constitutes an error, and this type of error, when a false null hypothesis is incorrectly accepted as true, is called a *type II error*. Note under approach A, the type I error could also be called a false alarm; i.e., we raise the alarm that we are failing to meet emission standards when in fact we are meeting them.

Note that we could develop the completely parallel decision error diagram for approach B. All the components would be reversed, and error types would be reversed. In particular, a false alarm in terms of emissions being declared above standards when they are not would now be a type II error. Either approach can be appropriate. It depends on the focus of the application and the degree of conservatism to be applied. Remember the DQO process? Its application should address these types of issues. In fact, the final steps of the DQO process address the content of the decision error box and acceptable probabilities of committing both type I and type II errors, but the formulation is done in a way that minimizes the statistical terminology and tries to make the development look not quite so "statistical."

The type I error rate or probability, typically denoted by α, is controlled at a low level such as 0.10, 0.05, or 0.01. The error rates represent a tradeoff since the smaller the type I error is held, the larger the type II error, typically denoted by β, will be. For a given α level, the larger the sample size, the smaller β will be. When the null hypothesis is false, we make either a type II error or a correct decision. This correct decision is called the *power* of the test and is therefore equal to $1 - \beta$. For a given α level, the larger the sample size, the greater the power of the test, i.e., the greater the ability to correctly reject a false null hypothesis. This is the mechanism used in the final step of the DQO process to determine required sample sizes given estimates of variability and tolerable error rates. Such determinations typically require the assistance of a statistician.

Returning to our example, we are assuming we are in compliance ($\mu \leq 50.0$) unless the data suggest otherwise. Thus we have

Null hypothesis	H_0: $\mu \leq 50.0$
Alternative hypothesis	H_a: $\mu > 50.0$

We will accept the null hypothesis unless sufficient statistical evidence shows the alternative is true. We maintain a type I error rate of, say, $\alpha = 0.05$. Recall that if the underlying distribution is normal, or if the sample size is sufficiently large, under H_0, $t = (\bar{x} - 50.0)/(s/\sqrt{n})$ can be treated as having a t distribution with $n - 1$ degrees of freedom. Thus the $\alpha = 0.05$ criterion will be satisfied if we use $t > t_{0.05,n-1}$ *as a rejection region*; i.e., if values of $\bar{x}$ are so large that the resulting t distribution value occurs in the upper 5 percent of its possible values, we conclude that the alternative hypothesis $\mu > 50.0$ is true instead. This gives us our type I error rate as 0.05. If in fact the true underlying mean is greater than 50.0, then we are even more likely to have the computed t value fall in the rejection region.

This is illustrated by the power curve in Fig. 4.4.12. Assume the true underlying mean lies somewhere along the horizontal axis. If $\mu = 50.0$, our rejection region was set

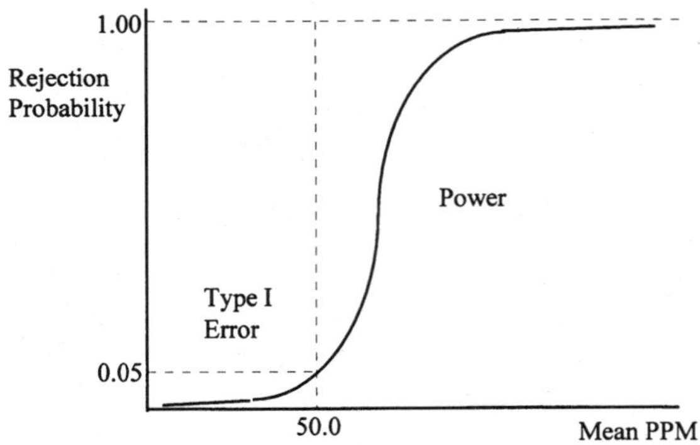

FIGURE 4.4.12 Power curve.

up so the probability of rejection is 0.05, as indicated on the plot. This is our type I error rate. If μ is in fact smaller, the rejection probability, and thus the type I error rate, will be even less. If $\mu > 50.0$, the alternative hypothesis is true, and we either have a type II error or a correct decision in rejecting the null hypothesis. Thus the rejection probability represented by the height of the curve in the figure is then the power of the test. The power always exceeds the type I error of 0.05 and gets large (close to 1.0) as μ increases. How fast the curve climbs to 1.0 will be a function of the estimate of variability and the sample size. The smaller the sample standard deviation s, or the larger the sample size, the faster the power will increase.

Say, for the example, that a sample size of 12 gives $\bar{x}$ and s, respectively, as 57.0 and 16.3. The computed t value is then $(57.0 - 50.0)/(16.3/\sqrt{12}) = 1.488$. From t distribution tables, $t_{0.05,\ 11\ \text{df}} = 1.796$. Thus we do not reject the null hypothesis since our computed t value does not exceed the rejection region t value. A sample mean of 57.09 is not sufficient relative to the estimated variability and the sample size to prove with potential type I error less than 0.05 that the true underlying mean is greater than 50.0.

The lack of conservatism of this approach is thus demonstrated. Even greater levels of emissions would need to be observed to determine that we were out of compliance with the standard value of 50.0. More likely we would set up the approach in the reverse, more conservative approach of assuming under the null hypothesis that we are out of compliance until the statistical evidence shows otherwise. For comparable sample size and variability to those in our computation, this would require sample means in the range of 40.0. Thus we would consider ourselves out of compliance with the standard value of 50.0 unless the sample mean were as small as 40.0, depending on the estimated variability and the sample size.

In our example, the *level of significance* of the test, i.e., the acceptable type I error probability, was preset at 0.05. This determines a rejection region of values of the test statistic that will result in the rejection of the null hypothesis in favor of the alternative. Most statistical hypothesis tests as computed by software do not use a preset significance level, but rather print the associated *p value* of the test statistic. This p value is the probability that a value as extreme as that obtained for the statistic will be obtained if the null hypothesis is true. Thus if this p value is "too small," in particular smaller than a level of significance of interest, then the value of the statistic seems relatively unusual for the null hypothesis and suggests instead that the alternative may be true. The p values such as 0.0001, 0.0207, or 0.05 would lead to rejection of the null hypothesis. The p values such as 0.2, 0.457, or 0.9999 would fail to reject the null hypothesis, suggesting it may be appropriate. The p values in the approximate range of 0.05 to 0.15 are considered something of a gray area where the suggestion is that the null hypothesis may not be true, but the evidence is not very strong. How a p value might appear on a computer output is illustrated in the following example of a two-sample t test.

Hypothesis testing is used for many statistical applications. A hypothesis regarding an unknown parameter of interest is made; typically what is desired to be proved is placed in the alternative hypothesis so rejection of the null accomplishes the goal. A common application is the comparison of two samples, one each from two underlying populations. The most common environmental example of this is comparisons to background. Some level of performance can be shown for unaffected background regions by sampling and generating the "background" data. "Site" data are then similarly generated for regions thought to potentially possess a negative characteristic due to operations or former use. The site data are then compared to the background data in a two-sample problem.

Once again what is to be compared is an essential component of the DQO planning process. Many common applications involve comparing the mean of the two populations, so a *two-sample t test* is appropriate, once again if the required conditions for normality of the sample means are met. Note that the original hypothesis-testing example discussed is a one-sample t test. In the two-sample case the hypotheses might be

$$\text{Null } H_0\text{: } \mu_s \geq \mu_b \qquad \text{versus} \qquad \text{Alternative } H_a\text{: } \mu_s < \mu_b$$

where μ_s and μ_b are, respectively, the underlying unknown population means for the site and background. Note that this represents the more conservative approach of assuming under the null hypothesis that the site mean may exceed the background mean. We assume this until the statistical evidence shows otherwise. Again the reverse approach could be used.

The two-sample t statistic is based on the site and background sample sizes, sample means, and sample standard deviations. Its specific form varies depending on whether the underlying standard deviations σ_s and σ_b can reasonably be assumed equal. Standard statistical tests contain descriptions of the associated methodology. For our purposes, consider the following computer output from the Statistical Analysis System [SAS (1989)] software:

The SAS System 1

TEST PROCEDURE

Variable: X

GROUP	N	Mean	Std Dev	Std Error	Minimum	Maximum
bkgrd	12	6.41666667	2.60971379	0.75335948	3.00000000	12.00000000
site	15	9.26666667	3.10452818	0.80158573	4.00000000	15.00000000

Variances	T	DF	Prob>|T|
Unequal	−2.5908	24.9	0.0158
Equal	−2.5399	25.0	0.0177

For H_0: Variances are equal, F′ = 1.42 DF = (14,11) Prob>F′ = 0.5695

The output gives the results when a variable X is measured for 12 background and 15 site samples. The respective sample means for the background and site were approximately 6.41 and 9.27, with standard deviations of 2.61 and 3.10. (Note that the standard error is $s/\sqrt{n}$, the variability of the sample mean.) When one is testing for differences in means, differences in variances can exert an influence. Thus the last line of the output gives the results for the test of equal population variances based on the estimated standard deviations 2.61 and 3.10. The value $F' = 1.42$ represents the test statistic based on the ratios of the variances that follows an F distribution with 14 and 11 degrees of freedom (see standard statistical texts for a description of this test procedure). And Prob>F' = 0.5695 gives the p value of the test. It is two-sided, meaning we would have been interested in detecting if the site variance either exceeded or were less than the background variance. Since the p value is large, certainly greater than a potential 0.05 significance level, we would not reject the null hypothesis that the variances are equal. Thus using the equal-variance two-sample t test to test differences in means is appropriate.

The results of both the equal- and unequal-variance t tests are given on the example output just prior to the F test results for the variances. The two lines for equal and unequal variances give results very close to each other, since assumption of equal variances is reasonable. If the test for equal variances had led to rejection of equal variances, the results on the two lines for the t test could be quite different; the preferred result would be that for unequal variances.

For the equal-variance t test, note the t value of −2.5399 with 25 degrees of freedom and associated p value 0.177. The notation Prob>|T| indicates that this is also the p value for a two-sided test. That is not what applies here since we are only interested in detecting the situation $\mu_s < \mu_b$. Remember, we are assuming the site exceeds background unless we can demonstrate statistically that this is not likely. Most software packages give you the option of specifying that you want a one-sided test and the direction of the test. Others simply give you the two-sided result. Consider a two-sided p

value of 0.2000. This arises from a result that has 0.1000 in one tail, and this is doubled due to the two-sided nature of the criterion. If, in fact, the one-sided result is needed, the direction of the results must be considered. If the sample results have $\bar{x}_s > \bar{x}_b$, then the appropriate one-sided p value that corresponds to the two-sided 0.2000 is 0.1000, that is, one-half of the two-sided value. If, on the other hand, $\bar{x}_s > \bar{x}_b$, then the appropriate one-sided p value is 0.9000, that is, 1 minus one-half the two-sided value. This illustrates the caution that should be applied when using p values generated by statistical software and one-sided hypothesis tests are of interest.

In the example output, the two-sided p value would be halved since $\bar{x}_s > \bar{x}_b$. Thus the actual p value of interest is 0.00885. This is strong evidence that the null hypothesis is not true. Thus the magnitude at which the site sample mean exceeds the background sample mean, relative to their estimated variabilities, is sufficient to conclude that the underlying site mean does exceed the underlying background mean.

Another type of t test that should be distinguished from the two "independent" sample t tests just discussed is the *paired t test.* If data are inherently paired, the approach to comparing means changes. Say that the same test subjects are used in odor testing, and emissions results are to be compared to control results. An odor result is obtained from each test subject for both an emissions sample and a control sample. In a comparison of means, the pairing information should be used by simply using the differences in the two results for each subject. The paired differences for each subject are computed first, and then inference is based on the observed mean and standard deviation of the differences (see standard introductory statistics texts). This is in contrast to the independent-samples test in which differences in sample means are compared to sample standard deviations. The two approaches are appropriate for their own types of applications, and they should not be confused with each other.

Standard statistical texts contain thorough discussions of hypothesis tests and their many types of applications. Another point that deserves mention here is the testing of whether the underlying population of values can reasonably be assumed to be normal. The probability plotting approach has already been mentioned, but its shortcomings are the subjectivity of its interpretation and lack of control of the decision errors that is possible in a more formal testing approach. A formal hypothesis *test for normality* has the null hypothesis that the underlying population is normally distributed, while the alternative hypothesis is that it is not normally distributed. Note that relatively large sample sizes are required to adequately detect patterns in the data that suggest nonnormal behavior.

Two commonly applied normality tests are the Wilk-Shapiro and the Kolmogorov-Smirnov tests (see standard statistical texts). Both are computationally difficult and would typically be applied through the use of a software package. In such cases a "small" p value indicates again that the null hypothesis should be rejected, meaning the underlying distribution should not be considered normal. Note in such cases that data transformations, such as a log transformation for lognormal data, can result in data that better conform to a normal distribution. Testing the transformed data for normality is a common application, but care should be exercised in interpreting resulting inference on the transformed data back in the untransformed data.

RELATIONSHIP OF CONFIDENCE INTERVALS AND HYPOTHESIS TESTS

The two major inference tools we have discussed, confidence intervals and hypothesis testing, are closely related. In fact, they are actually equivalent and simply display sample results in a different manner. Consider comparison of a mean to a standard as an example. We assume we are out of compliance (our mean is greater than the standard) unless the statistical evidence suggests otherwise. In hypothesis testing we have H_0: $\mu \geq$ standard and H_a: $\mu <$ standard. Relatively small values of the sample mean, somewhat smaller than 50.0, will cause us to reject H_0 in favor of H_a, thus concluding we are in compliance. The actual *rejection region* of sample mean values will depend on the level of significance of the test, say, $\alpha = 0.05$.

Consider also a confidence interval approach. We conclude we are in compliance with the standard only if the upper confidence limit for the mean is shown to be less than the standard. Say our confidence level is $1 - \alpha = 0.95$, so $\alpha = 0.05$. Exactly the same set of sample mean values that led to upper confidence limits less than the standard will also lead to the rejection of the null hypothesis in the testing approach. Thus the two approaches are exactly equivalent. The similar relationship between the two inference approaches holds for other applications as well.

DATA PROBLEMS

In the applications discussed to this point, we have not dealt with commonly encountered problems with data. One of the primary ones is the presence of *nondetects* in the data. These are sample results that are recorded only as "less than the detection or quantitation limit." The actual sample value is thus only observed to be less than a specified quantity. To make matters even worse, in many data sets this detection limit may change so the data contain nondetects at multiple detection limits. Larger numbers of nondetects, especially if some are at relatively large nondetect values relative to other detected values, result in an extremely complicated data analysis situation.

This complication depends, of course, on the estimation goals. If some type of inference is being made using the median or a larger percentile such as a 95th percentile, then the analysis may not be affected at all provided that the percentage of nondetects is not too large. This is because for percentiles the actual values of all the data are not used. This unfortunately is not the case for other common estimators. Any quantity that uses all the sample values, such as the mean, standard deviation, and thus confidence limits and tolerance limits, is directly affected by the way the nondetects are treated.

The simple fix is to replace nondetects by some alternative value. If that value is taken to be the detection limit, the true unobserved value is obviously being overestimated. This approach will build biases into estimators. Sample means will tend to be larger than the true underlying population means, and sample standard deviations will tend to be smaller than the true underlying population standard deviations. An alternative approach is to replace all nondetects by zero. This has the opposite effect of underestimating means and overestimating standard deviations.

The most commonly used replacement approach that splits the difference between the two approaches just mentioned is to replace nondetects by one-half their detection limits. This is certainly more intuitively appealing and would reduce estimation biases relative to the other replacement conventions. This is not to say that this is a "good" approach. If relatively small numbers of nondetects are present, say, less than 15 to 20 percent, and the detection limits are relatively small values compared to detected values, then such replacement will probably have minimal impact on inference decisions.

On the other hand, if many nondetects are present, and worse, if some of the largest values in the data are nondetects, as can easily happen, then even using one-half the detection limit will have a significant impact on the summary measures. Some of the larger values in the resulting data will in fact be the result of nondetects. In this latter case, no really good answer for an estimation approach is available. Up-front planning should ensure that such data quality problems are avoided. Detection limits should be well below the typical values of interest.

Note that more sophisticated nondetect replacement methods have been proposed that use assumptions regarding underlying population distributions. However, how useful they are has been questioned. The benefits derived may not be worth the added complexity.

Two-sample comparison problems are somewhat different with respect to nondetects since in comparing the behavior of two samples, the usual sample estimates may not be required. This is the case with *nonparametric* methods, discussed in greater detail in a following section. Generally in such methods the relative magnitude of sample values generates ranks, and this type of ranking may still be possible even in the presence of nondetects. Thus two-sample comparisons can still be made in a meaningful way. See the section on nonparametric methods for more information.

Another common data problem is occasional large outlying values. They can have a tremendous influence on summary statistics, and they likely bring into question assumptions about underlying distributions. Statistical tests that "label" such values as "outliers" are available. Many times such values are then summarily dismissed and omitted. In fact, iterative applications of the tests can result in several large values being deleted from a data set. This cannot be recommended as a general practice. Unless some special cause is identified that is not expected to recur, such outlying values should probably be retained in some type of decision-making criterion.

As an example, say you start recording your commuting time to work on various days. One day you get 23 min, another day 18 min, then 19, 24, 76, 22, and 19 min. The value 76 happens due to a road closure due to a bad accident. Should 76 be included in estimating the average commute time? It changes the estimate from 20.8 min to 28.7 min. This latter value exceeds all your recorded commute times except one. Is that really a good measure of your average time? Well, yes it is. Certainly another accident could result in another large commute time tomorrow. The value would certainly be identified by any statistical criterion as an outlier, but should it be simply deleted? Probably not. Again the best treatment of the value depends on the decision you are trying to make. Analysis methods are available that are less influenced by such extreme values. Maybe inference based on medians should be used instead of inference based on means (the median is 22 min with the large value, 20.5 without it). Maybe inference should be based on ranks. Again these are the approaches of nonparametric methods where extreme values will tend to have much less impact.

Another difficulty, as has been suggested earlier, is data which do not appear to have come from an underlying normal population. While inference regarding means can invoke the central limit theorem for large sample sizes to ensure the normality of the sample means, small-sample applications for the mean, or other applications such as tolerance limits, depend heavily on the underlying distribution. In some cases nonnormal data can be transformed to achieve approximate normality, but in others little else can be done to justify the use of the standard statistical approaches. The solution is again nonparametric methods. The use of these tools does not depend on underlying population distributions. Examples of such methods are given in the following section.

NONPARAMETRIC METHODS

Nonparametric statistical methods constitute a wide family of tools that are appropriate for use regardless of underlying population distributions. Their validity does not depend on underlying distributions, as does the validity of the methods discussed so far. The price paid for such broader applicability would be larger uncertainties of estimates or tests with less power, but this is relevant only when the underlying assumptions hold to make the parametric methods applicable. In addition to being valid for any underlying distributions, nonparametric methods can typically better handle data problems such as nondetects or outliers.

The reader is referred to a text on nonparametric methods such as that by Conover (1971) to see the variety of nonparametric methods available. Such methods typically involve proportions or ranks of values. The most commonly used nonparametric method is the *Wilcoxon rank-sum test* which is used analogously to a two-sample t test in order to compare two underlying populations. In particular, the method is often used in site-to-background comparisons.

Say the following two samples are obtained from underlying site and background populations. To obtain the associated ranks, the two samples are combined and ranked from smallest to largest, i.e., from 1 to 15 since there are 15 values altogether.

Site values:	12.8	16.0	17.2	22.8	33.5	34.0	34.0	45.8
Ranks:	*3*	*5*	*6.5*	*8*	*12*	*13.5*	*13.5*	*15*
Background values:	7.0	8.3	15.7	17.2	24.3	26.3	27.9	
Ranks:	*1*	*2*	*4*	*6.5*	*9*	*10*	*11*	

The background values 7.0 and 8.3 are the smallest, so they receive ranks 1 and 2. The next-smallest value is the site value 12, so it receives rank 3. All values are ranked in this manner. Note that when values are tied, average ranks are assigned, as is the case for the values 17.2 which are ranked as 6.5 since the two 17.2 values are actually the 6th and 7th smallest. Similarly, both 34.0 values get rank 13.5. Larger groups of tied values would also receive the average rank for the group in this way.

The test statistic is the sum of the ranks for one of the samples. For environmental applications the ranks for the site values are typically summed, and if the sum of these ranks is "too large," a null hypothesis of no difference between the site and background populations is rejected in favor of the alternative hypothesis that the site is "more contaminated" than background. What *too large* means depends on the tolerable type I error rate, i.e., the false-alarm rate as discussed earlier.

The least that the sum of such ranks could be in our example is the sum of ranks 1 through 8, which occurs if the eight site values are all smaller than the background values. The maximum is the sum of ranks 8 through 15 if the seven background values are all smaller than the site values. Under the null hypothesis of no differences in the populations, the ordering of the ranks among the site and background values is completely random whereas if the site really tends to have higher values, the sum of its ranks will tend to be larger. This fact is used in small-sample problems to table "rejection regions" of size α as a function of the sample sizes for the Wilcoxon test. For larger-sample problems, the probability distribution of the sum of the ranks can be well approximated by the normal distribution. See Conover (1971) for details.

Again the obvious advantage of the Wilcoxon test is that underlying distributions do not affect the distribution of the sum of the ranks, thus it is universally applicable. Even if underlying populations are normal so that the two-sample t test is appropriate, the t test is only marginally more powerful at detecting differences between populations than the Wilcoxon test. Thus the Wilcoxon test is often used for environmental applications. Note also that data transformations are not needed, since such transformation would result in the same rankings of the samples anyway.

Recall the other advantage of a nonparametric approach is that it is less sensitive to outliers and nondetects. Outliers will simply receive the larger ranks, and the magnitude of extreme outliers will not significantly affect results, as it can in methods that depend on the computation of the mean and standard deviation.

Similarly, if all nondetects are less than all the detects, the nondetects would simply receive the average of the lower ranks as their average ranks, and the Wilcoxon test could still be applied. One does have to be concerned, however, about too many "tied" values, since the distributional properties of the sum of the ranks will begin to become significantly influenced, but appropriate adjustments are available in such cases. Again nondetects are less of an issue since the computation of means and standard deviations of the sample values is not required.

Note that if multiple detection limits are involved with a mix in the magnitude of detects and nondetects, even the Wilcoxon test is not appropriate. However, modifications have been developed that can handle this situation as well. In them, e.g., a nondetect at 10.0 is considered tied with a detect at 8.0 since it is unknown which represents the larger value. However, a nondetect at 8.0 and a detect at 10.0 would clearly give a larger ranking to the detect at 10.0 since we know it is greater then the value that led to a nondetect at 8.0 regardless of what the true value actually is. Such methods are discussed in Millard and Deveral (1988). In particular, the *Gehan test* is analogous to the Wilcoxon test. It is appropriate for multiple detection limits, and if it is used with only a single detection limit, or for that matter, with all detection limits less than the minimum detect, it reduces to the Wilcoxon test.

Most parametric statistical methods have nonparametric counterparts. While the Wilcoxon method is pretty much universally available on computer software packages, even on spreadsheet applications, the generalization for multiple detection limits is not. Similarly, some of the more involved nonparametric applications can only be found on somewhat more statistically oriented software. A resource for locating such software is given later in this section.

REGRESSION AND CORRELATION

Another commonly applied statistical tool is *least-squares regression.* Paired data are observed, and a predictive relationship between the two variables is of interest. Given the paired sample values $(x_1, y_1), (x_2, y_2), \ldots, (x_n, y_n)$, drawn from an underlying population of possible paired values, we seek the linear relationship

$$y = a + bx$$

The values a (the y intercept) and b (slope) that result from least-squares regression are those which minimize the sum of the squared vertical deviations from the resulting line to each point. Since $y = a + bx$ is a straight line, this is simple linear regression. If a square term is included, $y = a + bx + cx^2$, then we have quadratic regression. If a variable y is related to multiple variables, as in $y = ax + bu + cv$, we have multiple regression. Introductory statistics texts discuss these methods. For a more rigorous discussion of regression topics, see Draper and Smith (1966).

Given the paired data for (x, y) as mentioned above, r is the *correlation coefficient* that estimates the underlying population correlation coefficient ρ. And r^2 is called the *coefficient of determination*; it is the proportion of variation in the y values that can be related to the changes in the x values; i.e., it measures the strength of the linear relationship. When x and y are "strongly correlated," r is close to either 1 or -1 with the sign being the same as that of the slope, and r^2 is thus also close to 1.0. When x and y have little or no relationship, r and r^2 are both close to zero. This is illustrated in the examples in Fig. 4.4.13. Note in the three examples based on the r^2 values, approximately 0, 94, and 20 percent, respectively, of the variability in y is explained by its linear relationship with x.

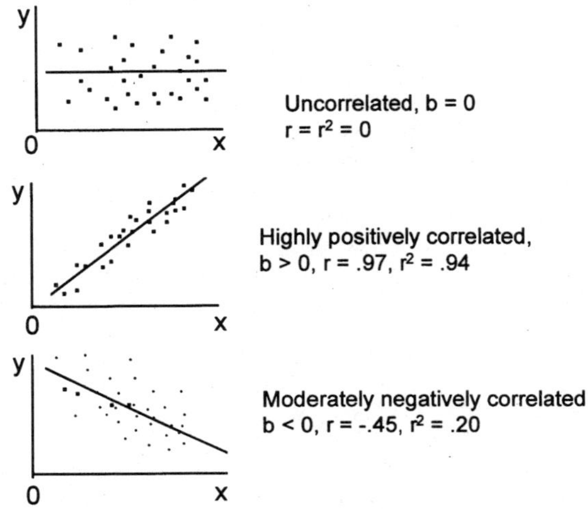

FIGURE 4.4.13 Regression and correlation examples.

No assumptions are being made regarding the underlying joint distribution of x and y in order to compute the estimates of slope, y intercept, and correlation. However, when one uses hypothesis testing to measure the significance of these quantities, then assumptions are being made. Given a value of x, the possible values of y that can then be realized are assumed to follow a normal distribution with mean $a + bx$ and standard deviation σ. Since σ does not involve x, an assumption of equal variability in y for any value x is being made. Methods are available to relax this final restriction, in particular, weighted least-squares regression; see Draper and Smith (1966).

Most software programs will compute the necessary quantities given the paired data. For example, the following output is from SAS (1989) software for the seven (x, y) paired values (1,2), (2,2), (2,3), (3,2), (3,3), (4,6), (5,6).

The SAS System

Model: MODEL1
Dependent Variable: Y

Analysis of Variance

Source	DF	Sum of Squares	Mean Square	F Value	Prob>F
Model	1	14.227	14.227	12.965	0.0155
Error	5	5.486	1.097		
C Total	6	19.714			

Root MSE	1.048	R-square	0.7217
Dep Mean	3.429	Adj R-sq	0.6660
C.V.	30.554		

Variable	DF	Parameter Estimate	Standard Error	T for H_0: Parameter = 0	Prob>|T|
INTERCEP	1	0.158	0.991	0.159	0.8796
X	1	1.145	0.318	3.601	0.0155

The 0.0155 value at both the top right and bottom right in the output is the p value associated with testing whether the slope is 0.0, that is, x and y are not correlated. Note that the parameter estimate of the slope is represented in the last line by the value 1.145. The result is significant; the p value is small (less than 0.05), so we conclude that x and y are significantly correlated. The r^2 value 0.7217 suggests that about 72 percent of the variability in the y's can be associated with the linear relationship with x. Note in the second-to-last line that the estimate of the y intercept is 0.158, and a statistical test of whether this is significantly different from zero has a p value of 0.8796, which suggests that is not the case. The *root mean square error* (root MSE), given on the output as 1.048, is the estimate of σ, the underlying standard deviation of y given a value x.

Regression can be a useful tool for estimating or predicting expected emission constituents as a function of process parameters such as speed, time, pressure, temperature, or the chemical composition of process inputs. By modeling emissions as a function of process parameters, one could estimate total emissions by tracking process operations. The uncertainty of such an estimate would then be based on the RMSE and possibly the variability in the input parameters, since they, too, are actually estimates and not necessarily exact values. This would not be a trivial approach to emission estimation.

EXPERIMENT DESIGN

Statistical *experiment design* deals with developing a set of experiment trials that efficiently generate information in an application. For example, in the previous discussion, the idea of modeling emissions as a function of process parameters was mentioned. Say you want to do that, but you want to determine an optimum way to run the process within an acceptable operating envelope that minimizes emissions. Several parameters may potentially have a significant impact, and experiment design addresses what combination of factors should be run to determine the influence of the parameters on emissions.

A common engineering approach is to hold all other parameters constant while varying only one at a time. This may not accomplish the goal of minimizing emissions, in particular, if the parameters have significant interaction, i.e., if their combined impact on emissions is different from what would be expected based on their individual impacts. This type of interaction is common, and it interferes with the one-at-a-time approach.

Designing a multiparameter experiment that uses a mathematically based set of trials is a more efficient approach. A detailed discussion of this statistical area is beyond the scope of this section. More elementary design topics are discussed in standard introductory statistics texts, and they can be implemented with most statistical software packages. They generally consist of running all combinations of preselected parameter levels. These are called *factorial* designs. For several parameters, or several levels, this approach quickly becomes impractical due to the large number of test runs required. Then more sophisticated statistical designs are required to efficiently draw appropriate conclusions. Such designs are typically called *fractional,* and several mathematically justified approaches are available. A text such as that by Kirk (1982) contains these more advanced topics.

STATISTICAL SOFTWARE

An extremely broad array of statistical software has become available in recent years. The capability and cost of such packages can vary dramatically. Selection of what is appropriate for you will depend on the type of applications you expect to encounter and what you are willing to spend. Obviously, the greater the cost, the greater the expected sophistication of the available statistical procedures within the capability of the software. For an example listing of software costs and capabilities, see http://www.scitechint.com/stats.HTM.

COMMON EMISSIONS APPLICATIONS

Odor Testing

In odor testing, subjects are presented with successively less diluted subsamples from an original sample of interest. The dilution at which an odor is first detected is the response. Several test subjects are used to generate data on which to base conclusions. What constitutes too high an odor level? Recall the DQO process of defining the problem and identifying associated decisions, decision criteria, and acceptable error rates. These should be points of discussion with the regulator.

Many sampling issues are pertinent. What are the boundaries of the area of interest, and how and where in this area will samples be taken? What time frame is of interest, and how will sampling activities account for this? Can compositing of several samples be used to come up with an "average result" over a broader area or over a time interval, in order to save measurement costs? Will comparisons be made to a standard, or to background samples using the same or different subjects? If background samples are used, are the sampling methods used in a similar fashion? If not, adjustments to statistical comparisons may be needed. Are stakeholders willing to assume the associated process is in compliance with odor standards unless statistical evidence shows it is not, or must the process be assumed out of compliance until statistical evidence shows that it is in compliance? Answers to these and many other questions will dictate the appropriate statistical methodology.

Recall that if comparisons are to be made to a standard, then point estimates such as sample means or percentiles can have no associated decision error rates. If a mean or percentile is simply compared to the standard, no measure of error probabilities is available. If, instead, confidence limits for the mean, upper tolerance limits (confidence limits for percentiles), hypothesis tests for the mean, or hypothesis tests for a percentile are used, decisions error rates can be quantified and controlled, as discussed in earlier sections. These are all reasonable approaches for comparison to standards.

If, instead, odor results are to be compared to those for background control samples, two-sample or paired-sample methods are appropriate. If the same subjects are used for both types of samples, a paired t test may be appropriate. If different groups of subjects are used for the two types of samples, the two-independent-sample t test may be appropriate. If sample sizes are small, these test approaches require underlying normal distributions, or possibly a transformation of the data needs to be normal. Recall that tests for normality are not really practical for small sample sizes, so the use of t test approaches may be difficult to justify. Then the analogous nonparametric approaches are useful.

It is conceivable that one might try to model odor test results as a function of process parameters, but since generating a data value is a rather involved process, it is unlikely that sufficient data would be available. Let us save that approach for emissions.

The logic in the previous paragraphs is the type that must be applied to decide on optimal statistical approaches. For some types of applications, such logic and the associated statistical methods may be straightforward. For others they will not be, and qualified statistical support should be sought. It is the responsibility of the investigators to know their statistical capabilities and to know when they are "getting in over their heads."

Emissions Compliance

An even wider variety of approaches could conceivably be taken to investigate emissions compliance. Again the up-front planning of the DQO process with regulator buy-in to an approach is essential. The simplest approach may be to base a decision on a set of emission samples where concentrations per volume of analytes of interest are measured. Then the problem is essentially the same as that initially discussed previously for odor detection. The first four paragraphs of that discussion apply to this emissions application as well.

Outliers and nondetects are quite likely for emissions data. How will they be handled? Recall that nonparametric methods are less sensitive to these data problems and can readily handle a broader set of applications. For substantial nondetects with multiple detection limits in comparisons to background, the nonparametric tests in Millard and Deveral (1988) are about the only technically defensible approach.

What if the compliance standard is based on total material released in a time interval, as is commonly the case? We then have much more complicated sampling and estimation issues. How does one estimate such a quantity, and given an estimation approach, how good are the estimates? If the uncertainty in the estimate is not quantified, no discussion of decision error rates can occur. The approach may have little power in detecting true violators, or it may readily falsely flag emitters that are in fact in compliance.

Total material released requires the estimation of at least two quantities. For a specified time interval, possibly a subset of the entire interval of interest, an emission concentration rate for an analyte and the total air volume released are required. Both quantities could be based on samples taken throughout the time interval. Both quantities then have measurement and sampling error and thus uncertainties in their estimates. These uncertainties can be combined to get the uncertainty in their product which is an estimate of the analyte emission total for the time interval. For example, if X and Y are independent, and Var indicates variance, then $\text{Var}\,(XY) = X^2\,\text{Var}\,(Y) + Y^2\,\text{Var}\,(X)$. The variance of products and other functions of random variables is included in discussions of *uncertainty, error,* or *variance propagation.* See *error propagation* in Kotz and Johnson (1986).

Averaging samples taken over a time interval may be the best approach when the emissions sampling point is a conglomeration of outputs from many processes being run on different schedules and at different levels. However, if instead the emissions are from a single, well-controlled process, a modeling approach may be more useful. The relationship between the process operation levels and emissions is characterized along with uncertainties through an approach such as regression. Then an operation log could be used to estimate total analyte emission and its uncertainty through the model that was derived. Efficient experimental design would facilitate such modeling activities. Again regulator buy-in would be essential. Obviously the statistical complexity of such an approach has increased as will the difficulty of defending the approach to stakeholders.

REMINDERS

Here are two final reminders of concepts involving statistical applications mentioned earlier. Most importantly, plan the data acquisition and analysis activities in advance, and obtain regulator buy-in *prior to* the collection activities. Use a planning process such as data quality objectives. Also realize the limitations on how far your statistical capability can take you. Applications outside of textbooks typically get complicated very quickly, so do not get in the uncomfortable position of being in over your head. Hopefully others with greater statistical expertise and experience are available to help you out in such situations.

REFERENCES

Cochran, W. G. (1977). *Sampling Techniques,* Wiley, New York.

Conover, W. J. (1971). *Practical Nonparametric Statistics,* Wiley, New York.

Draper, N. R., and H. Smith (1966). *Applied Regression Analysis,* Wiley, New York.

Gilbert, R. O. (1987). *Statistical Methods for Environmental Pollution Monitoring,* Van Nostrand Reinhold, New York.

Kirk, R. E. (1982). *Experimental Design,* Brooks/Cole, Monterey, Calif.

Kotz, S., and N. L. Johnson, eds. (1986). *Encyclopedia of Statistical Sciences,* Wiley, New York.

Millard, S. P., and S. J. Deveral (1988). Nonparametric Statistical Methods for Comparing Two Sites Based on Data with Multiple Nondetect Limits, *Water Resources Research,* 24: 2087–2098.

Montgomery, D. C. (1991). *Introduction to Statistical Quality Control,* Wiley, New York.

Odeh, R. E., and D. B. Owen (1980). *Tables for Normal Tolerance Limits, Sampling Plans, and Screening,* Marcel Dekker, New York.

SAS (1989). *SAS/STAT User's Guide,* Version 6, 4th ed., SAS Institute Inc., Cary, N.C.

Thompson, S. K. (1992). *Sampling,* Wiley, New York.

CHAPTER 5
MODELING

SECTION 5.1

EMISSION MODELING—METEOROLOGY, TOPOGRAPHY, AND DISPERSION

Kalliat T. Valsaraj
Department of Chemical Engineering
Louisiana State University
Baton Rouge, Louisiana

METEOROLOGY AND TOPOGRAPHY 5.2
EMISSIONS FROM AREA SOURCES 5.3
EMISSIONS FROM STACKS AND POINT SOURCES 5.3
ESTIMATION OF EMISSION RATE *Q* FOR DIFFERENT SCENARIOS 5.7
EMISSIONS FROM LINE SOURCES 5.7
COMPUTER MODELS FOR AIR DISPERSION CALCULATIONS 5.8
Industrial Source Complex Models 5.8
COMPLEX I Model 5.8
Rough Terrain Diffusion Model 5.8
AN EXAMPLE CALCULATION 5.9
REFERENCES 5.9

A volatile organic compound (VOC) released to the air from a manufacturing process, a transportation accident, or an inadvertent release can undergo a variety of transformations and find special niches in the atmosphere, where its effects can be felt. Figure 5.1.1 is a schematic of the fate of a VOC released from a stationary source (a power plant), from a mobile source (an automobile), or of natural origin. In the lower atmosphere the VOC undergoes initial mixing, thereby diluting it in the air, while at the same time undergoing both physical transformations and chemical reactions. Further the VOC and its reaction products are scavenged to the ground by dry particles (dry deposition) and by snow, rain, fog, or mist (wet deposition). The deposition may occur on land or water where they cause pollution. What we are concerned with in this section is how the air pollutants are transported by diffusion and dispersion and how their concentrations at some receptor locations downwind from the source can be ascertained. The aspects of chemical reactions are not considered in this chapter.

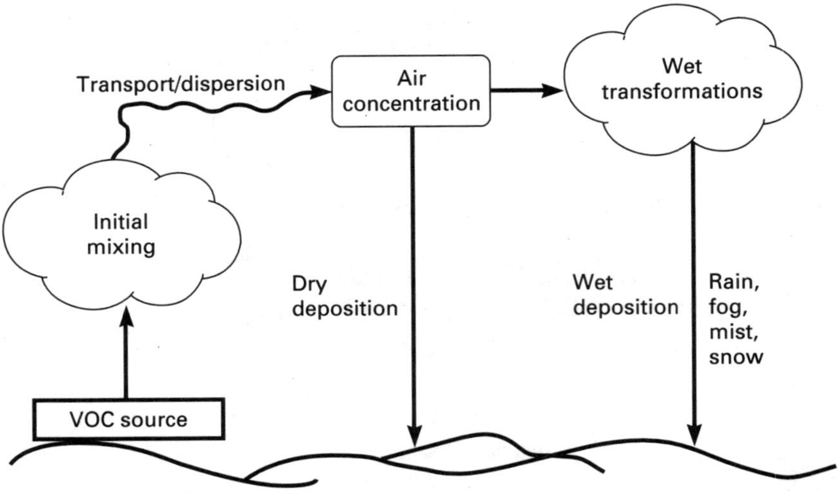

FIGURE 5.1.1 Schematic of the fate of a VOC released from a stationary source, a mobile source, or of natural origin.

METEOROLOGY AND TOPOGRAPHY

Two important meteorological parameters that affect the dispersion of pollutants in the atmosphere are temperature and wind velocity. The variation in temperature with height determines the extent of buoyancy-driven mixing within the atmosphere. This factor is ascertained in terms of the ideal case where a parcel of air is allowed to rise adiabatically through the atmosphere. The air parcel then experiences a decrease in temperature as it expands in response to decreasing temperature with increasing elevation. This is called the *adiabatic lapse rate* Γ ($= dT/dz$) and is a standard value of 1°C/100 m for dry air. For a saturated air, this value is slightly smaller (0.6°C/100 m). If the prevailing atmosphere has an *environmental lapse rate* $\Gamma_{env} > \Gamma$, the atmosphere is considered unstable and rapid mixing occurs, thereby diluting the air pollutant. If $\Gamma_{env} < \Gamma$, the atmosphere is stable and little or no mixing occurs.

The second important parameter is the wind velocity. In general, the velocity of wind above the surface follows a profile given by Deacon's power law

$$\frac{u}{u_1} = \left(\frac{z}{z_1}\right)^p \tag{5.1.1}$$

where u = wind velocity at height z
u_1 = wind velocity at height z_1
p = positive exponent (between 0 and 1)

The value of p in the above equation varies with atmospheric stability.

The above effects of temperature and wind velocity are introduced in the dispersion models through the Pasquill stability criteria. However, we are still left with one additional issue—that of the terrain itself. Obviously the wind velocity near the surface is going to be strongly influenced by the type of terrain. Rough surfaces or irregular terrain (buildings and other structures) will give rise to varying wind directions and velocities near the surface. Open, wide terrains do not provide that much wind resistance.

EMISSIONS FROM AREA SOURCES

Very large-area sources such as a large city experiencing air pollution problems can be analyzed by considering the source to be distributed over a wide area, as in Fig. 5.1.2. The volume of air above the city can be considered to be completely mixed and of uniform

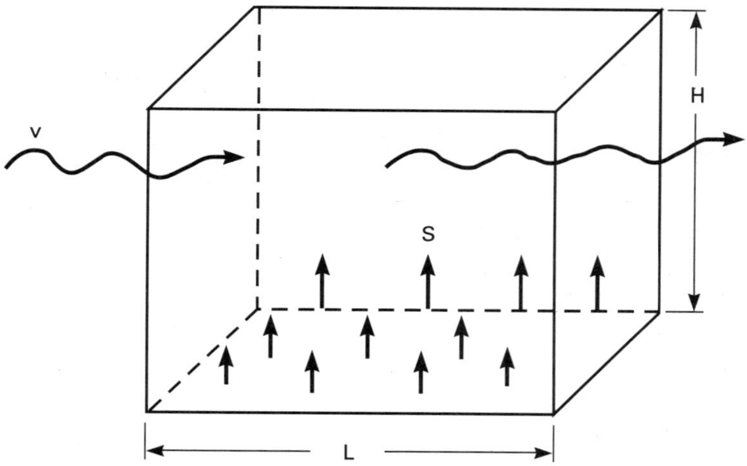

FIGURE 5.1.2 Box model for area sources.

concentration up to a mixing height H (m). This is called a *box model*. The pollutant concentration within the box is inversely proportional to the air velocity v (m/s) and directly proportional to the rate of emission from all sources S [$\mu g/(m^2 \cdot s)$] and is given by

$$C(t) = \frac{S \cdot L}{v \cdot H}\left[1 - \exp\left(-\frac{v}{L} \cdot t\right)\right] \tag{5.1.2}$$

Here L (m) is the length of the box in the direction of the wind, and $C(t)$ is the concentration in micrograms per cubic meter at any time t within the atmosphere. As time progresses, the concentration within the air increases from an initial value of zero and reaches a maximum value of $SL/(vH)$ after a long time. This maximum value of concentration is called the *steady-state concentration.* Obviously this is an oversimplified scenario; rarely does the concentration in the atmosphere vary so smoothly over time. In particular, both wind velocity and emission will fluctuate with time, and hence the value of C shows a complex behavior depending on wind direction, velocity, and time of day. However, the above equation is useful in ascertaining the factors that determine the air pollutant concentrations over very large-area sources.

EMISSIONS FROM STACKS AND POINT SOURCES

Dispersion from point sources and dispersion from small-area sources are handled somewhat differently. As stated earlier, the dispersion of VOCs in the atmosphere is influenced by the degree of atmospheric stability, which in turn is determined by the temperature profile in the atmosphere, wind direction, wind velocity, and surface roughness. These factors are considered in arriving at equations to determine the dispersion of pollutants in a prevailing atmosphere. Let us consider the emission of a plume of air pollutant from a tall chimney, as in Fig. 5.1.3. The air leaving the chimney is often at a higher temperature than the ambient air. The buoyancy of the plume causes a rise in

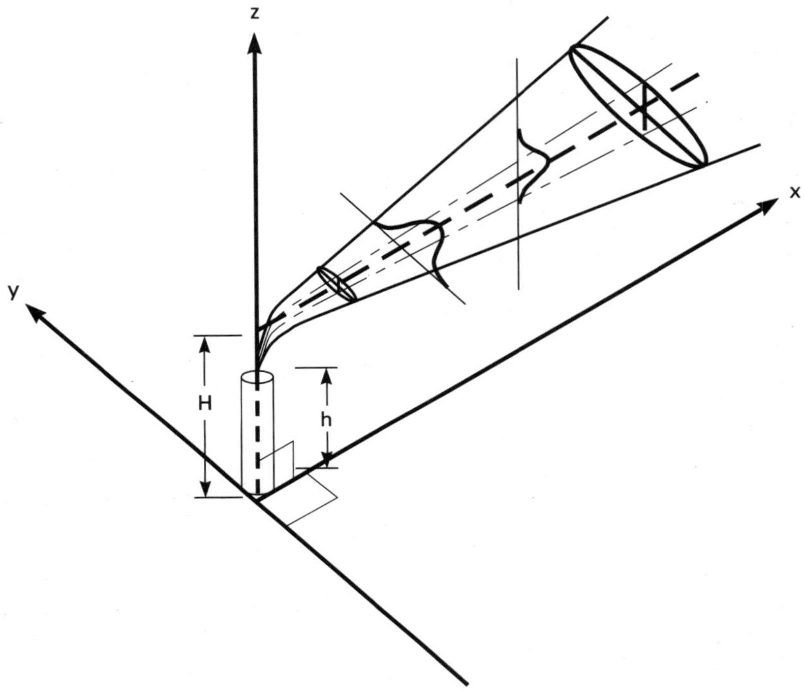

FIGURE 5.1.3 Emission of a plume of air pollutant from a tall chimney.

the air parcel before it achieves a more or less horizontal travel path. The concentration of VOC in air at any point with coordinates (x, y, z) from a plume of effective height H (m) is given by (see Turner, 1994)

$$C(x, y, z; H) = \frac{Q}{2\pi u \sigma_y \sigma_z} \left[\exp\left(- \frac{(H - z)^2}{2\sigma_z^2} \right) + \exp\left(- \frac{(H + z)^2}{2\sigma_z^2} \right) \right] \cdot \exp\left(\frac{-y^2}{2\sigma_y^2} \right) \quad (5.1.3)$$

where Q = emission rate at stack, g/s
u = wind speed at release point, m/s
σ_y = standard deviation of concentration distribution in crosswind direction at a distance x, m
σ_z = standard deviation of concentration distribution in vertical direction at a distance x, m
H = effective height of plume, m

The concentration C is given in grams per cubic meter which can be converted to the more common units of micrograms per cubic meter or ppmv using appropriate conversion factors.

TABLE 5.1.1 Pasquill Stability Categories

Wind speed at 10-m height, m/s	Day			Night	
	Incoming solar radiation			Cloud cover	
	Strong	Moderate	Slight	Mostly overcast	Mostly clear
<2	A	A-B	B	E	F
2–3	A-B	B	C	E	F
3–5	B	B-C	C	D	E
5–6	C	C-D	D	D	D
>7	C	D	D	D	D

Note: Class A is the most unstable, and class F is the most stable. Class B is moderately unstable, and class E is slightly stable. Class C is slightly unstable. Class D is neutral stability and should be used for overcast conditions during day or night.

The standard deviations σ_y and σ_z have been correlated to both mechanical turbulence (wind speed at 10-m height) and buoyant turbulence (time of day). These give rise to the Pasquill stability categories given in Table 5.1.1. Based on the stability class, Fig. 5.1.4 gives appropriate σ_y and σ_z, for any distance x downwind of the source.

The effective height of the stack H (m) is the sum of the actual stack height h (m) and the plume rise Δh (m). The plume rise can be estimated by using a number of different expressions (Wark and Warner, 1981). Of these, the Holland formula, given below, is one that is commonly used:

$$\Delta h = \frac{u_s d}{u}\left[1.5 + 2.68 \times 10^{-3}\, Pd\left(\frac{T_s - T_a}{T_s}\right)\right] \tag{5.1.4}$$

where u_s = stack gas exit velocity, m/s
d = stack exit diameter, m
P = pressure, mbar
T_s = gas temperature at exit of stack, K
T_a = ambient air temperature at stack height, K

The equations of Carson and Moses and the Concawe formulas are also frequently used to estimate the plume rise Δh, the details of which are given in standard air pollution texts (see, e.g., Wark and Warner, 1981).

One parameter that is often sought in air pollution modeling is the so-called ground-level centerline concentration $C(x, 0, 0; H)$ obtained from Eq. (5.1.3)

$$C(x, 0, 0; H) = \frac{Q}{\pi u \sigma_y \sigma_z} \exp\left(\frac{-H^2}{2\sigma_z^2}\right) \tag{5.1.5}$$

The maximum concentration at a receptor along the ground-level centerline is found by substituting $H^2 = 2\sigma_z^2$:

$$C_{\max} = \frac{0.1171 Q}{u \sigma_y \sigma_z} \tag{5.1.6}$$

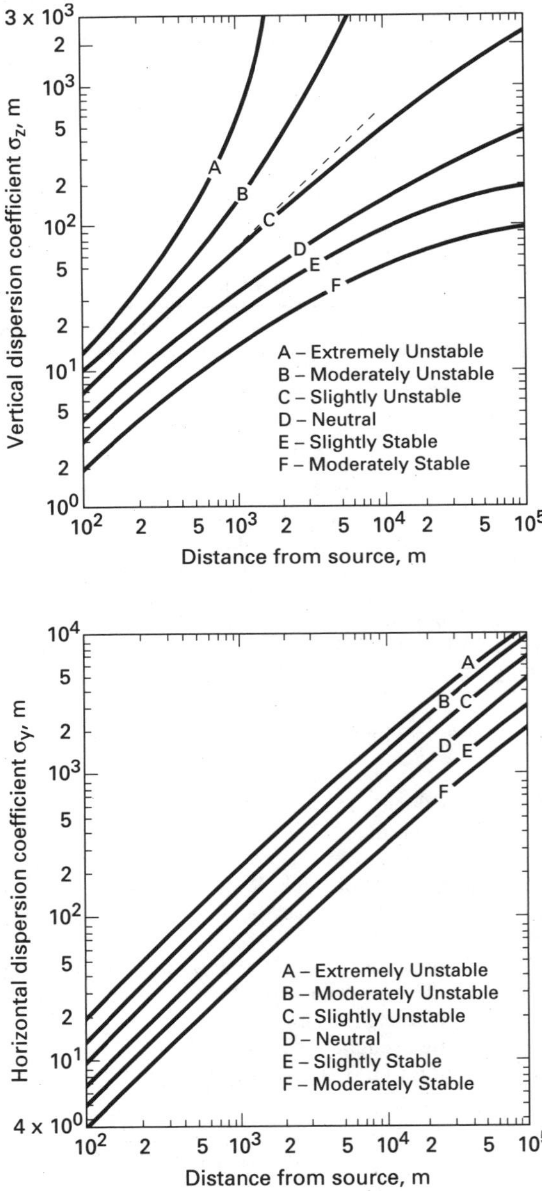

FIGURE 5.1.4 Pasquill-Gifford horizontal and vertical dispersion parameters as functions of stability class and downwind distance from the source.

ESTIMATION OF EMISSION RATE Q FOR DIFFERENT SCENARIOS

As far as emission from a chimney stack is concerned, the value of Q can be estimated if the process is from an industrial or a chemical manufacturing facility. All that one needs is the overall material balance on the reactors from which the emissions result.

In the case of spills on the ground resulting from a leak or an accident, one has to consider three different cases depending on whether it is a liquid or gas and the degree of volatility if it is a liquid (Kumar, 1996). In all the cases given below, Q is in grams per second.

Case 1: Low-Volatile Liquid Spilled on the Ground. The rate of liquid evaporation is given by

$$Q = 5.1 \times 10^{-6} u^{0.78} P^0 M^{0.67} A_p^{0.94} \tag{5.1.7}$$

where P^0 = vapor pressure of liquid, N/m^2
M = molecular weight of liquid
A_p = pool area, m^2

Case 2: Highly Volatile Liquids Spilled on the Ground. Consider a liquid of height L (m) contained in a tank. Upon rupture it forms a pool of area A_r (m^2). Since the evaporation from a pool of highly volatile liquid is almost instantaneous, one can obtain the following expression for the release rate:

$$Q = 700 A_r \rho_1 \left(19.6\,\mathrm{L} + 2\,\frac{P_t - P_a}{\rho_1} \right)^{1/2} \tag{5.1.8}$$

where ρ_1 = liquid density, kg/m^3
P_t = tank pressure, N/m^2
P_a = atmospheric pressure, N/m^2

Case 3: Spill of a Heavier-Than-Air Gas. Generally, a denser-than-air gas when spilled will form a small gas pool close to the ground which in time will dissipate by entrainment in the air by surface winds. The rate of release can be approximated using the following equation (Briggs et al., 1990)

$$Q = 0.05 u^3 \frac{d_p}{g_i} \frac{\rho_g}{\rho_g/\rho_a - 1} \tag{5.1.9}$$

where d_p = diameter of pool, m
g_i = gravitational acceleration, m/s^2
ρ_g = density of gas, kg/m^3
ρ_a = density of air, kg/m^3

EMISSIONS FROM LINE SOURCES

For a continuously emitting infinite line source, the concentration downwind when the wind direction is at an angle ϕ to the line is given by

$$C(x, y, 0; H) = \frac{2Q'}{\sin \phi \sqrt{2\pi}\, \sigma_z u} \exp\left(-\frac{H^2}{2\sigma_z^2}\right) \tag{5.1.10}$$

where Q' is the source strength per unit distance [g/(m•s)]. The above equation is invalid if $\phi < 45°$. The above equation can be manipulated to obtain the concentration resulting from a finite line source as well (Turner, 1994).

COMPUTER MODELS FOR AIR DISPERSION CALCULATIONS

The models described above are based on the gaussian plume formulation. They contain only limited provisions for including complicated atmospheric conditions such as VOC deposition and flow in complex terrains. There are some computer models recommended by the U.S. Environmental Protection Agency (EPA) that contain some information on these complex phenomena (EPA, 1986). Examples of available models are Industrial Source Complex (ISC), Complex Terrain (COMPLEX I), and Rough Terrain Diffusion (RTD) models.

Industrial Source Complex Models

For most flat or moderately rolling terrains, EPA recommends using the ISC models (either Short-Term ST or Long-Term LT version). Both of these are gaussian plume models. Their estimates of concentrations are subject to considerable uncertainty as they substitute the stack elevation for the actual elevation of the point source. The deposition of VOCs is handled via surface reflection coefficients and gravitational settling velocities.

COMPLEX I Model

This is an EPA-recommended screening-level model for use in complex terrains. As in the case of the ISC models, this is also a gaussian plume-based model. In fact, for terrains below the stack height, the algorithm in this model is the same as that for the ISCST model. For the terrain 400 m above the stack height, a terrain correction factor is incorporated per the recommendations of Turner (1986), while for terrains greater than this height, the VOC concentration is set equal to zero. There is, however, one major drawback to the COMPLEX I model, and that is its inability to include atmospheric deposition of pollutants. Modifications to include depositional events from the plume have been investigated (Yambert et al., 1991).

Rough Terrain Diffusion Model

This is another EPA-recommended model used for third-level screening in complex terrains. It differs from the COMPLEX I model in two respects. First, it accounts for the terrain effects by defining a critical plume height, which is the height above which air passes over the terrain and below which air impacts the terrain. Second, it accounts for sloping terrain and plume reflection from these locations. Whereas conventional gaussian plume models use virtual image sources to consider plume reelection which give unrealistically high VOC concentrations, the RTD model overcomes these inaccuracies by using a reelection factor *R*, based on terrain slope and plume growth. However, the RTD model also suffers from the disadvantage that it fails to adequately model VOC deposition rates. Yambert et al. (1991) suggest some modifications to take this into account.

In spite of these improvements to the gaussian plume models, the current EPA-recommended air dispersion models need further improvement to highly complex terrains by realistically simulating the actual flows in and around complex objects. The existing model simply modifies the gaussian plume algorithms to modify plume height as a means of accounting for complex terrains downwind of the source.

AN EXAMPLE CALCULATION

Problem. Nitrogen dioxide is being emitted from a stack. The stack diameter at the top is 1.5 m. The rate of emission of nitrogen dioxide is estimated to be 1500 g/s with an exit velocity at the stack of 15 m/s, an exit gas temperature of 642 K, and a pressure of 90 kPa. It is desired to estimate the design stack height required so that the ground-level centerline concentration 2 km from the source during slightly unstable atmospheric conditions does not exceed 1 ppmv.

Solution. Slightly unstable condition is represented by stability class C in Table 5.1.1. Consider an average wind velocity of 5.5 m/s. From Table 5.1.1 and Fig. 5.1.4, the values of σ_y and σ_z are 200 and 100 m, respectively.

Equation (5.1.3) pertains to this problem. Since we are interested in the ground-level centerline concentration, we set $z = 0$ and $y = 0$ in the equation to obtain the desired concentration

$$C(2\text{ km}, 0, 0; H) = \frac{Q}{\mu\,\sigma_y\sigma_z u}\exp\left(-\frac{1}{2}\frac{H^2}{\sigma_z^2}\right) \tag{5.1.11}$$

The target concentration C is 1 ppmv which is equivalent to 0.0019 g/m^3. Substituting the values of C, Q, u, σ_y, and σ_z in Eq. (5.1.11), we obtain $H = 128$ m. This represents an effective height of the stack that includes the physical height h and plume rise Δh. Now, using Eq. (5.1.4), we obtain the plume rise $\Delta h = 13$ m. Therefore, the actual stack height should be 115 m.

REFERENCES

Briggs, G. A., R. S. Thompson, and W. H. Snyder (1990). Dense gas removal from a valley by crosswinds, *Journal of Hazardous Materials,* 24:1–38.

EPA (1986). *Guidelines on Air Quality Models* (rev.), EPA-450/2-78-072R, Environmental Protection Agency, Washington.

Kumar, A. (1996). Emission estimation rates from Haz Mat releases, *Pollution Engineering,* February, pp. 52–53.

Turner, D. B. (1994). *Workbook of Atmospheric Dispersion Estimates,* 2d ed., Lewis Publishers Inc., Boca Raton, Fla.

Turner, D. B. (1986). Fortran Computer Code/User's Guide for COMPLEX I Version 86064: An Air Quality Dispersion Model in Section 4. Additional Models for Regulatory Use. Source File 31 Containing UNAMAP (Version 6), NTIS PB 96-222361/AS.

Wark, K., and C. F. Warner (1981). *Air Pollution,* 2d ed., HarperCollins Publishers, New York.

Yambert, M. W., G. D. Belcher, and C. C. Travis (1991). Evaluation of flat versus complex terrain models in estimating pollutant transport and deposition in complex terrain. In *Municipal Waste Incineration and Risk Assessment,* C. C. Travis (Ed.), Plenum Press, New York, pp. 1–21.

SECTION 5.2

EMISSION MODELS

Richard L. Corsi

David A. Olson

Department of Civil Engineering
University of Texas at Austin

FUNDAMENTAL PRINCIPLES 5.11
EMISSION MODELS: WASTEWATER COLLECTION SYSTEMS 5.14
General Considerations 5.14
CORAL+ 5.16
WATER8 5.16
TOXCHEM+ 5.17
naUTilus 5.17
Chloroform Mass Balance 5.18
EMISSION MODELS: WASTEWATER TREATMENT SYSTEMS 5.18
General Considerations 5.18
BASTE 5.20
TOXCHEM+ 5.21
WATER8 5.21
Comparison with Field Data 5.21
SUMMARY 5.22
REFERENCES 5.22

This section involves a discussion of models that are used for estimating volatile organic compound (VOC) emissions from wastewater. The intent is to provide a general overview of emissions estimation methods and available computer models. A brief review of the important principles which form the basis for many of those methods and models is also provided.

Computational models for estimating emissions from either wastewater collection systems or wastewater treatment systems are described. These sections focus on several of the most commonly used models for emissions estimation (BASTE, CORAL+, TOXCHEM+, and WATER8), although reference is also made to other models in the final section. Where appropriate, the strengths and limitations of each model are described.

FUNDAMENTAL PRINCIPLES

An important assumption of most VOC fate models is that entire wastewater treatment processes, or well-defined subsections of those processes, can be treated as well-mixed reactors, i.e., homogeneous contaminant concentrations. Within such reactors, contaminants may accumulate, pass through to the effluent stream, or be affected by one of five primary removal pathways: volatilization across the exposed wastewater-atmosphere interface, stripping to diffused air bubbles, adsorption to solid particles or biomass, absorption to immiscible liquids, and biodegradation. These pathways are illustrated in Fig. 5.2.1.

Assuming a constant reactor volume, a generalized mass balance on a well-mixed reactor volume leads to

$$V_1 \frac{dC}{dt} = QC_0 - QC + R_v + R_s + R_{\text{ad}} + R_{\text{ab}} + R_b \qquad (5.2.1)$$

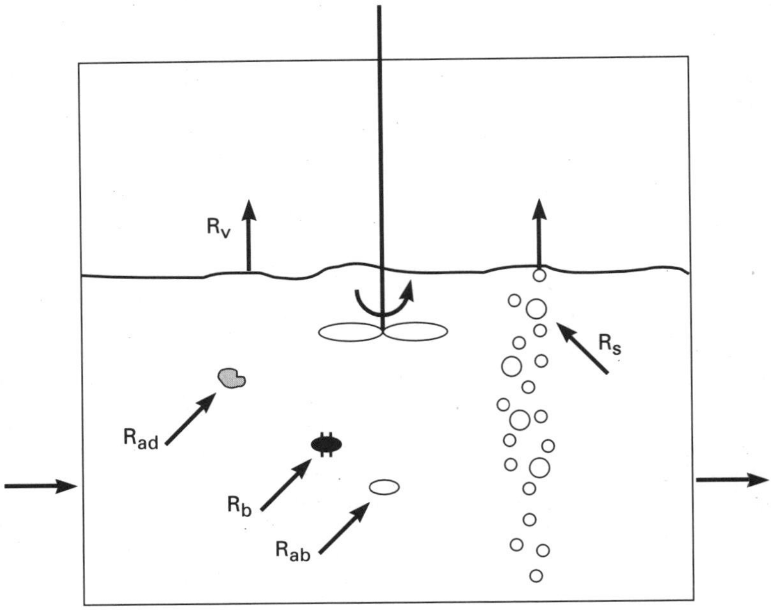

FIGURE 5.2.1 Removal pathways.

where C_0 and C are dissolved contaminant concentrations entering and within (leaving) the reactor (mg/m^3), respectively, V_1 is the reactor volume (m^3), and Q is the volumetric flow rate of wastewater entering the reactor (m^3/s). The variables R_v, R_s, R_{ad}, R_{ab}, and R_b represent rates of contaminant removal (mg/s) by *volatilization, stripping, adsorption, absorption,* and *biodegradation,* respectively.

The *rate of volatilization* is typically modeled as

$$R_v = -K_L\left(C - \frac{C_g}{H_c}\right)A \tag{5.2.2}$$

where K_L is an overall mass transfer coefficient (m/s), A is the interfacial area over which mass transfer occurs (m^2), C_g is the contaminant concentration in the gas phase adjacent to the well-mixed liquid (mg/m^3), and H_c is Henry's law constant (m_{liq}^3/m_{gas}^3) for the chemical of interest. In accordance with two-film theory, the overall mass-transfer coefficient can be represented as

$$\frac{1}{K_L} = \frac{1}{k_l} + \frac{1}{k_g H_c} \tag{5.2.3}$$

where k_l and k_g are liquid- and gas-phase mass-transfer coefficients (m/s), respectively. Existing models differ significantly with respect to how k_l and k_g are estimated, or as to whether k_g is even accounted for in source estimates. The first and second terms on the right-hand side of Eq. (5.2.3) are often referred to as liquid- and gas-phase resistances to mass transfer, respectively. These sum to the overall resistance to mass transfer ($1/K_L$).

Methods for estimating k_l and k_g for a chemical i often rely on information related to a specific contaminant which has been rigorously studied. For gas-liquid mass transfer in wastewater, the most common of these chemicals is oxygen, although it is generally assumed to be dominated by liquid-phase resistance to mass transfer. The relationships between liquid-phase and gas-phase mass-transfer coefficients for two chemicals i and j are given by

$$\psi_l = \frac{k_{li}}{k_{lj}} = \left(\frac{D_{li}}{D_{lj}}\right)^n \tag{5.2.4}$$

$$\psi_g = \frac{k_{gi}}{k_{gj}} = \left(\frac{D_{gi}}{D_{gj}}\right)^m \tag{5.2.5}$$

where ψ_l and ψ_g are relational parameters (dimensionless), D_l and D_g represent liquid- and gas-phase diffusion coefficients (m^2/s) for the system of interest, subscripts i and j denote the two chemicals of interest, and n and m vary between 1.0 (two-film theory) and 0.5 (penetration and surface renewal theories).

A major assumption (problem) with most emissions models is that ψ_l represents the ratio of overall mass-transfer coefficients for i and j. In fact, this is only a good assumption when gas-phase resistance to mass transfer is small for each chemical. Otherwise, the following theoretical expression should be used (Corsi, 1997):

$$\psi_T = \frac{K_{Li}}{K_{Lj}} = \psi_l \psi_g \frac{H_{ci}}{H_{cj}} \frac{1 + (k_{gj}/k_{lj})H_{cj}}{\psi_l + \psi_g H_{ci}(k_{gj}/k_{lj})} \tag{5.2.6}$$

The *rate of stripping* by dissolved air bubbles can be modeled as

$$R_s = -Q_b C_b = -Q_b C H_c \gamma \tag{5.2.7}$$

where R_s is the rate of stripping (mg/s), Q_b is the bubble volumetric flow rate (m^3/s), and γ is a variable which represents the degree of saturation (chemical equilibrium) achieved by a bubble by the time it reaches the exposed wastewater-air interface. Roberts et al. (1984) described a method for estimating γ based on a mass balance on a rising air bubble. However, many models simply assume a value of unity for γ. Such an assumption is generally valid for low- to middle-volatility VOCs, but may be significantly violated for VOCs with relatively high Henry's law constants, e.g., cyclohexane or 1,3-butadiene.

Equations (5.2.2) through (5.2.7) all relate to emissions of chemicals from wastewater to the atmosphere. In fact, such emissions are effectively reduced through the competitive mechanisms of absorption (to an oily phase), adsorption (to solids and biomass), and biodegradation. The *rate of absorption* is generally a complex process that is not well understood. Absorption is not considered in models intended for applications involving municipal wastewater, but may be important for industrial wastewater systems. A simplified, but not verified, absorption algorithm is included in one model (WATER8), but is not considered further here.

The rate of contaminant removal by *adsorption* processes is generally assumed to be very rapid such that equilibrium is achieved instantaneously. The rate of contaminant removal in primary or secondary sludge streams is modeled as

$$R_{\text{ad}} = -Q_w K_p C C_s \tag{5.2.8}$$

where R_{ad} is the rate of contaminant removal by *adsorption* (mg/s), Q_w is the volumetric sludge wastage rate (m^3/s), C_s is the solids or biomass concentration (mg/m^3), and K_p is a linear liquid-solid partition coefficient (m^3/mg). The variable K_p is often correlated with a chemical's octanol-water partition coefficient and fractional organic matter content of solids.

The *rate of biodegradation* is often assumed to follow Monod kinetics. However, at the low contaminant concentrations typical of most wastewaters, Monod kinetics are often assumed to simplify to a first-order process such that the *rate of biodegradation* is

$$R_{\text{bio}} = -k_b C X V_l \tag{5.2.9}$$

where R_{bio} is the rate of contaminant removal by biodegradation (mg/s), k_b is a first-order biodegradation rate constant [$m^3/(mg \cdot s)$], and X is the active biomass concentration (mg/m^3). For activated sludge systems, X is often assumed to equal the mixed-liquor volatile suspended solids (MLVSS) concentration. The biodegradation rate constant is generally chemical- and system-specific, and it is therefore difficult to estimate or extrapolate from laboratory-scale studies. Biodegradation rate constants have been determined experimentally for different operating systems and have been reported in the literature (Pitter, 1976; Kincannon et al., 1982; Petrasek et al., 1983; Hannah et al., 1986).

For steady-state conditions and open processes ($C_g = 0$), substitution of Eqs. (5.2.2) and (5.2.7) through (5.2.9) into Eq. (5.2.1) yields

$$\frac{C}{C_0} = \frac{1}{1 + K_L A/Q + (Q_b/Q)\gamma H_c + (Q_w/Q)K_p C_s + k_b X/Q} \tag{5.2.10}$$

Knowledge of the inlet concentration (C_0) and all variables in the denominator of Eq. (5.2.10) allows an estimation of the effluent concentration C. This concentration can then be substituted into Eqs. (5.2.2) and (5.2.7) through (5.2.9) to determine pathway-specific removal rates. If a treatment process is covered, a separate mass balance must be completed on the gas phase, with the term C_g/H_c retained in Eq. (5.2.2). The liquid- and gas-phase mass balances must then be solved simultaneously for C and C_g, with emissions equaling the product of C_g and the headspace ventilation (air exchange) rate.

Equation (5.2.10) serves to illustrate differences in existing models for estimating VOC emissions from wastewater. Some models account for all the terms in the denominator of Eq. (5.2.10). Others do not. In addition, models differ in how they estimate or prescribe parameters such as K_L (or k_l and k_g), H_c, γ, K_p, and k_b. Specific emission models are described in the remainder of this section. Where appropriate, differences in the formulations of these models are described.

EMISSION MODELS: WASTEWATER COLLECTION SYSTEMS

General Considerations

Both municipal and industrial sewers typically consist of reaches, junction boxes, and drop structures. Industrial wastewater generally originates from specific process units and enters a collection system through process drains, which for a large industrial facility can number several thousand (EPA, 1985). Few existing models include estimates of emissions from process drains based on mechanistic mass-transfer principles, although recent research has been completed for determination of mass-transfer coefficients from such systems (Enviromega, 1993; Shepherd and Corsi, 1996; Stubbe and Corsi, 1997).

Sewer reaches are used to convey wastewater to the point of discharge or end-of-pipe treatment. Existing sewer models can generally be divided into two categories: equilibrium-based or kinetics-based. An assumption of chemical equilibrium is most appropriate under conditions of low compound volatility (low Henry's law constant) and/or low rates of air exchange (ventilation rate) between sewer and ambient atmospheres. A kinetics-based approach involves estimation of mass-transfer coefficients, typically based on oxygen transfer correlations by Parkhurst and Pomeroy (1972) or Owens et al. (1964) with theoretical extrapolation to VOCs (Whitmore and Corsi, 1994). Chemical tracer releases in municipal sewers have shown reasonable consistency between predicted (using an adjusted Parkhurst-Pomeroy model) and measured values for a wide range of operating conditions (Corsi et al., 1992; Whitmore and Corsi, 1994; Jensen and Hvitved-Jacobsen, 1991).

Junction boxes serve as points of confluence when more than one sewer reach intersects. Existing models assume that instantaneous mixing occurs at junction boxes, and that mass transfer associated with mixing inlet streams is negligible. Corsi and Quigley (1996) determined stripping efficiencies for a wide range of chemical tracers in a pilot junction box and drop structure. However, existing models for estimating emissions from junction boxes are highly simplified, often treating the junction box as if it contained a pool with a surface aerator.

Drop structures are also locations where chemical emissions can occur, particularly for high drop heights and ventilation rates. Emissions from drop structures may occur via several mechanisms, including splashing at the tailwater surface and air entrainment in the tailwater pool. However, the combined effects of those mechanisms are typically lumped into an oxygen transfer parameter referred to as an *oxygen deficit ratio:*

$$r_{O_2} = \frac{C_{so} - C_{1o}}{C_{so} - C_{2o}} \tag{5.2.11}$$

where rO_2 is the deficit ratio for oxygen, C_{so} is the dissolved oxygen concentration at saturation (mg/L), C_{1o} is the dissolved oxygen concentration upstream of a drop (mg/L), and C_{2o} is the dissolved oxygen concentration downstream of a drop (mg/L).

The oxygen deficit ratio can be related to a similar term for VOCs:

$$r_{\text{VOC}} = r_{O_2}{}^{\psi_T} = \frac{C_g/H_c - C_{1v}}{C_g/H_c - C_{2v}} \tag{5.2.12}$$

where r_{VOC} is the deficit ratio for VOC, ψ_T is defined in Eq. (5.2.6), C_g is the chemical concentration in the gas phase (mg/m^3), H_c is Henry's law constant for the VOC ($m_{liq}{}^3/m_{gas}{}^3$), C_{1v} is the chemical concentration in the liquid phase upstream of a drop (mg/m^3), and C_{2v} is the chemical concentration in the liquid phase downstream of a drop (mg/m^3). The derivation of Eq. (5.2.11) has been described elsewhere (Corsi and Quigley, 1995). A major limitation includes the assumption that the gas concentration remains constant from the inlet to the outlet of the drop structure. This is only true if no mass transfer occurs, and it leads to significant uncertainties if air entrainment is the dominant mass-transfer mechanism.

The major drawback to existing VOC and sewer models is that the ventilation rate is typically a user-prescribed input. Previous research has shown that emissions estimates are highly sensitive to the ventilation rate, particularly for lower-volatility compounds (Whitmore and Corsi, 1994). Recent research has been completed to quantify ventilation rates from industrial sewers based on experimental work completed on a laboratory-scale sewer simulator. Ventilation caused by ambient wind speeds and buoyancy-driven ventilation caused by the discharge of hot effluent streams were determined to be the dominant mechanisms for air exchange (Olson et al., 1996; Olson et al., 1997a; and Varma et al., 1996).

CORAL+

The CORAL+ model was the first commercially available model to estimate VOC emissions from sewer reaches (Corsi et al., 1992). It allows for estimation of VOC emissions from a single sewer reach and drop structure. The CORAL+ model can be used to estimate emissions for both continuous and dynamic discharges of VOCs. It also has a sensitivity analysis component which allows for determination of single-parameter effects on chemical emissions. The CORAL+ model is menu-driven and has a chemical database containing a large number of VOCs and their relevant physicochemical properties.

The CORAL+ model uses a kinetics-based approach involving simultaneous solution of gas- and liquid-phase mass balances. Liquid-phase mass-transfer coefficients are based on an oxygen transfer correlation (Parkhurst and Pomeroy, 1972) with adjustments to account for differences in liquid-phase diffusion coefficients between oxygen and VOCs, i.e., Eq. (5.2.6) (Whitmore and Corsi, 1994). As little research has been completed to experimentally quantify gas-phase mass-transfer coefficients from sewer reaches, values of k_g are based on a surface impoundment correlation (MacKay and Yeun, 1983). Hydraulic conditions are calculated based on uniform flow hydraulics and corresponding use of Manning's equation (White, 1986). A uniform or cocurrent ventilation pattern may be prescribed by the user, as well as the ratio of gas-to-liquid flow rates.

Emissions from drop structures are estimated by empirical correlations of oxygen deficit ratios developed by Nakasone (1986). A k_g/k_l ratio of 3.0 is used for drop structures. This ratio is consistent with those observed for diffused aeration systems (Hsieh et al., 1992).

CORAL+ is the one model which has been confirmed experimentally using field data. Corsi et al. (1992) conducted field experiments for two operating collection systems. Deuterated chloroform ($CDCl_3$) was used as a volatile tracer for measurement of stripping efficiencies along each sewer reach. Predicted (CORAL+) and measured chloroform concentrations were generally within 30 percent of one another for four different sampling periods.

Although it is the only emissions model for sewers that has been evaluated using field data, CORAL+ does have some limitations. It is only capable of estimating emissions from single-sewer reaches or single-reach/drop combinations. Multiple-reach simulations require successive model executions with updating of relevant boundary conditions. The CORAL+ model does not account for competing fate pathways, i.e., adsorption and biodegradation, in sewers, although the latter is likely to be insignificant relative to volatilization of VOCs (Swanston, 1994). In addition, the user is required to prescribe information related to air exchange rates, values which have not been well defined and which are not easily determined for sewers. Finally, CORAL+ does not allow for estimates of hydrogen sulfide formation, dissociation, and emissions.

WATER8

WATER8 is a fate and transport model for both wastewater treatment and collection systems and was developed by Research Triangle Institute for the Office of Air Quality Planning and Standards of the Environmental Protection Agency (EPA, 1994). Unlike CORAL+, the WATER8 model is intended for estimating VOC emissions from industrial wastewater. Emissions from sewers are determined based on an assumption that chemical equilibrium exists between liquid and adjacent air. Although chemical equilibrium is the default method for determining emissions, the WATER8 model does allow for kinetics-based mass-transfer estimates along reaches. An oxygen-transfer coefficient is extrapolated to VOCs using Eqs. (5.2.4) and (5.2.5). Unlike CORAL+, an oxygen-transfer model developed by Owens et al. (1964) is used. The mass-transfer models employed by CORAL+ and WATER8 can lead to significantly different emissions estimates for sewer reaches.

Emissions from process drains are estimated for both sealed and unsealed cases. Emissions from unsealed drains are estimated based on an assumption of chemical equilibrium. Emissions from sealed and inactive (no flowing water) drains are estimated assuming that volatilization occurs only by molecular diffusion and that there is negligible chemical accumulation in the gas phase. For sealed drains with wastewater flow, liquid- and gas-phase mass-transfer coefficients are estimated from empirical correlations derived from pilot data (Enviromega, 1993).

For drop structures, liquid-phase mass-transfer coefficients are estimated from oxygen-transfer correlations by Nakasone (1986). A gas-phase mass-transfer coefficient of 0.05 m/s is chosen for benzene, which was noted to be equal in magnitude with gas-phase mass-transfer coefficients for mechanically aerated systems. The gas-phase mass-transfer coefficient is related to other chemicals using Eq. (5.2.5) with $m = \frac{2}{3}$.

Like CORAL+, the WATER8 model requires the user to prescribe air exchange rates. Another limitation is the assumption of chemical equilibrium between wastewater and adjacent air. This assumption is generally conservative, with the potential for significant overestimation of emissions for highly volatile chemicals and/or high air exchange rates. The WATER8 model does not account for adsorption or biodegradation pathways. However, it does attempt to account for absorption to, and emissions from, an oily phase in industrial sewers. It is assumed that equilibrium conditions exist between oil and wastewater. This approach will tend to overestimate emissions if oil is only located at the surface, and underestimate emissions if oil is only dispersed in the liquid and not located at the surface.

TOXCHEM+

The TOXCHEM+ model is intended for the prediction of the fate and transport of contaminants in wastewater treatment plants and was developed by Enviromega, Ltd. (Campellville, Ontario) in conjunction with Environment Canada (Burlington, Ontario). Like WATER8, TOXCHEM+ also allows for estimates of emissions from the collection system which conveys wastewater to a downstream treatment plant.

Emissions from sewer reaches and drop structures are estimated using an algorithm similar to CORAL+, but without the option of dynamic discharges of chemicals. Emissions from process drains are estimated in a manner similar to WATER8.

A significant advantage of TOXCHEM+ is its ability to simulate multiple sewer reaches. It also has a large chemical database, parameter sensitivity algorithm, and user-friendly interface. As with other models, TOXCHEM+ is limited in terms of requiring a user-prescribed air exchange rate, and it does not account for competing removal pathways in sewers.

naUTilus

Research has recently been completed to experimentally quantify emissions of VOCs from specific sewer components such as reaches, drop structures, and industrial process drains (Whitmore and Corsi, 1994; Shepherd and Corsi, 1996; Stubbe and Corsi, 1997; and Corsi and Quigley, 1996). These studies have led to the development of a model for estimating VOC emissions from entire industrial sewer networks. The model—naUTilus—is based on fundamental mass-transfer principles for estimating gas-liquid mass transfer and fluid mechanics and heat-transfer principles for determining air exchange rates between sewer and ambient atmospheres. Submodels for specific mass-transfer parameters have been determined based on experiments using a novel process drain simulator, and pilot systems for sewer reaches and drop structures. The naUTilus model can be used to provide a detailed inventory and mechanistic analysis for an entire sewer system, including the process drains and connecting pipes contained within specific process units and the larger sewer laterals and interceptors which connect such units. Although originally developed for industrial sewers, naUTilus can also be applied to municipal sewer networks. A detailed description of naUTilus is provided elsewhere (Olson and Corsi, 1997).

A major advantage of naUTilus is that the ventilation rate is calculated based on theoretical and experimental research findings rather than on a user-prescribed value. Experimental studies were recently completed on laboratory- and pilot-scale sewer systems to quantify several mechanisms which affect ventilation (Varma et al., 1996; Olson et al., 1996; Olson et al., 1997a). Theoretical models based on fundamental fluid mechanics and heat-transfer kinetics were generally within 30 percent of measured values (Olson et al., 1997a). The naUTilus model also includes an improved mass-transfer algorithm which substantially reduces errors associated with commonly used equilibrium approaches (Olson et al., 1997b).

Chloroform Mass Balance

A novel approach was recently proposed for estimating VOC emissions from entire municipal sewer systems (Koziel and Corsi, 1996). The *chloroform mass-balance* (CMB) approach is based on the use of chloroform as an in situ tracer. The CMB method requires knowledge of the chloroform concentration in the influent water supply, water usage rates, chloroform concentration entering a wastewater treatment plant, VOC concentration in the wastewater entering the treatment plant, and wastewater discharge rates. These data are readily available for most municipalities.

Chloroform losses across the entire sewer are attributed to stripping and emissions from sewers. The resultant stripping efficiency is used, along with mass-transfer coefficients estimated in a manner similar to those in CORAL+, to determine systemwide air exchange rates. These are then used in conjunction with mass loading rates in wastewater influent streams to back-calculate systemwide emissions for other VOCs. Variations in hypothetical system parameters result in probabilistic distributions for VOC emissions.

The major advantage of the CMB approach is that it allows for systemwide estimates of VOC emissions from sewers. The utility of this approach has been illustrated for the city of Dallas (Koziel and Corsi, 1996). Limitations include an assumption of *uniformity of discharge,* i.e., that chemical discharges are spatially uniform within a municipality. While this assumption is reasonable for chloroform, it may be violated for chemicals which are released from a small number of dischargers, particularly if those dischargers are located in the near vicinity of a treatment plant.

EMISSION MODELS: WASTEWATER TREATMENT SYSTEMS

General Considerations

Wastewater treatment plants typically consist of several different treatment units that can be grouped into the following categories: headworks, primary treatment, secondary treatment, advanced treatment, sludge treatment, and industrial processes. The headworks involves preliminary processes in the treatment plant and may include bar screens, grit chambers, and equalization tanks. Bar screens are used to remove large objects from influent wastewater streams and are included in most municipal treatment plants. Little research has been completed to address the effects of emissions from bar screens, and as such, they are generally omitted from emission estimates.

Grit chambers are used to remove inorganic solids from municipal wastewater. Equalization basins are used to dampen fluctuations in influent flow rate and are generally characterized by relatively long residence times. Both processes may be aerated or nonaerated. Aerated grit chambers or equalization tanks are typically modeled as either bubble aeration or mechanical aeration. For bubble aeration, bubbles may be assumed to be at equilibrium [$\gamma = 1$ in Eq. (5.2.7)] or an appropriate $k_l a$ can be related to a given VOC using oxygen-transfer correlations (Metcalf & Eddy, Inc. and Tchobanoglous, 1981). Oxygen-transfer correlations have been developed for mechanical surface aerators, and are typically expressed as a function of the power input and oxygen-transfer rate of the aerator (Thibodeaux, 1979). Nonaerated grit chambers or equalization basins are typically estimated using mass-transfer correlations for open surfaces.

Primary clarifiers are used to remove solids from the wastewater prior to biological treatment. Emissions of VOCs from primary treatment processes include *volatilization* from primary clarifiers and clarifier weirs. Contaminant removal in sludge streams is generally estimated using Eq. (5.2.8). Emissions from clarifier weirs are typically modeled in a manner similar to that of drop structures, i.e., by employing Eqs. (5.2.6), (5.2.11), and (5.2.12) and an empirical correlation to estimate an oxygen deficit ratio.

Secondary treatment processes include units which degrade constituents in wastewater biologically. Previous research has focused attention on activated sludge processes (ASPs). As such, ASPs have been the most commonly modeled unit for secondary treatment. The major removal mechanisms for this process are *biodegradation,* surface *volatilization, stripping* from diffused aeration, and *adsorption* (Fig. 5.2.1). Most models assume that uniform mixing and steady-state conditions exist, and that nonaqueous liquids are not present [Eq. (5.2.10)].

Advanced treatment processes may be used for nutrient removal and solids polishing. As most VOCs are typically removed from wastewater prior to this stage, most fate models omit advanced treatment processes.

Sludge treatment involves the stabilization and dewatering of solids that are removed from wastewater. Relevant processes include aerobic and anaerobic digesters, filter process, centrifuges, drying beds, etc. Compounds which are removed in the aqueous or sorbed phase of sludge may be biodegraded or stripped during these processes. The TOXCHEM+ model is unique in terms of its ability to address several sludge treatment processes.

Industrial plants often include additional treatment units for specific removal tasks. These industrial processes include gravity oil separators, dissolved air flotation units, and cooling towers.

Gravity oil separators are holding tanks that remove oil from wastewater using baffles and skimming devices. The most rigorous modeling approach would include analysis of mass-transfer kinetics between all three phases: water and oil, water and air, and oil and air. It is generally assumed that water and oil are completely mixed as separate phases, and that equilibrium conditions exist between water and oil phases. The latter assumption is reasonable given the high partitioning of most contaminants into the oily phase. The octanol-water partition coefficient is generally used to estimate partitioning between oil and water phases.

Dissolved air flotation (DAF) units are used to separate particles from wastewater by attaching those particles to rising air bubbles. Attachment of air bubbles to oil particles lowers the effective density of the oil agglomerate, thus allowing the oil to more rapidly rise to the surface where it can be removed by a skimming device. Dissolved VOCs may be transferred from oil or water to air bubbles upon contact, and are then released to the atmosphere or gas-phase control device as the bubbles reach the DAF surface. Estimates of VOC emissions generally incorporate assumptions similar to those for gravity oil separators, i.e., that water and oil are completely mixed as separate phases and that equilibrium conditions exist between water and oil.

Some industrial plants employ cooling towers to dampen high-temperature discharges to biological treatment units. Although extensive research has been completed regarding heat-transfer kinetics in cooling towers, little work has been completed to characterize VOC emissions from such systems. The TOXCHEM+ model includes an algorithm for estimating VOC emissions from cooling towers.

BASTE

The BASTE (Bay Area Sewage Toxics Emissions) model was developed for the San Francisco Bay Area Air Toxics (BAAT) group (Corsi and Card, 1991). It is a general fate model which uses a building-block approach to estimate removal of VOCs from individual treatment units. Building blocks are connected in series or parallel by links (which can also be modeled), thus allowing for flexibility in evaluating different treatment configurations. The BASTE model includes five building-block options: (1) series of continuous-flow stirred-tank reactors, (2) quiescent surfaces, (3) drop structures, (4) packed-media systems, and (5) measured emissions or emission factors. Emission estimates for any treatment process can be made by using the appropriate building blocks. For example, a primary clarifier can be modeled as a quiescent tank followed by a drop structure.

The BASTE model accounts for losses due to *adsorption, biodegradation, volatilization,* and *stripping*. Equilibrium conditions are assumed to exist for sorption to solids, where liquid-solid partition coefficients are based on research by Dobbs et al. (1989). Biodegradation is assumed to be first-order, where biological rate constants are estimated from literature values. BASTE also allows for user-prescribed biodegradation rate constants, which is important since rate constants may vary significantly between systems. Emissions are calculated as the sum of three different components: nonagitated surfaces, mechanical surface aeration, and rising air bubbles. Emissions from bubble aeration are modeled using Eq. (5.2.7), with estimation of γ using a method first published by Roberts et al. (1984).

Emissions from drop structures are estimated using Eqs. (5.2.4), (5.2.5), (5.2.11), and (5.2.12) with oxygen deficit ratios approximated using correlations by Nakasone (1986) or Pincince (1989) for primary and secondary clarifier weirs. Alternatively, BASTE allows for the determination of oxygen deficit ratios using dissolved oxygen measurements collected immediately upstream and downstream of specific drop structures.

For packed-media systems, e.g., trickling filters, *volatilization* is assumed to be the only removal mechanism for VOCs. This assumption tends to overpredict the *rate of volatilization* for degradable compounds, but may be appropriate for less degradable or recalcitrant compounds. These systems are modeled with either cocurrent or countercurrent gas and liquid flows, where in both cases it is assumed that equilibrium conditions exist between effluent gas and adjacent liquid.

The fifth building-block option in the BASTE model allows for inclusion of emission factors or measured emission rates. This option is particularly useful for treatment processes otherwise not accounted for by the BASTE model.

TOXCHEM+

The TOXCHEM+ model, described previously for sewers, was originally developed for wastewater treatment plants. Removal of VOCs from headworks is assumed to be due entirely to *volatilization* and *adsorption.* For headworks that are nonaerated, *volatilization rates* are estimated using correlations for surface impoundments (MacKay and Yeun, 1983). Oxygen-transfer correlations are used for estimation of liquid-phase mass-transfer coefficients for aerated systems. These are based on mechanical surface aerator data (Thibodeaux, 1979) or Eq. (5.2.7). A k_g/k_l ratio of 40 is assumed for aerated systems, based on work by Munz and Roberts (1989).

Removal from primary treatment is assumed to be due to adsorption and volatilization. *Volatilization* from clarifier weirs is estimated using Eqs. (5.2.6), (5.2.11), and (5.2.12), and oxygen-transfer correlations by Pincince (1989).

Biodegradation, stripping, volatilization, and *sorption* are included as removal mechanisms during secondary treatment, i.e., Eq. (5.2.10). Values of K_p are based on a correlation attributed to Dobbs et al. (1989). Values of k_b are based, in large part, on several years of novel experiments using a large pilot-activated sludge plant.

The TOXCHEM+ model also includes algorithms for predicting emissions during sludge treatment and dewatering processes, as well as cooling towers. The model consists of a large database of chemical properties, including those for VOCs, semivolatile organic compounds, and metals. Several of the model's subroutines are based on pilot experiments for specific processes and ranges of tracer chemicals, with corresponding physical, chemical, and biological parameter estimation.

WATER8

The WATER8 model described previously also has the capability of estimating emissions from wastewater treatment plants. For headworks treated as quiescent surfaces, emissions are estimated using liquid- and gas-phase mass-transfer coefficients by Springer et al. (1984) and MacKay and Matasuga (in Hwang, 1982), respectively. For mechanically aerated surfaces, liquid- and gas-phase mass-transfer coefficients are based on work by Thibodeaux (in Hwang, 1982) and Reinhardt (1985), respectively. For diffused aeration, rising air bubbles are assumed to be in chemical equilibrium with adjacent wastewater [$\gamma = 1$ in Eq. (5.2.7)]. *Volatilization* from clarifier weirs is estimated using Eqs. (5.2.4), (5.2.11), and (5.2.12), with oxygen-transfer correlations by Pincince (1989).

Removal mechanisms for secondary treatment include *biodegradation, stripping,* and *volatilization.* Removal rates from *stripping* and *volatilization* have been described previously. *Biodegradation rates* are estimated using Eq. (5.2.9) with rate constants based on literature values (Coburn et al., 1988) and quantitative structure activity relationships (QSARs).

Comparison with Field Data

Schroy (1997) compared simulated air emissions using several models with measured data from 15 different field studies. He compared the predicted and measured air emissions using WATER7, an earlier version of WATER8. Total predicted VOC emissions using WATER7 ranged from −100 to +100 percent of measured values for 14 of the field studies. All models generally had relative errors of greater than ±100 percent for the other field study, a covered activated sludge system. Total predicted VOC emissions using TOXCHEM ranged from −100 to +95 percent of measured values for 14 of the field studies, although most model simulations were within ±40 percent of measured values. Total predicted VOC emissions using BASTE were only evaluated for four sampling periods during one field study. Predicted values were within −37, −37, 0, and 2 percent of measured data.

Soszynski et al. (1997) compared predictions using WATER8, TOXCHEM+, and BASTE with measured data from both aerated grit chambers and aeration basins at two municipal wastewater treatment plants in Chicago. Predicted values using WATER8 were an average of 1386 and 3575 percent greater than measured values for the two aerated grit chambers, and within 22.7 and −46.7 percent for the aeration basins. Predicted values using TOXCHEM+ were an average of −17.4 and 56.1 percent within measured values for the aerated grit chambers, and within 11.9 and −74.0 percent for the aeration basins. Predicted values using BASTE were an average of −10.8 and 85.6 percent within measured values for the aerated grit chambers, and within −23.0 and −69.1 percent for the aeration basins.

SUMMARY

Several computational models are available for estimation of emissions from either wastewater collection systems or wastewater treatment systems. The focus of this section has been on the most commonly used models (BASTE, CORAL+, TOXCHEM+, and WATER8). A summary of collection and treatment models is provided in Table 5.2.1. Major limitations of existing models include the lack of ventilation algorithms and system-specific nature of biodegradation kinetics for many VOCs.

REFERENCES

Coburn, J., C. Allen, D. Green, and K. Leese (1988). Site Visits of Aerated and Nonaerated Impoundments, revised draft summary report, Environmental Protection Agency, project 68-03-3253.

Corsi, R. L. (1997). "VOC Emissions from Municipal Sewers: New Research Findings," *Proceedings of Control of Odors and VOC Emissions, Water Environment Federation Specialty Conference Series,* Houston, Tex.

Corsi, R. L., and T. R. Card (1991). "Estimation of VOC Emissions Using the BASTE Model," *Environmental Progress,* **10**(4): 290.

Corsi, R. L., D. P. Y. Chang, and E. D. Schroeder (1992). "A Modeling Approach for VOC Emissions from Sewers," *Water Environment Research,* **64**(5): 734.

Corsi, R. L., and C. J. Quigley (1995). "VOC Emissions from Sewer Junction Boxes and Drop Structures: Estimation Methods and Experimental Results," *Proceedings of the 88th Annual Meeting of the Air & Waste Management Association,* San Antonio, Tex.

Corsi, R. L., and C. J. Quigley (1996). "VOC Emissions from Sewer Junction Boxes and Drop Structures: Estimation Methods and Experimental Results," *Journal of the Air & Waste Management Association,* **45**(5): 395.

Dobbs, R. A., L. Wang, and R. Govind (1989). "Sorption of Toxic Organic Compounds on Wastewater Solids: Correlation with Fundamental Properties," *Environmental Science & Technology,* **23**(9): 1092.

Enviromega, Ltd. (1993). *Measurement of Hazardous Air Pollutant Emissions from Wastewater Collection System Components,* vol. 2: *Process Drains,* report to the Chemical Manufacturers Association, Hamilton, Ontario.

TABLE 5.2.1 Summary of Collection and Treatment Models

Name	Developer	Description
BASTE (Bay Area Sewage Toxics Emissions)	Richard L. Corsi of the University of Texas at Austin, and CH2M-Hill, Oakland, Calif. (for the BAAT group)	Fate model for treatment systems using building-block approach to model emissions
WATER8	Research Triangle Institute (for EPA OAQPS)	Treatment and collection model for aerated and nonaerated processes and impoundments
CINCI (EPA-Cincinnati model)	Richard Dobbs of Lamar University and Rakesh Govind of the University of Cincinnati with support from the EPA Risk Reduction Engineering Laboratory (Cincinnati, Ohio)	Models the following treatment systems: (1) primary treatment followed by secondary treatment with sludge recycle, (2) secondary treatment with recycle, (3) secondary treatment without recycle
CORAL+ (Collection System Organic Release Algorithm)	Richard L. Corsi of the University of Texas at Austin and Enviromega Ltd.	Models collection system reaches and drops based on gas-liquid mass transfer
EPA FATE (Fate and Treatability Estimator)	ABB Environmental for EPA Office of Water Regulations and Standards	Spreadsheet version of model developed by Namkung and Rittman (1987)
NOCEPM (NCASI Organic Compound Elimination Pathway Model)	D.A. Barton of the National Council of the Paper Industry for Air and Stream Improvement (NCASI), New York	Models aeration tanks and secondary treatment with removal from stripping, sorption, and biodegradation based on literature models
PAVE (Program to Assess Volatile Emissions)	Jerry Schroy of Monsanto and Chemical Manufacturers Association	Models treatment units based on literature values; also models emissions from chemical spills
SIMS (Surface Impoundment Modeling System—1990)	Radian (for EPA OAQPS)	Uses WATER8 and other EPA documents
naUTilus	David A. Olson, Sunil Varma, and Richard L. Corsi of the University of Texas at Austin	Kinetics-based model for entire wastewater collection systems; only model which calculates air exchange rates
TOXCHEM+ (Toxic Chemical Modeling Program for Water Pollution Control Plants)	Environment Canada's Wastewater Technology Centre and Enviromega Ltd.	Fate model for treatment units based on literature; only treatment model with unsteady-state option

EPA (1985). VOC Emissions from Petroleum Refinery Wastewater Systems—Background Information for Proposed Standards. EPA-450/3-85-001a.

EPA (1994). Air Emissions Models for Waste and Wastewater. EPA-453/R-94-080A.

Hannah, S., B. Austern, A. Eralp, and R. Wise (1986). "Comparative Removal of Toxic Pollutants by Six Wastewater Treatment Processes," *Journal of the Water Pollution Control Federation,* **58**(1): 30.

Hsieh, C. C., R. W. Babcock, and M. K. Stenstrom (1992). "Estimating Semi-Volatile Organic Compound Emission Rates and Oxygen Transfer Coefficients in Diffused Aeration," *Proceedings of the 65th Annual Conference of the Water Environment Federation,* New Orleans, La.

Hwang, S. T. (1982). "Toxic Emissions from Land Disposal Facilities," *Environmental Progress,* **1:** 46.

Jensen, N. A., and T. Hvitved-Jacobsen (1991). "Method for Measurement of Reaeration in Gravity Sewers Using Radioactive Tracers," *Research Journal of the Water Pollution Control Federation,* **63:** 758.

Kincannon, D., A. Weinert, R. Padorr, and E. L. Stover (1982). "Predicting Treatability of Multiple Organic Priority Pollutant Wastewater from Single Treatability Studies," *Proceedings of the 37th Purdue Industrial Waste Conference,* West Lafayette, Ind.

Koziel, J., and R. L. Corsi (1996). "A Novel Approach for Estimating VOC Emissions from Municipal Sewers: Methodology, Applications, and Implications," *Proceedings of the 89th Annual Meeting of the Air & Waste Management Association,* Nashville, Tenn.

MacKay, D., and A. Yeun (1983). "Mass Transfer Coefficient Correlations for Volatilization of Organic Solutes from Water," *Environmental Science & Technology,* **7:** 211.

Metcalf & Eddy Inc., and G. Tchobanoglous (1981). *Wastewater Engineering: Collection and Pumping of Wastewater*, McGraw-Hill, New York.

Munz, C., and P. V. Roberts (1989). "Gas- and Liquid-Phase Mass Transfer Resistances of Organic Compounds during Mechanical Surface Aeration," *Water Research,* **23:** 589.

Nakasone, H. (1986). "Study of Aeration at Weirs and Cascades," *Journal of Environmental Engineering, ASCE,* **113**(1): 64.

Namkung, E., and B. E. Rittman (1987). "Estimating Volatile Organic Compound Emissions from Publicly Owned Treatment Works," *Journal of the Water Pollution Control Federation,* **59**(7): 670.

Olson, D. A., and R. L. Corsi (1997). "naUTilus: A Model for Predicting HAP Emissions from Industrial Sewers," *Proceedings of the 90th Annual Meeting of the Air & Waste Management Association,* Toronto, Canada.

Olson, D. A., R. L. Corsi, and S. Rajagopalan (1996). "Buoyancy-Induced Ventilation between Industrial Process Drains and the Ambient Atmosphere," *Proceedings of the 89th Annual Meeting of the Air & Waste Management Association,* Nashville, Tenn.

Olson, D. A., S. Rajagopalan, and R. L. Corsi (1997a). "Ventilation of Industrial Process Drains: Mechanisms and Effects on VOC Emissions," *Journal of Environmental Engineering, ASCE,* 123(9): 939.

Olson, D., S. Varma, and R. L. Corsi (1997b). "A New Approach for Estimating VOC Emissions from Industrial Sewers: Methodology and Associated Errors," *Water Environment Research,* awaiting publication.

Owens, M., R. W. Edwards, and J. W. Gibbs (1964). "Some Reaeration Studies in Streams," *International Journal of Air and Water Pollution,* **8:** 469.

Parkhurst, J. D., and R. D. Pomeroy (1972). "Oxygen Absorption in Streams," *Journal of the Sanitary Engineering Division, ASCE,* **98**(SA1): 101.

Petrasek, A., B. Austern, and T. Neiheisel (1983). "Removal and Partitioning of Volatile Organic Priority Pollutants in Wastewater Treatment," *Proceedings of the 9th U.S.-Japan Conference on Sewage Treatment Technology,* Tokyo, Japan.

Pincince, A. B. (1989). "Transfer of Oxygen and Emissions of Volatile Organic Compounds at Clarifier Weirs," *Research Journal of the Water Pollution Control Federation,* **63**(2): 114.

Pitter, P. (1976). "Determination of Biological Degradability of Organic Substances," *Water Research,* **10:** 231.

Reinhardt, J. R. (1985). GCA Corporation. Emissions Data and Model Review for Wastewater Treatment Operations, draft technical note, Environmental Protection Agency, contract 68-01-6871.

Roberts, P. V., P. Dandliker, and C. Matter-Muller (1984). "Volatilization of Organic Pollutants in Wastewater Treatment—Model Studies," EPA-600/2-84-047.

Schroy, J. M. (1997). "Round Robin Comparison of Air Emission Models for Wastewater Treatment Systems," *Proceedings of Control of Odors and VOC Emissions, Water Environment Federation Specialty Conference Series,* Houston, Tex.

Shepherd, J. L., and R. L. Corsi (1996). "Mass Transfer of Volatile Organic Compounds from Industrial Process Drains," *Proceedings of the 89th Annual Meeting of the Air & Waste Management Association,* Nashville, Tenn.

Soszynski, S., D. T. Lordi, P. Tata, D. R. Zenz, and C. Lue-Hing (1997). "Comparison of Estimated and Measured Volatile Organic Compound Emissions from Large Publicly Owned Treatment Works," *Proceedings of Control of Odors and VOC Emissions, Water Environment Federation Specialty Conference Series,* Houston, Tex.

Springer, C., P. D. Lunney, and K. T. Valsaraj (1984). Emissions of Hazardous Chemicals from Surface and Near Surface Impoundments to Air. Environmental Protection Agency, Solid and Hazardous Waste Research Division, project 808161-02.

Stubbe, J., and R. L. Corsi (1997). "A Multizone Mechanistic Model for Estimating VOC Emissions from Industrial Process Drains," *Proceedings of the 90th Annual Meeting of the Air & Waste Management Association,* Toronto, Canada.

Swanston, M. (1994). "Biodegradation of Volatile Organic Compounds in Sewers," M.S. thesis, University of Guelph, Guelph, Ontario.

Thibodeaux, L. J. (1979). *Chemodynamics,* John Wiley & Sons, New York.

Varma, S., D. A. Olson, R. L. Corsi, and S. Rajagopalan (1996). "Air Exchange between Industrial Sewers and the Ambient Atmosphere," *Proceedings of the 89th Annual Meeting of the Air & Waste Management Association,* Nashville, Tenn.

White, F. M. (1986). *Fluid Mechanics,* McGraw-Hill, New York.

Whitmore, A., and R. L. Corsi (1994). "Measurement of Gas-Liquid Mass Transfer Coefficients for Volatile Organic Compounds in Sewers," *Environmental Progress,* **13**(2): 114–123.

CHAPTER 6
AIR POLLUTION PREVENTION

SECTION 6.1

INDUSTRIAL POLLUTION PREVENTION

Marcus J. Healey
Mark Walsh
Douglas Kretkowski
Jeffrey Lewis
New Jersey Technical Assistance Program for Industrial Pollution Prevention
New Jersey Institute of Technology
Newark, New Jersey

INTRODUCTION 6.1
- Overview 6.1
- Forces for Change 6.2
- The New Environmental Wave 6.3
- General Benefits of Pollution Prevention 6.3

PROCESS PROFILES 6.4
- Surface Coating and Painting 6.4
- Application of Pollution Prevention to the Surface Coating Process 6.5
- Additional Waste Minimization Strategies for the Surface Coating Process 6.11
- Solvent Cleaning and Degreasing 6.12

INDUSTRY PROFILES 6.16
- Textiles 6.16
- Paper and Pulp 6.22
- Printing and Publishing 6.26
- Secondary Metals (SIC Code 34) 6.29

REFERENCES 6.33

INTRODUCTION

Overview

Pollution prevention is a term that is both evocative and enigmatic. It evokes the idea of an environmental analog of preventive medicine; yet, it is puzzling to those who are not familiar with pollution prevention and its relationship to other, more mainstream areas of environmental science and engineering. The rationale for pollution prevention is often associated with its change in focus, its relationship to other environmental and manufacturing strategies, and its potential benefits.

Pollution prevention changes the focus from treatment to prevention, from single medium (i.e., air, water, or land) to multimedia (i.e., all media), from enforcement to encouragement, and from the end of the pipe to the "production" process in industry and in service providers. In so doing, a prevention ethic is inculcated into the way that people perform their business and lead their lives.

The switch from a focus on treatment to a focus on prevention represents a fundamental change in environmental thinking because waste is not accepted as a given. This change in thinking to one of prevention ultimately serves to shift slowly the focus from solvent usage and best-available control technology to rethinking the process, the product, and the overall role of volatile organic compounds (VOCs).

A multimedia approach recognizes that much of what was done in the past in terms of treatment constituted a "toxic shell game" whereby pollutants were transferred from one medium to another medium. Pollution prevention also signals a fundamental shift in compliance with VOC regulations. Pollution prevention's goal is zero emissions, or as close to zero emissions as possible, rather than switching from regulated solvents to unregulated solvents to stay one step ahead of the latest ban.

The movement from enforcement (i.e., "command and control") to encouragement recognizes that without industry-government-citizen cooperation, we as a nation are not going to fulfill, and sustain, our commitment to cleaning up the environment. End-of-pipe controls have been a necessary, effective, but insufficient means to move us away from the specter of wholesale environmental degradation. Pollution prevention serves, ultimately, to change our focus from viewing the waste that comes from the end of the pipe as necessary, to viewing it as conditional to what takes place upstream in the service or industrial process.

Forces for Change

Environmental management leaders in manufacturing, service, and government are unanimous in their collective belief that it is better to prevent pollution at its source than to deal with the attendant effects of releases, spills, and cleanups. Many people are finding economies and benefits in reducing pollution at the source.

Several present trends are propelling the movement toward pollution prevention:

- Environmental liabilities and risk management
- Total quality management, agile manufacturing, and International Standards Organization (ISO) 9000
- Environmental audits and anticipation of ISO 14000
- Prevention of accidental releases of chemicals
- Improved cost of doing business—cost avoidance and total cost accounting

A number of future environmental strategies also serve to promote pollution prevention:

- Strategic environmental management
- Life-cycle analysis
- Industrial ecology and sustainable development
- Clean manufacturing systems
- Actual bans of selected chemicals [e.g., chlorofluorocarbons (CFCs)]
- Proposed bans of selected chemicals (e.g., chlorine)

These converging forces catalyze the recognition of pollution prevention and its relationship to other forward-looking organizational strategies, such as total quality management (TQM). Service providers, manufacturers, academia, national governments, and the military are promoting pollution prevention because they all recognize the impetus to go beyond compliance toward stewardship, and they see the benefits associated with this change.

The New Environmental Wave

The waste minimization movement of the early 1980s was first advanced by industries which realized that the costs associated with environmental regulation and waste treatment were significant and spiraling upward. Most companies focused on means to reduce waste because they associate it with potential money savings and an opportunity to present a "green" public image. The end-of-the pipe focus on waste minimization or waste reduction gave way to pollution prevention in the late 1980s and early 1990s. Pollution prevention, with its up-the-pipe or in-process focus, represented a critical change in thinking for many because (1) waste was not accepted as a given and (2) the focus had changed from the waste treatment facilities to the production facilities, the *sanctum sanctorum* of the manufacturing firm.

General Benefits of Pollution Prevention

Managers in industry are often under pressure to achieve several objectives which are met by pollution prevention:

- Operate more economically, often with fewer resources (labor, capital, material, and energy)
- Continuously improve quality
- Meet increasingly stringent limits on quantities of pollutants released to the air, water, and soil

Pollution prevention is also a priority of many regulators and private organizations internationally (i.e., standards, trade, and environmental organizations). These parties will choose a pollution prevention course if it can be demonstrated to have an economic and environmental advantage over traditional options. The following benefits promote pollution prevention in the United States and internationally:

- Avoidance of future liability
- Cost avoidance (e.g., chemical purchases)
- Enhanced worker safety
- Safer transport of hazardous materials
- Easier compliance with environmental regulations
- More efficient operation of facilities
- "Green image" promotion

With a reduction in pollution, material costs often are reduced. In addition, manufacturing or service process may be rendered more economical inasmuch as the consumption of water and energy may also be diminished. A corporation will also reap benefits from an improved "green" public image. The same volume of product may still be produced, but remanufactured and reused, and/or recyclable materials may be used.

PROCESS PROFILES

Surface Coating and Painting

In terms of both the diversity of materials and the number of facilities, the surface coating and painting process is perhaps one of the most widely applied industrial processes. Surface coating and painting serve to protect multitudes of surfaces of almost any composition in an extremely wide range of commercial, industrial, and residential settings. The following list of applications of surface coatings highlights the wide variety of situations in which they are used: architectural (including bridges, buildings, and wooden structures), automobile finishes, aerospace applications, industrial surfaces (including chemical process vessels, reaction chambers, tank liners, piping, and ducts), marine vessels and structures, traffic paints, safety markings, food production equipment, computers and electronic devices, household appliances, wood products, and wood construction materials. These examples give a far from exhaustive glimpse of the almost limitless need for surface coatings and paints.

Primarily, paints and surface coatings are applied to inhibit corrosion and rusting, enhance weatherability and durability, and provide resistance to harsh environments, including abrasion, scratching, and staining. Also, and considerably important to many applications, the purposes include fortification against water infiltration, temperature extremes, exposure to damaging chemicals and solvents, and ultraviolet (uv) degradation. Perhaps the most significant applications for paints and surface coatings are the high-quality surface decoration and identification that consumers demand.

Role of Solvents in Painting and Coating. The traditional organic solvents, almost all of which are categorized as VOCs, have performed so well that they have dominated the surface coating process for decades. However, the common organic solvents listed in Table 6.1.1 serve almost identical roles in the surface coating and painting process.

TABLE 6.1.1 Typical VOCs Used in Surface Coating Applications

Ethyl benzene
Glycol ethers
Methanol
Methyl ethyl ketone (MEK)
Methyl isobutyl ketone (MIBK)
Methyl *t*-butyl ether
Toluene
Xylene

To better understand the role of organic solvents in this process, it is necessary to more fully understand their role as the coating's delivery vehicle. Coating materials and

One of the major advantages of waterborne coatings is that they are compatible with the same material substrates as are organic solvent-based coatings. Thus, the substitution process does not require a major overhaul of the coating equipment; but as will be discussed in the next section, minor modifications in the application equipment will be necessary when switching from a solvent-based to a water-based system.

Application Methods. Many different application techniques exist for liquid-based coatings. These include air spray, air-assisted spray, air-assisted airless spray, manual brush or roller, dip tanks, electrostatic spray, flow coating, high-volume low-pressure (HVLP) spray, and turbocharged spray techniques. All these application techniques may be used with the different waterborne coating technologies.

However, due to the increased viscosity and thixotropy of waterborne coatings, the modification or replacement of existing application equipment may be necessary. Furthermore, precautions against water-induced corrosion must be taken. In most instances this can be overcome by inserting stainless-steel parts wherever rusting is a possibility. In some critical parts of application equipment, such as pumps, piping, and spray nozzles, this may not be feasible, and therefore replacements may be necessary. In addition, spray arresters and spray booth filters will have to be modified in order to adapt to waterborne coatings. Technical advice and in-house experimentation will be necessary to determine the best configuration for your facility.

Other Considerations. Once they have been applied, it is commonly thought that the drying times for waterborne coatings are prohibitively long in comparison to those for organic-based coatings. Environmental conditions such as humidity and temperature play a critical role in the drying process, and when these factors are properly controlled, drying times are, in some cases, shorter for waterborne coatings. This fact allows applications in which multiple layers are required to be performed in a shorter time, with an increase in production rate.

Other considerations include these:

- Some waterborne coatings still contain solvents (VOCs), although usually less than high-solid formulations.
- Prior to fully curing, waterborne coatings are sensitive to degradation due to contact with water. Once the coating has fully cured, this is not a problem, however.
- Waterborne coatings are sensitive to humidity, requiring humidity control in the application and curing areas. Low humidity can cause those coatings to dry extremely fast, resulting in craters in the final film. High humidity can cause very slow drying times, resulting in sagging.
- Emulsion water-based coatings do not easily penetrate porous substrates. Of course, this characteristic is often beneficial, but precludes their use for some substrates.
- Some resins may degrade in water, limiting the shelf life for these formulations.

Pollution Prevention Benefits of Waterborne Coatings

- Waterborne coating technologies offer significant, but not total, reductions in VOC emissions to bring facilities into compliance with VOC regulations.
- Their use is compatible with almost all prevalent application systems, although equipment modifications are sometimes necessary.
- Tap water may be used for the majority of cleanups, although small quantities of solvent will be necessary to remove non-water-soluble coating components.

Source reduction typically results in important economic benefits for manufacturers in several ways, including decreased liability for waste, decreased waste of materials and energy (i.e., greater profits), decreased waste collection costs, decreased handling and treatment costs, and potential recycle and recovery value of waste.

A key to accomplishing pollution prevention with dyeing procedures is to have accurate dispensing systems, which meter given amounts of dyes and chemicals and feed them into machines at predetermined, programmable rates. For example, in fiber reactive dyeing, alkali added all at one time will produce a rapid exhaust of dye onto the fabric and reaction. Alkali, if added in small, discrete quantities, will produce a series of rapid strikes of dyes. However, a continual and controlled addition of alkali can cause a smooth and even exhaustion of dye onto the fabric. Similar kinds of systems are available for making print paste automatically. These systems deliver print paste from overhead tanks, automatically dispense into a print paste tank, mix and homogenize, and then deliver to the mix tank for use on the print machine.

Source reductions may be accomplished by several different techniques:

- Use of alternative methods
- Chemical and water conservation
- Raw material prescreening
- Conservation of chemicals
- Process modifications
- Improved maintenance
- Segregation of waste streams
- Waste recovery, recycle, and reuse
- Raw material quality control
- Optimization of chemical use
- Equipment modifications
- Improved housekeeping

Available Options. Material change strategies have focused on decreased toxicity, increased recoverability, and improved efficiency of reactions. Process change methods have focused on advanced dye bath techniques for aqueous processing, alternative application media and delivery, and improved washing efficiency. In addition, overall strategies such as combining individual process steps into one step offer a more systematic and holistic approach to minimizing waste. See Table 6.1.6.

It is important to note that in many cases, the overall waste situation is not improved by substitution. For example, substitution of certain surfactants can result in lower BODs but increased toxicity of treated wastes. In other cases, materials exhaust from the batch and become (temporarily or permanently) part of the product. Another example of tradeoffs is the use of mineral versus organic acids and alkali, which can substitute BOD for acidity or alkalinity.

Waste Reuse and Exchange. Various technologies exist to remove water from waste liquids. For example, the SAMSCO Water Evaporator is a hot evaporation system that is a natural-gas-fired, air-assisted evaporator for the removal of water from waste liquids. The purpose of the unit is to safely concentrate and contain liquid hazardous waste. Within this system the unit burner flue pipe is immersed in the waste liquid and directly heats and boils the water in the evaporator. The water vapor boils to the surface of the liquid in the evaporator and mixes with the ambient air and combusted gases in the premix exhaust flue stack.

TABLE 6.1.6 Material Change Strategies in Wet Processing

Option	Example	Benefit
Alternative desizing agents	Enzymes that degrade starch to ethanol	Reduced biological oxygen demand (BOD) of spent desizing bath
Substitute auxiliaries (all processes)	New washing agents	Increased wash efficiency, decreased water consumption, and improved fastness of reactives
Pad-patch dyeing	Transfer of dye to cotton, rayon, and blended goods through rollers (continuous method)	Reduction in water (2 gal/lb vs. 20 on becks), energy (2000 vs. 9000 Btu/lb) and chemical use
Foam technology	Application of dye through foam media (air dispersed in liquid) or other solvents for dyeing and printing	Reduced water and energy consumption, chemical waste, and time necessary for drying
Spraying technology	Application of finishes using sprays	Reduced energy and water usage and chemical waste (most finish remains on fabric)
Process consolidation	Single-pad steam-wash sequence using unique combinations of chemicals	Decreased water and energy usage, salt in effluent, and process times

Source: Freeman (1995).

The vapor, air, and flue gases are further mixed in the high-pressure radial-blade exhausting blower prior to discharge through the vertical exhaust stack.

When heavy metal contaminants are present in the waste liquid, they remain with the residual liquid concentrate inside the evaporator chamber. The heavy metals do not vaporize with the boiling water. The evaporator is also designed to contain heavy metal particulates within the system. Particulates are captured by the boiling liquid surface and further collected within the premix exhaust flue stack and high-pressure exhausting blower. The collected particulates drain back into the evaporator chamber.

Other Techniques and Possibilities

- Eliminate phosphates wherever possible.
- Explore solvent processing techniques.
- Use liquid-ammonia mercerization instead of caustic.
- Replace all chromates used for oxidation of vat dyes.
- Evaluate all maintenance chemicals for metals and toxicity (Noyes, 1993).

Paper and Pulp

The paper and pulp industry (SIC Code 26) can be classified into three types of mills: pulp, paper, and paperboard. Pulp mills manufacture pulp from wood or other materi-

paints typically consist of solutions of organic resins, organic and inorganic coloring pigments, stabilizers, and extenders. The organic solvents perform the critical role of transporting the components and giving the liquid the necessary flow characteristics, viscosity, surface tension, cross-linking between resins and pigments, and other chemical properties that aid in the consistent application of individual coating layers. The choice of organic solvent or solvents used depends on the particular requirements, including substrate composition and geometry, application equipment to be used, curing temperature and time range, and coating solids' characteristics. In this capacity, organic solvents, which evaporate relatively quickly under normal application conditions, and hence qualify as VOCs, do not make up a significant portion of the coating material after curing.

Organic Solvents Used and the Associated Pollution Problem. The fugitive evaporation during drying of the organic solvents listed in Table 6.1.1 constitutes the largest source of environmental problems linked with the coatings industry. In addition, other sources of VOC emissions contribute to the overall VOC output from the surface coating and painting process. These are coating materials which fail to bind to the substrate, wastes routinely generated during equipment cleaning and color changeovers, and the removal of coating material from defective or otherwise incorrectly coated parts. The next section will outline pollution prevention applications that address emissions from all the above sources.

Application of Pollution Prevention to the Surface Coating Process

The application of pollution prevention in the surface coating process offers the potential for significant reductions in, and in some cases elimination of, VOC content. Three primary pollution prevention avenues may be explored to decrease VOC emissions in the surface coating process. The rest of this section will provide an overview of these main approaches: substituting the coating material to reduce or eliminate completely the quantity of VOCs involved; improving the application of coating materials to minimize raw material usage; and maximizing the lifespan of virgin solvents already used. As will be discussed, these three differ in several ways, most notably in their effectiveness in abating VOCs, the degree of effort to enact them, capital costs, and modification of current practices.

All the pollution prevention approaches described and advocated here can be adopted individually or in conjunction with one another. Similar approaches to the ones described below will also be applicable in other processes. The goal now is to raise the awareness of pollution prevention and to enable facilities to customize pollution prevention aimed at reducing VOC emissions to their own needs. A key part of the pollution prevention strategy is for individual facilities to have the freedom to adopt these steps in their own particular situation.

Material Substitution. The surest way to meet the increasingly strict limits on VOC emissions imposed on surface coating facilities is to eliminate or reduce their use to the maximum degree possible. The substitution of VOCs with water-based solvents or, in the case of solid powder coatings, eliminating VOCs altogether allows manufacturers to reduce the total VOC composition in their coating materials. The emphasis on the substitution of reduced-VOC coatings has been growing for years, and numerous emerging technologies offer great potential to reduce VOC levels. Among these, powder and waterborne coatings are the best developed and have the longest successful histories. Therefore, they will form the basis for this discussion of pollution prevention through material substitution.

An Introduction to Powder Coatings. Powder-based coating technology uses specially engineered thermoplastic or thermoset dry resins for coating substrates. Numerous different categories of powder resins exist, and each possesses different hardness, durability, protection against corrosion and chemicals, and texture characteristics. The most common powder types are acrylics, epoxies, epoxy-urethane hybrids, nylon, and urethane polyesters. For each individual substrate, the appropriate powder resin must be selected to accommodate the desired coating. In general, powder coatings compare favorably with solvent-based coatings in terms of durability, protection, and quality. Please consult with vendor data and other appropriate sources of specific information to determine if powder coating will work for you.

The mechanics of powder coating technology differs greatly from that of traditional solvent-based liquid coating systems. The dry, fine coating powder is applied to the substrate using one of two dominant methods: electrostatic spray gun or fluidized-bed tank. These two application methods bind the solid coating powder to the substrate. The piece is heated, and the powder uniformly melts onto the substrate. This initiates the cross-polymerization that permanently binds the pigment to the substrate and forms the necessary high-quality finish.

Due to the need for heat curing of powder coatings, the technology is currently limited in the range of substrates to which it can be applied. The temperature to which the coating must be heated to ensure adequate coverage varies from resin to resin, with typical thermoplastic resins requiring temperatures between 325 and 400°F for between 8 and 20 min (EPA, *Manual: Pollution Prevention in the Paints and Coating Industry,* 1996). Currently, materials suitable for powder coating include aluminum, brass, galvanized steel, magnesium, most electroplated substrates, sheet steel, and other metals that can withstand elevated temperature for several minutes. Delicate substrates that cannot withstand the elevated temperatures are not currently compatible with powder coating technology. These include heat-sensitive plastics, rubber, paper, and wood. Efforts to produce resins with decreased melting temperatures promise to spread the use of the technology.

As is often the case, process modifications complement material substitutions to provide very effective pollution prevention solutions. The combined approach of powder coating substitution and one of the two application methods offers a dramatic decrease in VOC emissions and manifold cost savings.

Certainly not exclusive to powder coating systems, electrostatic spray guns are an effective pollution prevention complement to powder coating technology. Applying a powder coating with an electrostatic spray gun involves placing opposite, attractive charges between the powder coating material and the substrate. An electrode supplies a high-voltage, low-amperage current that ionizes the air surrounding the spray nozzle. As the powder particles are forced from the nozzle by air pressure, they acquire a negative electrostatic charge. The powder particles follow the electrostatic field that is formed between the ionized nozzle and the grounded substrate. Due to the need to carry an electric charge, the substrate must either be electrically conductive or be treated with a conductive primer.

Because the coating molecules are electrostatically guided to the substrate, a higher material transfer efficiency is attained than with traditional spray coating. Furthermore, the nonlinear electrostatic attraction forces the powder to wrap around corners to provide a superior coating in small niches as well as along edges. Electrostatic spraying also increases the rate of coating: only one pass is required as the coating thickness can be determined by the current applied to the powder particles. The electrostatic attraction between the coating and the substrate is maintained as the product is transferred to a curing oven, where the powder melts and permanently binds to the substrate.

The second main delivery technique for powder coating is a fluidized bed. Similar to a dip tank, the fluidized bed is an enclosed apparatus with a porous bottom grate through which dry powder resin is circulated by low-pressure compressed air. The heated substrate is placed into the container as the powder distributes evenly in a fluidlike state across its surface. As the swirling powder particles contact the substrate surface, which is at a higher temperature than the melting point of the powder resin, they instantly adhere to form a thick, even coating.

Other Considerations

- To ensure proper powder adhesion, workpieces must be preheated, often to temperatures greater than 500°F (EPA, *Manual: Pollution Prevention in the Paints and Coatings Industry,* 1996). Therefore, only substrates that can withstand the elevated temperature may use this system. Advances in powder resin technology may decrease the melting points of future resins.
- The size and geometry of the substrates to be coated must be compatible with the dimensions of both the fluidized-bed apparatus and the curing oven.
- When only partial coating of substrate surfaces is desired, the segment that is not to be coated must be masked to prevent powder adhesion. Properly masking the substrate can be tricky in some instances, and as a result, fluidized-bed coating is usually used when pieces are to be uniformly coated.
- Changing colors requires the powder reservoir as well as the powder recycling unit to be evacuated using compressed air. While relatively simple, these color changes are often time-consuming; consequently, fluidized-bed systems are most amenable to operations that require few color changes.

Pollution Prevention Benefits of Powder Coatings

- Powder coating formulations themselves contain almost no VOCs, thereby eliminating VOC problems at the source.
- The nearly 100 percent transfer efficiency maximizes the use of coating materials.
- Since only the resin particles that touch the heated substrate adhere to it, those that fail to contact the surface remain unaltered and may be easily recycled back into the process by using compressed air.
- The ease of cleaning greatly reduces the need for typically VOC-laden cleanup solvents.

Additional Benefits of Powder Coatings

- Since VOC emissions are almost entirely eliminated by powder coatings, facilities enjoy large cost savings by averting the need for expensive VOC control equipment.
- The improved transfer efficiency and the resulting savings in raw material costs reduce the unit cost of coating.
- By altering the current in electrostatic powder sprayers, experienced operators can provide a thicker coating in a single application, thus saving labor costs where thicker layers are required.

Despite the fact that powder coating materials are typically more expensive per mil-square foot than traditional solvent fluid coatings and that the one-time capital costs for powder application, powder reclamation, and high-temperature curing units are higher than for organic solvent coating systems, the three factors listed above make powder coatings a lower-cost alternative for a similar, high-quality coating.

An Introduction to Waterborne Coatings. While waterborne coating technology actually represents several different resin and delivery systems, they may be classified together by their substitution for all or a portion of the organic solvent carrier with water. As will be discussed momentarily, some coatings offer complete substitution of VOC solvents with water, as opposed to other coatings that possess a smaller, yet significant quantity of solvents.

One important factor in determining the waterborne coating's compliance with applicable VOC regulations is the prevailing method of calculating VOC content. Most regulations ignore the presence of water as a cosolvent. For the purposes of these rules, the VOC weight per gallon of coating is calculated for a concentrated mixture of VOCs and solids that would fit into a 1-gal container if the sizable water component were removed. The effect of this method of calculation is an artificial increase in the calculated VOC content. In reality, the approximate ratio between the amount of water and organic solvent for many formulations is 80:20. Although VOC contents vary with specific formulations, most waterborne coatings have VOC contents below 2.0 lb/gal (without water) (EPA, *Manual: Pollution Prevention in the Paints and Coatings Industry,* 1996).

The substitution of waterborne coatings for those carried by organic solvents is gaining in popularity. In fact, almost all types of major resin categories are currently available in a waterborne version. These include resins that fall into the following categories: acrylics, alkyds, amine solubilizers, epoxies, polyesters, styrene-butadiene acrylics, urethanes, and vinyl acetate and vinyl acrylics. Although the resin, pigment, stabilizer, and extender portions of the various waterborne coating systems vary significantly, the coating's important physical properties (viscosity, surface tension, and other physical flow characteristics) are largely determined by the type of solvent carrier, whether pure water or a mix of water and organic solvent. As is the case with organic-based coating materials, each of the different variants of waterborne coatings provides specific levels of durability, corrosion, stain and abrasion resistance, protection against temperature extremes, and appearance. Below is a simple introduction to the three main classes of waterborne coatings: water-soluble, colloidal, and latex coatings.

Water-soluble coatings are so named because the coating resins must be hydrophilic and water-soluble. Solvents may play a role in dissolving and transporting certain pigments and resins; however, water serves as the primary resin transport vehicle and determines the flow characteristics of the coating. The following coating formulations are compatible with water-soluble coating systems: acrylics, alkyds, polybutadienes, polyesters, and water-soluble oils.

Representing an intermediate between water-soluble and emulsion coating chemistries, colloidal coating resins are composed of water-soluble resin droplets dispersed in the water-solvent mixture. In colloidal coating systems, it is typical to have both water-soluble and emulsified resins present. Colloidal coating dispersions are used largely to cover porous materials such as architectural concrete and leather.

Emulsion coatings use water as the primary solvent to provide adequate flow characteristics, but use emulsifying and stabilizing agents to suspend the specific pigments and resins. Typically, pigments and resins need not be water-soluble. In general, the flow characteristics possessed by emulsified coating materials may differ significantly from those of the other two classes of waterborne coating systems.

Because the chemical characteristics of the three classes of waterborne coating systems may vary significantly according to the presence of organic solvents, emulsifiers, thickeners, stabilizers, and various compositions of resins and pigments, those interested in substituting waterborne coating systems for traditional organic-based coating systems have a large number of options from which to choose. In fact, the surface protection properties of most waterborne coating systems match the high-quality finishes provided by organic-based coating systems.

One of the major advantages of waterborne coatings is that they are compatible with the same material substrates as are organic solvent-based coatings. Thus, the substitution process does not require a major overhaul of the coating equipment; but as will be discussed in the next section, minor modifications in the application equipment will be necessary when switching from a solvent-based to a water-based system.

Application Methods. Many different application techniques exist for liquid-based coatings. These include air spray, air-assisted spray, air-assisted airless spray, manual brush or roller, dip tanks, electrostatic spray, flow coating, high-volume low-pressure (HVLP) spray, and turbocharged spray techniques. All these application techniques may be used with the different waterborne coating technologies.

However, due to the increased viscosity and thixotropy of waterborne coatings, the modification or replacement of existing application equipment may be necessary. Furthermore, precautions against water-induced corrosion must be taken. In most instances this can be overcome by inserting stainless-steel parts wherever rusting is a possibility. In some critical parts of application equipment, such as pumps, piping, and spray nozzles, this may not be feasible, and therefore replacements may be necessary. In addition, spray arresters and spray booth filters will have to be modified in order to adapt to waterborne coatings. Technical advice and in-house experimentation will be necessary to determine the best configuration for your facility.

Other Considerations. Once they have been applied, it is commonly thought that the drying times for waterborne coatings are prohibitively long in comparison to those for organic-based coatings. Environmental conditions such as humidity and temperature play a critical role in the drying process, and when these factors are properly controlled, drying times are, in some cases, shorter for waterborne coatings. This fact allows applications in which multiple layers are required to be performed in a shorter time, with an increase in production rate.

Other considerations include these:

- Some waterborne coatings still contain solvents (VOCs), although usually less than high-solid formulations.
- Prior to fully curing, waterborne coatings are sensitive to degradation due to contact with water. Once the coating has fully cured, this is not a problem, however.
- Waterborne coatings are sensitive to humidity, requiring humidity control in the application and curing areas. Low humidity can cause those coatings to dry extremely fast, resulting in craters in the final film. High humidity can cause very slow drying times, resulting in sagging.
- Emulsion water-based coatings do not easily penetrate porous substrates. Of course, this characteristic is often beneficial, but precludes their use for some substrates.
- Some resins may degrade in water, limiting the shelf life for these formulations.

Pollution Prevention Benefits of Waterborne Coatings

- Waterborne coating technologies offer significant, but not total, reductions in VOC emissions to bring facilities into compliance with VOC regulations.
- Their use is compatible with almost all prevalent application systems, although equipment modifications are sometimes necessary.
- Tap water may be used for the majority of cleanups, although small quantities of solvent will be necessary to remove non-water-soluble coating components.

Additional Benefits Attributed to Waterborne Coatings

- The costs of coating fluid preparation, application, cleanup, and disposal are similar for water-based and conventional coatings.
- Although water-based coatings typically contain some solvents, they are less likely to require costly VOC control equipment.
- The costs of storage are lower because waterborne coatings are nonhazardous and do not require storage in flameproof enclosures. The costs associated with installing ventilation systems in the workplace are also lower or, in some cases, are completely eliminated.
- Because the cost of water is less than that of organic solvents, waterborne coatings generally present an economic advantage.

Improvements in the Application of Coating Materials. Another potent application of pollution prevention is the improvement of the transfer efficiency of surface coating materials. Defined by the ratio of the quantity of coating that actually adheres to the surface to the total quantity expended, transfer efficiency gains conserve resources, while reducing VOC and other hazardous and nonhazardous emissions. Obviously increases in transfer efficiency do not preclude other pollution prevention measures.

At the heart of pollution prevention approaches to increasing transfer efficiency is the desire to coat an area with less waste, thereby reducing the need for virgin materials. As will be detailed further in this section, these improvements can be achieved by the following actions: upgrading or modifying application equipment for optimum performance, improved worker training in proper application techniques, and changes in the application environment.

Of course, efficiency improvements certainly can be enjoyed by all users of the surface coating processes, but perhaps the greatest appeal is for facilities for which other VOC control and pollution prevention measures are neither technically practical nor economically feasible. This is so because the pollution prevention steps outlined in this section are implementable for almost all current coating technologies in a relatively short time frame.

The manifold benefits of improving transfer efficiency include

- A reduction in VOC emissions
- Savings in raw material purchases
- A decrease in hazardous and nonhazardous waste generation and disposal costs
- Increased profitability

Relatively large gains in transfer efficiency can be reaped by relatively small alterations in worker practices. The suggestions below form the basis for a training program that, for minimal cost, can reduce VOC emissions and material costs while increasing profitability.

When spray coating equipment is used, a certain portion of the coating material is wasted after it is atomized and is transmitted from the spray gun to the substrate surface. This atomization creates an inverse relationship between the transfer efficiency and the distance from the gun to the substrate. By moving the gun closer to the target, effects of air currents and overspray are diminished. The spray pattern may need to be altered to maintain adequate spray coverage. However, unless the spray ejection pressure is decreased, increased bounce-off of coating from the target and excess coating application can occur. Experimentation can be used to determine the appropriate spray pattern and spray velocity to compensate for the decreased separation.

Matching the nozzle pattern on the spray gun to the geometry of the workpiece can greatly increase the transfer efficiency. Projecting a wide, fan spray field onto a narrow, elongated object such as tubing minimizes coating overspray.

Also, in situations where a narrow strip remains after previous coating passes have covered the majority of the surface, altering the spray pattern will increase transfer efficiency. Rather than overlap the previously coated area, operators often project, into the air, the portion of the spray field that does not cover the uncoated remainder. By reducing the spray pattern size so that adequate coverage and overlap are maintained, further increases in transfer efficiency can be attained.

In a similar effort to minimize coating overspray, operators of spray guns should be taught to align the initiation and termination of the spray stream as closely as possible with the leading and trailing edges of the workpiece. This pollution prevention technique can be applied to both manual and automated spray guns. For finishes that demand a high-quality appearance, sufficient leading and trailing edges will be necessary. However, for less exacting applications, this technique can contribute to a significant increase in transfer efficiency.

Changing the angle between the spray gun and the surface can reduce the tendency of the coating to bounce off the surface. Rather than maintain a low angle to the substrate surface, spray gun operators must maintain the spray field perpendicular to the surface. Minimizing bounce-back may be more difficult when the surface geometry is concave or convex, but experimentation can help increase the transfer efficiency.

Scheduling of paint spray jobs from light to dark colors can reduce wastes generated in equipment cleanup. Scheduling to permit the end-of-day cleanup to coincide with the end of a job can eliminate a cleanup procedure and eliminate the waste from one cleanup.

Additional Waste Minimization Strategies for the Surface Coating Process

- Increase precoating inspection of workpieces so paint operators can minimize unacceptable quality and paint waste. Not coating an otherwise defective workpiece avoids the need to strip and recoat the piece later.
- Mix coating batches according to specific job or jobs. This avoids producing unnecessary coating material that is likely to be wasted.
- Ensure that all filters are intact and fit properly. This prevents leaks and other sources of contamination, prolongs equipment life, reduces the need for equipment cleaning, and minimizes coating waste.
- Routinely clean production equipment, such as racks, that comes in contact with workpieces to reduce possible contamination.
- Remove coating solids from reclaimed coating material using filtration, centrifugation, or decantation techniques. Include collected solvent-coating mixtures in the next compatible coating formulation to reduce the need for virgin material. This is most applicable in applications where some solvent contamination can be tolerated.
- Utilize a computerized inventory control and tracking system to avoid overpurchasing of coating materials already in stock. An improved inventory system can assist you in taking advantage of bulk ordering.

A number of storage and transfer practices should be practiced to minimize waste, including these:

- Regularly inspect storage drums for leaks.

- Minimize contamination and the chance of accidents by storing drums in secure areas that see little traffic.
- Use metered material delivery spigots and/or pumps to automatically dispense new materials, thereby avoiding leaks and spills. Similarly, maintain dedicated funnels and other transfer equipment near storage drums.
- Install tight-fitting lids on all storage vessels to reduce evaporation.

Solvent Cleaning and Degreasing

Solvent cleaning and degreasing is the process of using an organic solvent to remove unwanted grease, oils, and other organic films from surfaces. Solvents are used for cleaning, degreasing, and rinsing in vehicle maintenance operations; in the manufacture of aircraft, motor vehicles, metal furniture, tools, appliances, paint, ink, organic and agricultural chemicals, adhesives, petrochemicals, plastics, rubber, vegetable oil, textiles, and other products; and in photographic processing, coating, printing, and dry cleaning.

In metal finishing industries, as an example, cleaning usually follows machining and precedes other surface-finishing steps such as rust inhibition or electroplating. During machining, a variety of oil emulsions or synthetic fluids are sprayed on the workpiece for lubrication or cooling. These fluids must be removed before final finishing. The slightest amount of oil or other residue can render the final surface finish ineffective. Pollutants generated from degreasing include liquid solvents containing unwanted film material and air emissions containing volatile solvents.

Organic Solvents Used in the Cleaning and Degreasing Process. Vapor degreasing and other solvent cleaning frequently require the use of chlorinated solvents. Chlorinated solvents were adopted by industry because of their ability to dissolve a wide range of organic contaminants, their low flammability, and their high vapor pressure, which allow them to evaporate at room temperature and leave a residue-free surface. Organic, chlorinated solvents' large role in the cleaning and degreasing process is due largely to the very effective cleaning results achieved by the solvents. Prior to 1995, popular chlorinated solvents included 1,1,2-trichloro-1,2,2-trifluoroethane (CFC-113); 1,1,1-trichloroethane (TCA, also called methyl chloroform); trichloroethylene (TCE); tetrachloroethylene (also called perchloroethylene, or Perc); and dichloromethane (or Meth for methylene chloride).

Concerns about the toxicity and environmental effects (ozone depletion in the upper atmosphere) of continued use of chlorinated solvents have led chemical companies to search for replacement solvents. For these reasons, pollution prevention methods focusing on reducing or eliminating the use of these solvents are high priorities.

Vapor Degreasing Description. Many industries use vapor degreasing. It involves using nonflammable, chlorinated hydrocarbons in a vapor phase to clean metallic and other suitable surfaces. The special apparatus, the degreaser, consists of a tank that is one-tenth full of solvent. The solvent is heated to its boiling point, usually by steam coils, producing solvent-saturated vapor in the upper portion of the tank. The item to be cleaned is inserted either manually or automatically into the vapor region. Hot solvent vapor immediately condenses onto the surface of the item. The condensed solvent then drips back into the liquid bath and takes with it the removed dirt and grease. Solvent vapor is prevented from escaping to the atmosphere by use of a refrigerated section on the upper part of the tank. The solvent condenses on the walls of this section and returns to the liquid sump.

Cold cleaning and spot cleaning are two other solvent cleaning techniques. In cold cleaning, the part to be cleaned is immersed in a tank of solvent and typically agitated mechanically or ultrasonically. In this instance, the cleaning performance degrades with use because the solvent bath becomes loaded with contaminants—a distinct disadvantage compared to vapor degreasing, because solvent vapors remain relatively pure. The typical method of spot cleaning is to use a lintfree cloth saturated in solvent to clean a specific area or a part or workpiece, often as a part of a repair operation.

Applications of Pollution Prevention in the Degreasing Process. When attempting to reduce the use of hazardous cleaning solvents, a facility should first fully understand the cleaning process being practiced. Does the workpiece in question need to be cleaned with a chlorinated solvent, or does it even need to be cleaned at all? Can modifications be made to a process prior to the cleaning step that could obviate the cleaning step altogether? Reevaluating the need for extensive, and possibly unnecessary, cleaning of noncritical parts can serve as an important first step in a successful pollution prevention program. Once it is established that cleaning is unavoidable, alternative cleaning methods need to be considered.

Below are several pollution prevention alternatives that may be used in your chlorinated solvent cleaning and degreasing process. The options—aqueous substitution for organic solvents, ultrasonic agitation, solvent recycling and reuse, and modifications to existing vapor degreaser units—are listed roughly in their order of impact on reducing VOC emissions. They represent different levels of process reorganization with respect to pollution prevention and may be enacted individually or in conjunction with one or more other techniques.

Aqueous Cleaning. Cleaning using water-based solutions and various detergents (*aqueous cleaning*) is an alternative to conventional chlorinated solvent cleaning. The reduction of VOC use and other environmentally friendly aspects of aqueous cleaning make it an increasingly popular alternative to cleaning with organic solvents. Aqueous cleaning systems can include alkaline solutions and detergents to enhance their soil removal capabilities or consist solely of hot water washes.

Water-based cleaners have few or no volatile components, which means cleaning must take place in the liquid phase. Therefore, immersion tanks are most commonly used in these applications. These systems are often used with pressurized sprays, agitation, heat treatment, or some other physical process to further provide effective cleaning. Agitation can be accomplished either with ultrasonics or by mechanically rotating the part and circulating the system.

An aqueous system usually requires many tanks (often more floor space than vapor degreasers) and the addition of rinsing and drying steps after cleaning to accomplish results comparable to those of solvent degreasing. In a common aqueous cleaning operation, parts travel through a series of chambers, each with different concentrations of cleaning and finishing steps. Most systems are *closed-loop* systems, whereby the used solutions are not disposed of daily, but can be recirculated for a week of relatively continuous use. This minimizes what otherwise could be a potentially large wastewater problem.

Unlike some chlorinated solvents, which effectively clean almost any surface, the aqueous cleaning system chemistry needs to be specifically tailored for particular cleaning surfaces. For this reason, converting to an aqueous system can be expensive in the short term, but cost-effective in the long run. In some cases, cleaning suppliers will formulate cleaners for a client's particular soils. The combination of additives chosen for an aqueous cleaning system alters the foaming, wetting, and soil removal properties of the solution. Different additives include synthetic detergents and organic surfactants, saponifiers, acids and alkalines, and corrosion inhibitors. Some alkaline cleaners, due to their high pH, need to be tested for potential corrosion characteristics prior to use on sensitive materials. A rinsing step is required to remove any residual films left by these additives in the cleaning process.

Pollution Prevention Benefits of Aqueous Solvents

- Switching to aqueous solvents greatly reduces or eliminates the need for organic, chlorinated solvents.
- Aqueous cleaning units offer comparable cleaning effectiveness when compared with solvent degreasers.

Ultrasonic Cleaning. Ultrasonic cleaning uses high-frequency sound waves in an immersion tank to improve the efficiency of aqueous and semiaqueous cleaners. By generating zones of high and low pressures in the liquid, the sound waves create microscopic vacuum bubbles that implode as the sound wave moves and the zone changes from negative to positive pressure. This process, called *cavitation,* exerts enormous localized pressures (approximately 10,000 lb/in^2) and temperatures (approximately 20,000°F) on a microscopic scale that loosen contaminants and actually scrub the workpiece. Ultrasonic energy is applied to a solution by means of a transducer that converts electric into mechanical energy. The use of ultrasonics as part of an aqueous system can be particularly effective at reaching into crevices and small holes.

Ultrasonic cleaning technology is compatible with organic solvents and semiaqueous and aqueous solutions. Typically, aqueous systems are used in conjunction with ultrasonic agitation, because the high effectiveness decreases the need for powerful VOC-laden solvents.

In general, ultrasonic cleaning can be used with almost any substrate, including numerous substrates which are otherwise more difficult to clean, such as plastic, glass, and ceramics. Obviously, workpieces must be submersible in liquid solutions.

Workpieces must be placed carefully within the ultrasonic cleaners in order to clean them effectively. Contact between substrate surfaces and the liquid medium is extremely significant. Workpieces must be placed so that receded holes and niches are placed upright so that the liquid medium fills the inner surfaces. Otherwise, ultrasonic cavitation will not reach those surfaces to scour the grease and dirt.

Successful cleaning also requires that enough ultrasonic energy reach each portion of the substrate to be cleaned. The cleaning of small parts within baskets may present a problem. When the baskets are overloaded, the outermost layers of workpieces are likely to absorb the majority of the ultrasonic energy. Therefore, it is extremely important that baskets and racks not be so full that they hinder the cleaning of the interior pieces.

Pollution Prevention Benefits of Ultrasonic Cleaning

- Ultrasonic agitation is compatible with organic, semiaqueous, and aqueous cleaning technologies.
- Due to the increased effectiveness of ultrasonic agitation, a decreased amount of cleaning solvents (organic or aqueous) is needed to achieve a comparable level of cleanliness.
- Use of ultrasonic agitation may increase workpiece processing speeds.

Reusing and Recycling Solvents. The fact that large quantities of solvents are used, but not consumed, during the cleaning and degreasing process creates an opportunity for reusing and recycling already present materials and thereby minimizing virgin material use.

Solvents can be reclaimed by recycling and distillation. Frequently, high-quality solvents are used once for high-precision cleaning and are then disposed of, either on a scheduled basis or because they do not meet their original specifications. These solvents can be reused for applications that do not require such purity. Cascading used solvents downward for use in applications that require a less stringent level of purity can greatly extend the bath life of a solvent. This pollution prevention application is commonly referred to as *countercurrent rinsing* and is perfectly suited to aqueous cleaning and degreasing. Also suitable for many other dipping or immersion processes, it requires that contaminated bath solutions be rotated in the direction opposite to the movement of the parts to be cleaned. This ensures that during initial cleaning the part encounters the least clean bath and during the final rinse it encounters a virgin cleaning solution.

Solvents may be purified by distillation so that they can be reused for their original purpose. It involves heating a solvent until it vaporizes, then condensing the vapor, which is reused. Aside from purchasing a recycling system outright, on-site recycling services are available in many areas. Off-site recycling is also an option; however, on-site distillation equipment is economically favored over commercial off-site recovery. Off-site recovery incurs costs such as transportation, the risk of spills, and the potential liability for improper disposal of still bottoms. Commercial systems today are reasonably efficient and recover 70 to 90 percent of solvent input.

Pollution Prevention Benefits of Recycling

- Solvent quantities are used over and again, thereby reducing the need for virgin materials.
- Limiting the amount of hazardous substances that need to be either disposed or destroyed decreases legal liability, overall costs, and human health exposure.

Vapor Degreaser Modifications. Conventional open-top vapor degreasers allow for quick and thorough cleaning, but produce significant air emissions and solvent losses. Emissions from vapor degreasers are specifically due to workpiece entry and removal, dragout, idling, and degreaser shutdown. The following tips can cut down emissions from traditional open-top vapor degreasers:

- Be sure to place a cover on the tank opening during idling and shutdown. A rolling, retractable cover reduces air turbulence, which may dissipate vapor during use.
- Increase the freeboard height (the height of the tank walls above the vapor layer) to better contain the solvent vapors. Increasing the ratio between the freeboard height and the width of the vapor degreaser—the freeboard ratio—from 0.75 to 1.0 can reduce air emissions by up to 20 percent. Increasing the freeboard ratio from 1.0 to 1.25 can reduce air emissions a further 5 to 10 percent (EPA, *Guide to Cleaner Technologies: Organic Coating Replacements,* 1994).
- Convert water-cooled coils to refrigerated coils to better contain solvent vapors within the degreaser by lowering the temperature of the zone around the freeboard. Emissions that occur during degreasing operation can be reduced by 20 to 50 percent for above-freezing coils and by 30 to 80 percent for below-freezing coils (EPA, *Guide to Cleaner Technologies: Organic Coating Replacements,* 1994).
- Move workpieces slowly and steadily into and out of the degreaser to avoid turbulence that can disperse solvents.
- Retain workpieces over the tank to cut down on dragout while removing workpieces.
- Rack workpieces so that the longest axis is parallel to the floor, thereby minimizing dragout.

- Minimize air turbulence around the degreaser unit in order to avoid removing vapor from the degreaser.

Pollution Prevention Benefits of Modifying Current Vapor Degreasers

- Modifying existing equipment reduces worker exposure to potentially harmful vapors.
- Efficiently using already existing inventories of organic solvents or phased-out solvents reduces the need to purchase new stock.
- Modifying existing equipment is typically less expensive than purchasing new equipment and also may result in faster compliance with VOC regulations.

INDUSTRY PROFILES

Textiles

Markets for textiles are most frequently divided into apparel (clothing), domestic (household items, such as towels and bed coverings), industrial (automotive components, such as tire cords, filters, and carpets), and recreational equipment (such as backpacks and parachutes). Of the total textile products, cotton fiber has the highest demand, and the apparel market accounts for the majority of consumption.

The textile industry includes fiber, fabric and apparel production, as well as retail sales. Generally, production includes the following steps:

1. Generation of a usable fiber from either a natural source (i.e., cotton, wool, and silk) or a manufactured source (polyester, rayon, and nylon)
2. Production of yarn for knitted and woven fabrics
3. Construction of product through knitting, weaving, and nonwoven methods
4. Preparation, dyeing, printing, and finishing to produce finished fabric
5. Fabrication to produce final product (i.e., garment, household item, industrial, or specialty product)

Raw Materials Used. Textile fabrics are made from a wide range of products, including a wide variety of fiber types. Table 6.1.2 lists some different fibers and their sources.

The number, amounts, and types of chemical raw materials used in textile facilities vary. Many chemical raw materials are "specialty" chemicals of a proprietary composition. The actual amount used depends on some very mill-specific factors, including product type, quality goals, and equipment. Some of the more common types of processing assistants are listed in Table 6.1.3.

Various Textile Processes. Because there is such a diverse product and application range of textiles, the type of processing used is highly variable. Textile processing depends on site-specific manufacturing practices, type of fiber used, and final physical and chemical properties desired (such as tensile strength, flexibility, uniformity, and luster). Wet processing varies depending on whether the fabric is cotton, wool, synthetic, or a cotton-synthetic fiber blend. Natural fibers receive more processing than manufactured ones to achieve the same product performance. Even for a constant product type, no two textile mills use exactly the same methods of production.

TABLE 6.1.2 Fibers and Their Sources

Fiber type	Resource
Natural	
Cellulosic	Cotton, flax (linen)
Animal hair	Wool, mohair, angora, camel's hair, musk ox, cashmere
Silk	"Sericulture": cultivation of *Bombyx mori* silkworm
Synthetic	
Rayon	Wood pulp
Acetate	Purified cellulose from wood pulp + acetic acid + catalyst
Nylon	Polyamide production from various monomers
Polyester	Dicarboxylic acid + dihydric alcohol
Olefin	Polymerization of ethylene; polypropylene
Acrylic	Polymerization of acrylonitrile
Spandex	Polyester

Source: Freeman (1995).

TABLE 6.1.3 Processing Assistants in Textile Processes

Preparation	Finishing
Wetting	Antimigrant
Scouring (solvent-free)	Resin
Desizing auxiliary	Catalyst
Chelates and sequesterants	Softener
Emulsifiers	Builder
Dyeing	
Dispersant	Softener
Compatibilizer	Leveler
Carrier	Lubricant
Buffer	Emulsifier
Acid	Wetters
Alkali	Retarder
Antimigrant	Accelerant
Defoamer	Fixative

The common systems used for processes include batch, continuous, and semicontinuous equipment setups. In the pad batch method, prepared fabric is impregnated with liquor containing premixed fiber-reactive dyestuff and alkali. Excess liquid is squeezed out on the mangle. Then the fabric is batched onto rolls or in boxes, is covered with plastic film to prevent carbon dioxide absorption from air and evaporation, and is stored for 2 to 12 h. After the reaction is finished, the substrate is removed, and chemical solutions are discharged. Any subsequent auxiliary processes (rinsing, washing, etc.) occur in the same vessel.

In continuous systems, chemical solution is placed in the machine, and fabric is moved through it without interruption. Because batch systems inherently tend to generate more waste than continuous systems, batch systems are generally used for treating (e.g., dyeing) small quantities of a particular type of product, while continuous systems are more amenable to processing large yardages.

Preparation processes include several steps. The minimum aqueous processing for cotton-containing, woven-fabric preparation prior to dyeing includes desizing, scouring, bleaching, and mercerizing. Desizing is done by the pad batch method, which consists of applying enzyme from a standard padder, batching on rolls at room temperature, then washing off. Since the desizing process is usually done continuously, waste streams can be easily neutralized by the addition of acid as produced.

Scouring and bleaching are done on large continuous ranges. Scouring and bleaching of synthetics should be done using minimal amounts of alkali; however, scouring of cotton requires much larger amounts of alkali. Typical concentrations of alkali for cotton scouring are 2 to 4 percent of caustic (50 percent) based on the weight of the fabric processed. When cotton scouring or bleaching is done in a continuous process unit, there is potential for neutralizing or reuse of waste streams.

After preparation, color is applied to fabric through dyeing or printing. Stripping, or the removal of colorants from textile substrates, is used in essentially all commercial dyeing operations for repair purposes. Basic operations include

Preparation

- Desizing
- Scouring
- Bleaching
- Mercerizing

Dyeing

- Beck, jig, beam, jet, and continuous dyeing equipment
- A multitude of chemical commodities and specialties

Finishing

- Chemical and mechanical processes
- Resinous cross-linking finishes
- Nonresinous, nonreactive materials (e.g., softeners)

Environmental Impacts of Textile Processes. Traditionally, the textile industry is very energy-, water-, and chemical-intensive. Within the industry, approximately 60 percent of the total energy consumed is within the wet processing stages. Most wet processing involves treatment with chemical baths, which often require washing, rinsing, and drying between key treatment steps. Consequently, large volumes of wastewater are generated with a diverse range of contaminants that must be treated prior to disposal. Much energy is consumed to heat and cool chemical baths and washwater and to dry fabrics or yarns. Additionally, there are a few problems of air pollution by chemicals, lint, etc., but these are minor.

When the complete textile product life cycle is considered, it is apparent that chemical suppliers (for fibers, dyes, and auxiliaries) have a tremendous influence on the environmental impact of textile manufacturing. In fact, there is increasing pressure from the textile mills on their suppliers to improve product performance via increased exhaustion and efficiency of dyes, less toxic and fewer auxiliary chemicals needed, and low-volatility and permanent finishes.

The greatest potential source of alkali in preparation is the process of mercerization of cotton. This uses caustic concentrations of about 20 percent. Dyeing processes vary in acidity and alkalinity depending on the substrate and dye class.

Wastewater. Wastes in textile mill effluents can be divided into the following categories:

1. Naturally occurring dirts, salts, oils, and greases on cotton and wool
2. Chemicals added or removed during the many different process operations
3. Fibers removed by chemical and mechanical action during processing

One of the primary environmental concerns now under study is the discharge of toxic materials from point sources. Types of materials that may be expected to predominate in textile wastewater include

1. Metals
2. Nondegradable surfactants
3. Toxic organic materials, such as phenols, aromatic solvents, methylene chloride, perchloroethylene (Perc), and oxalic acid

Air Emissions. Air emissions emanate from several places in typical textile operations:

1. Hot air driers
2. Dyeing machines
3. Storage tanks
4. Warehouse area
5. Fugitive (general ventilation)

Air emissions from finishing in high-temperature drying and curing ovens contain varying amounts of any volatile residues left in fabric from prior processing. Two materials are almost certainly present in all textile operations: acetic acid and formaldehyde. Acetic acid could be emitted from bulk storage tanks as well as from dyeing machines and driers. This would occur through vents during filling and due to breathing losses. Formaldehyde may be emitted from bulk resin storage tanks, finished-fabric warehouses, driers, and curing ovens. Table 6.1.4 lists additional chemicals which may be present in specific textile operations.

TABLE 6.1.4 Chemicals Used in Textile Operations

Chemical	Potential source
Chlorine	Shop, water treatment
Ethylene oxide	Drier stacks (wetting agents)
Hydrochloric acid	Drier stacks (catalyst)
Methylene chloride	Shop, paint stripper
Perchloroethylene	Dry cleaner, scour, carrier
Toluene	Becks, driers (carrier/scour)
Xylene	Becks, driers (carrier/scour)
Ammonia	Shop, storage tank
Tetrachloroethane	Shop, inspection (spot remover)
Trichloroethylene	Shop, inspection (spot remover)

Some of the solvents used in the textile manufacturing which are of concern include

- Perchloroethylene (or tetrachloroethylene)
- Trichloroethylene (TCE)
- Methylene chloride (or dichloromethane)
- 1,1,1-Trichloroethane
- Toluene (or toluol)
- Benzene (or benzol)
- Xylene (or xylol)
- Ethylene dichloride
- Varsol (or naphtha)

These solvents are typically used in the drycleaning of 100 percent synthetic fiber knit fabrics, woven and wool fabrics; in specialty operations such as tricot and solvent scouring; in dyeing operations; and in some finishing operations for impregnation or coating of textile fibers. In addition, solvents are used to clean machinery, such as rollers and spinning machines, used in textile manufacturing.

General Approaches to Textile Waste Reduction. Some ideas which have been used commercially for pollution source management are listed in Table 6.1.5.

TABLE 6.1.5 Pollution Source Management in the Textile Industry

Preparation	General
Recovery systems	Waste characterization
Waste stream reuse	Raw materials
Chemical substitutions	The fate of processing chemicals
Alternative processes	Equalization
Finishing	**Dyeing**
Reuse	Reconstitution or reuse of dye baths
Substitution	Chemical substitutions
Alternative processes	Alternative processes

Pollution Prevention Opportunities. Substantial reductions in textile facility waste can be realized through good operating practices, without major investments in new technology. Nonprocess strategies include activities such as improved equipment maintenance, valves, pumps, material handling, and production scheduling.

The primary focus for achieving pollution prevention in textile manufacturing is on reductions in water, chemicals, and energy use and the utilization of less hazardous, more efficient, and more recoverable compounds. These goals are attained by material substitution, process changes, and reuse strategies.

Long-term solutions to waste generation may involve fiber-specific product reformulation, such as increased reactivity and resistance to abrasion during weaving. However, it is unlikely that drastic changes in the fiber types will occur.

als, such as rags, linters, wastepaper, and straw. Paper mills may manufacture pulp and paper, as long as the primary product is paper. Paperboard is made from wood and other fiber pulp or converted paperboard products.

Secondary fibers are recycled material from previous paper products and have almost become the industry standard. Low-quality fibers may be used in corrugated papers while high-end fibers are used primarily as newsprint. Ultimately, the level of deinking, washing, and bleaching is determined by the desired final product.

Wood Yard Operations. The pulp and paper industry is dominated by the Kraft process. It begins with wood, which may be in the form of logs, chips, or sawmill wastes and secondary fiber. Debarking of logs makes the largest amount of solid waste, which can be burned for power generation. Once the logs are chipped, chips are screened to remove large, nonuniform pieces and are stored in large piles ready for digestion. Waste products from screening may be dirt, trash, and fine particles, which are burned for their energy value.

Pulping. Pulping can be achieved chemically, thermally, mechanically, or by a combination of the three methods. Wood consists of many different materials, which can be classified into four major categories: cellulose, hemicellulose, lignin, and extractives. Lignin is the chemical which acts as a binding agent for wood fibers. The pulping process removes the lignin and extractives, freeing up the fibrous material for paper production.

Wood chips are mixed with white liquor, by batch or continuously, to promote digestion. The digestion process is typically carried out by heating to 350°F and adding sodium hydroxide and sodium sulfide. The process yields two materials, a dark brown pulp and a black liquor, which are carefully separated through a process known as *brownstock washing.* After being washed, the pulp is screened, thickened, and washed.

Deknotting, Washing, and Screening. At this point, the fate of the pulp is determined by the desired final product. Clean, bleachable pulps are deknotted before washing, a process which involves removal of uncooked chips, overthick chips, and irregularly sized pieces by way of vibrating screens or completely enclosed, pressure-vibrated screens. Finally, the pulp is washed again and screened through a perforated barrier for acceptable fibers. The waste may be sent back into the digester or used as boiler fuel.

Chemical Recovery. The separation of the spent black liquor is the target of many pollution prevention efforts. The solution contains the initial chemicals, by-products, and organic material which can be recycled back into the process or used to generate heat for the pulping step.

The solution resulting from the brownstock washing, a weak black liquor, is concentrated to 60 to 80 percent solids and burned, forming an inorganic smelt. The smelt consists of sodium carbonate and sodium sulfide and is dissolved in water to produce green liquor. The liquor is then reacted with quicklime (CaO) to convert the sodium carbonate (Na_2CO_3) to sodium hydroxide (NaOH), which can be used in white liquor. Finally, the lime mud is burned to recover the quicklime.

During the various process steps, air and other noncondensible gases are released from the cooking tanks, boilers, and other process units. Turpentine is recovered from digestion vapors and sodium soaps converted from fatty and resin acids, alcohols, and other compounds which may be sold for profit.

Pulp Bleaching. The accepted measurement of the lignin content is the kappa (K) number. Softwoods contain greater amounts of lignin, when compared to hardwood, which directly affects system conditions. The brightness or whiteness of paper is measured by its ability to reflect monochromatic light. Unbleached Kraft pulp typically ranges from 15 to 30 in brightness, whereas mechanical bisulfite and sulfite (unbleached) pulps range from 50 to 65. The source of the brightness is the bleaching of the pulp with several chemicals or a combination of chemicals. The bleaching steps, chlorination, alkaline extraction, chlorine dioxide, oxygen, hypochlorite, peroxide, and ozone all remove the residual lignin from the pulp. These steps produce a significant portion of the total hazardous waste generated, releasing toxins to the air, water, and land. Much research has been done on the bleaching processes, focusing on optimizing parameters such as temperature, pressure, and residence time.

Pulp Drying and Papermaking. In order to lower transportation costs for off-site use, the pulp is first denatured (90 percent). Pulp drying methods include air float drying, flash drying, and paper machine adopted techniques. The water may be removed using circulating heated air, vaporization in a hot gas stream or vacuum, pressure, and heating. Surface sizing, coatings, and other treatments may be added to the product at this point.

Wastewater Treatment. Large volumes of wastewater are generated as a result of the pulping, bleaching, and associated processes. The water contains dissolved lignins, suspended solids, adsorbable organic halides (AOXs) and several other organic compounds. There are several well-known recycling techniques which have dramatically decreased the need for freshwater and loads on waste treatment facilities; they will be discussed in detail below.

Gaseous Emissions. Table 6.1.7 is a list of point and fugitive emission sources followed by emission control technologies.

Operational Modifications. There are several different operational modifications which can be made to improve both the plant efficiency and product quality. First, the quality of the chip size can greatly affect the digestion process. If the chips are too thick or large, Kraft liquor will not fully penetrate the particle. Introducing thickness and size screens and maintaining chipper machinery will increase the uniformity of chips, reducing bleach consumption and associated water and air emissions. Oversized material can be rechipped and recycled back into the process. Fewer chip rejects from the digestion process will lower emissions from boilers, which contribute 60 to 80 percent of total emissions.

Dry debarking methods will reduce water consumption and wastewater loads. Wet debarking utilizes 2000 to 3000 gal of water per ton of wood debarked, contributing to increased BOD and total suspended solids (TSS). Wastes generated from dry debarking operations can be used directly in energy recovery boilers, and do not require dewatering unit operations.

Diking chip piles and processing areas prevents chips and dust from entering runoff water and ultimately stormwater. The benefits of this practice include the avoidance of contaminated surface water and the need for separate stormwater permits.

Pulping and Chemical Recovery. Many pollution prevention methods can improve performance in the pulping and chemical recovery processes while reducing emissions. Extended delignification will remove more lignin and reduce the demand on bleaching processes. The use of polysulfides, known as polysulfide cooking, has shown significant benefits when compared to typical Kraft cooking. Pretreatment of pulp with nitrogen dioxide (NO_2) allows up to 80 percent lignin removal in the oxygen stage, which has normal limits of 50 percent. Demethylation is another pretreatment which enhances the

TABLE 6.1.7 Emission Sources and Control Technologies in the Paper and Pulp Industry

Wood preparation
Wood/chip pile; debarking; chipping
Pulping
Blow and foam tank vents; washer, decker, and knotter hood vent gases; filtrate tank vent gases; screening; fugitive emissions from washers, deckers, screens, and knotters; wastewater to open sewers
Chemical recovery
Evaporator vent gases; evaporator condensate to sewer; black liquor oxidation off gases; recovery furnace exhaust; smelt dissolving tank vent gases; causticizing tank emissions; tall oil/turpentine recovery; wastewater to sewer; turpentine off gases; tall oil acidulation off gases; lime kiln exhaust
Bleaching
Bleaching tower vent gases; bleaching stage wastewater effluent to sewer; bleaching washers; fugitives from washers; bleach filtrate tank vent (seal tank)
Papermaking
White-water emissions; mixing tank for additives; paper machine (Fourdrinier) table, press, and drier
Deinking
Pulper
Wastewater
Junction of combined acid/caustic bleach wastewater streams
Miscellaneous
Bleach scrubber vent gases; storage tanks; sludge drying and landfilling; incineration of sludge; power boiler; equipment leaks

delignification process by providing more active sites on the pulp. The reagents used are potassium thiophenoxide in diethylene glycol at 200°C.

Closed screen rooms improve the efficiency of the recovery process and lower the demand for chemicals in white liquor. Maintaining equipment such as pumps, valves, and tanks can reduce fugitive emissions as well as curb spills and leaks. Improved brownstock washing leads to increased chemical recovery and reduced VOC and odorous emissions. The use of appropriate defoamers and pitch dispersants can reduce the formation of chlorinated compounds and contaminant loading in wastewater sludges.

Steam stripping of foul condensates can greatly reduce VOC emissions while lowering BOD loads downstream. Stripper overheads can then be burned for energy recovery, and the bottoms can be recycled in brownstock washing. Installing controls for high-volume low-concentration and low-volume low-concentration vents can allow collection of emissions which can be used for their heating value. Conversion of recovery boilers to low-odor designs can improve boiler efficiency and reduce emissions by more than 90 percent. Weak washwater from the causticizing system can be recycled as a scrubber solution for smelt-dissolving-tank emissions, bleach plant and chlorine dioxide plant vents, lime slakers, and lime kilns. Installation of a lime mud diversion basin would improve resource recovery, reduce wastewater loading and treatment cost, and reduce sludge accumulation. Oxygen delignification results in bleach plant chemical savings and reduced wastewater contamination, i.e., reduced BOD, chlorinated compounds, AOX, and color. Anthraquinone can be used as an additive to the digesters to accelerate the pulping process and increase pulp yield.

There are a host of other large and small process or chemical modifications that can be employed to enhance efficiency and product quality.

Pollution Prevention in Pulp Bleaching. Pulp bleaching processes contribute significant amounts of hazardous waste to the environment as dioxins, furans, and chlorinated compounds. The following methods can be employed to promote pollution prevention. Special instrumentation can be used to detect and control chemical inputs in the first bleaching stage, which would control the kappa factor within tight constraints, minimizing chlorinated emissions. High shear mixing equipment can be used to promote uniform solutions and avoid localized overchlorination. This method will maximize bleaching efficiency and minimize wastewater and atmospheric loads. Furthermore, multiple additions of chlorine will promote chemical application and minimize overchlorination. Chlorinated compound formation can be dramatically lowered by substituting chlorine dioxide for elemental chlorine in the first bleaching stage, while maintaining a kappa factor of less than 0.15 through careful control.

Lignin extraction can be enhanced through the introduction of oxygen alone or in combination with peroxides or hypochlorites. This method can be used at relatively low cost while at the same time garnering the advantages of reduced BOD, chemical oxygen demand (COD), and color in the effluent. Hypochlorite can be replaced by chlorine dioxide in the bleaching process and is most effective when high-grade chlorine dioxide is used, which leads to reduced formation of chloroform.

The use of ozone as a bleaching agent, in place of chlorine, followed by oxygen delignification, eliminates the formation of chlorinated compounds. In addition, wastewater loadings of AOX, BOD, and color are reduced as a result of the substitution. Lastly, countercurrent, jump-stage washing decreases water demands, increases energy savings, and reduces treatment facility waste loads.

Pollution Prevention in Pulp Drying and Papermaking. A number of pollution prevention options apply to pulp drying and papermaking processes. Fibers can be recovered and recycled from white water or brown water by using settling tank, drum, flotation, or polydisk save-alls with an efficiency of over 90 percent. The water, depending on its quality with regard to TSS, can be recycled back into the main process and several side systems. As a result, freshwater consumption is reduced along with loads on the waste treatment plant. Water and energy demands can be further reduced by the use of cascade systems or recirculating water systems for the operation of water ring vacuum pumps. Also, the steam condensates from pulp driers and paper machines can be collected and recovered. Finally, some chemical substitutions can be made, including sizing agents, resins, fillers, dyes or pigments, flocculants, and brighteners.

Printing and Publishing

The printing and publishing industry (SIC Code 27) is defined to include establishments engaged in printing by one or more of the common processes, such as letterpress, lithography, gravure, or screen, and those establishments which perform services for the printing trade, such as bookbinding, typesetting, engraving, photoengraving, and electrotyping. This major group also includes establishments engaged in publishing newspapers, books, and periodicals, regardless of whether they do their own printing.

Printing Processes. The five most common printing processes, in order of their market share, are lithography, gravure, flexography, letterpress, and screen. Presses are also categorized by the form of paper or other substrate used.

1. *Lithography.* Used for printing books, posters, greeting cards, labels, packaging, advertising flyers and brochures, periodicals, and reproducing artwork.
2. *Gravure.* Used for large-volume runs for printing high-quality publications, magazines, catalogs, and advertising. It is also used to print polyolefin packaging, paperboard boxes, and labels.
3. *Flexography.* Used primarily for packaging, such as plastic wrappers, corrugated boxes, milk cartons, and for shower curtains, foil, and paper bags.
4. *Web letterpress.* Used for printing newspapers and magazines.
5. *Screen printing.* Used for specialty printing, T-shirts, posters, banners, decals, and wallpapers. It is also used to print patterns on electric circuit boards prior to etching (Noyes, 1993).

Environmental Impacts of Printing. The environmental issues facing the printing industry today have expanded considerably beyond traditional concerns. Wastewater, air emissions, potential soil and groundwater contamination, solid waste disposal, and employee health and safety have become increasingly important concerns. The management and disposal of hazardous substances, including both process-related wastes and residues from waste treatment, receive significant attention because of regulations and economics. Although the average small print shop is not a major source of pollution, the cumulative impact of all printing operations on the environment is substantial.

Lithographic printing uses potentially hazardous chemicals, requires energy, and generates wastes. Isopropyl alcohol (IPA) is used to decrease the surface tension of lithographic fountain solutions, allowing the solution to adhere more readily to the nonimage area of the plate cylinder. However, most of the IPA evaporates, contributing to ozone creation in the lower atmosphere. Emissions from heat-set presses can be reduced with an afterburner or regenerative thermal oxidizer. Newer, low-volatility fountain solutions use surfactants, glycols, and glycol ethers in place of IPA to cut surface tension. Many printers are changing over to non- and low-IPA fountain solutions.

Raw Materials Used. The principal raw materials used by the graphic arts industry are inks and substrates. A substrate is any material upon which ink is impressed, such as paper, plastic, wood, or metal. Other raw materials used in SIC 27 are gravure cylinders, photographic films, photoprocessing chemicals (developers, fixers, wash baths, reducers, intensifiers), printing plates, plate processing chemicals, lubricating oils, fountain solutions, cleaning solvents, and rags.

Waste Description. Commercial printers generate a variety of wastes during printing operations, the vast majority of which is paper. For example, almost 98 percent of the total waste generated by the lithographic industry is spoiled paper and paper wrap. Most of this paper is recycled, incinerated, or disposed of as trash. Other wastes that can be disposed of in trash include film, empty containers, used blankets, damaged products, and other items. These wastes result from image processing, platemaking, proofmaking, printing, and finishing processes. Wastewater results from image processing, platemaking, and printing processes. Equipment cleaning wastes and air emissions are other categories of waste.

Listed below are potential sources of wastes in the printing industry, categorized as hazardous wastes and VOCs (air emissions). Even though the list indicates the types of wastes that *may* be generated by a printing process, the actual wastes generated by a printer will depend on the technologies and material used and whether waste reduction measures are employed.

Hazardous Wastes

- Photographic wastes (intensifiers, scrap lithographic film)
- Spent solvents (1,1,1-trichloroethane, methylene chloride)
- Waste inks with solvents or heavy metals (inks and empty containers)
- Cleaning rags soaked with solvent or containing residues of heavy metals
- Strong acid or alkaline wastes (ammonium hydroxide, hydrochloric acid)
- Spent plating wastes (spent plating solutions and sludges)
- Ink sludge containing heavy metals (ink sludges with chromium, lead, or cadmium)
- Containers with hazardous residues of solvents, inks, or adhesives
- Miscellaneous (used oil from compressors, refrigeration tank filters)

VOCs

- Inks (xylene, ketones, alcohols)
- Fountain solutions (isopropyl alcohol)
- Adhesives (binding, labels)
- Cleanup (toluene, naphtha, kerosene, methanol)

Source Reduction. The types of inks used by a printer may have significant impacts on the shop's generation of hazardous wastes and the releases of VOCs from the printing and cleanup operation. The Toxics Use Reduction Institute (TURI) of the University of Massachusetts-Lowell has released a case study report "Alternatives to Solvent- and Petroleum-Based Inks" (Toxics Use Reduction Institute, 1992). Prepared by Boston-based Tellus Institute, the report examines seven categories of alternative inks for use in lithography, flexography, and gravure processes. These categories include soy-based inks, ultraviolet and electron beam radiation-curing inks, and water-based inks. The report finds that prospects for large-scale replacement of solvent- and petroleum-based inks with alternatives are good. Environmental benefits, chiefly in the form of reduced VOC emissions, are possible in all seven categories of alternatives. It was also reported that product quality for four or five of the alternatives was either as good as or better than that achieved by conventional inks.

According to the TURI report, most printers using flexography or gravure have substituted water-based inks for alcohol-based inks due to concerns for worker health and safety, flammability, and high insurance costs. A number of printers have reported sizable dollar savings through this substitution. The successful changeover from solvent-based to water-based inks improves safety, aids compliance with air pollution regulations, and reduces costs.

Waste Minimization Options. Depending on the specific process and company, different waste minimization and source reduction options are possible. Some of the general waste minimization methods for the commercial printing industry are listed in Table 6.1.8.

Recommendations. There is considerable uncertainty as to which inks are considered hazardous within the printing industry. Due to this uncertainty, ink manufacturers should conduct the tests necessary to determine which waste inks should be classified "hazardous" with guidance from the Printing Industries Association (PIA). This information should then be recorded and transferred to all facilities which handle hazardous and nonhazardous inks. In accordance with these tests, a technology transfer program for printers should be implemented and made widely available to the commercial printing industry with guidance from the Department of Health Services (DHS).

TABLE 6.1.8 Waste Minimization Methods for the Commercial Printing Industry

Operation	Waste minimization methods
Material handling and storage	Material preinspection Proper storage of materials Restrict traffic through area Inventory control Purchase quantities according to needs Recycle empty containers Recycle photographic film and paper
Image processing	Electronic imaging and laser platemaking Material substitution Extend bath life Use of squeegees Employ countercurrent washing Recover silver and recycle chemicals
Plate processing	Reduce solution loss Replace metal etching/plating operations Use nonhazardous developers and finishers
Make-ready	Ink-water ratio sensor Computerized registration Automatic scanners Automatic ink key setting system
Printing	Store ink properly Collect and reuse solvent Alternative printing techniques Recycle lube oils and waste ink Use less hazardous inks Use alternative fountain solutions Use alternative cleaning solvents Reduce the need to clean
Finishing	Reduce paper use and recycle waste paper

Source: Noyes (1993).

Secondary Metals (SIC Code 34)

The fabricated metal products industry comprises facilities that generally perform one of two functions: forming metal shapes (metal fabrication) and performing metal finishing operations, including surface preparation. Standard Industrial Classification (SIC) Code 34 is composed of facilities that fabricate ferrous and nonferrous metal products and those that perform electroplating, plating, polishing, anodizing, coloring, and coating operations on metals.

Raw Material Inputs and Pollution Outputs in the Production Line. The material inputs and pollution outputs resulting from metal fabrication, surface preparation, and metal finishing processes by media can be found in Table 6.1.9.

Each of the metal-shaping processes can result in wastes containing chemicals of concern. Clearly, the application of solvents to metal and machinery results in air emissions. Additionally, wastewater containing acidic or alkaline wastes and waste oils, and

TABLE 6.1.9 Process Material Inputs and Outputs for the Metal Processing Industry

Process	Material input	Air emission	Process wastewater	Solid waste
		Metal shaping		
Metal cutting and/or forming	Cutting oils, degreasing and cleaning solvents, acids, alkalies, and heavy metals	Solvent wastes (for example, 1,1,1-trichloroethane, acetone, xylene, toluene)	Waste oils (e.g., ethylene glycol) and acid (e.g., hydrochloric, sulfuric, nitric), alkaline, and solvent wastes	Metal chips (e.g., scrap steel and aluminum), metal-bearing cutting fluid sludges, and solvent still-bottom wastes
		Surface preparation		
Solvent degreasing and emulsion, alkaline and acid cleaning	Solvents, emulsifying agents, alkalies, and acids	Solvents (associated with solvent degreasing and emulsion cleaning only)	Solvent, alkaline, and acid wastes	Ignitable wastes, solvent wastes, and still bottoms
		Surface finishing		
Anodizing	Acids	Metal-ion-bearing mists and acid mists	Acid wastes	Spent solutions, wastewater treatment sludges, and base metals
Chemical conversion coating	Metals and acids	Metal-ion-bearing mists and acid mists	Metal salts, acid, and base wastes	Spent solutions, wastewater treatment sludges, and base metals
Electroplating	Acid/alkaline solutions, heavy metal bearing solutions, and cyanide bearing solutions	Metal-ion-bearing mists and acid mists	Acid/alkaline, cyanide, and metal wastes	Metal and reactive wastes
Plating	Metals (e.g., salts), complexing agents, and alkalies	Metal-ion-bearing mists	Cyanide and metal wastes	Cyanide and metal wastes
Painting	Solvents and paints	Solvents (xylene, toluene, MEK, MIBK)	Solvent wastes	Still bottoms, sludges, paint solvents, and metals
Other metal finishing techniques (including polishing, hot dip coating, and etching)	Metals and acids	Metal fumes and acid fumes	Metal and acid wastes	Polishing sludges, hot dip tank dross, and etching sludges

solid wastes, such as metals and solvent-saturated waste, can be generated during these processes. Metal fabrication facilities are major users of solvents for degreasing during pre- and postcleanup operations. In cases where solvents are used solely in degreasing, records of the amount and frequency of purchases provide information to calculate emission rates.

Metalworking fluids are applied to either the tool or the metal being tooled to facilitate the shaping operation. Metalworking fluid is used to

- Control and reduce the temperature of tools and aid lubrication
- Control and reduce the temperature of workpieces and aid lubrication
- Provide a good finish
- Wash away chips and metal debris
- Inhibit corrosion or surface oxidation

Preparation. Metal surface preparation typically results in air emissions, contaminated wastewater, and solid wastes. The primary air emissions are due to the evaporation of chemicals from solvent degreasing and emulsion cleaning processes. These emissions may result through volatilization of solvents during bulk storage, fugitive losses during use, and direct ventilation of fumes. Solvent-containing wastes are typically pretreated to comply with any applicable National Pollutant Discharge Elimination System (NPDES) permits and then are sent off-site, while aqueous wastes from alkaline and acid cleaning, which do not contain solvents, are usually treated on-site.

Anodizing and electroplating operations produce air emissions, contaminated wastewaters, and solid wastes as well. Mists and gas bubbles arising from heated fluids are a source of air emissions, which may contain metals or other substances present in the bath. Industry has begun adding fume suppressants to electroplating baths to reduce air emissions of chromium, one of the most frequently electroplated metals. Fume suppressants lower the surface tension of the bath, preventing hydrogen bubbles in the bath from bursting and producing a chromium-laden mist. Fume suppressants are highly effective when used in decorative plating, but less effective when used in hard-chromium plating.

Painting operations also result in air emissions, contaminated wastewaters, and the generation of liquid and solid wastes. Air emissions consist primarily of the organic solvents used as carriers for the paint. Emissions also result from bulk paint storage, mixing, application, and drying. In addition, cleanup processes can result in the release of organic solvents used to clean equipment and painting areas. Sources of solid- and liquid-phase wastes include

- Paint application emissions control devices (e.g., paint booth collection systems, ventilation filters)
- Equipment washing
- Disposal of materials used to contain paint and overspray
- Excess paints discarded upon completion of a painting operation or after expiration of the paint shelf life

These solid and liquid wastes may contain metals from paint pigments and organic solvents, such as paint solvents and cleaning solvents. Still bottoms also contain solvent wastes. The cleaning solvents used on painting equipment and spray booths may also contribute organic solid waste.

Pollution Prevention Options. The following list of numerous pollution prevention techniques may be useful to facilities specializing in metal fabrication and finishing operations. Many suggestions mentioned in the Solvent Cleaning and Degreasing section also apply to the metal finishing industry.

Plating Operations

- Install atmospheric evaporator to reduce metal concentrations.
- Install process (for example, CALFRAN) to reduce pressure to vaporize water at cooler temperatures and recycle water by condensing the vapors in another container, thus concentrating and precipitating solutes out.
- Replace methylene chloride, 1,1,1-trichloroethane, and perchloroethylene (solvent-based photochemical coatings) with aqueous-base coating of 1 percent sodium carbonate.

Finishing Operations

- Always use proper spraying techniques.
- Improved paint quality and work efficiency and lower vapor emissions can be attained by formal training of operators.
- Avoid buying excess finishing material at one time due to its short shelf life.
- Use the correct spray gun for particular applications: conventional air spray gun for thin-film build requirements, airless gun for heavy-film application, and air-assisted airless spray gun for a wide range of fluid output.
- Preinspect parts to prevent painting of obvious rejects.
- Ensure the spray gun air supply is free of water, oil, and dirt.
- Replace galvanizing processes requiring high temperature and flux with one that is low-temperature and does not require flux.
- Investigate use of transfer methods that reduce material loss such as dip and flow coating and electrostatic spraying and electrodeposition.
- Change from conventional air spray to an electrostatic finishing system.
- Use solvent recovery or incineration to reduce the emissions of volatile organics from curing ovens.
- Use alternative coatings for solvent-based paints to reduce VOC use and emissions, such as high solids coatings (this may require modifying the painting process including high-speed/high-pressure equipment, a paint distributing system, and paint heaters), water-based coatings, and powder coatings.
- Segregate nonhazardous paint solids from hazardous paint solvents and thinners.
- Use activated carbon to recover solvent vapors, then recover the solvent from the carbon by steam stripping, and distill the resulting water-solvent mixture.

Paint Cleanup

- Reduce equipment cleaning by painting with lighter colors before darker ones.
- Reuse cleaning solvents for the same resin system by first allowing solids to settle out of solution.
- Flush equipment first with dirty solvent before final cleaning with virgin solvent.
- Use virgin solvents for final equipment cleaning, then as paint thinner.
- Use pressurized air mixed with a mist of solvent to clean equipment.

- Replace water-based paint booth filters with dry filters. Dry filters will double paint booth life and allow more efficient treatment of wastewater.
- To prevent spray gun leakage, submerge only the front end (or fluid control) of the gun in the cleaning solvent.
- Solvent waste streams should be kept segregated and free from water contamination. Solvent recovery units can be used to recycle spent solvents generated in flushing operations. Install a recovery system for solvents contained in air emissions. Use batch distillation to recover isopropyl acetate generated during equipment cleanup. Use batch distillation to recover xylene from paint equipment cleanup. Use a small solvent recovery still to recover spent paint thinner from spray gun cleanups and excess paint batches. Install a methyl ethyl ketone solvent recovery system to recover and reuse waste solvents.
- Make agreements with other small companies to jointly recycle cleaning wastes.

REFERENCES

This list includes both works cited in the chapter and those which are not.

Brown, L., et al., "Chemical Substitution for 1,1,1-Trichloroethane and Methanol in Manufacturing Operations," *16th Annual Hazardous Waste Research Symposium,* April 1990, EPA/600/9-90/037.

"Development Document for Best Conventional Pollutant Control Technology Effluent Limitations Guidelines for the Pulp, Paper and Paperboard and the Builders' Paper and Board Mills Point Source Categories," EPA 440/1-86/025, Environmental Protection Agency, December 1986.

"Development Document for Effluent Limitations: Guidelines and Standards for the Textile Mills," Environmental Protection Agency, EPA 440/1-79/0226, October 1979.

"Development Document for Effluent Limitations: Guidelines and Standards for the Metal Finishing Industry," EPA-440-1-80-091A, Environmental Protection Agency, Office of Water Regulations and Standards, 1980.

DHS, "Waste Audit Study: Fabricated Metal Products Industry," prepared by Jacobs Engineering Group for Alternative Technology Section, Toxic Substances Control Division, California Dept. of Health Services, 1989.

Durney, L. J., ed., *Electroplating Engineering Handbook* 4th ed., Van Nostrand Reinhold, New York, 1984.

Edde, Howard, *Environmental Control for Pulp and Paper Mills,* Noyes Publications, Park Ridge, N.J., 1984.

"The Effect of Textile Dyeing Effluents on the Environment," ADMI-EPA joint publication, 1979.

Environmental Protection Agency, *Guide to Cleaner Technologies: Alternatives to Chlorinated Solvents for Cleaning and Degreasing,* Office of Research and Development, Washington, 1994.

Environmental Protection Agency, *Manual: Pollution Prevention in the Paints and Coatings Industry,* Office of Research and Development, Washington, 1996.

Environmental Protection Agency, *Manual: Best Management Practices for Pollution Prevention in the Textile Industry,* Office of Research and Development, Washington, 1996.

Environmental Protection Agency, *Guides to Pollution Prevention: The Commercial Printing Industry,* Office of Research and Development, Washington, 1990.

Environmental Protection Agency, *Guide to Cleaner Technologies: Cleaning and Degreasing Process Changes,* Office of Research and Development, Washington, 1994.

Environmental Protection Agency, *Guides to Pollution Prevention: The Paint Manufacturing Industry,* Office of Research and Development, Washington, 1990.

Environmental Protection Agency, *Guide to Cleaner Technologies: Organic Coating Replacements,* Office of Research and Development, Washington, 1994.

Environmental Protection Agency, *Seminar Publication—Solvent Waste Reduction Alternatives,* Center for Environmental Research Information, Cincinnati, Ohio, 1989.

Environmental Protection Agency, *Waste Minimization in Metals Parts Cleaning,* Office of Solid Waste and Emergency Response, Washington, 1989.

Environmental Protection Agency, *Guide to Cleaner Technologies: Alternatives to Chlorinated Solvents for Cleaning and Degreasing,* Office of Research and Development, Washington, 1994.

Environmental Protection Agency, *Profile of the Lumber and Wood Products Industry,* Office of Enforcement and Compliance Assurance, Washington, 1995.

Environmental Protection Agency, *Profile of the Pulp and Paper Industry,* Office of Enforcement and Compliance Assurance, Washington, 1995.

Freeman, H.M., *Industrial Pollution Prevention Handbook,* McGraw-Hill, New York, 1995.

Galecki, G., *Proceedings of Waste Stream Minimization and Utilization Innovative Concepts—An Experimental Technology Exchange,* Battelle Memorial Institute, PNL-SA-19337, April 1991.

"Guides to Pollution Prevention: The Commercial Printing Industry," prepared by Risk Reduction Engineering Laboratory and Center for Environmental Research Information, EPA/625/7-90/008, August 1990.

Hawkins, Robert F., "Use of Water-Based Inks to Meet EPA Requirements," Paper 87-48.4, 80th Annual Air Pollution Control Association Meeting and Exhibition, New York City, June 21–26, 1987.

Hazelwood, D., et al., "Program toward Waste Minimization in Painting Operations in the Defense Community," International Conference on Pollution Prevention, June 1990, EPA/600/9-90/039.

Jacobs Engineering Group, "Case Studies of Minimization of Cyanide Waste from Electroplating Operations," U.S. Environmental Protection Agency Contract No. 68-01-7053, Dec. 1986.

Jones, H., *Pollution Control and Chemical Recovery in the Pulp and Paper Industry,* Noyes Data, 1973.

Knox, J. H., ed., *High Performance Liquid Chromatography,* Edinburgh University Press, Edinburgh, 1977, pp. 1–4, 34–39, 52–58, 68–73, 175–179.

Kohl, J., P. Moses, and B. Triplett, "Management Recycling Solvents: North Carolina Practices, Facilities, and Regulations," North Carolina State University, Raleigh, 1984.

Kulube, Herbert, "Residual Carrier Components in Exhausted Textile Dyebaths," North Carolina State University, 1987, Master's thesis.

Licis, L., "Industrial Pollution Prevention Opportunities for the 1990s," EPA/600/8/-91/052, Environmental Protection Agency, August 1991.

Mertens, J., "CFCs: In Search of a Clean Solution," *Env. Prot.,* September 1991.

Noyes, Robert, *Pollution Prevention Technology Handbook,* Noyes, Park Ridge, N.J., 1993, pp. 1–4, 611–621.

Operations, Environmental Protection Agency, contract no. 68-01-7053, task 46, December 1986.

Pferdehirt, Wayne, "Roll the Presses but Hold the Wastes: P2 and the Printing Industry," *Pollution Prevention Review,* Autumn 1993.

Pinkerton, H. L., and K. A. Graham, "Rinsing," *Electroplating Engineering Handbook,* 4th ed., Durney, L. J., ed., Van Nostrand Reinhold, New York, 1984, pp. 691–709.

Pollution Prevention Opportunities in Paper Manufacturing, Environmental Protection Agency, Philadelphia, October 1990.

Report to Congress. *Waste Minimization Issues and Options,* vol. 2, App. B: "Process Studies," prepared by Jacobs Engineering Group for EPA, Office of Solid Waste, Waste Treatment Branch, EPA/530-SW-86-042, 1986.

Schaffer, G., "Recycling Coolant Reduces Cost," *American Machinist,* February 1978.

Schlesinger, Herbert A., Emil F. Dul, and Thomas A. Friday, Jr., "Pollution Control in Textile Mills," chap. 15, pp. 15.1–15.30.

Schwartz, S. I., "Recycling of Hazardous Waste Solvents: Economic and Policy Aspects," *Solvent Waste Reduction Alternatives Symposia Conference Proceedings,* Los Angeles, sponsored by California Department of Health Services, 1986.

Sluhan, W. A., "The Application of High Speed, Disc Bowl Centrifuges to Water Miscible Cutting and Grinding Fluids," Master Chemical Corporation, Perrysburg, Ohio.

Smith, Brent, "Identification and Reduction of Pollution Sources in Textile Wet Processing," North Carolina Division of Environmental Management, 1986.

Smith, Brent, and James Rucker, "Water and Textile Wet Processing," *American Dyestuff Reporter,* July 1987, p. 15.

Smook, Gary A., *Handbook for Pulp and Paper Technologists,* 2d ed., Angus Wilde Publications, Vancouver, B.C., 1992.

Sokolowska-Gajda, Jolanta, Harold S. Freeman, and Abraham Reife, "Synthetic Dyes Based on Environmental Considerations," reprinted from *Textile Research Journal,* vol. 64, no.7, July 1994.

Solvent Substitution: The First Annual International Workshop on Solvent Substitution, December 4–7, 1990, Noyes Data, 1992.

"Strategy for the Regulation of Discharges of PHPDs and PHDFs from Pulp and Paper Mills to Waters of the United States," Environmental Protection Agency, May 21, 1990.

"Success through Waste Reduction," Washington State Department of Ecology, Publication 90-22, 1990.

"Summary of Technologies for the Control and Reduction of Chlorinated Organics from the Bleached Chemical Pulping Subcategories of the Pulp and Paper Industry," Environmental Protection Agency, April 27, 1990.

Tedder, Steve, "Aquatic Bioassay Toxicological Summary," March 31, 1986 (unpublished). Waste Minimization Opportunity Assessment Manual, EPA; Hazardous Waste Engineering Research.

Toxics Use Reduction Institute, *Alternatives to Solvent- and Petroleum-Based Inks,* Tellus Institute, Boston, 1992.

Willson, Edgerley, "Recycling/Waste Minimization—One Mill's Success Story," *Technical Association of the Pulp and Paper Industry Proceedings,* 1991 Environmental Conference, TAPPI Press, Atlanta, Ga., 1991.

Yazdani, A., "Regulatory Pressures on Chlorinated Solvents—How Can Users Cope," *AESF/EPA Proceedings,* January 27–29, 1992.

SECTION 6.2

WASTEWATER

Gayle P. Van Durme
Black & Veatch
Kansas City, Missouri

CONVEYANCE SYSTEMS 6.36
Emission Sources 6.37
System Design to Reduce Emissions 6.38
Treatment Methods 6.38
WASTEWATER TREATMENT FACILITIES 6.43
Liquid Treatment Processes 6.45
Solids Handling Processes 6.50
REFERENCES 6.55

When volatile compounds are emitted to the atmosphere, they cause air pollution. The most effective and reliable approach to air pollution control is to prevent these emissions. In some cases, air pollution can be significantly reduced by merely changing operations or modifying a production process. When those options are not feasible, other source control measures may be applied to minimize emissions. The objective of these control techniques is to stop air pollution before it occurs.

The first subsection of this section examines air pollution prevention methods for municipal wastewater conveyance systems and treatment facilities. The second subsection discusses industrial air pollution prevention techniques.

Odors and volatile organic compound (VOC) emissions from wastewater can be a significant concern to public health and well-being. Some odorous compounds and VOCs are present in the raw wastewater, while others can be generated during transport in a wastewater conveyance system or during processing at a wastewater treatment facility. Once generated, these compounds can be released from the wastewater and cause nuisance odors, impact air quality, and cause corrosion of equipment and materials.

In an early paper on wastewater odor reduction, Morris M. Cohn reflected that "'An ounce of prevention is worth a pound of cure'—yes, and a pound of prevention costs less than an ounce of cure." Cohn's insight still holds true today (Cohn, 1929). This section includes techniques for minimizing odors and VOCs from wastewater by preventing their generation or by reducing their release from the liquid phase. Many of the techniques are relatively low-cost operational changes that, in some cases, may eliminate the need for more expensive containment and control equipment.

CONVEYANCE SYSTEMS

Wastewater conveyance systems are complex networks of gravity sewers and pressurized force mains that transport sewage from residential and industrial sources to treatment facilities. In conveyance systems that serve large areas, the wastewater may travel many miles and be pumped several times before reaching its final destination. Long detention times and low sewer velocities encourage the generation of hydrogen sulfide.

Emission Sources

Under anaerobic conditions, sulfate-reducing bacteria in the slime layer on the wetted sewer wall use sulfate (SO_4) as an oxygen source and reduce it to sulfide ($S^=$), which is emitted as gaseous hydrogen sulfide (H_2S). The H_2S gas is utilized by another type of bacteria residing above the liquid surface, particularly on the crown of the sewer. A by-product of these bacteria is sulfuric acid, which causes the rapid corrosion of unprotected concrete. In some municipalities this has led to the complete collapse of a section of the sewer line, as shown in Figure 6.2.1. Also H_2S directly corrodes exposed metal surfaces. Concrete can be protected with a corrosion-resistant coating or lining, but H_2S remains a problem, because high concentrations are toxic and pose a safety hazard for maintenance personnel. If sewer gases containing H_2S are vented to the air, a severe odor nuisance may result.

Due to the serious problems associated with H_2S in conveyance systems, it is of prime importance to identify effective methods to reduce its generation or inhibit its release. Recent research (Quigley, 1995) has shown that, in addition to H_2S, significant amounts of VOCs are present in wastewater and a large fraction are emitted during transport in the conveyance system.

Before we address effective control methods, it is helpful to examine the conditions that promote sulfide generation. The main factors that influence sulfide production are the biological oxygen demand (BOD), sulfate concentration, dissolved oxygen, pH, detention time, temperature, surface area, and stream velocity. In most municipal wastewater, sulfate is abundant, so soluble BOD typically governs the rate of sulfide production. When dissolved oxygen is present, sulfate is not reduced and H_2S production is limited. Sulfate-reducing bacteria tolerate a pH of 5.5 to 9, but are optimal at 7.5 to 8. As pH decreases, there is a greater potential for release of H_2S gas. Long detention times

FIGURE 6.2.1 H_2S crown corrosion can lead to pipe collapses.

favor the generation of sulfide, because the organic matter becomes more solubilized. Higher wastewater temperatures encourage the metabolic activity of sulfate-reducing bacteria. The greater the amount of wetted surface area available, the more slime growth will develop. Higher stream velocities reduce solids deposition and decrease the sulfide concentration by increasing reaeration, which oxidizes sulfide in the wastewater before it escapes.

System Design to Reduce Emissions

Once the mechanisms of H_2S generation and release are understood, it is possible to design new conveyance systems or modify existing systems to minimize problems. Equations for predicting sulfide buildup and corrosion in conveyance systems were initially developed in the 1970s by Pomeroy and Parkhurst (1977). A thorough discussion of these equations is found in design manuals by the Environmental Protection Agency (1985), American Society of Civil Engineers (1989), and the American Concrete Pipe Association (1984). Computer modeling programs incorporating these equations have been developed that allow the rapid analysis of an entire conveyance system. Some states now require that sulfide modeling be performed for all new conveyance systems. The HS program (American Concrete Association, 1991) and the Sulfide Works program (Microcomp Systems, 1991) are modeling programs that predict where sulfide generation is likely to occur. The designer can vary pipe slopes and diameters to minimize sulfide generation as much as possible. In areas where sulfide generation is unavoidable because topography or subsurface features limit pipe slope, the designer can use corrosion-resistant materials or plan for appropriate treatment.

In the design of conveyance systems, sulfide can be minimized by providing adequate flow velocity to prevent solids deposition. The often cited minimum velocity of 2 ft/s (0.6 m/s) regardless of pipe diameter is not sufficient for large pipes. The critical shear stress approach (Yao, 1976) yields increasing minimum velocities as pipe size increases. In addition to pipe slope and diameter, there are several other structural elements in a conveyance system that influence sulfide generation. Critical structures in a conveyance system are locations that either encourage sulfide generation or bring about its release. A variety of critical structures are shown in Table 6.2.1, along with their sulfide impacts and design considerations.

There are a wide variety of corrosion-resistant materials available for use in conveyance systems. These materials protect the pipe and structures from corrosion, but do nothing to slow the formation of sulfide and the release of odors in the conveyance system and downstream at the treatment plant. Because they do not prevent air pollution, details of specific types of corrosion-resistant materials are not presented here. Extensive information on corrosion-resistant materials and corrosion protection coatings is given in a 1992 EPA manual (Environmental Protection Agency, 1992).

Treatment Methods

There are many chemical products and treatment techniques that can be applied in conveyance systems to inhibit sulfide formation or react with sulfide in the aqueous phase before it is emitted. These treatment methods protect the conveyance system infrastructure from deterioration and at the same time minimize odors. However, these methods may not completely remove sulfide and may not reduce VOCs, so some locations may require additional vapor-phase treatment. Conveyance system treatment methods discussed include mechanical and chemical oxidation, nitrate addition, metabolism modi-

TABLE 6.2.1 Critical Conveyance System Structures, Impacts, and Design Considerations

Structure	Sulfide impacts	Design considerations
Drop manhole	High turbulence at drops provides reaeration, which is beneficial in maintaining dissolved oxygen and discouraging sulfide production. However, drops release H_2S already formed.	When sulfide is present, corrosion-resistant material is needed, along with treatment to prevent odors. Vortex drop structures dissipate energy, but H_2S release is still a problem.
Junction structure	Turbulence results from a change in grade within structure or difference in flow or velocity of the incoming streams.	Minimize turbulence by design, so that flows converge smoothly and velocity changes occur gradually.
Force main	Full pipe flow increases slime surface area. Low velocity or intermittent flow can cause high sulfide generation.	Install air release valves and use corrosion-resistant materials where air pockets occur. Corrosion-resistant materials are needed at force main discharge into gravity sewer.
Siphon	Siphons remain full even at low flows and generate substantial sulfide. Wastewater dropping into the siphon inlet releases H_2S. Solids deposition occurs.	Multiple siphon lines can be used to maintain velocity. Periodic cleaning is needed to remove solids. Air jump piping conveys air to the end of the siphon, but transfers H_2S downstream.
Lift station wet well	Solids deposition occurs in wet well. High turbulence results from a drop influent. Wet wells commonly have a long detention time at low flow.	Design pump suction and wet well to prevent solids accumulation. Clean the wet well periodically. Submerge the influent pipe to reduce turbulence. Use corrosion-resistant wet well liner.

fication, iron precipitation, pipe cleaning, and pH elevation. A chemical treatment to inhibit crown corrosion is also discussed.

Mechanical Oxidation. There are numerous techniques to mechanically inject air or oxygen into wastewater. These methods are most applicable to force mains and siphons because pressure increases the transfer of oxygen into the wastewater. The least costly mechanical aeration method is to use a venturi aspirator. Wastewater flowing under pressure through a narrow constriction allows the introduction of air through an opening at the neck.

Another mechanical aeration method involves injection of compressed air directly into a pressurized pipe, or into a gravity sewer after the air is predissolved. This method requires a compressor and appurtenances. One source (Laughlin, 1964) suggests 0.1 to 0.3 scfm of air per gallon of wastewater (0.75 to 2.25 m^3/m^3), but system hydraulics and wastewater characteristics affect air requirements. One problem with air injection in pressure lines is that it can cause the formation of large air pockets at some locations on the line. High-purity oxygen is 5 times more soluble than air, and since less oxygen gas is required, the potential for air pocket formation in force mains is reduced. High-purity oxygen systems include a liquid storage vessel, vaporizer, pressure regulator, oxygen injection system, and controls. Most of the components can be leased from an oxygen supplier. Oxygen can be injected directly into a force main or added to a pressurized sidestream, or U-tube system.

A U tube is basically a means to increase oxygen transfer by increasing the pressure. A U tube consists of a U-shaped pipe arrangement added to a pressure pipe section in which oxygen is introduced at the top of a downflow leg, and returns back to grade through an upflow leg. The increased pressure at the bottom of the U increases the oxygen transfer. U tubes can be applied for both air and high-purity oxygen. A U tube system on a 2-mi force main in Keene, New Hampshire (Bowker et al., 1995), decreased dissolved sulfide from 2.4 mg/L to below the detection limit and increased dissolved oxygen from 0.2 to 9.5 mg/L.

Chemical Oxidation. Chemical oxidants react with the organics in wastewater as well as dissolved sulfide, so their dosage is several times greater than the stoichiometric requirement to oxidize sulfide. Because oxidants are highly reactive, their effects are relatively short-lived and multiple injection stations may be required to treat long reaches of the conveyance system. However, because they are so reactive, chemical oxidants will treat a wide range of odorous compounds in addition to sulfide.

Chlorine will oxidize sulfide to sulfate or elemental sulfur depending on pH. Chlorine can be added directly as a gas or as an aqueous solution. Chlorine gas is a hazardous material, and appropriate safety measures must be observed when it is used. Injection of chlorine gas may be appropriate for application at a wastewater facility, but is not well suited for installation at conveyance system structures located throughout the community. Liquid solutions of hypochlorite are more applicable for use in remote areas in a conveyance system. Sodium hypochlorite (NaOCl) is much less expensive than calcium hypochlorite, so it is most commonly used. Because of side reactions, the recommended NaOCl-to-sulfide dosage rate is 12 lb/lb. At a typical bulk cost for NaOCl, the unit cost is \$6.36 per pound (\$14 per kilogram) of sulfide treated.

Hydrogen peroxide (H_2O_2) is a highly reactive oxidant that has been used extensively in conveyance systems. For wastewater applications peroxide is typically used in 35% or 50% solutions, and protective gear must be worn when handling. Some municipalities report peroxide-to-sulfide dosage rates of less than 2 lb/lb. However, many others have found that effective treatment requires a peroxide-to-sulfide dosage of 4 lb/lb. At typical peroxide costs, the unit cost is \$4 per pound (\$8.80 per kilogram) of sulfide removed.

Potassium permanganate ($KMnO_4$) has been employed for sulfide control in sludge dewatering applications. It has also been applied in conveyance systems and pump station wet wells. Permanganate-to-sulfide dosage rates as high as 17 lb/lb have been reported. A typical permanganate unit cost is \$30 per pound (\$66 per kilogram) of sulfide. The relatively high cost of this chemical makes it unattractive for continuously treating large wastewater flows or high sulfide concentrations.

Nitrate Addition. Unlike oxidants, which react with sulfide already formed, the addition of nitrate acts to suppress sulfide generation. This occurs because sulfate-reducing bacteria preferentially use nitrate rather than sulfate as an oxygen source. Bioxide is a proprietary nitrate formulation that has been used extensively in conveyance systems throughout the United States. Factors that affect the dosage include wastewater strength, detention time, dissolved oxygen, and dissolved sulfide.

An evaluation at the city of Los Angeles (Hagekhalil et al., 1994) showed calcium nitrate dosage ranges from 1.5 to 3 gal of 65% calcium nitrate solution per pound of sulfide removed (12.5 to 22.5 L/kg). The unit cost range is $2 to $4 per pound of sulfide removed ($4.4 to $8.8 per kilogram). This product is nonhazardous as well as economical and has been shown to be effective in many applications. Calcium nitrate suppresses sulfide generation, and in some applications existing sulfide was also decreased, but in other cases this was not evident. To obtain systemwide sulfide control, it may be necessary to use this product in conjunction with another chemical that treats existing sulfide.

Metabolism Modification. A product called Aquit or Sulf Control with the active ingredient of anthraquinone inhibits sulfide generation by interfering with the metabolism of sulfate-reducing bacteria. Sulf Control is manually applied directly to the sewer at 2-week intervals. The dosage is related to the wetted surface area in pipes and wet wells, and the manufacturer's guideline is 50 to 100 lb per mgd [2500–5000 kg/($m^3 \cdot s$)]. For effective treatment Sulf Control must be applied so that it contacts the slime on the pipe walls. This product has been most effective where complete wall coverage can be obtained such as in a force main. Some success has been reported with gravity lines. Because the dosage is dependent on the pipe wall surface area and not on the concentration, it becomes most cost-effective for lines with higher sulfide concentrations, such as force mains with long detention times.

In Tampa, Florida, a dosage of 50 lb (110 kg) of Sulf Control was applied to several small pump stations located upstream of a 1 mgd (0.044 m^3/s) pump station (Ballinger et al., 1994). Over the ensuing 14 days, average dissolved sulfide decreased from 11 to 0.7 mg/L at the pump station. At $20 per pound ($44 per kilogram) of Sulf Control, this equated to slightly less than $1 per pound ($2.2 per kilogram) of sulfide removed. If downstream reductions in H_2S are attributed to Sulf Control, the product appears to be even more cost-effective. Because it is manually applied, the product requires no capital expenditures for storage tanks or pumps.

Pipe Cleaning. Pipe cleaning can be an effective means of controlling sulfide generation in conveyance systems. Various methods have been developed to clean large gravity sewers. Flushing systems have been used to increase flow and scour away deposits. This method has limitations because it is difficult to store enough water to provide a significant flush. Gates can be added within sewers to hold the wastewater and create a flushing flow. There is a large variety of water jetting equipment available. The type of jet nozzle, pressure, and flow rate vary depending on the type of deposits and the particular application. The Sielwolf and Live Line water jetting systems use wastewater for the jetting flow, so a clean water supply is not needed.

A recent study in Council Bluffs, Iowa (Beck, 1995), concluded that cleaning was the most cost-effective means to reduce H_2S in 10,000 lin ft (3408 m) of 36-in (91-cm) interceptor sewer carrying 2 mgd (0.088 m^3/s). The analysis compared doing nothing (which required interceptor replacement), feeding chlorine, and pipe cleaning. For this particular conveyance system, it was determined that the areas of high corrosion correlated with solids deposition in the line. It was theorized that deposited solids slow wastewater flow and increase detention time, and that the solids serve as a food source for sulfate-reducing bacteria, so that H_2S production increases. Pipe cleaning was estimated to cost $6 per foot ($1.80 per meter) of 36-in (91-cm) diameter pipe. A present-worth evaluation, which assumed cleaning every 3 years, found that pipe cleaning costs about one-half as much as chemical addition for this application.

Shock dosing with caustic has been used to remove biological slime from the pipe invert and thus inhibit sulfide generation. In a study at the city of Los Angeles, sodium hydroxide (NaOH) was used to shock-dose a 126-in (320-cm) trunk sewer and a 30-in

(76-cm) branch line (Hamilton et al., 1994). For the 150 mgd (6.6 m^3/s) trunk sewer, 3700 gal (14,000 L) of 50% NaOH was applied to raise downstream pH to 11.4 for 52 min. Headspace H_2S remained at 100 ppm, but dissolved sulfide decreased from 10.5 to 7 mg/L for the next 50 h. Higher doses were not tested on the trunk sewer, because of high costs and concern about high pH at the treatment plant. At the 4.6 mgd (0.2 m^3/s) branch line, 3000 gal (11,350 L) of NaOH was applied to raise the downstream pH to 13.2 for 30 min. Headspace H_2S decreased from 98 to 7 ppm initially, but increased to 49 ppm in 3 days. Dissolved sulfide decreased from 4.5 mg/L to zero, and increased to 0.25 mg/L after 3 days. It was concluded that NaOH shock treatments 3 times per week in the branch line would cost $5300 per week. The annual cost would be $60,000 per million gallons ($15.85 per cubic meter), and this method was not deemed cost-effective. Other municipalities have found shock dosing to be cost-effective for smaller flows with treatment required at 2-week intervals.

Continuous pH Elevation. Unlike shock dosing, continuous addition of caustic is intended not to inactivate the slime layer and inhibit sulfide generation, but rather to hold sulfide in solution and thereby prevent H_2S volatilization. If wastewater pH is raised to 9, less than 1 percent of dissolved sulfide is present as H_2S.

Hydrated lime [$Ca(OH)_2$] and caustic (NaOH) are the least expensive chemicals to raise the wastewater pH. To elevate pH from 7 to 9, with a wastewater alkalinity of 200 mg/L as $CaCO_3$, requires 390 lb hydrated lime per mgd [37.8 kg/(m^3 • s)], or 420 lb of NaOH per mgd [40.7 kg/(m^3 • s)]. For 80% hydrated lime, the annual cost is about $18,250 per million gallons ($4.80 per cubic meter), and for 25% NaOH, the annual cost is about $63,900 per million gallons ($16.88 per cubic meter). This becomes expensive for large wastewater flows and simply delays the release of the H_2S. Unless treated further, H_2S will be released when the pH of the wastewater is lowered through dilution in the conveyance system or at the wastewater treatment facilities.

Iron Precipitation. Salts of iron and zinc will react with sulfide to form insoluble precipitates. Iron salts are less expensive than zinc salts, and wastewater sludge disposal regulations limit zinc concentrations, so zinc salts are not recommended. Iron salts are widely used in the United States for control of sulfide in conveyance systems. Generally, ferrous chloride ($FeCl_2$) is the least expensive type of iron salt, and it can be readily obtained from local chemical suppliers. Ferric chloride ($FeCl_3$) also provides effective treatment, but is typically more expensive. The lowest treatment costs are usually obtained by purchasing the product from a local supplier to minimize transportation costs. Effective treatment has been obtained with an iron-to-sulfide dosage rate of 3.5 lb/lb or a ferrous chloride–to–sulfide rate of 8 lb/lb (Van Durme, 1989).

One advantage of iron salts over oxidation chemicals is that the salts do not react with wastewater organics and are fairly sulfide-specific. Ferrous chloride pilot testing in Yuma, Arizona (see Fig. 6.2.2), verified that iron salts could be overdosed in the upstream portion of a conveyance system and used to treat long reaches. Because they are sulfide-specific, this means that iron salts do not treat other odorous VOCs. Bulk purchase costs have been as low as $0.12 per pound of $FeCl_2$ ($0.26 per kilogram) in some areas, which yields a cost of about $1.00 per pound of sulfide ($2.20 per kilogram).

Crown Treatment. An innovative pipe treatment has been evaluated by the Los Angeles County Sanitation District (LACSD) (Sydney et al., 1996). The method involves spraying the pipe crown with a magnesium hydroxide solution. A 50% magnesium hydroxide slurry provided the longest-lasting treatment because it had better adherence than the other caustic solutions investigated. The LACSD predicted that annual reapplication would be sufficient. It is theorized that the alkaline material neu-

FIGURE 6.2.2 Testing in Yuma, Arizona, confirmed that ferrous chloride could be overdosed upstream to treat long reaches.

tralizes existing surface sulfuric acid, inhibits further bacteria-generated sulfuric acid, and provides some buffering capacity. Sodium hydroxide and sodium carbonate solutions were also tested by LACSD, but were not as effective.

Crown spraying may not be as successful in conveyance systems that experience more frequent high flows, because the crown coating is washed off and must be reapplied if the sewer is surcharged, such as after a storm event. For LACSD, crown treatment was found to be more cost-effective than traditional corrosion control measures already tested by the district. The amortized capital and operational cost for treating 85 mi (137 km) of conveyance system was estimated to be $500,000 per year.

The advantages and disadvantages of the various control methods discussed are summarized in Table 6.2.2.

WASTEWATER TREATMENT FACILITIES

Wastewater treatment facilities consist of a wide variety of liquid treatment processes that remove suspended solids, dissolved organics, and nutrients from the liquid before discharging it to a receiving water. In addition to the liquid treatment units, a variety of solids handling processes are applied to dewater and stabilize the sludge before disposal or beneficial reuse. Some of the liquid and solids treatment processes induce turbulence, which strips sulfide and VOCs from the wastewater, while other processes have detention times that may encourage sulfide generation. Sulfide control is important for health and safety reasons as well as odor and corrosion control. This subsection focuses on operational techniques and low-cost aqueous-phase treatment methods that can be employed to minimize H_2S and VOC emissions from liquid treatment and solids handling processes.

TABLE 6.2.2 Conveyance System Treatment Techniques

Control measure	Advantages	Disadvantages
	Mechanical oxidation	
Venturi aspirator Compressed air Oxygen injection U tube Sidestream injection	No hazardous chemicals on site. Reduces sulfide formation by maintaining aerobic conditions.	High power costs. Strips sulfides already formed. Numerous application points in gravity sewers. Gas pockets in force mains.
	Chemical oxidation	
Chlorine gas Sodium hypochlorite Hydrogen peroxide Permanganate	Applicable for sewers and force mains. Can reduce VOCs as well as sulfide.	Chlorine gas not applicable for remote sites due to public safety concerns. Multiple injection sites required. Chemical handling concerns.
	Nitrate addition	
Sodium nitrate Bioxide Calcium nitrate	Nonhazardous chemical. Relatively low capital and operating costs.	Requires 1 to 2 h of detention time. May not treat sulfides already formed. Treats sulfide only.
	Metabolism modification	
Sulf control Anthraquinone	Successful where detention time allows long surface contact. Manual application, negligible capital costs.	Variable results. Requires good surface contact, works better in force mains than sewers. Reapplication every 2 weeks.
	Pipe cleaning	
Manual scrubbing High-pressure water Pipe pigging Caustic shock dose	Minimal capital cost for small lines. Improves sewer flow by removing deposits.	Short-lived effects, may require frequent retreatment. Shock cleaning requires hazardous chemical handling.
	Continuous pH increase	
Hydrated lime Sodium hydroxide	Side benefit of alkalinity addition, if required at wastewater treatment plant.	Costly for high flow. May require neutralization. Sulfide emitted when pH drops. Hazardous chemical handling.
	Iron precipitation	
Ferrous chloride Ferric chloride	Sulfide-specific, so it can be overdosed to treat long sewer reaches. Low operating cost.	Treats sulfide only, not VOCs. Adds solids to wastewater.
	Crown treatment	
Magnesium hydroxide Sodium hydroxide Sodium carbonate	Relatively low cost. Magnesium hydroxide has best adherence and longest life.	Does not reduce wastewater sulfide. Coating may wash off in storm events. Sodium hydroxide and sodium carbonate must be reapplied frequently.

Liquid Treatment Processes

Air pollution prevention techniques are discussed for liquid treatment processes in the general order that they occur in the process flow. Preliminary treatment processes are addressed first, followed by primary, secondary, disinfection, and advanced treatment.

Preliminary Treatment. Influent wastewater entering wastewater treatment facilities from the conveyance system may contain high concentrations of sulfide and VOCs. Therefore, the preliminary treatment processes are often major sources of emissions at wastewater facilities. Because of their configuration, preliminary treatment processes such as coarse bar screens, rotary fine screens, and Parshall flumes are sources of turbulence that induce stripping of H_2S and VOCs. Influent screw pumps are also a major source of emissions. Decreasing turbulence may not be a viable control option when it is inherent to the process and cannot be reduced without adversely affecting performance. One method of reducing turbulence is to use an alternative type of flow-measuring device, such as an ultrasonic or magnetic flowmeter, rather than a Parshall flume. Turbulence can also be reduced by submerging the inlets on recycle streams that are routed to the head of the plant.

Liquid-phase chemical treatment can be used to reduce emissions at preliminary treatment units, but the chemicals should be injected far enough ahead of the units to provide adequate reaction time. Chemical feed piping can be routed to an upstream manhole to achieve the required time. In many cases, chemical treatment alone is not sufficient to provide adequate control of emissions, so containment and vapor-phase treatment may also be required. At preliminary treatment processes, ferrous chloride is a good choice for sulfide control because it can be overdosed at the head of the plant and can provide treatment through subsequent processes.

Chlorine gas is often used for prechlorination of the influent flow to reduce sulfide and odor. Chlorine reacts with a wide variety of compounds as well as sulfide, but produces volatile chlorinated by-products as well (Water Pollution Control Federation, 1990). Chlorine emissions are also corrosive to equipment, so care must be taken to avoid overdosing. Chlorine gas in ton containers is less expensive than liquid sodium hypochlorite, so it is widely used. However, many states are imposing tighter restrictions on the use of chlorine gas, which may require comprehensive risk management plans and emergency scrubbers. Many facilities have stopped using chlorine gas, in favor of liquid chemicals or ultraviolet disinfection. Hydrogen peroxide has been used with success for treating preliminary processes, but it will be consumed by organics, so it will not treat downstream processes. Hydrogen peroxide is relatively expensive, but it will treat both sulfides and VOCs without chlorinated by-products.

Grit removal processes are often a source of high emissions. Aerated grit chambers use air diffusers to separate the heavier particles from the lighter organics. Because of the intense aeration, these units have significant emissions, but it is not possible to reduce the aeration without affecting the process, so liquid-phase treatment or vapor-phase control may be needed. Other types of grit removal processes such as vortex removal systems and horizontal flow systems have less turbulence, so emissions are reduced.

One method of treating influent sulfides is to recycle return-activated sludge (RAS) to the head of the plant (Joyce et al., 1995). This treatment technique is reported to work because microbes in the RAS convert dissolved sulfide to nonodorous sulfur species. Reaeration of the RAS is reported to enhance sulfide removal. This process is most applicable to plants without primary clarifiers, but is not well suited to plants with primaries, because it may cause septic conditions in the sludge.

Primary Treatment. Much of the sulfide and VOCs in the influent wastewater have been stripped by the time the flow reaches the primary settling process. In the settling process, the flow is held in a large basin under quiescent conditions for several hours to allow suspended particles to collect at the bottom of the basin, where they are collected and removed. During detention in a primary basin, under anaerobic conditions, additional sulfide formation is likely to occur. Sulfide production is usually highest during the summer when temperatures are warmer and microbial metabolism is at a peak. Sulfide formation is further encouraged by the practice of holding solids for long periods to obtain a thicker sludge. The practice of cosettling secondary solids in the primary clarifier can also accelerate sulfide generation when active microorganisms are mixed with primary organics under anaerobic conditions. Solids sloughed from fixed film processes such as trickling filters and rotating biological contactors may be largely inactive, but solids from suspended growth processes can be highly active and can lead to rapid generation of sulfides.

There are several operational controls that can be used to minimize sulfide production during primary settling. Some holding time is necessary to achieve good suspended solids removal, but many facilities have excessive detention times of more than 3 h. This situation frequently occurs when the settling basins have been sized to treat future flows or seasonal peak flows. The problem can be easily remedied by removing basins from operation to match the flow. Operational changes can be made to prevent excessive holding of sludge by increasing the frequency of sludge pumping to maintain a shallow sludge blanket in the basin. This practice results in a thinner primary sludge, but it may significantly reduce sulfide generation. The practice of cosettling sludges in the primary basins, particularly for suspended growth processes, should be avoided in the design of new facilities.

Approximately 90 percent of the emissions from a primary clarifier occur at the overflow weirs. The higher the drop from the weir to the water surface in the flume, the greater the emissions. One technique for reducing emissions is to reduce the height of the drop by raising the water surface of the flume. This can be accomplished by installing stop plates in the effluent discharge box, as shown in Fig. 6.2.3. Reducing emissions at the primary basin weirs may shift the problem to another location depending on the type of secondary process used.

When chemical treatment is needed, iron salts are a good choice because they will react with sulfide as it is formed. Iron salts can be overdosed in the influent to provide treatment through the primary process. At facilities which require nutrient removal, iron salts can be used to enhance primary treatment to remove phosphorus and significantly increase solids removal, while at the same time controlling sulfide.

Flow equalization basins may be located at a remote site on the conveyance system or at the plant site. They are intended to temporarily hold high flows for future processing when the flow subsides. Although not intended for solids removal, deposition of solids is an unavoidable part of their operation. In equalization basins that hold flows for long periods, sulfide generation is a problem and chemical addition may be required. To reduce odorous emissions from residual solids that remain after draining, the basins should be flushed with high-pressure hoses to remove the solids. In flow equalization basins that are used frequently to dampen diurnal flows, sludge collection and removal equipment should be provided similar to a primary clarifier.

Secondary Treatment. Secondary treatment systems are generally thought to be less odorous than primary treatment processes, but they can have severe odors when overloaded or operated improperly. Trickling filters can be a source of H_2S emissions, when there is sulfide in the liquid from primary treatment. The rotary distributors or fixed nozzles that spread the wastewater over the media create turbulence and cause volatilization. Trickling filters generate sulfide when they are overloaded or lack adequate ventilation to keep the process aerobic. Atmospheric conditions impact trickling

FIGURE 6.2.3 Stop plates added at primary clarifier outlet decrease the drop at weirs and reduce emissions.

filter odors, because natural draft filters can stagnate and create oxygen-limited conditions. Primary clarifier operation should be improved or chemical treatment provided to reduce emissions at the distributors. Installing blowers to provide forced air ventilation will prevent stagnation of the airflow that occurs in natural draft systems when the air temperature equals the wastewater temperature. Slowing the distributor arm increases the flow intensity and yields a thinner biofilm, which discourages anaerobic activity in the lower layers of deeper films. Rotational speed can be decreased by installing an electric drive, adding a hydraulic brake, or simply opening ports on the front side of the distributor arm, as shown in Fig. 6.2.4. In some trickling filters and rotating biological contactors, it may be necessary to decrease the organic loading to within the aerobic limits of the units.

Activated sludge processes require aeration, which results in the stripping of VOCs to the atmosphere. However, strong odors are not common at aeration basins, as long as sufficient oxygen is supplied to keep the secondary process aerobic. Because of the high turbulence, surface aeration strips more VOCs than diffused aeration. Research also indicates that coarse bubble diffusers strip more than fine bubble diffusers (Parker et al.,

FIGURE 6.2.4 At trickling filters, the rotational speed of the distribution arm can be slowed by opening front ports.

1993). By comparison, the pure oxygen process has much lower VOC emissions (Joint Powers Agencies, 1990). If VOC emissions are a concern, consideration should be given to converting the existing aeration system to one that produces fewer emissions.

The release of VOCs during aeration is unavoidable, but operational measures can be taken to reduce emissions. Decreasing aeration rates at the head of the tank where the mixed liquor is returned will encourage biological uptake of volatile compounds before they are released. Increasing the aeration rates in smaller, covered processes, such as preaeration basins or aerated grit chambers, will volatilize compounds at those locations and reduce them in subsequent treatment processes.

Secondary clarifiers are usually not significant emission sources. The secondary solids have little remaining organics, and VOCs have been thoroughly degraded or stripped out of solution in the secondary treatment process.

Disinfection. As previously noted, the use of chlorination for wastewater disinfection produces volatile chlorinated by-products (WPCF, 1990). One significant chlorinated by-product is trichloromethane. Emissions could be minimized by using automatic controls to pace the chlorine dosage to the flow to avoid adding excess chlorine. In cases where plant VOC emissions must be reduced, the use of ultraviolet disinfection should be considered.

Advanced Treatment. Generally, advanced liquid treatment processes will not be a major source of odor or VOCs. The use of chlorine during backwashing of effluent filters for cleaning and algae control may be a minor source of emissions. Effluent filter emissions can be minimized by reducing the frequency of backwashing. The use of a cover or screen over the filters will block sunlight and help control algae growth and may reduce backwashing requirements.

A summary can be found in Table 6.2.3.

TABLE 6.2.3 Liquid Process Emissions Control

Process	Problems	Control measures
	Preliminary treatment	
Coarse bar screens Fine bar screens	Influent sulfide and VOCs are stripped by turbulence inherent to these processes.	Upstream chemical addition. Recycle RAS to headworks.
Parshall flumes	Turbulence causes stripping.	Use ultrasonic or magnetic flow measurement.
Grit basins	Aerated grit basins induce stripping.	Vortex and horizontal flow grit basins are less turbulent.
	Primary treatment	
Primary clarifiers	Sulfide formed in basins during holding. Sulfide and VOCs stripped at weirs. Sulfide forms in settled solids.	Remove unneeded basins from service. Raise water level in flume to decrease weir drop. Pump sludge more often. Avoid cosettling of sludge. Add iron salts directly or upstream.
Flow equalization basins	Odor from residual solids in flow equalization basins.	Provide collection and removal equipment or flush solids with high-pressure hoses.
	Secondary treatment	
Trickling filters Rotating biological contactors (RBCs)	Influent sulfides stripped at distributors. Sulfide formed when overloaded or oxygen deficient.	Add iron salts upstream. Limit loading. Provide power ventilation. Slow distributors or increase wetting rate to maintain a thin aerobic biofilm.
Aeration basins	Influent sulfide and VOCs stripped at head of basin. Sulfides form when oxygen deficient.	Decrease aeration at head of basin. Fine bubble diffusers cause less stripping than coarse bubble. Pure oxygen has lowest emissions.
	Disinfection	
Chlorination	Volatile chlorinated by-products formed during disinfection.	Use automatic controls to pace chlorine dosage to flow. Convert to ultraviolet disinfection.
	Advanced treatment	
Effluent filters	VOCs formed by chlorine used during backwashing to clean and control algae.	Minimize filter backwashing. Cover filters to block sunlight and control algae.

Solids Handling Processes

Thickening. Sludge thickening facilities can be significant sources of emissions. One process that is a common source of significant emissions and odor is gravity thickening (see Fig. 6.2.5). The long retention times and anaerobic conditions in gravity thickeners

FIGURE 6.2.5 H_2S emissions at gravity thickeners are reduced by separate thickening of primary and secondary sludges.

can result in sulfide generation. When biological sludge is cothickened with primary sludge, the combination of a hungry population of organisms and a surplus of food in an oxygen-deficient environment has the potential to create severe odors. The solution to this problem is to thicken primary and biological sludges in separate tanks, if possible. If only one tank is available, it may be necessary to add chemicals along with the influent sludge to treat sulfide as it is formed.

Dissolved air flotation (DAF) thickeners are typically used for thickening waste-activated sludge. These units use a high aeration rate to float solids, so any sulfide or odors present in the secondary solids will be stripped out in the DAF thickener. As with gravity thickeners, chemical pretreatment may be needed to control odorous emissions.

Dewatering. Belt presses are major sources of odor because the physical process of pressing the sludge to remove water also forces out odors and VOCs. Although ventilation hoods are often installed over belt presses, they are typically not effective at capturing emissions before their escape into the room. Belt press rooms are often found to have excessive H_2S concentrations, which are a cause of concern for operator safety.

One of the most effective methods for controlling belt press emissions is the use of potassium permanganate to pretreat the feed sludge. Compared to other chemicals, potassium permanganate is relatively expensive for large wastewater flows, but the cost difference is not as significant with smaller sludge flows. It offers an advantage over iron addition because it is an oxidant that reacts with both H_2S and other odorous compounds. It has the added benefit of reducing polymer requirements during dewatering, which decreases the overall cost. Another chemical treatment option is hydrogen peroxide, which also treats a wide range of compounds. Permanganate is generally preferred to peroxide because it is safer to store and handle. In sensitive areas, where vapor-phase control is needed to treat ventilation air from the belt press room, chemical pretreatment can reduce operational costs of the vapor-phase treatment.

Stabilization. Anaerobic digestion is one of the most widely used sludge stabilization methods at large municipal wastewater facilities. The process provides a high degree of volatile solids destruction and is highly effective at inactivating pathogens. The process also yields methane gas, which can be used to heat buildings or generate electricity. The H_2S in anaerobic digester biogas is a major concern because it causes corrosion and can significantly reduce the life of cogeneration equipment. The air quality impact is also an issue, because H_2S is converted to sulfur dioxide (SO_2) during combustion. Depending on the region, air permitting agencies may require reduction of SO_2 emissions. When internal-combustion engines are used, carbon monoxide (CO) emissions may be above the major source threshold. An oxidation catalyst can be installed to reduce carbon monoxide emissions, but proper catalyst operation requires treatment to reduce H_2S in the biogas to 25 ppm or less.

Proper operation of anaerobic digesters, to maintain a pH and temperature favoring the methane-forming bacteria over the sulfide-forming bacteria, will maximize biogas production and thereby decrease H_2S concentrations. However, due to the nature of the anaerobic process, moderately high H_2S concentrations will always be present. Biogas H_2S concentrations usually range from 1000 to 3000 ppm. If biogas utilization is planned, then some form of treatment may be required. One method for reducing H_2S is to add iron salts to treat sulfide in the liquid phase before it is released to the biogas. Otherwise H_2S must be removed in the vapor phase by using wet or dry scrubbing.

Based on the experiences at numerous wastewater facilities, the application of iron salts effectively reduces H_2S in digester biogas. To avoid scaling and plugging due to vivanite formation, overdosing of the influent wastewater or direct addition into the digester is preferable to injection into sludge pipes. When iron salts are added to the influent, more iron is consumed by sulfide and other compounds in other processes. Therefore, less iron will be required for controlling H_2S in the digester biogas, if it is added directly to the digester tank. However, adding the iron salts at the plant influent also provides odor control at preliminary and primary treatment processes.

Ferrous chloride has been the most frequently used iron salt, because it is less expensive than ferric chloride. Ferrous chloride costs about $0.04 per pound ($0.09 per kilogram), which is one-half the cost of ferric chloride. The amount of iron added must be in excess of stoichiometric requirements to account for reactions with compounds other than sulfide, such as phosphates. Generally, a ferrous chloride–to–sulfide dosage rate of 12 lb/lb provides effective control, but exact dosages are site-specific, due to interferences with other compounds. Based on sulfate concentration, the ferrous chloride–to–sulfate dosage rate is about 4 lb/lb. If the sulfate content of the feed sludge is not known, the dosage can be estimated based on sludge solids, with a ferrous chloride–to–solids dosage rate range of 7 to 15 lb/1000 lb total suspended solids. It appears practical to reduce digester biogas H_2S down to 100 ppm using ferrous chloride, but it may be difficult to maintain lower concentrations, so vapor-phase treatment may be required when carbon monoxide reduction is needed.

Aerobic digesters can also be a source of odor generation when the process is overloaded and becomes oxygen-deficient. Many times mixing is inadequate to thoroughly aerate the sludge. Operators should take care to provide adequate mixing and aeration to prevent anaerobic conditions.

Lime stabilization involves adding quicklime or hydrated lime to sludge to increase the pH and inactivate pathogens. During lime stabilization, significant amounts of ammonia are released due to the high pH. High concentrations of ammonia may pose a health problem for workers if the process exhaust is not contained and vented outdoors. Generally, ammonia does not pose an odor problem, because it is rapidly dispersed to below-detectable levels. However, if the ammonia concentrations are extremely high or the plant is next to a sensitive residential area, vapor-phase treatment may be required.

The composting process generates heat and produces a wide variety of reduced sulfur and amine compounds, which have highly objectionable odors. Several operational practices are known to affect odor generation in the composting process. In general, well-digested sludge has much less odor-generating potential than raw sludge. Feed sludge that is too wet has a greater potential for odor generation, so a feed solids of at least 15 percent should be maintained. Aerated compost piles produce less odorous compounds than nonaerated piles; however, surface emissions may be higher in an aerated pile unless the air is drawn down through the pile and captured for treatment before discharge.

The use of wood ash as an amendment to the compost pile will reduce emissions of reduced sulfur compounds such as dimethyl sulfide (DMS) and dimethyl disulfide (DMDS). There are also various chemical products on the market which purport to reduce emissions from composting when sprayed on the surface of the piles.

Storage. The short-term storage of combined biological and primary sludge prior to stabilization can cause sulfide generation problems. The best operational practice is to maintain separate storage of the sludges. If combined sludges must be stored, it may be necessary to add iron salts directly to the storage tank to treat sulfide. If mixers have been installed in the sludge storage tank, continuous mixing of the sludge may reduce sulfide generation.

Long-term sludge storage facilities are a potential source of odor. It was commonly thought that lime stabilization completely arrested microbial activity, and that odor production would cease. However, recent experience with lime-treated liquid sludge shows that stratification of the sludge occurs, and that pH decreases during long-term storage.

FIGURE 6.2.6 Emissions can be reduced during pile breakdown by limiting the amount of open face exposed.

When the pH drops, microbial activity begins and odor generation proceeds. This problem can be avoided by providing mixing to prevent stratification of the stored sludge. Supplemental lime addition may be necessary to ensure that pH is maintained.

Long-term storage of lime-stabilized sludge cake in piles has also resulted in odor problems during breakdown of the piles for transport and application. The breakdown of active compost and curing piles for relocation or disposal can also be a source of odors. One means of reducing emissions during pile breakdown is to minimize the amount of open face that is exposed (see Fig. 6.2.6). It is also a good operational practice to cover the open face when the breakdown activity is stopped overnight or on weekends. Another potential operational strategy is to limit work on the piles to favorable wind and weather conditions to minimize the transport of odors off the site.

Land Application. During land application, biosolids are transported to the application site and spread over a large area. Nuisance odors are sometimes released during the transport and application of biosolids. One option to lessen the odor impact is to wait and apply biosolids during ideal wind and weather conditions, but this is not always possible. In many cases, land application must be completed rapidly to allow timely planting of a crop. One operational practice to minimize the odor nuisance is to apply highly odorous biosolids in remote areas of the field. The type of application method also has an impact on odor. The use of subsurface injection of liquid sludge may result in less odors than the surface spreading of sludge cake.

See the summary in Table 6.2.4.

TABLE 6.2.4 Solids Handling Emissions Control

Process	Problems	Control measures
	Thickening	
Gravity thickeners	Cothickening biological and primary sludges causes sulfide generation. Long detention under anaerobic conditions is problematic.	Avoid cothickening, if multiple basins are available. Use direct chemical treatment to reduce sulfide formed during thickening.
Dissolved air flotation	Aeration strips sulfide and odors from sludge.	Use chemical pretreatment to remove sulfide from sludge before processing.
	Dewatering	
Belt presses	Pressing strips sulfide and VOCs from feed sludge into belt press room.	Potassium permanganate or hydrogen peroxide will treat sulfide and other odorous VOCs.
	Stabilization	
Anaerobic digesters	The H_2S formed in the process corrodes combustion equipment. Air quality is a concern, because H_2S is converted to sulfur dioxide during combustion.	Maintain proper temperature and pH in the process. Add iron salts directly to the digester, at headworks, or at primaries.
Aerobic digesters	Odorous compounds form when process is overloaded or oxygen-deficient.	Provide adequate aeration and mixing to maintain aerobic conditions.
Lime stabilization	Ammonia is released due to high pH.	Vent ammonia to outside unless concentrations are very high or the site is in a sensitive area.
Composting	Decomposition of organics at high temperatures produces a wide variety of odorous compounds.	Aerate piles. Draw air down through pile and treat. Avoid too wet sludge. Use wood ash as an amendment. Apply pile surface treatment chemical.
	Storage	
Short-term	Storage of combined biological and primary sludge generates sulfide.	Avoid storing combined sludge. Provide mixing to maintain aerobic conditions. Add iron salts directly to storage tanks to treat sulfide formed.
Long-term	Lime-stabilized biosolids may become odorous in long-term storage. Breakdown of storage piles releases odor.	Add supplemental lime to maintain pH during long-term storage. Limit open face exposed, and cover when possible. Break down piles under favorable weather conditions.
Land application	Odor is released during application of biosolids to a large land area.	Apply highly odorous biosolids in remote areas of field. Use subsurface injection to minimize odor.

REFERENCES

American Concrete Pipe Association, *Design Manual for Sulfide and Corrosion Prediction and Control,* Vienna, Va., 1984.

American Concrete Pipe Association, HS Program, University of Gainesville, Fla., 1991.

American Society of Civil Engineers, *Sulfide in Wastewater Collection and Treatment Systems,* ASCE Manuals No. 69, New York, 1989.

Ballinger, K. E., Knaver, R., and Daily, J., Controlling H_2S generation in wastewater treatment systems by metabolic modification of sulfate reducing bacteria, *Proceedings 1994 WEF Specialty Conference on Odor and VOC Emissions Control,* Alexandria, Va., 1994, p. 7-37.

Beck, G. S., Deposition and pipe corrosion, *Water Environment & Technology,* Feb. 1995, p. 46.

Bowker, R., King, A., and Holcomb, G., U-tube oxygen dissolver controls odors, *Water Environment & Technology,* Jan. 1995, p. 20.

Cohn, Morris M., Sewage works and their odor control, *Sewage Works Journal,* Oct. 1929, **1**(5): 568.

Environmental Protection Agency, *Design Manual for Odor and Corrosion Control in Sanitary Sewerage Systems and Treatment Plants,* EPA/625/1-85/018, Office of Research and Development, Cincinnati, Ohio, 1985.

Environmental Protection Agency, *Detection, Control, and Correction of Hydrogen Sulfide Corrosion in Existing Wastewater Systems,* EPA 832-R-92-001, Washington, D.C., 1992.

Hagekhalil, A., et al., Control of H_2S and related odor and corrosion problems in the City of L.A. wastewater collection system utilizing nitrate addition. *Proceedings 1994 WEF Specialty Conference on Odor and VOC Emissions Control,* Alexandria, Va., p. 6-13.

Hamilton, J., et al., H_2S abatement in the City of L.A. wastewater collection system, *Proceedings 1994 WEF Specialty Conference on Odor and VOC Emissions Control,* Alexandria, Va., 1994, p. 6-25.

Joint Powers Agencies, *Pooled Emissions Estimating Program Final Report,* James M. Montgomery Consulting Engineers, 1990.

Joyce, J. et al., H_2S biological oxidation, *Water Environment & Technology,* Mar. 1995, p. 40.

Laughlin J. A., Studies in force main aeration, *Journal ASCE Sanitary Engineering Division,* Dec. 1964, **90**(SA 6): 13.

Microcomp Systems, Sulfide Works, 1991, Contact: Kienow Associates, Redlands, Calif.

Parker, W. J., et al., Fate of volatile organic compounds in municipal activated sludge plants, *Water Environment Research,* Jan./Feb. 1993, **65**(1): 58.

Pomeroy, R. D., and Parkhurst, J. D., The forecasting of sulfide buildup rates in sewers, *Progress in Water Technology,* 1977, **9**(3): 621.

Quigley, C. J., and Corsi, R. L., Emissions of VOCs from a municipal sewer, *Journal Air & Waste Management Association,* May 1995, **45:** 395.

Sydney. R., Esfandi, E., and Surapaneni, S., Control concrete sewer corrosion via the crown spray process, *Water Environment Research*, May/June 1996, **68**(3): 338.

Van Durme, G., and Berkenpas, K., Comparing sulfide control products, *Operations Forum,* Feb. 1989, **6**(2): 12.

Water Pollution Control Federation, *Vapor Phase Control Technology Assessment,* WPCF Research Foundation Report 90-2, Alexandria, Va., 1990.

Yao, K. M., Functional design of sanitary sewers, *Journal Water Pollution Control Federation,* July 1976, **48**(7): 1772.

CHAPTER 7
FOUL AIR COLLECTION AND CONVEYANCE

Phillip Wolsteinholme
Brown & Caldwell
Seattle, Washington

BUILDING PRESSURIZATION 7.2
TANK AND CHANNEL COVERS 7.4
Air Movement Theory 7.4
Empirical Testing 7.5
Cover Construction 7.5
Cover Costs 7.8
HOODS AND ENCLOSURES 7.9
FOUL AIR DUCTWORK 7.9
FOUL AIR FANS 7.12
Centrifugal Fans 7.13
Axial Fans 7.14
Fan Sizing 7.14
Fan Laws and Performance 7.14
Fan Material 7.14
REGULATIONS 7.14
SUMMARY 7.15
RECOMMENDED FURTHER READING 7.15
REFERENCES 7.16

The trend in recent years toward improved odor and volatile organic compound (VOC) control at municipal and industrial facilities, in particular wastewater treatment plants, has significantly increased the availability and performance of foul air treatment systems. A wealth of expertise is available to help. Engineers relish designing the systems, there is no shortage of vendors to supply them, and operators are becoming increasingly proficient at running them. While foul air treatment systems have generated a lot of interest, foul air collection and conveyance are often neglected. Yet foul air cannot be treated until it has first been collected and conveyed to a treatment structure or facility.

This chapter discusses the theory and practice of designing foul air collection systems—with particular emphasis on preventing fugitive odor release through effective design of building pressurization, tank and channel covers, hoods, and ductwork. The chapter also describes some of the empirical testing performed by engineers and suppliers and suggests design criteria for foul air containment and collection equipment. It should also be remembered that, in some cases, preventing the release of odors (such as reducing sulfide generation in sewers or turbulence in process units) can be simpler and more cost-effective than installing and operating a foul air treatment system.

The collection and conveyance of odorous air are based on establishing positive or negative pressures. Pressurization of a space is established by creating an airflow into or out of that space. The airflow is created by fans, the workhorses of all contaminated air collection systems.

Smoke testing is always a useful tool to directly observe air movement and to determine where leaks or fugitive emissions are occurring. Smoke testing can be used in buildings and in most of the other situations described in this chapter.

BUILDING PRESSURIZATION

Buildings must be placed under a positive or negative pressure in order to prevent fugitive odors from entering or exiting a building or room. Figure 7.1 demonstrates the typical concept of establishing a negative pressure in an odorous area, such as a belt press room. The exhaust airflow rate is significantly greater than the supply airflow rate, which causes an inflow of outside air into the room through gaps in doors, windows, and pipe penetrations. As a rule of thumb, the differential in supply and exhaust airflow rates is usually 10 percent for a typical building. However, the differential airflow rate required can be lower or higher and should be more accurately calculated, particularly for critical odor or VOC situations and for large-volume airflows, where power costs may be significant.

The differential pressurization of a space can be expressed as

$$\Delta p = \frac{\rho}{2}\left(\frac{Q}{C_f A}\right)^2 \tag{7.1}$$

where Δp = differential pressure across flow path
ρ = density of air entering flow path
Q = volumetric flow rate
C_f = flow coefficient
A = area of airflow path (through openings)

The term $(\rho/2)(Q/A)^2$ is the velocity pressure. For typical building openings, the flow coefficient C_f is normally 0.6 to 0.7. Using 0.65 as the coefficient and an air density of 1.2 kg/m^3 (0.075 lb/ft^3), we see that Q becomes

	SI units	English units
	$Q = 0.839\ A\ \Delta p^{1/2}$	$Q = 2610\ A\ \Delta p^{1/2}$
where units for Q, A, and Δp are as follows		
Q	m^3/s	cfm
A	m^2	ft^2
Δp	Pa	in, WC

The area of flow path can be easily calculated for large areas such as open windows and louvers, but is impossible to measure or calculate for gaps in doors and windows, etc. Therefore, tables must be used for overall leakage areas of buildings. The American Society of Heating, Refrigerating, and Air-Conditioning Engineers' (ASHRAE's) *HVAC Applications* (1991, chap. 47) provides typical leakage areas for commercial buildings as a ratio of total wall area. The ratio ranges from 0.13×10^{-2} for very loose construction to 0.70×10^{-4} for tight construction external walls. The ratio is 0.42×10^{-3} for a loosely constructed external wall. For example, a 50 m × 30 m × 5 m belt press room would have a wall area of 800 m^2 and, therefore, a leakage area of $800 \times 0.42 \times 10^{-3}$, or 0.32 m^2.

The lowest measurable and sustainable pressure differential for a building, and a quantity used by various regulatory authorities, is 25 Pa (or 0.1 inWC). Lower levels down to 12 Pa may be suggested but are difficult to sustain and measure for most buildings.

The required pressurization flow rate, to pressurize the hypothetical belt press room described above to 25 Pa (0.1 inWC), can therefore be estimated from

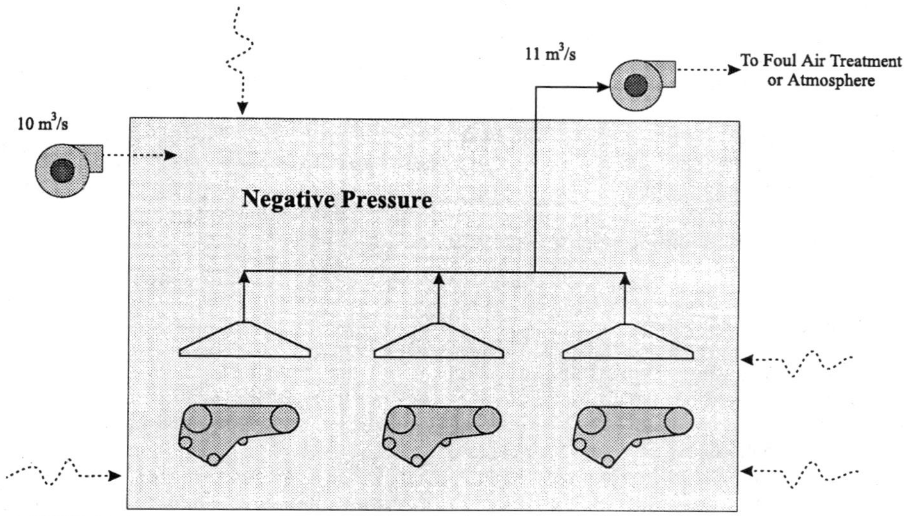

FIGURE 7.1 Building ventilation.

$$Q = 0.839 \times 0.32 \times 25^{1/2} \tag{7.2}$$
$$= 1.34 \text{ m}^3/\text{s}$$

If the belt press room were to be ventilated at a typical 12 air changes per hour, the air exchange flow rate would be

$$(50 \times 30 \times 5) \text{ m}^3 \times 12/\text{h} = 90{,}000 \text{ m}^3/\text{h}$$
$$= 25 \text{ m}^3/\text{s}$$

In this case, then, the pressurization rate of 1.34 m³/s is only 5 percent of the air exchange flow rate, and the room would require a supply airflow of 25 m³/s and an exhaust airflow of 26.34 m³/s (56,000 cfm) to create a 25-Pa (0.1-inWC) negative pressure.

Similar calculations should be performed if, e.g., an electrical control room is to be positively pressurized to prevent infiltration of corrosive gases.

Roof or floor area openings such as skylights or pipe penetrations must be included in leakage area calculations.

Whenever there are large openings in buildings (such as roll-up doors), it is impractical to supply or exhaust sufficient air to maintain a positive or negative pressure. For instance, a 3-m (approximately 10-ft) square opening would require a flow rate of about 80,000 cfm to maintain a measurable differential pressure, even under still weather conditions. Indeed, the effect of wind passing across the opening would overwhelm any prospect of building pressurization. In such cases, when doors need to remain open for long periods, either plastic curtains or air curtains can be used to reduce the wind effect. When doors are left open, the indoor atmosphere may appear improved, but such actions can destroy delicate air balances inside the building and create outdoor impacts which would otherwise be contained.

TANK AND CHANNEL COVERS

The prevention of fugitive odor release from tanks and channels is achieved by the use of covers to contain foul air and the same pressurization concept described above for buildings. Studies, such as that at the East Bay Municipal Utility District in Oakland, California (Witherspoon et al., 1994), have shown that covers themselves are de facto major contributors to air emission control. They achieve this control by increasing the partial pressure at the liquid-air interface, by reducing convection across the liquid surfaces, and by promoting condensation of vapors.

Air Movement Theory

Figure 7.2 illustrates a typical situation at an aerated tank or channel. A fan exhausts air from the enclosed headspace and creates an infiltration of air through openings in the cover. The quantity of outside air infiltrating into the headspace *I* is equal to the difference between exhaust air *E* and process air *P*.

The outside air infiltration and exhaust quantities have traditionally been based on an air exchange rate principle. This is still the governing parameter when frequent operator access to the enclosed space, such as a dome-covered vessel, is required. However, a typically required exchange rate of 12 air changes per hour can result in enormous quantities of air flowing through a tank and the downstream foul air treatment system. If frequent operator access is not required, substantial cost savings can be achieved by reducing the quantity of air to be moved. It may be argued that lowering the outside airflow reduces the purging effect in the headspace and accelerates corrosion. However, the initial cost of concrete protection is likely to be significantly less than the alternative—many years of high-volume air purging. For temporary occupation of such a space, the code-recommended rate of ventilation could be obtained by use of a push-pull system: supply fan(s) and exhaust fan(s).

The negative pressure developed at a headspace cover is a function of velocity through the cover openings. As was the case with buildings, Eq. (7.1) is again used to calculate the pressure established at the cover:

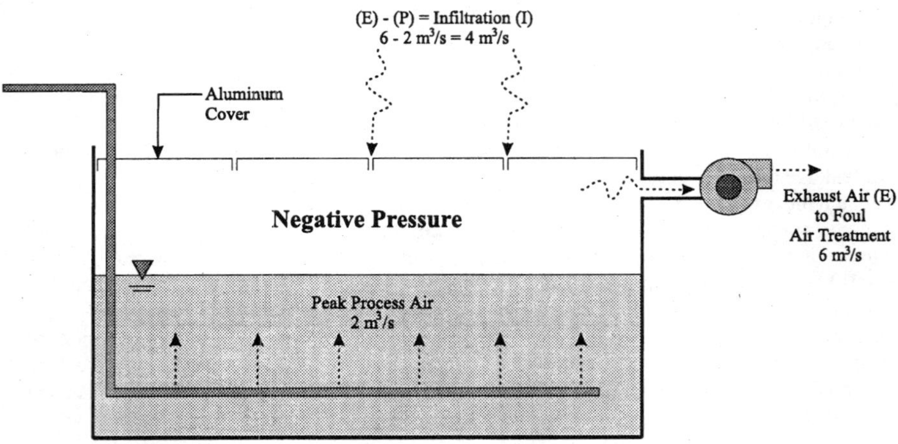

FIGURE 7.2 Covered channels and tanks.

$$\Delta p = \frac{\rho}{2}\left(\frac{Q}{C_f A}\right)^2$$

This can alternatively be expressed as a typical heating, ventilating, and air conditioning (HVAC) pressure loss equation:

$$\Delta p = C_L V_p$$

where C_L = loss coefficient, which is dimensionless, and V_p = velocity pressure.

The American Conference of Governmental Hygienists' 1980 handbook recommends, based on empirically established data, a value of 1.78 for the loss coefficient; thus

$$\Delta p = 1.78 V_p$$

For fixed solid or permanently sealed covers that are not frequently removed and have few penetrations or joints, outside air inlets can be furnished to provide the calculated cross-sectional area. This will create a known velocity and establish the required pressure. An example would be concrete covered tanks with sealed access hatches. However, for typical removable covers with many joints and openings, determining velocity pressure by measuring or calculating the area of cracks and openings in the cover presents an impossible challenge. The relationship between flow and consequent pressure must currently be determined empirically.

Empirical Testing

During the mid-1990s, tests were performed at several facilities. One test was performed (Brown and Caldwell, 1993) on flat aluminum covers at King County's East Section Reclamation Plant in Renton, Washington, to determine what negative pressure could be established under covers and to establish a correlation between negative pressure and airflow rate in terms of airflow per cover area.

Initial pilot testing of temporary, flat, aluminum, ungasketed covers over an unaerated, primary channel showed that an exhaust airflow rate of 0.0076 m^3/s per square meter (1.5 cfm per square foot) of cover area was required to establish a measurable negative static pressure in the headspace. Negative static pressure of 12 Pa (0.05 inWC) was achieved for unsealed flat covers. This increased to 25 Pa (0.1 inWC) when penetrations and joints were caulked.

Full-scale testing in 1995 on installed flat, aluminum, ungasketed covers later demonstrated that an airflow rate of 0.0061 m^3/s per square meter (1.2-cfm per square foot) produced a negative static pressure of 50 Pa (0.2 inWC) (Pacific Inspection and Research Laboratory, 1995). A summary of these and test results on covers at other wastewater treatment plants (Brown and Caldwell, 1996) is presented in Table 7.1. The 1995 King County tests (by isolating areas of the cover) revealed that approximately one-half of the exhaust airflow was infiltrating through the cover surface and one-half from the perimeter connection between the aluminum and concrete.

Cover Construction

Cover construction factors to be considered are

Permanency	Fixed or removable
Accessibility/visibility	Hatches, clear panels, hinged, counterbalanced

TABLE 7.1 Pressures Established under Flat Aluminum Covers

Test location*	Air flux		Pressure established		Notes
	$m^3/(s \cdot m^2)$	cfm/ft^2	Pa	inWC	
1	0.0076	1.5	12	0.05	Uncaulked
2	0.0076	1.5	25	0.10	Caulked at perimeter
3	0.0061	1.2	50	0.20	Gasketed, uncaulked
4	0.0051	1.0	25	0.1	
5	0.0061	1.2	10	0.04	Ungasketed
6	0.0051	1.0	2	0.01	Ungasketed, many penetrations

*Test locations: 1, 2: pilot test, King County's East Section wastewater treatment plant (WWTP), 1993; 3: installed covers test, King County's East Section WWTP, 1995; 4: Sioux City, Iowa, WWTP; 5, 6: city of Edmonton, 1996.

Aesthetics	Sun reflection, surface oxidation
Sealing	Gasketed, caulked, permanently sealed
Air inlet openings	Designed or through gaps

Tank or channel covers have been constructed from concrete, aluminum, fiberglass, and fabric materials. A greater proportion of concrete reduces air infiltration but decreases accessibility and increases the need for corrosion protection. Aluminum is probably the most widely used cover material with availability and competition from several established national vendors as well as local aluminum fabricators. Figure 7.3

FIGURE 7.3 Aluminum covers on circular tanks.

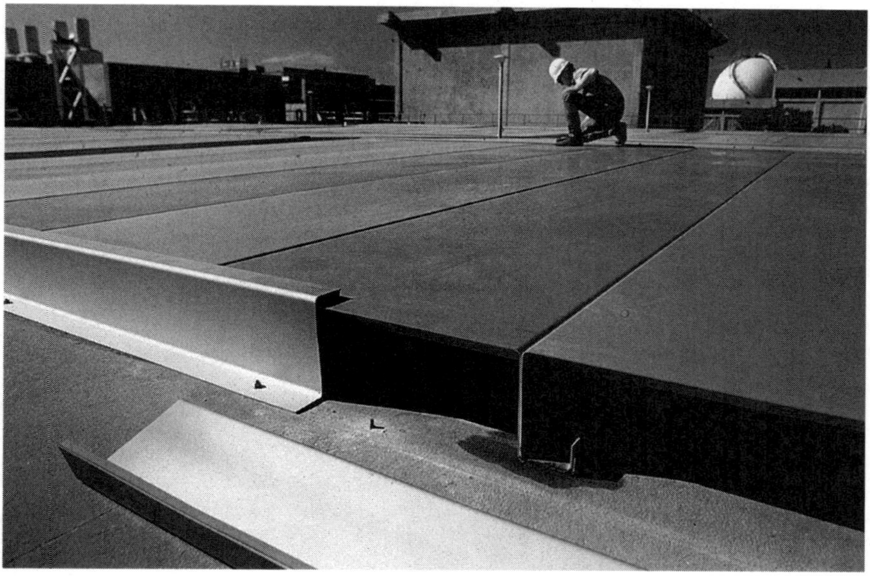

FIGURE 7.4 Aluminum cover on rectangular tank. (*Photo courtesy of Temcor.*)

shows a flat, aluminum cover on a circular tank, and Fig. 7.4 shows aluminum covers on a rectangular tank. Fiberglass offers excellent corrosion resistance, but its life span is likely less than that of aluminum. Structural-fabric covers are a more recent material option. The covers can be fabricated for all shapes of tank and are frequently designed to accommodate normal snow loads. A fabric cover is shown in Fig. 7.5.

Removal of covers may be performed by hoist or crane, or manually. In the last case, the maximum cover weight must meet Occupational Safety and Health Administration (OSHA) standards, which currently require the weight to be less than 45 kg (100 lb) to allow two people to lift the cover.

Handrailing around the tank must be provided if removable covers are installed on otherwise open tanks. Handrailing may either be permanent or temporary. In the same way as leaving open the doors of buildings destroys the potential for creating a negative pressure, open hatches on covers ensure that no negative pressure can be achieved in the vicinity of the hatch. Thus hatches should be easily opened and closed, counterbalanced if large, or replaced by transparent hatches. Grease and condensation accumulation on the inside of transparent hatches can be frustrating to the observing operator, although wiper blades have been successfully installed to maintain transparency, at least for several years.

Operator-accessible covers should have an integral nonskid surface and minimal deflection for the safety and security of operators who routinely walk on the covers. Structural design criteria should be based on the cover function. However, when covers are designed for typical live loads that only accommodate one or two people, warning signs to that effect should be posted to prevent overloading and failure.

To reduce the expense of covering large liquid surfaces such as found at clarifiers, it may be cost-effective to cover only the area of the tank where emissions are highest. Areas such as aerated zones, weirs, and scum collection troughs can be covered to collect the majority of the emissions from the process vessel.

FIGURE 7.5 Structural-fabric covers on circular tanks.

Certain process units, despite their potential for odorous emissions, can often be controlled without covers. Trickling filters can be designed to operate with airflow in the downdraft (top-to-bottom) mode such that the upper surface of the trickling filter medium is under slightly negative pressure. The air quantity required to achieve a negative pressure under windy conditions may be considerable, and it will normally require foul air treatment. The second option for trickling filters is to cover the filter, continuously recirculate the process air, and exhaust only a portion of the process air to the treatment system. This smaller quantity of foul air directed to the treatment process should be replaced by fresh outside air of sufficient quantity to satisfy the oxygen demand of the process. The cost of the trickling filter cover in the second option should be compared to the cost of conveying and treating the higher airflow in the first option.

Trickling filters are themselves often used to treat foul air from other sources, particularly when the trickling filter is lightly loaded and/or in the nitrifying mode (Lutz et al., 1994).

Cover Costs

Typical costs for rectangular, flat, aluminum covers over wastewater channels are in the range of \$130 to \$200 per square meter (\$11 to \$18 per square foot). The cost for circular, flat, aluminum covers over clarifiers will be up to twice that of the narrower, rectangular covers. Aluminum domes (e.g., over trickling filters) are likely to cost \$270 to \$350 per square meter (\$25 to \$33 per square foot). Fiberglass covers are similar in cost to aluminum and fabric covers and cost \$100 to \$200 per square meter (\$10 to \$20 per square foot) (Kepets, 1996). It may be more cost-effective if local sheet-metal fabricators or plant staff provide custom covers for small, irregularly shaped areas than to modify standard covers.

HOODS AND ENCLOSURES

The function of hoods (when used for foul air collection) is to capture the contaminated air as close to the point of release as possible, with as low an airflow as possible. Ideally, for maximum capture, the hood should be connected to the piece of equipment. This is not always possible, as a hood may impair operator visibility or access. For instance, belt presses and gravity belt thickeners are often equipped with rectangular canopy hoods over the equipment to capture released vapors and odors. In such cases, hoods should be located as low as possible, yet allow operator access. The air velocity to capture the odorous vapor emissions *at the point of release* should be in the range of 0.25 to 0.50 m/s (50 to 100 fpm) (ASHRAE, chap. 27, 1991). If the hood face were at the point of contaminant release, the hood face velocity would be the capture velocity. As the hood face is elevated some distance above the release point, the flow rate needs to be increased according to the equation

$$Q_n = Q_c \frac{10X^2 + A}{A}$$

where Q_n = hood flow rate, m^3/s
Q_c = capture flow rate, m^3/s
A = hood face area, m^2
X = vertical distance from hood face to point of release, m

For example, consider a gravity belt thickener hood with a face dimension of 8 m × 3 m (26 ft×10 ft) with the hood 1.2 m (4 ft) above the point of release. Even at the low end (0.25 m/s) of the capture velocity range, this would result in a required hood velocity of 0.4 m/s and a hood airflow rate of 9.6 m^3/s (20,800 cfm).

Airflow rates can be significantly reduced by collecting the contaminated air locally at the point of release. For example, by installing exhaust ducts at a gravity thickener feed tank, a belt press cake hopper, or a conveyor transfer point, or by totally enclosing the equipment item, a lower airflow rate can be used to capture foul air. Often it is not practical to enclose or even cover equipment or other odorous sources. In that case, alternatives may include adding odor-reducing chemicals to the feedstock or locating the equipment item in a room within a room. Figure 7.6 illustrates the example of sludge-thickening equipment located in a glass and plastic enclosure ventilated at a higher air exchange rate than that of the remainder of the surrounding room.

If air is collected for treatment from the points of highest concentration, it may be possible to exhaust the general room air to atmosphere without treatment, thus considerably reducing the size and cost of an otherwise expensive treatment system. Gases can accumulate near ceilings or floors, depending on their density and temperature, and at least some of the exhaust collection points should be located accordingly.

Hoods should be constructed of the same material as their downstream exhaust ducts.

FOUL AIR DUCTWORK

Ductwork is used to convey air from the point of contaminant release and collection to the treatment system. Its cost can be significant, often on a par with that of the treatment system.

Notwithstanding site-specific criteria that are noted below, ductwork is generally designed to balance construction and operating cost. Construction cost is a function of type and quantity of material, while operating cost is a function of the energy required

FIGURE 7.6 Full enclosure around gravity belt thickeners.

to transfer the air through the system. Ductwork is constructed of a wide variety of materials, including galvanized steel, aluminum, stainless steel, fiberglass, and plastic. The material choice is dependent on both the external environment and the corrosiveness of the conveyed air. The most cost-effective ductwork material that meets corrosion demands should be used. Duct sizing is based on velocity (to reduce noise) and air friction loss (to conserve fan energy). If facilities are used intermittently, then smaller ducts and slightly higher than normal friction losses will be cost-effective. The three accepted methods of ductwork design are equal function, static regain, and T method. The T method was developed (Tsal, 1990) to optimize the complete fan and duct system. Nowadays, computer software is available to assist engineers, particularly when designing larger, more complex systems. Nonetheless, even the most accurately designed systems should be fitted with dampers at all major branches. The inclusion of dampers allows for some tolerance in design, permits minor modifications during construction, and greatly assists the balancing contractor.

Accurate and regular air balancing is important because variations in airflow rates can upset the pressure balance, resulting in leaks and fugitive emissions. A balancing contractor should always be specified to be a member of the Associated Air Balance Council or the National Environmental Balancing Bureau and should perform the balancing to the respective standard methods.

Pressure losses in ductwork can be reduced by using gradual rather than abrupt transitions. Figure 7.7 shows duct-fitting options that provide superior flow characteristics, resulting in up to 5 times lower frictional pressure drop, less noise, and lower energy cost. Other components that significantly reduce frictional pressure drop and produce overall cost savings are turning vanes to smooth out airflow through elbows, using extractors to direct air to and from registers, and straightening vanes to reduce turbulence, particularly at fan inlets.

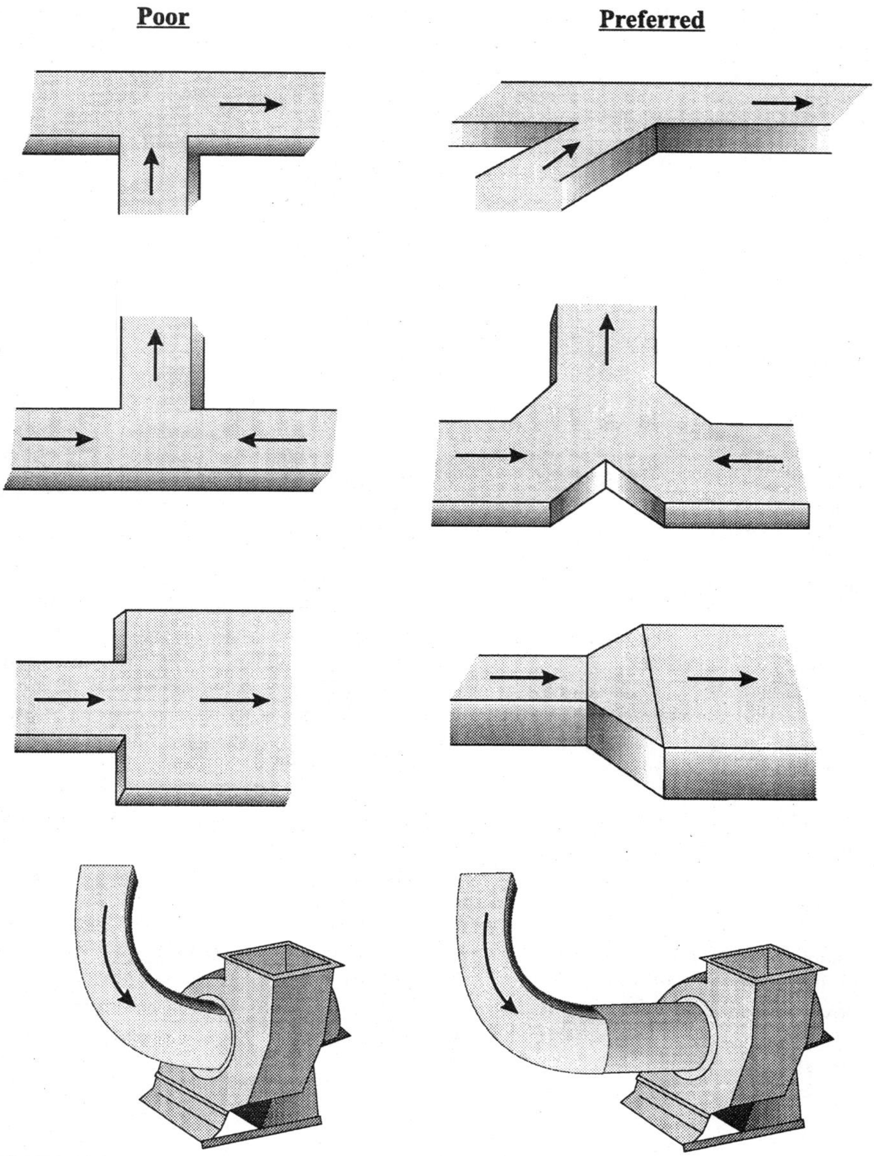

FIGURE 7.7 Alternative duct fittings.

Duct wall thickness is determined by the anticipated positive or negative air pressure. Ductwork should be selected according to the rating classification for the anticipated pressure.

Suggested design criteria for odor and VOC ductwork are presented in Table 7.2. Higher pressure may be experienced around control devices, for instance, bulk media biofilters, where static pressures up to 4 kPa (16 inWC) can be generated.

Site-specific criteria that will influence ductwork design are as follows:

TABLE 7.2 Ductwork Design Criteria

Air velocity	<13 m/s <2500 fpm
Friction loss	8–12 Pa/m 0.10–0.15 inWC/100 ft
Stack exit velocity	15 m/s 3000 fpm
Typical internal pressure rating	500–2300 Pa 2–10 inWC

Noise. Ductwork can be oversized to reduce velocity, or where this is impractical, noise or sound attenuation devices may be added.

Particulates. Minimum velocities in ductwork are required to maintain particles in suspension (see ASHRAE, 1991, chap. 27). Ductwork is then designed for constant velocity.

Designers should be aware that the velocity required to resuspend settled particles may be higher than the velocity needed to convey those same particles. An additional concern when handling particulate streams is the potential for explosion. When such a potentially dangerous situation is anticipated, experts in the field should be consulted.

A comparison of alternative ductwork materials is presented in Table 7.3. In the design of fiberglass and plastic ductwork, care should be taken to accommodate high linear expansion, which is several times that of metal duct. The majority of foul air ductwork experiences internal condensation when, having collected air from warm processes or wastewater, it is routed outside or underground through a cooler environment. Provision should be made for trapped drainage from those ducts.

Fire Control. Ductwork can be fire-rated, fire and smoke dampers included, and smoke detectors strategically located. National Fire Protection Association (NFPA) Standard 90A addresses fire protection in ductwork. Particular care should be taken in designing nonmetallic, potentially flammable ductwork.

Space Allocation. Rectangular ductwork aspect ratios (width to height) may be adjusted beyond the normally preferred limit of 3:1 to allow construction in space-constricted locations. Rectangular metal ductwork is slightly more costly than round ductwork. However, rectangular fiberglass ductwork can be up to 5 times as costly as round fiberglass ductwork.

FOUL AIR FANS

The fan is the air equivalent of a pump in a water system. Although the terms *fan* and *blower* are often used synonymously, a blower, in theory, has most of its pressure resistance on the downstream side. The construction of fans and blowers is identical. Fans are classified as centrifugal or axial. The impeller of a centrifugal fan turns the airstream through 90°, whereas the airstream does not change direction with an axial fan.

TABLE 7.3 Alternative Ductwork Materials

Ductwork Material	*Corrosion Resistence*	*Installed Cost*	*Available Cross-Sections*	*Comments*
Galvanized Steel	◕	○	Round, Rectangular	Used in a non-corrosive environment
Aluminum	◐	◔	Round, Rectangular	Normal alloy 3003-H14. Has moderate sulfide resistance.
Fiberglass	○	◐	Round, Rectangular	Vinyl ester preferred. Rectangular very costly. Indoors needs fire protection. Quality control required.
Stainless Steel	◔	◐	Round, Rectangular	Use grade 316 when acids are present.
PVC	○	◔	Round	Not available in larger diameters.
High-Density Polyethylene	○	◔	Round	Good support needed. Often used for underground application.

Key:
○ Most favorable
● Least favorable

Key:
◯ Round
□ Rectangular

Centrifugal Fans

Centrifugal fans are generally more efficient than axial fans, are used for higher-pressure applications, and are consequently more expensive. There are four types of centrifugal fans, as described by their impeller shapes: air foil, backward-inclined, radial, and forward-curved. The first two types are typically used in odor and VOC applications due to their stability and efficiency. Radial impellers are used for particulate, abrasive conditions, and forward-curved are used in low-pressure HVAC applications.

Axial Fans

Axial fans are generally used for high-volume, low- to medium-pressure applications. The pressure capability of a simple axial propeller fan is increased by enclosing the impeller in a tube and further increased by adding guide vanes in the tube to straighten out the airflow.

Both centrifugal and axial fans are used as roof ventilators. The typical mushroom-cap roof exhauster should be specified in an upblast configuration when air must be dispersed vertically.

Fan Sizing

Similar to pumps, fans should be sized so their fan curve matches the system curve. Operating points should almost always be to the right of the fan curve peak. For a single operating point, a steep point on the fan curve should be selected. Always review the fan curve when a selection is made. Some fan curves, particularly those for axial fans, may have an irregular shape.

Fan Laws and Performance

The fan laws show the effect on a fan of changing such variables as speed and air density. They can also be used to relate the performance of dynamically similar fans. At a given speed, the fan is a constant-volume device. For a given fan in constant air conditions, the dependence of parameters on rotational speed (rpm) can be expressed as follows:

$$\text{Airflow} \propto \text{rpm}$$

$$\text{Developed pressure} \propto (\text{rpm})^2$$

$$\text{Fan power} \propto (\text{rpm})^3$$

A full explanation of the fan laws can be found in the ASHRAE *Systems and Equipment Handbook.*

Fans should be tested in accordance with Air Moving and Conditioning Association (AMCA) and ASHRAE standards. Except for the smallest of axial fans, all fans should be specified as a minimum to meet AMCA Standard 210, the AMCA's Test Code and Certified Ratings Program.

Fan Material

Fan material should be at least as corrosion-resistant as the connected ductwork. In wastewater facilities, fiberglass fans have become widely accepted for handling all but very low-concentration corrosive airstreams, in which case aluminum can be specified. Outside air supply fan material is normally steel, but at wastewater plants with a potential for hydrogen sulfide in ambient air, it is often specified as aluminum.

REGULATIONS

Air emissions from the majority of wastewater treatment plants are still generally regulated in terms of odor rather than hazardous air pollutants (HAPs), as defined by federal clean air acts. Only a handful of North American wastewater treatment plants are currently large enough to be required to possess permits for HAP emissions. Nevertheless, the same principles of collection apply to both odorous and HAP emissions.

The National Fire Protection Association (1995) publication is probably the most pertinent guideline to ventilation and odor control at wastewater plants. For instance, it recommends ventilation rates for enclosed vessels and channels. In many process units, ventilation rate thresholds of 6 or 12 air changes per hour will allow a reduction of the electrical classification from division 1 to division 2, or from division 2 to unclassified.

The Occupational Safety and Health Administration (OSHA) specifies ventilation rates and general procedures for employee entry into confined spaces. In particular, Regulation 29 CFR, Part 1910.146, should be referenced for design and operational criteria. It should be remembered that adding covers can create a confined-space entry condition. Covers can also produce an atmosphere of explosive or hazardous gas.

Another code which should be referenced when designing air collection systems is the mechanical code governing the location of the plant.

SUMMARY

The containment and collection of foul air prior to treatment are accomplished by several design methodologies. For buildings, a differential pressure is established by supplying and exhausting disparate air quantities. The base airflow rate is calculated on the exchange rate principle. The differential quantity is calculated by adding or subtracting the differential pressurization quantity according to the established ASHRAE equation, Eq. (7.1).

For covered tanks and channels, the concept is similar, but the required flow rates continue to be established empirically. Recent test experience indicates that a flow rate of 0.005 $m^3/(s \cdot m^2)$ (1.0 cfm/ft^2) can be expected to establish a measurable pressure. Using well-gasketed covers will enable that flow rate to be reduced, but documented testing is needed to substantiate the effectiveness of lower rates.

For localized capture of odors from equipment using, e.g., hoods, configuring the exhaust pickup point as close as possible to the release point is the key to minimizing airflow.

RECOMMENDED FURTHER READING

The bible of the air handling profession is the ASHRAE handbook series. ASHRAE publishes four volumes of its handbook: *Fundamentals, HVAC Systems and Equipment, HVAC Applications,* and *Refrigeration.* Each handbook is available in an inch-pound and an SI edition.

Other sources of information and the subject areas addressed by their publications include

- American Conference of Governmental Hygienists: industrial ventilation guidelines, especially contaminant capture.
- Sheet Metal and Air Conditioning Contractors National Association, Inc. (SMACNA): sheet-metal ductwork design and construction.
- Air Moving and Control Association International, Inc. (AMCA): fan performance standards.
- National Fire Protection Association (NFPA): building ventilation guidelines. NFPA 820 addresses wastewater facilities.

- Occupational Safety and Health Administration (OSHA).
- American Society of Heating, Refrigerating and Air Conditioning Engineers (ASHRAE).

REFERENCES

American Conference of Governmental Hygienists, *Industrial Ventilation, A Manual of Recommended Practice,* Cincinnati, Ohio, 1980.

American Society of Heating, Refrigerating, and Air Conditioning Engineers, *HVAC Applications Handbook,* Atlanta, Ga., 1991.

Brown and Caldwell, internal memorandum, Seattle, Wash., June 4, 1993.

Brown and Caldwell, internal memorandum, Denver, Colo., Dec. 1996.

Kepets, A., "Containing odor control costs," *Industrial Wastewater,* Nov./Dec. 1996.

Lutz, M., S. Davidson, and D. Stowe, "Emission Control," *Proceedings of the Water Environment Federation Specialty Conference*, April 1994.

National Fire Protection Association, *Fire Protection in Wastewater Treatment and Collection Facilities,* NFPA 820, Quincy, Mass., 1995.

Pacific Inspection and Research Laboratory, Inc., letter report, Seattle, Wash., Mar. 27, 1995.

Tsal, R. J., H.F. Behls, and R. Mangel, "T-method duct design," *Proceedings of the 1990 Annual Meeting of the American Society of Heating, Refrigerating and Air Conditioning Engineers,* St. Louis, Mo., 1990, pp. 3–31.

Witherspoon, J.R., et al., "Can covers contain air toxics?" *Water Environment and Technology,* Apr. 1994, pp. 45–49.

CHAPTER 8
EMISSION CONTROL TECHNOLOGY

Editor's Note. This chapter of the book is a critical part in that it explains the equipment, processes, and options available to solve odor or volatile organic compound (VOC) emission problems. Many of the contributors for this chapter are actively involved in the specific technology about which they write. This leads to an active, timely, and informative discussion; however, it also runs the risk of allowing the author's preferences to be shown. As an editor, I have aimed for a fair and balanced presentation.

While the chapter aims to be broad, covering the major process options, it cannot mention every variety of equipment or all the manufacturers—this is not a buyer's guide listing. After having read the chapter and obtained guidance toward possible solutions, the reader should find current buyers' guides in major trade magazines and professional society publications, and obtain technical literature from a wide range of writings.

Some of the contributors may refer to brand names for processes or products; these are not the *only* products available. In no way is any mention in this book meant to be an endorsement of any product. The market is very competitive, and there are usually similar products manufactured by many manufacturers.

This chapter could not be totally comprehensive, and some technologies have a limited coverage. Examples are ozone, which 20 years ago may have been popular for odor control, and the use of zeolites for adsorption, which has been treated in combination with thermal oxidation, with which it is often linked.

Technological progress is ongoing. There must be a balance, when evaluating newer technologies, to review the advantages of these technologies, and the improvements they make over existing systems, though field experience may be limited. Readers should be open to such changes and progress, and while it is commendable to rely upon the tried and true processes, I urge readers to give consideration to newer technologies. Unless good equipment developments gain acceptance, there will be no meaningful equipment development by the industry. We recommend that readers consider the fundamentals, understand how the equipment functions, consider their individual special needs, and then review all equipment, new and old.

SECTION 8.1

THE ADDITION OF CHEMICALS TO LIQUID TO CONTROL ODORS

Edward A. Bonanni
Eaglebrook, Inc.
Edgemoor, Delaware

INTRODUCTION 8.2
Hydrogen Sulfide Equilibrium 8.2
COMMON CHEMICAL ODOR CONTROL METHODS 8.3
Control of H_2S and Other Odors—General 8.3
Control of Principally H_2S Odors—General 8.4
Control of H_2S and Other Odors—Continued 8.4
Control of Principally H_2S Odors—Continued 8.11
REFERENCES 8.16

INTRODUCTION

One method of controlling odors is by reaction of the odorous component with a chemical treating reagent. The chemical treating reagent alters the concentration of the odor ous component in the aqueous phase and hence lowers the emission of the component. For example, a common odorous component in wastewater is hydrogen sulfide (H_2S). Chemical addition can alter the oxygen balance in the wastewater by (1) oxidizing sulfides, (2) precipitating dissolved sulfides, or (3) changing the ability of the sulfate- or organic sulfides-reducing organisms to generate sulfides. Several different chemicals, when added to the liquid phase, will help control odors. Some chemicals control H_2S formation and other odor components while other chemicals control just H_2S.

The common chemical addition techniques will be discussed. Their advantages and disadvantages will be addressed along with a discussion of general cost considerations. In addition, other factors, such as other reactions of the chemicals with other components in the liquid (e.g., how oxidants can be wasted when added to wastewater) or reaction of precipitants with wastewater to form metallic salt sludges (e.g., iron sulfides or iron phosphates) will be discussed.

Hydrogen Sulfide Equilibrium

Since the control of hydrogen sulfide, an inorganic sulfide, is probably the most common wastewater odor concern, it is very important for a plant manager to understand how and why H_2S forms. Hydrogen sulfide is detectable at 0.02 ppm and can also dull the olfactory senses at higher concentrations. Therefore, prolonged exposure to H_2S diminishes its odor intensity. It is readily soluble in both acidic and basic solutions and is a weak acid.

Sulfates (SO_4^{2-}) in the presence of organic matter are reduced to sulfides (S^{2-}) by sulfate-reducing bacteria at low oxidation-reduction potential (ORP) usually found in septic (anaerobic) wastewater. Since there is no oxygen present (anaerobic conditions), the organic matter uses sulfate as an oxygen source. The H_2S emission into the atmosphere is a function of various chemical equilibria. The protonation of sulfides (S^{2-}) to

HS^- and subsequently H_2S in the aqueous phase will eventually lead to airborne hydrogen sulfide when the aqueous phase becomes saturated.

Several methods can be used to control this equilibrium and subsequently H_2S emissions. All work by eliminating the precursors (hydrogen ions and sulfide ions) of the aqueous hydrogen sulfide and, therefore, never allow the liquid phase to accumulate a significant concentration of H_2S. The first method uses pH control to the alkaline side (pH elevation) which lowers the concentration of protons (H^+) that form the aqueous hydrogen sulfide. Sulfide concentration may also be lowered by oxidation which prevents the formation of aqueous hydrogen sulfide. Another method uses metal salts, especially iron, to precipitate dissolved sulfides. When ferric salts are used, the sulfide precursor may be bound to the hydrated iron oxide matrix. When ferrous salts are used, the sulfide precursor is bound to the ferrous ion. Both prevent aqueous hydrogen sulfide formation. All these methods affect H_2S gas solution equilibrium. Therefore, the precursors to the hydrogen sulfide do not need to be completely eliminated, just reduced just enough to shift the equilibrium in favor of the aqueous phase.

Chemicals which remove hydrogen sulfide odors can also prevent corrosion in both plant structure and sewer lines. At the plant level, corrosion occurs when sulfuric acid collects on metal, electrical equipment, and concrete. Sulfuric acid is produced when H_2S gas is released to the air and reacts with moisture and oxygen in the air. In sewer pipes, H_2S gas reacts with water and oxygen to form sulfuric acid which not only corrodes the pipes, but also removes oxygen from wastewater, which retards the treatment process (Gonzalez, unpublished).

COMMON CHEMICAL ODOR CONTROL METHODS

Control of H_2S and Other Odors—General

Oxidation (Oxygen, Air, Chlorine, Sodium Hypochlorite, Potassium Permanganate, Hydrogen Peroxide, Ferric Salts). Oxidation is a process involving a loss of electrons while reduction involves a gain. The substance that gains electrons in an oxidation-reduction reaction is an oxidant; the substance that loses electrons is a reductant (Watt and Holmes, 1958). The common chemical oxidants are oxygen, air, chlorine, sodium hypochlorite, potassium permanganate, hydrogen peroxide, and ferric salts. These oxidants provide an odor control advantage because they typically replace dissolved oxygen in anoxic wastewater and increase its ORP which hinders the formation of both H_2S as well as other malodorous compounds. They are added to odorous wastewater which is normally "rich" in electrons. The oxidant gains electrons and is reduced while the malodorous reductants lose electrons and are oxidized.

Regarding H_2S, as the chemical oxidant is reduced, it takes electrons away from the sulfide ions which are present in wastewater. The wastewater has high levels of dissolved sulfides. The dissolved sulfides are oxidized to sulfur or sulfate ions. This prevents H_2S from being formed. Regarding other odors, this oxidation-reduction (transfer of electrons) can take place with other compounds and either prevent or cause odors, depending on the compound. Therefore, oxidation of odorous compounds (dissolved sulfides and others) in the wastewater, involving oxygen or air injection or addition of oxidizing chemicals such as chlorine (Cl_2), sodium hypochlorite (NaOCl), or potassium permanganate ($KMnO_4$), and hydrogen peroxide (H_2O_2), controls both H_2S and other odors. However, when adding these chemicals, plant managers may discover that they have little control over what odors get reduced, and they may add more chemical than is necessary, especially if they need H_2S reduction. Also, oxidants may oxidize non-odorous compounds and create odorous compounds. Ideally, the end products from oxidation are odorless compounds, that is, CO_2, H_2O, SO_4^{2-}, etc., which result from the complete mineralization of the odorous compounds in the wastestream. Unfortunately, complete oxidation does not always take place, and an intermediate can be generated which may be more odorous than the original compound.

Oxidation, Precipitation, and Adsorption (Ferric Salts). Oxidation, precipitation, and adsorption of odorous compounds (H_2S and other odors) with ferric chloride, $FeCl_3$, and ferric sulfate, $Fe_2(SO_4)_3$, occurs because these ferric salts are mild oxidants, good precipitants, and excellent coagulants. As mentioned, ferric salts oxidize wastewater and increase its ORP which hinders the formation of both H_2S and other odors. They are also inorganic coagulants which may further reduce H_2S and other odors by adsorbing the precursors that cause these odors. Adsorption is the physical adhesion of molecules or colloids to the surfaces of solids. They precipitate dissolved sulfides to produce an insoluble iron sulfide and control H_2S odors. This occurs when the ferric ion is reduced to ferrous. The ferrous ion precipitates dissolved sulfides.

Control of Principally H_2S Odors—General

Precipitation (Iron Salts). Precipitation occurs when two or more ions in solution react to exceed the solubility limit of the compound formed and settle out as a solid. The insoluble product of this process or reaction is called the *precipitate.* This type of chemical reaction is sometimes called *double replacement* because two compounds exchange atoms or groups of atoms (Smoot et al., 1995). The common chemical precipitants used for odor control are the ferrous salts—ferrous chloride, $FeCl_2$, and ferrous sulfate, $FeSO_4$. These precipitants control H_2S odors very well. Moreover, a full-scale test at San Diego Metropolitan Sewerage District showed that ferrous chloride also reduced mercaptans and dimethylsulfide gas-phase concentrations below their lower detection limits (Mohleji et al., 1992). The ferric salts—ferric chloride and ferric sulfate—also react with H_2S to form a precipitate and reduce H_2S and other odors, but they will be discussed separately.

pH Elevation (Caustic). Sodium hydroxide, NaOH, elevates wastewater pH which then inactivates the sulfate-reducing bacteria and prevents the formation of H_2S emissions. Hydrogen sulfide is much more soluble, in the ionized form, and at higher pH values it remains in solution.

Biochemical Treatment (Nitrites and Nitrates). Nitrite and nitrate addition to wastewater prevents sulfide generation because they provide an oxidant source to wastewater. Under anaerobic conditions, certain bacteria prefer to use nitrites and nitrates as an oxidant, reducing them to nitrogen gas (U.S. EPA, 1985). No sulfate reduction occurs until all the nitrites or nitrates have been reduced. Therefore, nitrite and nitrate addition to wastewater will inhibit H_2S production and odors.

Control of H_2S and Other Odors—Continued

Oxidation

General. Oxidants (oxygen, air, chlorine, sodium hypochlorite, potassium permanganate, hydrogen peroxide, and ferric salts) typically oxidize "oxygen-poor" wastewater and increase its ORP. This hinders the process of odorous compound formation. ORP is a convenient method of measuring odor formation potential. In general, most wastewater has a negative value, and when an oxidant is added to wastewater, the ORP is raised.

One way to predict the strength of an oxidant is to measure its *standard reduction potential* $E°$. Many chemistry texts have half-cell reactions listed for common oxidants together with their standard reduction potentials. The following table lists the half-cell reactions and the $E°$ values for the common oxidants used in wastewater. The higher the value; the stronger the oxidant (Pontius, 1990; Smoot et al., 1995).

Compound	Half-cell reaction	$E°$, V
Hydrogen peroxide	$H_2O_2 + 2H^+ + 2e^- \rightarrow 2H_2O$	1.78
Hydrogen peroxide	$HO_2^- + 2e^- + H_2O \rightarrow 2H_2O$	0.85
Potassium permanganate	$MnO_4^- + 4H^+ + 3e^- \rightarrow MnO_2 + 2H_2O$	1.68
Potassium permanganate	$MnO_4^- + 8H^+ + 5e^- \rightarrow Mn^{2+} + 4H_2O$	1.49
Chlorine	$Cl_2 + 2e^- \rightarrow 2Cl^-$	1.36
Oxygen	$O_2 + 4H^+ + 4e^- \rightarrow 2H_2O$	1.23
Ferric salts	$Fe^{3+} + e^- \rightarrow Fe^{2+}$	0.77

Oxygen (O_2). Oxygen injection to wastewater can improve oxygen balance and control H_2S and other odors. It has been used in Great Britain and Australia for many years, but less frequently in the United States (WEF and ASCE, 1995). The most common method used to improve wastewater oxygen balance acts by injecting oxygen in the pressure piping from the discharge of the wastewater treatment plant pumps, because the increased pressure in the force main permits greater dissolution of oxygen in the wastewater (ASCE, 1989). When added to wastewater, oxygen raises its ORP.

ADVANTAGES. Pure O_2 has a major advantage over air because there is 5 times more oxygen in an equal volume of gas. Therefore, smaller amounts of gas need to be added which should reduce gas pocket formation (ASCE, 1989). Also, when O_2 vaporizes, it is under pressure and compressors are not needed. Also BOD_5 may also be reduced along with odors (WEF and ASCE, 1995).

DISADVANTAGES. The improper application of oxygen can aggravate an odor problem. Oxygen is not very soluble in water, and when inadequate oxygen transfer occurs, the excess gas can strip hydrogen sulfide from the wastewater and odors may increase (ASCE, 1989).

On-site storage and application of oxygen require specialized equipment which must be either purchased or leased from oxygen suppliers. For example, liquid oxygen requires a specialized containment vessel to maintain the pressure required to keep it as a liquid (U.S. EPA, 1992).

COST FACTORS. Typical methods of oxygen injection are direct oxygen injection into pipes under pressure, single U-tube injection, multiple U-tube dissolution, pressurized sidestream injection, or hydraulic fall injection. If any of these methods are used, then equipment is required to introduce oxygen into the wastewater. Also, the design of O_2 requirements and injection systems is important. These items make up the majority of the cost to implement an oxygen injection system. Oxygen suppliers may lease oxygen injection equipment, and they should be contacted when evaluating this odor control option to determine the most cost-effective approach (ASCE, 1989).

Air (O_2/N_2). Aeration can be a cost-effective method for controlling odors, especially H_2S. Other odors may be difficult to control with air (U.S. EPA, 1992). Air can improve the dissolved oxygen (DO) concentration of the wastewater to prevent downstream formation of dissolved sulfides and can help oxidize existing dissolved sulfides to sulfur or sulfate. Usually, if at least 1.0 mg/L DO is present in the wastewater, then dissolved sulfides do not accumulate (U.S. EPA, 1985). When added to wastewater, air raises its ORP.

ADVANTAGES. Air is a readily available source of oxygen that can be added to wastewater to control odors. It is a low-cost option. It adds DO to wastewater to prevent further dissolved sulfide generation. Another advantage of air injection is that BOD_5 (biological oxygen demand, 5 days) may be reduced along with odors (U.S. EPA, 1992).

DISADVANTAGES. Air injection has seen limited use for odor control in the liquid phase. It has been limited primarily to force main application. Air does not completely dissolve in water at normal pressures. Therefore, residual gases are always released (air binding). So, there is strong potential that the odor problem may get worse because odorous compounds may be present at the air injection point, and these odors may diffuse in undissolved air and escape into the atmosphere (WEF and ASCE, 1995).

COST FACTORS. Typical methods of air addition are direct injection of compressed air into the force main, venturi aspirators, air lift pumps, U-tube aeration, or pressure tank air injection. If any of these methods are used, then equipment is normally required to introduce air into the wastewater. Also, the design of air requirements and control systems is important. These items make up the majority of the cost to implement an air injection system. The additional power cost should be factored into the decision to use air for odor control.

Chlorine (Cl_2). Chlorine is a greenish-yellow gas that is usually packaged as a liquid under pressure in special containers. In its liquid state, chlorine is amber-colored and 1.5 times as dense as water. Liquid chlorine vaporizes into a greenish-yellow gas about 2.5 times as dense as air. One volume of liquid chlorine, when vaporized, yields 460 volumes of gas (AWWA, 1992).

Chlorine can be added to the liquid phase to control the formation and emission of odorous compounds. It will oxidize dissolved sulfides and depress the generation of H_2S odors as long as a chlorine residual exists (ASCE, 1989). When added to wastewater, Cl_2 raises a negative ORP. It is a strong oxidant ($E° = +1.36$) (Pontius, 1990; Smoot et al., 1995). Chlorine reacts with water to form hypochlorous and hydrochloric acids and oxidizes hydrogen sulfide to produce hydrochloric and sulfuric acids (U.S. EPA, 1985).

Theoretically, 8.9 parts by weight of chlorine reacts with each part of sulfide under acid conditions and 2.2 parts under alkaline conditions. In field applications, 5 to 15 parts of chlorine are normally needed because chlorine oxidizes other compounds (WEF and ASCE, 1995).

ADVANTAGES. Chlorine is relatively inexpensive and widely available, depending upon the caustic/chlorine supply/demand balance. Also, chlorine equipment is relatively inexpensive and widely available. Chlorine (gas or liquid) is not explosive or flammable (AWWA, 1992).

DISADVANTAGES. Chlorine is a hazardous chemical, and employee and community safety must be considered before using it. Chlorine gas is a respiratory irritant. High concentrations in air cause coughing, labored breathing, and eye irritation. In extreme cases, breathing difficulty may cause death by suffocation. Liquid chlorine causes skin and eye burns on contact. Excellent safety information is available through Chlorine Institute, Inc., Suite 506, 2001 L Street NW, Washington, DC 20036 (AWWA, 1992).

Chlorine gas is very difficult to handle. Piping of chlorine around the plant poses serious leak hazards. Chlorine is very corrosive, and special care must be taken with all equipment and piping. Safety devices must be provided to protect operators in the event of chlorine leaks (AWWA, 1992). Chlorine also reacts chemically with many substances. It is very reactive (strongly corrosive) when moisture is present. It reacts spontaneously with iron, steel, or copper at elevated temperatures (AWWA, 1992).

COST FACTORS. With chlorine gas used for odor control, the proper equipment must be installed to safely introduce it into the wastewater. Also, control system design is important. These items make up the majority of the cost to implement chlorine gas addition.

Sodium Hypochlorite. Chlorine may be added to the liquid phase as sodium hypochlorite solution (NaOCl) to control odors. It is a clear, light-yellow liquid containing up to 160 g/L available chlorine (16 trade percent) (AWWA, 1987). It is manufactured by passing chlorine into caustic soda solution or into a caustic–soda ash mixture (AWWA, 1987). It, too, will oxidize dissolved sulfides and depress the generation of H_2S. It ionizes when added to water to yield the hypochlorite ion, OCl^-. The OCl^- ion establishes an equilibrium with hydrogen ions in the wastewater. The reactive component of chlorine in water is the hypochlorite ion, regardless of whether chlorine is a gas or a solution, and the same equilibrium occurs whether chlorine is added as a gas or as hypochlorite. However, chlorine gas dissolved in water is an acidic solution, while commercial hypochlorite solutions are basic due to residual caustic.

Chlorine and hypochlorite can be used interchangeably (ASCE, 1989). Chlorine in dilute solutions is a mixture of hypochlorite and hypochlorous acids. Because hypochlorite concentration is expressed in chlorine equivalents, it is important to understand how much chlorine is available in sodium hypochlorite solution. The following table (AWWA, 1987) shows how much Cl_2 is available in NaOCl:

Available Cl_2 (%)	16.0	15.0	12.5
lb Cl_2/gal NaOCl	1.30	1.25	1.04
gal NaOCl/lb Cl_2	0.76	0.80	1.00

Theoretically, 8.9 parts by weight of chlorine equivalent reacts with each part of sulfide under acid conditions and 2.2 parts under alkaline conditions. In field applications, 5 to 15 parts are normally needed (WEF and ASCE, 1995).

ADVANTAGES. Sodium hypochlorite is normally used instead of chlorine gas because it is more convenient for occasional applications, where dosages are small, or where, for safety reasons, chlorine gas systems are excluded from consideration (ASCE, 1989). Sodium hypochlorite addition is a very popular method for odor control.

DISADVANTAGES. Light, heat, organic matter, and certain heavy-metal cations accelerate the decomposition of NaOCl. Sodium hypochlorite solution is corrosive to some degree and will affect skin and eyes on contact (AWWA, 1987). Any affected areas should be washed with copious amounts of water. Proper safety equipment is recommended by all personnel handling NaOCl. Additionally, care must be taken to ensure that NaOCl solution does not inadvertently mix with acids or chemicals that behave as acids (i.e., ferric chloride) because chlorine gas can be released (AWWA, 1987). For example, this could possibly occur if a ferric chloride and sodium hypochlorite storage tank shared common containment.

COST FACTORS. With sodium hypochlorite use for odor control, the proper equipment must be installed to safely introduce it into the wastewater. Also, control system design is important. These items make up the majority of the cost to implement sodium hypochlorite addition.

Potassium Permanganate. Potassium permanganate is a very strong oxidant in acid or neutral solutions ($E° = 1.49$ or 1.68) (Pontius, 1990; Smoot et al., 1995) which will react with hydrogen sulfide and form elemental sulfur under acidic conditions and potassium sulfate under alkaline conditions. Since most wastewater pH is close to neutral (pH 7), the acid-neutral reaction predominates. When added to wastewater, $KMnO_4$ also raises a negative ORP. Theoretically, 3 parts of potassium permanganate reacts with each part of sulfide under acid conditions; 12 parts under alkaline and 6.6 parts under neutral conditions. In field applications, 6 parts is usually needed under acid conditions and 14 parts under alkaline. However, some field reports have shown that ratios between 6 and 7 parts are needed, and as low as 1 part may be enough for odor control where there is exposure to air. [See ASCE (1989); Peroxidation Systems (undated); Vella (1993); and WEF and ASCE (1995).]

ADVANTAGES. Unlike chlorine and sodium hypochlorite, $KMnO_4$ does not produce chlorinated reaction products. It is listed by the Environmental Protection Agency (EPA) as a substitute for chlorine as a preoxidant to control trihalomethanes (THMs).

If potassium permanganate is being used at a treatment plant for odor control, then, if needed, it can help the plant in other ways. It may reduce BOD, chemical oxygen demand (COD), and total organic carbon (TOC) when used as a pretreatment prior to biological treatment systems. The partial breakup of certain organic substrates can make them more susceptible to destruction by the biological system. Some compounds may be toxic to bacteria if not pretreated by $KMnO_4$ (Carus Chemical Company, Ref. c).

Potassium permanganate provides solution oxygen which is extremely efficient in aqueous oxidation reactions (Carus Chemical Company). Its use as an oxidant in wastewater treatment for odor control usually provides additional benefits such as improved dewatering, grease control, filamentous bulking control, and added dissolved oxygen (Carus Chemical Company, Ref. d).

Regarding hydrogen sulfide removal, $KMnO_4$ is its own reaction indicator. The permanganate ion is purple in solution. When $KMnO_4$ reacts with H_2S, manganese dioxide (MnO_2) forms which has a yellow-brown color. This is a fluffy, brown floc which is stable and settles out in the treatment plant process. As sulfides are eliminated, the purple turns to brown, indicating that H_2S odors are being eliminated.

Disadvantages. All $KMnO_4$ reactions to control H_2S odors produce manganese dioxide as a by-product which adds slightly to plant solids (WEF and ASCE, 1995).

In the United States, $KMnO_4$ has not been embraced widely as an odor control chemical for collection systems or in plant applications because it usually costs more to use than other competitive chemicals (WEF and ASCE, 1995). However, for biosolids odor control, it is extensively used and is cost-competitive (Carus Chemical Company, Ref. d). It is a dry crystalline, granular pellet and normally mixed with water to make a 4% solution (WEF and ASCE, 1995).

Like other oxidants, $KMnO_4$ is incompatible with acids, peroxides, and all combustible organic or readily oxidizable materials, including inorganic oxidizable materials and metal powders. When it is mixed with hydrochloric acid (HCl), chlorine gas is liberated. Although $KMnO_4$ is not combustible, it will support combustion and may decompose if exposed to extreme heat (302°F) (Carus Chemical Company, Ref. b).

When one is handling $KMnO_4$, respirators should be worn to avoid irritation to mucous membranes. Eye protection should also be worn when handling it as a solid or as a solution. In neutral and alkaline solutions, potassium permanganate is not corrosive to iron, mild steel, or stainless steel. However, chloride corrosion of metals may be accelerated when an oxidant such as permanganate is present in solution (Carus Chemical Company, Ref. b).

Making and handling permanganate solution require caution. If not carefully dissolved, $KMnO_4$ may plug nozzles and pumps in the feed system.

Cost Factors. Of domestic $KMnO_4$ dry crystals, 97 percent are marketed under the tradename Cairox potassium permanganate (Carus Chemical Company, Ref. c). Potassium permanganate is usually uneconomical for high-flow or high-dissolved-sulfide wastewaters. However, for biosolids odor control applications which range from small to intermittent size, it is usually cost-competitive. It can be applied using dry or liquid feed systems, depending on the application. Costs of these systems are available from $KMnO_4$ suppliers.

Hydrogen Peroxide. Depending upon wastewater pH, hydrogen peroxide is a strong to very strong oxidant ($E° = 0.85$ or 1.78) which reacts with hydrogen sulfide (Pontius, 1990; Smoot et al., 1995). When added to wastewater, H_2O_2 raises a negative ORP. In acidic and neutral solutions, hydrogen peroxide will react with hydrogen sulfide to form elemental sulfur. In alkaline solutions, with a pH greater than 8.5, hydrogen peroxide will react with hydrogen sulfide to produce sulfate. Theoretically, 1 part of H_2O_2 reacts with each part of sulfide under acid conditions; 4 parts under alkaline (ASCE, 1989). In field applications, usually 1.2 to 1.8 parts is needed under acid conditions (DuPont, Ref. b). Since most wastewater pH is close to neutral (pH 7), the acid-neutral reaction predominates to form elemental sulfur. Information from actual alkaline field applications is limited. A plant test would be appropriate. Like chlorine, sodium hypochlorite, and potassium permanganate, H_2O_2 oxidizes other compounds.

ADVANTAGES. In wastewater, hydrogen peroxide decomposes to oxygen and water—harmless by-products. This is accomplished by the enzyme catalase. Catalase is present in wastewater and is produced by living cells. So, hydrogen peroxide actually can act to supplement the oxygen supply in wastewater (ASCE, 1989). It also acts as a bactericide to the sulfate-reducing bacteria, which helps prevent the formation of H_2S odors.

When hydrogen peroxide is added to wastewater with pH less than 7.5 and reacts with dissolved sulfides, a brief yellow color appears and quickly changes to a white turbidity. This white turbidity is the formation of colloidal sulfur. At wastewater pH greater than 7.5, the yellow turns colorless which indicates the formation of sulfate. These color changes are good indicators that the chemical is reacting with dissolved sulfides (DuPont, Ref. b).

In the presence of ferrous and ferric ions, hydrogen peroxide decomposes to oxygen and water through a series of reactions involving hydroxyl and perhydroxyl radicals, which are very powerful oxidants. This is commonly referred to as *Fenton Chemistry*. A common source of the ferrous ion is ferrous sulfate while ferric chloride is a good source for the ferric ion (DuPont, Ref. b).

If hydrogen peroxide is being used at a treatment plant for odor control, then, if needed, it could help the plant in other ways. It can reduce BOD, COD, and TOC by adding DO for microbiological oxidation or by directly oxidizing organics using iron salts and Fenton chemistry (DuPont, Ref. b). It can help to control denitrification which causes "rising sludge" by adding DO to the sludge blanket.(DuPont, Ref. b) It can also help control filamentous bulking because two common filamentous bulking organisms, *Thiothrix* and *Sphaerotilus natans,* respond well to hydrogen peroxide treatment (DuPont, Ref. b).

Hydrogen peroxide is a more efficient source of oxygen than O_2 or air, and hydrogen peroxide solutions are easily dispersed in wastewater. Therefore, the mass transfer of oxygen is very efficient, and more latent oxygen from H_2O_2 can be used than from oxygen or air. The DO yield from H_2O_2 can be 90 to 95 percent, or 2.25 parts H_2O_2 give 1.0 part DO (DuPont, Ref. b).

DISADVANTAGES. Although H_2O_2 can be effective in controlling hydrogen sulfide generation and the rate of reaction is rapid, it is not instantaneous (DuPont, Ref. b). Therefore, peroxide should normally be added a minimum of 10 to 15 min upstream of where hydrogen sulfide control is desired (ASCE, 1989). It may also be necessary to have multiple addition points, depending upon the degree of odor control (U.S. EPA, 1992).

Hydrogen peroxide can react with high concentrations of some organic or inorganic wastes, and this reaction can be strongly exothermic. Amines, cyanides, formaldehyde, phenols, ferrous ion, or hypochlorite at levels greater than 1000 ppm have shown rapid temperature increases and possible splattering from gas evolution when treated with hydrogen peroxide. The pH changes can potentially release poisonous fumes (DuPont, Ref. b).

Equipment used for H_2O_2 addition must be made of corrosion-resistant material since hydrogen peroxide will react vigorously with contaminants such as iron and organic materials. Protective gear must be worn when handling hydrogen peroxide. Special storage tanks are required.

Hydrogen peroxide reacts strongly with organic materials including grease and oil. All wetted surfaces of a delivery system must be specially cleaned and prepared before hydrogen peroxide is introduced. This is called *passivation* and consists of rigorous cleaning with detergent, water, nitric acid, and water to avoid reactions with contaminants. This passivation removes catalytic impurities on metal surfaces and renders the metal surface inactive by forming an oxide coating (WEF and ASCE, 1995).

Proper safety equipment must be worn when handling H_2O_2 solutions and equipment, to prevent exposure to the skin, eye, or body. OSHA's airborne exposure limit for H_2O_2 is 1 ppm (DuPont, Ref. b). It can be a fire hazard if it comes into contact with combustible materials.

Less than 52% H_2O_2 is a strong oxidizer which causes irritation and may cause eye injury. However, effects of an injury from H_2O_2 which is less than 52% in strength may be delayed. Greater than 52% H_2O_2 is a powerful oxidizer which causes severe burns of the skin and eyes. Effects of an eye injury from H_2O_2 which is greater than 52% in strength may be delayed (DuPont, Ref. b).

COST FACTORS. Equipment required for hydrogen peroxide addition is relatively simple and easy to operate and maintain. However, the cost to purchase or lease storage and feed equipment must be included in a cost evaluation.

If H_2O_2 is not properly mixed into the wastewater, then its reaction rate will be slowed and its efficiency will be adversely affected. This will cause more H_2O_2 to be used than is needed for odor control. Improper mixing enhances the decomposition of H_2O_2 to O_2. This can prevent anaerobic (septic) conditions and sulfides from being formed downstream, but this does not oxidize sulfides that have already formed (DuPont, Ref. b).

Oxidation, Precipitation, and Adsorption

Ferric Salts. Ferric chloride and ferric sulfate will precipitate dissolved sulfides to produce an insoluble iron sulfide which is removed with the sludge. Ferric salts are mild oxidants ($E° = 0.77$) (Pontius, 1990; Smoot et al., 1995). When added to wastewater, a ferric salt reacts with sulfides and raises a negative ORP, and Fe^{2+} is produced as the ORP rises and can also react with sulfides. Therefore, during the redox reaction of the ferric salt with the reductants in the wastewater, ferrous ion is generated and can precipitate the ionized forms of sulfides. During the oxidation-reduction process, depending upon ORP, pH, and the wastewater sulfide concentration, the ferric ion either can oxidize the sulfide ion or can interfere with the metabolic pathways of anaerobic bacteria which are present in wastewater. Either process can stop hydrogen sulfide from being formed. So, ferric salts can act as both a precipitant and an oxidant during odor control (Niedenzu, 1991). Ferric chloride has been shown to control odors prior to centrifuging when added to a liquid-phase sludge stream (Prakasam et al., 1990). Also, when added to wastewater at the grit chamber effluent, ferric chloride has decreased levels of both inorganic (H_2S) and organic sulfur compounds (dimethylsulfide) (Niedenzu, 1990).

Ferric salts are also inorganic coagulants which may further reduce sulfide and other odors by adsorbing the precursors that cause these odors. Adsorption is the physical adhesion of molecules or colloids to the surfaces of solids. Odor precursors are adsorbed directly onto the floc and settle with the sludge. Chemically, the ferric ion (Fe^{3+}) participates in reactions with a suspended particle surface and/or complexes with other atoms in the surface structure so that the finer particles are agglomerated into larger particles via FeO(OH), iron hydroxide polymer. The adsorption prevents further protonation and eliminates H_2S formation (Niedenzu, 1991).

Theoretically, 3.2 parts of $FeCl_3$ (1.1 parts Fe) and 3.8 parts of $Fe_2(SO_4)_3$ (1.1 parts Fe) react with each part of sulfide. In field applications, usually 2 to 4 parts Fe is needed (Peroxidation Systems, undated).

A relatively new patented technology called Ferrisorb (Dunson, undated) may improve chemical odor control performance by treating odorous emissions with core particles of high interfacial area along with odor improvement activating agents which together catch surface-active oils before they can be emitted and carry odors downwind with steam droplets. The hydrated iron oxide which precipitates from ferric chloride use in buffered wastewater is an ideal core particle because it works well with odor improvement activating agents (DuPont, Ref. a). Other good core particles are oxidized starch and tannin glycoside solids (Dunson, undated).

ADVANTAGES. Ferric salts are much more selective in reducing sulfide odors than other, more powerful oxidants. They do not produce chlorinated reaction products. They can reduce trihalomethanes in the wastewater by coagulation and adsorption.

If a ferric salt is being used at a plant for odor control, then, if needed, it can help the plant in other ways. It can reduce BOD, COD, and TOC when used as a pretreatment chemical prior to biological treatment systems. It is also effective for phosphorus, trace metals, and color removal. It enhances sludge cake dewatering, reduces polymer consumption, increases sludge cake solids, and controls struvite formation (Bonanni, 1997).

DISADVANTAGES. Ferric salts oxidize, precipitate, and adsorb. Several competing reactions result when ferric salts are added to wastewater—ferric hydroxide coagulant is formed, sulfides and phosphates are precipitated and adsorbed. This normally increases the optimum odor control dose, especially if the plant manager just wants to control H_2S odors.

Ferric chloride storage tanks can be constructed of either fiberglass-reinforced polyester (FRP) or rubber-lined steel. Because of its low freezing point, −58°F at 30% concentration, tanks generally do not require insulation or heaters. Ferric chloride solution is corrosive to most metals except titanium and tantalum. Aluminum, brass, copper, and stainless steel are readily attacked and should never be used in ferric chloride service. Piping of fiberglass-reinforced polyester, polyvinyl chloride, polyvinylidene fluoride, and other resistant plastic materials subject to temperature limitations is satisfactory (Bonanni, 1997). Ferric chloride will also stain skin and clothing and cause deterioration of fabrics and leather. It will crystallize (fall out of solution) depending on its concentration and temperature (AWWA, 1993).

Ferric sulfate is fairly corrosive to mild steel. It can decompose into hazardous sulfur oxides which are combustible. Ferric sulfate is not compatible with bases or alkalies (Eaglebrook, 1996b).

COST FACTORS. Ferric chloride is supplied domestically usually as a 36% to 38% solution by weight. Ferric sulfate is usually 45% solution by weight. They should be evaluated on an anhydrous or dry weight basis or on a dry Fe (iron) basis. Iron makes up 34.4% of $FeCl_3$ and 27.9% of $Fe_2(SO_4)_3$. When comparing these two chemicals, their price should be adjusted to make sure that treatment costs are on a dry Fe basis.

Control of Principally H_2S Odors—Continued

Precipitation

Ferrous Salts. Ferrous chloride and ferrous sulfate both precipitate dissolved sulfides and control sulfide odors only. Unlike ferric salts and other oxidants, they do not increase the oxidation-reduction potential of the wastewater or form a coagulant. Ferrous salts are used primarily for H_2S and phosphorus removal in wastewater treatment applications. They precipitate the sulfides and phosphates in wastewater, and these iron sulfide and phosphate precipitates are removed with the biosolids. Theoretically, 4.5 parts of $FeSO_4$ (1.7 parts Fe) and 3.8 parts of $FeCl_2$ (1.7 parts Fe) are required to react with 1 part of sulfide (ASCE, 1989). In field applications, actual $FeCl_2$ dosages have been reported at 7.2 parts of $FeCl_2$ (3.2 parts Fe) (Peroxidation Systems, undated).

There may be an advantage to using ferrous chloride instead of ferrous sulfate for H_2S removal. If sulfide generation depletes available sulfate in the wastewater, then ferrous sulfate may allow more sulfide to be generated downstream. In anaerobic digesters, sulfate depletion is very common, and sulfate addition will most likely cause an increase in hydrogen sulfide concentrations in the digester gas (ASCE, 1989).

ADVANTAGES. Ferrous salts are one of the most selective ways to reduce dissolved sulfides compared to other chemicals. They should be seriously considered if a plant has only an H_2S problem and is interested in a chemical solution. They also do not produce any chlorinated reaction products. Ferrous salts are usually the least expensive chemical option. They are easy to implement.

DISADVANTAGES. Ferrous salts precipitate, and competing reactions result when they are added to wastewater. For example, ferrous hydroxide is formed and phosphates are precipitated. As with ferric salts, this increases the optimum odor control dose, especially if the plant manager just wants to control H_2S odors.

Ferrous chloride can be stored in either FRP or rubber-lined steel tanks. All fittings, piping, and valves should be of the same materials of construction as the tank. At standard concentrations, ferrous chloride freezes at −4°F. Since ferrous chloride attacks most metals, all wetted parts should be of compatible material of construction. Titanium or tantalum will not corrode (Eaglebrook, 1996c). Ferrous sulfate can be stored in stainless-steel, FRP, or rubber-lined steel tanks. All fittings, piping, and valves must be of the same materials of construction as the tank. At standard concentrations, ferrous sulfate freezes at 28°F.

COST FACTORS. The $FeSO_4$ and $FeCl_2$ pickle liquor solutions are available domestically depending upon where your plant is located. They are usually available within reasonable shipping distance of heavy-steel producers. Currently, $FeSO_4$ solution can be found on the east coast and $FeCl_2$ solution in the midwest. The $FeSO_4$ solution usually contains 8% to 13% $FeSO_4$ (3% to 5% Fe) and 3% to 5% free acid. Usually $FeCl_2$ solution contains 18% to 28% $FeCl_2$ (8% to 12% Fe) and 1% to 2% free acid (Eaglebrook, 1996c). It is also marketed under the tradename OdorLok.* And $FeSO_4$ contains 36.8% Fe; $FeCl_2$ contains 44.1% Fe.

These ferrous solutions are purchased directly from steel mills and usually vary in concentration and product consistency. This causes difficulty in dosing and in pricing. They are usually priced on a cost per gallon or cost per load basis which makes it difficult to understand the cost on a dry iron (Fe) basis. They also contain higher levels of free acid than other iron salts. This causes the treatment plant to neutralize this free acid and incur a cost penalty.

Ferrous sulfate crystal ($FeSO_4 \cdot 7H_2O$) is manufactured as a by-product of the sulfate titanium dioxide manufacturing process (AWWA, 1990). Most is imported and delivered as 50% solution. It is also marketed under the tradename Odophos.†

pH Elevation

Sodium Hydroxide. Sodium hydroxide, commonly known as caustic soda, can be added to wastewater to control sulfides. The addition of NaOH to wastewater elevates pH, reduces the concentration of protons (H^+) that form the aqueous hydrogen sulfide, and thus prevents aqueous hydrogen sulfide formation. Equal concentrations of dissolved hydrogen sulfide and hydrosulfide ion exist at equilibrium at a pH of 7. However, at a pH of 8 only about 10 percent of the dissolved sulfide exists as hydrogen sulfide (ASCE, 1989). Therefore, increasing the pH reduces H_2S odors by keeping the dissolved sulfides in the HS^- form (U.S. EPA, 1985).

Also, NaOH can be added periodically (slugging) to the liquid phase to inactivate or kill the biological slime layer and subsequently remove all forms of sulfide (WEF and ASCE, 1995). This layer is a collection of bacteria and microorganisms which reduce sulfate to sulfide. Exposing this layer to pH levels between 12.5 and 13 for 30 min will destroy it. Dosages are field-dependent.

ADVANTAGES. Caustic is not flammable or combustible. The 50% caustic solution has a very low corrosion rate, and iron and steel equipment is usually adequate if operated at moderate temperatures.

---

*OdorLok is a registered trademark of Eaglebrook, Inc., Schererville, Ind.

†Odophos is a registered trademark of Davis Water & Waste Industries, Tallevast, Fla.

Caustic soda is an excellent source of hydroxide radicals (OH^-). It also increases wastewater alkalinity along with pH. It is much easier to handle than other strong alkalies such as lime.

DISADVANTAGES. NaOH addition (pH elevation) may result in the formation of secondary odorants which may be even more odorous than hydrogen sulfide.

When NaOH slugging is used to destroy the slime layer and sulfate-reducing bacteria, a plant manager must realize that the layer will reform and H_2S odors will return unless retreated. Slugging should be done in the collection system and away from the treatment plant because it can affect treatment effluent pH (WEF and ASCE, 1995). Special handling of the caustic slug dose may be necessary (ASCE, 1989), especially if the effluent discharge pH is affected or the treatment plant process cannot tolerate wastewater pH>9 (ASCE, 1989). Although practiced in several communities, NaOH slugging has not been widely accepted because large amounts of caustic are usually required and this can be costly (WEF and ASCE, 1995). With NaOH slugging, it may also be necessary to mechanically scrape the slime layer before applying the first dose of caustic (U.S. EPA, 1992).

Continuous NaOH addition to maintain a high pH is usually not practical or economical because wastewater pH is easily depressed by the biochemical production of carbon dioxide and organic acids and by the tributary flows which are at normal pH levels. (U.S. EPA, 1985). If pH drops by 0.5 unit, then a significant increase in H_2S gas to the air can result (WEF and ASCE, 1995).

Caustic should not come in contact with water or acids because it may generate enough heat to ignite nearby combustible materials (ASCE, 1989). If it comes in contact with certain metals (aluminum, tin, zinc), flammable and combustible hydrogen gas will evolve. Caustic is corrosive to body tissue and will cause severe burns. Proper procedures and personal protective equipment are required.

50% caustic solution crystallizes at 54°F, 73% solution at 145°F, and 25% at 8°F (AWWA, 1988). Therefore, steps must be taken to prevent a storage problem. Tank trucks, tank cars, and storage tanks should be heated or the solution diluted to a concentration that will not crystallize (ASCE, 1989).

It is recommended that softened water be used to dilute NaOH solution. If hard water is used, then calcium carbonate will precipitate and clog feed equipment. Also, because considerable heat is produced during dilution, the rate of dilution and the methods of cooling need to be controlled to prevent boiling and splattering of liquid caustic (AWWA, 1988). Also, water must not be added to caustic soda, but caustic soda must always be added to water with agitation. If storage tank operating temperatures exceed 100°F, then 50% caustic solution is corrosive and corrosion-resistant materials of construction (nickel, plastic-lined steel) should be used. Caustic is not compatible with aluminum, tin, zinc, brass, or copper (ASCE, 1989).

COST FACTORS. Caustic is typically made by the electrolysis of sodium chloride brine, which produces chlorine and hydrogen in addition to caustic (ASCE, 1989). It is usually available as 25, 50, and 73% solution and is a solution of anhydrous caustic soda and water.

A City of Los Angeles pilot study determined that continuous NaOH treatment which removed a continuous dissolved sulfide concentration of at least 8 mg/L was economical, but that NaOH slugging was twice as expensive (WEF and ASCE, 1995).

Biochemical Treatment

Nitrites (NO_2^-) and Nitrates (NO_3^-). Nitrites and nitrates have been used successfully for H_2S odor control in wastewater systems (U.S. EPA, 1992). They provide an oxidant source in the wastewater. Facultative (aerobic and anaerobic) and obligate anaerobic bacteria prefer to reduce nitrites and nitrates before sulfate. No sulfate reduction occurs until all the nitrites and nitrates have been reduced. Bacteria will use these electron acceptors preferentially in this order: (1) dissolved oxygen, (2) nitrite, (3) nitrate, (4) sulfate (Eaglebrook, Ref. g). Therefore, nitrite and nitrate addition to wastewater will inhibit hydrogen sulfide production. However, since raw wastewater contains no nitrites or nitrates, they must be added artificially (WEF and ASCE, 1995).

A buffered nitrite/nitrate solution may provide superior H_2S control compared to a standard nitrate-only product (Eaglebrook, Ref. g). During the bacterial oxidation of organic matter in a wastewater treatment plant, dissolved oxygen is reduced initially. When all the oxygen has been reduced and the system becomes anaerobic, then other compounds must be reduced to continue the oxidation process. In the absence of oxygen, the best compounds for reduction are nitrites, nitrates, and finally sulfates. All the reduction reactions of these compounds are pH-dependent, and calculated standard reduction potentials over a pH range of 6 to 8 indicate that nitrite is more readily reduced than dissolved oxygen which is more readily reduced than nitrate. All are more readily reduced than sulfate if there is enough material present.

Generally, the nitrite concentration in a wastewater treatment system is insignificant and not normally considered. The sulfate will be reduced when all oxygen, nitrite, and nitrate have been depleted and ORP is between −50 and −300 mV (Eaglebrook, Ref. g).

Nitrites and nitrates are marketed under the tradenames OdorLok Super N and Bioxide (Eaglebrook, Ref. g; Davis, Ref. b). OdorLok Super N is a combination of sodium nitrate ($NaNO_2$) and sodium nitrite ($NaNO_3$). Bioxide is a proprietary calcium nitrate product that has been patented as a biochemical process solution which controls odors naturally rather than chemically. When calcium nitrate solution was tested on H_2S odors, a dose of 2.4 parts nitrate-oxygen to 1 part sulfide was shown to control existing sulfides in wastewater (WEF and ASCE, 1995).

Advantages. Because nitrate is very soluble, it appears to perform best when added to long-force or low-velocity gravity mains. Wastewater bacteria consume nitrate 15 times more slowly than DO (AWWA, 1992). Nitrate addition produces no additional chemical precipitates which increase solids production at wastewater treatment plants (AWWA, 1992). Nitrates are also nonhazardous and nonflammable, and a storage system requires no containment or safety devices (WEF and ASCE, 1995).

Bioxide addition can also reduce BOD (Davis, Ref. a). Bioxide contains no hazardous substances as defined by the EPA CERCLA list. Because this is a biochemical process, it is very cost-effective in applications where extended detention times produce septic conditions when compared to oxidizers (Davis, Ref. b). Unlike oxidants and precipitants, Bioxide addition is specific to H_2S odors. Unlike caustic addition, which can kill many helpful wastewater organisms, Bioxide enhances biological organism growth.

Disadvantages. Usually, before a reduction in H_2S is experienced, a population of nitrite/nitrate-reducing bacteria has to be developed in the wastewater system, and this may require several days. Therefore, a nitrite/nitrate feed interruption could be a concern because H_2S removal might be affected for several days. Also, if the nitrite/nitrate reaction is incomplete, this might cause nitrites/nitrates to be discharged into a plant's effluent and subsequently to a receiving body of water. This may cause this body of water to become eutrophic. Eutrophication is water enrichment causing excessive growth of aquatic plants which eventually chokes and deoxygenates the water. This would require the addition of dinitrification facilities. Also, if nitrites/nitrates are present in a plant's effluent, then they may anaerobically convert to ammonia, NH_3, another malodorant. Moreover, if a plant disinfects this ammonia-rich effluent with chlorine, then odorous chloramines can result.

OdorLok Super N solution is corrosive and an irritant. It can cause eye burns and can irritate the skin. Avoid body contact by wearing proper personal protective equipment. It can be stored in stainless-steel, FRP, or rubber-lined steel tanks. Storage tanks should be insulated, heated, or located in a heated building. OdorLok Super N solution crystallizes at <50°F. Tanks need to be diked to contain any spills (Eaglebrook, Ref. h). It is incompatible with acids and solutions containing reducing agents such as ferrous iron.

COST FACTORS. Nitrites/nitrates are commercially available in solution as sodium or calcium nitrate. Bioxide comes delivered with 2.0 to 3.5 lb/gal nitrate-oxygen.

Several odor control chemicals provide similar advantages and/or disadvantages. Table 8.1.1 has been developed to assist the plant manager in the evaluation.

TABLE 8.1.1 Common Advantages and Disadvantages of Odor Control Chemicals

Chemical	Advantages*	Disadvantages†
Chlorine	1	6, 7, 8
Sodium hypochlorite	1	6, 7, 8
Potassium permanganate	3	6, 10
Hydrogen peroxide	—	6
Ferric salts	2, 3, 4, 5	9, 10, 11, 12, 13, 14, 15, 16
Ferrous salts	2, 4, 5	9, 10, 11, 12, 13, 14, 15, 16
Caustic	—	—
Nitrites/nitrates	—	—

*Advantages:

1. This chemical may already be used at the wastewater treatment plant for disinfection, and it may make sense to use the existing chemical distribution system to treat plant odors.
2. The use of this chemical may add beneficial nutrients to a wastewater treatment plant's biosolids (Fe, P, N). This may assist the plant manager with biosolids management.
3. The addition of this chemical for biosolids odor control usually results in a reduction in polymer consumption for biosolids dewatering. Any polymer savings should be included in a cost evaluation.
4. If an iron salt is being considered for odor control, then a plant manager should consider purchasing equipment that would be compatible with a variety of inorganic salts. This gives the flexibility to convert to another chemical if needed because of price or performance.
5. If an iron salt is being used for odor control, then, if needed, a plant manager should consider using it to also reduce phosphorus. This would help control struvite (magnesium ammonium phosphate).

†Disadvantages:

6. This chemical's strong oxidizing capability can be a disadvantage in odor control. Its overall odor control effectiveness may be reduced because of competing reactions with other wastewater components. It will oxidize reduced wastewater compounds and create competing side reactions which require additional oxidant to make sure dissolved sulfide oxidation occurs and H_2S odors are eliminated (WEF and ASCE, 1995).
7. This chemical is a strong disinfectant. It acts as a bactericide and kills or inactivates many bacteria that cause odors. But because it is nonselective, it also kills organisms which help wastewater treatment plants. Application point and dose are critical to prevent this problem (WEF and ASCE, 1995).
8. This chemical may react with organic matter prevalent in wastewater to form halogenated products, trihalomethanes, and haloacetic acids. These may be toxic, carcinogenic, and/or regulated substances that have their own objectionable odors (U.S. EPA, 1985).
9. This chemical produces additional sludge that must be managed. The volume of excess solids produced depends upon the amount of sulfides that need to be treated and the iron salt dose (ASCE, 1989).
10. This chemical stains many materials. Special coatings are available to protect and to improve plant housekeeping.
11. This chemical contains trace metals which report to the biosolids. This may be important to the plant manager depending upon the biosolids management program. Recent EPA regulations place limits on the trace-metal levels in a wastewater treatment plant's biosolids.
12. Because iron salts may become tightly bound to sulfide and other radicals in wastewater, this may reduce their availability to form iron sulfides downstream from the point of application (ASCE, 1989). However, there have been applications in which iron addition has held dissolved sulfide concentrations to <1.0 mg/L for distances up to 20 mi (ASCE, 1989).

TABLE 8.1.1 Footnotes continued

13. The free mineral acid usually associated with this chemical can react with metals and produce hydrogen which is combustible. The chemical is not compatible with strong mineral acids, bases, alkalies, or sodium hypochlorite (Eaglebrook, 1996a, 1996d, 1996f).
14. Iron salt solutions may cause eye burns. Acid goggles should be worn when handling them. Brief skin contact may cause only slight irritation, but prolonged exposure could cause burns. To prevent contact with skin and clothing, the use of proper protective personal equipment is recommended. Safety showers, water hoses, and eyewash stations should be easily accessible from unloading, storage, and feeding locations, and should be clearly marked.
15. Iron salt solution spills should be neutralized with caustic, soda ash, limestone, or lime and the area flushed with water to a process sewer. Carbon dioxide may be generated, so additional ventilation may be required (Eaglebrook, 1996a, 1996b, 1996d, 1996f). Adequate diking and drainage should be provided at unloading and storage areas to contain spills or leaks from harming surrounding equipment and facilities (Bonanni, 1997). Tanks should be diked to contain any spills. Neutralize spills with lime, soda ash, limestone, or caustic (Eaglebrook, 1996e).
16. Since the specific gravity of iron salt solutions varies with concentration, a plant manager should request a table showing the correlation between these two parameters.

ACKNOWLEDGMENTS

The author wishes to thank Harold S. Rafson, Rafson Engineering; James B. Dunson, DuPont Co.; Dr. Phillip A. Vella, Carus Chemical Co.; Dr. Philip Niedenzu, DuPont Co.; Robert D. Harriger, Eaglebrook, Inc.; Bruce Evans, DuPont Co.; and Dr. Kenneth Denning, Eaglebrook, Inc. for reviewing, commenting on, and upgrading this section.

REFERENCES

American Society of Civil Engineers, *Sulfide in Wastewater Collection and Treatment Systems,* ASCE Manuals and Reports on Engineering Practice No. 69, 1989.

American Water Works Association, *AWWA Standard for Caustic Soda ANSI/AWWA B501-88,* Denver, Colo., 1988.

American Water Works Association, *AWWA Standard for Ferrous Sulfate ANSI/AWWA B402-90,* 1990.

American Water Works Association, *AWWA Standard for Hypochlorites ANSI/AWWA B300-87,* Denver, Colo., 1987.

American Water Works Association, *AWWA Standard for Liquid Chlorine ANSI/AWWA B301-92,* Denver, Colo., 1992.

American Water Works Association, *AWWA Standard for Liquid Ferric Chloride ANSI/AWWA B407-93,* 1993.

Bonanni, Edward A., *Eaglebrook's Ferric Chloride Booklet,* 1997.

Carus Chemical Company, *Cairox Potassium Permanganate for Industrial Water Treatment,* Peru, Ill. (a).

Carus Chemical Company, *Cairox Potassium Permanganate, Free Flowing Grade Data Sheet,* Peru, Ill. (b)

Carus Chemical Company, *Cairox Potassium Permanganate ($KMnO_4$), Fact Sheet, Industrial Waste Treatment.* Cairox is a registered trademark of Carus Chemical Company, Peru, Ill. (c).

Carus Chemical Company, *Cairox Potassium Permanganate, The Cairox Method,* Peru, Ill. (d).

Davis Water & Waste Industries, Inc., *Bioxide, Process Division Technical Bulletin No. A-401,* Tallevast, Fla., 1993. (a).

Davis Water & Waste Industries, Inc., *Bioxide, Process Division Technical Bulletin No. A-402,* Tallevast, Fla. Bioxide is a registered trademark of Davis Water & Waste Industries, 1993. (b).

Dunson, James B., *Control of Odors by Physical-Chemical Approaches,* unpublished Technical Report, DuPont Co., International Composting Research Symposium, Columbus, Ohio, 1992.

Dunson, James B., *Fundamentals of DuPont Ferrisorb Technology for Sludge Odor Abatement.* Ferrisorb is a registered trademark of the DuPont Company, Wilmington, Del.

DuPont Co., *DuPont Odor Control Qualification Package,* Wilmington, Del. (a).

DuPont Co., *Tysul WW Hydrogen Peroxide, Waste Treatment Information,* Wilmington, Del. (b).

Eaglebrook, Inc., *Ferric Chloride Solution MSDS,* Schererville, Ind., 1996a.

Eaglebrook, Inc., *Ferric Sulfate Solution MSDS,* Schererville, Ind., 1996b.

Eaglebrook, Inc., *Ferrous Chloride Solution Data Sheet,* Schererville, Ind., 1996c.

Eaglebrook, Inc., *Ferrous Chloride Solution MSDS,* Schererville, Ind., 1996d.

Eaglebrook, Inc., *Ferrous Sulfate Data Sheet,* Schererville, Ind., 1996e.

Eaglebrook, Inc., *Ferrous Sulfate Solution MSDS,* Schererville, Ind., 1996f.

Eaglebrook, Inc., *OdorLok Super N Application Brief, The Advantages of OdorLok Super N over Standard Nitrate Based Odor Control Products,* Schererville, Ind. OdorLok Super N is a registered trademark of Eaglebrook, Inc. (g).

Eaglebrook, Inc., *OdorLok Super N Data Sheet,* Schererville, Ind. (h).

Gonzalez, Jose M., *The Use of Ferric Chloride for Odor Control,* unpublished Technical Report, PVS Technologies, Inc. Detroit, Mich.

Mohleji, S. C., V. Y. Occiano, P. Wong, and P. L. Schafer, *Odor Characterization and Control by Ferrous Chloride Addition in Sewers (Full-Scale Field Test)*, presented at 65th WEF Annual Conference and Exposition, New Orleans, La., 1992.

Niedenzu, Philip, *Demonstration of Ferric Chloride Effectiveness in Sulfide and Phosphate Removal at Blue Plains WWTP,* unpublished Technical Report, DuPont Co., Wilmington, Del., 1991.

Niedenzu, Philip, *Northeast Philadelphia WPCP—Odor Abatement Using Ferric Chloride,* unpublished Technical Report, DuPont Co., Wilmington, Del., 1990.

Peroxidation Systems Inc., *Summary of Chemical Reactions for the Treatment of Hydrogen Sulfide.*

Pontius, F. W. (ed.), *Water Quality and Treatment, A Handbook of Community Water Supplies,* 4th ed., McGraw-Hill, New York, 1990.

Prakasam, T. B. S., D. R. Zenz, and C. Lue-Hing, *Pilot and Full-Scale Centrifuge Studies on Simultaneous Dewatering and Odor Control of Anaerobically Digested Sludge,* R&D Laboratory, Metropolitan Water Reclamation District of Greater Chicago, Cicero, Ill., 1990.

Smoot, R. C., R. G. Smith, J. Price, and T. Russo, *Merrill Chemistry,* Glencoe/McGraw-Hill, New York, 1995.

U.S. Environmental Protection Agency, *Design Manual, Odor and Corrosion Control in Sanitary Sewerage Systems and Treatment Plants,* Center for Environmental Research Information, Office of Research and Development, Cincinnati, Ohio, 1985.

U.S. Environmental Protection Agency, *Detection, Control, and Correction of Hydrogen Sulfide Corrosion in Existing Wastewater Systems,* Office of Wastewater Enforcement and Compliance, Washington, 1992.

Vella, P. A., *Odor Control: Determining the Best Solution,* presented at WEF Specialty Conference on Collection Systems: Operation and Maintenance, Tucson, Ariz., 1993.

Water Environment Federation and American Society of Civil Engineers, *Odor Control in Wastewater Treatment Plants.* WEF MOP No. 22, ASCE Manuals and Reports on Engineering Practice No. 82, 1995.

Watt, G. W., and J. R. Holmes, *Basic Concepts in Chemistry,* McGraw-Hill, New York, 1958.

SECTION 8.2

MASKING AND ODOR NEUTRALIZATION

Timothy W. Planker
Hinselblon Labs.
Cape Coral, Florida

INTRODUCTION 8.18
HOW OUR SENSE OF SMELL WORKS 8.18
MASKING 8.19
ODOR NEUTRALIZATION 8.19
ESSENTIAL OILS 8.20
TRACE ELEMENTS 8.20
MANUFACTURING 8.22
FOGGING TECHNOLOGIES 8.22
Application Equipment Costs 8.23
SAFETY 8.24
CONCLUSION 8.24
REFERENCES 8.24

INTRODUCTION

The odor treatment technique of fogging an odor, modifying or neutralizing product in the air, is effective and misunderstood. The proper use of essential oil products provides a positive solution for nuisance odors in many municipal and industrial applications. The misunderstandings by some specifying and operating personnel are generally due to grouping all aerosol experiences together. Installations that employ inappropriate formulations and inadequate application equipment can add to the odor problem instead of reduce it.

Most nuisance odor complaints are from the general public, the neighbor to the municipal or industrial plant. Interestingly, the general public accepts an odor modification as a means of odor control in all levels of everyday life. To demonstrate this, think of every personal care product we use.

Most things we purchase for the bath, kitchen, home, and workplace include a fragrance to enhance its pleasantness. Odor modifiers added to a product, food, room, or workplace influence our decision to use the product or our ability to enjoy our surroundings. In the same way, an odor modification program at a municipal or industrial facility can be acceptable to the general public. The understanding and implementation of a proper aerosol modifier or neutralizer are the key to a successful solution to a nuisance odor problem.

HOW OUR SENSE OF SMELL WORKS

Our sense of smell was developed early in our evolution to allow us to experience our environment. We use our sense of smell to perceive different odors and classify them in our brains. This process is our *olfactory perception,* which we use to identify and characterize the odors we experience.

Odor is a property of certain gases and vapors only. Solids and liquids do not have an odor. The odor of a solid or liquid is perceived only when molecules detach from the surface or volatilize.

Human olfactory sensitivity is fairly great. Substances such as hydrogen sulfide can be detected at concentrations as low as 1 part per billion (ppb). This is largely due to the way the olfactory system operates. The nasal cavities contain thousands of tiny hairs called *cilia.* These cilia have tiny receptor depressions to receive odor molecules. The cilia are covered with a thin film of mucus. When we breathe, the odors in the air are drawn over the mucus-covered cilia and caught in the mucous layer. These odorous molecules are dissolved in the mucus and attach to the receptors at specific sites on the cilia. This theory of odor stimuli was discovered by J. E. Amoore in 1962. The receptors then transmit the information to the brain through nerve impulses. The human olfactory mucous membrane region is approximately 1 in^2 and contains 10 million olfactory nerve cells.

The olfactory nerves must process many identical odorous molecules to perceive and identify an odor. The perception process is sensitive to not only the odorous molecule but also the concentration. The brain responds differently with concentration: A pleasant perception is experienced in low levels and unpleasant at high concentration. We also have the ability to desensitize to certain odors over time through acclimatization. This phenomenon, known as *olfactory fatigue,* explains why workers at a wastewater or industrial facility do not experience the odors, but the neighbors do.

MASKING

The process of masking involves overpowering the odorous molecules with a stronger, pleasanter molecule. Masking compounds frequently utilize only one essential oil, usually vanilla, citrus, pine, or floral to permeate the area with its aroma. Masking usually does not modify or neutralize the odorous molecules because there is usually only one essential oil in play. The human nose can identify both the masking agent and the odorous molecule.

ODOR NEUTRALIZATION

Much work has been done to quantify and identify the olfactory response to odorous substances. The use of a gas chromatograph–mass spectrometer can isolate individual molecules. These odorous substances have been identified and their odor thresholds measured using odor panel analysis. But the olfactory reaction to a blend of odorous molecules at varying concentrations is very different.

What is known is that the olfactory perception of odorous molecules paired with a modifying odorous molecule can result in significantly lower odor perception. This decrease in the olfactory sensation has been studied for years in order to formulate aromatic blends specifically designed for this purpose. Many experiments have been conducted to determine what inhibitors and synergists possess properties that modify the character of the odor. Certain pairs of odors have a neutralizing effect on each other (e.g., rancid butter and juniper oil; rank tobacco and oil of wintergreen). Each has a recognizable odor on its own, but when combined in the vapor state, both become unrecognizable by a canceling effect. Generally known as *Zwaademaker pairing,* this phenomenon can be used quite effectively by prudent selection of known "opposites" to malodorous molecules. Why two odorous molecules produce a lower intensity of odor is not totally understood. It is thought that the phenomenon occurs upon entry of the molecules to the olfactory sensory system. The two signals result in *olfactory confusion*—the inability of the brain to accept and categorize the signals it receives from the olfactory sensors.

Olfactory chemists have isolated the aromatic substances which intensify and modify these perceptions. The phenomenon of odor modifying or neutralization is a science primarily centered in the fragrance and aromatherapy industries. These industries work primarily with essential oils and organic solvents. The mechanisms through which essential oils may work to modify the odors in the air are adsorption, absorption, and combination.

- *Adsorption.* Molecules of odorous vapors may selectively adhere or attach to the essential oil molecule. The attached pair is not perceived in our olfactory process as individual components.
- *Absorption.* Molecules of odorous vapors may dissolve in essential oil molecules. Malodors are volatile organic compounds in a vapor or gaseous state. Each molecule has a specific chemical which determines its solubility in other materials. Many malodors are soluble in certain organic oils and are no longer recognized by the nose as a malodor.
- *Combination.* Many chemicals have the ability to react with each other to form totally different compounds. These new compounds have not only different chemical structures, but also different physical and chemical properties, including odor. Organic chemicals can change odor very dramatically when they combine to form new compounds. As examples, organic alcohols and acids, which are not particularly pleasant, can combine to form organic esters (acetates) which are less harsh.

These reactions may proceed more rapidly in the vapor state than in the liquid state. In applications where heat is added, the reactions can take place very quickly and effect some very dramatic results. Many malodors are of the type that can chemically react to form compounds of lower intensity, or pleasanter odor.

ESSENTIAL OILS

Odor-modifying and -neutralizing products include essential oils. Essential oils provide a natural solution for the treatment of numerous odors. They are extremely complex organic compounds consisting of hundreds of components and trace elements.

Essential oils are liquids extracted from plant material by pressing, solvent extraction, or steam distillation. These organic, naturally occurring essential oils are extracted from the seeds, bark, roots, leaves, flowers, wood, balsam, resin, and fruit of plants. The oils are then redistilled or rectified to remove any unwanted materials. Essential oils easily evaporate, infusing the air without leaving an oily residue behind.

The versatility of essential oils for odor control has made them an acceptable and effective technique. The correct blending and selection of the oils are a science itself. The chemical and pharmaceutical industries have developed synthetic substitutions for naturally produced essential oils and have isolated the main component from aromatic chemicals. These products are referred to as *natural identical* but act differently from real oils. Although they may have a similar fragrance, they do not possess the complex trace elements in naturally produced essential oils.

TRACE ELEMENTS

Essential oils contain many constituents. The predominant components are terpenes and esters, but a large number of trace elements are also present. It is these trace elements that give the real essential oil its character and enhance its ability to blend with the other oils.

An example of the many trace components is folded orange peel oil. Orange peel oil is a concentrated product obtained by high-vacuum distillation of the winterized peel oil of fresh oranges. The folding process removes much of the terpene hydrocarbon, *d*-limonene, while retaining the highly aromatic components. Clarified (or folded) oils have several times the potency of the original oil while maintaining the best odor-modifying properties. The Valencia orange oil, e.g., contains more than 200 trace elements which have a synergistic or controlling effect on the desired potency. Trace elements exist in all natural essential oils at concentrations of less than 1%. Some trace elements exist at down to a few parts per trillion, that is, ⅕ of a teaspoon (1 g) in 200 million gal of water. These trace elements provide the delicate, extremely complex nuances that give the natural oil its unique odor-neutralizing quality.

The synergistic effect of trace elements enhances the blending and performance of the odor-neutralizing product. The blended product actually is more effective than the oils individually.

The efficacy of a product to modify and neutralize odors is dependent upon the correct formulation. Indiscriminate blending of synthetic substitutions renders the product ineffective.

In general, essential oils consist of chemical compounds that have hydrogen, carbon, and oxygen as their building blocks. These can be subdivided into two groups: the hydrocarbons, which are made up almost exclusively of terpenes (monoterpenes, sesquiterpenes, and diterpenes), and the oxygenated compounds, mainly esters, aldehydes, ketones, alcohols, phenols and oxides. Acids, lactones, sulfur, and nitrogen compounds are sometimes also present.

Aldehydes. Citral, citronellal, and neural are important aldehydes found notably in lemon-scented oils such as melissa, lemongrass, lemon verbena, lemon-scented eucalyptus, and citronella. Aldehydes in general have a sedative effect; citral has been found to have specifically antiseptic properties. Other aldehydes include benzaldehyde, cinnamic aldehyde, cuminic aldehyde, and perillaldehyde.

Phenols. These tend to have a bactericidal and strongly stimulating effect, but can be skin irritants. Common phenols include eugenol (found in clove and West Indian bay), thymol (found in thyme), and carvacrol (found in oregano and savory); methyl eugenol, methyl chavicol, anethole, safrole, myristicin, and apiol among others.

Terpenes. Common terpene hydrocarbons include limonene (antiviral, found in 90 percent of citrus oils) and pinene (antiseptic, found in high proportions in pine and turpentine oils) as well as camphene, cadinene, caryophyllene, cedrene, dipentene, phellandrene, terpinene, sabinene, and myrcene. Some sesquiterpenes, such as chamazulene and farnesol (both found in chamomile oil), have been the object of great interest recently because of their medical uses and outstanding anti-inflammatory and bactericidal properties.

Ketones. Some of the most common toxic constituents are ketones, such as thujone found in mugwort, tansy, sage, and wormwood; and pulegone found in pennyroyal and buchu. But this does not mean that all ketones are dangerous. Nontoxic ketones include jasmone, found in jasmine, and fenchone in fennel oil. Generally considered to ease congestion and aid the flow of mucus, ketones are often found in plants that are used for upper respiratory complaints, such as hyssop and sage. Other ketones include camphor, carvone, methone, methyl nonyl ketone, and pinocamphone.

Oxides. By far the most important oxide is cineol (or eucalyptol), which has an expectorant effect, well known as the principal constituent of eucalyptus oil. It is also found in a wide range of other oils, especially those of a camphoraceous nature such as rosemary, bay laurel, tea tree, and cajeput. Other oxides include linalol oxide found in hyssop (decumbent variety), ascaridol, bisabolol oxide, and bisabolone oxide.

Esters. Esters are probably the most widespread group of compounds found in essential oils, and they include linalyl acetate (found in bergamot, clary sage, and lavender) and geranyl acetate (found in sweet marjoram). They are characteristically fungicidal and sedative, often having a fruity aroma. Other esters include bornyl acetate, eugenyl acetate, and lavendulyl acetate.

Alcohols. One of the most useful groups of compounds, alcohols tend to have good antiseptic and antiviral properties with an uplifting quality; they are also generally nontoxic. Some of the most common terpene alcohols include linalol (found in rosewood, lineloe, and lavender), citronellol (found in rosewood, lineloe, and lavender), citronellol (found in rose, lemon, eucalyptus, and geranium) and geraniol (found in palmarosa); also borneol, methol, nerol, terpineol, farnesol, vetiverol, benzyl alcohol, and cedrol among others.

MANUFACTURING

Atomization and water-based aerosol fogging technology use essential oil products that are prediluted with water. Essential oils do not mix with water to form a stable solution. There are two manufacturing processes to prevent the oils from coalescing. The particles of oil can be reduced in size through a microemulsion process with an acoustic blender. Surfactants and solubilizers can be added. Some oils require both manufacturing processes. All essential-oil odor-modifying products are prediluted with water during the manufacturing process.

FOGGING TECHNOLOGIES

There are three fogging technologies which use essential-oil odor-modifying and neutralization products and are based on the amount of water involvement. The first method (water-based aerosol) mixes the product with water for maximum water involvement, the second method (atomization) for minimal water involvement, and the third method (direct vaporization) with no water involvement.

Many odor treatment applications are required only for peak, occasional, or seasonal loads. The significantly lower-capital-cost fogging systems are ideal for these applications where a much more costly scrubber system would not be fully utilized.

Fogging systems are much easier to maintain and operate. They usually require no sophisticated electrical controllers. The systems are often manufactured with component parts readily available from local supply houses. Plant personnel can operate and maintain the equipment with minimal effort. Winterization and freeze protection also are significantly less complicated and costly compared to packed-bed scrubber systems. Existing carbon systems can be converted to fogging and can thus eliminate the hazardous waste disposal problem that exists with spent carbon.

Fogging systems are the only adequate means of odor treatment for very large exposed areas such as lagoons, landfills, sludge drying beds, and aeration basins. To treat such large areas, a fogging nozzle system is installed either along the prevailing downwind side or around the entire area.

Which fogging system is utilized depends upon a number of considerations. Cost should not be the most significant because the capital costs of these systems are relatively low compared with those of other alternatives.

Application Equipment and Costs

Drum-Top Foggers. These units are small, self-contained fractional-horsepower fans that screw into the top of a 55-gal drum. The product being used must be diluted with water manually. The units cost between $250 and $1000. The drum-top units only last 3 to 6 months and must be replaced due to a high motor burnout rate. There is a significant rainout with these units as they produce a spray of large particles. The drum top has a propulsion range of 5 to 20 ft.

Atomizing Fans. The better atomizing fans dispense the diluted liquid off the fan blades. There are fan systems that have liquid nozzles placed in the fan discharge, but they are not as efficient. The product used must be diluted manually. The units range from ¼ to 1 hp, available in wash-down duty and explosionproof motors. The units cost between $1000 and $2000. These units have an expected life of 2 to 5 years. Wall- or stand-mounted units with oscillators are available. If they are operated from ½ to 1 gal/h, there should be no rainout. They have a propulsion range of 20 to 40 ft.

High-Pressure Liquid. These systems utilize a 600- to 1200-psi 1-hp high-pressure pump. There is no fan or air source required. The systems usually dilute the product with water automatically. The system fogs through very small orifice nozzles placed every 2 to 4 ft. The systems cost between $5000 and $20,000 to cover a 400- to 1500-foot-long line of nozzles. These systems have a 2- to 5-year expected life. The nozzles have a propulsion range of 2 to 6 ft. Properly operated, they should not experience any rainout. They do experience nozzle clogging and dripping. The system requires both 5- and 10-μm filters to reduce the clogging.

Compressed Air Systems. These systems require both a compressed air and diluted liquid line to service the nozzles. The systems usually dilute the products with water automatically. The nozzles are placed every 5 to 15 ft apart or directly over the odorous source. The system costs between $3000 and $5000 and utilizes a facility's available compressed air. The expected life of the system is 2 to 5 years. The nozzles have a propulsion range of 5 to 10 ft. Properly operated at less than 1 gal/h per nozzle, there should be limited rainout.

Diffused Air Systems. These systems have both air and liquid line servicing nozzles for both atomization and water-based aerosol operation. Most of these systems perform the product dilution process automatically and have a 1-hp blower. The nozzles are placed every 10 to 35 ft apart. The system costs between $5000 and $20,000 and has an expected life of 3 to 5 years. The systems should utilize nonclogging, large-orifice, external mixing nozzles. Nozzles have a propulsion range of 15 to 35 ft and liquid flow adjustment. These systems take advantage of existing low-pressure air sources (diffused air) used at wastewater plants for creation or easily generated with regenerative blowers. This air source, as opposed to high-pressure compressed air, is used in volume not pressure, thus carrying the liquid particle farther, covering a wider total volume area. The diffused air nozzles generate a fine mist type of particle fog that disperses in the atmosphere with little rainout.

Direct Vaporization Systems. These systems use an evaporator to convert a superconcentrated liquid essential-oil product to vapor. The vapor is then distributed through an air delivery system via fans or nozzles. The system costs between $5000 and $20,000 and has a life expectancy of 3 to 5 years.

The direct vaporization system allows winter operation and operates without water. The essential oils have a freezing point well below −100°F, so there is no concern of freezing. The direct vaporization systems allow the municipal or industrial facility to operate with the essential oils directly, eliminating heat tracing, storing, handling, and shipping a factory product that is prediluted with water.

Operating Costs. There are a number of essential-oil odor-neutralizing and -modifying products on the market. They range in cost from $10 to $35 per gallon. These products can be used directly for atomization or diluted 10 to 100 times with water and fogged as a water-based aerosol. The majority of the systems have a 1-hp blower or fan and are used to fog an area up to 1000 ft^2. These systems will use 1 to 2 gal of odor-neutralizing product per day. Common 1-hp odor control applications would be in wet wells, headworks, screw pump, grit chambers, sludge bins, digesters, covered tanks, and belt press rooms.

Larger applications for landfills, aeration basins, and sludge drying pads require approximately 1 hp per 100 ft of linear coverage. These larger systems use approximately 1 gal/day of odor-neutralizing product per 100 ft of linear coverage.

SAFETY

Odor modification and neutralization techniques are effective in controlling low levels of odors. This provides a safety benefit where the odorant can be toxic at high concentrations. For example, hydrogen sulfide is toxic at more than 10 parts per million (ppm) and is very recognizable. Essential-oil odor control products are not very effective at levels over 10 ppm, thus eliminating the concern of exposing people to toxic levels without their ability to recognize it. To ensure safe use, the products should be EPA/FIFRA class IV, biodegradable, nontoxic, nonflammable, and nonhazardous and have a minimum acute bioassay LC_{50} rating of 100.

CONCLUSION

The fogging of essential-oil odor modification and neutralization products provides a safe, effective, and economical solution for controlling nuisance odors. The natural essential oils derived from plant sources have the complex trace elements that provide a synergistic and controlling effect, making the vapor extremely effective in modifying and neutralizing odors. Considering that odors are perceived only in the vapor or gaseous state, an odor control system that provides a vapor-phase fogging treatment is the most effective.

REFERENCES

The H&R Book of Perfume, Gloss Verlag, Hamburg, Germany, 1992.

Lawless, Julia: *The Illustrated Encyclopedia of Essential Oils,* Barnes & Noble Books, New York, 1995.

Williams, David G.: *The Chemistry of Essential Oils,* Michelle Press, Dorset, England, 1996.

SECTION 8.3

CONDENSATION

Richard E. Waldrop
Edwards Engineering Corp.
Pompton Plains, New Jersey

INTRODUCTION 8.25
STORAGE/TRANSFER SYSTEMS 8.26
Safety 8.27
Backpressure 8.27
Efficiency 8.27
PROCESS VENT SYSTEMS 8.27
TEMPERATURES 8.28
PRECOOLER 8.29
LOW-TEMPERATURE COILS 8.29
DEFROSTING 8.29
SAFETY 8.30
LIFE EXPECTANCY 8.30
CONDENSATION SYSTEM ADVANTAGES 8.30
CONCLUSION 8.30

Editor's Note. The following section was written concerning the condensation of hydrocarbon vapors, with a strong emphasis on petroleum applications. While this is certainly a major application for condensation technology, it does not cover the wider scope of this handbook, and, therefore, I have added this preface.

The design and application of condensation technologies depend on the condensing temperatures of the compounds in the vapor stream. The more volatile the compound, the more complete the removal required, the lower will be the design temperature of the condenser. This is discussed in the following section concerning hydrocarbons, and all relate to refrigeration temperatures. However, many odors and volatile organic compounds (VOCs) are emitted at higher temperatures from driers, cookers, and process vessels and are often associated with odors. Condensation temperatures are often achieved by ambient water or air. A typical example is an animal rendering facility, where the vaporous emissions from the cooker are condensed either by a finned-tube condenser with forced airflow blown over it, or a shell-and-tube water condenser, where the heat is discharged in a cooling tower and the water recirculated. During such condensation, a large proportion of the gaseous odor is condensed, or dissolved in the condensing water vapor.

The use of water, or air condenser, is applicable in many cases to inexpensively take off the load from a subsequent colder condensation step, such as chilled water, or brine, or a refrigerant.

The equipment that is used for such condensation is conventional chemical process equipment for which there are many suppliers. Care must be taken in the selection of such equipment to avoid plugging on the vapor side of the condenser. Where the vapor contains a great deal of particulates, a separation step should be performed before the condenser, or a condenser design selected for easy cleaning.

INTRODUCTION

When solvents are used in a manufacturing process, they must be transferred numerous times from transports to holding tanks and finally to the manufacturing process. All these actions cause vapors to be generated.

Utilizing the Rankine cycle of refrigeration, a refrigerant-based unit uses compressors to cool vapors by way of a secondary heat exchanger. This is commonly known as the *refrigeration-* or *condensation*-based vapor recovery system. Liquid nitrogen also can be used as a refrigerant to cool vapors. When combined with the Rankine cycle refrigeration process, this becomes a two-stage mechanical cryogenic vapor recovery package.

The condensation method of vapor recovery allows vapors to be converted back to liquid and reused. This process eliminates potential contamination of any filtering or scrubbing media. This lack of contamination eliminates the need to handle and dispose of filtering materials as potentially hazardous waste. Two major applications of solvent vapor recovery relate to storage/transfer and process.

STORAGE/TRANSFER SYSTEMS

For a single tank application (Fig. 8.3.1), the simplest and most economical approach is to mount a vent condensing coil on the tank itself and to mount the refrigeration equipment on ground level near the tank. In general, by cooling these vapors to −20°F (−28.9°C), 90 to 95 percent recovery of gasoline vapors can be obtained. For higher efficiencies, lower condenser temperatures may be required. The condensed vapors are gravity-drained back into the tank. For multiple-tank applications where the liquid being stored is the same in all tanks, it is most economical to pipe the vent lines to a common recovery system and to pump the condensed liquid to a single point for reuse.

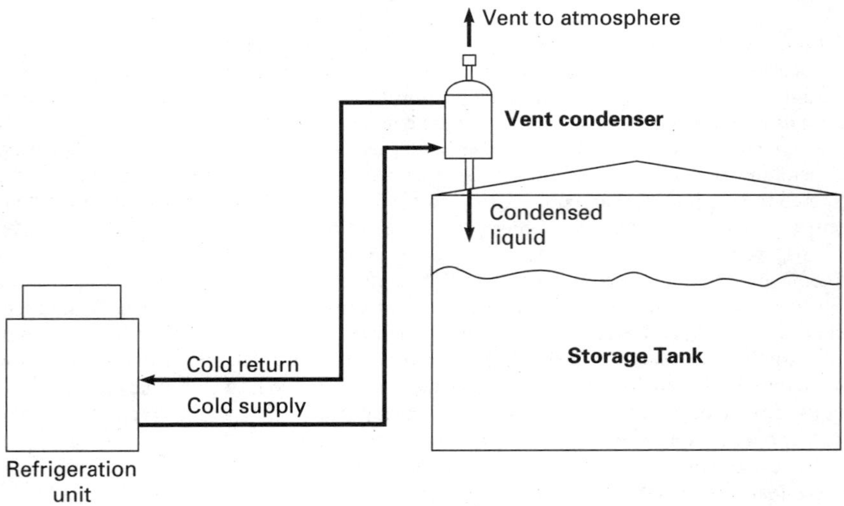

FIGURE 8.3.1 Single-tank-mounted vent condenser.

For multiple-tank applications where the liquids being stored are of different types, it is usually desirable to keep the condensates separate. For this reason the best approach is to mount a vent condensing coil on each tank. These vent condensing coils are typically supplied with a cold heat-transfer fluid, such as ethylene glycol, from a central refrigeration system.

In any of the above applications, when tank construction precludes mounting vent condensing coils on the tank, they can be mounted at ground level. The condensate is then pumped back into the tank.

Storage tanks with fixed roofs which vent to the atmosphere will contain water molecules in the vapor space that will form hydrates on the vent condensing coil. These hydrates will return to the storage tank after defrosting, which may be undesirable. If a nitrogen blanketing system is used, it will prevent ambient air and water molecules from entering the vapor space. This will eliminate the problem of water being condensed and returning to the storage tank. Keeping water from returning will eliminate the need for a defrosting system.

The above noted refrigeration units use less than 1 kWh of power per 1000 gal loaded for gasoline.

There are a number of concerns in vapor recovery, regardless of application. Three of the important ones are safety, backpressure, and efficiency.

Safety

Condensation is generally considered safe. The vapors are passed over a cold metal surface and condensed. Marine loading regulations require either enrichment of vapors above the upper explosion limit (UEL) or dilution below the lower explosion limit (LEL) prior to transfer from a barge or ship to the shore. Condensation works the best for gasoline by using the enrichment technique so as not to introduce additional cooling load or require a lower condensation temperature. This enrichment can be done by passing the vapor stream through the product to saturate the stream.

Backpressure

Backpressure is of major concern due to vapor leakage from a barge or ship or loading truck design point. Condensation systems typically have backpressure in the range of 1 to 2 in water column. This low backpressure means that, in most locations, no additional vapor blowers are necessary. Addition of flame arresters and length of piping will increase backpressure and may require the use of a supplementary blower system.

Efficiency

Early units were designed to emit no more than 80 mg/L of gasoline vapor effluent based on the original Environmental Protection Agency (EPA) standards. Second-generation systems reduced this to 35 mg/L. The addition of cryogenic systems meets EPA requirements of 10 mg/L and can potentially reduce to levels of 1 mg/L and less.

PROCESS VENT SYSTEMS

Process vent systems generally fall within four categories:

1. *Reactors.* Solvents are used in reactors and are displaced through vent systems. Multiple-reactor systems may vent to a common recovery unit.
2. *Mixers.* Materials are blended in rotary-type mixers, and the heat of mixing evaporates solvent to a vent system.

3. *Pan Coaters.* Materials are coated with a spray solution, and the solvent is evaporated and exhausted to a vent system.
4. *Spray Driers/Granulators.* When solvents are used for agglomeration of particles and powders they are subsequently evaporated and exhausted to atmosphere.

All the aforementioned processes use solvents which generally require temperatures from −60°F (−51.1°C) to −120°F (−83.3°C) to attain 90 to 95 percent recovery. For higher efficiencies, lower vapor temperatures down to −300°F may be required.

The reactor vents generally are *one-pass* systems in which the vapors are exhausted from the reactor, pass through the recovery unit, are cooled, and are expelled to the atmosphere. In the mixing, coating, and drying/granulating operations, a more efficient and economical method of recovery is to use a closed-loop system. In either case, one important point to consider is the installation of filters so that dust and other particles are not carried over to the condensing coils.

A closed-loop system is best utilized on a batch process system. The closed-loop system has the advantage of having no emissions during operation, which in effect gives 100 percent recovery of the solvent. The temperature required can be as warm as desired, 35°F (1.7°C) to 60°F (15.55°C) and is only limited by the maximum allowable concentration in the return stream or LEL concentrations at the end of the cycle. The only emission occurs when the process is opened to empty and refill, or when there is leakage.

For some processes that are fed continuously or cannot be made vaportight, a modified closed-loop system may be required. In this case a sidestream of air and vapor is pulled out of the closed-loop system after cooling, so that any leakage will be into the system. The solvent vapor present in this sidestream determines the overall recovery efficiency and the required condensation temperature.

In sizing a condensation unit, the primary concern is the characteristics of the process stream introduced into the vapor recovery unit so that sufficient refrigeration and heat transfer capacity can be calculated.

In order to make proper selection of unit size, the following information is required:

1. The components of the vapor stream
2. The concentration of the various components
3. The volume flow rate of the stream
4. Inlet temperature
5. Desired recovery
6. Hours of operation

The VOCs are recovered as a liquid. If the inlet vapor stream contains more than one condensable vapor, the recovered liquid will contain the same mixture of components, depending on the outlet vapor temperature. If the VOCs are immiscible with water, condensate can be gravity-separated and the water-free product can be returned to storage. If these liquids are miscible with water, the mixture must be separated, or disposed of.

TEMPERATURES

The temperatures required to recover 95 or 99 percent of a given vapor depend primarily on the type of vapor involved. Most chlorinated hydrocarbons encountered in tank breathing can be recovered with 95 percent efficiency by cooling to −20°F (−28.9°C).

Different applications may require lower temperatures, such as reactor vent recovery which generally requires temperatures of −60°F (−15.55°C) to −300°F (−181.1°C). Typical working ranges of refrigeration systems are described as follows:

Single	Cascade	Mechanical/liquid nitrogen
R-22/AZ-50*	AZ-50/R-23*	AZ-50/R-23*/LN2
+50 to −30°F	−10 to −100°F	+50 to −300°F

*Or equivalent new refrigerants

All the above refrigerants are available for the vapor recovery effect desired, and in many applications, combinations can be used to cool the vapor stream in stages. All the above refrigerants are considered to be environmentally safe and meet or exceed regulations as established by the Montreal Protocol.

PRECOOLER

A precooler system reduces power costs, minimizes subsequent hydrate formation, and standardizes the condition of the vapor entering the low-temperature condensers. The precooler refrigeration system is of a standard glycol-and-water chiller set to operate just above 35°F. The precooler refrigeration system operates as a single-stage R-22 system equipped with either an air-cooled or water-cooled condenser.

LOW-TEMPERATURE COILS

After leaving the precooler, the vapors pass to the low-temperature coils. When the vapors first enter the low-temperature coils, a hydrate is formed on the finned surfaces. The hydrate may contain from 50 to 80 percent hydrocarbons. The coils are constructed to accommodate a considerable column of hydrate without seriously affecting the heat-transfer rate of the system or increasing the pressure drop.

New regulations mandate a recovery rate of at least 99 percent, which can be achieved by vapor recovery coils using liquid nitrogen. By cooling vapors to as low as −300°F (−181.1°C), enhanced mechanical vapor recovery systems using liquid nitrogen make it possible to achieve high recovery rates.

DEFROSTING

Once every 24 h, or as required, the condensing coils should be defrosted by passing a warm fluid through a second set of tubes in the condensing coils. Heat for the defrost system is collected from the compressors by a gas-to-liquid heat exchanger. The defrost liquid is held in a storage reservoir for use as needed. Defrosting takes about 1 h and can be scheduled to take place during periods when loading or tank breathing is not normally occurring. The defrosting shutdown period can be eliminated by installing two complete condensing coils in parallel. This approach is regularly employed. The LN_2 section is defrosted using warm air recirculation and an indirect heat source.

SAFETY

Safety from explosions or fire is of primary concern to all operators. The low-temperature refrigeration vapor recovery condensers are essentially refrigerated flame arrestors of great depth. The vapors pass through 10 to 12 ft of finned, frost-covered surfaces chilled as low as −300°F (−181.1°C). The vapors pass through the explosive range at greatly reduced temperatures. The tubes are made of copper or stainless steel. The refrigerant is nonflammable. The refrigeration vapor recovery system is quite safe. The finned coils contain only a small quantity of hydrocarbons in the form of a frozen hydrate.

LIFE EXPECTANCY

Condensation vapor recovery machines have been in operation for more than 20 years. So long as the oil return to the refrigeration compressors is properly maintained, an indefinite operating life is expected. Large users of refrigerating machines, as in supermarkets, expect 20 years of operation, with occasional replacement of controls as required. The vapor recovery condenser section contains extended surface tubing requiring little or no maintenance.

CONDENSATION SYSTEM ADVANTAGES

1. Low capital cost—for the proper application.
2. Low power cost. Units operate on demand only.
3. Low mainenance cost. Only preventative maintenance is required.
4. Availability of maintenance personnel—use of plant or local refrigeration personnel.
5. Safety—low probability of a fire or explosion.
6. Ease of performance monitoring. Outlet temperature read and correlated with vapor pressure chart (as accepted by EPA 40 CFR parts 9 and 63 dated 12/14/94).
7. Low-pressure drop.
8. Product recovered in a liquid state—easily measured, ready for immediate reuse.
9. No significant additional pollution problem created. The processing of ambient air only adds 1 lb of condensed water per 1000 ft^3 of air processed.

CONCLUSION

In conclusion, vapor recovery by direct condensation is the cleanest and, in many cases, the most energy-efficient method of control for hydrocarbons typically encountered in today's vapor recovery applications. Its flexibility in design and broad range of applicable processes will make it the preferred method of emissions control.

SECTION 8.4

THERMAL OXIDATION

Gene H. Anguil
Anguil Environmental Systems, Inc.
Milwaukee, Wisconsin

INTRODUCTION 8.31
CHARACTERIZING THE PROCESS STREAM 8.33
- **Process Flow** 8.35
- **Organic Loading** 8.36
- **Contaminants** 8.36
- **Operating Costs** 8.36
- **Destruction Efficiency** 8.37
- **Multiple Process Lines** 8.47

THERMAL OXIDATION TECHNOLOGIES 8.47
- **Catalytic Recuperative Oxidizer** 8.52
- **Thermal Recuperative Oxidizer** 8.53
- **Regenerative Thermal Oxidizer** 8.57
- **Regenerative Catalytic Oxidizer** 8.61
- **Rotor Concentrator/Thermal Oxidizer** 8.62
- **Flare/Burners** 8.65
- **Ceramic Filter** 8.65

INTRODUCTION

The objective of this section is to present various volatile organic compound (VOC) thermal oxidation technologies, their strengths and weaknesses, and the elements required to determine what is the preferred technology for a given application.

In 1970 the environmental authorities started to identify various industries which had to bring their emissions within boundary limits. At the beginning of 1997, there were more than 6000 systems operating throughout the world controlling VOC emissions. The majority of the equipment is in the United States although systems have proliferated in Canada, Europe, Taiwan, Korea, and Japan. The major industries that have employed thermal oxidation technology are shown in Table 8.4.1. Since the early 1990s the baking industry, the chemical process industry, pharmaceutical industry, fabric coating, rubber extrusion, electronic components, soil remediation, metal coating, woodworking, formaldehyde, and sterilizers have become the specific industries being targeted by the environmental authorities.

The air pollution control solutions that are available are normally dictated by the volume of air that is to be processed. The volume of airflow, measured in cubic feet per minute, is designated as acfm, for *actual* cubic feet per minute or scfm where s stands for *standard* cubic feet per minute, at 70°F, sea level, and 1 atm.

Thermal oxidizer sizes range from 100 up to 100,000 scfm. Each industry has operations that dictate the exhaust flow that must be processed. Table 8.4.2 shows various industries and typical airflows from their operations.

Most thermal oxidizers are preengineered; i.e., the equipment is designed on the principle that in order for the equipment to be competitive in the marketplace, a series of products of fundamentally standard design are tailored to the application by changing some of the parameters as dictated by the requirements. Since thermal oxidation equipment has a burner, the designs require controls for safety and operation. Controls are termed *user-friendly* with control logic being relay- or computer-based.

Particular hydrocarbons affect health in several ways. Some affect the respiratory system, while others are air toxins. The Environmental Protection Agency (EPA) issued

TABLE 8.4.1 Major Industrial Sources of VOCs Emitted from Stationary Sources

Industry	VOCs, typical solvents, and other off-gases
Acetate finish coating	Silicate solution
Alcohol synthesis	C1, C2, C3, C6 hydrocarbons
Automobile coating	Ketones, xylene, toluene, phenols
Bakery ovens	Ethanol
Can coating	Ketones, alcohols, aromatic hydrocarbons, ethers
Chemical processing	
Coffee roasting	Heavy oils from coffee beans
Coil coating	Phosphates, solvesso, cyclohexanol, alcohols, carbitols, hydrocarbons
Electronic components	Butyl acetate, xylene, MEK, cellosolve
Fiberglass coating	Teflon emulsion, fiberglass, synthetics, styrene
Flexographic printing	Flexographic ink derivatives, alcohols, glycol
Formaldehyde	Formaldehyde, methanol, CO
Lithographic print/paint	Butyl cellosolve, ciacetone alcohol, solvesso, cellosolve acetate, xylene, MIBK
Metal coating	Alcohols, cellosolve acetate, phthalates, solvesso
Paper coating	High-boiling organics, latexes
Pharmaceuticals	Isopropanol, toluene, hydrocarbons
Phthalic anhydride mfg.	Organic acids
Resin plant	Formaldehyde, phenols, phthalic anhydride
Rubber processing	Formaldehyde, phenols, butyl benzyl phthalate
Soil remediation	Benzene, toluene, ethylene, xylene
Sterilizers	Ethylene oxide
Vinyl surgical gloves	Polyvinyl chloride, phthalate
Wire enameling	Cellosolve acetate

TABLE 8.4.2 Industrial Process Volumes

Industry	scfm
Soil remediation	100–5,000
Vent treatment	100–1,000
Automobile coating	10,000–100,000
Semiconductor	20,000–50,000
Flexographic printing	5,000–15,000
Rotogravure printing	10,000–30,000

new rules and regulations in 1990 regarding VOCs and air toxins. Subsequent controls or regulations have been issued which have identified that air toxins are particularly dangerous to industrial workers. One of the major air toxins is benzene, and benzene is one of the principal components of gasoline. The remediation of benzene from underground storage tanks and leaky underground petroleum bases is a significant source of air toxins. Carbon tetrachloroethylene, hydrogen cyanide, and ketones are all air toxins being controlled today or being regulated, since they are carcinogenic and may create an immediate or delayed problem.

CHARACTERIZING THE PROCESS STREAM

In determining the most appropriate technology to use to control a process stream, it is necessary to characterize the airstream. The first step is to establish what organics and other contaminants need to be destroyed or controlled.

Standard forms called Application Data Sheets (see Table 8.4.3) are available in which a series of questions are asked to determine the possible technologies to control the VOCs being generated. The sizing of the equipment required is dependent on the airflow and the organic loading. Figure 8.4.1 is a chart with percent of lower explosive limit (LEL) on the vertical axis and the process flow in 1000 scfm on the horizontal axis. In general, for a given flow and concentration, preferred oxidation technology exists. In this figure, four general oxidation technologies—thermal recuperative, thermal regenerative, concentrator oxidizer, and catalytic—are identified. As can be seen from the chart, low-flow, low-concentration streams are best handled by a catalytic recuperative oxidizer. When the concentration of the stream is between 15 and 20 percent LEL, then a catalytic recuperative or thermal recuperative is the best technology. For process streams between 20 and 25 percent LEL, thermal recuperative is the preferred solution.

Regenerative thermal oxidizer (RTO), regenerative catalytic oxidizer (RCO), and rotor concentrator systems are typically higher in capital costs, but the operating-cost savings on high-volume, low-concentration streams make these technologies attractive in this region. Subsequently the capital and operating costs of the equipment and the cost of the installation of that equipment are related to the sizing. Since oxidation equipment requires fuel and electricity, the operating costs will also depend on the hours of operation per day, per week, and per year. In addition, the costs will depend on the quantity of organics to be processed, in pounds per hour (lb/h) or in parts per million by volume (ppmv). A critical consideration is whether the organic loading is steady-state or has peaks or spikes.

The volume of the process stream to be treated must be determined. Since most companies have not needed to buy pollution control equipment in the past, their approach to eliminating pollution has been: the more exhaust, the better. This is particularly true in the low-margin industries. The more sophisticated industries, such as the chemical industry, have normally been concerned about the air exhausted from plants and have paid attention to the exhaust volume by process control. The industries that have not paid attention to exhaust volume are not sufficiently aware when buying pollution control equipment that the cost of that equipment is going to depend on two elements:

1. The air volume that must be processed
2. The pounds per hour of organics that must be processed

TABLE 8.4.3 Application Data Sheet

FAX TO: ______________________ FROM: ______________________

SUBJECT: ______________________ DATE: ______________________

QUOTE: ESTIMATE: __________ FORMAL: ________ NEEDED BY: __________

Name ______________________ Title ______________________

Company ____

Address ______________________ City__________ State _____ Zip _____

Phone ______________________ Fax ______________________

How did you hear about our company? ______________________

What type of process is producing the emission? ______________________

Hours/Day ______________ Days/Week ______________ Weeks/Year ___

Waste Stream flow rate: ______________________ scfm, acfm, nm^3/h, kg/h

Is recirculation or flow rate reduction possible? Yes __________ No __________

Waste stream temperature: ____________ F° ____________C°

VOC Type	Usage or Concentration: lb/h, ppm, mg/m^3, vol %, wt %
______________________	______________________
______________________	______________________
______________________	______________________
______________________	______________________

Does the process emit any of the following potential catalyst masking agents or deactivators?

Yes _____ No _____

Halogen ___ Silicone ___ Phosphorous compound ___ Heavy metal ___ Sulfur compound ___ Particulate ___

If any of the above items are present, what is the concentration in the exhaust? __________

Fuel available_____ Fuel cost_____$/million Btu Power_____V/____Hz/____Phase Electric Cost_____$/kWh

Can excess heat be utilized? __________ Comfort heat __________ Process Heat ________

Project stage: Immediate __________ 3–6 months __________ 6–12 months __________

Is project EPA-motivated? Yes _____ No _____

What percent destruction of VOC is required? ______________________%

Location: Concrete pad ______________ Structural steel platform ______________

Estimated total length of ducting: ______________________________

COMMENTS: ____

____

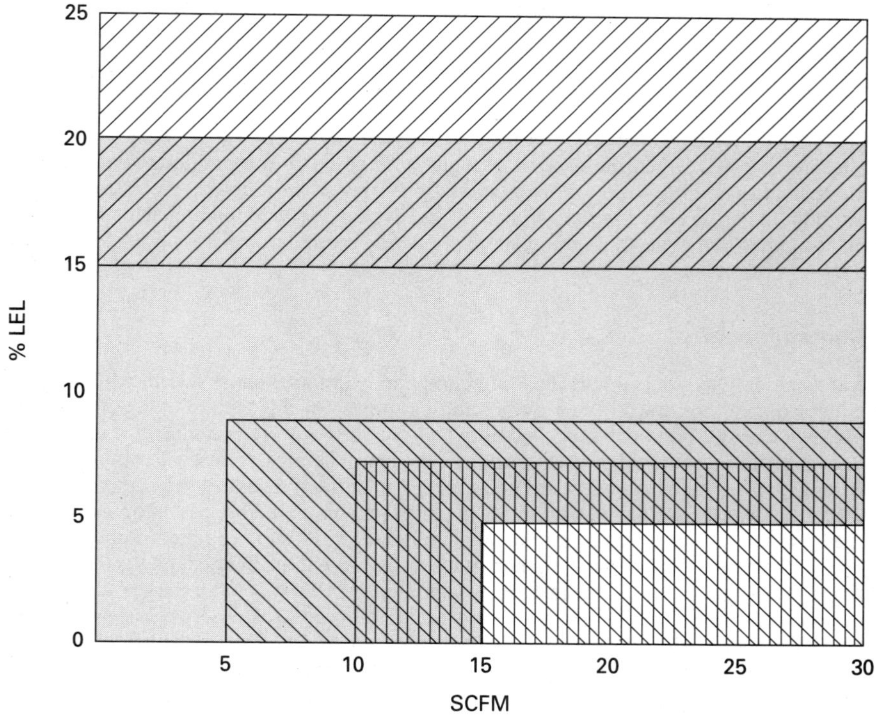

Type of system	SCFM	% LEL
Thermal recuperative	0–30,000	15–25%
Thermal regenerative	5,000–30,000	0–8%

Type of system	SCFM	% LEL
Concentrator oxidizer	10,000–30,000	0–6%
Catalytic recuperative	0–15,000 15,000–30,000	0–20% 5-20%

Thermal recuperative

Thermal regenerative

Concentrator oxidizer

Catalytic recuperative

FIGURE 8.4.1 Percent of LEL versus process flow—oxidizer technology identification.

Process Flow

Process flow reduction should be targeted, and this must be done in such a way that it will not affect the operation of the process or plant. The usual first response by a plant engineer is that airflow reduction cannot be achieved. However, subsequent investigation has shown that most of the time airflow reduction can be achieved.

Organic Loading

The determination of the concentration of the organics is often difficult. However, in an operation where solvents or other organics are purchased, the number of pounds or tons of organics bought in the past year is known, since this information must also be supplied to the environmental authorities. Hence, the first step is done through the purchasing records. Given the tons of organics purchased in the previous years and the number of operating hours per year, the ratio of those two numbers yields the pounds per hour, a conservative initial estimate of the average organic loading. This methodology assumes that everything that is purchased will be emitted up the stacks.

Contaminants

It is necessary to determine if there are compounds in the exhaust stream which can be deleterious to the catalyst, thereby precluding the use of this technology. As an example, when automotive catalytic mufflers and converters were introduced a number of years ago, the automobile industry required the petrochemical industry to eliminate lead from gasoline since lead degraded and reduced the effectiveness of the catalyst and the destruction of the gasoline. One set of industrial compounds that can harm catalysts is the halogens, a family of compounds which include chlorine, bromine, iodine, and fluorine. Bromine, while not prevalent in industry, is present in chemical plants. Freons are fluorine compounds. Silicone is another compound which is deleterious to catalyst. It is used as a slip agent or a lubricant in many industrial processes. Phosphorus, heavy metals (zinc, lead), sulfur compounds, and any particulate can shorten the life of the catalyst. It is necessary to estimate the volume or the amount of each of those contaminants, to assess the viability of catalytic technologies for the application.

Operating Costs

Since all oxidation equipment has to operate at an elevated temperature, it is necessary to know the type of auxiliary energy or fuel available. All oxidizers can operate on electric heating, natural gas, or liquefied petroleum gas (LPG), whereas thermal oxidizers can operate with sulfur-based fuels such as no. 2 or no. 6 fuel oil. Electric heat is only suitable for small airflows.

Since a continuing cost of air pollution control is the operating cost, all thermal oxidizers should have some form of heat recovery. Available heat-recovery efficiencies vary depending on the type of thermal oxidation technology selected. The specific heat recovery efficiency selected for an application depends on the organic concentration. Table 8.4.4 shows a series of different economic analyses for a particular airstream.

A process stream of 10,000 scfm with an inlet temperature of 70°F was utilized in the analysis shown in Table 8.4.4. Toluene was chosen as the contaminant in this example. Toluene has a calorific content of 16,720 Btu/lb and an LEL of 1.10 percent by volume. For each of the technologies, an economic analysis was performed at a toluene loading of 1 percent LEL and at 10 percent LEL. The results of this analysis are reported as operational costs in dollars per hour. The operational cost is the sum of the fuel use and the electricity needed to run the system fans. All the technologies shown have been assumed to be running on natural gas at a cost of $4.50 per TBtu. The electricity price has been taken at $0.06 per kilowatthour.

TABLE 8.4.4 Operating-Cost Comparisons

Process stream 10,000 scfm percent LEL	Operational costs, $/hr				
	Catalytic recuperative oxidizer	Thermal recuperative oxidizer	Regenerative catalytic oxidizer	Regenerative thermal oxidizer	Rotor concentrator with thermal oxidizer
1% LEL	$11.00/h	$28.37/h	$2.50/h	$5.65/h	$0.30/h
10% LEL	$ 3.70/h	$16.72/h	$1.65/h†	$1.60/h†	N/A

1. Assumes a 10,000 scfm process stream with an inlet temperature of 70°F.
2. Assumes a process mixture of 16,720 Btu/lb with a mixture LEL of 1.10%.
3. Assumes an electrical cost of $0.06 per kWh.
4. Assumes a natural gas cost of $4.50 per TBtu.
5. The catalytic recuperative oxidizer assumes a 65% efficient heat exchanger.
6. The thermal recuperative oxidizer assumes a 65% efficient heat exchanger.
7. The regenerative catalytic oxidizer assumes a 95% efficient heat exchanger.
8. The regenerative thermal oxidizer assumes a 95% efficient heat exchanger.
9. The rotor concentrator wheel assumes a 6:1 concentration ratio.

*Refer to Figs. 8.4.2 to 8.4.10 for detailed operating-cost information.
†Excess heat has been produced at this loading, and a hot-side heat-exchanger bypass is required.

From Table 8.4.4 one can see the operating-cost advantages of one technology over another at two loading conditions. A catalytic recuperative oxidizer, with a 65 percent effective heat exchanger, is more economical to operate at the 1 to 10 percent LEL range than a thermal recuperative oxidizer with a 65 percent effective heat exchanger. The thermal recuperative oxidizer will only begin to show a significant reduction in operational costs around the 15 to 25 percent LEL range. Both of the regenerative oxidizers utilize a 95 percent effective heat exchanger. These technologies are ideal for high-airflow, low-loading process streams. As shown in Table 8.4.4, both of the regenerative oxidizers begin to show high-temperature bypass at the 10 percent LEL loading. This means that the loading point at which the oxidizers would require zero supplemental fuel has been exceeded. Excess heat is now being produced in the oxidizer. This heat must be able to escape from the oxidizer by way of a high-temperature bypass. The table also shows that the rotor concentrator wheel is another technology suited for high-airflow, low-loading process streams. The rotor concentrator will concentrate the process stream from anywhere from 6:1 to 13:1 concentration ratio. Thus, the airflow will be reduced 6 times, while the contaminant loading will be increased 6 times. The rotor concentrator technology is not suited for process streams at a 10 percent LEL loading.

Destruction Efficiency

Typically throughout the United States, 95 percent destruction of VOCs is required, and when toxic chemicals are present, the destruction efficiency is raised to 99 percent. Characterization of the VOC-laden stream assists in selecting the appropriate technology to achieve the required destruction efficiency.

INPUT PARAMETERS		Assumes 460/60/3	Dilution Air Rate 0
Heat Exchanger Efficiency	65 %	Electricity Price	$0.06 per KWH
Process Exhaust Flow	10000 SCFM	Fuel Price	$4.50 per MMBTU
Process Exhaust Temp.	70 Deg. F	LEL of Mixture	1.10% by Volume
Burner Setting	600 Deg. F	Avg Vapor Density	0.24 Lb/cuft
Pressure drop (max)	17 in. H20	Pressure Drop(Min)	17 in. H20
Solvent Loading	15.70 LB/Hour	Percent LEL	1.00% LEL
Solvent Calorific EQUIV.	16,721 BTU/LB	Compound Average	110 PPMV

SYSTEM POINT	TEMP	Deg. F		Flow	SCFM
System Inlet	T(1)	70		F(1)	10000
Heat Exchanger Cold Outlet	T(A)	430		F(A)	10000
Heat Exchanger Bypass	T(B)	70		F(B)	0
Burner Inlet	T(2)	430		F(2)	10000
Reactor Inlet	T(3)	600		F(3)	10000
Reactor Outlet	T(4)	623	Catalytic Range	F(4)	10000
System Outlet	T(5)	264		F(5)	10000

Results		Economics	1.13 Gross/avail ratio
Firing Rate Pilot	13,000 BTU/Hour	Fuel Cost	$9.42 per Hour
Firing Rate, HHV	2,080,034 BTU/Hour	Electric Cost	$1.60 per Hour
Dilution/purge Blower Reqd (FD)	36 HP		
Combustion Air Required	516 SCFM	Total Operating Cost	$11.01 per Hour
Percent Bypass HX	0 %		

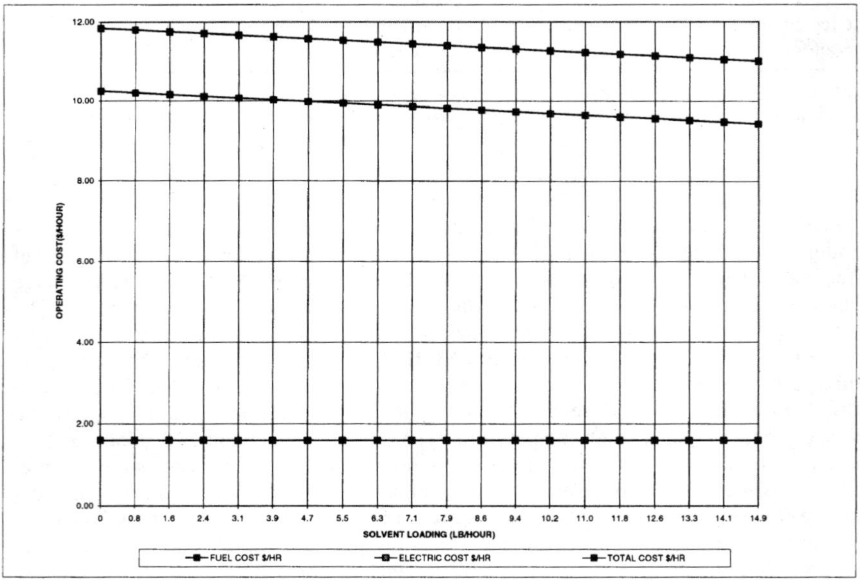

FIGURE 8.4.2 Catalytic recuperative oxidizer, operating costs—1 percent LEL.

INPUT PARAMETERS		Assumes 460/60/3	Dilution Air Rate 0
Heat Exchanger Efficiency	65 %	Electricity Price	$0.06 per KWH
Process Exhaust Flow	10000 SCFM	Fuel Price	$4.50 per MMBTU
Process Exhaust Temp.	70 Deg. F	LEL of Mixture	1.10% by Volume
Burner Setting	600 Deg. F	Avg Vapor Density	0.24 Lb/cuft
Pressure drop (max)	17 in. H20	Pressure Drop(Min)	17 in. H20
Solvent Loading	157.40 LB/Hour	Percent LEL	10.01% LEL
Solvent Calorific EQUIV.	16,721 BTU/LB	Compound Average	1101 PPMV

SYSTEM POINT	TEMP	Deg. F	Flow	SCFM
System Inlet	T(1)	70	F(1)	10000
Heat Exchanger Cold Outlet	T(A)	563	F(A)	10000
Heat Exchanger Bypass	T(B)	70	F(B)	0
Burner Inlet	T(2)	563	F(2)	10000
Reactor Inlet	T(3)	600	F(3)	10000
Reactor Outlet	T(4)	828 Catalytic Range	F(4)	10000
System Outlet	T(5)	335	F(5)	10000

Results		Economics	1.13 Gross/avail ratio
Firing Rate Pilot	13,000 BTU/Hour	Fuel Cost	$2.11 per Hour
Firing Rate, HHV	455,362 BTU/Hour	Electric Cost	$1.60 per Hour
Dilution/purge Blower Reqd (FD)	36 HP		
Combustion Air Required	191 SCFM	Total Operating Cost	$3.70 per Hour
Percent Bypass HX	0 %		

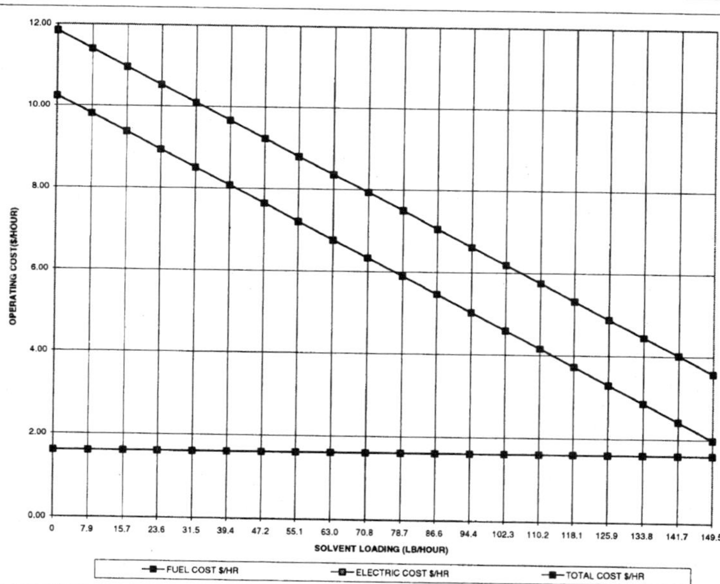

FIGURE 8.4.3 Catalytic recuperative oxidizer, operating costs—10 percent LEL.

INPUT PARAMETERS		Assumes 460/60/3	Dilution Air Rate 0
Heat Exchanger Efficiency	65 %	Electricity Price	$0.06 per KWH
Process Exhaust Flow	10000 SCFM	Fuel Price	$4.50 per MMBTU
Process Exhaust Temp.	70 Deg. F	LEL of Mixture	1.10% by Volume
Burner Setting	1379 Deg. F	Avg Vapor Density	0.24 Lb/cuft
Pressure drop (max)	14 in. H20	Pressure Drop(Min)	14 in. H20
Solvent Loading	15.70 LB/Hour	Percent LEL	1.00% LEL
Solvent Calorific EQUIV.	16,721 BTU/LB	Compound Average	110 PPMV

SYSTEM POINT	TEMP	Deg. F		Flow	SCFM
System Inlet	T(1)	70		F(1)	10000
Heat Exchanger Cold Outlet	T(A)	935		F(A)	10000
Heat Exchanger Bypass	T(B)	70		F(B)	0
Burner Inlet	T(2)	935		F(2)	10000
Reactor Inlet	T(3)	1379		F(3)	10000
Reactor Outlet	T(4)	1400	Thermal Range	F(4)	10000
System Outlet	T(5)	536		F(5)	10000

Results		Economics	1.25 Gross/avail ratio
Firing Rate Pilot	13,000 BTU/Hour	Fuel Cost	$27.05 per Hour
Firing Rate, HHV	5,998,745 BTU/Hour	Electric Cost	$1.31 per Hour
Dilution/purge Blower Reqd (FD)	29 HP		
Combustion Air Required	1300 SCFM	Total Operating Cost	$28.37 per Hour
Percent Bypass HX	0 %		

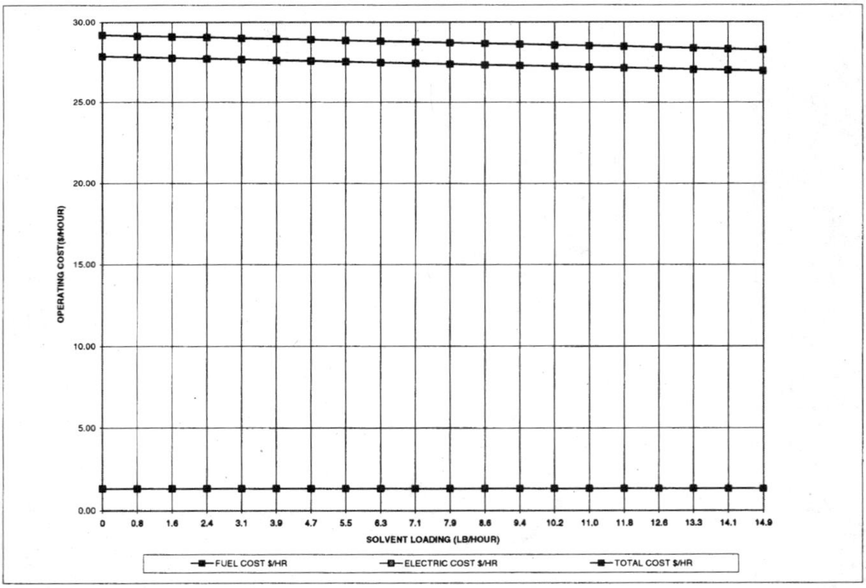

FIGURE 8.4.4 Thermal recuperative oxidizer, operating costs—1 percent LEL.

INPUT PARAMETERS		Assumes 460/60/3	Dilution Air Rate 0
Heat Exchanger Efficiency	65 %	Electricity Price	$0.06 per KWH
Process Exhaust Flow	10000 SCFM	Fuel Price	$4.50 per MMBTU
Process Exhaust Temp.	70 Deg. F	LEL of Mixture	1.10% by Volume
Burner Setting	1187 Deg. F	Avg Vapor Density	0.24 Lb/cuft
Pressure drop (max)	14 in. H20	Pressure Drop(Min)	14 in. H20
Solvent Loading	157.40 LB/Hour	Percent LEL	10.01% LEL
Solvent Calorific EQUIV.	16,721 BTU/LB	Compound Average	1101 PPMV

SYSTEM POINT	TEMP	Deg. F	Flow	SCFM
System Inlet	T(1)	70	F(1)	10000
Heat Exchanger Cold Outlet	T(A)	934	F(A)	10000
Heat Exchanger Bypass	T(B)	70	F(B)	0
Burner Inlet	T(2)	934	F(2)	10000
Reactor Inlet	T(3)	1187	F(3)	10000
Reactor Outlet	T(4)	1400 Thermal Range	F(4)	10000
System Outlet	T(5)	535	F(5)	10000

Results		Economics	1.25 Gross/avail ratio
Firing Rate Pilot	13,000 BTU/Hour	Fuel Cost	$15.40 per Hour
Firing Rate, HHV	3,410,184 BTU/Hour	Electric Cost	$1.31 per Hour
Dilution/purge Blower Reqd (FD)	29 HP		
Combustion Air Required	782 SCFM	Total Operating Cost	$16.72 per Hour
Percent Bypass HX	0 %		

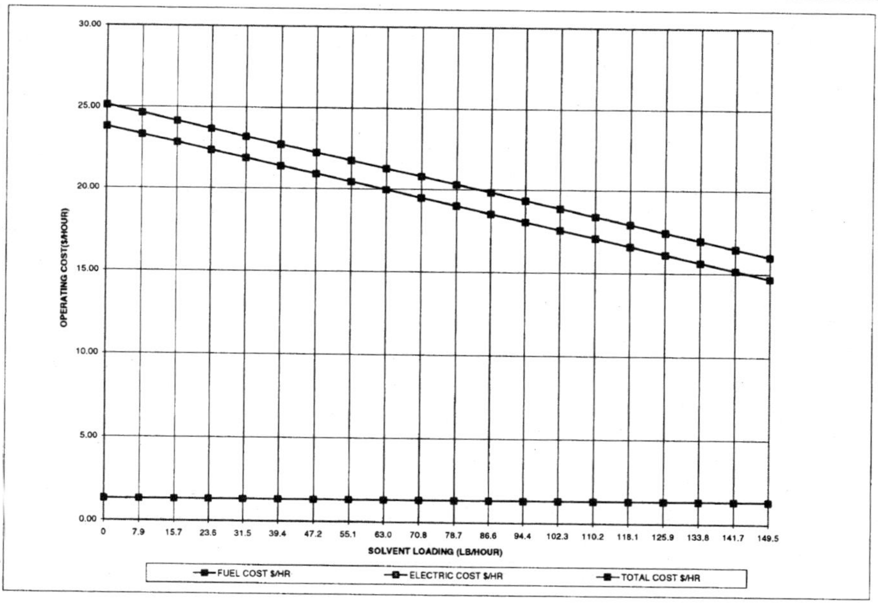

FIGURE 8.4.5 Thermal recuperative oxidizer, operating costs—10 percent LEL.

INPUT PARAMETERS		Assumes 460/60/3	Dilution Air Rate 0
Heat Exchanger Efficiency	95 %	Electricity Price	$0.06 per KWH
Process Exhaust Flow	10000 SCFM	Fuel Price	$4.50 per MMBTU
Process Exhaust Temp.	70 Deg. F	LEL of Mixture	1.10% by Volume
Burner Setting	600 Deg. F	Avg Vapor Density	0.24 Lb/cuft
Pressure drop (max)	17 in. H20	Pressure Drop(Min)	17 in. H20
Solvent Loading	15.70 LB/Hour	Percent LEL	1.00% LEL
Solvent Calorific EQUIV.	16,721 BTU/LB	Compound Average	110 PPMV

SYSTEM POINT	TEMP	Deg. F	Flow	SCFM
System Inlet	T(1)	70	F(1)	10000
Heat Exchanger Cold Outlet	T(A)	584	F(A)	10000
Heat Exchanger Bypass	T(B)	70	F(B)	0
Burner Inlet	T(2)	584	F(2)	10000
Reactor Inlet	T(3)	600	F(3)	10000
Reactor Outlet	T(4)	623 Catalytic Range	F(4)	10000
System Outlet	T(5)	109	F(5)	10000

Results		Economics	1.13 Gross/avail ratio
Firing Rate Pilot	13,000 BTU/Hour	Fuel Cost	$0.91 per Hour
Firing Rate, HHV	189,782 BTU/Hour	Electric Cost	$1.60 per Hour
Dilution/purge Blower Reqd (FD)	36 HP		
Combustion Air Required	138 SCFM	Total Operating Cost	$2.51 per Hour
Percent Bypass HX	0 %		

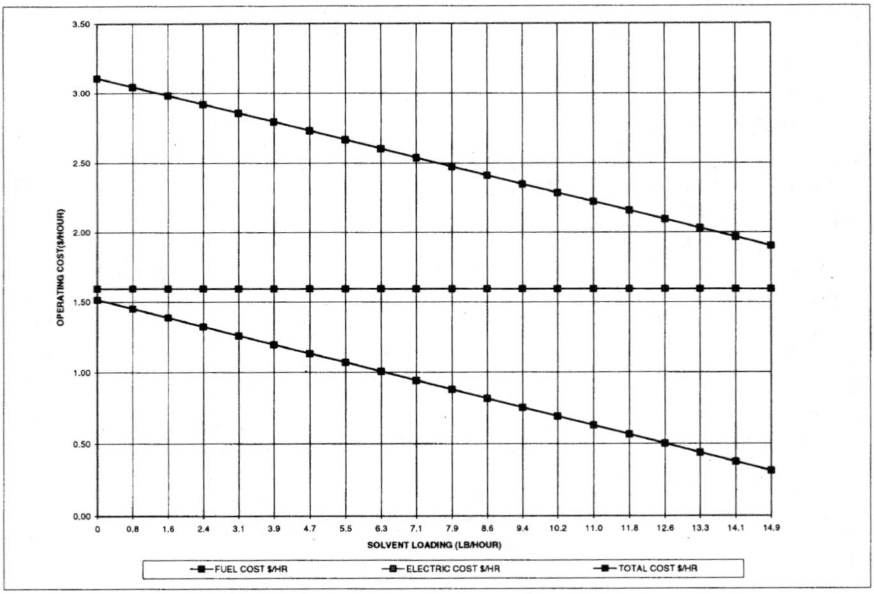

FIGURE 8.4.6 Regenerative catalytic oxidizer, operating costs—1 percent LEL.

INPUT PARAMETERS		Assumes 460/60/3	Dilution Air Rate 0
Heat Exchanger Efficiency	95 %	Electricity Price	$0.06 per KWH
Process Exhaust Flow	10000 SCFM	Fuel Price	$4.50 per MMBTU
Process Exhaust Temp.	70 Deg. F	LEL of Mixture	1.10% by Volume
Burner Setting	600 Deg. F	Avg Vapor Density	0.24 Lb/cuft
Pressure drop (max)	17 in. H20	Pressure Drop(Min)	17 in. H20
Solvent Loading	157.40 LB/Hour	Percent LEL	10.01% LEL
Solvent Calorific EQUIV.	16,721 BTU/LB	Compound Average	1101 PPMV

SYSTEM POINT	TEMP	Deg. F	Flow	SCFM
System Inlet	T(1)	70	F(1)	10000
Heat Exchanger Cold Outlet	T(A)	775	F(A)	7519
Heat Exchanger Bypass	T(B)	70	F(B)	2481
Burner Inlet	T(2)	600	F(2)	10000
Reactor Inlet	T(3)	600	F(3)	10000
Reactor Outlet	T(4)	828 Catalytic Range	F(4)	10000
System Outlet	T(5)	298	F(5)	10000

Results		Economics	1.13 Gross/avail ratio
Firing Rate Pilot	13,000 BTU/Hour	Fuel Cost	$0.06 per Hour
Firing Rate, HHV	0 BTU/Hour	Electric Cost	$1.60 per Hour
Dilution/purge Blower Reqd (FD)	36 HP		
Combustion Air Required	100 SCFM	Total Operating Cost	$1.65 per Hour
Percent Bypass HX	25 %		

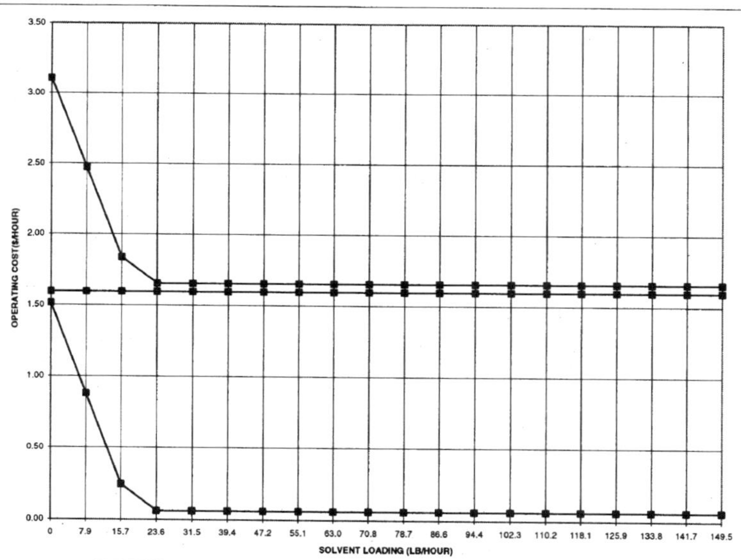

FIGURE 8.4.7 Regenerative catalytic oxidizer, operating costs—10 percent LEL.

INPUT PARAMETERS		Assumes 460/60/3	Dilution Air Rate 0
Heat Exchanger Efficiency	95 %	Electricity Price	$0.06 per KWH
Process Exhaust Flow	10000 SCFM	Fuel Price	$4.50 per MMBTU
Process Exhaust Temp.	70 Deg. F	LEL of Mixture	1.10% by Volume
Burner Setting	1479 Deg. F	Avg Vapor Density	0.24 Lb/cuft
Pressure drop (max)	14 in. H20	Pressure Drop(Min)	14 in. H20
Solvent Loading	15.70 LB/Hour	Percent LEL	1.00% LEL
Solvent Calorific EQUIV.	16,721 BTU/LB	Compound Average	110 PPMV

SYSTEM POINT	TEMP	Deg. F		Flow	SCFM
System Inlet	T(1)	70		F(1)	10000
Heat Exchanger Cold Outlet	T(A)	1400		F(A)	10000
Heat Exchanger Bypass	T(B)	70		F(B)	0
Burner Inlet	T(2)	1400		F(2)	10000
Reactor Inlet	T(3)	1479		F(3)	10000
Reactor Outlet	T(4)	1500	Thermal Range	F(4)	10000
System Outlet	T(5)	170		F(5)	10000

Results		Economics	1.13 Gross/avail ratio
Firing Rate Pilot	0 BTU/Hour	Fuel Cost	$4.34 per Hour
Firing Rate, HHV	964,126 BTU/Hour	Electric Cost	$1.31 per Hour
Dilution/purge Blower Reqd (FD)	29 HP		
Combustion Air Required	293 SCFM	Total Operating Cost	$5.65 per Hour
Percent Bypass HX	0 %		

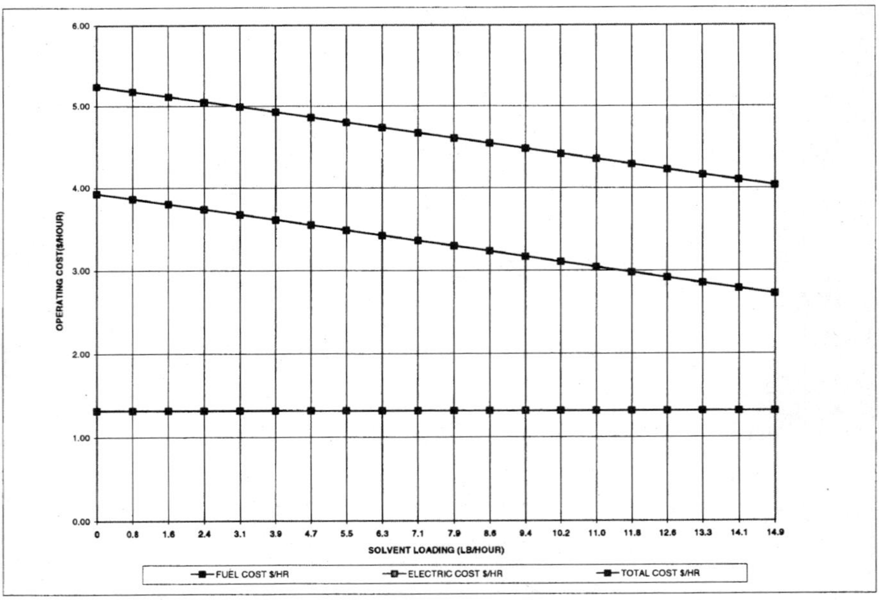

FIGURE 8.4.8 Regenerative thermal oxidizer, operating costs—1 percent LEL.

INPUT PARAMETERS		Assumes 460/60/3	Dilution Air Rate 0
Heat Exchanger Efficiency	95 %	Electricity Price	$0.06 per KWH
Process Exhaust Flow	10000 SCFM	Fuel Price	$4.50 per MMBTU
Process Exhaust Temp.	70 Deg. F	LEL of Mixture	1.10% by Volume
Burner Setting	1289 Deg. F	Avg Vapor Density	0.24 Lb/cuft
Pressure drop (max)	17 in. H20	Pressure Drop(Min)	17 in. H20
Solvent Loading	157.40 LB/Hour	Percent LEL	10.01% LEL
Solvent Calorific EQUIV.	16,721 BTU/LB	Compound Average	1101 PPMV

SYSTEM POINT	TEMP	Deg. F	Flow	SCFM
System Inlet	T(1)	70	F(1)	10000
Heat Exchanger Cold Outlet	T(A)	1400	F(A)	9165
Heat Exchanger Bypass	T(B)	70	F(B)	835
Burner Inlet	T(2)	1289	F(2)	10000
Reactor Inlet	T(3)	1289	F(3)	10000
Reactor Outlet	T(4)	1500 Thermal Range	F(4)	10000
System Outlet	T(5)	281	F(5)	10000

Results		Economics	1.13 Gross/avail ratio
Firing Rate Pilot	0 BTU/Hour	Fuel Cost	$0.00 per Hour
Firing Rate, HHV	0 BTU/Hour	Electric Cost	$1.60 per Hour
Dilution/purge Blower Reqd (FD)	36 HP		
Combustion Air Required	100 SCFM	Total Operating Cost	$1.60 per Hour
Percent Bypass HX	8 %		

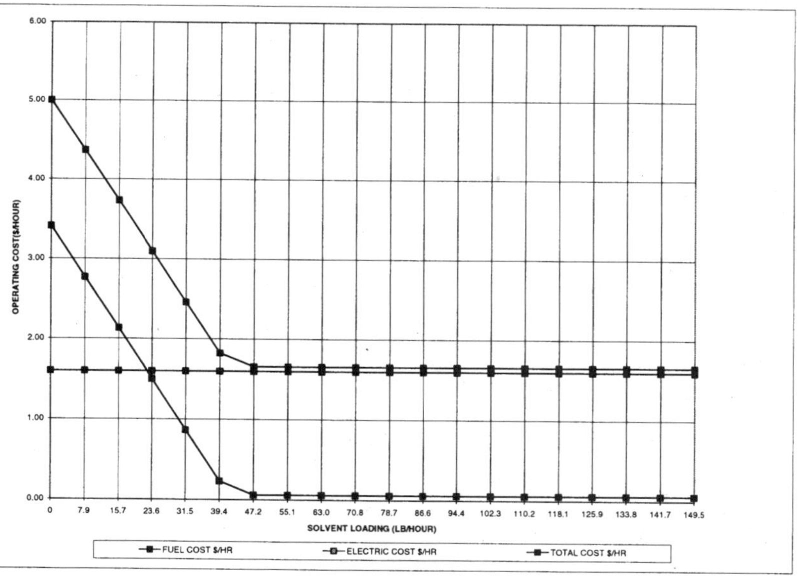

FIGURE 8.4.9 Regenerative thermal oxidizer, operating costs—10 percent LEL.

INPUT PARAMETERS		Assumes 460/60/3	Dilution Air Rate 1676
Heat Exchanger Efficiency	65 %	Electricity Price	$0.06 per KWH
Process Exhaust Flow	1683 SCFM	Fuel Price	$4.50 per MMBTU
Process Exhaust Temp.	125 Deg. F	LEL of Mixture	1.10% by Volume
Burner Setting	744 Deg. F	Avg Vapor Density	0.24 Lb/cuft
Pressure drop (max)	14 in. H20	Pressure Drop(Min)	14 in. H20
Solvent Loading	94.20 LB/Hour	Percent LEL	35.61% LEL
Solvent Calorific EQUIV.	16,721 BTU/LB	Compound Average	3917 PPMV

SYSTEM POINT	TEMP	Deg. F		Flow	SCFM
System Inlet	T(1)	125		F(1)	1683
Heat Exchanger Cold Outlet	T(A)	1018		F(A)	1166
Heat Exchanger Bypass	T(B)	125		F(B)	517
Burner Inlet	T(2)	744		F(2)	1683
Reactor Inlet	T(3)	744		F(3)	1683
Reactor Outlet	T(4)	1500	Thermal Range	F(4)	1683
System Outlet	T(5)	880		F(5)	1683

Results		Economics	1.37 Gross/avail ratio
Firing Rate Pilot	13,000 BTU/Hour	Fuel Cost	$0.06 per Hour
Firing Rate, HHV	0 BTU/Hour	Electric Cost	$0.24 per Hour
Dilution/purge Blower Reqd (FD)	5 HP		
Combustion Air Required	17 SCFM	Total Operating Cost	$0.30 per Hour
Percent Bypass HX	31 %		

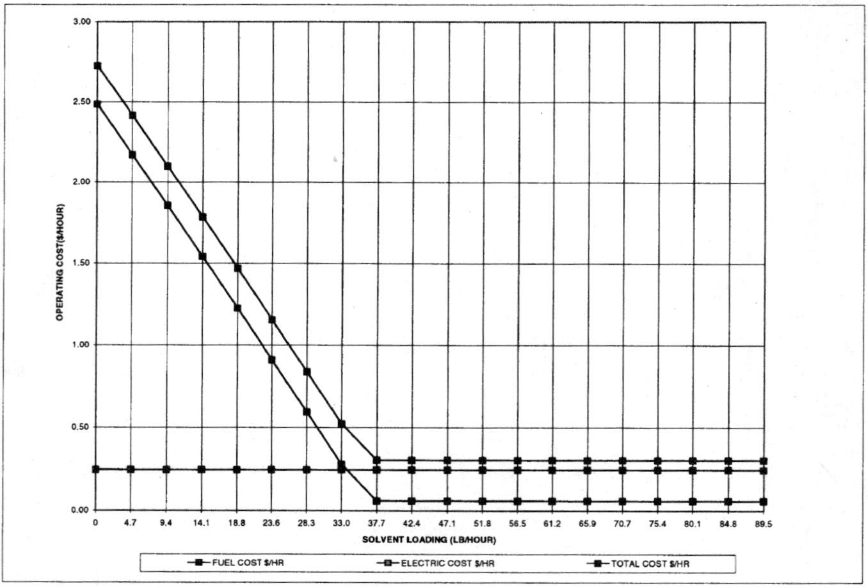

FIGURE 8.4.10 Thermal oxidizer with rotor concentrator—1 percent LEL.

Multiple Process Lines

If multiple process lines exist, a means to combine the airstream and to control the volume to the oxidizer must be present. For example, several process exhausts feed into the collection plenum, go through a roughing filter, and then are drawn through a system fan. The pressure drop across that filter is monitored to determine when to clean the filters. The amount of flow going into the system fan is controlled by a vortex damper or alternatively, a variable-frequency drive (VFD). The amount of air from multiple process lines into this plenum can vary. The pressure in the plenum is sensed by a pressure transducer, and this signal is used to vary the speed of the fan via the VFD or position the vortex damper to control the amount of volume drawn from the plenum. Consequently, the fan will slow down or speed up until the amount of air drawn from the plenum is the same as the amount of air entering. This pressure control loop is standard on all systems that have multiple process lines. It is not required when a single process line exists, but is beneficial to minimize air volume variations drawn from the process. When a fixed-speed fan is used, variation in organic loading causes differing pressure drops across the oxidizer. The differing pressure drops result in varying flow from the process unless a pressure control loop is utilized.

THERMAL OXIDATION TECHNOLOGIES

Thermal oxidation, by definition, converts a hydrocarbon, in the presence of oxygen and heat, to carbon dioxide and water vapor. A general equation showing this relationship is

$$C_nH_{2m} + \left(n + \frac{m}{2}\right)O_2 \Rightarrow nCO_2 + H_2O + \text{heat}$$

The particular n and m subscripts define the number of carbon atoms and hydrogen atoms. The number of oxygen atoms converts to n molecules of carbon dioxide and m molecules of water vapor and heat which is given off in the exothermic reactor.

In the section Characterizing the Process Stream, it was stated that process air volume and organic loading were important variables in identifying the appropriate oxidation technology. One measure of the organic loading or the organic concentration of the process stream is the *lower explosive limit,* designated as the LEL, or the *lower flammable limit,* designated as the LFL. The LEL of a stream is the lowest organic concentration in a stream that would, in the presence of an ignition source, yield a combustible mixture. At this concentration, the stream is said to be at 100 percent LEL. As examples, if a steam has toluene or alcohol and it is at 100 percent LEL, this means that the concentration within that stream is explosive if a spark or ignition source is present. Characteristics of some common organics are shown in Table 8.4.5, where they are divided into five general types: alcohols, aliphatics, aeromatics, esters, and ketones.

For organic benzene, 12,000 ppm is 100 percent LEL. The National Fire Protection Association (NFPA) states that equipment can operate, without LEL monitors or controls, if the LEL is less than 25 percent. For benzene, then, 25 percent LEL is equal to 3000 ppm. This upper boundary becomes a dictating factor in the selection and design of the oxidation equipment. If the concentration is higher than 25 percent LEL, the NFPA requirements state that an LEL monitor is required. Using an LEL monitor, NFPA guidelines allow operation up to 50 percent LEL (a 2:1 safety factor). In summary, 100 percent LEL is explosive; if the stream is at 25 percent, a safety factor of 4 exists.

In addition to the explosive aspects of the LEL, another issue is the heat energy given off during oxidation. An estimate of the exotherm is that there will be a 25°F rise per 1

TABLE 8.4.5 Organic Compound Characteristics

Gas/Vapor	Chemical formula	LEL	Calorific value (Btu/lb)
	(1) Alcohols		
Methanol	CH_3OH	3.4	9,033
Isopropanol	$(CH_3)_2CHOH$	1.1	12,960
N-Butyl alcohol	$CH_3(CH_2)_2CH_2OH$	2.7	14,230
N-Propyl alcohol	$CH_3CH_2CH_2OH$	1.2	12,960
	(2) Aliphatics		
Naphtha	No chemical description	—	16,960
Methyl cellosolve	$CH_3OCH_2CH_2OH$	1.5	9,274
N-Hexane	$CH_3(CH_2)_4CH_3$	2.8	19,391
N-Heptane	$CH_3(CH_2)_5CH_3$	2.4	19,314
N-Pentane	$CH_3(CH_2)_3CH_3$	—	19,493
Dimethyl formamide	$(CH_3)_2NCHO$	0.8	11,280
Cyclohexane	$CH_2(CH_2)_2CH_2$	2.0	18,826
	(3) Aromatics		
Benzene	C_6H_6	3.2	17,443
Toluene	$C_6H_5CH_3$	1.7	17,599
Tetrahydrofuran	$OCH_2CH_2CH_2CH_2$	—	14,990
Xylene	$C_6H_4(CH_3)_2$	1.1	17,725
	(4) Esters		
Isopropyl acetate	$(CH_3)_2CHOOCCH_3$	12.5	10,960
Isobutyl acetate	$CH_3COOCH_2CH(CH_3)_2$	6.5	13,000
Ethyl acetate	$CH_3COOC_2H_5$	Gas	10,235
N-Butyl acetate	$CH_3COOC_4H_9$	3.0	12,460
Propyl acetate	$C_3H_7OOCCH_3$	10.0	10,525
	(5) Ketones		
Acetone	CH_3COCH_3	2.5	12,250
Methyl ethyl ketone	$CH_3COC_2H_5$	1.0	13,729
Methyl butyl ketone	$CH_3CO(CH_2)_3CH_3$	2.2	16,100
Methyl isobutyl ketone	$(CH_3)_2CHCH_2COCH_3$	1.4	10,400

percent LEL in the stream. Hence, if the process air enters the oxidizer at a given temperature, and if the stream has a concentration of 2 percent LEL, then a 50°F rise in process stream temperature is expected after oxidation. If the process stream were running at a 10 percent LEL, then a 250°F temperature rise would be predicted. A maximum LEL of 25 percent yields a 625°F temperature rise of the process stream.

The concentration that is required to yield 100 percent LEL varies with the characteristics of the identified organics. For example, ethanol has a 20,000 ppmv concentration whereas benzene has a 12,000 ppmv.

Not only do different organics have different LELs, but also every time a pound of a different organic is oxidized, a different amount of heat will be given off. An example that demonstrates this energy release is the catalytic muffler in an automobile. When mufflers were initially installed, the high level of unburned gasoline which went through the muffler caused excessive heating and subsequently fires if the automobile had been parked on leaves. The reason for the excessive heat was that for each 1 lb of gasoline being oxidized 20,000 Btu was being released.

Chlorinated organics are hydrocarbons that have one or many chlorine atoms. Oxidation of chlorinated hydrocarbons yields CO_2, water vapor, and hydrogen chloride (HCl) gas. Some typical chlorinated organics are TCE and PCE. These organics have calorific values as low as 5000 Btu/lb.

Catalyst. The characteristics of oxidation catalysts are many and varied. Fundamentally, if an airstream containing organics is heated and passed across catalyst, the organics will be converted to carbon dioxide and water vapor. However, the conversion happens at different temperatures for different organics and for different catalysts. (See Figs. 8.4.11 and 8.4.12.)

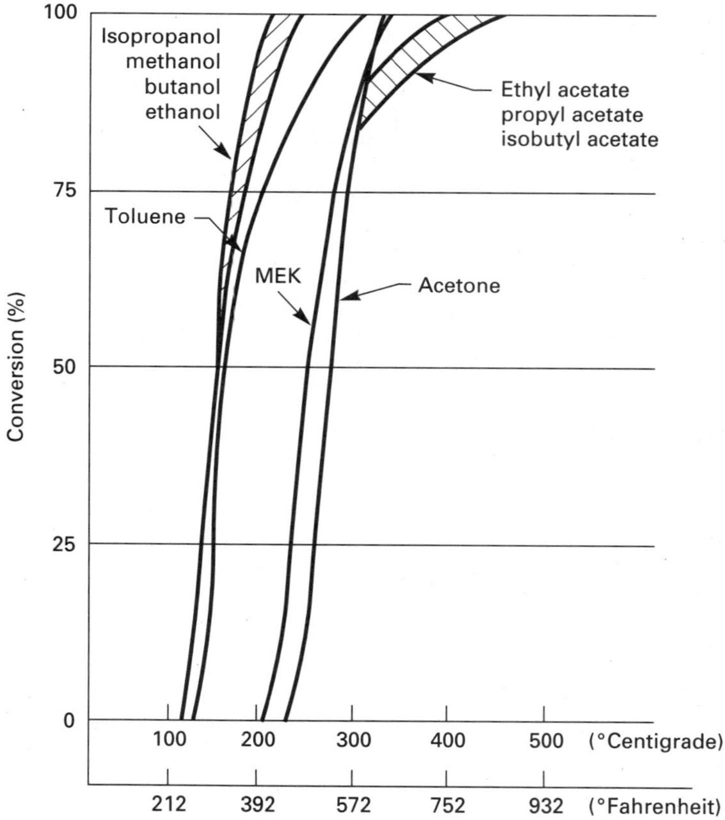

FIGURE 8.4.11 Catalyst destruction versus temperature—alcohols, esters, ketones, and aromatics.

Consider the compound toluene. To destroy 25 percent of the toluene in an airstream, enter the chart on the *y* axis at 25 percent, come over to the toluene line and down to the

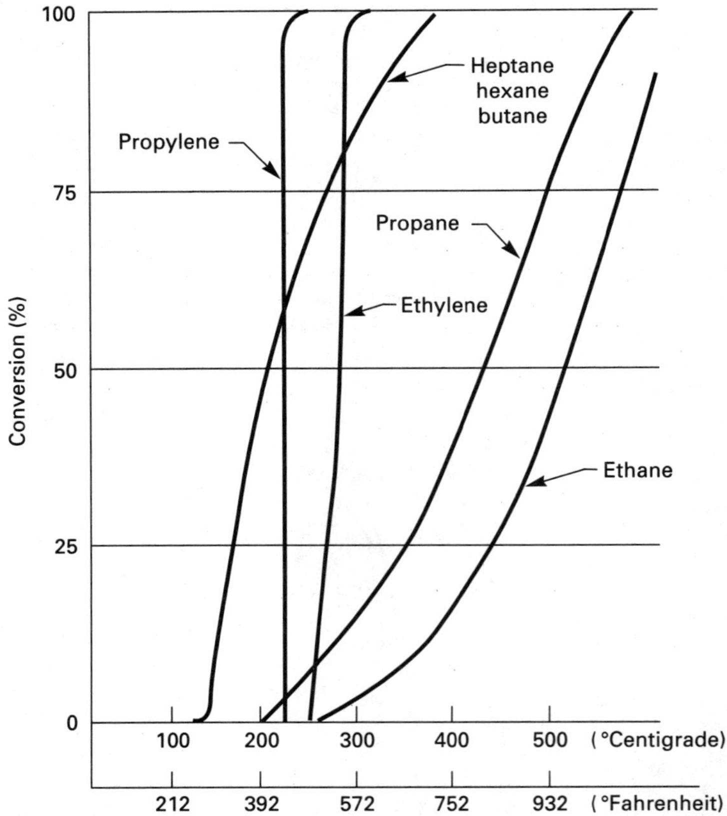

FIGURE 8.4.12 Catalyst destruction versus temperature—aliphatics.

Fahrenheit line, and it is seen that a temperature of about 300°F is required. However, the EPA does not require 25 percent destruction, but 95 to 99 percent destruction. To destroy 95 percent of the toluene, enter the graph at 95 percent destruction, come over to the toluene line and down vertically, and it is seen that 500 to 550°F is required to destroy the toluene. Some organics require higher temperatures to be destroyed than others, catalytically. Alcohols, isopropyl alcohol and ethanol, can be destroyed relatively simply whereas the acetates, particularly the ethyl acetates and propyl acetates, may require temperatures near 750°F to achieve adequate destruction. Depending on the process stream, either a single organic may be present, as found in the chemical industry, or, as in printing operations, a multiplicity of organics exists. Having a multiplicity of organics imposes the requirement of focusing on the ability to destroy the most difficult organic.

Some organics cannot be effectively destroyed by catalyst. For example, heptane and hexane can be destroyed at temperatures of 600 to 700°F, whereas propane, ethane, and methane require temperatures beyond a reasonable temperature range. Since methane is not a smog-producing organic, a guarantee to destroy 95 percent of the organics means that the methane is not considered and is removed from that stream in the computation process. However, if propane is the auxiliary fuel, i.e., if the burners are being driven by

FIGURE 8.4.13 Catalyst media.

LPG or by propane directly, it means that there will be contribution to the VOC at the end of the stack.

Catalyst samples are shown in Fig. 8.4.13. Some catalyst is deposited on a ceramic substrate. These ceramics are extruded in a malleable state and then fired in ovens. The process consists of starting with a ceramic and depositing an aluminum oxide coating. The aluminum oxide makes the ceramic, which is fairly smooth, have a number of bumps. On those bumps a noble metal catalyst, such as platinum, palladium, or rubidium, is deposited. The active site, wherever the noble metal is deposited, is where the conversion will actually take place.

An alternate to the ceramic substrate is a metallic substrate. In this process, the aluminum oxide is deposited on the metallic substrate to give the wavy contour. The precious metal is then deposited onto the aluminum oxide. Both forms of catalyst are called *monoliths.*

An alternate form of catalyst is pellets. The pellets are available in various diameters or extruded forms. The pellets can have an aluminum oxide coating with a noble metal deposited as the catalyst. The beads are placed in a tray or bed and have a depth of anywhere from 6 to 10 in. The larger the bead ($\frac{1}{4}$ versus $\frac{1}{8}$ in), the less the pressure drop through the catalyst bed. However, the larger the bead, the less surface area is present in the same volume, which translates to less destruction efficiency. Higher pressure drop translates to higher horsepower required for the oxidation system. The noble metal monoliths have a relatively low pressure drop and are typically more expensive than the pellets for the same application.

Base-Metal Catalyst. An alternate to a noble metal catalyst is a base metal catalyst. A base metal catalyst can be deposited on a monolithic substrate or is available as a pellet. These pellets are normally extruded and hence are 100 percent catalyst rather than deposition on a substrate. A benefit of base metal extruded catalyst is that if any poisons are present in the process stream, a deposition of the poisons on the surface of the catalyst occurs. Depending on the type of contaminant, it can frequently be washed with water. When it is washed, abraded, or atritted, the outer surface is removed and subsequently new catalyst is exposed. Hence, the catalyst can be regenerated. Noble metal catalyst can also be regenerated, but the process is more expensive and often does not work effectively.

A noble metal catalyst, depending on the operation, will typically last around 30,000 hours. As a rule of thumb, a single shift operation of 40 hours/week, 50 weeks per year results in a total of 2000 hours/year. Hence, the catalyst might have a 15-year life. It may degrade, it may need some washing, it may need some other operations on it, but it should last that long. If a plant operates three shifts per week, or 6000 hours/year, then the expected catalyst life is about 5 years. From a cost factor, a typical rule of thumb catalyst might be 10 to 15 percent of the overall capital cost of the equipment.

There are seven fundamental oxidizer technologies that achieve the oxidation of organics in various ways:

- Catalytic recuperative
- Thermal recuperative
- Catalytic regenerative
- Thermal regenerative
- Rotor concentrators
- Flare/burner
- Ceramic filter

Catalytic Recuperative Oxidizer

A catalytic recuperative oxidizer consists of several main elements, as shown in Fig. 8.4.14, and classified as

1. System fan
2. Heat exchanger
3. Burner
4. Reactor
5. Catalyst
6. Exhaust stack

Figure 8.4.15 shows a typical process and instrumentation diagram (P&ID) for such a system. The process stream is ducted to the oxidizer and enters a system fan. The system fan is driven by a motor, and the process stream is forced under positive pressure through a heat exchanger. The heat exchanger shown is a cross-flow heat exchanger of the shell and tube type. Plate type heat exchangers can be used in this application. Normally, the process stream enters through the tube side of the heat exchanger to clean the tubes. As the air enters and goes through the heat exchanger, it is heated and then exits into the section called the *reactor.* As it comes into the reactor, the process stream will be further heated by a burner, controlled by a thermocouple measuring the temperature of the air and a temperature controller regulating the burner firing to bring the process stream up to the catalyzing temperature of 300 to 700°F. The catalyzing tem-

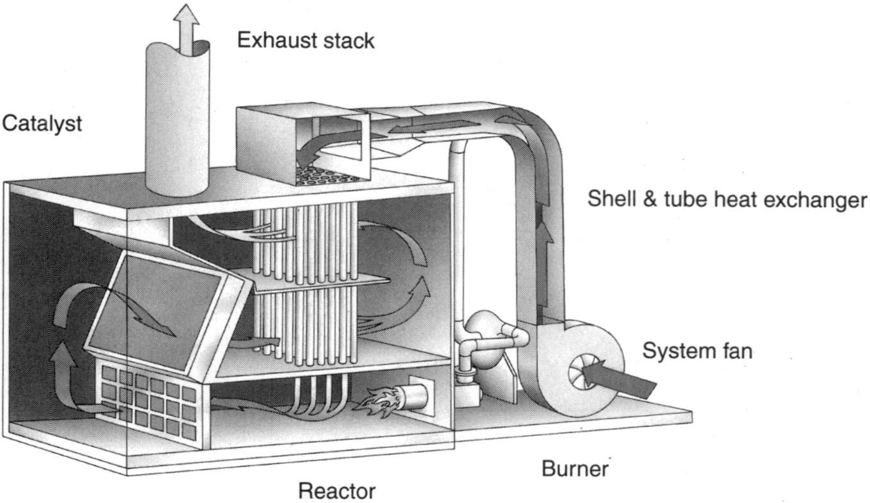

FIGURE 8.4.14 Catalytic recuperative oxidizer.

perature depends on the organic, the requirement for the destruction of the organic, and the type and volume of catalyst. At the catalyzing temperature, the process stream will pass through a series of beds, having catalyst in them. As the air containing organics comes across the catalyst, the organic is converted to CO_2 and water vapor and an exothermic reaction occurs. This exotherm will raise the temperature of the stream exiting the catalyst bed. Hence the catalyst outlet temperature will be higher than the temperature going into the catalyst bed. The process stream is then directed though the shell side of the heat exchanger where it preheats the incoming air and is then exhausted to the atmosphere.

Thermal Recuperative Oxidizer

The theory of operation for thermal recuperative oxidation is called the three Ts of combustion:

- Time
- Turbulence
- Temperature

Time, as used in thermal oxidizers, refers to the retention time or residence time, which is the length of time that an organic is at the appropriate oxidation temperature. Typically, if a 95 percent destruction efficiency is required, a residence time of 0.5 second is adequate. That is, the organic is brought up to a temperature of about 1400°F, and maintained at 1400°F for a retention time of 0.5 second. Both retention time and turbulence must be achieved; turbulence is a term which implies an adequate level of mixing. If the turbulence occurs and if the 1400°F for 0.5 second is achieved, with adequate oxygen, then 95 percent of the organics will be destroyed. In some cases, particularly in the chemical indus-

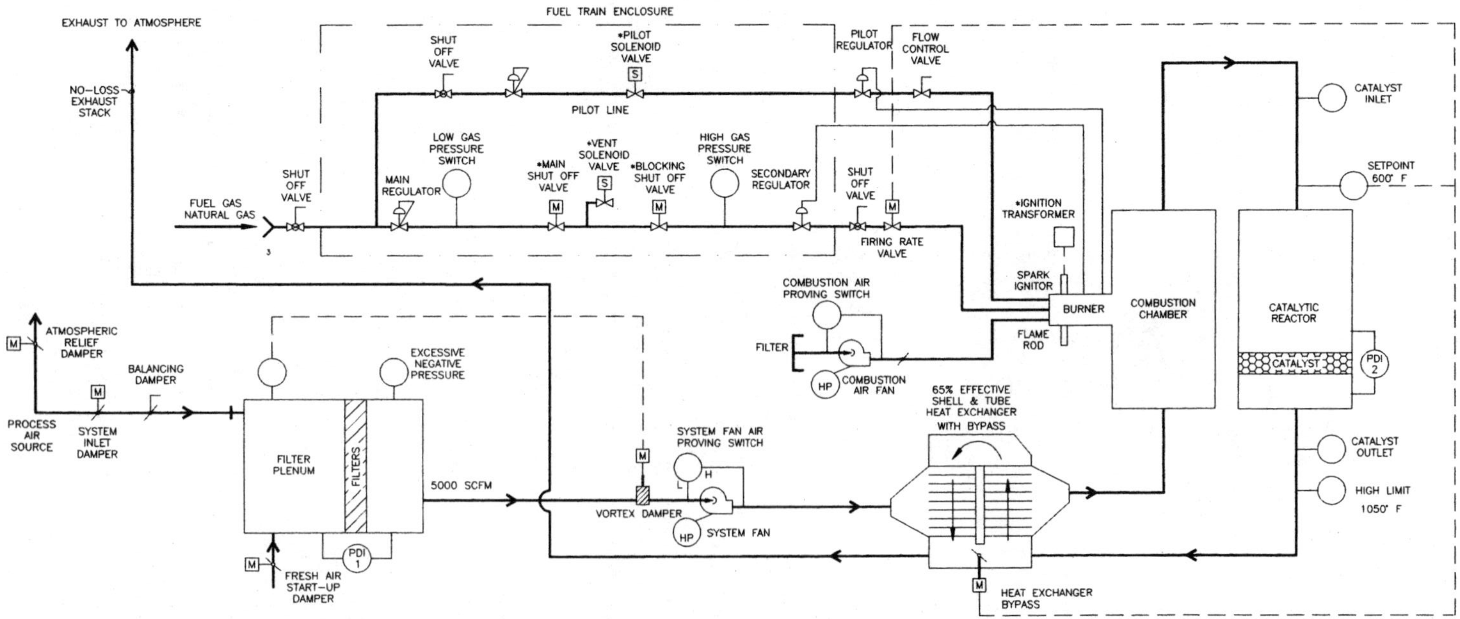

FIGURE 8.4.15 P&ID for catalytic recuperative oxidizer.

try, oxygen must be added to a process stream. For each organic, a specific temperature and a specific retention time are required to achieve the desired level of destruction. Table 8.4.6 illustrates these points for several organics, some being chlorinated.

TABLE 8.4.6 Temperatures and Residence Times

	0.5-s Retention time		1.0-s Retention time	
	95% DE	99% DE	95% DE	99% DE
Toluene	1351	1372	1317	1338
Benzene	1489	1640	1415	1640
Styrene	1445	1480	1391	1424
PCE	1843	1921	1726	1797
TCE	1592	1675	1972	1545

PCE: Perchloroethane
TCE: Trichloroethane
DE: Destruction efficiency

- **Toluene.** The 95 percent destruction efficiencies are achieved with 0.5-second residence time if the temperature is 1351°F. At 1-second residence time, the temperature can be reduced slightly to 1317°F. To achieve 99 percent destruction with 0.5-second residence time, the operating temperature would have to be 1372°F. But if a 1-second residence time existed, then 99 percent destruction efficiency could be achieved at 1338°F. Increased temperature gives better destruction efficiency but increases operating costs. Increased retention time gives better destruction efficiency but increases capital cost. The balance between increased capital cost and lowered operating cost must be weighed.
- **Benzene.** Benzene, which is carcinogenic, requires a higher destruction temperature than toluene. Rather than operating at 1351°F, operating temperatures of 1489°F are required at 0.5-second residence time.
- **PCE.** The chlorinated organics require higher temperatures. Rather than operating at the 1400°F range, the oxidizer must operate around 1800°F, for a longer residence time allows a decreased temperature to about 1700°F; an increase to a 99 percent destruction requires an increased temperature to about 1900°F.

One of the concerns that exists in thermally processing chlorinated organics is that the by-products of combustion that are formed may be more harmful than the initial product to be destroyed. A by-product of combustion is a compound formed, either catalytically or thermally, when most of the organic compound is converted to CO_2 and H_2O. For example, when 95 percent of toluene is destroyed, what happens to the other 5 percent? Does it remain as toluene, or are new products formed? Frequently, "by-products of combustion" are formed, and it is not possible to predict what those by-products are. They could be a mixture of the original compounds, or they could be more harmful. The concern that people have in processing chlorinated organics thermally at these high temperatures is that dioxins may be generated. Dioxins are in effect much more harmful than the organic that will cause VOCs or smog in the air.

Oxidizer Fabrication. Thermal oxidizers must be built to provide the residence time and temperatures to achieve the destruction of the organics. Therefore, thermal oxidizers are comparatively larger than catalytic oxidizers since their residence time is 2 to 4 times greater. Thermal oxidizers used to be manufactured with carbon steel for the outer shell and castable refractory or brick as the thermal liner. A refractory is like a cement, which is put on the inside of the reactor shell to act as a thermal insulation barrier. New thermal oxidizers, offered in the marketplace today, are designed and built using ceramic fiber insulation on the inside. This is a rather lightweight material and has a relatively long life, whereas the old refractory would tend to fail over a period of years by attrition of expansion and contraction.

Figure 8.4.16 shows a forced-draft thermal recuperative oxidizer, where the system fan forces the process airstream through a heat exchanger. As the process flow exits the heat exchanger, the burner fires and the process stream is brought up to temperature. The reactor chamber, which is lined with a high-temperative ceramic fiber, is designed for the required retention time. A shell-and-tube heat exchanger is normally used for thermal oxidizers because of expansion concerns. The material selection for the shell-and-tube heat exchanger needs to be a high-grade stainless, adequate for expected steady-state and peak temperatures.

The thermal oxidizer in Fig. 8.4.16 and its associated process and instrumentation diagram in Fig. 8.4.17 illustrate a forced-draft system, but induced-draft systems also exist. The induced-draft system is slightly more expensive but is recommended when particulates or organic oils are present. Particulates impact upon a forced-draft fan and will have a negative effect on the system performance. If the process stream is clean, a forced-draft system is appropriate.

The burners employed are modulating burners; i.e., if the organic is preheated to an adequate temperature, the burners will modulate down to zero so that there is no energy required for the continued oxidation.

One of the benefits of the thermal recuperative oxidizer is that it is possible to process organics that may be a poison or otherwise detrimental to the catalyst. In addi-

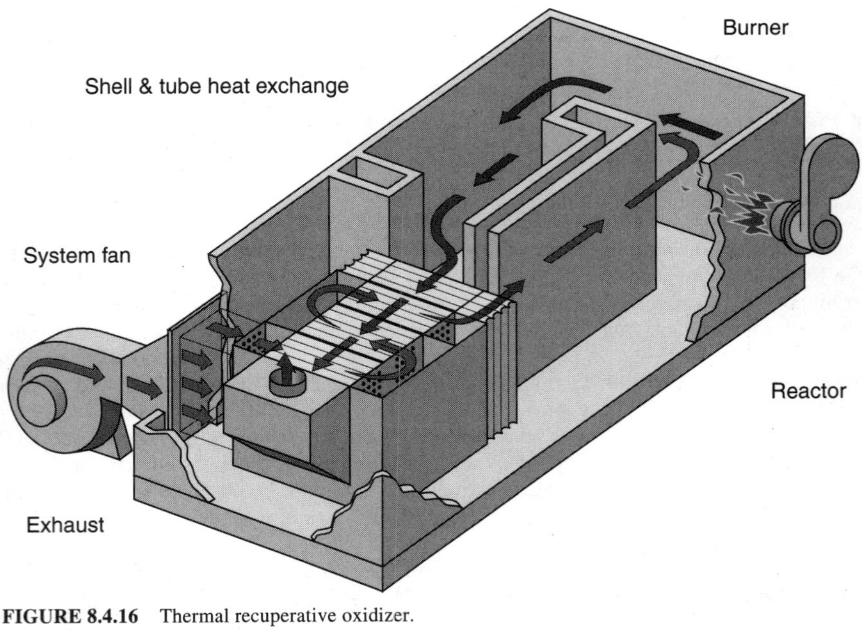

FIGURE 8.4.16 Thermal recuperative oxidizer.

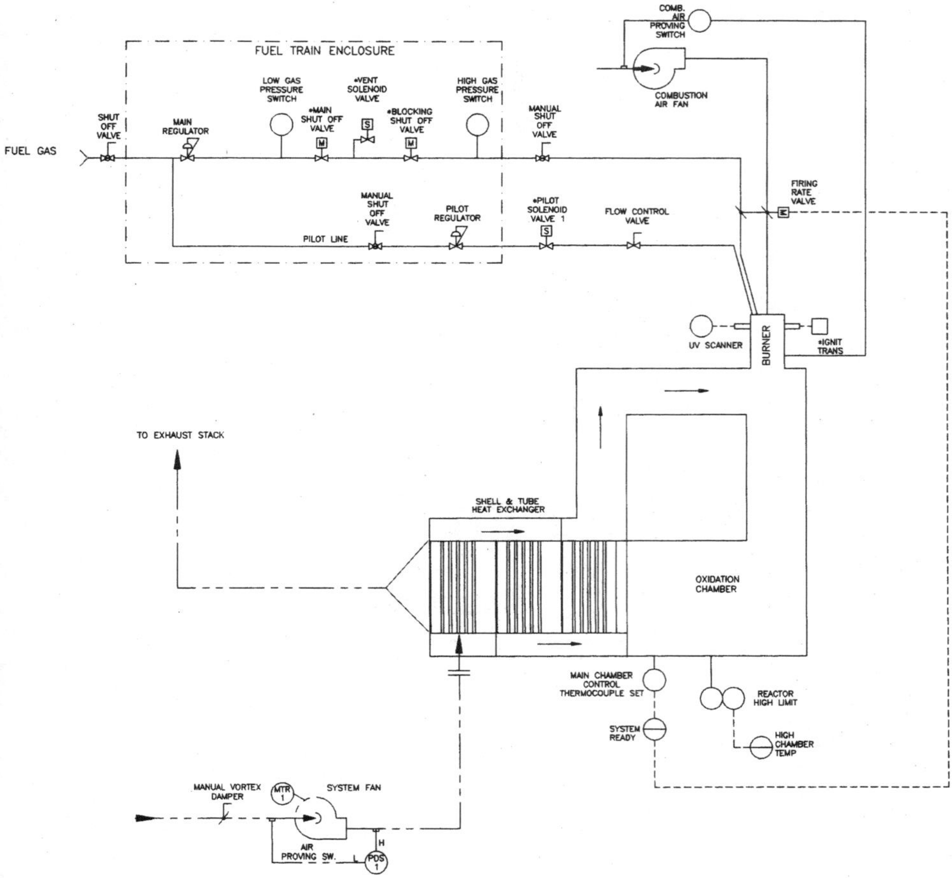

FIGURE 8.4.17 P&ID for thermal recuperative oxidizer.

tion, if the organic concentration is very high, e.g., the organic level is in the 20 to 25 percent LEL range, then thermal recuperative oxidation is the appropriate technology.

For relatively large-volume and low-concentration streams, the problem with the thermal recuperative design is that the metallic heat exchanger only recovers 70 percent of the heat and operating costs increase dramatically. The level of heat-exchanger efficiency is limited by material cost to increase heat-exchanger efficiency and the fact that autoignition, where organics ignite due to temperature alone, becomes a problem with the life of metallic heat exchangers.

Regenerative Thermal Oxidizer

The regenerative thermal oxidizer is a thermal oxidizer consisting of two or more ceramic heat-transfer beds, which act as heat exchangers, and a purification chamber or retention chamber where the organics are oxidized and converted to CO_2 and H_2O vapor.

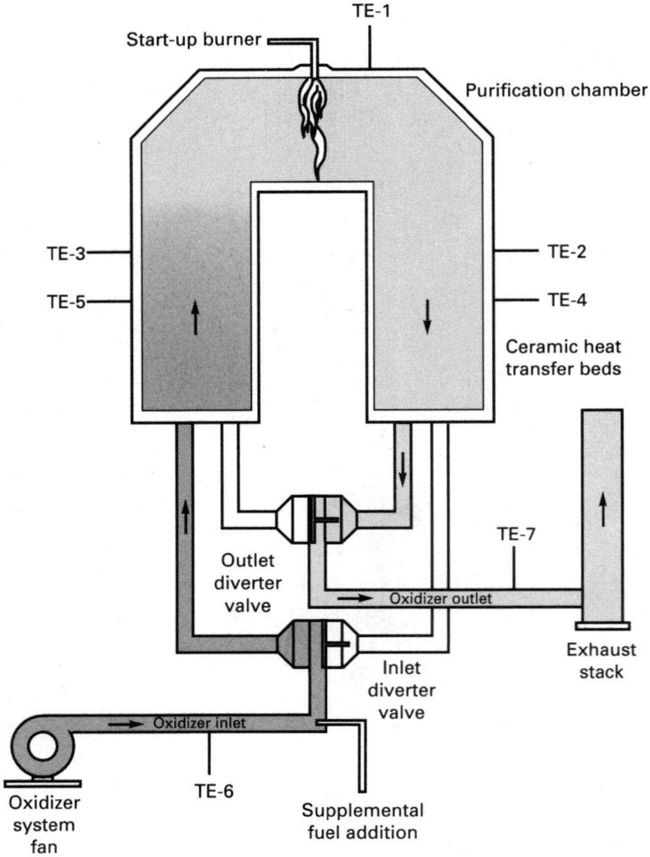

FIGURE 8.4.18 Twin-bed RTO—valve arrangement. Shown in clockwise cycle.

The operation of a twin-bed regenerative thermal oxidizer is shown in Figs. 8.4.18, 8.4.19, and 8.4.20 (its associated process and instrumentation diagram). This thermal oxidizer requires the initial preheating of the ceramic heat-transfer beds to a temperature of 1500°F during the start-up mode. This is accomplished by operation of a fuel-fired burner located in the purification chamber. To equalize the preheating of the ceramic heat-transfer beds, the air is directed into and out of the ceramic heat-transfer beds by operation of two pneumatic diverter valves located under each ceramic heat-transfer bed. During initial start-up, outside air is supplied to the oxidizer through the makeup air damper tee located on the inlet side of the process air fan. A programmable logic controller (PLC) monitors and controls the direction of the airflow.

After the ceramic heat-transfer beds have reached an operating temperature of 1500°F, the unit is ready for the process airstream. As the process airstream enters the ceramic heat-transfer beds, the heated ceramic media preheat the process airstream to its oxidation temperature. Oxidation of the airstream occurs when the autoignition of the hydrocarbon is reached. At this point the heat released by the oxidation of the process hydrocarbons is partially absorbed by the inlet ceramic heat-transfer bed. The heated air passes through the retention chamber, and the heat is absorbed by the outlet ceramic

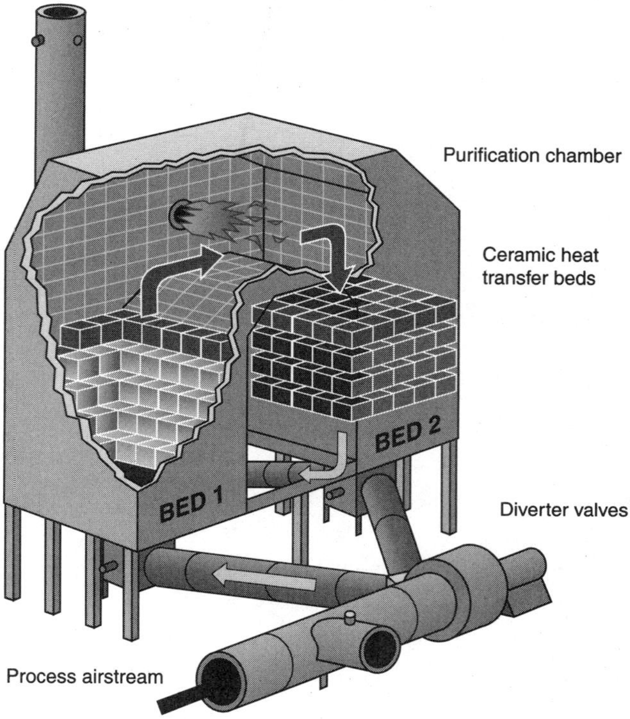

FIGURE 8.4.19 Twin-bed RTO—component identification.

heat-transfer bed. If the oxidizer is self-sustaining, the net increase in temperature (inlet to outlet) is 100°F.

During the normal mode of operation of the system (see Fig. 8.4.16), the process air enters the RTO system fan and passes through the inlet diverter valve, where the process air is forced into the bottom of the left ceramic heat-transfer bed. As the process air rises through the ceramic heat-transfer bed, the temperature of the process stream will rise. The tops of the beds are controlled to a temperature of 1500°F. The bottom of the beds will vary depending upon the temperature of the air that is coming in. If it is assumed that the process air is at ambient conditions or 70°F, then as the air enters the bottom of the bed, the bottom of the bed will approach the inlet air temperature of 70°F. The entering air is heated, and the medium is cooled. As the air exits the ceramic medium, it will approach 1500°F. The process air then enters the second bed at 1500°F, and now the ceramic medium recovers the heat from the air and increases in temperature. At a fixed time interval of 4 to 5 min, or based on thermocouple control, the diverter valves switch and the process air is directed to enter the bed on the right and exits the bed on the left. Prior to valve switching, the air heated the right bed and now this bed is being cooled. The cooling starts at the bottom and continues upward because the medium is hot and the energy is transferred. The process air then goes through the purification chamber and exits through the second bed.

Destruction Efficiency. When the valves are switched, whatever organics had not been destroyed prior to the flow reversal are then exhausted out of the stack. In addition, the rapidity of switching or closure of the valves is critical to minimize the bypass of

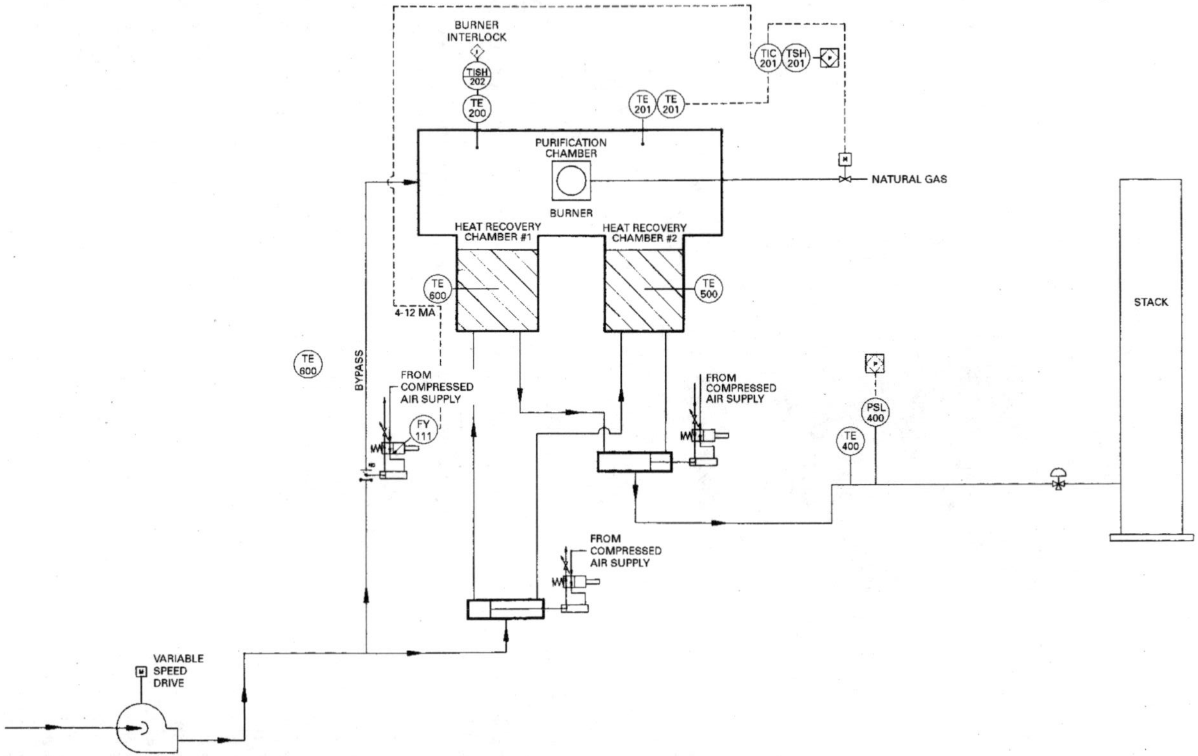

FIGURE 8.4.20 P&ID for RTO.

unoxidized organics. If emissions versus time were plotted, the graph would reflect a very low exhaust concentration level, say, 1 ppm; but whenever the diverter valve switches, an organic pulse occurs in the exhaust stream. Since the valves shift every 4 min, these pulses reduce the overall destruction efficiency of the organics. Several methods of processing the pulse exist in order to achieve higher destruction efficiencies.

Oxidizer Fabrication. The retention chamber and heat-transfer chamber are fabricated of reinforced carbon steel exterior and ceramic fiber lining. The thickness of the ceramic fiber lining is based on the required destruction temperature of the organics and the desired outside shell temperature.

The ceramic heat-exchange medium can be of various types including ceramic saddles, tye pacs, or structured packing. The ceramic structured packing is a recent development in the industry reflecting lower pressure drops for equivalent heat transfer (see Fig. 8.4.19).

A reinforced carbon-steel structure is provided to support the load of the oxidizer chambers, the structured packing support grid, and the wind and/or earthquake loads. The packing support grid is fabricated of stainless steel and is designed to support the structured packing.

Diverter Valves. The oxidizer has a process air diverter valve located under each energy recovery chamber. The process air diverter valves control the process airflow into and out of the energy recovery beds. The valves can be fabricated of carbon steel or stainless steel depending on the application. The valves are operated by pneumatic actuators and controlled by a PLC to maintain optimum energy efficiency.

A number of different types of valves are used. All valves must be highly reliable and have the ability to seal effectively. Many systems use heavy-duty case butterfly valves, whereas newer systems utilize popit-type valves.

Oxidizer Bake-out. If organic particulates in the process exhaust build up on the "cold" surfaces at the bottom of the oxidizer, the process must be shut down and a volatilization of these organics or a *bake-out* is required. When bake-out is activated, the flow diverter valves will stay in one position until the exhaust air temperature from the outlet bed reaches 850°F. At this temperature, most organic oils will volatilize, as in a self-cleaning oven. When the first outlet bed reaches 850°F, the flow diverter valves will switch and will stay in position until the outlet temperature of the second bed reaches 850°F.

Supplemental Fuel Injection. The supplemental fuel system is designed for natural gas operation, but allows the oxidizer to be operated with natural gas, propane, or butane for any condition requiring energy input to initiate or sustain operation. Supplemental fuel may be injected via the combustion burner in the purification chamber or by injection of natural gas properly mixed with the process stream. As the natural gas mixture rises through the ceramic medium, it reaches autoignition temperature and the exothermic reaction takes place, thereby providing heat to the process stream.

The benefit of natural gas injection into the process stream rather than through the burner results in utilization of 1000 Btu/ft^3 of natural gas versus 600 Btu/ft^3 of gas when a burner is employed.

Regenerative Catalytic Oxidizer

Operating-Temperature Comparisons. The oxidation temperature for catalytic or thermal oxidation is shown in Table 8.4.7. Lower temperature translates to lower fuel

TABLE 8.4.7 Oxidation Temperatures, Degrees Fahrenheit, 99 Percent Destruction

	Catalytic	Thermal
Benzene	440	1460
Carbon TET	610	1430
MEK	600	1780
Cyanide	480	1800

costs, assuming the same heat-exchanger efficiency. It takes energy to take the airstream from whatever temperature at which it enters up to its operating temperature. Even though a heat exchanger will help reduce net fuel costs, the lower the operating temperature, the lower the fuel costs.

Catalytic oxidizer design uses the term *gas hourly space velocity* (GHSV) which relates the amount of air to be processed and the volume of catalyst required. The GHSV is the inverse of the residence time. As shown in Table 8.4.8, for benzene destruction, the residence time in a catalytic unit is less than 0.25 s.

TABLE 8.4.8 Residence Time, Seconds, 99 Percent Destruction

	Catalytic	Thermal
Benzene	0.18	1.0
Carbon TET	0.18	1.0
MEK	0.18	1.0

Rotor Concentrator/Thermal Oxidizer

For high-volume airstreams with relatively low organic concentration, the energy required to bring the process stream to the thermal oxidation temperature can be significant. A technology developed uses a rotor concentrator in a two-step process, and it has proved to be effective in high-volume, low-concentration airstreams. The overall flow of a rotor concentrator system is shown in Fig. 8.4.21 and its process and instrumentation diagram is shown in Fig. 8.4.22.

The first part of that system consists of a slowly rotating concentrator wheel, utilizing zeolites or carbon deposited on a substrate, which adsorbs the organics as they are exhausted from the original process and passed through the wheel. A sector of the concentrator wheel is partitioned off from the main section of the rotor, and clean, heated air is passed through this section to desorb the organics. The air volume of clean heated air varies for applications, but is roughly 10 percent of the original airstream volume. As the clean, heated air passes through the rotor section, the organics are desorbed, resulting in a process stream that is 10 percent in volume of the original process stream and 10 times the original concentration. For example, if the process airstream is 50,000 cfm with 1 percent concentration, then the desorption stream will be 5000 cfm and 10 percent LEL. The desorption airstream will drive off all the organics in that particular

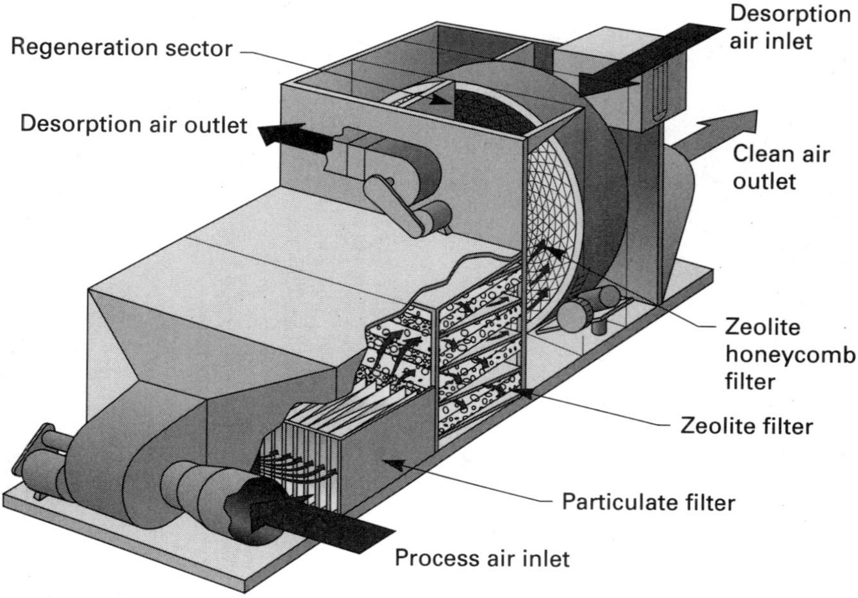

FIGURE 8.4.21 Rotor concentrator.

sector of the wheel, and as the wheel continues to turn, the entire wheel is desorbed. As the concentrated-organic, low-volume airstream exits, in the second part of the system it is processed through a thermal recuperative oxidizer. Consequently, what has occurred is that a large airstream with low concentration, which requires considerable energy to heat up, has been reduced to a low airstream with a high concentration, which is significantly less costly to process.

The concentrator wheel is shown in Fig. 8.4.21 with the exhaust passing through. A review of the P&ID shows that a 10 percent air volume slipstream from the original process stream is drawn through the rotor on the cooling pass. A secondary fan draws this air and forces this volume through the secondary heat exchanger, where the reduced air volume temperature is raised to the required desorption temperature. The preheated air is then used to desorb the air in the pie-shaped sector of the wheel. As the air exits the desorption section, the organic concentration is approximately 10 times the concentration of the original process stream. This low-volume, higher-concentration stream then enters the induced-draft section of a catalytic or thermal recuperative oxidizer, where the organics are destroyed.

Destruction Efficiency. Note that the total hydrocarbon reduction efficiency for the rotor/concentrator is the adsorption efficiency of the rotor/concentrator times the destruction efficiency of the oxidizer. Hence, if the rotor can adsorb 98 percent of the organics in the airstream and the oxidizer can destroy 99 percent of the organics that pass through it, then the overall system organic reduction is

$$\text{Destruction efficiency} = 0.98 \times 0.99 = 0.97 = 97\%$$

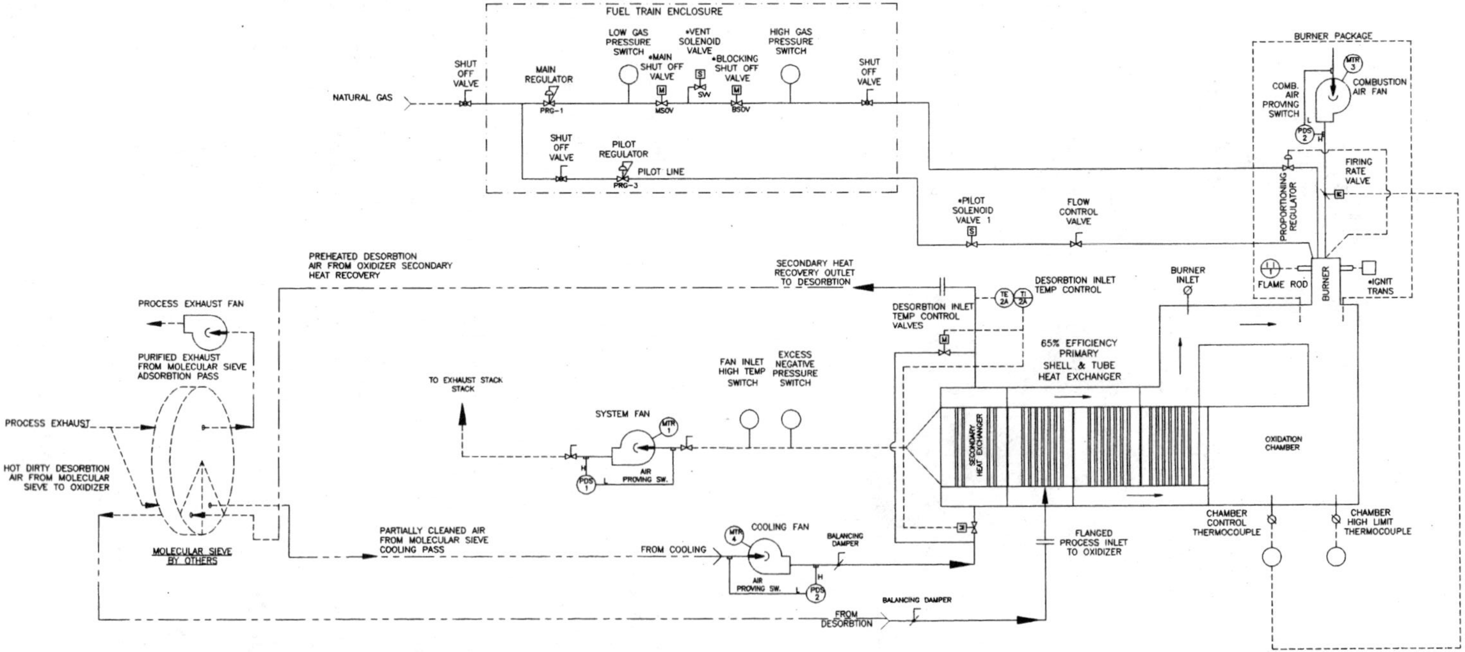

FIGURE 8.4.22 P&ID for rotor concentrator.

Flare/Burners

The flare is an oxidation technology that continues to be used both domestically and internationally. They are used in the petroleum, petrochemical, and other industries that require the disposal of waste gases of high concentration on both a continuous and intermittent basis. As in other thermal oxidation technologies, the three T's of combustion—time, temperature, and turbulence—are necessary to achieve adequate emission control.

Flares ideally burn waste gas completely and smokelessly. Two types of flares are normally employed. The first is called the *open flare,* and the second is called the *enclosed flare.* The major components of a flare consist of the burner, stack, water seal, controls, pilot burner, and ignition system. Flares required to process variable air volumes and concentrations are equipped with automatic pilot ignition systems, temperature sensors, and air and combustion controls.

Open flares have a flare tip with no restriction to flow, the flare tip being the same diameter as the stack. Open flares are effectively a burner in a tube. Combustion and mixing of air and gas take place above the flare with the flame being fully combusted outside of the stack.

Enclosed flares are composed of multiple gas burner heads placed at ground level in a stacklike enclosure that is usually refractory- or ceramic-lined. Many flares are equipped with automatic damper controls that regulate the supply of combustion air depending on temperature, which is monitored upstream of the mixing, but inside the stack. This class of flare is becoming the standard in the industry due to its ability to more effectively control emissions.

Requirements on emissions include carbon monoxide limits and minimal residence time and temperature. Exhaust gas temperatures may vary from 1000 to 2000°F.

Ceramic Filter

The ceramic filter collects, volatilizes, and destroys particulate and condensible organics emitted from industrial process streams, such as paint spray, lost foam casting, condensible organics, tenter frames, and cured rubber operations. It may be used independently as a hot or cold filter medium or coupled with an oxidization module for total odor and VOC control.

The ceramic filter is designed based on exhaust airflow volume, type of contaminant, and desired collection efficiency. During operation, the process stream, containing hot or cold particulate-laden air, condensible organics, or VOCs, is drawn into the ceramic filter. The process stream passes over a ceramic matrix selected on particle size and collection efficiency considerations.

The ceramic matrix is periodically heated by a natural gas burner, whereby any organics collected on the ceramic matrix are volatilized. Any nonorganics collected on the ceramic matrix are converted to inorganic ash and retained in the unit dropout chamber. If required, the volatilized organics can then be processed through a catalyst bed. As in the catalytic oxidizer, these volatiles are converted to CO_2 and H_2O.

SECTION 8.5.1

ADSORPTION: CARBON

Charles Parmele
Science Applications International Corporation
Oak Ridge, Tennessee

Teresa Kovalcson
TN & Associates, Inc.
Oak Ridge, Tennessee

INTRODUCTION 8.66
ROLE OF ADSORPTION 8.66
CONVENTIONAL ADSORPTION SYSTEM 8.67
- **Adsorption Performance** 8.68
- **Choosing an Adsorbent** 8.70
- **Adsorber Operation** 8.72
- **Regeneration** 8.76

ALTERNATIVE ADSORBER DESIGNS 8.80
FULL-SCALE DESIGN 8.83
- **Design Data and Scale-up** 8.84
- **Safety Considerations** 8.86
- **Cost Factors** 8.87

ACHIEVING HIGH REMOVAL EFFICIENCIES 8.88
ACKNOWLEDGMENT 8.90
REFERENCE 8.91

INTRODUCTION

Volatile organic compounds (VOCs) are often used as solvents in the manufacture of countless products. Some prominent examples of these products include crystallized organic chemicals, printed materials, paints and coatings, and dry-cleaning fluids, to name just a few.

One consequence of these manufacturing operations is that large amounts of organics are emitted to the atmosphere each year. Stricter air pollution laws have dramatized the environmental significance of these difficult-to-control emissions. At the same time, the rising costs of volatile materials have made them economically significant. Furthermore, higher costs for heating and cooling the ventilation air in a work area make it economically attractive to reuse the air. But before the air can be recycled, the concentration of organics present must be reduced to limits set by Occupational Safety and Health Administration (OSHA) guidelines or fire protection codes.

VOCs can also be released as secondary emissions that are generated during activities to manage wastes containing volatile materials or to remediate disposal or leaks of volatile materials into the environment. These secondary emissions can originate as exhausts from air strippers or soil vapor extraction (SVE) systems to remove VOCs from contaminated groundwater or soil. They can also be created when vacuum trucks are used to empty tanks, pits, or sumps containing volatile materials.

ROLE OF ADSORPTION

Carbon adsorption is one versatile technology that has been successfully applied to controlling these emissions of VOCs. Vapor-phase activated-carbon adsorption has also gained favor as a method of recovering valuable solvents from atmospheric emissions (Anderson et al., 1977). This is especially true for applications where other techniques are unable to produce the required low outlet concentrations at reasonable cost.

Condensers—another nondestructive pollution control option—are also effective in reducing heavy emissions, but they sometimes cannot achieve low enough discharge levels. Condenser outlet concentrations are often limited to 10,000 to 20,000 ppm due to the vapor pressures of the organics at the temperatures of the cooling water. Better recoveries can be achieved with refrigerated condensers, but this option is usually limited to applications where water is not present to foul the condenser with ice.

Scrubbers using nonvolatile organics as the scrubbing medium are another option. However, they are usually limited technically and economically to applications in which the spent scrubbing fluid can be reused directly without the absorbed organics having to be removed. Desorption requirements are often prohibitively expensive because the residual organic concentration in the liquor must be extremely low for it to be suitable for reuse.

In some cases the emissions from large solvent storage tanks have been controlled by scrubbing with the subcooled solvent itself. Package systems for this patented technology (U.S. Patent 3,981,156) have been offered by Ecology Control, Inc., Houston, Texas.

Although condensers and scrubbers are not effective in controlling dilute organic emissions, fume incineration is. However, it is usually a more expensive control method than recovery technologies because of higher energy costs and the value of the organic that is destroyed. Fume incineration can be competitive with methods that require additional separation equipment to effect a suitable recovery.

CONVENTIONAL ADSORPTION SYSTEM

Adsorption is usually a batch operation involving multiple beds. At least one bed remains on-line while the other(s) is(are) being regenerated.

Most of the systems already installed pass the vapor-laden airstream downward through a fixed carbon bed. Upflow design is generally avoided because carbon particles can become entrained in the exhaust at higher superficial velocities. Granular carbon is usually favored for vapor-phase applications because it is not so easily entrained as powdered carbon, and offers lower pressure drop across the bed.

Spent carbon is usually regenerated with upflow low-pressure steam. This drives off the organic, which is usually recovered by condensing the vapors and separating the organic from the water by either decantation or distillation. The countercurrent upflow pattern favors high removal efficiency and effluent quality.

Figure 8.5.1 shows a flow sheet for a typical fixed-bed system. Feed to the adsorber is pretreated to remove solids (dust or lint), liquids (droplets or aerosols), or vapors (high inlet concentrations or high boiling components) since these can hamper performance.

Accumulations of solid materials can cause an excessively high pressure drop across the adsorbers and may eventually blind them. To prevent an excessive temperature rise across the bed due to the heat of adsorption, inlet concentrations of organics are usually limited to about 0.5 to 1 percent (sometimes by addition of dilution air). This also explains why droplets or aerosols should be removed upstream of the adsorbers. Because operating costs are directly proportional to the weight of organic compounds to be adsorbed and desorbed, pretreatment with condensers is advantageous in reducing inlet concentrations.

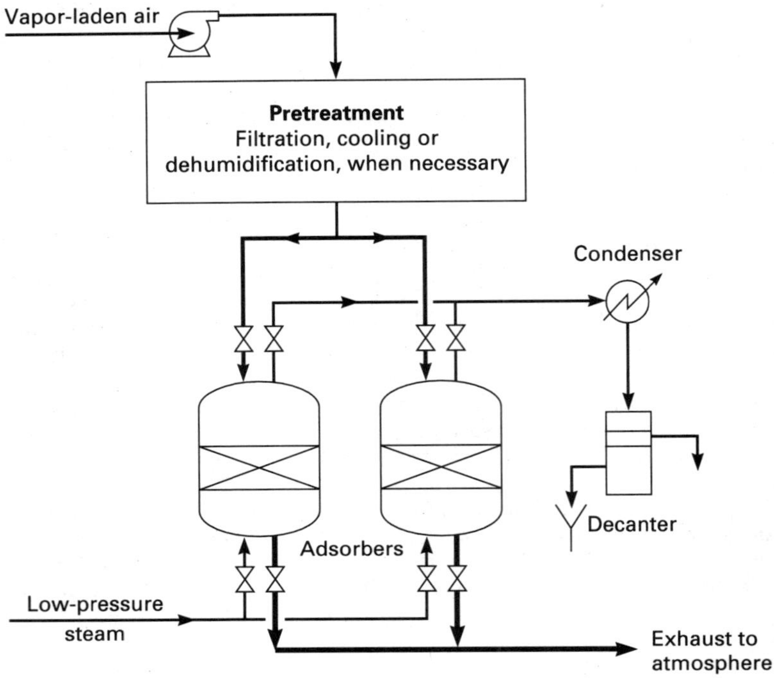

FIGURE 8.5.1 Process flow sheet for a conventional vapor-phase adsorption system.

Figure 8.5.2 traces the inlet and outlet concentrations of a fixed-bed adsorber through one complete cycle of adsorption and regeneration. The following objectives are necessary to obtain high removal efficiency:

1. Low baseline effluent concentrations. These are usually less than 10 ppm, as shown in zone 1 on Fig. 8.5.2.
2. Containment of breakthrough emissions (zone 2). This is done by avoiding premature breakthrough, or by sending material emitted during breakthrough to another vessel hooked up in series.
3. Efficient recovery of the desorbed organics (zone 3). This is usually achieved by condensing the vapors and recycling the noncondensables from the condenser back to the inlet of the on-line adsorber.
4. Containment of organics exhausted during the cooling and drying cycle (zone 4). This cycle prepares the carbon bed for renewed service.

In order to achieve the optimum interaction between adsorption and regeneration, it is important to understand the factors that affect the performance of activated carbon during these two operations.

Adsorption Performance

As yet there is no method that satisfactorily and accurately predicts the performance of adsorption systems from theory. This chapter will touch only briefly on the theoretical

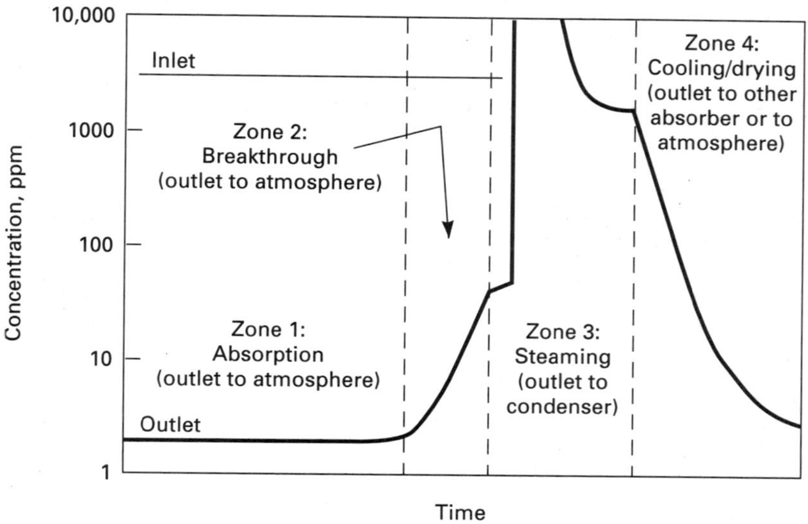

FIGURE 8.5.2 How the concentration in the outlet changes during the adsorption cycle.

aspects. Readers who want to consider these aspects further should consult other references (EPA, 1973; Kovach, 1978; and Smisek, 1970).

The adsorption of vapors onto activated carbon is controlled by the properties of both the carbon and the adsorbate, and by the conditions under which they are contacted. The phenomenon is generally believed to result from the diffusion of vapor molecules into the surface of the carbon. These molecules are retained at the surface in the liquid state because of intermolecular or van der Waals forces.

As the temperature falls, or as the partial pressure of the vapor above the carbon rises, the average time that a molecule resides on the surface increases. So does the fraction of the available surface covered by the adsorbate. In this early stage, the adsorption can be described by the following equation, developed by Freundlich (Kovach, 1978):

$$V = kP^{1/n} \tag{8.5.1}$$

where k and n are Freundlich constants. A log-log plot of V, the liquid volume adsorbed, as a function of P, the partial pressure of the vapor, yields a straight line.

However, the surface of the carbon is not uniform and consists of sites whose activities vary. More active sites will become occupied first, and as the activities of the remaining available sites decrease, the adsorption energy will change. This is why experimentally determined isotherms will deviate from Eq. (8.5.1).

The Freundlich equation also applies to chemisorption, where the adsorbed material undergoes a chemical reaction on the surface. This phenomenon is not common in air pollution control applications, except when an impregnated carbon is used.

Equation (8.5.1) rarely applies to systems in which the carbon is regenerated in place. Economics dictate that most regeneration processes do not drive off enough adsorbate to reach the region where the equation is valid.

The most accepted theory describing adsorption in the smallest pores is Polanyi's potential theory of adsorption (Smisek, 1970). This theory states that for a fixed amount of a material adsorbed, the free energy of adsorption ΔF is a constant and is a function of the relative vapor pressure P/P_o:

$$\Delta F = -RT \ln \frac{P}{P_o} \tag{8.5.2}$$

where T = absolute temperature
P = partial pressure of vapor
P_o = vapor pressure
R = ideal gas constant

This relationship makes it possible to calculate from one measured isotherm the relative vapor pressures of a material in equilibrium with carbon at a variety of temperatures. These data are usually presented as a series of isotherms (Fig. 8.5.3). Isotherms can also be predicted by comparisons with homologous materials, as well as by modification of Eq. (8.5.2).

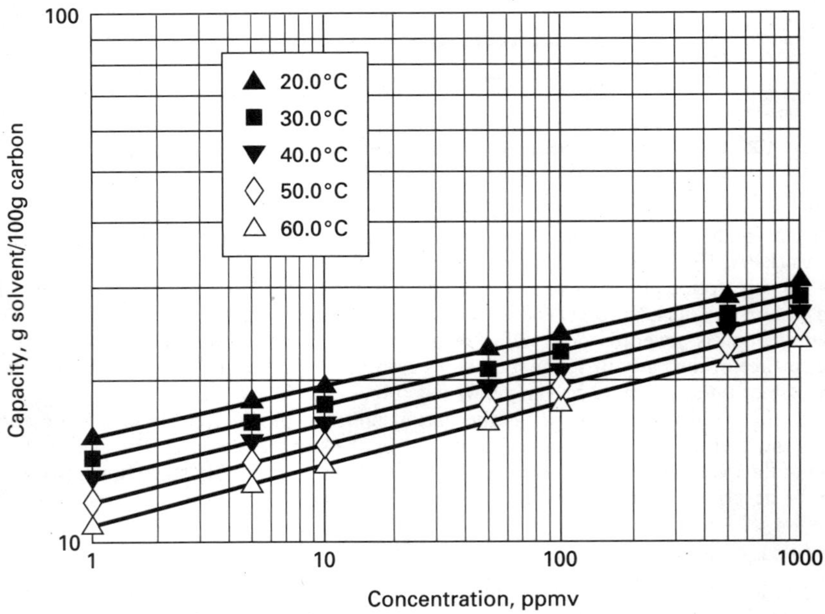

FIGURE 8.5.3 Isotherms make it possible to estimate equilibria at various temperatures. Vapor phase isotherm results for toluene at 1 atm. (*Calgon Carbon Corporation.*)

Choosing an Adsorbent

The physical structure of activated carbon is not known in detail, but it is probably composed of randomly distributed carbon pores, between which lies a complex network of irregular interconnected passages. These pores range in diameter from a few micrometers down to a few angstroms (Å) and provide a surface area of around 1000 m^2/g of carbon. The volume of pores at each diameter is an important variable that affects carbon performance (Fig. 8.5.4).

Since adsorption takes place at the carbon-gas interface, the surface area of the carbon is probably the most important factor to consider. Generally, the greater the surface

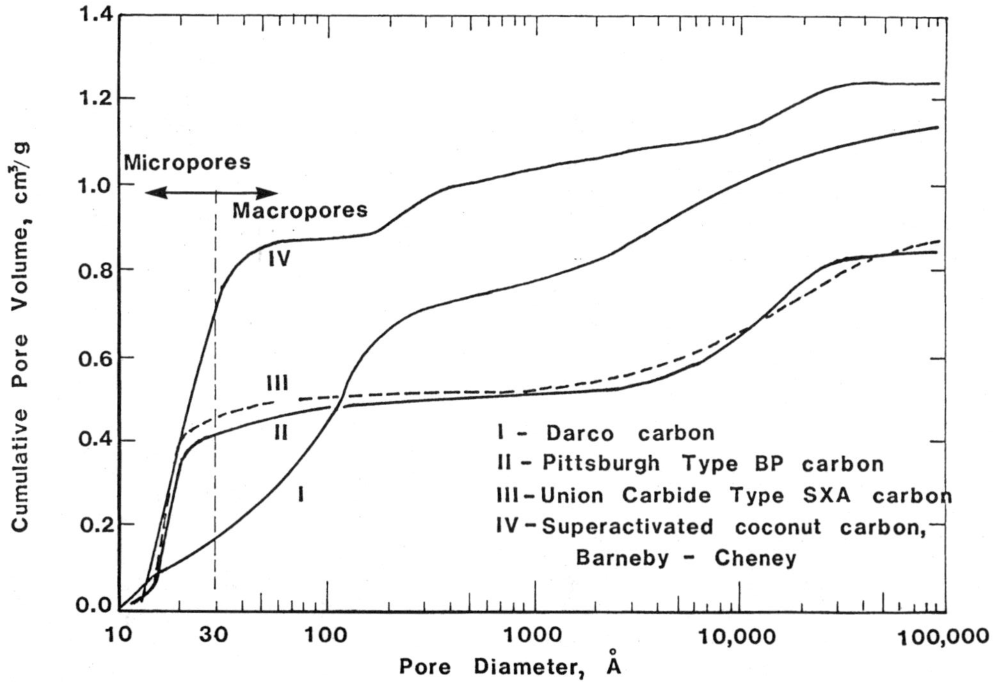

FIGURE 8.5.4 The distribution of pore sizes strongly influences carbon performance. [*From Parmele et al. (1979).*]

area, the greater the adsorption capacity for all compounds. However, the surface area has to be available in the proper range of pore sizes. If too much of the surface area is available in pores smaller than 5 Å, many solvent molecules will be unable to penetrate the pores. That area of the carbon will be essentially unavailable for adsorption.

For most pollution control applications, the surface area of pores whose diameters range between 5 and 50 Å yields good efficiency, because the relative pressure P/P_0 of the vapor is usually too low for the larger pores to be filled. At high relative pressures, however, the total pore volume becomes important because the macropores also become active.

The adsorption of benzene, e.g., has been shown to be affected by pore size distribution (Kovach, 1978). At high benzene concentrations, carbons in which large pores predominate have higher capacities than those in which medium or small pores predominate. But at low concentrations, the large-pore carbon has a lower capacity.

Although the larger pores are active only at high partial pressures, they serve as passageways, leading the vapor to the smaller pores in the interior of the carbon. This phenomenon is most important during regeneration, when the molecular flux is highest.

In general, the nonpolar surface of activated carbon makes it more efficient for nonpolar organics than for polar ones. Silica gel, alumina, and molecular sieves, however, have polar surfaces and adsorb polar materials more efficiently. The polar adsorbents are rarely used for air pollution control because of their unfavorable competition for water.

Recently, it has become possible to alter the type of oxide formed on the carbon surface, in order to make it somewhat more efficient in capturing polar materials at low loadings. Some carbons, especially those of vegetable origin, are more efficient for polar materials because their surface is more hydrophilic.

Adsorber Operation

Figure 8.5.5 shows a profile of the adsorbate vapor concentration along the axis of a fixed bed. Note that a portion of the bed L_s is saturated and that there is a relatively short active transfer zone L_z where most of the adsorption is taking place; also note that a portion of the bed is unused.

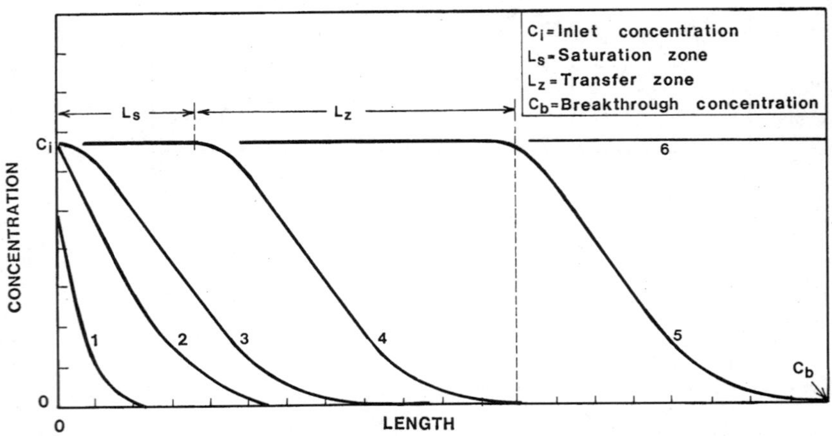

FIGURE 8.5.5 Concentration profile along the axis of a fixed-bed during the adsorption step.

The transfer zone moves through the bed until breakthrough occurs, at which time the air pollutant starts to exit the bed. A bed that is shorter than the transfer zone will have essentially no capacity. Similarly, the longer the bed, the smaller the fractional capacity lost.

The applicability of activated carbon for solvent recovery or air pollution control is determined to a large extent by the characteristics of the adsorbate—principally its volatility and polarity. From Eq. (8.5.2), the capacity of carbon is proportional to the natural log of P/P_0, the relative saturation or vapor pressure, and therefore falls off considerably at low solvent concentrations.

Since standard in-place regeneration always leaves some solvent on the carbon, even a freshly regenerated and cooled bed will have an equilibrium concentration of vapors that will not be adsorbed. If the inlet concentrations are low and vary widely, the bed will actually add organics to the air.

This difficulty has been confirmed by data taken from full-scale adsorbers used by a client to recover carbon tetrachloride from a product dryer exhaust (Fig. 8.5.6). The operation of this installation differs from that elsewhere, in that one bed, A, was in service during the first 30 min of the drier cycle and was regenerated after each batch; the second bed, B, was in service during the last 100 min or so of the cycle, but was regenerated daily or after three drier cycles.

Therefore, the inlet concentration in bed A is always high, as is shown by the data for three successive adsorption cycles (Fig. 8.5.6*a*). The inlet concentration at bed B varies more widely (Fig. 8.5.6*b*), because of the decreasing amount of solvent left in the batch. In fact, the inlet concentration at bed B actually drops below the outlet concentration at the end of the second batch.

These data show that greater than 99 percent removal is achieved by bed A. In bed B, because the outlet concentration is more a function of operating conditions in the carbon, the removal efficiency of the bed is directly related to the inlet concentrations.

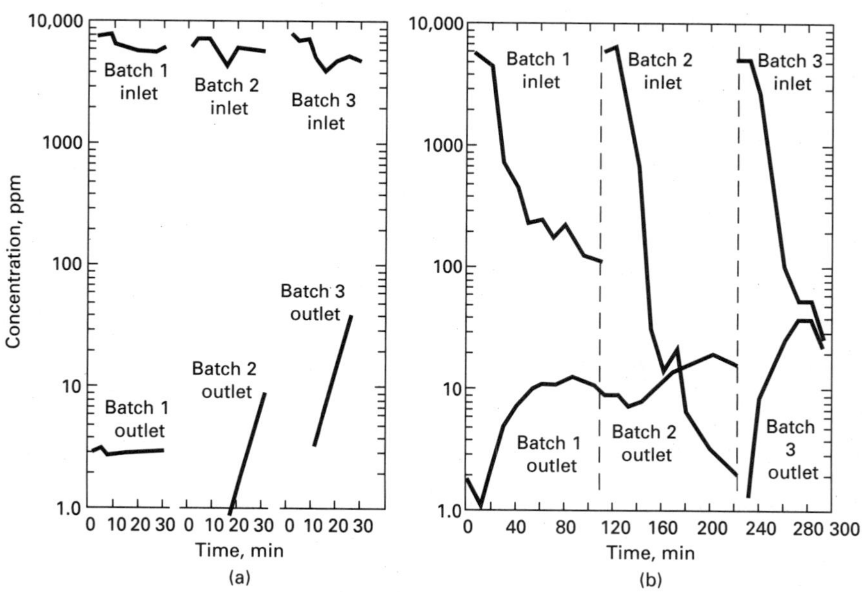

FIGURE 8.5.6 Performance of system adsorbing CCl_4 shows how heel on carbon can add VOC to the outlet. (*a*) Bed A performance, regeneration each batch; (*b*) bed B performance, regeneration every third batch.

If flammable vapors are present, the upper concentration is frequently set so that the vapor content does not exceed 25 percent of the lower explosion limit. Concentration may also be limited because of a reduction in capacity or because of safety problems posed by high bed temperatures produced by the heat of adsorption. The maximum practical concentration is usually about 10,000 ppm, and higher concentrations are frequently reduced by condensation or dilution before the adsorption step.

Heat dissipation from the interior of the bed through the walls of the container is inhibited because granular carbon is a good thermal insulator. The heat of adsorption (which is approximately the same as the heat of condensation) can usually be kept from building up by allowing the temperature of the airstream to go up. The extent of the temperature rise can be estimated by assuming adiabatic thermal equilibrium.

Frequently, the carbon is left partly wet with condensate after the regeneration step. Subsequent evaporation of the water dissipates some of the heat of adsorption and limits the rise in bed temperature. This tactic is helpful when the inlet concentration is high. However, it does impose a greater resistance to the collection of the organic and may cause problems when a very high capture efficiency (of, say, 99 percent) is required.

The influence of adsorbate volatility on performance is indicated by the term ln (P/P_0) in Eq. (8.5.2). In general, at the normal commercial operating temperature of 20 to 55°C, carbon has insufficient capacity for compounds that boil below 0°C to make adsorption of these materials practical.

Carbon's capacity to adsorb chemicals is frequently reduced when competing water vapor is present at high relative-saturation pressures. Figure 8.5.7 shows that whereas

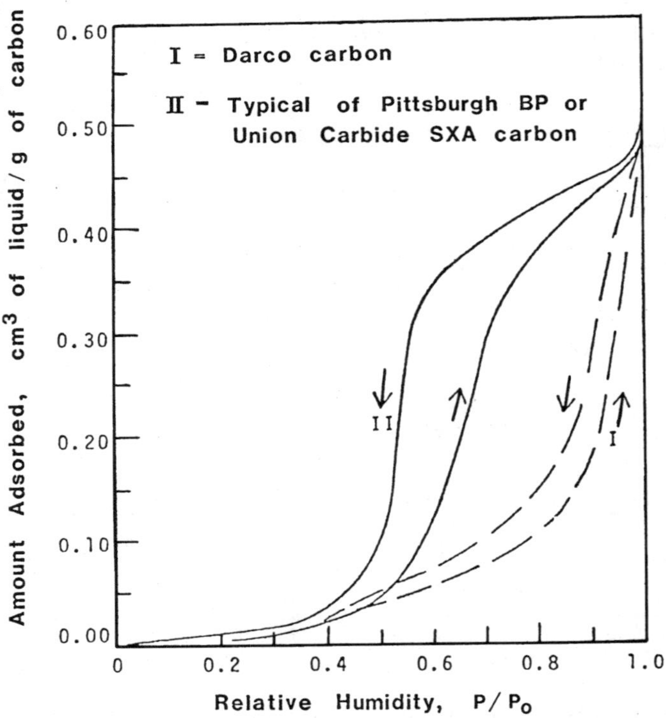

FIGURE 8.5.7 Water competes with other organics and shows a hysteresis between adsorption and desorption. [*From Parmele et al. (1979).*]

water is only weakly adsorbed by carbon, it can effectively compete with marginally adsorbable organics at higher relative humidities. (Rectilinear axes are used to show water loading at adsorbate concentrations and loadings near zero.)

Figure 8.5.7 also illustrates that the carbon loading of water can be a function of the type of carbon used, and that water exhibits a hysteresis of desorption. Thus, if a carbon bed is exposed to a high water concentration, satisfactory performance cannot be achieved until the bed is dried to the point where its exhaust has a relative humidity lower than 40 to 50 percent.

This has become an increasingly important circumstance in three types of applications of vapor-phase adsorption with high relative humidity in the 1980s and 1990s:

1. Increasing the control efficiencies of adsorption systems used to control the emissions of volatile chemicals
2. Controlling the emissions from groundwater treatment and soil remediation systems (such as air strippers and SVE systems)
3. Using respirators to remove organics from air breathed by workers at remediation sites

Several articles published in the 1980s (Werner, 1985; Underhill, 1987; Adams et al., 1988; and Noll et al., 1988) presented laboratory measurements for one adsorption cycle for several organics that showed the effect of water on the adsorption of organics. The data showed that water reduced adsorption capacities at low concentrations of organics, especially when the relative humidities were above 50 percent. Note that this is important for remediation applications where the exhausts from SVE or air-stripping systems have high relative humidities and very low concentrations of organics.

These articles also showed that it is difficult to precisely predict the impact of water on the adsorption of organics without making measurements in the expected service because the impact is the result of a complex interaction of several parameters. Werner (1985) found that these parameters included

1. Adsorbate's concentration
2. Type of adsorbate
3. Type of adsorbent
4. Preconditioning of the adsorbent

Most of the measurements in the references above were made in laboratory settings that simulated conditions in the field. Some measurements were made on small columns 2.54 cm in diameter by 5 to 10 cm deep in which the removal of the target organic was monitored on-line by gas chromatography at the outlet of the carbon column. One article (Noll et al., 1988) was based on gravimetric measurements of the uptake of two organics (toluene and trichloroethylene) and water on a single layer of carbon particles suspended in a Teflon screen in equilibrium with the test conditions.

Foster also evaluated the impact of relative humidity on loading capacity in a parametric study during the operation of a larger pilot plant system designed, constructed, and operated under contract to the Environmental Protection Agency (EPA). This study was more representative of full-scale adsorbers because the carbon beds were 2 ft in diameter and 3 ft high and operated automatically 24 h/day by a computer that programmed the adsorbers through adsorption, regeneration, and cooling/drying, when used. Regeneration with steam was initiated upon detection of breakthrough of the VOC in the exhaust of the on-line adsorber by an on-line analyzer. The results showed that at constant conditions of 90°F, a steam regeneration ratio of 0.5 lb steam/lb carbon, and cooling/drying cycle time of 60 min, increasing the relative humidity from 30 to 90 percent decreased the loading of toluene from 0.08 to 0.04 lb/lb carbon (Foster, 1985).

Less volatile materials are more strongly adsorbed on carbon. They can usually be removed from air to very low levels. However, the affinity of the carbon for the organic may make these materials difficult to remove during regeneration. Again, experience has shown that it is impractical to remove materials whose boiling points are above 200°C by in-place regeneration using hot gas or steam.

Regeneration

Conventional Regeneration. The key to the effectiveness and utility of activated carbon is its regenerability. Short adsorption cycles (often less than 12 h) dictate that in-place nondestructive regeneration be used in order to avoid the carbon losses that are characteristic of thermal techniques. Therefore, systems using throwaway carbon or thermal regeneration are restricted to low inlet concentrations or to odor control applications that have extended adsorption cycles.

Regeneration, or desorption, is usually accomplished by changing the conditions in the adsorber to bring about a lower equilibrium loading capacity. Increasing the temperature or decreasing the partial pressures enhances desorption.

Low equilibrium loadings can also be achieved by introducing a more strongly adsorbed material to displace the previously adsorbed materials, or by eluting the carbon with a suitable solvent.

Vapor-phase adsorption systems are usually regenerated using steam, hot gas, or a combination of vacuum and hot gas. Solvent regeneration is usually limited to the desorption of nonvolatile organics collected from aqueous phases.

The choice of regeneration method depends on the downstream use for the desorbed organics. The adsorption step itself serves only to concentrate the organics for subsequent collection or destruction by other techniques. Although condensing them from the regenerating fluid is the customary approach, incinerating them in the regenerant can be a cost-effective alternative to direct incineration of dilute fumes—if the savings in capital and energy that result from using a smaller fume incinerator offset the expense of the adsorption system.

Steam Regeneration. Direct steaming of the carbon bed is the most widely used regeneration technique, because it is cheap and simple. Steam is very effective in raising the bed temperature quickly and is a more concentrated source of heat than hot gas. It is a readily available, condensable fluid from which many organics can be easily separated and recovered. And, as previously mentioned, the high concentration of water promotes desorption of the organics.

Figure 8.5.8 illustrates how desorption profiles change as a function of the amount of steam or other regenerant used. Note that these profiles are not simply a reversal of the progression shown in Fig. 8.5.5.

To begin with, the individual curves do not represent equal time intervals or equal amounts of regenerating agent. The desorption requirements to achieve profiles 1 and 2 are relatively small. Increasingly larger amounts of regenerating agent are required to produce profiles 3, 4, and 5. Regeneration of a bed is usually halted once profile 4 or 5 is achieved. A "heel" is always left on the carbon bed because complete desorption is technically difficult to achieve and economically impractical.

The amount of steam required for regeneration depends on how much material is to be desorbed from the bed. A fixed amount is required to raise the bed to its regeneration temperature and provide the heat of desorption. A certain flow is also required to reduce

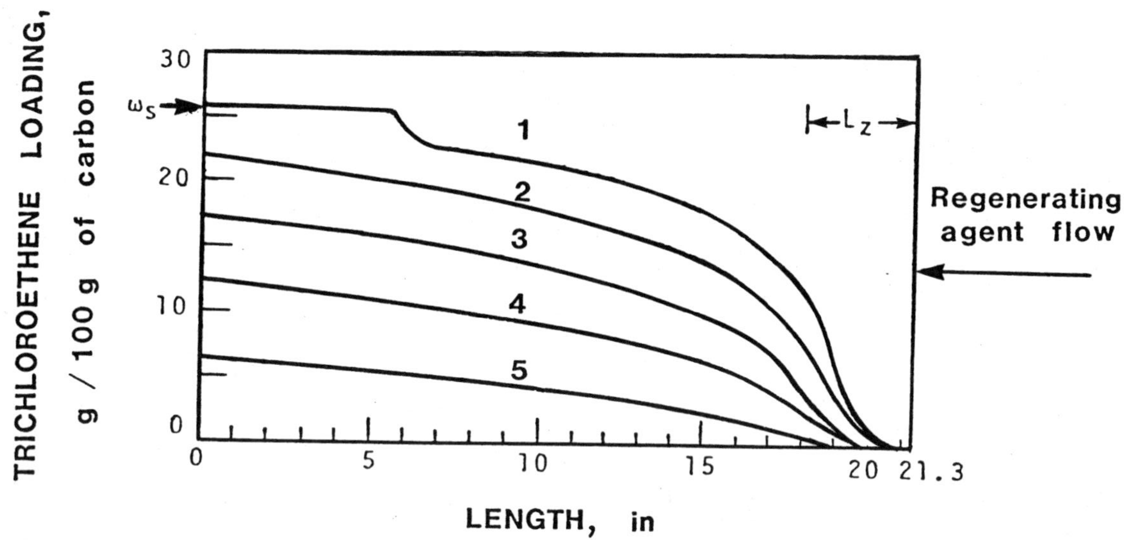

FIGURE 8.5.8 Complete desorption is difficult to achieve and economically impractical.

the partial pressure of the adsorbate and to carry the organics out of the bed. Passing steam through the bed for a long time produces an equilibrium between the steam and the heel.

Optimizing Steam Consumption. Figure 8.5.9 shows a typical relationship between steam consumption and desorption efficiency (relative to what would be desorbed with

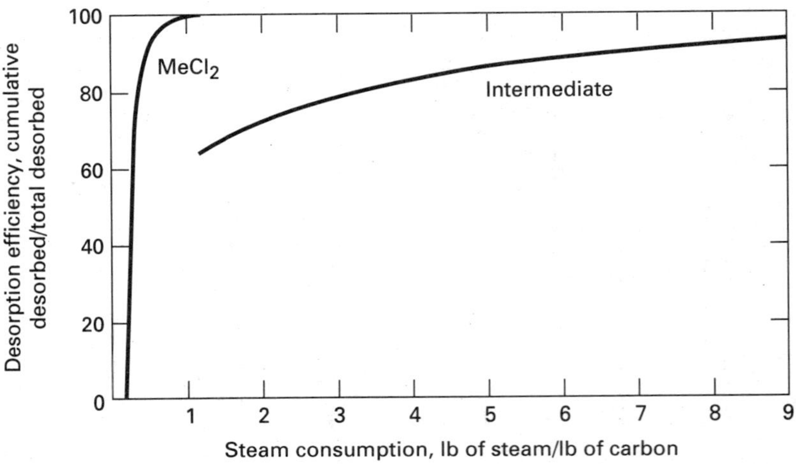

FIGURE 8.5.9 Desorption efficiency and steam consumption must be carefully optimized.

exhaustive steaming). The curves were developed for methylene chloride and an intermediate (boiling point, 110°C) produced by a Hoffmann-La Roche, Inc. pharmaceutical process.

The data show that during the initial heat-up period, nothing comes out of the adsorber. After this period, substantial amounts of the adsorbed material are released, until a plateau is attained. This occurs when an equilibrium is reached between the remaining adsorbed organic and the prevailing operating conditions.

It is not cost-effective to try for 100 percent desorption. Acceptable working capacities can be achieved without consuming so much steam. For solvent recovery systems, a requirement of 0.25 to 0.35 lb steam/lb carbon usually has been specified. This requirement represents an optimization between the cost of steam and the costs of bigger beds to compensate for reduced working capacity.

The shape of the desorption efficiency curve depends upon what is adsorbed, the loading capacity, and the regeneration conditions. Saturated steam at pressures of up to 15 psig is usually used for regeneration, although pressures of up to 50 psig are feasible. Superheated steam is not usually advantageous, because the adsorbed organics may react at elevated temperatures. Often, these reactions decrease the life of the carbon bed. Also, the extra cost of superheated steam is not usually justified by the marginal improvement in desorption.

The rate of steaming should be limited to 1 to 4 lb/(min • ft^2) to prevent the carbon from being shifted in the bed. Such shifts can cause flow distribution problems.

The heel that is left on the carbon influences the equilibrium and dynamics of the next adsorption cycle. Any optimization of steam requirements will have to take into account the impact of effluent concentration during the next adsorption cycle. Data are generally unavailable, but the concentrations will probably take the qualitative form shown in Fig. 8.5.10, in which no values for steam usage have been assigned to the abscissa.

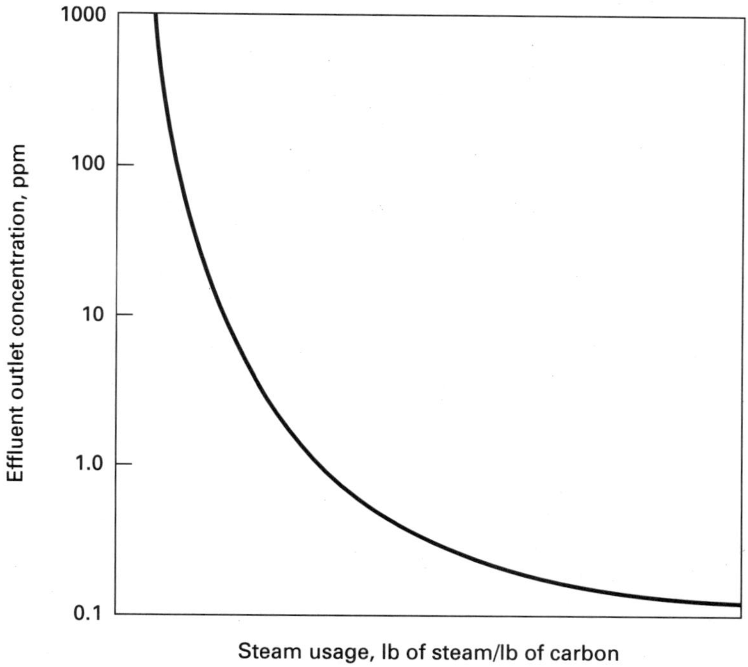

FIGURE 8.5.10 Qualitative impact of steam use on effluent concentration.

Importance of Cooling and Drying. After regeneration, a hot, wet carbon bed will not remove organics from air effectively because, as previously discussed, high temperature and humidity do not favor complete adsorption. If a separate cooling/drying cycle is not included, the concentration at the outlet will at first nearly equal that at the inlet, then fall to the baseline concentration as the bed is cooled and dried by the incoming gas.

The laboratory data in Fig. 8.5.11 show that, when air with a velocity of 40 ft/min is used for cooling, a well-insulated, 18-in-deep bed will cool in approximately 10 min.

The amount of organics emerging from the outlet during cooling depends on the amount of heel, which in turn depends upon how much steam was used. To reach high removal efficiencies required in the 1990s, it is necessary to control these organics, usually by recycling the exhaust to that bed currently in the adsorption stage (Paresse, 1997).

To do this, the beds and blower must be sized to provide for the additional time and flow capacity required. To route the cooling and drying air to the other bed, additional valves and piping should be installed so that the beds can be operated in series. The piping necessary for recycling of the cooling/drying exhaust was not provided in many older existing systems. The added capital costs usually can be justified only when high capture efficiencies are required, which is often the case in the 1990s.

A properly designed and operated adsorption system should be able to handle this recycling. Carbon requirements will not be greatly affected, since the amount of organ-

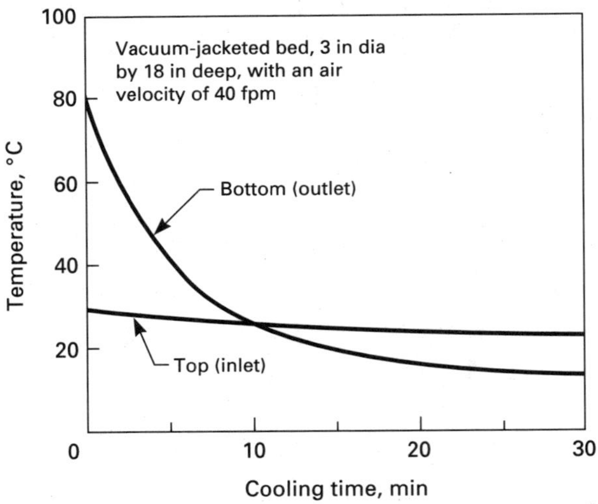

FIGURE 8.5.11 Regenerated bed must be cooled and dried to achieve efficient adsorption.

ics recycled from one adsorber to another is usually no more than 3 to 5 percent of the total adsorber duty. The additional load of water and organics will generally not exceed normal process variations. However, the additional water may pose a problem if the relative humidity of the vapor-laden airflow is already too high.

ALTERNATIVE ADSORBER DESIGNS

Conventional steam-regenerated fixed-bed systems have encountered problems in some applications, requiring development of alternative systems.

One problem has been the failure of the adsorber vessel due to corrosion. Wet carbon in contact with a metal fosters formation of galvanic cells, which accelerate the corrosion of some metals, especially carbon steel. Although stainless steel is an acceptable alternative for many applications, it may sustain crevice corrosion when exposed to chlorinated solvents because many chlorinated solvents hydrolyze slightly to generate HCl.

Steel adsorbers with inexpensive linings have often not performed acceptably in this service because of pinhole leaks in the lining. Another difficulty with liners is that their coefficients of thermal expansion differ from that of the substrate, making for poor bonding in services that have frequent thermal cycles.

Another potential area for improvement is the amount of energy required for regeneration. Not all the energy supplied is used to supply the heat of desorption. The bulk of the steam is used either to heat up the vessel and the bed, or to lower the partial pressure of the adsorbate by sweeping it out to the condenser.

To circumvent these problems, other approaches are being developed. For example, one firm, Met-Pro Systems Div. (Harleysville, Pennsylvania), has developed a vacuum-regenerated system that changes the regeneration conditions to reduce corrosion. Direct injection of steam into a bed has been eliminated, and the regeneration temperature has been reduced.

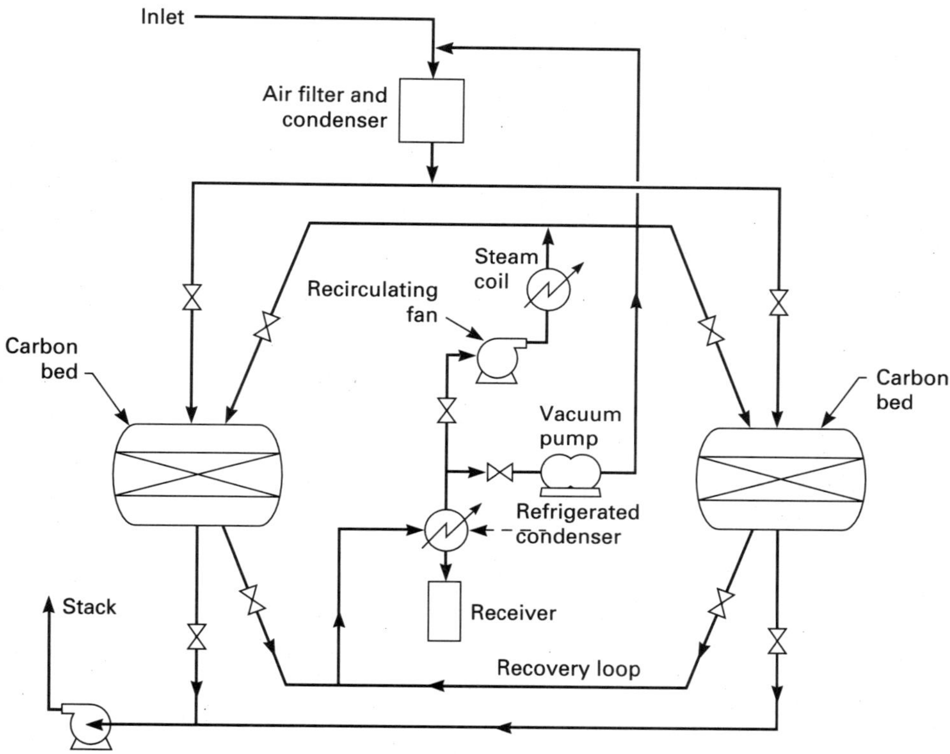

FIGURE 8.5.12 Vacuum regeneration can avoid the corrosion problems associated with direct steam injection. [*From Parmele et al. (1979).*]

The system shown in Fig. 8.5.12 controls emissions from a pill-coating operation. Regeneration is accomplished by a combination of convective heating to approximately 65°C (using a preheated purge stream) and conductive heating from a heat exchanger or electric element embedded in the carbon. The vessel is then evacuated to 1 mmHg absolute to promote desorption. The desorbed organics are removed from the adsorber by a closed-loop purge stream, and they are then separated by a refrigerated condenser. Noncondensables are recirculated back to the adsorber so that the organics will not be emitted.

Vacuum regeneration can offer a cost-effective alternative when direct injection of steam demands the use of more expensive materials of construction. It may also be called for when steam is not effective because the boiling point of the adsorbate is too high or because polymerization of the adsorbate is promoted by temperature and/or water (as with vinyl chloride monomer emissions from polyvinyl chloride plants).

The decision to use vacuum regeneration is determined by the tradeoff between costs for materials of construction and related steam system requirements versus the added equipment costs for the vacuum pump, refrigeration unit, and the slightly stronger adsorber vessel design.

The CADRE VOC abatement process is an alternate design that addresses the disposal of mixed VOCs desorbed from the carbon (SenGupta, 1997). The process, shown in Fig. 8.5.13, was developed by VARA International, a division of Calgon Carbon

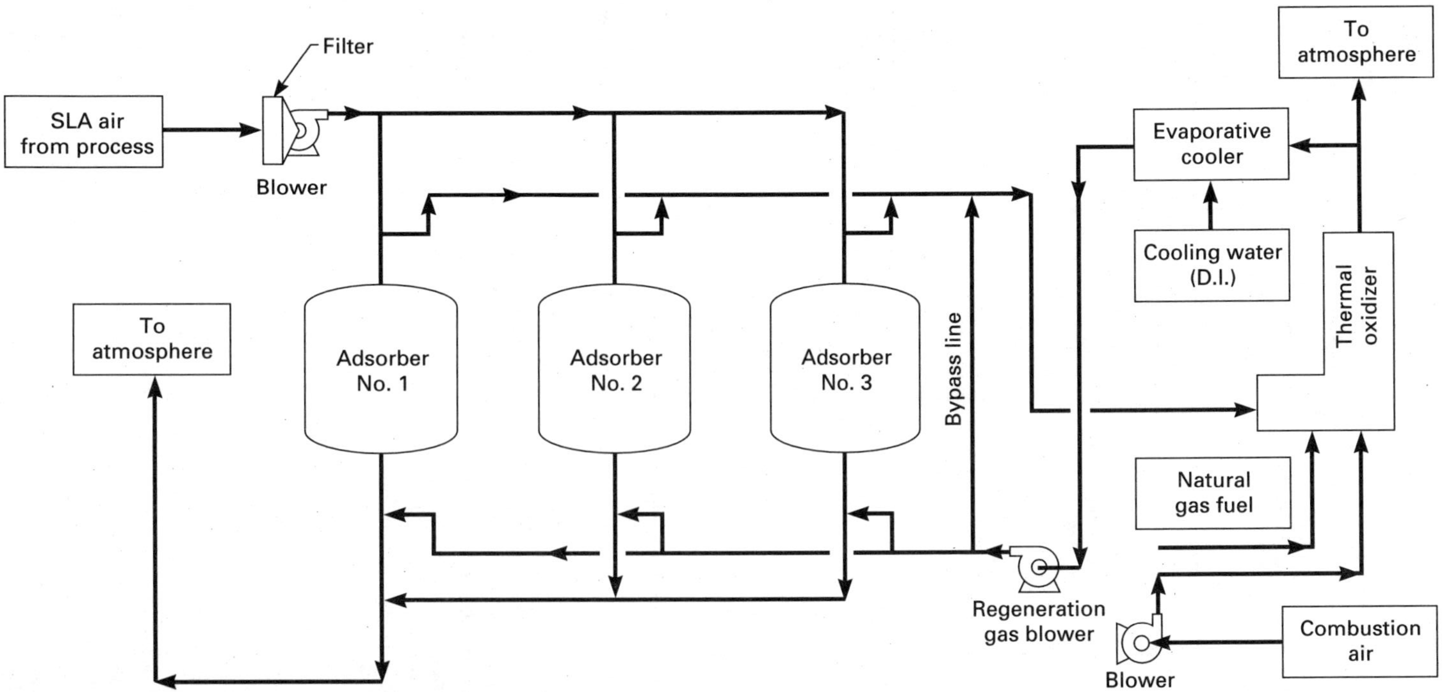

FIGURE 8.5.13 The CADRE process regenerates the carbon with a recirculated hot gas stream from a thermal oxidizer that destroys the VOC. (*VARA Internationsl, Calgon Carbon Corporation.*)

Corporation. VOCs are removed from solvent-ladened air in one or more adsorbers, as in other systems. The unique feature is the regeneration of a spent adsorber by recirculating a hot gas stream between the adsorber and a thermal oxidizer where the desorbed VOCs are combusted.

FULL-SCALE DESIGN

A crucial aspect of designing a full-scale system is an accurate definition of the adsorber duty. The total flow rate and the amount of emitted organic must be predicted as accurately as possible, because the size of the adsorbers will be determined from these factors.

Overall material balances based on monthly inventories will often yield only an order-of-magnitude estimate. Measuring the weight lost when a batch of material is dried is another way of determining emissions, but the assumption that all the weight lost represents volatile organics should be verified. Water, entrained or vaporized product, and lint are only three examples of other sources of misleading weight loss. Such losses would give high estimates of the amount of material to be adsorbed.

There is usually some flexibility in sizing an adsorber, since two factors will be balanced:

1. There must be enough carbon for an adequate and effective adsorption cycle.
2. Cross-sectional area must be sufficient to prevent excessive pressure drop across the bed.

Thus, adsorber sizes are limited either by the mass flow rate of the organics and the working capacity, or by the volumetric flow rate and the depth of the bed.

Applications limited by adsorption capacity are usually characterized by high inlet concentrations. Capacity must be large enough to allow adequate time for the spent adsorber to be desorbed, cooled, and dried. Although systems with adsorption cycles of 1 h have been installed, they provide little flexibility for operational changes that improve performance or increase throughput. If possible, new systems should be sized to contain enough carbon for at least 2 h of adsorption, normally an adequate time for the carbon to be regenerated.

Pressure drop considerations usually govern the selection of bed height and cross-sectional area. Figure 8.5.14 shows how pressure drop is influenced by linear velocity and bed depth. If a centrifugal fan is used to move the air, the allowable pressure drop is limited to 10 to 20 in of water, and cross-sectional areas are designed for linear velocities of 50 to 100 ft/min. Bed depths then usually range from 12 to 36 in. The appropriate vessel size can be determined directly from the volume of carbon, linear velocity, bed depth, and pressure drop.

If inlet concentrations are high, the cross-sectional area dictated by the volume of carbon needed usually results in a linear velocity at the low end of the range of 50 to 100 ft/min. Vertical cylindrical vessels are favored for small flows. For large flows, horizontal vessels offer a preferred alternative to fabricating short, large-diameter vertical vessels.

Some additional design flexibility is available if the process stream to be treated is under pressure before discharge to an adsorber. This is the case in many oxidation reactions that use air as the source of oxygen. Higher linear velocities and/or deeper beds are possible when a higher pressure drop can be taken across the adsorber. The advantage is a smaller adsorption system, since more of the volume inside the vessel is filled with carbon.

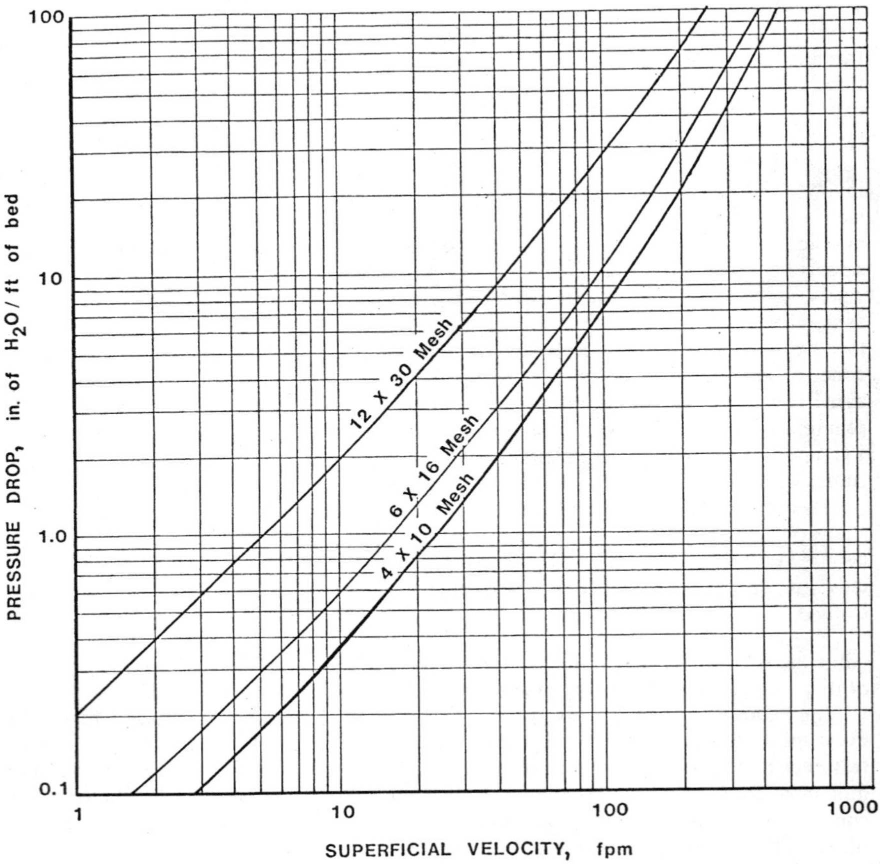

FIGURE 8.5.14 Influence of bed depth and linear velocity on pressure drop for a particular carbon. [*From Parmele et al. (1979)*.]

Flow rate poses a restriction when it is high and the inlet concentrations are low (usually below 1000 ppm). This situation often occurs when ventilation is needed inside the equipment to keep the organic concentrations below the lower explosion limits mandated by fire and/or insurance codes. Dilute emissions also occur in odor control, and in installations specified to meet workplace concentrations set by OSHA regulations.

In these cases, adsorbers are sized primarily for handling a large flow at a low pressure drop. If the linear velocities must be limited to 100 ft/min, the vessel is usually much larger than that indicated by the adsorption duty. Even though bed depths of 6 in or less will yield an adsorption cycle of 2 h or more, deeper beds are often used if the pressure drop will allow it. The incremental cost of adding more carbon to a large vessel is more than offset by the added flexibility that a longer adsorption cycle gives.

Design Data and Scale-up

One of the most important design parameters is the working capacity of the carbon for a particular service.

Isotherms for the more common volatile organics (such as Fig. 8.5.3) are available from the literature, vendors, or carbon manufacturers. It should be remembered that these curves describe the equilibrium capacity that was measured under specific conditions of temperature, partial pressure, carbon type, and relative humidity. Working capacities are always less than isotherm capacities, because the isotherm conditions are rarely duplicated in the real world.

Fortunately, significant scale-up experience exists for straightforward applications involving the more common organics. A general rule of thumb is to use one-half the isotherm loading at the design temperature and partial pressure. If the isotherm data are not available in the anticipated concentration ranges, laboratory evaluations are necessary.

Isotherms can be measured for single-component systems, but are not as easy to do as for liquid-phase applications. An isotherm is the measurement of the equilibrium loading (in milligrams per gram of carbon) as a function of the equilibrium, vapor-phase concentration. Isotherms are measured by placing a known volume of gas (usually air) along with known quantities of carbon and the VOC in a vapor-tight, sealed, gas-sampling bag. It is important to add just enough carbon to remove some, but not all, of the VOC because a value of zero cannot be plotted on a logarithmic scale for the equilibrium concentration.

The bag and its contents are shaken for enough time to reach equilibrium at a constant temperature—often overnight. The gas is analyzed to determine the equilibrium concentration of the VOC in the vapor. The equilibrium loading is determined by calculating the mass of VOC that is not in the vapor by difference and assuming that it has been adsorbed by the carbon.

The accuracy of the isotherm measurement will be determined by two factors: (1) the validity of the assumption that the missing VOC was adsorbed and (2) the thoroughness of the contact between the carbon particles and the VOC molecules.

The loading capacity will be overstated if the adsorption is not the only mechanism by which the VOC concentration decreased (e.g., some of the VOC decomposed or leaked out of the bag). The loading capacity will be understated if the molecules do not thoroughly contact the carbon particles. It is especially important to be sure of thorough contact when measuring isotherms at low concentrations because of the reduced quantities of VOC and carbon in the bag.

If the adsorption performance of a mixture is in question, one should make a series of column runs to measure the working capacity of the carbon under the anticipated scale-up conditions. Laboratory runs should be made in a carbon bed that is 18 to 24 in deep—the normal depth of full-scale units. The diameter should be large enough in relation to the particle size to eliminate wall effects. The system should be designed to minimize losses of volatile materials. It must be large enough that the loss of several grams of material would not greatly affect material balance calculations.

As in any miniplant or pilot plant evaluation, conditions in the laboratory must accurately reflect the anticipated full-scale conditions. Accurate characterization of the emission that will be controlled is essential. Obviously, the flow and concentration of primary material to be adsorbed should be known. But other materials that might be present should also be identified so that a full understanding of the service can be reached.

The importance of this understanding was demonstrated during the abovementioned Hoffmann-La Roche project, whose purpose was to control emissions bearing methylene chloride. Analysis of the emissions revealed the presence of a volatile intermediate. As indicated by data in Fig. 8.5.9, an uneconomical amount of steam would be needed to effectively desorb it. Consequently, process changes were made to minimize the amount of intermediate emitted. This tactic forestalled adverse effects on the adsorber's performance in removing methylene chloride.

The influence of water vapor on adsorption has already been noted. When design data are generated in the laboratory, it is important to measure the relative humidity so that differences between the laboratory conditions and plant conditions are fully defined. One should recognize that plant emissions rarely have the low relative humidity (~20 percent) that is characteristic of the laboratory compressed air used to approximate an emission. If high-humidity conditions are anticipated, the emission synthesized in the laboratory should be humidified.

Safety Considerations

Activated carbon or its impurities can catalyze the decomposition of some organics, such as ketones, aldehydes, and possibly organic acids and esters. These exothermic decomposition reactions can generate localized hot spots and/or bed fires within an adsorber if the heat is allowed to build up. These hazards will crop up if the flow is low and/or the inlet concentrations are high, or if an adsorber is left dormant without being completely regenerated.

Bed fires have occurred especially in the application of vapor-phase adsorption to recover solvents used in coating processes. Engineering reviews by the technical staffs and consultants have found that the bed fires result from the complex chemical and physical processes during the adsorption, desorption, and cooling/drying cycles. These reviews (Lang, 1979; Miller et al., 1987; and Naujokas, 1985) reveal that the significant factors which can lead to bed fires include

1. An activated carbon with a high activity or ash content
2. The type of solvents being adsorbed on the carbon
3. The moisture content of the carbon throughout the adsorption, desorption, and cooling/drying cycles
4. The airflow distribution throughout the carbon bed

A fire hazard can also exist during the virgin operating cycle, because there is little water present on the carbon to act as a heat sink and because more material is adsorbed during this cycle than during other cycles. One such situation is the use of vapor-phase carbon to control the emissions of VOCs during the initial operation of a soil-vapor extraction (SVE) system when the concentrations of VOCs are often very high. The adsorption of relatively large amounts of VOCs onto dry, virgin carbon has released enough heat to raise the temperature of the vapor going through the adsorber from 100°F or more (due to the energy of compression from the vacuum pump) to >250°F. This high temperature has melted plastic piping after the adsorber. This is a condition that should be avoided—especially if the vapors contain flammable or combustible materials.

These considerations do not imply that vapor-phase adsorption should not be selected for these potentially hazardous applications, but rather that safety and performance require consideration for hazards during adsorber design and operation without incident. To reduce the hazard of bed fires, the following procedures (Miller et al., 1987; Naujokas, 1985; Zanitsch, 1979) are usually recommended:

1. The bed should never be left dormant unless it has been thoroughly regenerated.
2. A deluge system should be provided above the bed to water-cool the bed in the event of an excursion.
3. Strong consideration should be given to rendering an idle bed inert with nitrogen or deluging it with water during shutdown.

4. Design parameters should be set so as to avoid high inlet concentrations and low flow.
5. Adsorbers should be designed to maintain good flow distribution and avoid areas where the air velocity is less than 40 ft/min.
6. Instrumentation, including alarms, should be installed to monitor the temperature change across the adsorber bed and the outlet CO/CO_2 concentrations. The instrumentation should signal the first signs of decomposition, so that any acceleration leading to a bed fire can be forestalled.
7. Superheated steam should not be used for regeneration in systems recovering oxidizable organics because it can dry out the bed during successive adsorption-desorption cycles. Superheated steam can also elevate the temperatures in the bed.
8. A virgin bed should be steamed before the first adsorption cycle. Residual condensate will remove heat generated during the virgin cycle.

Cost Factors

The example of a fixed-bed system is used here to illustrate the associated costs. The system is designed to handle 20,000-cfm of air with a pollutant concentration of 1000 ppm (a 25:75 mixture of toluene and hexane). Major items include a blower, two or more carbon beds, a condenser, a decanter, provisions for cooling and drying a regenerated adsorber, interconnecting piping and valves, and automatic controls.

Instrumentation that automatically switches the beds between the adsorption and regeneration modes will either follow a timed cycle or detect organic breakthrough at the outlet. An adsorption cycle based on time is most effective when the emission profile is predictable. When the emission profile is variable, a breakthrough monitor becomes cost-effective because steam is not wasted in regenerating a partly loaded carbon bed.

The highest-cost items of an adsorption system are the adsorbers, blowers, and automatic valves, which are all dependent on the feed-gas flow rate.

The amount of carbon required for the beds will also have an effect on capital costs in terms of blower pressure drop, bed size, and condenser size. Vendors report that the purchase price of an adsorption system ranges from \$20 to \$25 per cfm (Paresse, 1997). Installation cost ranges from 50 to 100 percent of the purchase price, assuming that there is already some ductwork to the adsorber and that utility hookups are available at the battery limits. Skid-mounted units in the 5000-cfm range or below are available off the shelf, and can provide a saving over custom-built systems.

Materials of construction strongly influence the system cost. The purchase costs cited above do not apply to systems that require protection against the presence of HCl generated, e.g., by hydrolysis of a chlorinated solvent. Metals that corrode at a high overall rate or pit rapidly or sustain crevice corrosion are not acceptable in some of these applications. Materials should be selected with prudence, especially if failure of the emission control system would force a significant loss of production.

Valve selection is very important for applications requiring high capture efficiency, because the valves isolate the outlet from the inlet or regenerant streams. A small leak will result in unacceptable outlet concentrations even when adsorption is satisfactory. Systems in excess of 10,000 cfm need large valves to handle the flow, and frequently require more than two adsorbers. These two factors can make the cost of valves significant. However, one should avoid the temptation to reduce installation costs by using inexpensive valves, which will not be leaktight.

The annual costs for a system include the fixed costs arising from capital requirements and the operating costs for steam, electricity, cooling water, and carbon replacement. The annual capital charges are normally about 24 percent of the installed cost, which includes depreciation and interest at 15 percent; taxes, insurance, and administrative charges at 4 percent; and maintenance, labor, and material at 5 percent.

The installed cost of the 20,000-cfm system in this example will be estimated at \$800,000, based on a purchase price of \$20 per cfm and an installation factor of 2. The steam consumption for such a system is around 0.3 lb/lb carbon. It will require about 18,000 lb of carbon per bed, based on 3-ft-deep beds whose design superficial velocities are 100 ft/min. The carbon capacity is assumed to be 0.0625 lb organic/lb carbon, which means that the adsorption cycle will be 4.5 h.

The amount of electricity used is estimated from the horsepower requirements of the blower. Cooling water usage depends on the amount of steam used for regeneration; it was calculated to be 4.2 gal/lb steam.

Carbon replacement costs are based on a 4-year lifetime. Actual lifetimes have ranged from 1 year to 10 years, depending on how effectively materials other than the primary organic were excluded from the bed.

Total annual cost for this carbon adsorption system is \$325,700 (Table 8.5.1). This cost is equivalent to \$0.15 per pound of organic emitted in the process exhaust, or about 60 percent of the purchase price of the organic.

Three other cases are given in Table 8.5.1 to illustrate the impact that a reduction in total flow or organic flow will have on operating costs. Given the assumptions noted above, decreasing the flow rate by 50 percent (and maintaining the same mass flow rate of the VOCs) has a larger impact on the annual costs than reducing the mass flow rate of organics by 50 percent and maintaining the same flow. Cost savings of 50 percent are projected if both flow rate and mass flow rate of organics are cut by 50 percent.

ACHIEVING HIGH REMOVAL EFFICIENCIES

The concentration profile in Fig. 8.5.2 shows that high removal efficiencies require good effluent quality and adequate adsorption capacity. Enough capacity must be provided to prevent premature breakthrough and to limit emissions from a hot, wet adsorber during the cooling/drying period.

Maintenance of good effluent quality requires effective desorption and the close control of materials (such as other organics and water) and conditions (such as temperature) that impair performance.

Prevention of premature breakthrough requires a favorable balance between the adsorption capacity dictated by the mass flow rate of organics and the speed and effectiveness with which the adsorber can be desorbed, cooled, and dried.

To avert the discharge of organics from a hot, wet adsorber, piping must be arranged so that exhaust can be recycled to an on-line adsorber. There must be sufficient time in the total adsorption cycle for the bed to be cooled and dried.

Removal efficiencies of 95 to 99 percent can be achieved, but it must be recognized that the system is comprised of three closely related process operations: adsorption, desorption, and cooling/drying. The idea that adsorption systems are black boxes to be installed on the roof and run with no supervision must give way to an understanding of the factors that affect operation.

When designing new systems, one should size the adsorbers large enough to allow sufficient time for effective desorption and cooling/drying. Take into account increases in flow and/or inlet concentration so that they will not adversely affect performance. Make provisions so that more steam can be furnished by increasing the steaming time and/or the steam flow rate.

TABLE 8.5.1 Annual Costs for Carbon Adsorption System

Flow rate: Feed concentration:	20,000 cfm 1,000 ppm	10,000 cfm 2,000 ppm	20,000 cfm 500 ppm	10,000 cfm 1,000 ppm
Capital charges at 24%	$192,000	$ 96,000	$192,000	$ 96,000
Utilities				
Steam at 0.3 lb/lb carbon and at $6/1000 lb	63,200	63,200	31,600	31,600
Electricity at 5 hp/1000 cfm and at $0.08/kWh	47,200	23,600	47,200	23,600
Cooling water at 4.2 gal/lb of steam and at $0.12/1000 gal	5,300	5,300	2,650	2,650
Carbon replacement at 4-year lifetime and at $2.00/lb	18,000	9,000	18,000	9,000
Total annual costs	$325,700	$197,100	$291,450	$162,850

Improving the performance of existing systems not designed for high efficiencies will often prove more difficult than designing a new system. The most important step is to verify that the adsorption section is operating properly. If it is not, take steps to define clearly the cause of the poor effluent quality or the premature breakthrough.

If there is insufficient adsorption capacity, look for ways to reduce the mass flow rate of organics to the adsorbers, as well as ways to improve the desorption efficiency. The mass flow rate of organics can be reduced by minimizing the airflow through vessels that contain volatile materials, by installing efficient ventilating hoods to collect fugitive emissions, and by verifying the effectiveness of such pretreatment equipment as condensers, filters, scrubbers, and coolers.

For many existing systems, the opportunity to improve the desorption efficiency may exist. If the amount of steam originally specified was dictated solely by economic considerations, it may have been insufficient for attaining high removal efficiencies. In some cases, the flow of steam supplied for the regeneration cycle may have fallen even lower due to the failure of steam valves or instrumentation.

Unless desorption efficiency data similar to those in Fig. 8.5.9 show that enough steam is already being used, cast about for ways to furnish more. Increasing the length of the steam cycle (if time permits) and increasing the steam flow rate (if the size of the condenser permits) are the two simplest means. If the desorption cycle is initiated by a timer, consider changing to a system that initiates regeneration when breakthrough is detected.

If the existing system does not have a cooling/drying cycle, evaluate methods for including one. Although the total adsorption cycle will have to be changed, the main task will be to install the piping, and possibly a second blower, to move ambient air through the cooling/drying bed and into the inlet of the on-line adsorber vessel.

If additional steam has proved sufficient to reduce the residual organic content, it may be cost-effective to install the equipment necessary to recycle the exhaust from cooling and drying.

When the effluent quality remains poor, despite apparently adequate desorption and cooling/drying, it can usually be traced to an accumulation of other organics on the carbon. The accumulation is often undetected until performance can be improved only by changing the carbon. Sometimes, more steam will improve the desorption efficiency enough to extend the life of the carbon. However, if the extra steam cost is greater than the cost of changing the carbon, an optimum can be reached between the cost of steam and the cost of changing carbon.

Finally, do not overlook the possibility of modifying the emission source as an option for improving performance. Often, a solution to the problem is unnecessarily restricted by an arbitrary decision that the process cannot be changed. It has been the authors' experience that the most cost-effective solution is usually a combination of process changes that reduce required capacity and modifications that upgrade operating performance.

ACKNOWLEDGMENT

This section is an updated version of the article "Vapor-Phase Adsorption Cuts Pollution, Recovers Solvents" (Parmele et al., 1979). Copyright 1979 by McGraw-Hill, Inc., New York, NY 10020. It is reprinted by special permission from *Chemical Engineering*.

REFERENCES

Adams, L. B., C. R. Hall, R. J. Holmes, and R. A. Newton, "An Examination of How Exposure to Humid Air Can Result in Changes in the Adsorption Properties of Activated Carbons," *Carbon,* vol. 26, no. 4, pp. 451–459, 1988.

Anderson, R., "Solvent Recovery Pays Off," *Business Week,* Apr. 18, 1977, p. 138.

Environmental Protection Agency, EPA Package Sorption Device System Study, document no. PB-221-138, April 1973.

Foster, M. L., "Evaluation of the Parameters Affecting Activated Carbon Performance of a Solvent-Ladened Air Stream," presented at the 78th Annual Meeting of the Air Pollution Control Association, Detroit, Mich., 1985.

Kovach, J. L., "Gas-Phase Adsorption and Air Purification," chap. 9 in *Carbon Adsorption Handbook,* edited by P. N. Cheremisinoff and F. Ellerbusch, Ann Arbor Science Publishers, Ann Arbor, Mich., 1978.

Lang, W., General Tire and Rubber Company, personal communication with C. Parmele, Oct. 5, 1979.

Miller, K. J., C. R. Noddings, and R. C. Nattkemper, "Preventing Bed Fires in Carbon Adsorption Systems," presented at 80th Annual Meeting of Air Pollution Control Association, New York, June 1987.

Naujokas, A. A., "Spontaneous Combustion of Carbon Beds," *Plant/Operations Progress,* vol. 4, no. 2, April 1985.

Noll, K. E., D. Wang, and T. Shen, "The Effect of Moisture on the Vapor Phase Adsorption of Toluene and Trichloroethylene," presented at the 81st Annual Meeting of Air Pollution Control Association, Dallas, Tex., June 1988.

Paresse, T., Met-Pro Systems Div., personal communication with C. Parmele, October 1997.

Parmele, C. S., H. S. Basdekis, and W. L. O'Connell, "Vapor-Phase Adsorption Cuts Pollution, Recovers Solvents," *Chemical Engineering,* Dec. 31, 1979, pp. 58–70.

SenGupta, P., VARA International Division of Calgon Carbon Corporation, personal communication with C. Parmele, July 1997.

Smisek, M., and S. Cerny, *Active Carbon,* Elsevier, New York, 1970.

Underhill, D. W., "Calculation of the Performances of Activated Carbons at High Relative Humidities," *Journal of Industrial Hygiene Association,* vol. 48, no. 11, pp. 909–913, Nov. 1987.

Werner, M. D., "The Effects of Relative Humidity on the Vapor Phase Adsorption of Trichloroethylene by Activated Carbon," *Journal of Industrial Hygiene Association,* vol. 46, no. 10, pp. 585–590, Oct. 1985.

Zanitsch, R., "Solvent Recovery from Low Concentration Emissions," Seminar on Emission Control for Publication, Packaging and Specialty Gravure Printers, sponsored jointly by Gravure Research Institute and Gravure Technical Association, Louisville, Ky., Sept. 26, 1979.

SECTION 8.5.2

ADSORPTION: FOG

Robert Rafson, P.E.
Rafson Engineering, Inc.
Chicago, Illinois

BACKGROUND 8.92
MECHANISM OF ABSORPTION BY FOG DROPLETS 8.92
MECHANISM OF ADSORPTION BY FOG DROPLETS 8.96
REACTIONS OF ABSORBED AND ADSORBED COMPOUNDS 8.98
RESULTS OF TESTING 8.99
APPLICATIONS 8.99
REFERENCES 8.101

BACKGROUND

Fog scrubbing technology was developed in 1975 and has been applied, in almost all cases, to odor control applications. This is discussed in Sec. 8.6.2. Although most odorous compounds are water-soluble, and reactable, not all are. The exhausts from some odorous sources contain other compounds, such as benzene, toluene, perchlorethylene, and a wide variety of other compounds.

This is illustrated in Tables 8.5.2 and 8.5.3, analyses of exhaust from a wastewater pumping station and from a composting operation. Wastewater contains a wide variety of compounds which are dumped down the drain and find their way to the water treatment process, where a portion is emitted into the air and thus will need to be treated. It was observed that a large proportion of these compounds, even those which were non-water-soluble and nonreactable, were removed by fog scrubbing.

MECHANISM OF ABSORPTION BY FOG DROPLETS

It is necessary to understand the different mechanisms at work when one considers removing water-soluble and water-insoluble components from the gas stream for fog droplets. As discussed in Sec. 8.6.2, the water-soluble compounds are removed by diffusions into the droplet. The fog-sized droplets, of less than 10 μm, are extraordinarily effective in such removals because of high surface area and contact time, allowing for diffusion into the droplet. This diffusion is further enhanced by controlled chemical addition to the atomized liquid stream, which reacts with the absorbed compounds, encouraging further diffusion into the droplet.

Volatile organic conpounds (VOCs) represent a wide range of compounds; some are readily water-soluble, others moderately so, and still others are very water-insoluble. The removals by fog droplets are affected by these characteristics. VOCs which are water-soluble are removed, as are many water-soluble inorganic odorous compounds. The insoluble compounds are removed by adsorption on the fog droplet, as discussed in greater detail in the following sections. Removals of VOCs which are moderately water-soluble can be enhanced by the addition of certain surface active agents, or by reactants

TABLE 8.5.2 Wastewater Pumping Station—Inlet to Odor Control System

GC/MS analysis of volatile organics associated with odorous components

Spec. no.		Compound	Peak area (counts)
159	CCl_3F	Trichlorofluoromethane	380,379
166	$C_4H_{10}O$	Methoxypropane	10,526
176	$C_2Cl_3F_3$	Trichlorotrifluoroethane	5,403,005
197	$C_2H_2Cl_2$	Vinylidene chloride	21,082
205	C_2H_6S	Dimethylsulfide	18,927
235	C_3H_6O	Propanol	6,532
255	C_3H_6O	Acetone	429,152
312	C_4H_8O	Tetrahydrofuran	32,711
329	$C_2H_3Cl_3$	Methylchloroform	36,689
346	C_4H_8O	Methylethylketone	52,998
351	CH_4O	Methanol	38,170
368	CH_2Cl_2	Dichloromethane	157,454
386	C_6H_6	Benzene	80,025
397	C_2H_6O	Ethanol	594,755
447	C_2HCl_3	Trichloroethylene	26,130
466	$CHCl_3$	Chloroform	735,438
496	C_2Cl_4	Tetrachloroethylene	132,604
499	C_7H_8	Toluene	289,900
503	C_3H_8O	1-Propanol	20,006
919	C_7H_6O	Benzaldehyde	85,205
923	$C_{10}H_{20}O$	Decanal	62,149
1176	$C_8H_{16}O_2$	Methyl propanoic acid, butyl ester	4,418
1183	$C_{11}H_{20}O$	Undecanal	37,009

which render the compounds more water-soluble. Examples of these kinds of compounds are vegetable oils, in which increased removals are obtained by the addition of soap to break the water-oil barrier, or by reactants which enhance solubility in the droplet, such as caustic.

Another factor, when one is considering solubility and the vapor-pressure relationship between the contaminant and the fog droplet, is the absolute pressure. In most conventional scrubbers, the absolute pressure within the droplet is generally very close to 1 atm. In a fog scrubber, the surface tension of the very small droplets maintains a solution at a higher absolute pressure within the droplet. At a droplet size of less than 10 μm, this pressure difference is significant and can exceed 5 atm. This enhances the solubility of some compounds in the droplet and encourages transfer into, and retention within, the droplet. Removal of water-soluble compounds may also be improved at increasing pressure within the droplet.

Transfer into the droplets can be further enhanced by some types of surface active agents and reactants. Surface active agents function in two ways. First, they reduce the surface tension of the liquid; second, since the droplets are more easily formed by atomization, smaller droplets are generated and increase the total surface area.

Compounds which are less soluble are still removable, by another mechanism, which is adsorption on the surface of the droplet.

TABLE 8.5.3 Compost Odor Emissions—Inlet and Exhaust from Odor Control System

GC/MS analysis of volatile organics

Ret. no.		Compound	Inlet	Outlet	% Removal
		Sulfur compounds			
505	CS_2	Carbon disulfide	1.00	0.00	100
543	C_2H_6S	Dimethyl sulfide	22.00	0.00	100
851	$C_2H_6S_2$	Dimethyl disulfide	106.00	0.07	100
1190	$C_2H_6S_3$	Dimethyl trisulfide	3.00	0.00	100
		Aldehydes			
626	C_4H_6O	2-methyl 2-propanol	3.00	0.00	100
685	$C_5H_{10}O$	2-methyl-butanal	5.00	0.00	100
691	$C_5H_{10}O$	3-methyl-butanal	8.00	0.00	100
788	$C_7H_{14}O$	Dimethyl pentanal (tent)	3.00	0.00	100
1014	$C_{10}H_{18}O$	Methanol + $C_{11}H_{18}$ cyclical	8.00	0.00	100
1192	$C_7H_{14}O$	Methyl hexanal	5.00	0.00	100
1240	$C_5H_4O_2$	Furfural	1.00	0.08	92
1241	$C_{10}H_{20}O$	Decanal	4.00	0.21	95
1300	C_7H_6O	Benzaldehyde	3.00	0.33	89
		Ketones			
607	C_3H_6O	Acetone	114.00	3.37	97
632	$C_6H_{10}O$	Hexenone (tent)	4.00	0.00	100
683	C_4H_8O	Methylethylketone	24.00	0.27	99
763	$C_6H_{10}O$	Hexenone (tent)	3.00	0.00	100
765	$C_8H_{14}O$	Cyclohexylethanone	7.00	0.00	100
966	$C_7H_{12}O$	Triethylcyclobutanone	3.00	0.00	100
1095	$C_{10}H_{16}O$	Triethylbicycloheptanone	3.00	0.00	100
1421	$C_8H_{18}O$	Trimethylpentanone	2.00	0.00	100
		Alcohols			
676	CH_4O	Methanol	2.00	0.34	83
735	C_2H_6O	Ethanol	3.00	0.62	79
874	$C_9H_{20}O$	Methylpropylpentanol	10.00	0.00	100
1063	$C_8H_{16}O$	Octanol	2.00	0.13	94
1281	$C_8H_{18}O$	Ethylhexanol	7.00	0.00	100
1344	$C_8H_{18}O$	Isooctanol	1.00	0.10	90
1505	$C_{10}H_{22}O$	Decanol	4.00	0.04	99
1506	$C_9H_{12}O$	Alpha-dimethylbenzylalcohol	2.00	0.00	100
		Ring compounds			
710	C_6H_6	Benzene	5.00	0.14	97
900	C_8H_{10}	Ethylbenzene	1.00	0.13	87
914	C_8H_{10}	Xylene	2.00	0.22	89
959	C_8H_{10}	Xylene	7.00	0.09	99
961	C_8H_{10}	Xylene	3.00	0.00	100
964	$C_{11}H_{20}$	Decahydromethylnaphthalene	7.00	0.00	100
1026B	C_8H_8	Styrene	2.00	0.12	94
1081	$C_9H_{12}O$	Methylbenzeneethanol	2.00	0.00	100
1110	C_9H_{10}	Alpha-methylstyrene	2.00	0.00	100
1200	$C_6H_4O_{12}$	Dichlorobenzene	2.00	0.11	95
1588	C_7H_5NS	Benzothiazole	1.00	0.01	99
1600	C_6H_6O	Phenol	1.00	0.21	79

TABLE 8.5.3 Compost Odor Emissions—Inlet and Exhaust from Odor Control System (*Continued*)
GC/MS analysis of volatile organics

Ret. no.		Compound	Inlet	Outlet	% Removal
		Esters			
1077	$C_6H_{12}O_3$	2-ethoxyethylacetate	2.00	0.00	100
		Furans			
595	C_4H_4O	Furan	1.00	0.00	100
619	C_5H_6O	Methylfuran	1.00	0.00	100
638	C_5H_6O	Methylfuran	1.00	0.00	100
968	$C_9H_{14}O$	2-Pentylfuran	2.00	0.00	100
1500	$C_{10}H_8$	Naphthalene	1.00	0.04	96
		Terpenes			
785	$C_{10}H_{16}$	4-Carene	2.00	0.00	100
800	$C_{10}H_{16}$	Alpha-pinene + C_2Cl_4	74.00	0.00	100
825	$C_{10}H_{18}O$	*p*-Menthol + $C_{10}H_{16}$	2.00	0.00	100
834	$C_{10}H_{16}$	Camphene	10.00	0.00	100
885	$C_{10}H_{16}$	Beta-pinene	5.00	0.00	100
960	$C_{10}H_{18}$	Decahydronaphthalene	14.00	0.00	100
962B	$C_{10}H_{16}$	Limonene	29.00	0.03	100
1425	C_8H_8O	Acetophenone	5.00	0.21	96
1517	$C_{11}H_{10}$	Methylnaphthalene	2.00	0.00	100
		Alkyl benzenes			
991	C_9H_{12}	Alkylbenzene	5.00	0.08	99
1033	$C_{10}H_{14}$	Alkylbenzene + $C_{11}H_{24}$alkan	7.00	0.00	100
1052B	C_9H_{12}	Alkylbenzene	5.00	0.09	98
1104	$C_{10}H_{14}$	Alkylbenzene + $C_{14}H_{26}$alkan	4.00	0.00	100
1107	C_9H_{12}	Alkylbenzene	2.00	0.04	98
1125	$C_{10}H_{14}$	Alkylbenzene	2.00	0.00	100
1157	$C_{10}H_{14}$	Alkylbenzene + $C_{11}H_{16}$	3.00	0.00	100
1178	$C_{10}H_{14}$	Alkylbenzene	5.00	0.00	100
1193	$C_{11}H_{16}$	Alkylbenzene	4.00	0.00	100
1198	$C_{10}H_{12}$	Methylpropenylbenzene	3.00	0.00	100
		Ethers			
490	C_4H_8O	Vinylethylether	31.00	0.48	99
		Chlorinated compounds			
491	CCl_3F	Trichlorofluoromethane	22.00	0.23	99
700	CH_2Cl_2	Dichloromethane	1.00	1.00	90
782	C_2HCl_3	Trichloroethylene	2.00	0.00	100
		Others			
962A	$C_9H_{15}N$	Methyloctanenitride	9.00	0.00	100
1375	C_7H_5N	Benzonitrile	2.00	0.18	91

**Note:* Analysis of compounds with less than 1-in inlet have been omitted.

MECHANISM OF ADSORPTION BY FOG DROPLETS

This process became understandable through literature research of work done in the field of atmospheric chemistry. The standard approach to thinking about mass transfer is based on Henry's law, which concerns the proportion of vapor pressure in liquid and gas. It was determined that transfer of VOCs to small droplets was far greater than would have been predicted by Henry's law. Part of this increased removal ratio is referred to as an enrichment factor, resulting from adsorption to fog-sized droplets.

A wide range of organic compounds have been found concentrated in fogs, mists, generated mists, clouds, rain, snow, and aerosol particles. Aneja (1993) and Capel (1991) provide extensive bibliographies of studies identifying a wide range of atmospheric contaminants.

Capel (1991) describes the first detailed examination of hydrophobic organic chemicals in fogs. His study examines the contaminants in fog, rain, and air in Switzerland. He discusses the air/fog transfer mechanism of Henry's law and the enhancement of Henry's law observed, and he posits some explanations for this enhancement. Some of his observations are as follows:

> All the non-alkanes ($C_{19\text{-}35}$) were observed at concentrations many orders of magnitude greater than their calculated water solubility in most of the fog samples.
>
> The consistent occurrence of organic chemicals, particularly hydrophobic organic compounds in fog suggests that it is an excellent scavenger of organic compounds from the atmosphere. (Refer to Table 8.5.4.)

TABLE 8.5.4 Organic Chemicals in Fog and Rain in Dubenborf

Compound	Fog/rain enrichment factor
N-Alkanes	1.3–130
Total PAH	50–280
Total PCB	4,000–6,000
	2,000–22,000
Chlorinated dioxins and furans	3–100
Phenol	2–20
2-Nitrophenol	0.1–0.3

Source: Capel et al., 1991.

> One mechanism of the transfer of VOCs to fog droplets is Henry's Law of the exchange between vapor and dissolved phases.…To illustrate the relative acceptance of this mechanism for a particular compound, its dimensionless Henry's Law coefficient, α, can be compared to the overall washout ratio (W). W is the ratio of the chemical's concentration in a precipitation event. If the concentration of the chemical in the droplet is due only to air/water exchange, then W should equal α. When median fog concentrations are divided by median atmospheric concentrations, W_S of the order of 10^5–10^7 are obtained for the hydrophobic compounds.
>
> The enrichment, as compared to simple air/water exchange, of the hydrophobic organic chemicals in the fog droplet is exactly the same result that was reported by Glotfelty (1987)

> for a number of pesticides simultaneously measured in fog and air. Based on the comparison of an estimated *W*, simple Henry's Law exchange, although contributing to the chemical loading of the fog, cannot account for the concentration of hydrophobic organic compounds in fog.

Capel (1991) then suggests that this enrichment may be partly due to particle scavenging and dissolved organic compounds. He continued:

> The wide discrepancy between droplet/vapor washout ratios of rain and fog may indicate a third important mechanism. This mechanism is probably based on the effects of dissolved, and/or colloidal organic carbon. Studies have shown that the dissolved organic carbon decreases the surface tension of the water, which indicates that a portion of the dissolved organic material is surface active. This could yield a vapor/organic film exchange mechanism which would probably be efficient in scavenging of hydrophobic chemicals from the vapor phase and incorporating them into the fog droplet.
>
> Another way that the dissolved/colloidal organic carbon can act to increase the flux of hydrophobic chemicals is through partitioning. As the hydrophobic chemical enters the fog droplet through simple air/water exchange, it can become associated with the dissolved and/or particulate organic matter, thus decreasing the concentration of the truly dissolved chemical concentration and increasing the flux due to Henry's Law exchange.
>
> The relative contribution of these three mechanisms, which act to incorporate inorganic chemicals in the fog droplet, cannot be assessed from these data although an estimation of the reaction of each of the organic chemicals in the truly dissolved phase and associated with particles at the time of collection, can be made.

Other authors comment on these and other mechanisms, and this is the primary focus of this literature. Capel continues:

> The relatively large concentrations of organic chemicals present in fog droplets comprise a significant mass of the total atmospheric burden of these chemicals during a fog event.…Only about $1 \times 10^{-10}\%$ of the total volume is composed of fog droplets. In contrast to this, typical 5–100% of the mass hydrophobic organic compounds are contained in the fog droplet. This again reflects the tremendous scavenging efficiency of hydrophobic compounds by the fog.

Valsaraj (1992) in a personal communication goes beyond the studies of Capel, and all the prior literature references, in correlating the data from prior studies, which are found to be in good agreement when temperature adjustments are considered. Valsaraj suggests three mechanisms to explain the enhancements: the temperature correction, colloidal organic material, and the effect of large air-water interfacial area for adsorption of hydrophobic organics. Valsaraj's conclusion is that the air-water interfacial adsorption is of utmost significance in explaining the uptake of organics in the fog droplet. Valsaraj also indicates that compounds with increasing hydrophobicity are more readily removed. He notes that the enrichment in fog droplets appears to be universal, and not characteristic of a particular location.

Valsaraj, in evaluating the mechanisms, concludes that colloidal matter within the droplet cannot adequately explain the 1000-fold or larger enrichment found in fog water. Valsaraj, referencing numerous prior studies, concludes that there is considerable evidence to suggest that surface adsorption of hydrophobic compounds on fog droplets is an important factor in fog-water enrichment. He then presents a mathematical analysis that supports this hypothesis. Further, the mathematical analysis conforms with observations that there is decreased enrichment by larger droplets and by compounds with decreased hydrophobicity. This explains why raindrops have 1 to 2 orders of magnitude less enrichment than fog droplets.

The literature is quite extensive and quite recent, and while there are always individual perspectives, the studies all point in the same direction (Capel, 1991):

> Since both fog and rain derive their chemical concentrations from the atmosphere, a direct comparison can be made....The main differences are the surface areas, water volumes, and lifetimes in contact with a particular parcel of air. The raindrop is one or two orders of magnitude greater in size than the fog droplet. The rain falls through the atmosphere quickly,...while the fog droplet moves slowly....Both the size (increase in surface area) and direction of the fog droplet tend to increase its chemical concentration....The hydrophobic compounds are enriched tens to thousands of times that of rain.

Glofelty et al. (1987, 1990) investigated pesticides in early morning mists above the ground, paralleled Capel's later discussions, and found enrichments up to 4 orders of magnitude in mist size droplets.

Aneja (1993) extensively reviewed the literature and was concerned with enrichments that occur in clouds, which he sampled in peaks of the Blue Ridge Mountains. He refers not only to Capel, but also to Perona (1992), who "has shown, based on thermodynamic treatment, that the solubility of hydrophobic compounds dissolved in small droplets (i.e., systems with large surface-volume ratios) can be significantly higher than values obtained with solutions having small surface-to-volume ratio and/or planar surface."

Aneja parallels Capel's earlier work, suggesting the same mechanisms for removal, but supplementing and reconfirming the results by his study of clouds (Table 8.5.5). He emphasizes the significance of small droplets.

TABLE 8.5.5 Overall Washout Ratios

Compound	$W_{\text{-rain}}$	$W_{\text{-cloud}}$
Acetone	0.0034	31.25
Chloroform	202.5	2,025.0
Ethylbenzene	26.8	13,402.0
Hexane	22.3	223.2
Mesitylene	10.4	61.2
Methylene chloride	716.0	7,161.0
Methyl furan	30.0	2.2E8
o-Xylene	35.0	349.9
Toluene	23.4	159.7
Trichloroethene	3.6	897.9
Trimethylbenzenes	22.3	252.6
2-Butane	57.9	579.2

Source: Aneja, 1993.

REACTIONS OF ABSORBED AND ADSORBED COMPOUNDS

As long as the reaction increases the driving force for a compound to diffuse into a droplet, or to decrease its likelihood to escape, removal will be enhanced. This is clearly demonstrated by reaction of water-soluble compounds, with the addition of agents that enhance the diffusion. An acid gas will be removed much more effectively by a droplet of basic pH than one with water which becomes more acid as acid compounds diffuse into it. This is also true with moderately soluble compounds. The mechanism of holding insoluble compounds is adsorption forces on the surface of the droplets. All these mechanisms accomplish the same goals of enhanced removals from the gas stream. What is not expected is that this adsorption mechanism is effective and useful for removing many water-insoluble compounds from the gas stream by using water droplets. The effectiveness of removal depends on the chemical, the structure of the gaseous compounds, and is demonstrated by the atmospheric chemistry work which has shown the effectiveness of fog droplets to cleanse the atmosphere.

In a fog scrubber, the conditions are much more controllable than in the atmosphere. It is possible to establish the density of the fog, the contact time, and the addition of reactive chemicals and surface active agents which enhance removals. This has been shown to be effective, as discussed in the following section on field test results.

RESULTS OF TESTING

For over 20 years, mist scrubbing has been treating odors and VOCs, but only in the last 10 years has extensive testing been done on VOCs. In field pilot plant tests, the results of removals are summarized in Table 8.5.6. From these several simulated studies it is apparent, as described by the above-referenced works, that removal results will vary for different compounds. However, it is also very clear that significant removals are obtainable.

Removals were also reported in 1989 by Tom Card in a study that compared a mist scrubber and packed column. Very good results were reported with approximately 94 percent removals of VOCs obtained. Packed columns, which are recirculating scrubbers, achieved much lower levels.

In Water Environment Research Foundation–supported research work, a team of researchers reported no effective removals of VOC, and increases of chlorinated hydrocarbons in packed columns. The results of tests on mist scrubbers were confusing, and recommendations were made for further studies. The conclusion drawn, however, was that mist/fog scrubbers offered a good potential for VOC removal at wastewater facilities.

The Washington Suburban Sanitary Commission (WSSC) composting process has generated and controlled significant odors. WSSC conducted extensive testing and upgraded and expanded its mist scrubbing systems. Many tests have demonstrated the effectiveness of removals and functionality of the controls. The plant presently adjusts the chemistry of the removal system by daily testing to maximize VOC removal.

APPLICATIONS

The kinds of VOC compounds that are best removed are those which are highly water-insoluble. Those that are moderately soluble are somewhat less removed. Those that are readily soluble are removed by the mechanism of diffusion and solution.

The removals obtained fall into three categories:

1. Removals are adequate to meet environmental objectives.
2. Removals are good, but not good enough to meet emission objectives. In this case, the fog scrubber can be followed by a carbon adsorption system, biofilter, or some other added treatment.

TABLE 8.5.6 Mist Scrubber—Performance Tests, Various Operating Conditions*

	Benzene		Toluene		Unidentified		Total	
	Inlet	% Removal	Inlet	% Removal	Inlet	% Removal	Inlet	% Removal
Wastewater								
Grit removal	20,000	80	40,000	82	140,000	74	200,000	75
Aeration	15	100	70	85	4,915	93	5,000	93
Dewatering	140	100	460	93	6,400	100	7,000	99
Foundry								
Casting line	2,000–13,000	100	—	—	—	—	—	—

	Formaldehyde		Phenol		Unidentified		Total	
	Inlet	% Removal	Inlet	% Removal	Inlet	% Removal	Inlet	% Removal
Wood								
Chip drier	194,000	87	30,200	79	—	—	224,200	87
Press exhaust	169,400	92	21,100	93	—	—	190,500	93
Press exhaust—peak	—	—	—	—	—	—	420,000	99
Veneer drier	—	—	—	—	—	—	2,110,000	90

*All values in parts per billion.

3. Removals are not good on the contaminant compound, and some other technology is preferable.

The fog scrubber, which has low liquid flow rates, acts as a concentrator of the contaminant. The final disposal of the liquid effluent is, therefore, simplified and made more economical. There are several mechanisms which can be effective.

1. Strip the contaminant from the water stream with a small flow of air. This is particularly effective when the airstream can then be disposed of by burning in a plant boiler, or drier.

2. Adsorb the contaminant on carbon, and dispose of it, either as a solid waste or by incineration.

3. Treat the liquid stream with biological treatment. Because of the low liquid flow this makes the bioreaction design efficient.

4. No posttreatment at all is required. It has been our observation that the concentration of contaminant is often low enough that the liquid stream can be disposed of with normal wastewater treatment technologies.

Actual applications have been in operation for many years treating VOCs simultaneously with treating odors. These applications have included paint resin cooker emissions, brake shoe lining oven emissions, compost emissions, and wastewater pumping stations.

REFERENCES

Aneja, V. P.: "Organic Compounds in Cloud Water and Their Deposition at a Remote Continental Site," *Air & Waste Management Assoc.,* vol. 43, pp. 1238–1244, Sept. 1993.

Capel, P., C. Leunberger, and W. Giger: "Hydrophobic Organic Chemicals in Urban Fog," *Atmospheric Environment,* vol. 25A, no. 7, pp. 1335–1346, 1991.

Card, T.: "Volatile Organic Removal in Packed Towers and Atomized Mist Odor Scrubbing Systems," Presented at the Annual Water Pollution Control Federation Conference, San Francisco, Oct. 1989.

Glotfelty, D. E., J. N. Seiber, and L. A. Lilijedahl: "Pesticides in Fog," *Nature,* vol. 325, Feb. 12, 1987.

Glotfelty, D. E., M. Majewski, and J. Seiber: "Distribution of Several Organophosphorus Insecticides and Their Oxygen Analogues in a Foggy Atmosphere," *Environment Science Technology,* vol. 24, Nov. 3, 1990.

Valsaraj, K. T.: Personal communication, Nov. 1992.

SECTION 8.6.1

CHEMICAL SCRUBBING: PACKED COLUMNS

Charles M. McGinley, P.E.
McGinley Associates, P.A.
Stillwater, Minnesota

MASS TRANSFER 8.102
PACKED-COLUMN MEDIA 8.104
BASIC DESIGNS 8.106
DESIGN EXAMPLE 8.108
REFERENCES 8.110

When the environmental industry addresses odorous and volatile organic compound (VOC) emissions, packed-column scrubbers appear on the list as proven control technology. Packed columns by other names include packed-tower scrubbers, packed-bed scrubbers, and chemical packed beds. These common names, often interchangeable, suggest the versatility of packed-column configurations. Application versatility for packed-column chemical scrubbing is derived from absorption technology. Understanding basic absorption technology leads to understanding the advantages, features, and benefits of packed-column chemical scrubbing.

MASS TRANSFER

When one or more components (odorous or VOC) of a gas transfer to a liquid, because of physical solubility, absorption occurs. The transfer phenomenon of a component from the gas to the liquid is known as *mass transfer.* Chemical reactions that may occur in the liquid phase may affect the overall mass transfer of the component from the gas. Physical absorption phenomena can be understood by using models to describe the mass transfer that occurs from the gas phase to the liquid phase. The two models that provide simple conceptual understanding of the absorption phenomena are the *two-film model* and the *surface renewal model.*

The two-film model, proposed by Whitman (1923), defines an interface between the gas phase and the liquid phase. At the interface, two boundary layers exist. These boundary layers, called *films,* represent the zones of mass transfer from the bulk gas phase, through the interface, to the bulk liquid phase. The two-film model presumes that one of the films, the gas film or the liquid film, controls the rate of mass transfer from the gas to the liquid. An absorption process can be either gas-film controlling (gas-film limiting) or liquid-film controlling (liquid-film limiting). Figure 8.6.1 illustrates the elements of the model: bulk gas phase, gas film, interface, liquid film, and bulk liquid phase.

Two common systems represent the two choices for packed-column, mass-transfer modeling. The air-water system represents the gas-film controlling model. The carbon dioxide-air-water system represents the liquid-film controlling model. Performance data for packed columns are readily available for these two systems. The selection of one model over the other will depend on the specific application under consideration for the packed-column scrubber.

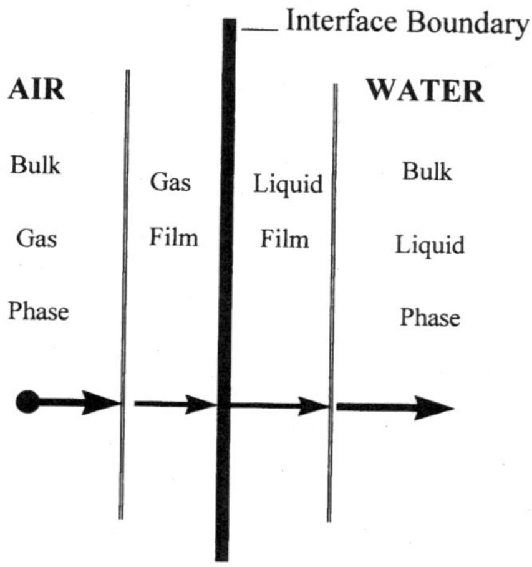

FIGURE 8.6.1 Two-film model of absorption mass transfer.

The air-water system, which is gas-film controlling (gas-film limiting), presumes that the liquid film is less limiting. In the case of simple water evaporation, from a pure water surface, the water molecules are readily available at the interface boundary's liquid film. The water molecules do not encounter a barrier or resistance in the liquid film at the interface boundary. Therefore, the only "barrier" or "resistance" is the gas film at the interface boundary.

In a specific example of odorous molecules in the gas phase, if the odorous molecules are highly soluble in the liquid phase (i.e., water), the gas film will be limiting and controlling the overall mass transfer from the bulk gas phase to the bulk liquid phase. The air-water system performance data for packed columns can then be used as a starting point for predicting or designing a packed-column scrubber for highly soluble odorous molecules.

The carbon dioxide-air-water system, which is liquid-film controlling (liquid-film limiting), presumes that the gas film is less limiting. In the case of simple carbon dioxide in air, the carbon dioxide molecules are readily available at the interface boundary's gas film. The carbon dioxide molecules do not encounter a barrier or resistance in the gas film at the interface boundary, compared to the "barrier" or "resistance" encountered at the liquid film. Carbon dioxide is slightly soluble in water, and as carbon dioxide enters water, the properties of the water change (that is, pH decrease). The changes in the water properties that occur further inhibit the solubility of the carbon dioxide. Therefore, the only barrier or resistance is the liquid film at the interface boundary.

In another example of odorous molecules in the gas phase, if the odorous molecules are in high concentration, but have low solubility in the liquid phase (i.e., water), the liquid film will be limiting. In this example, the liquid film will control the overall mass transfer from the bulk gas phase to the bulk liquid phase. The carbon dioxide-air-water system performance data for packed columns can then be used as a starting point for predicting or designing a packed-column scrubber.

The examples presented above, to illustrate the two-film model of absorption phenomena, help in the basic understanding of how packed columns work. Extreme caution needs to be applied when using the basic models and when using the performance data from the two common systems, air-water and carbon dioxide-air-water. The relative solubility of the odorous molecules or VOC molecules is a primary factor in the absorption process. Other primary factors include the concentration of the components in the gas phase, chemical reactions in the liquid phase, temperature, active surface area of the packed column, and turbulence within the system. These factors are discussed further in this section.

The second model that provides a simple conceptual understanding of the absorption phenomenon is the *surface renewal model.* The surface renewal model, developed by Danckwerts in 1951, assumes that the bulk liquid phase consists of small "packets" of liquid that are brought into contact with the bulk gas phase. Danckwerts' model was a modification of a model proposed by Higbie in 1935. The surface renewal model calculates an "average exposure time" for gas and liquid contact. The average exposure time of the surface renewal model translates to a minimum contact time requirement (for example, 1.5 s) for scrubbing a gas with a liquid in a packed column.

Mass-transfer coefficients, experimentally determined, provide the most direct method for predicting packed-column performance and for designing packed-column scrubbers. An overall coefficient incorporates gas- and liquid-film mass-transfer coefficients. Overall mass-transfer coefficients are system-specific. The specifics of the system include the temperature, gas components (i.e., specific odorous or VOC compounds), liquid composition (i.e., chemistry of the scrubbing solution), liquid flow rates, and packed-column type. The mass-transfer coefficient provides a method to calculate the required total volume of the packed-column scrubber, especially when accompanied by correction factors (i.e., temperature, liquid rates, and liquid composition concentrations).

Estimating mass-transfer coefficients for systems with limited or no experimental data leads to predictions with uncertainty in the packed-column design. The designer, aware of the estimating uncertainties, must recognize that the predictions may be an order of magnitude only with errors up to 50 percent. Therefore, the careful use of overall mass-transfer coefficients must be preceded with an extensive search and verification for the most applicable experimental data.

PACKED-COLUMN MEDIA

Packed-column media for gas-liquid contact in chemical scrubbing have been made in myriad shapes and forms in all types of materials including wood, rock, ceramic, metal, plastic, and woven filaments. Packed-column media may be divided into three categories: random packing, structured packing, and mesh packing. Usually packed columns are filled with random packing media, consisting of individual pieces of packing that are randomly dumped in the column. However, structured packing represents an alternative choice to random packing. Structured packing provides an ordered, predictable arrangement of the packing surfaces in a column. The third category of packing media, known as mesh packing, consists of woven or random filaments in a mesh pad.

The chemical resistance of the various thermoplastic fabrication materials has made *plastic* the predominant packed-column media material. Plastic technical bulletins, available from suppliers, detail complete chemical service and temperature conditions. The most common thermoplastic materials used as packed-column media are polypropylene, polyethylene, polyvinyl chloride, polystyrene, Kynar, and Teflon. Specific recommendations can be made by packing suppliers when complete service and temperature conditions are specified.

All packing media suppliers developed the shape and form of their packing with the goal to fill the basic requirements for an *ideal* packing media:

1. Low weight per unit volume
2. Low cost per unit volume
3. Large active surface area per unit volume
4. Large free volume (void space)
5. Low liquid hold-up (or high liquid hold-up)
6. High strength, durability, and temperature stability
7. High chemical resistance
8. High mixing properties of the gas and the liquid
9. Low pressure drop (low gas flow resistance) with high flooding velocity
10. High mass-transfer coefficients
11. Minimum side thrust on column walls
12. Good liquid redistribution properties
13. Ease of cleaning

Each packing media type and brand have achieved some or most of these requirements. The selection of a specific packing media for an application will depend on a combination of overall performance, project costs, and other commercial factors, i.e., warranties and availability. However, one of three performance factors generally predominates the selection criteria for packed-column media.

The first performance factor that might predominate for packed-column media selection is *mass-transfer coefficients.* Packing medium suppliers and research organizations have developed performance data for specific applications and simplified methods to apply mass-transfer coefficients to new applications. However, for most odor and VOC applications, the mass-transfer coefficients for a specific packed medium will be only slightly applicable. The principal value of mass-transfer coefficients is for preliminary screening of packing media choices, i.e., packing type, shape, and size. Odor and VOC applications may be emulated by the two-film mass-transfer model: gas-film controlling or liquid-film controlling. If the odorous or VOC molecules are in low concentration and the solubility of the molecules in the liquid phase (i.e., water) is slight to moderate (or if solubility is enhanced with chemical reactions), then gas-film mass-transfer coefficients might be suitable for predictions and design. However, if the odorous or VOC molecules exhibit very low solubility, liquid-film mass-transfer coefficients might be prescribed for predictions and design work.

The second factor that might predominate packed-column media selection is *liquid distribution and redistribution characteristics.* Uniform liquid distribution at the top of a packed column is essential. Liquid distribution at the top of a packed column can be accomplished with spray nozzles, perforated pipes, and weir troughs. Sprays that are directed at the packing surface and that avoid liquid spray on the walls provide optimum initial liquid distribution. Perforated pipes (i.e., holes in header and lateral pipes) and weir troughs require a minimum of four points of liquid distribution per square foot.

Liquid redistribution may be required, because liquid channeling can occur in packed columns. For small-diameter columns, i.e., less than 2 ft, liquid redistributors should be considered every 7 column diameters. Although packed-column beds have been constructed up to 30 ft deep without redistribution, packing suppliers generally recommend that a depth of 20 ft be considered a maximum without a liquid redistributor.

The third performance factor that might predominate packed-column media selection is *pressure drop* and the related issue of flooding. The resistance to gas flow through a packed column is influenced by both the gas flow rate and the liquid flow rate. The gas phase flowing through a packed column is generally turbulent. The liquid phase flowing through the packing column covers the surface of the packing media and flows from one surface to the next, i.e., trickles. Increases in gas flow rate or liquid flow rate would increase turbulence and would increase liquid on the media surface with a resulting increase in gas flow resistance, i.e., pressure drop. The condition of maximum practical gas and liquid flow is called *flooding.* At flooding conditions, the pressure drop rises rapidly as the liquid flow through the packed column prohibits uniform gas flow. Packed-media suppliers and research organizations have developed pressure drop data and methods for estimating pressure drop through packed columns. The majority of these methods provide excellent predictions of operating pressure drop and flooding conditions.

Random packed-media comparisons are typically based on one of four characteristics of the packing: *size, packing factor, height of a transfer unit,* and *surface area.* The *size number* of random packing media is typically the minor dimension of the individual packing piece. For example, a Raschig ring's minor dimension is its diameter, a Berl saddle's minor dimension is its width, and a round packing's minor dimension is typically its radius. The *packing factor* reported for packed media closely relates to pressure drop performance. A high packing factor generally corresponds to a high pressure drop and vice versa. The *height of a transfer unit* (HTU) is one measure of a packing media's mass-transfer performance. HTUs can be used successfully in scrubber design equations and in comparisons of packing types and sizes. The *surface area* of a packing media is generally reported as an estimated *geometric surface area* per unit volume of the packed column rather than the calculated *actual surface area* of the individual random packing piece.

BASIC DESIGNS

Packed-column scrubbers for odor and VOC control are typically configured in vertical towers or cross-flow vessels. Vertical tower designs primarily utilize a countercurrent flow scheme, as illustrated in Fig. 8.6.2. The countercurrent flow scheme uses gravity to carry the liquid phase from the point of liquid distribution to the tower reservoir. The gas phase flows up through the packed-column media. The optimum contact and mass transfer occur in this flow scheme as compared to the lower turbulent and generally less efficient concurrent flow configuration.

Figure 8.6.2 presents the basic components of a packed-column scrubber tower. A typical packed column consists of a cylindrical vessel shell containing a packed-bed support plate, packing media, liquid distribution device (i.e., spray nozzles on a pipe header), mist eliminator mesh pad, reservoir for the liquid, recirculation pump with associated piping, chemical addition system, water addition and level controls, instrumentation, and gas inlet and outlet apertures. Each component and subsystem must be selected to meet chemical, mechanical, structural, and hydraulic requirements. The components and subsystems must also fit the overall system requirements for performance (i.e., odor and VOC removal efficiency or emission limits).

The cross-flow packed-column scrubber configuration, as illustrated in Fig. 8.6.3, provides an alternative physical layout of the standard components and subsystems found in a vertical tower configuration. The cross-flow configuration of a packed column is a low-profile scrubber for applications with limited ceiling height.

Packed-column scrubbers, either tower vessels or cross-flow vessels, can be configured in multiple stages. The vertical configuration of a packed column can be staged as side-by-side units or can be staged as stacked units. The side-by-side units would have liquid reservoirs in the base of each vessel. The stacked units would have liquid collec-

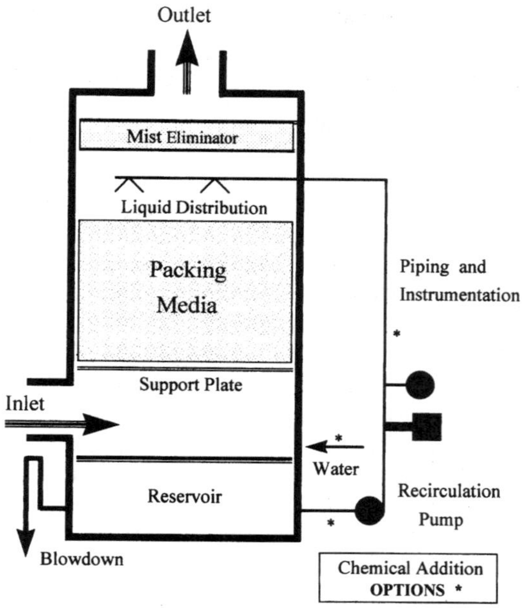

FIGURE 8.6.2 Packed-column scrubber tower (schematic).

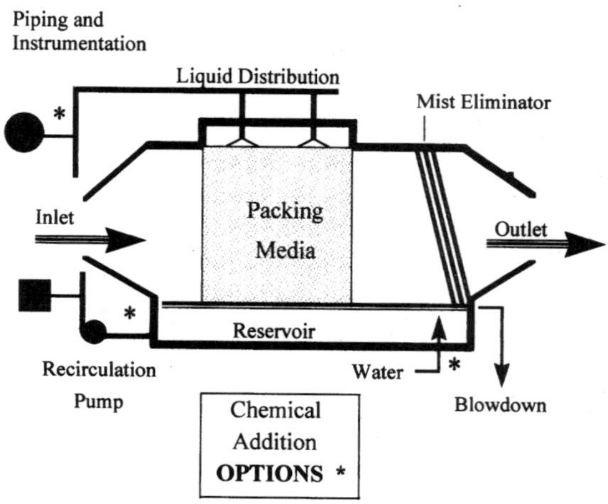

FIGURE 8.6.3 Cross-flow scrubber (schematic).

tion devices in between the stacked stages. The cross-flow configuration of a packed column allows for the combining of more than one stage within a single vessel. The process advantage of multiple staging of packed columns is the capability to use specific scrubber liquids and chemicals in a preferred order.

An example of an application that would benefit from a multiple-stage, packed-column scrubber system is an industrial wastewater treatment system with odorous emission. Sulfur gases produced from anaerobic activity would most likely include hydrogen sulfide and methyl mercaptan. A single packed-column scrubber might be selected for the combined odor. However, the multiple sulfur gases in this example require attention to the properties and chemical reactions of the individual sulfur gas components.

The chemistry of sulfur gases, specifically hydrogen sulfide and methyl mercaptan, is well known. However, the use of chlorine compounds can lead to the adverse side reactions of chlorine with methyl mercaptan (Jensen et al., 1966). Methyl mercaptan reacts with chlorine in water to form dimethyl disulfide. A single-stage packed-column scrubber operating with sodium hydroxide and sodium hypochlorite would remove the hydrogen sulfide in this example, but would also convert some of the methyl mercaptan to dimethyl disulfide as a by-product. The by-product dimethyl disulfide could then be stripped from the scrubbing liquid during the recirculation to the distribution spray at the top of the tower.

Therefore, the alternative to a single-stage scrubber configuration is a multiple-stage packed-column system with two stages. The first stage would utilize sodium hydroxide as the chemical to modify the water (increase pH) and enhance the absorption of hydrogen sulfide and methyl mercaptan. The sodium hydroxide would retain the sulfur compounds and effectively would increase the solubility of the compounds in the scrubber liquid. Gas-film mass-transfer coefficients would be used for the design calculation (i.e., the absorption being gas-film controlling). The second stage of the packed-column system would utilize sodium hypochlorite for the oxidation of the remaining hydrogen sulfide.

DESIGN EXAMPLE

The sizing of packed-column tower vessels is conducted following industry standards and guidelines for diameter selection and liquid flow rate selection. Gas-phase loading is typically selected in the range of 300 to 500 ft^3/min per square foot of tower cross section or 1300 to 2200 lb/h of gas per square foot. Liquid loading is typically selected in the range of 5 to 10 gal/min per square foot of tower cross section or 2500 to 5000 lb/ft^2 of liquid. The water and chemical requirements would be determined though calculations of the mass loading of the components.

An example application is an odorous exhaust from a wastewater treatment process. An exhaust of 10,000 ft^3/min from a headworks building might contain 5 ppm hydrogen sulfide. Using the vessel diameter guideline above, the (10,000 ft^3/min exhaust would require a vessel diameter of 6 ft with 28.3-ft^2 cross section [350 $ft^3/(min \cdot ft^2)$ or 1600 $lb/(h \cdot ft^2)$]. Using the scrubber liquid loading guideline above, the system would require a 150 gal/min recirculation rate [5.3 $gal/(min \cdot ft^2)$ or 2650 $lb/(h \cdot ft^2)$].

This application would be considered a gas-film controlling system, because the gas component, hydrogen sulfide, is dilute and moderately soluble in the liquid phase. The solubility of hydrogen sulfide in the liquid phase would be ensured and enhanced with the use of sodium hydroxide to increase the pH.

The height of the packed-column medium, for this gas-film controlling system, would be calculated using a simplified method. The method assumes that the gas components are dilute and that the components have "unlimited" solubility in the liquid phase. This is a reasonable assumption when sodium hydroxide is used alone or with a chemical, such as sodium hypochlorite, to oxidize the hydrogen sulfide. The number of overall gas-film transfer units N_{OG} is calculated using the hydrogen sulfide concentrations of 5-ppm inlet and a desired 0.01-ppm outlet:

$$N_{OG} = \ln\left(\frac{\text{inlet}}{\text{outlet}}\right) = \ln\frac{5}{0.01} = 6.2$$

The height of a gas-film transfer unit H_{OG} would be found in a packing supplier's technical data. In this example, the packing selected is a No. 1 sized saddle with an H_{OG} of 1.0. The height of the packed-column media is then calculated:

$$Z = N_{OG}H_{OG} = 6.2 \times 1.0 = 6.2 \text{ ft}$$

A designer of packed-column systems might routinely add a safety factor to the calculated bed depth, say, 25 percent, resulting in a selected packed-column media depth of 8 ft for this application.

The management of water chemistry addresses most of the operating and maintenance challenges of packed-column scrubbers. Proper water chemistry will enhance absorption of the gas components, prevent scaling on the packing medium, and minimize release of the gas components or release of reaction by-products from the recirculating scrubber liquid. Water chemistry can be monitored using pH, oxidation reduction potential (ORP), conductivity, or total dissolved solids and specific chemical concentrations (i.e., chlorine). A common parameter used to maintain water chemistry is the blowdown rate from the scrubber vessel reservoir. The blowdown rate is often controlled using a method of continuous freshwater makeup to the reservoir that causes a reservoir overflow or blowdown. A second method used in the maintenance of water chemistry is the batch discharge of the scrubber liquid reservoir. Either method of water maintenance can prevent scaling and will minimize precipitate accumulation and the release of volatile components from the liquid phase (VOCs) at the top of the packed column (i.e., liquid distribution).

This example application of a 10,000 ft^3/min packed-column scrubber would have seven categories of operating and maintenance costs:

1. Power consumption of the fan (blower), dependent on the collection ductwork and the pressure resistance (pressure drop) of the packing medium
2. Power consumption of the recirculation pump
3. Chemical costs
4. Water costs, including water treatment of blowdown
5. Cleaning costs associated with any descaling chemicals required
6. Fan and pump maintenance (i.e., belts, seals, bearings)
7. Instrumentation maintenance (pH probe replacement, calibration)

The design of the packed-column scrubber system would address each of these cost categories for proper planning and preventive maintenance.

The design of packed-column scrubbers for odor and VOC control is based heavily on empirical data, pilot tests, and application experience. Therefore, a designer must use caution when applying basic theory and design guidelines. Each application is unique with regard to gas-phase components, performance requirements, and economic factors. Successful odor and VOC control can be achieved with packed-column scrubbers that utilize absorption technology and engineering judgment.

REFERENCES

Danckwerts, P. V., *Ind. Eng. Chem.,* 43: 1460 (1951).

Higbie, R., *Trans. AIChE,* 31: 365 (1935).

Jensen, George A., Donald F. Adams, and Harry Stern, *JAPCA,* 16(5): 248 (May 1966).

Whitman, W. G., *Chem. Met. Eng.,* 24: 147 (1923).

SECTION 8.6.2

CHEMICAL SCRUBBING: MIST SCRUBBING

Harold J. Rafson
Environmental Engineering Consultant
Highland Park, Illinois

INTRODUCTION 8.111
GENERAL DESCRIPTION 8.111
GAS STREAMS TO BE TREATED 8.112
MAJOR DESIGN DECISIONS 8.112
CRITICAL DESIGN ELEMENTS OF FOG SCRUBBERS 8.113
Atomization 8.113
CONTACT TIME 8.115
MIXING OF DROPLET AND CONTAMINATED AIR 8.115
CHEMICAL FEED AND CONTROLS 8.116
CHARACTERIZING OF GAS STREAM TO BE TREATED 8.116
DETAILED SYSTEM DESCRIPTION 8.116
Contact Chambers 8.117
Atomization System 8.118
Chemical Storage System 8.118
Chemical Metering System 8.119
Chemical Controls, Including Sensors 8.119
Fans 8.121
Air Compressor, Receiver, Controls 8.121
Water Softener 8.121
Spare Parts 8.122
Start-up Services 8.122
Training 8.122
O&M Manuals 8.122
EASE OF OPERATION 8.123
PRACTICALITY OF OPERATION 8.123
COMPARISONS TO ALTERNATIVE TECHNOLOGIES 8.123
Criteria for Comparisons 8.123
COMPARISONS OF FOG SCRUBBING TO ALTERNATE TECHNOLOGIES 8.124
To Chemical Scrubbers of Other Designs 8.124
Critical Design Differences 8.125
Other Factors of Comparison 8.126
Comparison to Other Chemical Scrubbers 8.126
Comparison to Thermal Oxidizers 8.127
Comparison to Adsorption 8.127
Comparison to Biological Treatment 8.127
Application of Fog Scrubbing 8.127
Applications 8.128
PERFORMANCE RESULTS 8.128
HYBRID SYSTEMS 8.129
TESTING 8.129
CONCLUSION 8.130
REFERENCES 8.130

INTRODUCTION

Mist or fog scrubbing is a gas-liquid contacting system, which was invented in 1975 and subsequently has found a range of successful applications. However, it was marketed by a few small companies, which did not broaden the application of the technology or educate the wider engineering community. Therefore, the technology to this point, while having numerous advantages, has still not been widely applied.

GENERAL DESCRIPTION

Fog or mist scrubbing achieves gas-liquid contact by creating a contact between the gas to be treated and a fog of droplets, which are predominantly less than 10 μm in size.

This creates an enormous amount of contact surface *S*, compared to traditional scrubber designs. Further, the settling rate of these droplets is very low, and these systems are designed with a much greater contact time *T* than conventional scrubbers. Since the mist scrubber has large liquid surface area and long retention times, the scrubbing solution is almost entirely consumed in one pass through the scrubber. Thus reactivity *R* of the scrubbing solution is high, since it is fresh. Also, as a result of nonrecirculation, the solution is controllable. The efficiency of removal of any scrubbing system is determined by the mass-transfer equation

$$\text{Efficiency} = f \log STR$$

In all cases a fog scrubber has significantly higher numbers (for *S, T,* and *R*) than conventional scrubbers, and these are multiplied by each other; therefore, the efficiency of removal of a fog scrubber is high. Each of these factors is discussed in detail below.

There is a second mode of operation of a fog scrubber in which nonreactive, hydrophobic compounds are removed from the gas stream, not by diffusion, but by adsorption on the surface of the droplets. This use of fog scrubbing is discussed in Sec. 8.5.2.

GAS STREAMS TO BE TREATED

Fog scrubbing is best applied to compounds which are soluble in water and are reactable, especially where low outlet levels must be achieved, which is the case for odor control. Fog scrubbing is applicable to low concentrations and high volumes of air because the cost of treating high volumes of gas is less than with conventional scrubber technologies. Fog scrubbing is applicable to treat cases where neighbors are located near the process exhaust because of its high rate of removal of odors. Fog scrubbers are also applicable to high concentrations of inlet odors, and to complex mixtures, requiring multiple stages of reactants because each stage of reaction can be controlled accurately. Further, fog scrubbing treats variable inlets and can provide unattended operation, through the use of automatic controls.

The most significant factor in design is to characterize the gas stream to be treated, and to have a clear understanding of the goal of the design, the level to which the contaminants must be reduced. With this information, the major design decisions can be made.

MAJOR DESIGN DECISIONS

Because of the wide range of applications, numerous examples are given to illustrate design decisions.

Single Stage, Single Scrubbing Chemical. In a simple example, virtually all the odors are one compound, e.g., ammonia, and this is removed by an acid solution, in a single-stage design. This design will usually include automatic pH control of the rate of the acid being fed.

Single Stage, Multiple Chemicals. Slightly more complex odors are anaerobic sewerage odors which are high in hydrogen sulfide and other reduced organics. Here a single stage of reaction is adequate, using a mixture of two compatible chemicals—sodium hydroxide and sodium hypochlorite. This case will be discussed later regarding the amount of chemical usage.

Multiple Stages, Multiple Chemicals. An example of complex odors is the odors from composting sludge. The breakdown of organic material reveals hundreds of compounds including ammonia, organic nitrogen compounds, and organic sulfur compounds, which are generated in aerobic or partially aerobic conditions. Further, the sludge had adsorbed many volatile organic compounds (VOCs) from wastewater, and these are volatilized in the warm composting process. Clearly, not all these compounds can be treated with one chemical solution; therefore, there must be at least two stages—a first acidic stage to treat ammonia and a second stage to oxidize the odorous compounds. This has been adequate for some users; however, others have gone further, adding a second oxidant. In addition, numerous VOCs are removed in the first stage by the addition of a surfactant. Designs for this complex application illustrate the ability of a fog system to be fine-tuned to achieve adequately low levels of emissions from complex air contaminants.

Concentration-Reaction Time. To adequately treat 50,000 cfm of air is quite a different design challenge from treating 500 cfm, because of the limits of the sizes of the practical contact chambers. Luckily, high airflows are usually associated with low concentrations and low airflows with high concentrations; therefore, realistic systems can be designed.

System Configuration. Typical issues must be dealt with concerning configuration: Must equipment be located indoors? Wat are the height limits? Is there a cluttered site and limited space available? Since the design is based on contact time and airflow rate, this determines the volume required in the contacting chamber. The design problem is to fit this volume into the space available, considering the location of the source of odors and the location of the exhaust.

Removal Efficiency. The level to which odors must be reduced after treatment is a critical issue. That is the purpose for which the equipment is designed and constructed. It is a greater challenge to design a system where the nearest neighbor is at the property limit, versus 1 mi away. With distant neighbors, there is the luxury of additional dilution as the exhaust travels to the receptor. The acceptable outlet odor emission level (or VOC emission level) is critical to design and performance testing, and will recur in the following discussions.

In some cases, hybrid systems are considered, where a fog scrubber is followed by a further technology, carbon adsorption, biofilter, etc., and the issues of the coworking of the complementary systems must be considered.

CRITICAL DESIGN ELEMENTS OF FOG SCRUBBERS

There are certain critical elements in the design of a fog scrubber system, to ensure that the system will work at optimal levels.

Atomization

The atomization system must generate very fine droplets (the predominant number of droplets no larger than 10 μm), at the design liquid flow rate.

There are many different designs of atomizers that will generate fog droplets. High-frequency nebulizers are designed for very low-flow volumes (rated in milliliters per minute) and therefore are impractical for most industrial pollution control applications. At least 1 gal/min of liquid must be atomized to <10-μm droplets to treat 10,000 cfm. In many cases, and especially for higher-intensity odors, the droplet density (liquid flow rate per cfm treated) will be higher.

Specialized two-fluid nozzles (water and compressed air) have been developed to operate at these higher liquid flow rates. Two major designs have been used commercially: a sonic atomizer design and an opposed-jet design. In the sonic atomizer design, the spray is directed away from the nozzle, as it exits. In a typical design, the compressed air (the power for atomization) has the liquid introduced to it at 90° to the airflow. The compressed airflow is at sonic velocities, which creates an extremely turbulent atmosphere, and the liquid is shattered into many very small droplets. Because corrosive chemicals are being atomized, the nozzles are usually made of titanium. This nozzle design requires fine filtering of the liquid stream, prior to the nozzle, to avoid plugging.

The opposed-jet design is based on the idea of dividing a compressed airstream into two parts, adding liquid to each part, and then directing the airstreams toward each other. The opposing streams of compressed air and entrained droplets smash against each other, creating fine droplets. A major concern is that the nozzle configuration is much like the letter C, and when the two streams collide, the spray hits the back parts of the C, resulting in rapid corrosion and a shortened life of such nozzles, and a portion of the spray is eliminated.

There are several other designs from spray nozzle manufacturers, but none achieve the fine droplets of the above noted designs, at these high flow rates. The term *fog,* or *mist,* is widely misused and is often used to describe nozzles which produce much larger droplet size distributions.

As a practical measure, the fog generated must be a dense, opaque, dry fog. You should not be able to see your hand, in daylight, passed through the middle of the fog generated by the nozzle, and your hand should not come out wet or feel the impact of droplets. If you can see spray droplets or can see through the spray, or if your hand comes out dripping, then the droplet size is too large and the nozzle will not perform well in a fog scrubber.

The following example explains the importance of small droplets to the performance of a fog scrubber. Let us assume that we compare the surface generated in a fog scrubber with a conventional scrubber. The droplets in a fog scrubber are approximately 10 μm, and that in a packed column, or coarse spray scrubber, is about 1000 μm; the ratio of the diameters is 100. The surface generated is as the square of that ratio; or for the same amount of liquid, a fog scrubber will generate 10,000 times the amount of surface. The number of droplets is as the cube of that ratio; or, for the same amount of liquid, the fog scrubber will generate 1 million times the number of droplets.

Fog scrubbers do not use the same amount of liquid as conventional scrubbers, typically 100 times less. This still means that there is 100 times the amount of surface for the contaminant to diffuse into, or 10,000 times the number of droplets to collide with.

If one makes a comparison between a fog scrubber with an excellent atomization system that generates 10 μm, with a system fitted with less efficient nozzles that generate 50-μm droplets, then for the same amount of liquid, the fog scrubber will have 25 times the amount of surface and 125 times the number of droplets.

There are two ways that one can envision a contaminant gas molecule contacting the liquid reactant: passage through the droplet surface or collision with a droplet.

It can be seen that performance is affected by design of atomizers and by the maintenance of the nozzle to ensure no impairment of good atomization. To simplify maintenance, good filters must be provided and soft water supplied to decrease plugging of the nozzle. The nozzle should be removable from the outside of the chamber without interruption of the system. Disassembly, cleaning, and replacement should be easy and simple, and then they will be performed regularly. In many cases, this is a short 10-min cleaning required once each month.

Another issue concerning atomization nozzles is the power used for atomization. There must be some power consumption simply to overcome the surface tension as droplets are generated. However, this theoretical amount is lower than the actual usage rate of nozzles. It is difficult to make a power usage comparison between nozzles since it should be made when approximately the same droplet distribution is being atomized.

To summarize, the proper design of an atomization system is critical in fog scrubbing design.

CONTACT TIME

Contact time in a fog scrubber is much the same as the number of transfer units in a conventional design. The longer the contaminated gas and liquid droplets are in contact, the more diffusion, or collisions of contaminant and droplets, will occur and the more effective the removal. The question is, How much contact time is required? This depends on the inlet concentration, the design outlet concentration, and the ease of diffusion and reaction of the contaminant. Let us take a *hypothetical* example, that the inlet concentration of a compound (expressed in dilutions to threshold *D/T*) is 7500 *D/T* and that it must be reduced to an outlet *D/T* of 75. Let us assume that there is a 50 percent decrease in 1 s of contact. In the first second, 7500 is reduced to 3750; in the next, 1870; etc.; 7 s is needed to get below 75. If the inlet had been 75,000, it would have taken 10 s. If the outlet were acceptable at 150, then from 7500 *D/T* it would have taken 6 s.

It is not quite so simple. This example assumes that a decrease in concentration is directly related (with a slope of 1) to the decrease in odor detectability. But that is not generally true. In many cases the relationship between odor intensity and concentration has a butanol intensity slope of 0.2 to 0.6. Let us assume that it has a slope of 0.2 (the odor does not dissipate easily). Then in our example,

7,500→75	will take	11 s
75,000→75	will take	16 s
7,500→150	will take	9 s

This means that to design a chamber for a compound that does not dissipate easily, that is pervasive, will require a chamber about 60 percent larger.

Design should be based on as much knowledge as possible of the compounds to be treated and their concentrations, and knowhow of what has proved to be successful, to arrive at proper design sizing and operating conditions.

MIXING OF DROPLET AND CONTAMINATED AIR

This factor is part of the equipment design and is very important. For the contaminant to be removed from the gas stream by the droplets, there must be good mixing between gas and droplets. Even though there are an enormous number of droplets generated, without good mixing, some part of the contaminated air may pass through with little treatment. Different designs are used to obtain good mixing, and design factors include the number and placement of nozzles, airflow velocity and trajectory, and baffles.

CHEMICAL FEED AND CONTROLS

As with any liquid-gas contacting scrubber, it is critical to have the right reactants, or else little or nothing will happen. Once the correct reactants are selected, it is important that the chemical metering system and the controls of the feed of those chemicals be properly designed.

CHARACTERIZING OF GAS STREAM TO BE TREATED

In other chapters of this book, you will find explanations of sampling and test methods to assist in characterizing inlet odors or contaminants. In various sections on control technologies the need to characterize the inlet contaminants is often emphasized. This is particularly true for chemical reaction systems, and for odor control systems that aim to achieve low odor residuals in the exhaust. An equipment manufacturer that is asked to design a system for excellent performance and to provide performance warranties must be given good information about the exhaust to be treated.

Often an engineer does not have this information, and because it may be difficult or require extra effort to obtain, the engineer will make educated guesses about the contaminants, concentrations, variability, required efficiencies of removal, and other matters such as temperatures, relative humidity, dispersion conditions, plant process, and operations. Although the engineer aims to write a clear description of process performance, he or she must realize the difference between assumptions and actual conditions, and the effect that imprecise assumptions have on the equipment designer. In wastewater odor control applications, the engineer frequently will specify H_2S removal, rather than odor removal, because H_2S is one of the major odor contaminants at the front end of the wastewater plant (prior to oxidation) and is easy to test, and a lot has been written about H_2S. However, removing all the H_2S will remove some, but not all, of the odors. The engineer may not accomplish the primary objective of odor removal because he or she defined the problem inadequately. Providing imprecise information, and requiring performance testing based on these improper assumptions, has resulted in a great deal of inadequate equipment currently in use. If the initial and final conditions are better defined, the equipment manufacturers will be able to design better equipment and can be held to more meaningful standards.

The information required to characterize emissions and provide information to designers includes components, type and concentration, temperature, particulate content, variability of contaminants, airflow rates, space constraints, and control objectives.

DETAILED SYSTEM DESCRIPTION

The odor control system removes odors by spraying very fine mist droplets of controlled diluted chemical solution into an odorous stream that passes through a hollow, cylindrical, fiberglass reaction chamber. This is illustrated in a general flow diagram, Fig. 8.6.4. A spray nozzle is situated inside the chamber so that the mist of fine droplets

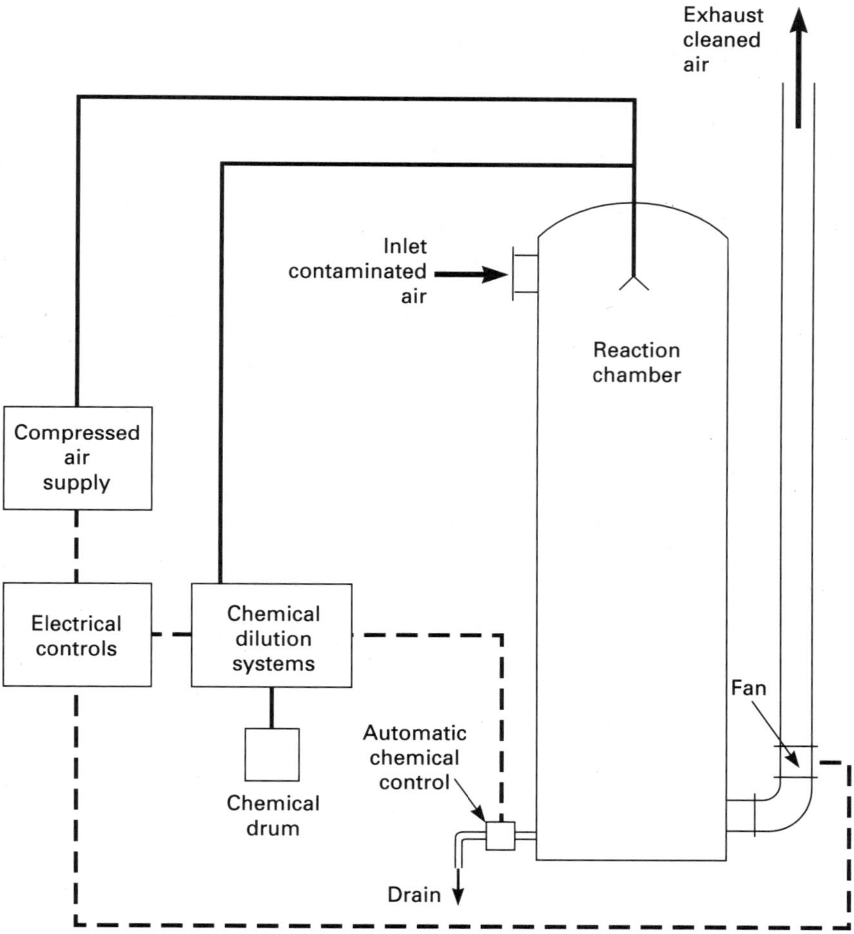

FIGURE 8.6.4 Odor control system flow diagram.

mixes with the incoming odorous air. Mist and odors flow cocurrently in each reaction chamber. The odorous air is treated with the mist of different chemical solutions, as is required to treat the specific contaminations and concentration contained within the airstream. Cleaned air is exhausted through the exhaust stack to the atmosphere. The fan may be located either following or before the contact chamber. Collected spent scrubbing solution in the chamber bottom flows out of the chamber drain. The spent solutions are monitored for residual pH and oxidation redirection potential (ORP). The pH, in the drain, controls the flow of acid or caustic required to maintain the set point pH. The bleach or other oxidant is controlled to maintain a set point ORP.

Contact Chambers

The contact chambers are usually fabricated of fiberglass-reinforced plastic, though chambers have been constructed of stainless steel and polyvinyl chloride (PVC). The resins selected for fabrication must be resistant to the chemicals used since they are highly reactive. Since systems are expected to last 20 years or more, it is a good idea to be conservative in the construction. It is possible to design chambers where the inner layer (a corrosion barrier) is thin and is only resistant to the oxidizing chemicals, and the major body of the wall is constructed of less costly resins, which will deteriorate with oxidants. It is more conservative to construct the entire wall of the premium resin, so that even if there is a breakthrough of the corrosion barrier, the wall will not deteriorate.

The wall strength is designed to withstand the maximum wind and seismic loads. This is usually for 100 to 120 mi/h winds, with 150 mi/h as the basis in hurricane areas. Similarly the seismic zone is considered in the design.

The exterior of the fiberglass should be treated with an opaque ultraviolet-resistant resin coating such as gel coat, although ultraviolet protection agents may also be added to the resins. When in use after many years, and loose fiberglass fibers become apparent on the surface, due to decomposition of some surface resins in the sunlight, an extra coat of gel coat should be applied.

Designs with tie-down bolts to be set in concrete are significantly less flexible and more difficult to install than bolts drilled in place and set in epoxy.

Where a chamber is on the pressure side of the fan, there is little risk of collapsing. Where the chamber is on the suction side of the fan, care should be exercised that the chamber wall is adequately strong for the maximum suction pressure conditions. Traffic stanchions should be installed where the chambers are near the road, or passageway used by trucks or forklifts.

Ladders and platforms should be provided of weather-resistant materials, and for taller chambers they should be self-supporting. Access platforms should be located near nozzle ports for convenient use by operating and maintenance personnel. The sidewall manway should be designed for easy access to the chamber and for easy removal of the manway cover.

The drain should be designed to *fully* drain the bottom of the chamber. The less liquid retained in the bottom, the better, since that affects the response time of the control system. It is desirable to use flanged or well-made clamp duct connections to avoid dripping from leaky connections. There is usually a flexible duct connection between the fan and the chamber to decrease vibration transmission to the chamber.

The configuration of the chamber depends upon the locations of inlet duct and exhaust stack, the height of buildings or ceiling limits or other height restrictions, and ease of access. A mist system can be designed to place those elements requiring maintenance, at ground level. Since a mist scrubber can be designed for vertical flow downward, or upward, or horizontal flow, there are many design configurations possible.

Atomization System

The properly designed atomizer, which generates 10-μm or less droplets, should be made of noncorrosive or nonerosive materials and located for easy access for maintenance. It should also be located to give good mixing between the inlet contaminated airflow and the droplets.

Chemical Storage System

The chemical storage system should be fabricated of noncorrosive materials and for the calculated storage requirements. It should be located in secondary containment enclosures. Two chemicals that react with each other should be in separate containment enclosures. The chemical storage tanks usually will have level controls and volume markers. Outdoor sodium hydroxide tanks should be heated in all northern areas (50 percent NaOH crystallizes at 55°F), or sodium hydroxide can be purchased at 25 percent which crystalizes at much lower temperatures. Sodium hypochlorite loses its oxidizing potential with heat, sunlight, and time. Sodium hypochlorite should not be stored for more than 1 month and should be stored indoors, if possible. All storage tanks should have vents and overflows. Where hookups are provided for fill from a chemical supply truck, an overflow basin should be provided as well as a high-level alarm to warn of overflowing, Special care must be taken with tanks for hydrogen peroxide.

Chemical Metering System

Every part of the chemical metering system should be corrosion-proof. The parts should be located for easy access and replacement. Metering pumps should be properly sized, and if greatly oversized, this will affect the responsiveness of control systems as well as the effectiveness of odor removal. A drawing of a typical metering panel is seen in Fig. 8.6.5 and illustrates the major components. The filter element should be a 50-μm filter to avoid nozzle plugging due to salts from the chemicals. The electrical panel design in some systems provides a timed shutdown cycle, so that the chemical shuts down first, the water continues to run and flushes the line, and then water, compressed air, and fan shut down. Not only does this avoid corrosion during shutdown periods, but also when maintenance is to be done, e.g., replacement of the filter, there is only water present in the lines, which is much safer for the workers.

Chemical Feed. It is desirable to have a chemical metering pump capacity not more than about twice the expected chemical usage. This should result in effective chemical concentrations with less wastage. However, it is not uncommon to find situations where the range of chemical containment concentrations exceeds 2 times the average. One approach to solving this is to have multiple pumps, which activate when the first pump cannot keep up with the feed requirements. This is a better design solution than one oversized pump, but requires a knowledge of the process emission variation.

After the chemical is fed into the chemical feed panel header by the metering pump, it is necessary to have enough volume in the line that mixing occurs, and that the solution fed to the nozzle will not fluctuate with every pulse of the metering pump.

If a chemical feed is automatically controlled, and the metering pumps are greatly oversized, then there will be slugging of chemical, overdosing, which then "tells" the automatic control to cut back the feed, and then the chemical will be underfed. Reaction will be inadequately achieved, odors will pass through. The controls will "hunt," and part of the time odor removal may have inadequate chemical treatment and part of the time chemical may have been wasted.

Chemical Controls, Including Sensors

The major cost of operation is usually chemical usage, and proper chemical feed is a major factor affecting odor removal effectiveness. Therefore, it is critical that the chemical control system be properly designed, and only high-quality controls and sensors be used. It is also important that systems be readily calibrated.

The purpose of the chemical feed control is to have enough chemical to complete reaction, but not to have much excess. That would be a waste of chemical and would

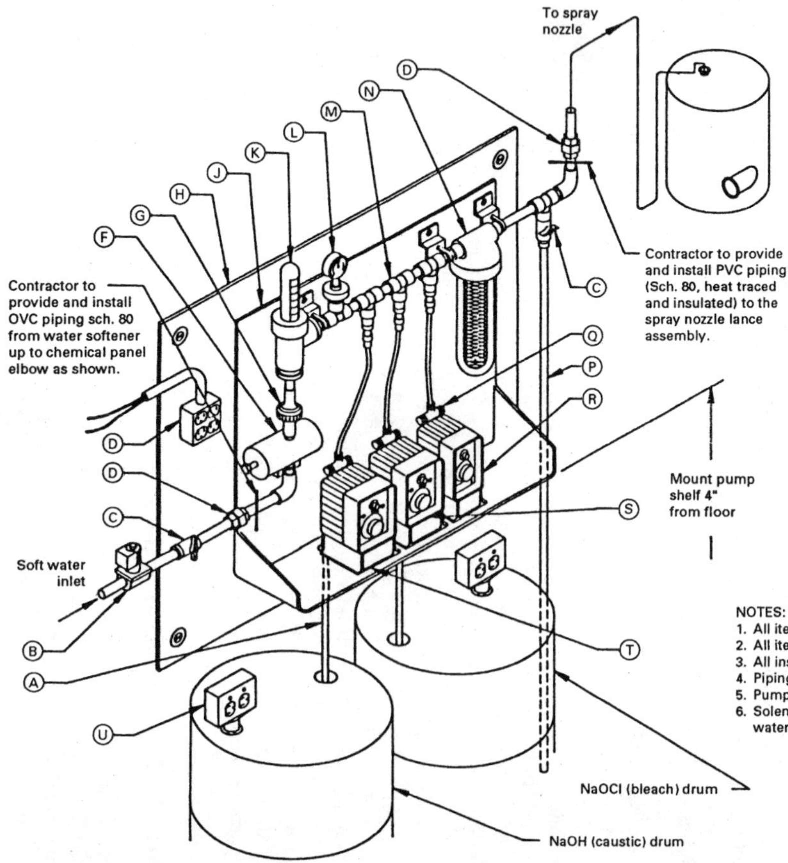

Legend:

- (A) * ø 1/2" polyethylene tubing
- (B) * Solenoid valve
- (C) 1/2" PVC ball valve
- (D) 1/2" PVC union
- (E) Four gang electrical receptacle (by electrical contractor)
- (F) * Water pressure regulator
- (G) * Diaphragm check velve
- (H) * Fiberglass back panel 40"H x 48"W
- (J) * Fiberglass mounting panel and pump shelf
- (K) * Water flowmeter – gpm
- (L) * Pressure gauge
- (M) * Injection check velve
- (N) * Liquid filter – 50 μm
- (P) * PVC drain line
- (Q) * Four function valve
- (R) * manual acid pump – (use for nozzle wash only)
- (S) * Sodium hypochlorite (bleach) pump – manual
- (T) * Sodium hydroxide (caustic) pump – automatic
- (U) * Drum level switch

NOTES:

1. All items marked with * are supplied.
2. All items not marked with * are supplied by installing contractor.
3. All installation, assembly, wiring and piping is by installing contractor.
4. Piping to and from chemical panel to be 1/2" PVC schedule 80 pipe.
5. Pumps are shipped separate to be installed on pump shelf as shown.
6. Solenoid valve shipped separate to be installed in 1/2" PVC line between water softener unit and chemical panel.

FIGURE 8.6.5 Typical chemical metering panel.

raise operating costs, and might generate chemical odors. Let us take, as an example, treating H_2S with sodium hydroxide (controlled by pH measurement in the effluent) and sodium hypochlorite (controlled by ORP in the effluent). The control points are usually a small amount of residual hypochlorite and a slightly basic pH. It should not be strongly basic, or highly oxidative, or this will waste chemical. Recognize that when H_2S oxidizes, it forms an acid, and the pH falls. If it falls too low, it will deter the oxidative reaction and also the solubility of H_2S from the gas stream into the droplet. Therefore, the amount of caustic needed is to be controlled to maintain the droplet slightly on the basic side. This operational mode is a very significant difference between fog scrubbing and recirculating scrubbers. The impacts of these controls, and set point decisions, are such that far less chemical can be used to accomplish effective removals.

The control equipment that is selected should be of high quality, and training should be provided to operators concerning maintenance and operation of controls. Recent equipment designs have included programmable controllers which have simplified the operation and calibration of controls.

The signals provided by the control sensors, and the control set points, are combined with control systems that allow adjustment of the timing and rate of response. Along with the sizing of the metering pumps this allows for a close tracking of the effluent conditions, and control to set point objectives.

Fans

It is important that fans be made of corrosion-resistant materials. This is a necessity when the fan is located on the exhaust of the chamber, and it is often desirable when the fan is located on an inlet. The correct resins should be used in fiberglass fans; driveshafts should be stainless steel or better; bearings should be removed from the gas stream, or protected. All fans should have drains, and for larger fans, access hatches. Fans should be mounted with vibration isolators from other equipment and flexible joints between fan and ducting installed.

It is desirable to oversize fans and run them at lower speeds. This usually saves power and is less noisy. Fan motors should be resistant to the chemical atmosphere and should be protected from the weather by covers.

Air Compressor, Receiver, Controls

Since odor control systems run continuously and use compressed air for atomization, compressors must run continuously. Rotary screw compressors are recommended, as they are designed for full-time use with heavy loading, unlike some other types of compressors. Rotary screw compressors are also quite small and require very low maintenance.

Compressed air receivers are necessary to allow the compressor longer idle time and to remove the condensed water, and the receivers should preferably be vertical and include an automatic drain. Materials of construction of air controls should be selected based upon the amount of contaminated gas in the area.

Water Softener

In many applications, caustic and bleach are added to the solution to treat the contaminated gases. This usually precipitates water hardness from the water and forms scale on the nozzles, equipment, and whatever else it contacts. In fog systems which use low volumes of water, it is preferable to remove the water hardness and avoid nozzle plugging and scale buildup. This is far less expensive than the maintenance required to remove scale. It is not desirable to greatly oversize the water softener, as this may cause channeling in the ion-exchange resin, and allow residual salts to pass through. It is desirable to use a dual system so that softened water supply is uninterrupted. Usually regeneration is activated by a measure of liquid flow rate, or a timer.

Spare Parts

Since odor control systems and VOC control systems most often operate continuously and must meet regulatory requirements, there is little tolerance for long equipment breakdowns. Therefore, the maintenance department should stock parts for all equipment with moving parts and should have available the manufacturer's recommended spare parts. In addition, even for equipment that does not move, critical parts such as filters, or nozzles, should have a replacement on hand. A system supplier can usually provide a recommended spare-parts list for a total system.

Start-up Services

Start-up services are often critical, to catch errors of installation and to perform proper adjustment and testing. It is very important to have experienced personnel do this, whether it is from the equipment supplier or an independent consultant. There is a great deal of knowhow needed to adjust equipment properly. A plant installing new equipment may not have that experience and must rely on others. In complex odor situations, more than the knowhow of a competent mechanic is required, as issues may involve complex chemical reactions and controls. Care must be taken since some equipment can be damaged if improperly started up. This is a very important part of the total project.

Training

Again, an experienced individual must be used for training, whether from the equipment supplier or an independent consultant. Emission control equipment is only a tool to remove gaseous contaminants, and that tool must be operated and maintained by skilled workers.

Any training program should include training of both operations and maintenance personnel, and comprise both classroom and hands-on training. The training should be comprehensive, providing a detailed explanation of every part of the system—its adjustment, maintenance, and safety. Operators must understand what they are doing if they are to respond knowledgeably to changed conditions in plant operations.

O&M Manuals

A system supplier should provide an O&M manual which includes detailed equipment description and maintenance information on every component of the system, as well as complete information on the operation, adjustment, and testing of the system.

EASE OF OPERATION

Certain features must be attended to in order to have a system which is easy to operate and maintain. The chemical feed system should be simplified, with all parts accessible and easily replaceable. All materials of construction should be selected to withstand the chemicals and temperatures encountered. The nozzle should be designed for ease of cleaning and long life. All parts requiring maintenance within the reaction chamber should be removable from the outside, and access ports conveniently located. Most mist systems have generally been trouble-free operating systems which require a low and very reasonable amount of care.

PRACTICALITY OF OPERATION

The fog scrubber is understandable to the operator. It reacts very much as a titration works—add a measured amount of chemical, and a measured amount of contaminant is reacted. Maintain the control set points, keep a clean nozzle for good atomization, and the system works well. It is a continuous system. It is easy to train operators so that they understand the process and can knowledgeably adjust for changing conditions. All component parts are simple and accessible and require very limited maintenance. Equipment can be adjusted easily and quickly. Therefore, the system is very practical to operate to continuously remove odors.

COMPARISONS TO ALTERNATIVE TECHNOLOGIES

Criteria for Comparisons

The following are the major issues for comparisons to other technologies:

Efficiency of Removal of Contaminant from the Gas Stream. Removal efficiency is the reason why the gaseous emission control equipment is purchased. If it does not do this adequately, all other issues are secondary. Further, in making comparisons consideration should be given to one system achieving higher levels of removal than others.

Life-Cycle Costs—Initial Capital, Continuing Operating Costs. It is easy to compare initial purchase prices. Capital costs seldom reflect the true cost of a system. One should not forget to factor in costs for equipment with shorter life spans or that require replacement, or the cost of added equipment because redundant equipment is necessary due to long shutdown periods. It is not easy to arrive at realistic operating costs for comparison. The information provided by the equipment suppliers often reflects the best-case conditions, and should be carefully evaluated. Ongoing operating costs include operating labor, maintenance labor, power, chemicals, replacement of parts, replacement of media, and regeneration of components. There is a more detailed discussion of life-cycle costing in Chap. 9.

Continuity of Operation at Design Removal Rates. Odor control systems are installed to eliminate odor complaints. Critical systems that are shut down for even short periods will receive renewed complaints. Systems that require lengthy maintenance or major changes to continue to function (replacement of media for adsorbers, dry chemical, biofilters) all must be evaluated. Similarly, VOC removal systems are designed to remove a weight of contaminants, and batch systems with limited capacities will require shutdown.

Practicality of Equipment and Ease of Operation and Maintenance. Some systems appear to work well in the textbooks, laboratories, and manufacturer's literature, but fail in full use for a variety of reasons. They are often too hard to maintain and require frequent replacement of media or components; have very high operating costs; have few meaningful controls or monitoring; are too difficult to operate; have system components that fail.

Ability to Treat Variable Sources. A great many emission sources are highly variable, varying with the hour, the day, and the season. Systems must be able to treat peak concentrations adequately. The way in which a system accomplishes this should not compromise the operation at other, lower levels of contamination. Critical systems are meant to remove contaminants. A system that, at times, is a collector and, at other times, an emitter may fail the objectives of odor and VOC control.

Flexibility for Change. Control systems are designed to operate for 15 to 20 years. Frequently a process and emissions will change considerably within that period. It is important that any system can, without too much difficulty or cost, be modified to treat the new conditions.

If system criteria have been inadequately stated and system designs are limited, this inflexibility can pose a problem, possibly requiring expensive modification.

Disposal of Waste. All systems remove compounds and discharge these compounds or their reaction products as some form of gas, liquid, or solid. Those disposal mechanisms must be acceptable and environmentally satisfactory, comply with regulations, and not contribute new and costly problems. Therefore, the gas emitted by a thermal oxidizer, the liquid effluent from a scrubber, and the solid disposal of an adsorber must be considered as part of the system evaluation.

COMPARISONS OF FOG SCRUBBING TO ALTERNATE TECHNOLOGIES

To Chemical Scrubbers of Other Designs

The most direct comparison of most fog scrubbing technology to other gas emission control systems is to other chemical scrubbers, which take all different approaches to gas-liquid contacting. It is desirable to clarify functional differences between fog scrubbers and other scrubbers. There are many different designs of gas-liquid contactors, and we will only consider comparison to one other major design. However, similar reasoning will apply to other mechanical designs.

Packed Columns. Packed columns are the most common design for chemical scrubbers, and they have been in use since the beginnings of the chemical industry in the 19th century. The liquid contacting surface is created by a downward flow over the packing, which creates thick films and large drops. Liquid flow rates are high to increase surface area. In most cases, the scrubbing solution is recirculated, and due to the large reservoir of chemical the response to gas emission changes is slow; but for consistent emissions this is satisfactory. Pressure drop is moderate. Operation is simple. The system is not easily variable, so chemical concentrations are maintained at high levels. (Also refer to Sec. 8.6.1.)

The comparison between a packed column and a fog scrubber highlights the radical differences in the functioning of these two technologies, even though these systems both fall in the category of chemical scrubber. In the following we discuss the principles of these differences. This discussion will not be repeated except by reference, for other recirculating scrubber designs or for other systems which have other means of generating the liquid surface.

Critical Design Differences

Chemistry and System Design. There are significant differences in the way that the same chemicals react with the same contaminant gases, in a packed column scrubber and a fog scrubber. Let us take an example using H_2S, and the reactant chemicals are sodium hypochlorite and sodium hydroxide.

In a packed column, the flow rate of liquid is high, the concentrations of the scrubbing chemicals are almost always fairly high and are slow to change as the solution is recirculated, and a small portion is discharged during each recirculation.

In a fog scrubber, liquid quantities are much less, the chemical concentrations are much lower and are varied, and there is no recirculation—the system is "once-through."

Let us visualize the fate of the H_2S molecule in those two chemical scrubber types. In a packed column (we are assuming the recirculating chemical is at equilibrium), the H_2S encounters far less surface, for a far shorter period than in a fog scrubber. The liquid, in a packed column, contains a higher concentration of chemical and by-products of reaction due to recirculation, and usually salts from the use of hard water. In a fog scrubber, the solution has a lower concentration of controlled fresh chemical, and no salts from the water.

In both cases, there is some reaction of this H_2S which encourages further diffusion of H_2S. But what happens after solution is very different in the two cases. Once in the liquid in a packed column, the H_2S is in contact with excess hypochlorite and sodium hydroxide, for a long period, due to recirculation. If the recirculation rate is 300 gal/min and the blowdown is 3 gal/min, one could estimate that the same chemical is in the packed column for 100 min. Therefore the H_2S reaction will go to full oxidation, and one could expect the formation of sulfates, using 4 mol of oxidant for every mole of H_2S. This is what is claimed by packed column companies. Actually, usage must be higher to account for fresh chemical lost in the blowdown, which is a necessity to remove accumulating salts.

This is not what occurs in the fog droplet. The only amount of chemical in the droplet is the amount that has been fed into it. That amount is controlled by monitoring the residual chemical (by some measure of oxidant ORP, or pH for caustic). When the H_2S enters the droplet, the S^{2-} oxidizes to S^0. If there is no more oxidant present, it cannot oxidize further. Therefore, one would expect the reaction to call for 1 mol of oxidant for 1 mol of H_2S. In practice, in wastewater plants, over years of use and many repeated tests, we have found chemical usage of approximately 1.4 mol/mol H_2S because the additional oxidant is being used to treat other compounds, and there is a need to retain some residual. The important fact is that the H_2S is reduced in the gas stream to 0, and the chemical usage is not 4 mol, but something a little over 1 mol of oxidant per mole of H_2S.

As explained in a previous example, a fog scrubber typically has greater surface area. Diffusion is also effected by gas-liquid contact time. Packed columns, such as used in the wastewater industry, seldom are designed for more than 1-s contact time, not considering the volume of the downflowing liquid. When the displacement of the liquid is considered, the contact time is closer to 0.6 s. Depending on the inlet concentration, the fog scrubber is designed anywhere from 5 s to 120 s or more of contact time. Let us say that there were moderate concentrations of odors and that the fog scrubber was designed for 15-s contact time. It is obvious that there are two different design concepts at work, with the fog scrubber designing for significantly greater contact.

The above discussion is meant to illustrate that chemical scrubbers using the same chemicals can have very different chemical uses, and different efficiencies of treatment of the gas stream.

There are two additional issues, not discussed in the above simplified example: the *additional* wastage in the blowdown and possible recontamination from the gas stream.

Wastage. The recirculating solution in a packed column is a relatively steady concentration, and it is necessary to discharge some blowdown to eliminate salts. The amount of the discharge is often a decision made by each plant engineer after considering cost and odors which build up in the recirculating solution (some may be exhausted). Whatever the wastage is, some unused chemical is lost, and this adds to the chemicals which are consumed in the scrubber.

Recirculation. The recirculating system results in revolatilization of some compounds. Some may be odorous, and these decrease the overall effectiveness of odor removal by recirculating scrubbers. Some may be VOCs, and work by the Water Environment Research Foundation (WERF) has shown packed columns to be ineffective for VOC removal.

Other Factors of Comparison

Packed columns have a relatively lower capital cost. There are many fabricators and equipment suppliers. This is well known and familiar to many engineers. They are simple to operate. Packed columns can be heavy, requiring strong foundations. Pressure drop is higher, adding to continual power costs. Running higher concentrations of chemical often requires greater maintenance of recirculating pumps. For certain applications in which two reactors are required with noncompatible chemicals, two separate packed towers are needed. Generally there is low maintenance when running, and more maintenance at shutdown periods.

Fog scrubbers consume lots of space. Being designed for more contact time, the systems are larger, and this may be a limitation for some sites. They are generally somewhat more expensive than packed columns in initial cost, although usually they have lower life-cycle costs. Operators and engineers are not as familiar with them, and there is more ongoing maintenance to keep performance high. There are only a few experienced suppliers.

Comparison to Other Chemical Scrubbers

The major other scrubber designs are conventional spray scrubbers, whether vertical or horizontal. All these designs recirculate chemical and use flow rates equal to or exceeding the flow rates of packed columns. Therefore, much of the discussion of packed column performance is relevant. However, typically these scrubbers have even shorter contact times than packed columns due to the falling rate of large droplets or other design features.

Comparison to Thermal Oxidizers

Thermal oxidation is rarely used for odor control, except where the odors are very concentrated and in a small stream, or where a boiler can be used, because of the high costs of energy to incinerate these gases.

VOCs are an excellent application of thermal oxidizers, except where they are dilute and in high volumes of air. Some concentrating mechanism, however, can extend the range of applicability. Fog scrubbers used in this way are described in Sec. 8.5.2.

Comparison to Adsorption

Carbon adsorption is a batch system. Odors are generated continuously. As long as the odor level is low, the carbon lasts for a long time before breaking through. For large quantities of odor, carbon is impractical when compared to the much lower cost of chemical scrubbing. With fog scrubbers and carbon as a polisher, there is a complementary process in which the low-cost scrubbing process greatly extends the life of the carbon, and the carbon decreases emission levels, resulting in a useful hybrid combination. Carbon adsorption is discussed further in Sec. 8.5.1.

For VOCs, carbon adsorption has considerable application, particularly where the recovered VOC has value.

Comparison to Biological Treatment

Biological treatment has gained increasing acceptance only in recent years, and more field experience and knowledge of costs, operating benefits, and problems are required. As with any newer field, rapid changes and improvements are made as field experience accumulates. Further discussion of biofilters can be found in Sec. 8.8.1.

Application of Fog Scrubbing

Fog scrubbing has a wide range of possible applications, far wider than has been used to date. Current applications are discussed based upon design characteristics, types of odors, by industry and field experience.

By Design. Let us survey the various range of applications encountered. This table is a broad generalization:

Contaminant concentrations	Airflow	Fog scrubbing applicability
Low	Low	Poor—too costly.
High	Low	Good—because of low outlets achieved, may be costly.
Low	Medium	Good, but many other scrubber designs also may work.
High	Medium	Best—low cost and better performance.
Low	High	Best—low cost and better performance.
High	High	Good—low cost and better performance, and alternates and hybrids could be considered.
For odors that have complexity		Best—may have advantages over alternate designs.
For odors that have variability		Good.

By Types of Contaminants to Be Treated

Reactable	Good
Nonreactable	Poor
Water-soluble	Good
Non-water-soluble	Variable applicability (see Sec. 8.5.2)

It should be understood that when *reactable* is listed, this includes a wide range of compounds. A very reactable contaminant, such as acid, is easily treated. More slowly reactable contaminants, such as mercaptans or dimethyl sulfide, are treated with different levels of effectiveness and require care in system design and control. More slowly reactable compounds such as esters require added design considerations.

A single-component odorous mixture is easiest to treat for any scrubber design. A complex mixture requires better contact, time, control, and possibly multiple stages.

Applications

In all cases, these are successful applications; the number of installations listed is for all suppliers of mist scrubbers and is an approximate number.

	Volume	Conc.	Complex odors
Animal rendering (35)			
Animal by-products, beef, pork, chicken, fish, noncondensable gases	Low	Very high	Very complex
Room ventilation	High	Low	Complex
Blood drier	Medium/high	High	Very complex
Wastewater (>200)			
Pumping station and headworks	Low/high	Low/high	Very complex
Aerobic digestion	Medium/high	Low	Complex
Anaerobic digestion	Low	High	Complex
Wastewater—Solids Handling (>100)			
Dewatering	High	Low	Complex
Sludge storage	Low	High	Complex
Drying	Medium	High	Complex
Lime stabilization	Low	High	Complex
Composting	Medium to high	High	Very complex
Automotive (1)			
Brake shoe ovens	High	High	Complex
Paint (1)			
Reactors	Low	High	Complex
Flavor houses (5)			
Driers and mixers	Medium	High	Very complex

PERFORMANCE RESULTS

Virtually all the installations of mist scrubbers have included performance testing. Most of the performance testing was based on odor panel tests; some was based on H_2S tests.

In many cases H_2S performance criteria were defined as 99 percent removal, and odor tests at 75 or 100 ED_{50}. Several papers cited in References have reported on performance results written by independent consultants, as well as equipment supplier technical personnel, and some included comparisons to alternate technologies.

HYBRID SYSTEMS

A fog scrubber works well in conjunction with other technologies, where that is determined to be advantageous.

1. *With biofilters.* Preceding a biofilter, the fog scrubber humidifies the air and can also remove an undesirable component such as ammonia or particulates, which will affect the biofilter operation.
2. *With carbon.* It precedes a carbon system, in which carbon, or some other adsorbent, preferentially may remove a compound and polish the exhaust from the fog scrubber. At the same time, the scrubber has removed the bulk of the contaminants, which extends the useful life of the carbon system.
3. *With thermal oxidation.* The scrubber is used to remove components which will be highly corrosive when heated in a fume incinerator, such as a concentrator, particularly for VOCs as described in Sec. 8.5.2.
4. *With dilution and dispersion.* Odors can be reduced to very low levels in a fog scrubber; however, there are situations where it is necessary to disperse or dilute the odors additionally.
5. *With conventional scrubbers.* Sometimes, in retrofit situations, existing scrubbers may not achieve adequate removals. This can be followed by a fog scrubber to achieve higher levels of removal of the residual contaminants.

TESTING

Fog scrubbing is a continuous process. Equilibrium is established almost immediately, and exhaust samples that are drawn are representative of the treatment of the gas stream as it occurs.

This is not true of recirculating scrubbers which have a sump of chemicals, which take time to come to equilibrium and can volatilize from the solution as well as absorb. To obtain an accurate test result, one must test after the recirculating solution has come to equilibrium. Start-up conditions do not reflect continuous operation.

Carbon systems change continuously as the batch system fills up, and where revolatilization is common in treating a variable inlet. Testing of carbon systems should be done somewhere near the carbon's half-life. Biofilters have an enormous bed which adsorbs and releases. Biofilters should also be tested around the system's half-life.

Regenerative incineration can release contamination between the heat recovery and oxidation stages. Flares and many thermal oxidizers work as continuous flow devices, and can be tested once they come up to temperature.

These comments were only made to emphasize the problem of comparing different technologies. Many require time to come to stable operations, even assuming a uniform inlet. With a variable inlet, a stable outlet is an elusive goal. But quick-response, continuous flow systems, like fog scrubbers, give realistic measurements all the time. This must be borne in mind as tests are performed in evaluating and comparing systems.

CONCLUSION

A fog scrubber is a relatively new technology which is gaining acceptance. Its advantages over conventional technologies are higher efficiencies of gaseous treatment, lower operating costs, and ease of operation.

REFERENCES

Card, T.: Volatile Organic Compound Removal in Packed Towers and Atomized Mist Odor Scrubbing Systems, presented at Water Pollution Control Federation Conference, San Francisco, Oct. 1989.

McRedmond, E.: New Odor Control System Saves Operating Dollars, *Meat Processing,* 1977.

Patterson, R., R. Jain, and S. Robinson: Odor Controls for Sewage Treatment Facilities, Air Pollution Control Association, presentation, June 1984.

Rafson, H.: Mist Scrubbing Technology, Recent Developments and Current Practices in Odor Regulations, Control and Technology, *Transactions Air & Waste Management Association,* Oct. 1989.

———: Latest Developments in Removal of Applications of Mist Scrubbing Technology to VOC Removal, *Proceedings of USEPA/A&WMA Intl. Specialty Conference, Tropospheric Ozone, Non-Attainment and Design Value Issues*, Oct. 1992.

———: Recent Developments in Removal of Hazardous and Toxic Chemicals Using Moist Scrubbing Technology. *Proceedings of Odor and Volatile Compound Emission Control for Municipal and Industrial Treatment Facilities,* Water Environment Federation (WEF) Specialty Conference, April 1994.

Rustoleum Installs Novel Odor Treatment System, *American Paint and Coatings Journal,* Oct. 20, 1986, p. 52.

Waltrip, D., and E. Snyder: Elimination of Odor at Six Major Wastewater Treatment Plants, *Journal WPCF,* vol. 57, no. 10, pp. 1027–1032, Oct. 1995.

Water Environment Research Foundation (WERF): Control and Production of Toxic Air Emissions by Public Owned Treatment Works, Odor Control Equipment, project 91-VOC-2, Alexandria, Va., 1991.

Weber, R., and W. Stinson: Elimination of 99% of Odors Saves $300,000 in Equipment Cost, *Food Processing,* Sept. 1977.

SECTION 8.6.3

CHEMICAL SCRUBBING: OTHER DESIGNS

Robert P. G. Bowker
Bowker & Associates, Inc.
Portland, Maine

GAS-LIQUID CONTACTING METHODS 8.131
Spray Towers 8.131
Ejector Venturi Scrubbers 8.131
Plate Scrubbers 8.132
ALTERNATIVE CHEMICAL TECHNOLOGIES 8.132
Wet Scrubber with Electrolytic Hypochlorite Generation 8.132
Wet Scrubbers with Catalyst Reactor 8.135
Mobile-Bed Absorber Using Chelated Iron Solution 8.135
APPLICATION 8.137
PERFORMANCE 8.137
REFERENCES 8.138

GAS-LIQUID CONTACTING METHODS

Although packed-bed and mist scrubbers are the most common wet scrubber technologies used for odor control, variations of packed-bed scrubbers or other types of scrubbers are sometimes employed. Since there are a great many different mechanical designs for gas-liquid contacting devices, this section will treat only major designs. These include spray towers, ejector venturi scrubbers, and plate scrubbers.

Spray Towers

Spray towers collect particles or gases on droplets of liquid formed by atomizing the liquid through a spray nozzle. The droplet size is larger than that used in a mist scrubber. Spray towers can be designed for cocurrent or countercurrent operation. The relative velocity between the droplets and the gas establishes the terminal settling velocity of the droplets. Spray towers have low pressure drop and high scrubbant flow rate, and disposal of effluent should be considered. They are inexpensive to construct and operate, but are likely to be less efficient than packed-tower or mist scrubbers for removal of odorous gases due to the lower mass-transfer efficiency. They are generally used for quenching of high-temperature gases or for removal of particulates greater than 5 μm in diameter, gases of high solubility and reaction rate, or liquid aerosols. As with any other scrubber with recirculating liquid, they find little application for removing volatile organic compounds (VOCs) (Cheremisinoff and Young, 1976).

Ejector Venturi Scrubbers

Ejector venturi scrubbers are sometimes used for simpler odor control applications or where the gas stream contains particulates. In these scrubbers, the airflow is induced by the vacuum created as a result of the high velocity of the spray of scrubbant through the throat. A diagram of an ejector venturi is shown in Fig. 8.6.6. Scrubbant liquid is recirculated through the nozzle using a pump, and the gas flows cocurrently with the atomized liquid. A separation chamber allows liquid and particles to be removed before the cleaned gas is discharged. Collection efficiency is high for particulates greater than 1 μm in diameter or for soluble gases. Such scrubbers have high pressure drops. Operating cost is dependent on the operating pressure drop and scrubbant liquid flow rate. Such scrubbers have relatively low capital costs, are compact, and are easy to install (Cheremisinoff and Young, 1974; Buonicore and Davis, 1992).

Plate Scrubbers

Plate scrubbers consist of a vertical column with a series of horizontal trays mounted in the interior. The plates contain perforations, slots, or other openings through which contaminated gas flows upward. The scrubbant liquid is introduced at the top and flows downward countercurrent to the gas flow. Mass transfer is effected by the contact between the gas and the liquid. Plate scrubbers are typically characterized by the mechanical design of the plate. Two common designs are bubble cap plates and sieve or perforated plates. With bubble cap plates, the gas moves upward through "caps" or risers and through foam and bubbles created in the scrubbant liquid (see Fig. 8.6.7*a*). Plates are normally about 24 in apart, and the number of plates can be varied depending upon the application and the desired efficiency. In sieve plates the contaminated gas rises upward through perforations in the plate (typically ⅛ to 1 in in diameter) and through the froth of the scrubbant liquid. In some cases the plates are equipped with weirs to form a shallow pond of liquid on each plate to improve gas-liquid contact. Spacing of sieve plates is usually 12 to 15 in. A schematic diagram of a sieve plate scrubber is shown in Fig. 8.6.7*b* (Cheremisinoff and Young, 1974; Buonicore and Davis, 1992).

ALTERNATIVE CHEMICAL TECHNOLOGIES

Another group of wet chemical scrubbers falls into the class where unique and usually proprietary methods are used for generating, or regenerating, chemicals used for the gaseous treatment. Three examples are presented: a packed-bed scrubber coupled with an electrolytic generator of sodium hypochlorite, a packed-bed scrubber with a fixed-bed catalytic reactor, and a mobile-bed scrubber using chemical chelating agents.

Wet Scrubber with Electrolytic Hypochlorite Generation

Sodium hypochlorite is frequently used as an oxidizing agent to treat oxidizable gaseous chemical contaminants. Sodium hypochlorite can be generated from saltwater and electricity by using an electrolytic cell. Such cells have been in use for many years in marine applications, where saltwater is free and transport of chemicals is costly. When such cells are used to generate hypochlorite to replace commercially available sodium hypochlorite, a careful cost comparison should be made. The benefits of electrolytic generation will vary with the cost of locally available chemicals, the cost of electricity, and the capital and maintenance costs of the electrolytic cells.

Several manufacturers have coupled the electrolytic cell with packed-column gas-liquid contacting to save on chemical costs. There are limits created by this linkage since the best solution for operating conditions for an electrolytic cell may not be the best to treat the odorous gases. To solve such problems, two-stage packed towers have been used, where one stage may be a caustic solution. As there is a recirculating solution, there is negligible VOC removal.

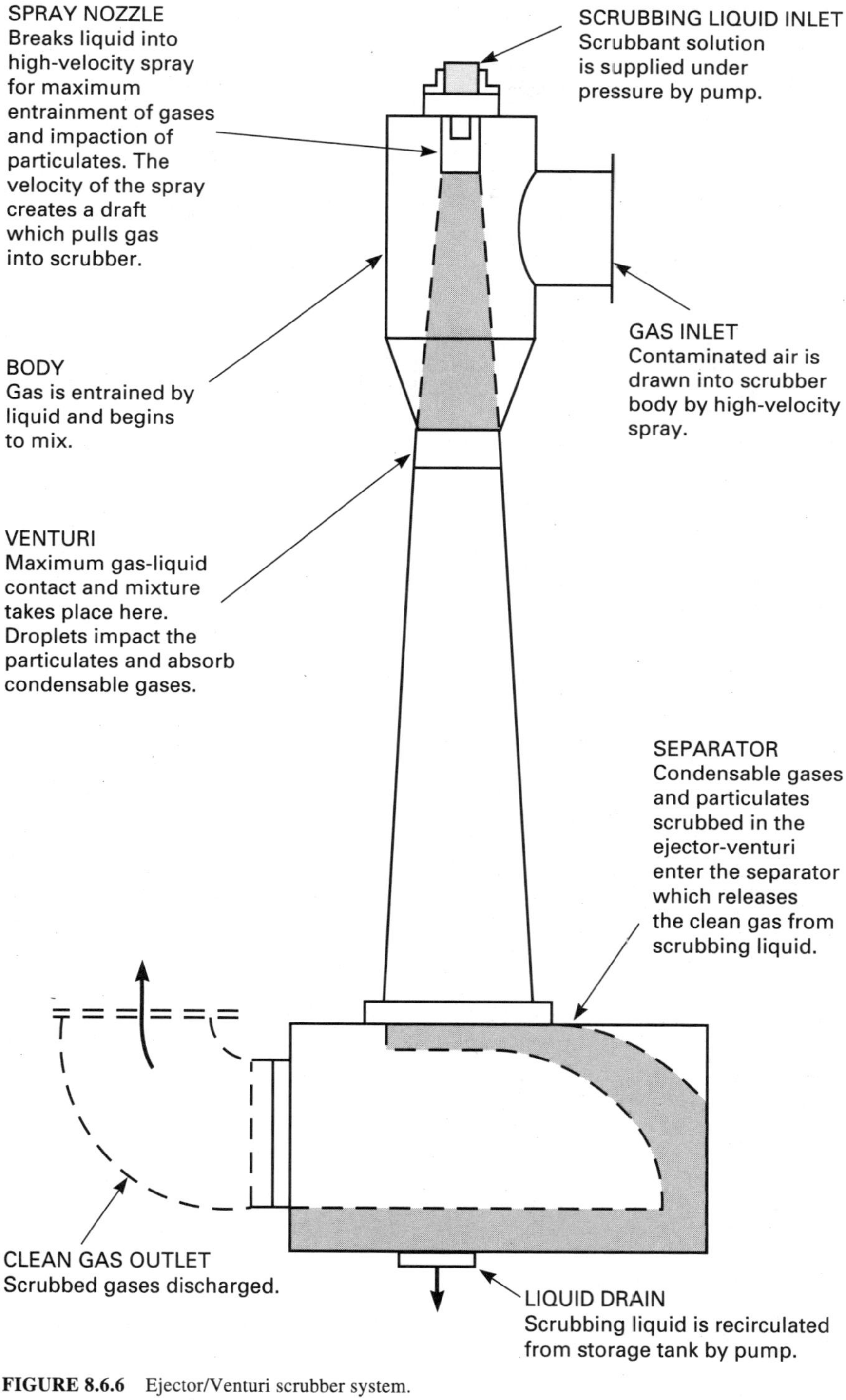

FIGURE 8.6.6 Ejector/Venturi scrubber system.

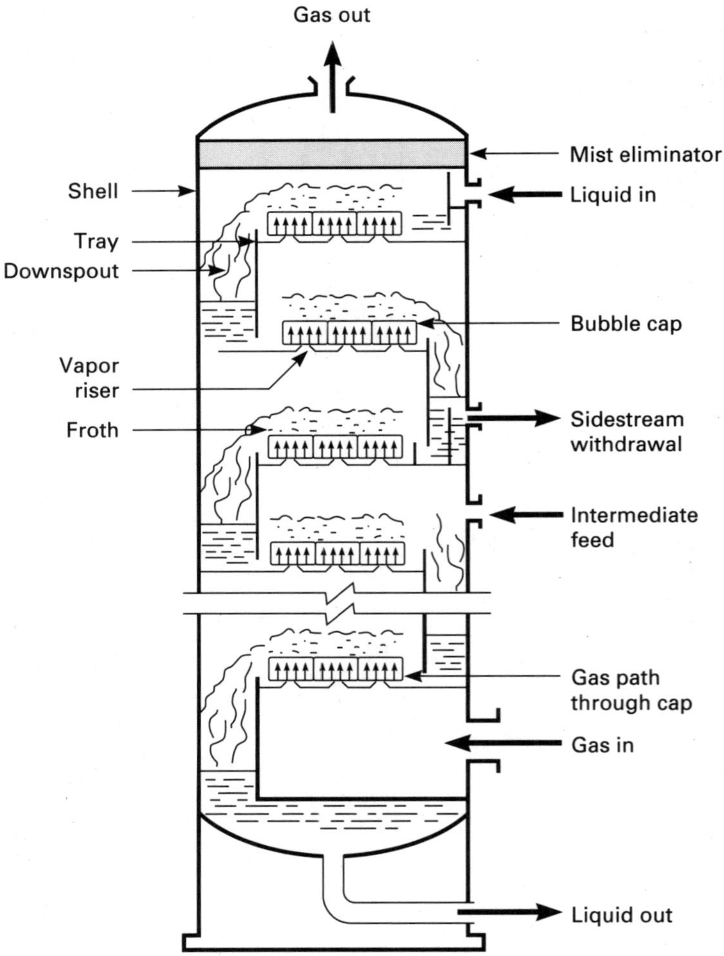

FIGURE 8.6.7(*a*) Typical bubble-cap plate column.

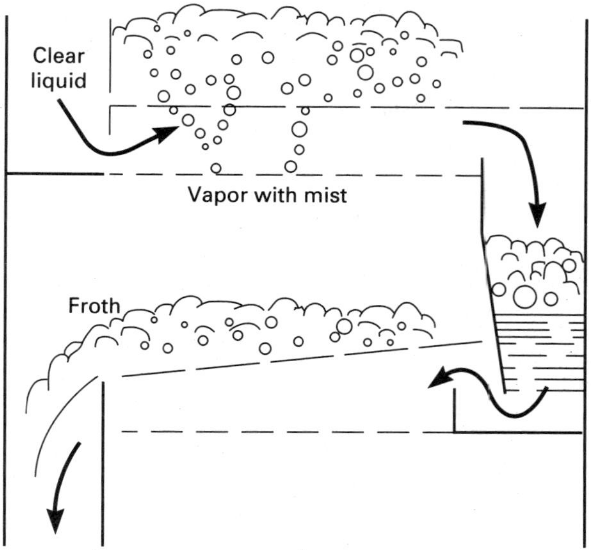

FIGURE 8.6.7(*b*) Sieve or perforated plate with downcomer.

Wet Scrubbers with Catalyst Reactor

A proprietary scrubber system developed in the United Kingdom recirculates the scrubbant liquid through a fixed-bed catalyst reactor. The catalytic reactor contains nickel which catalyzes the oxidation reaction between sodium hypochlorite and odorous compounds such as hydrogen sulfide, organic sulfides, mercaptans, and amines. A schematic diagram of the system is shown in Fig. 8.6.8. The wet scrubber is a modified packed tower. The claimed advantages of the catalytic reactor system include elimination of multiple scrubbing stages to achieve the same performance, decreased NaOCl concentrations and release of free chlorine, and lower capital and operating costs (Stitt and Fakley, 1996; Stitt et al., 1996a; Stitt et al., 1996b). The catalytic reactor used for the scrubbant solution can be retrofitted to an existing wet scrubber system.

Mobile-Bed Absorber Using Chelated Iron Solution

Another proprietary scrubber system employs a mobile-bed absorber and a chelated ferric ion scrubbing solution to oxidize hydrogen sulfide to elemental sulfur. Packing consist of smooth, spherical plastic balls that remain fluidized and agitated, preventing the elemental sulfur from adhering to the packing. The scrubbing water contains particles of elemental sulfur that, depending on the application, may be continuously filtered out and the solution recirculated. A diagram of the process is shown in Fig. 8.6.9. In some applications, a venturi is used for initial contact of the gas with the ferric ion solution. The process is typically used for applications involving high levels of H_2S. Other reduced-sulfur compounds are oxidized to a lesser extent, with no effect on VOCs, organic material, and amine odors (Nagl, 1991; Nagl, 1996).

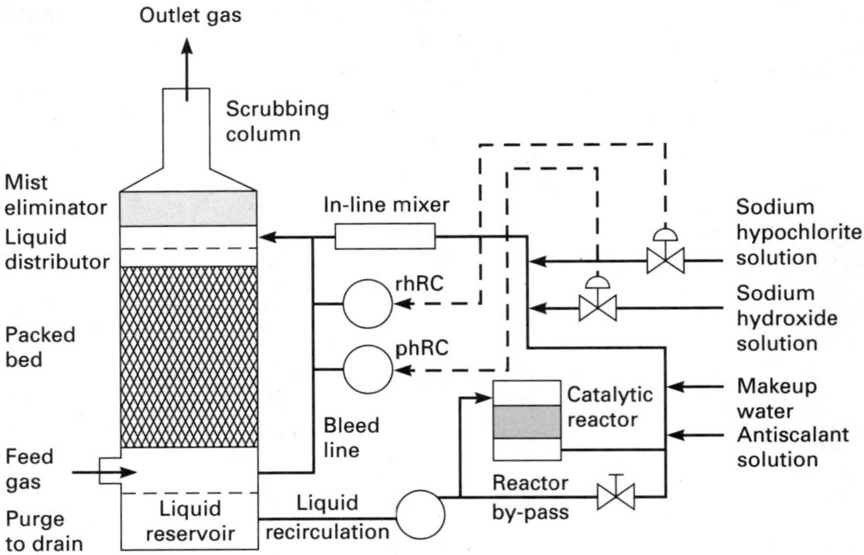

FIGURE 8.6.8 Flow diagram of wet scrubber with catalytic reactor.

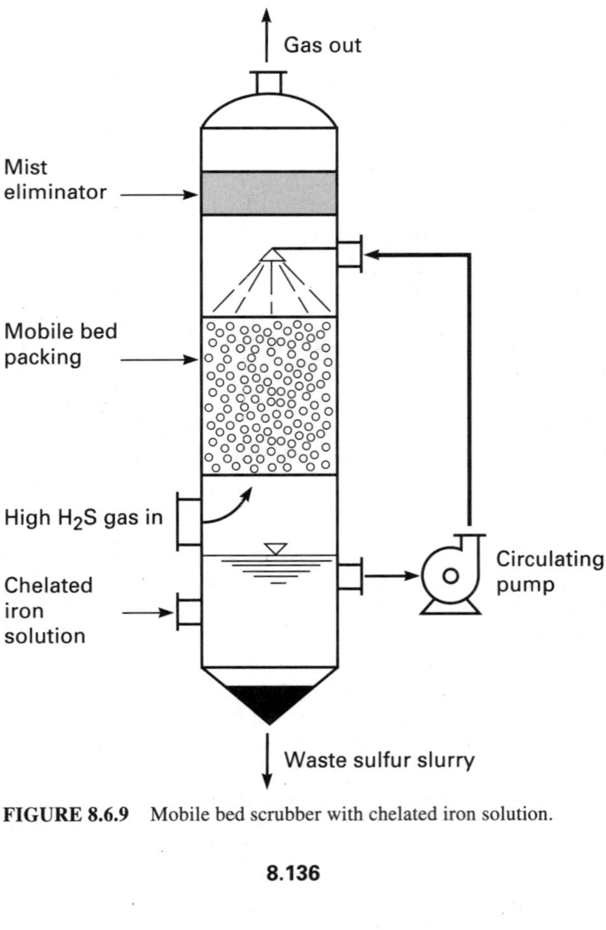

FIGURE 8.6.9 Mobile bed scrubber with chelated iron solution.

APPLICATION

Table 8.6.1 summarizes the applicability of these alternative scrubber systems for the control of odors and VOCs. Spray towers are useful for quenching of high-temperature gases or removal of particulates prior to some other odor control device, but have limited application as a technology for odor and VOC control. Ejector venturis and plate scrubbers are often applied to control highly soluble gases but are usually not used for complex, multicomponent gas streams. The packed-tower scrubber, coupled with the fixed-bed catalytic reactor, would appear to have a wide range of applications for soluble gas odor control. Full-scale industrial odor control applications include meat rendering, foundries, paint, resin manufacturing, and leather tanning. Finally, the mobile-bed scrubber with chelated iron scrubbant is best suited for handling very high-H_2S gas streams in which the hydrogen sulfide is the only contaminant of concern.

PERFORMANCE

The performance of these various control devices can vary widely depending on the application, the type and concentration of contaminants, and other factors. Since spray towers are normally not applied as an odor control process, no performance data are provided. Ejector venturis and plate scrubbers can be expected to provide 95 percent removal of soluble gases such as SO_2, but data from their use as odor control devices are limited. For the removal of H_2S, the mobile-bed scrubber with chelated iron scrubbant can be expected to remove 98 percent of the inlet H_2S (Nagl, 1996). However, removal efficiencies for other sulfur-based odorants or VOCs are likely to be minimal.

Table 8.6.2 provides H_2S and odor detectability data from pilot testing of the proprietary catalytic scrubber system. Table 8.6.3 provides inlet and outlet concentrations for an airstream containing a mixture of odorous compounds. The wet scrubber coupled with the catalytic scrubbant reactor appears to be very effective for a wide range of odorous compounds.

TABLE 8.6.1 Applicability of Miscellaneous Wet Scrubber Systems

	Contaminants to be controlled				
Scrubber type	Highly soluble non-H_2S gases	Particulates	Hydrogen sulfide	Other reduced sulfur compounds	Low-level VOCs
Spray towers	High	High	Low	Low	Low
Ejector venturis	High	High	Medium	Low	Low
Plate scrubbers	High	Medium	Medium	Low	Low
Packed tower with catalytic reactor	Low	Low	High	High	Medium-high*
Mobile bed with chelated iron scrubbant	Low	Low	High	Medium	Low

*Dependent on type and concentration of VOCs.

TABLE 8.6.2 Pilot Test of Catalytic Reactor Unit, Odor and H_2S Data

Sample	Odor in, OU/m³	Odor out, OU/m³	Reduction, %	H_2S in, mg/m³	H_2S out, mg/m³	Reduction, %
1	183,000	4,400	98	49,000	44	99.9
2	477,000	2,900	99	2,000	19	99.1
3	546,000	7,500	99	10,000	39	99.6
4	55,000	300	99	88,000	82	99.9

Gas rate = 3600 m³/h; Liquid rate = 50 m³/h; OU = odor units.
Source: Stitt et al. (1996a).

TABLE 8.6.3 Pilot Test of Catalytic Scrubber Unit, GC-MS Data

Pollutant	Concentration in, μg/m³	Concentration out, μg/m³	Reduction, %
Methyl mercaptan	45	0.1	99.8
Dimethyl sulfide	26	nd	>99.9
Carbon disulfide	35	3.1	91.1
Acetic acid	150	nd	>99.9
Dimethyl disulfide	1,730	3.4	99.8
Triethyl amine	170	24	85.9
Dimethyl trisulfide	560	nd	>99.9
Limonene	390	nd	>99.9
Dimethyl tetrasulfide	96	nd	>99.9
Methyl naphthalenes	260	20	92.1
C9 to C12 alkanes	2,170	383	82.4

Gas rate = 3600 m³/h; liquid rate = 25 m³/h; pH = 9.2; rH = 720 mV; nd = not detectable.
Source: Stitt and Fakley (1995).

REFERENCES

Buonicore, A. J., and W. T. Davis, eds., *Air Pollution Engineering Manual,* Air & Waste Management Association, Van Nostrand Reinhold, New York, 1992.

Cheremisinoff, P. N., and R. A. Young, "Wet Scrubbers—A Special Report," *Pollution Engineering,* May 1974.

Nagl, G., "Case History: Odor Control at a Water Treatment Plant," *National Environmental Journal,* Mar./Apr. 1996.

Nagl, G., "Economic Removal of Sulfides in Anaerobic Wastewater Treatment," presented at 64th Annual Conference, Water Pollution Control Federation, Toronto, Oct. 7–10, 1991.

Stitt, E. H., and M. E. Fakley, "New Process for the Abatement of Odours and Low Level VOCs," presented at American Institute of Chemical Engineers national spring meeting, Houston, Tex, Mar. 20, 1995.

Stitt, E. H., K. Kelly, A. R. Elgood, and M. E. Fakley, "Guarding against Odors," *Environmental Protection,* January 1996a.

Stitt, E. H., F. J. Taylor, and K. Kelly, "Odour and Low Level VOC Abatement Using Catalytically Enhanced Wet Scrubbing," presented at 69th Annual Conference, Water Environment Federation, Houston, Tex, Oct., 1996b.

SECTION 8.7

DRY CHEMICAL SCRUBBERS

Richard L. Davie
Environmental Systems Division
Purafil, Inc.
Oconomowoc, Wisconsin

DRY CHEMICAL SCRUBBING (GAS-PHASE FILTRATION) 8.139
MEDIA 8.141
Physical Properties 8.141
Advantages and Disadvantages 8.141
Specific Gas Applications 8.142
Chemical Reactions 8.142
Use of Multiple Media 8.143
TYPICAL SCRUBBERS AND ACCESSORIES 8.144
SCRUBBER DESIGN PARAMETERS 8.145
APPLICATIONS 8.146
Wastewater Odor and VOC Control 8.146
Polishing Wet Chemical Scrubber and Biofilter Discharge 8.147
Ammonia Removal 8.148
Polishing Carbon Scrubbers 8.148
ECONOMICS 8.148
Initial Cost 8.148
Operating Cost 8.149
DISPOSAL OF SPENT MEDIA 8.149
REFERENCES 8.149

DRY CHEMICAL SCRUBBING (GAS-PHASE FILTRATION)

A highly efficient means to remove trace gases from an airstream is through dry chemical scrubbing. This process is also referred to as *gas-phase filtration.*

Originally conceived and developed as a means to prevent severe corrosion of electronic components of computers and other electric equipment, the process has evolved to apply to a broader range of contaminant gas problems for odor control, volatile organic compound (VOC) removal, archival protection, preservation of indoor air quality, and scavenging of accidentally released toxic gases.

The heart of this process is manufactured, pelletized, porous media which are impregnated with one or more reacting chemicals. Media are manufactured from a mix of powdered substrate materials, such as activated alumina (Al_2O_3), powdered activated carbon, and sodium bicarbonate ($NaHCO_3$).

Mixed with the impregnating chemicals, these constituents are rolled into spherical pellets, dried, and screened to various particle sizes, typically from $\frac{1}{16}$ to $\frac{1}{4}$ in in diameter.

Other forms of media are produced by soaking an existing substrate pellet with the impregnate, after the porous pellet is formed. Such media using both alumina and zeolite substrates are available from various manufacturers. The latter are normally impregnated with $KMnO_4$. Typical media are shown in Fig. 8.7.1.

Media are arranged in an airtight housing where contaminated air is drawn or forced through and brought in contact with the media. As gas-contaminated air flows through the media bed, the contaminant gas(es) is(are) absorbed onto the pellet and chemically reacted with the impregnate to form, usually, solid reaction products. This irreversible process is referred to as *chemisorption.* The reaction products stay in the media. In con-

FIGURE 8.7.1 Photograph of media.

trast with adsorption on activated carbon, the contaminant gas cannot desorb, or reappear, in the airstream.

Other side processes also occur in the media. Certain impregnates also are thought to serve as catalysts to promote the in situ oxidation of the gas molecule (such as H_2S) by using oxygen in the airstream.

Physical adsorption of the gas also applies, to a degree, as both activated alumina and constituent activated carbon, when present, contain micro surface-area adsorption sites.

Evidence of additional physical and/or chemical activity, other than the classic chemical reaction, can be verified through a simple chemical reaction analysis, compared to actual media capacity test results.

The Al_2O_3 media, when impregnated with 4 percent by weight $KMnO_4$, demonstrates a removal capacity of 7 to 10 percent of the media weight for H_2S before breakthrough. This is more than 30 times the stoichiometric chemical reaction of H_2S with the $KMnO_4$ present in the media.

Dry chemical scrubbing is especially useful for removal of low-molecular-weight compounds such as H_2S, SO_2, and NH_3. These compounds are not readily adsorbed on activated carbon.

Multiple media may be used to address a variety of contaminant gases, either mixed or arrayed in series in separate beds.

Dry chemical scrubbers, when applied with deep media beds, offer two principal advantages when compared to other methods:

1. Gaseous discharge concentrations consistently less than 10 ppb, usually less than 3 ppb
2. Simple, unattended operation

Various scrubber configurations are presented in a later section titled "Typical Scrubbers and Accessories."

MEDIA

Physical Properties

Media are formed to provide substantial porosity for gas access, but must also be resistant to abrasion and crush-resistant to permit handling and resistance to the compressive stress when arrayed in deep vertical beds.

Bulk density is generally 30 percent greater than that for activated carbon, due to the use of Al_2O_3 as the principal substrate. Typical physical properties of manufactured media are listed in Table 8.7.1.

TABLE 8.7.1 Typical Physical Properties—Dry Chemical Scrubbing Media

Moisture content	20–30%
Crush strength	35–60%
Average abrasion	4.5% maximum
Bulk density*	45–55 lb/ft^3

**Note:* Some zeolite-based media approach 70 lb/ft^3 bulk density.

Advantages and Disadvantages

Dry chemical scrubber media offer specific advantages and some disadvantages in gas removal applications, compared to activated carbon. Advantages include the following:

- Performance is unaffected by high relative humidity.
- There is a 30 percent higher bulk density.

- They are nonflammable (usually UL class 1).
- There is inherent buffering of media pH when applied to acid gases.
- They are readily disposable in common landfills (nontoxic, nonhazardous near neutral pH).
- Resistance to airflow is one-half to one-third of typical granular activated carbon (reduced blower energy). Manufactured media do not exhibit change, or increases in airflow resistance with time. This is attributed to the greater porosity, which allows reaction products to fill the void space within the pellet, rather than collecting on the pellet's exterior surface.

Disadvantages are

- They are not regenerable.
- There is limited removal capacity for many organics.

Specific Gas Applications

Most dry chemical media are designed for removal of specific gases. Table 8.7.2 lists media supplied by one manufacturer (Purafil, Inc.), showing the chemical constituents and specific gas applications.

Chemical Reactions

Muller (1995) has compiled a useful reference of certain key chemical reactions for various media. A few of these reactions will illustrate the principles of media performance and specific capacities for gas removal. Refer to Table 8.7.2 for media composition.

Hydrogen Sulfide Reactions

Media A. The active ingredient in this media is potassium hydroxide (KOH) which reacts with H_2S to form a nontoxic salt—potassium sulfate (K_2SO_4)—and elemental sulfur (S), both of which remain bound to the media. This media can remove a weight of H_2S of approximately 16 to 18 percent of the original media weight. This is the normal method for expressing the media's capacity for a given gas.

Media B. The active ingredient for media B is potassium permanganate. The main reaction product formed upon exposure to H_2S is also potassium sulfate. The reaction mechanism most commonly cited is as follows:

TABLE 8.7.2 Media Composition versus Specific Gases

Product	Principal gas applications	Primary materials
A	H_2S, Cl_2, SO_2, mercaptans, hydrocarbons	Powdered activated carbon, activated alumina, KOH
B	Most organics plus H_2S, SO_2	Activated alumina, $KMnO_4$
C	NH_3	Powdered activated carbon, activated alumina, acid
D	Cl_2	Powdered activated alumina, sodium thiosulfate

$$3H_2S + 8KMnO_4 + 4H_2O \rightarrow 8MnO_2\downarrow + 3K_2SO_4\downarrow + 2KOH\downarrow + 6H_2O\updownarrow$$

This media contains 4 percent, by weight, $KMnO_4$ and has demonstrated a removal capacity of approximately 7 to 8 percent by weight for H_2S. A more recent development using 8 percent $KMnO_4$ has shown greater than 50 percent increase in gas removal capacity.

In general, $KMnO_4$ (potassium permanganate) media is effective in removing oxides of sulfur, formaldehyde, nitric acid, hydrogen sulfide, and other lower-molecular-weight aldehydes and organic acids. Activated carbon is not particularly effective in removing many of these gases.

The $KMnO_4$ media does not remove compounds such as acetone, *n*-butanol, MEK, benzene, or toluene, to list a few.

Use of Multiple Media

In most odor control and VOC applications, the use of multiple media has provided maximum removal efficiency against a multitude of odor-causing compounds. Typical wastewater odor control systems include, as a minimum, caustic-impregnated media (type A) with a polishing bed of $KMnO_4$ media (type B). This combination offers the higher capacity for H_2S in the caustic media, with the polishing capability to chemically react bypassed or desorbed compounds with $KMnO_4$ in the polishing step, thus preventing their reentry in the discharge airstream.

This principle is best illustrated in the performance data obtained from a combined scrubber installation at the Monterey Regional Water Pollution Control Agency's Sewage Treatment Plant in Monterey, California.

In this example, $KMnO_4$ media is employed in a separate polishing scrubber to treat the discharge from a conventional caustic-impregnated carbon scrubber.

With reference to Fig. 8.7.2, it is seen that the H_2S discharge from the carbon scrubber varies from 10 to over 400 ppb v/v over a 24-h period. The polishing $KMnO_4$ media

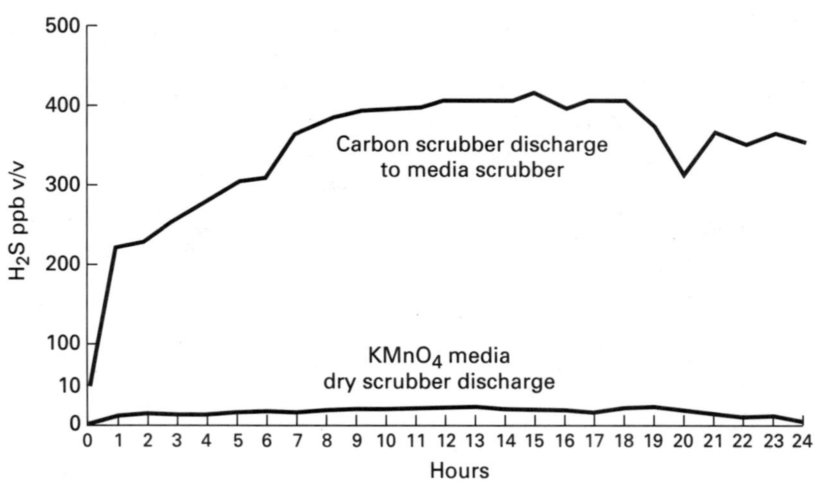

FIGURE 8.7.2 Monterey data—discharge from carbon and $KMnO_4$ polish scrubber. (Average inlet H_2S concentration to caustic carbon scrubber: 29.16 ppm v/v.)

reduces these variations to a maximum discharge of 4.0 ppb v/v or a 24-h average discharge concentration of 2.6 ppb v/v.

The use of caustic-impregnated media, followed by a polishing stage of $KMnO_4$ media, is now widely used in single multistage multimedia scrubbers to achieve the demonstrated low ppb discharge concentrations shown in this example. This discharge concentration compares favorably with the widely reported odor detection threshold, for most people, in the range of 1 to 3 ppb H_2S.

Dry chemical media do have limitations in their capacity for removing a wide range of organics, many of which are odorous, such as methyl mercaptan (CH_3SH).

In applications where relatively high organic contaminants are expected, some vapor-phase activated carbon may be employed either in a separate media bed or mixed with the media. Comparative testing (Muller and England, 1995) has shown that separate, discrete media beds offer the highest removal efficiencies and capacity.

TYPICAL SCRUBBERS AND ACCESSORIES

Dry chemical scrubbers have evolved to apply to a wide range of airflows and contaminant loads. Due to the importance of the use of multimedia, the most common scrubber configuration for a wide range of odor control and VOC applications is designed with separate discrete media beds. Referred to as a multistage scrubber (MSS) or deep bed scrubber (DBS), this version is shown in cutaway form in a fiberglass reinforced plastic (FRP) housing in Fig. 8.7.3.

When applied to wastewater treatment odor control, this scrubber includes a mist eliminator-prefilter at the blower inlet, or installed integral to the scrubber housing. Media are filled into separate, usually 12-in-deep beds, at the top of the scrubber,

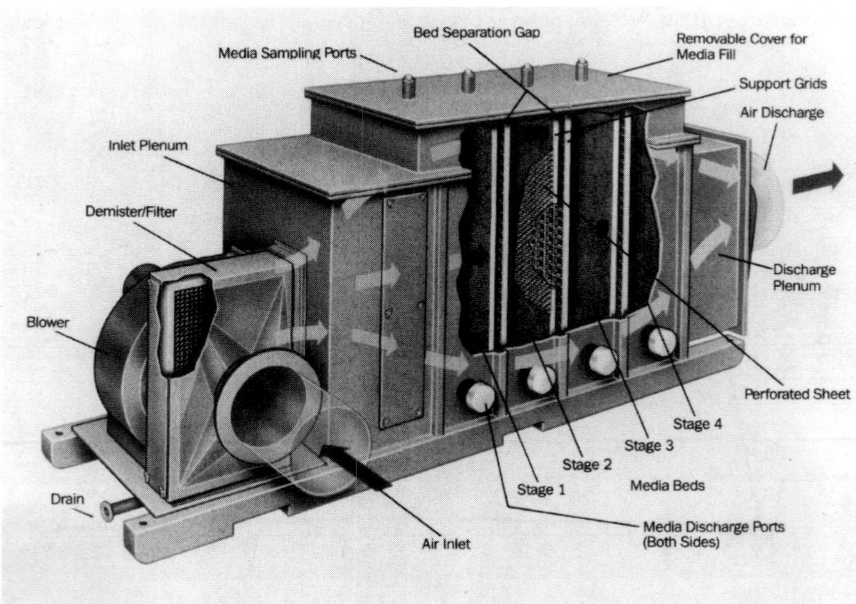

FIGURE 8.7.3 Artist's concept of multistage scrubber (cutaway).

through the covered fill chutes. The entire system is skid-mounted to include mist eliminator, blower, and scrubber housing.

Manufactured in aluminum, stainless steel, or FRP, these scrubbers usually contain a minimum of two 12-in beds, up to four and occasionally five beds (60-in bed depth). Media is shipped separately and loaded at the installation site. This configuration applies to airflow rates from 500 to 10,000 cfm.

Modular units are applied commonly for corrosion control, and occasionally for odor control of higher airflows and very light contaminant loads.

Replaceable media containing V-shaped modules achieve media beds of 1 to 3 in in depth. Rows of modules may be arranged in series to contain different media, and extended total bed depths.

Small drum units (55-gal and 200-gal) will handle 65 to 500 cfm. These units come with an integral top-mounted blower, or passive (no blower). These scrubbers are commonly applied to smaller pump station wet wells, tank vents, etc. A 100-cfm drum scrubber is shown in Fig. 8.7.4.

Higher airflows, to 25,000 cfm and more, can be achieved by parallel bed scrubbers (PBSs) and other configurations which maximize the media bed face area.

Dry chemical media may also be employed in conventional rectangular or cylindrical vessels, more common to activated-carbon adsorbers. Various manufacturers employ different, though similar, variations on these scrubber designs.

SCRUBBER DESIGN PARAMETERS

The major variables considered in the selection and sizing of any gas scrubber are the airflow rate and contaminant gas concentrations. With these, and the media capacity data, the media inventory is determined to achieve a design media life cycle.

FIGURE 8.7.4 A 100-cfm drum unit (Oconomowoc, Wisconsin).

Face velocity limits have been established to ensure adequate gas residence time and reasonable pressure drop through the scrubber. For deeper bed scrubbers (>12 in), face velocity from 80 to 125 ft/min may apply. Thin bed systems (modular units) may operate at up to 500 ft/min, depending on total bed depth.

With face velocity established, the required face area of the scrubber is determined, and that essentially completes the scrubber sizing process.

The key sizing parameter is the long-term average gas(es) concentration. Peak loads, if infrequent, do not enter into the process since a dry chemical scrubber will continue to discharge at very low concentrations, even during peak loading events. Air temperature and moisture content are less explicit considerations.

High air temperature is limited only by moisture evaporation from the media. Media requires a limited moisture content to function. Experience with 130°F saturated air has shown no apparent degradation in media performance. Temperatures much below 32°F may be of concern, although many scrubbers are known to operate effectively in cold climates with incoming air well below the freezing mark.

Gas loading rate and heat release are known to affect performance at air temperatures below freezing. Tests of one media removing chlorine, at very high loading rates associated with emergency gas scrubbers, have demonstrated acceptable performance at initial media temperature of −40°F.

APPLICATIONS

Dry chemical scrubbers have wide technical application to virtually any airborne gas removal problem. Application is limited primarily by operating cost, or economics, which is addressed specifically below.

Experience has shown the following applications to be especially viable:

- Electronic corrosion control
- Wastewater odor and VOC control
- Upgrading activated carbon scrubbers
  Media addition
  Separate polishing scrubber
  Replacement of carbon with dry chemical media
- Polishing wet chemical scrubber discharge
- Polishing biofilter discharge
- Emergency toxic gas removal (Cl_2, SO_2, NH_3)
- Indoor air quality (IAQ)

Wastewater Odor and VOC Control

Odor removal in wastewater facilities inherently includes VOC removal. Dry chemical media cannot distinguish between airborne gases as long as they can be chemically reacted or adsorbed.

Odorous compounds from wastewater include inorganics such as H_2S as well as sulfur- and nitrogen-based organics, most of which result from biological decomposition in the collection and treatment processes.

Dry scrubbers have been successfully applied to the following wastewater wet stream odor sources:

- Pump station wet wells
- Force main vents
- Plant headworks
- Primary clarifiers
- Aerobic digesters
- Sequencing batch reactors (SBRs)
- Septage transfer and storage

Applications also exist in certain sludge processing facilities, including

- Aerated sludge storage tanks
- Auto thermal aerobic digestion (ATAD) sludge digestion
- In-vessel sludge composting

In ATAD sludge digestion, dry scrubbers are combined with a wet scrubber to economically remove high ammonia concentrations.

Polishing Wet Chemical Scrubber and Biofilter Discharge

Wet scrubbers, particularly packed towers, frequently attain less than 99 percent removal of odorous gases. Ninety-nine percent removal of 20 ppm of H_2S allows a discharge concentration of 200 ppb.

With an odor detection threshold no greater for most people than 2 ppb, additional dilution of 100 times would be necessary to reduce discharge air to below the detection level. Dry chemical scrubbers are applied to reduce the discharge from the system to, usually, less than 3 ppb, in those cases where the discharge air must be essentially odor-free. Such an installation is shown in Fig. 8.7.5, where a two-stage (dual-media) dry

FIGURE 8.7.5 Photograph of Kent County, Delaware, installation, wet and dry scrubber.

scrubber is coupled with a packed tower chemical scrubber. This installation was required to reduce 200 ppm H_2S to <10 ppb in the discharge.

Table 8.7.3 reports field verification tests' average performance. Biofilter installations in Europe have also employed dry chemical polishing scrubbers for the same purpose. In both cases, the dry scrubber serves the additional purpose of removing peak loads, which may pass through the primary scrubber. In the case of chemical scrubbers, the use of a dry polishing stage permits chemical feed rates to be established for the average contaminant load, thus reducing chemical cost.

TABLE 8.7.3 Packed Tower Polishing Application—Field Test Data

Packed tower inlet H_2S	203 ppm
Packed tower discharge H_2S	600 ppb
Polishing dry scrubber discharge H_2S	<1.0 ppb

Overall, the use of wet chemical or biofilter scrubbers for "roughing" removals can minimize operating cost, while still achieving successful control of odors when followed by a dry chemical polishing stage.

Ammonia Removal

Although dry chemical media are available to remove ammonia (NH_3), this gas is somewhat problematical. Ammonia has an unusually high odor detection threshold, in the range of 20 to 40 ppm. Below that level, there is no need to remove it for odor control purposes.

In those sludge treatment processes which produce free NH_3, the levels are usually so high as to make dry media application uneconomical. Such applications will usually require the use of a water or wet chemical acid scrubber to selectively remove ammonia, followed by a dry scrubber to remove the remaining odorous gases.

Polishing Carbon Scrubbers

Caustic-impregnated activated carbon is commonly used for wastewater-related odor control. While it is usually capable of removing most organics and some reduced sulfur (H_2S, mercaptans), performance can be variable.

For those gases that are removed through adsorption on the carbon, desorption also occurs with changing load conditions, temperature, and moisture content.

Performance data presented in Fig. 8.7.2 illustrate the variability of H_2S discharge from a standard caustic carbon scrubber, with 3 ft of carbon bed depth. Performance is substantially improved when polished by a bed of $KMnO_4$ dry chemical media.

ECONOMICS

Initial Cost

Initial cost of dry chemical scrubbers will vary considerably with size and throughput. There is less economy of scale than there might be with a more complex wet chemical scrubber, where substantial cost is apportioned to controls and instrumentation, which do not change substantially with increased system size.

The principal costs of dry scrubbers are due to scrubber housings and media, which tend to increase proportionally with increased contaminant loads.

Where modular, thin media beds are applied to trace gas concentrations (usually less than 1 to 2 ppm), equipment costs may range as low as $5 per cfm. Smaller deep-bed, bulk-loaded scrubbers fall in the range of $25 to $30 per cfm. Higher airflow systems (15,000 to 20,000 cfm) will approach $8 to $10 per cfm. All these unit costs include scrubber, blower, mist eliminator, and media.

Operating Cost

Primary operating costs include electric energy for blower and the ongoing cost of replacement media. Operating blower energy costs for a dry scrubber should be less than for typical carbon adsorbers, due to lower pressure drop through the dry chemical media. Pressure drop, for deep-bed systems, is higher than that for conventional packed towers and will incur higher power costs.

By far, the major operating cost is for media replacement.

A useful way to compare operating costs for a scrubber is to reduce the cost of replacement media, or makeup chemicals, to the cost per unit of contaminant gas removed.

With reference to H_2S removal, the cost of makeup chemicals for a conventional, single-stage packed tower, using NaOH and NaOCl, can range from $6 to more than $8 per pound of H_2S removed.

Replacement dry chemical media and some impregnated carbons currently cost from $1.50 to more than $2.50 per pound. At $2.50 per pound cost, and end-of-life capacity for H_2S of 15 percent by weight, the replacement media cost equates to over $16 per pound of H_2S removed.

General operation and maintenance of dry chemical scrubbers are minimal, due to their simplicity. The additional cost of spent-media disposal is minor, consisting mainly of handling and transportation to a common landfill.

DISPOSAL OF SPENT MEDIA

When applied to odor control in most wastewater collection and treatment facilities, spent dry chemical media will not contain toxic or hazardous chemicals. Chemical content consists mainly of salts from the primary reactants.

Media used to remove reduced sulfur compounds frequently contain bicarbonate which further buffers pH, resulting in acceptable levels for common landfill application.

Media is usually removed from the bulk-loaded scrubbers by vacuuming through top-mounted fill hatches, or through specific vacuum connections at the bottom of the media bed.

For those systems which use media- containing modules, disposable modules are also available. Made of cardboard, the module structure is biodegradable when applied to the landfill.

REFERENCES

Muller, Christopher: Internal report, Purafil, Inc., 1995, Doraville, Ga.

Muller, Christopher, and William G. England: "Achieving Your Indoor Air Quality Goals: Which Filtration System Works Best?" *Ashrae Journal,* Feb. 1995.

SECTION 8.8.1

BIOLOGICAL SYSTEMS: BIOFILTERS

George J. Skladany
Director, Remediation Technology Testing Facility
Fluor Daniel GTI, Inc.

Marc A. Deshusses
Assistant Professor of Chemical Engineering
University of California, Riverside

Joseph S. Devinny
Professor of Civil and Environmental Engineering
University of Southern California

A. Paul Togna
Director, Systems Research and Development
Envirogen, Inc.

Todd S. Webster
Group Leader, California Projects
Envirogen, Inc.

OVERVIEW OF BIOFILTRATION 8.151
BIOFILTER REACTOR CONFIGURATIONS 8.155
- **Open Soil Bed Biofilters** 8.156
- **Open Single-Bed Biofilters** 8.157
- **Engineered Multiple Open-Bed Biofilters** 8.159
- **Closed Single- and Multiple-Bed Biofilters** 8.159

BIOFILTER DESIGN 8.162
- **Problem Definition** 8.162
- **Biofilter Sizing** 8.163
- **Engineering the Physical Biofilter System** 8.170

START-UP, OPERATIONS, AND MONITORING 8.178
- **Start-up of Biofilter Systems** 8.178
- **Biofilter Operations** 8.179
- **Monitoring Overall System Performance** 8.180

BIOFILTRATION MARKETPLACE 8.181
- **Odor Control Systems** 8.181
- **Hydrocarbon Treatment Systems** 8.184
- **VOC and HAP Control Systems** 8.185

SUMMARY 8.188
REFERENCES 8.188

Biofilters are the most widely used and accepted form of vapor-phase biological treatment system, and have been systematically applied in various forms throughout many parts of the world for more than 30 years. The *biofiltration process* relies on the beneficial action of microorganisms to metabolize volatile organic and inorganic contaminants passing through some form of organic media, also commonly referred to as packing or packing media. The packing material is usually held within some sort of engineered biofilter structure. By carefully monitoring and controlling the properties of the influent waste gas stream as well as the microbial growth environment within the biofilter, rapid and complete biodegradation of many compounds can be accomplished in a cost-effective manner. General information on biofilters and their application can be found in numerous publications (Webster and Devinny, 1998; Swanson and Loehr, 1997; Institute of Clean Air Companies, Inc., 1993; VDI Richtlinien 3477, 1991; Dharmavaram, 1991; Leson and Winer, 1991).

Biofilters have been used extensively to control odors generated at wastewater treatment and composting facilities. Much effort is currently underway to further expand the use of biofiltration technology to the treatment of volatile organic compounds (VOCs), hazardous air pollutants (HAPs), various petroleum hydrocarbons, and other organic and inorganic compounds. Future acceptance and use of biofilters depend on a more complete understanding of the benefits and limitations of the technology by professionals in the waste treatment field.

This section is designed to provide the reader with a basic understanding of the biofiltration process as well as the design and operation of biofilter systems. In the first part of the section, an overview of biofilters, the biofiltration process, and biofilter reactor configurations is presented. The basis for the design of a biofilter system follows, including determination of vapor contact time and packing volume, biofilter configuration, and selection and optimization of major system components. Next, information concerning the start-up, operation, and maintenance of biofilter systems is presented. Last, the biofiltration marketplace as it currently exists in the United States for the treatment of odors, petroleum hydrocarbons, and VOCs and HAPs is reviewed, along with a brief discussion of biofiltration capital and operating expenses.

OVERVIEW OF BIOFILTRATION

Figure 8.8.1 shows the major components of a typical biofilter system. The vapor to be treated first enters a pretreatment system, where the gas stream is adjusted as needed for

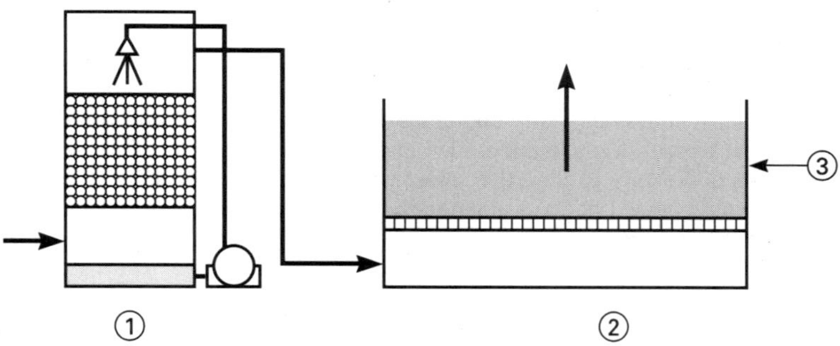

FIGURE 8.8.1 Cross-sectional drawing of the major components of a typical biofilter system: (1) air pretreatment system (using packed tower), (2) upflow open-bed biofilter, and (3) packing media. The microorganisms that catalyze the destruction of vapor-phase contaminants are found within the packing media.

moisture content, particulate load, and temperature. The conditioned vapor next passes into an air distribution system, where the gas enters the biofilter medium. Biofilter medium is made up of a porous bed of organic and/or inert material, typically less than

1 m in depth. A variety of media formulations are used, though each usually contains a mixture of peat or compost, bulking agents, inorganic nutrients, pH-buffering compounds, and possibly an organic carbon source. Indigenous or inoculated cultures of pollutant-degrading organisms are naturally immobilized on the media.

As the vapor passes through the filter bed, contaminants and nutrients are transferred to the microorganisms, where cellular metabolism breaks down the chemicals into simpler components. Biological metabolism provides the microorganisms with additional carbon and energy for growth and reproduction. Many different classes of compounds are amenable to treatment using biofiltration, including (1) aliphatic hydrocarbons, such as hexane; (2) aromatic hydrocarbons, such as benzene, toluene, ethylbenzene, xylenes, and styrene; (3) oxygen-containing compounds, such as alcohols (methanol, ethanol, and propanol), aldehydes (formaldehyde and acetaldehyde), ketones (acetone, methyl ethyl ketone, and methyl isobutyl ketone), carbonic acids (butyric acid), carbonic acid esters (ethylaldehyde and methyl methacrylate), and phenols; (4) sulfur-containing compounds, such as sulfides (hydrogen sulfide and dimethyl sulfide), thiocyanates, mercaptans (methyl mercaptan), and carbon disulfide; and (5) nitrogen-containing compounds, such as amides, amines (trimethylamine), nitrogen heterocyclics (pyridine), nitriles (acetonitrile and acrylonitrile), and ammonia (modified from VDI Richtlinien 3477, 1991).

Biofilters are most effective against single or complex mixtures of contaminants present at total carbon concentrations of 1500 ppmv or less. Removal efficiencies greater than 99 percent are attainable for many compounds. Typical vapor contact times for low-concentration odor control systems range from 10 to 40 s, compared to 20 to 90 s required for the treatment of higher concentrations of VOCs and HAPs.

Mineralization is the term used to describe the complete bioconversion of a target compound to carbon dioxide, water, and possibly inorganic salts or acids. Partial or incomplete biodegradation of the target compound(s) can result in the formation of various "daughter products," some of which may themselves be odorous and/or hazardous. While hydrocarbons and oxygenates are generally completely converted to carbon dioxide and water, the elimination of sulfur- or chlorine-containing pollutants results in the formation of sulfate and chloride ions, respectively. Over time, the formation of such acidic end products will cause acidification of the biofilter bed and a decrease in pollutant removal efficiency, unless steps are taken to readjust bed pH values or remove the acidic products. In general, successful biofiltration projects require that the pollutants be readily biodegradable and that their degradation not lead to the accumulation of hazardous or noxious end products.

The biofilter media is typically contained within some sort of constructed bioreactor. Biofilter reactor designs include open and closed systems of different shapes and sizes. Open-bed systems expose the surface of the biofilter packing directly to the atmosphere, while closed-bed systems totally enclose the biofilter beds within an airtight structure. Biofilter materials of construction include wood, concrete, polyvinyl chloride (PVC), fiberglass, metal, and fiberglass-reinforced epoxy resin. The construction materials chosen for a specific system depend on the characteristics of the gas to be treated, the anticipated generation of end products (such as acids), the degree of system control and sophistication desired, and the project budget.

The current state of the art in biofiltration knowledge and application is actively evolving. Many successful odor control systems have been installed without a detailed understanding of the underlying biological processes. The successful treatment of complex mixtures of organic compounds, especially those waste streams containing high concentrations of organics, requires a more complete understanding of the biological, physical, and chemical processes at work. In many situations, laboratory and/or field pilot testing can be used to provide detailed information. As the understanding of biofiltration processes increases, new opportunities will arise for the use of the technology.

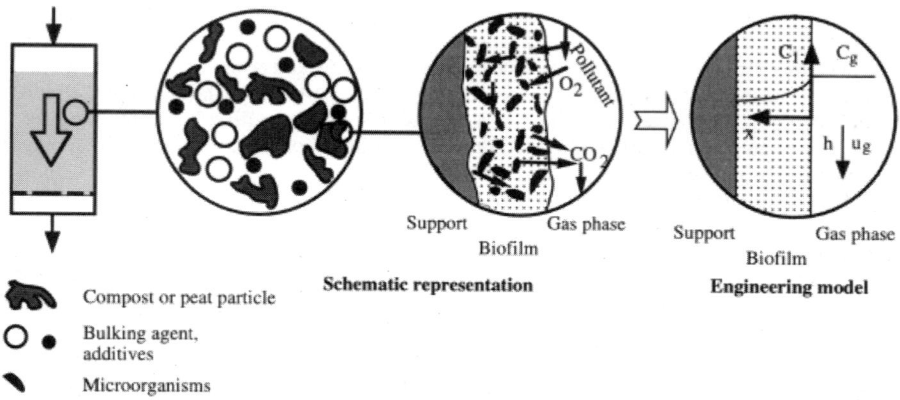

FIGURE 8.8.2 Schematic representation of the biofiltration process showing gross composition of the packing media, closeup of the biofilm layer with its major interactions, and major engineering parameters. (*Courtesy of Marc A. Deshusses.*)

As described in Fig. 8.8.2, the elimination of a gaseous pollutant in a biofilter is the result of a complex combination of different physical, chemical and biological phenomena. Initially, the vapor-phase contaminants and any nutrients needed for microbial growth (oxygen, nitrogen, and phosphorus, e.g.) must become dissolved in the water layer surrounding the biomass. The pollutant vapors and oxygen are transported in humid air through the medium by forced convection. Inorganic nutrients, such as nitrogen and phosphorus, may already be present in the medium (such as the compost fraction), or they may be added in liquid or gaseous forms as appropriate. Interphase mass transfer occurs, and provided that the biofilter bed particles are small, interfacial equilibrium is achieved so that gas-phase resistance can usually be neglected.

Next the dissolved contaminants and nutrients must be assimilated by the biomass. The biomass portion of the packing media is frequently envisioned as a separate phase called the *biofilm*. The biofilm is often thought of as "coating" the surface of the solid support media. This is a simplification. The actual structure of the biomass community solid support is more complex (Hugler et al., 1996; Moller et al., 1996; de Beer et al., 1994). However, for simplicity, the biofilm model will be used during further discussions. In the biofilm, simultaneous diffusion and biodegradation of the pollutants occur as a result of growing or resting microorganisms. Oxygen is also subject to diffusional resistance. Sorption of all chemical species involved is possible either after diffusion through the biofilm or directly from the gas-solid contact occurring on bare portions of particles.

The actual biodegradation of target contaminants occurs within the biofilm, and the formation of secondary metabolites, or daughter products, is possible. If formed, secondary metabolites will undergo the same simultaneous diffusion, biodegradation, and sorption processes. Acidic metabolites produced can be neutralized by limestone or other pH-buffering agents added to the biofilter packing. During treatment secondary metabolites are seldom observed. This suggests that they are degraded faster than the primary pollutants either by ancillary microbial strains in direct contact with the primary pollutant degraders (acting as a microbial consortia) or directly by the primary pollutant degraders themselves. The carbon dioxide resulting from the oxidation of primary and secondary metabolites diffuses through the biofilm and back to the free gas phase. Some of the carbon dioxide produced may also accumulate within the media as carbonate.

The biofilter media provides surface area on which the active microorganisms grow. In addition, the media serves as a reservoir for water, pollutants, and nutrients, which may accumulate by either adsorbing onto the packing matrix or dissolving into the pore water. The media is usually comprised of a fair amount of organic residues, which may be utilized as nutrients by the microorganisms present.

Biofilters rely on the metabolic action of mixed microbial cultures, or consortia, to catalyze the destruction of target compounds. Classical microbiological techniques have shown the presence of mixed populations of bacteria, yeasts, fungi, and higher organisms in biofilters. In most cases, these laboratory techniques are limited to the quantification and characterization of culturable organisms, i.e., only a small portion of the actual total microflora present (Hugler et al., 1996). However, this limitation has recently been overcome (Webster et al., 1997; de Castro et al., 1996). To date, little detailed information is known about biofilter process cultures, microbial interactions, and how the biological populations present are affected by operating conditions or by their direct environment. The active microorganisms are typically found attached to the matrix support as biofilms, and in most instances, microorganism emissions from biofilters can be neglected. Also, nutrient leaching occurs at a slow rate, so that biofilters can be considered as closed systems with respect to nutrient balance (Smet et al., 1996; Ottengraf and Konings, 1992). Microorganisms can obtain nutrients by recycling cellular materials from dead cells ("cryptic" growth) as well as through the action of higher organisms (such as protozoa) also present in biofilters (Mason et al., 1986).

In most cases, the biological reactions occurring involve the aerobic oxidation of the pollutant. In the presence of pollutant-degrading microorganisms, the pollutant(s) and oxygen are biocatalyzed into carbon dioxide, heat, and additional biomass.

In attempts to mathematically describe the biofiltration process, the pollutant diffusion flux at the interface and the differential change of pollutant concentration in the gas phase can be equated:

$$-\mu_g \cdot \frac{dC_g}{dh} = -A_s \cdot D_{\text{eff}} \cdot \left(\frac{dC_l}{dx}\right)_{x=0} \tag{8.8.1}$$

In the biofilm simultaneous diffusion and biological reaction occur, hence at steady state

$$D_{\text{eff}} \cdot \left(\frac{d^2C_l}{dx^2}\right) - k = 0 \tag{8.8.2}$$

The variables used in Eqs. (8.8.1) and (8.8.2) are defined below:

A_s = interfacial area, m^2/m^3 or ft^2/ft^3
C_l = concentration in liquid phase or in biofilm, kg/m^3 or lb/ft^3
C_g = concentration in gas phase, kg/m^3 or lb/ft^3
D_{eff} = effective diffusion coefficient, m^2/s or ft^2/min
h = height in bed measured from gas inlet side, m or ft
k = reaction rate in biofilm, $kg/(m^3/s)$ or $lb/(ft^3/min)$
μ_g = superficial gas velocity, m/s or ft/min
x = coordinate in biofilm, with 0 = gas-biofilm interphase, m or ft

Using Eqs. (8.8.1) and (8.8.2), the efficiency of a biofilter system can be predicted if the parameters of the process are known. The equations use a biokinetic reaction rate constant k to describe the mass of contaminant biologically treated per volume of media per time. Various forms of biological kinetics (such as first-order, zero-order, Monod, and Haldane) have been used to describe this rate constant (Deshusses et al., 1995; Shareefdeen et al., 1993; Ottengraf, 1986). In most instances, it can be reasonably assumed that the amount of active microorganisms, hence the kinetics, in a biofilter remains constant over time because nutrient availability is restricted. Even so, a nonlinear relationship for k with respect to concentration (predicted from use of the Monod or Haldane equations) in Eq. (8.8.2) will necessitate numerical integration to solve Eqs. (8.8.1) and (8.8.2), and therefore, usually first- or zero-order kinetics is used to estimate performance when these two equations are used. For first-order kinetics, the rate of degradation of the contaminant is linearly proportional to the concentration in the biofilm. For zero-order kinetics, the rate of degradation of the contaminant is a constant value, independent of concentration. Monod kinetics combines first- and zero-order behavior into a single equation. Monod kinetics predicts that the biodegradation rate is proportional to contaminant concentration at low concentrations (first-order behavior), and that the biodegradation rate is a constant value, independent of concentration, at high contaminant concentrations (zero-order behavior). The Haldane equation is often used to model microbial behavior where higher concentrations of contaminants will cause inhibition or toxicity to the microorganisms present. The Haldane equation predicts first-order kinetics at low contaminant concentrations, zero-order kinetics at intermediate contaminant concentrations, and decreased degradation at higher contaminant concentrations where microbial inhibition occurs.

No single biokinetic equation is appropriate for use with all contaminants and concentration ranges. The choice of which biokinetic equation to use will depend upon the physical properties of the contaminants (including solubility, Henry's law constant, vapor pressure, etc.), the concentrations of contaminants, as well as the nature of the microorganisms present.

BIOFILTER REACTOR CONFIGURATIONS

The full-scale implementation of vapor-phase biological treatment can be accomplished using any one of several bioreactor designs, including biofilters, biotrickling filters, and bioscrubbers. Each of these bioreactor designs shares key characteristics with one another, but differ in their level of contaminant applicability, process control, and attainable removal efficiencies.

On the simplest level, biological reactors for air pollution control can be classified on the basis of the presence or absence of a free liquid phase. Bioreactors with no free liquid phase are called *biofilters,* whereas those involving a free liquid phase are generally referred to as *biotrickling filters* and *bioscrubbers.* Biofilters work by passing a humidified polluted airstream through a porous organic-based packed bed on which pollutant-degrading organisms are immobilized. Biofilters are the mechanically simplest and usually the least expensive form of vapor-phase bioreactor available.

In contrast, biotrickling filters consist of pollutant-degrading organisms immobilized on a packing media (usually plastic or ceramic materials), and they employ continuous water recirculation over the media. Biotrickling filters offer enhanced opportunities for pH and process control over biofilters. Biotrickling filters can also be designed with smaller footprints than biofilters because the packing height within trickling filters can be significantly greater than the height of media within biofilters, since the synthetic trickling filter packing will not deteriorate and compact over time. Bioscrubbers initially absorb vapor-phase contaminants into a recirculating liquid. The dissolved contaminants then undergo a subsequent biological treatment step. Absorption and biodegradation can occur within a single vessel, or each treatment process can occur sequentially in separate structures. Bioscrubbers are also commonly referred to as *biowashers.*

Concentrating now on biofilters only, the successful biological treatment of vapor-phase contaminants relies on the survival and continued growth of metabolically suitable microorganisms within the biofilter media. Thus, different biofilter designs influence, to various degrees, the suitability and optimization of the microbial growth environment present. Biofilter designs offering limited process control correlate with a limited optimization of the microbial growth environment. While these system designs are relatively inexpensive from a capital cost perspective, they are more prone to upsets and typically have fluctuating removal efficiencies. On the other hand, biofilter designs with extensive process control can optimize and maintain favorable microbial growth conditions. These systems are typically able to maintain consistently high removal efficiencies, but at a greater capital cost. The major features, benefits, and limitations of different biofilter system configurations are described below.

Open Soil Bed Biofilters

The earliest biofilters developed from the recognition that soils, in reality the microorganisms present in soils, have the ability to destroy odors. Soil beds are the simplest and cheapest form of biofiltration, literally using soil as the packing media. Perforated pipes are laid in trenches and covered with soil, so that the discharged air passes up through the soil bed before it is released to the atmosphere. Figure 8.8.3 shows the major elements of a typical soil bed biofilter.

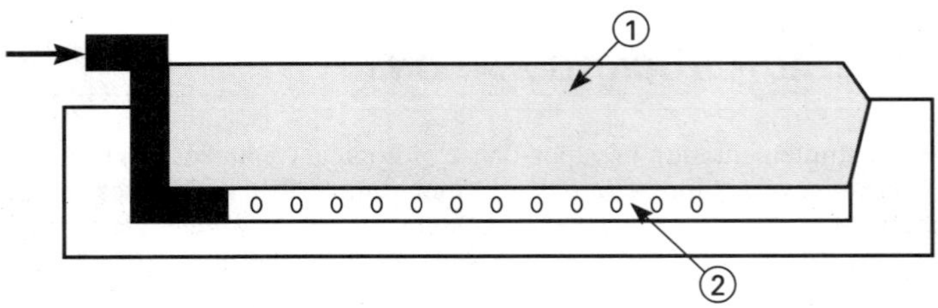

FIGURE 8.8.3 Cross-sectional drawing of a typical soil bed biofilter. The soil media (1) is placed above the air distribution system (2), which is located either below or on grade. A leachate collection system and/or secondary humidification system can also be used if needed.

Soil bed biofilters are still in use for many applications, but particularly for odor control, in which the contaminant loading on the biofilter is small. Their greatest advantages include mechanical simplicity and low cost. Trenching and piping can be installed quickly and easily. Soils are typically rich in microbial species capable of degrading organic compounds, so no inoculation is needed.

The success of a soil bed biofilter, defined here as the maximum contaminant removal efficiency attainable in combination with consistency of treatment, is dependent upon three major factors: (1) the quality of the soil, (2) the types and concentrations of contaminants present, and (3) the weather. The soil used as packing material for this biofilter design must have large, well-connected pores to provide the high permeability that will allow the air to pass through at a low pressure drop. Small amounts of organic material are also needed to provide a benign environment and nutrients for the microorganisms. Soils with a high clay content are not usable because they will pack tightly and block airflow. While it is possible to fill the trenches with soil from another site, this may lead to increased project costs.

Soil beds are not appropriate for high influent contaminant concentrations (high contaminant loads) because the relatively small particle size of soil makes the bed susceptible to clogging if the biomass grows rapidly. In addition, the biomass cannot be removed easily. Because natural soils are often nonhomogeneous and tend to settle and compact, airflow through the bed may short-circuit, overloading some areas of the bed while bypassing others. Soil beds may also suffer weather-related problems. Heavy rainfall may saturate the soil, interrupting airflow with resultant short-circuiting. Saturated soils may also become anaerobic. Soil beds are subject to freezing during the winter if the warm airflow is interrupted.

It is also important to note that since soil beds are typically installed without an impermeable liner beneath the air distribution piping and soil bed, leaching of partially biodegraded or nonbiodegraded compounds to the subsurface soils and groundwater is possible. This problem may occur during periods of heavy rain. The condition may also occur if the contaminants to be treated are unsuitable for rapid biodegradation because of high concentrations and/or the presence of difficult-to-degrade chemical structures.

Accurately measuring the performance of open-bed systems can be difficult. Influent contaminant concentrations may be obtained by analyzing vapor samples taken from the gases before they enter the air distribution system. Effluent concentrations may be monitored in a qualitative fashion by using a handheld detector, such as an organic vapor analyzer (OVA) or photoionization detector (PID), suitable for the contaminants of concern, as one walks over the surface of the biofilter. However, in open systems the biofilter effluent mixes with the atmosphere to some degree as soon as it leaves the filter bed. Air can be quantitatively sampled by placing a small collection hood on the surface of the bed (VDI Richtlinien 3477, 1991), but this may create a back pressure which disturbs the flow, raising doubt as to whether the sample is representative. Effluent vapors captured by the hood can be adsorbed onto a chemical trap for future desorption and analysis, or captured into gas sampling bags or evacuated metal canisters for later analysis. If the hood method of sampling is used, it is important to sample from several randomly selected points on the biofilter surface, since it is unlikely that treatment is completely uniform across the entire surface of the biofilter bed.

Open Single-Bed Biofilters

Open single-bed biofilters, in the form of either "in-ground" or engineered structures, are perhaps the most widely used form of biofilter design and include various improvements over the simplest soil bed biofilters. The construction of these systems is more elaborate, and the corresponding capital costs are also higher.

In-ground systems often include a liner to prevent the escape of contaminated vapor or leachate into the surrounding soil, as shown in Fig. 8.8.4. To prevent flooding from rainfall or inadvertent overwatering, these designs often include a drainage system. A layer of material such as gravel is installed around the air distribution piping. The large pores of the gravel help to evenly distribute the airflow and prevent clogging of the distribution piping by the biofilter media. The gravel layer also serves as a drainage pathway for water. The bottom of the biofilter is sloped to allow collection and removal of the water. A geotextile fabric may be spread on top of the gravel layer to prevent clogging of the airholes from pieces of the media, though it has been reported that the fabric itself may clog (Williams and Boyette, 1997). The packing material can simply be

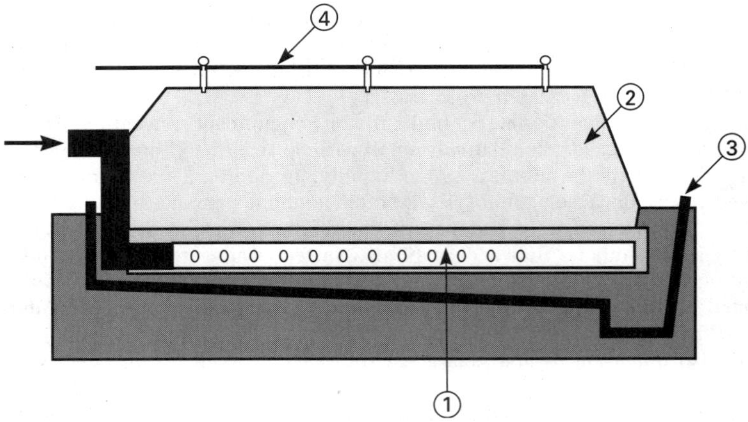

FIGURE 8.8.4 Cross-sectional drawing of an open single-bed biofilter. In this version of the design, the packing media is simply placed above the air distribution system, which is located either below or on grade. Major components include (1) air distribution system, (2) media, (3) optional impermeable liner, which can be installed to collect any leachate generated, and (4) optional secondary humidification system, shown here as sprinkler heads.

stacked on top of the gravel or fabric; or if a greater packing depth is required, sides for the biofilter can be constructed to contain the packing media. Wood and concrete are frequently used in wall construction.

Instead of soil, these biofilters contain packing material made from compost and other amendments, all chosen to maintain appropriate porosity and other desirable characteristics. Compost and other natural media often include particles of a wide variety of sizes, and the smaller particles may be gradually washed downward to accumulate in the drainage layer or the air distribution system. Careful choice of the media and flushing of the drainage system may alleviate potential problems. It is important to design the drainage piping for easy cleanout.

Engineered variations of the in-ground design use a concrete slab as a support for the biofilter packing media. In one design, concrete slabs with perforations to allow airflow are constructed on top of an air distribution system. These slabs can support the weight of heavy equipment used to load, mix, and unload the packing, and allow for the uniform distribution of air through the medium. As an even more engineered alternative, specialized concrete or plastic plates, containing an integral air distribution and leachate collection system, can also be installed (Leson et al., 1993).

Last, open-bed biofilters can be built entirely from concrete, wood, metal, plastics, or fiberglass. These systems typically have contiguous walls and floors, preventing the release of leachate and/or contaminated vapors from the bottom and sides of the unit, as shown in Fig. 8.8.5. Many open-bed systems come preengineered and/or prefabricated, ready for use. In other situations, they can be custom-designed with regard to size and shape for a particular application. Many of these systems also utilize secondary humidification systems in the form of sprinklers placed above the bed surface and/or irrigation hoses buried within the media.

Open-bed biofilters may suffer the same weather-related and monitoring problems previously described for soil bed systems. While heavy rainfall may saturate the pack-

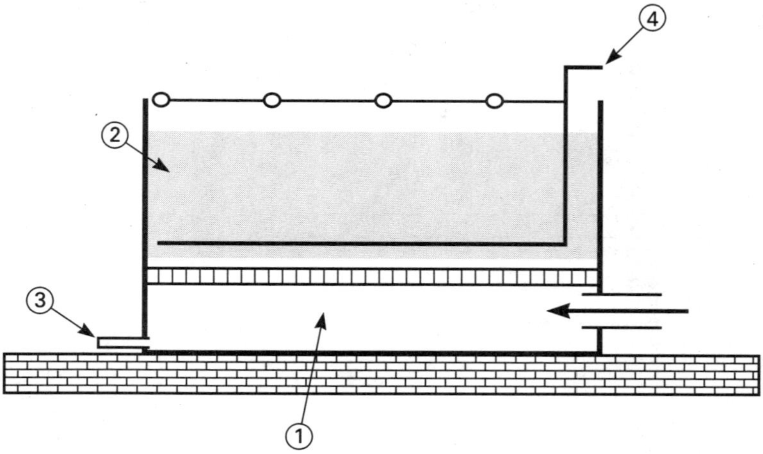

FIGURE 8.8.5 Cross-sectional drawing of an open single-bed biofilter. In this version of the design, the entire biofilter system is constructed above grade, and sidewalls are used to hold the packing media in place above the air distribution system. Major components include (1) air distribution system, (2) media, (3) leachate drain and collection, and (4) optional secondary humidification systems, shown here as sprinkler heads above the bed and also watering hoses buried within the packing medium.

ing, interrupting airflow with resultant short-circuiting and the formation of anaerobic zones, this problem can be minimized through the use of inexpensive covers.

Engineered Multiple Open-Bed Biofilters

Engineered open-bed biofilters can be operated as multiple units in parallel, as shown in Fig. 8.8.6. Often this is done simply to make use of limited space. Individual biofilter units may be stacked vertically, with side ports allowing the lower beds to discharge around the upper beds. Although this design increases costs, it offers the flexibility for one or more beds of the system to be shut down while the remaining beds continue in operation, providing for maintenance and media replacement without interruption of the total airflow.

Closed Single- and Multiple-Bed Biofilters

Closed-bed biofilters enclose one or more filter beds within an airtight engineered structure, as shown in Fig. 8.8.7. By shielding the beds from the outside weather, a more controlled (and ideally favorable) microbial growth environment can be maintained, resulting in higher removal efficiencies as well as more consistent treatment. For example, control of bed moisture content is improved because a closed-bed biofilter is protected from rainfall and direct sunlight. Other advantages of closed-bed systems include more consistent temperature control, less likelihood of bed freezing if gas flow is temporarily discontinued during winter months, and easier monitoring. In a closed system, representative effluent gas samples can be readily collected from the discharge ducting.

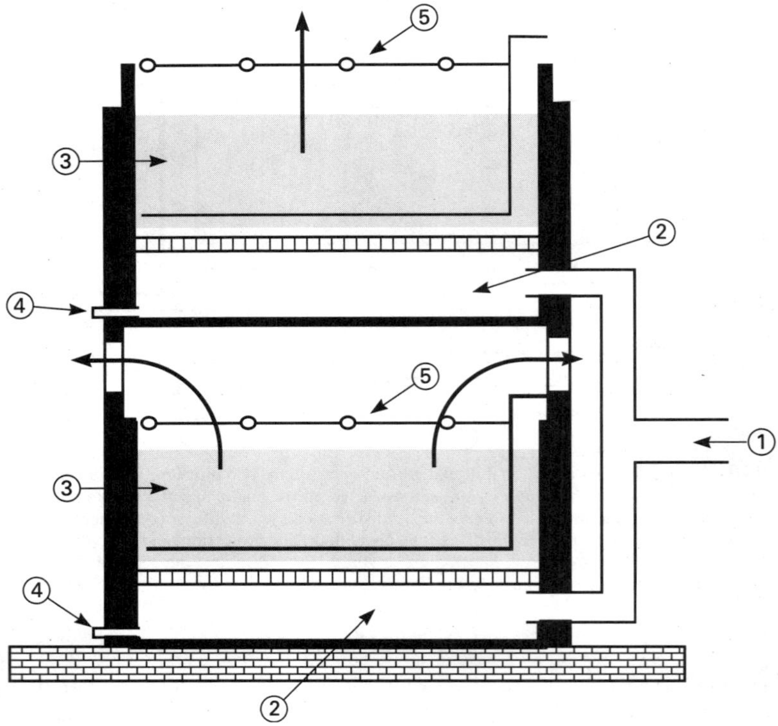

FIGURE 8.8.6 Cross-sectional drawing of a multiple open-bed biofilter. In this reactor design, several biofilter beds are constructed above one another to minimize the footprint of the system. The beds operate in parallel to one another, and the system can be designed to independently control the flow of waste gas to each bed. Such control allows for single-bed shutdown for maintenance purposes or for medium removal and replacement. Major components include (1) influent ducting, (2) air distribution systems, (3) media, (4) leachate drain and collection, and (5) optional secondary humidification system, shown here as a combination of sprinkler heads and buried watering hoses.

Real-time monitoring is also possible, and may be critical to regulatory approval of the project. In addition, if closed-bed systems are operated under negative airflow (fan downstream of the biofilter bed), any leaks in the biofilter structure will not result in the release of partially treated or untreated vapors to the atmosphere.

Closed-bed biofilter systems also allow upward or downward airflow through the filter bed(s). This flexibility increases the moisture control options within the system. In open-bed configurations, contaminated vapors must flow up through the media. The highest concentration of contaminants enters the bottom of the bed, and the greatest amount of biological activity and bed drying also occurs in this location. However, secondary humidification sprinklers typically add water to the top of the bed. Because the water must trickle down through the media, it is very difficult to remoisten the bottom of the bed uniformly. The water tends to collect in streams as it flows downward, flooding parts of the media and bypassing others. This problem can be addressed by using

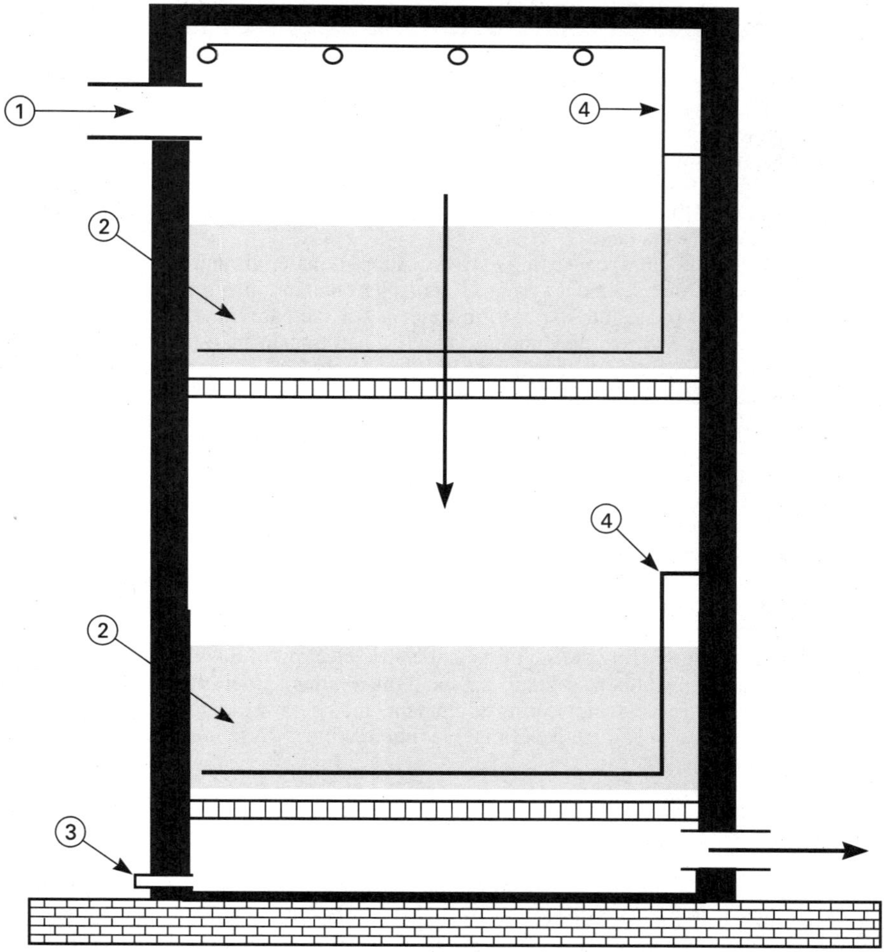

FIGURE 8.8.7 Cross-sectional drawing of a closed multiple-bed biofilter. In this reactor design, the flow of contaminated vapor can be routed either up or down through the medium. Downflow operation with beds operating in series is shown. Major components include (1) influent ducting, (2) biofilter beds, (3) leachate drain and collection, and (4) optional secondary humidification system, shown here as a combination of sprinkler heads and buried watering hoses. (*Courtesy of George J. Skladany.*)

irrigation hoses buried within the media to selectively control moisture, as shown in Fig. 8.8.7. With a downflow biofilter, supplemental water can be added directly to the top of the bed where the drying is most rapid; this option is also shown in Fig. 8.8.7.

Closed-bed biofilters have been constructed of insulated metal panels, concrete, and fiberglass. While their capital costs are in most cases higher than those of simple open-bed systems, the consistently high removal efficiencies attainable provide a return benefit for critical applications.

BIOFILTER DESIGN

Biofilter system design consists of two major tasks: (1) sizing the unit, which entails determining the type and volume of packing media required along with the appropriate vapor contact time, and (2) engineering the physical system, which includes specifying air pretreatment equipment, the reactor vessel, and ancillary process control equipment needed to create favorable environmental conditions within the media. The overall goal is to design a biofiltration system that provides environmental compliance in a consistent, cost-effective manner.

The biofilter medium, consisting of matrix support materials and biomass, forms the catalytic basis of the treatment system. Vapor-phase contaminants enter the media and pass from the gas phase into the biofilm water layer, then are transported into individual microorganisms where they undergo biochemical transformation. The composition of the media directly affects the flow of air through the system, gas-liquid mass transfer, microbial distribution and density, as well as the retention of water and nutrients. The metabolic capabilities of the microorganisms present determine what compounds can be degraded, how fast they are transformed, and what (if any) daughter products are formed. Thus, the proper selection of medium and supporting microbial activity determines the maximum potential effectiveness of biofiltration systems.

The function of the air pretreatment system, constructed bioreactor vessel, and supporting process control equipment is to provide the active biomass with the most favorable microbial growth conditions possible under the technical and financial constraints imposed on the project. Properly designed and operating hardware allows the biomass present to reach its full metabolic potential, resulting in the highest level of system effectiveness. Engineering design options include selection of the biofilter reactor configuration, air pretreatment system, air handling system, process control equipment, materials of construction, monitoring equipment, and operating conditions. Engineering efforts are made to maximize beneficial medium qualities while minimizing any corresponding deficiencies.

Problem Definition

The key to successful biofilter design is to have an accurate problem definition with which to work. For most projects, the end user supplies the consulting engineering firm and/or biofilter system vendor with a problem definition which includes the contaminants and concentrations to be treated; influent vapor flow rate, temperature, pressure, and relative humidity; particulate load; contaminant removal efficiency required; degree of instrumentation and control required; and budget. In many cases, the information supplied is inaccurate or only partially complete. Poor problem definition is the cause of many of the problems experienced by full-scale biofilter systems. This point cannot be overemphasized.

Depending upon the biofilter application and treatment levels required, the composition and concentration of the waste gas may require extensive assessment. Generally, when a biofilter is to be used at a specific site, the regulated target contaminants are known as a function of the process which generates them. However, depending upon the methods used to characterize the waste gas stream, not all the constituents present may be identified. Plant engineers and managers are usually able to provide basic information about the specific input chemicals which go into the process as well as the flow rate of the air to be treated. In many cases, information about the minor constituents and by-products formed is missing or unknown. This is true when one is treating vapors produced from batch operations, especially if many different products are manufactured, or when one is treating air generated from processes using ill-defined feedstock. For example, when vapors produced from the processing of wood, oil, or other similar organic materials are treated, a significant fraction of the constituents in the off-gas stream may be unknown.

If additional analytical evaluation is warranted, typical Environmental Protection Agency (EPA) gas sampling methods, such as EPA method TO-11 or TO-14 (for aldehydes, ketones, and chlorinated and nonchlorinated aromatics) can be used. Often, minor components which may be harmful to the biomass of the system may be present that can jeopardize the reactor performance if not accounted for.

Monitoring over time may be required to establish average and spike concentrations of the contaminants. For example, batch processes typically release very high concentrations of chemicals when the production cycle is complete and the reaction vessel is initially vented. As the venting process continues, vapor-phase concentrations usually decline significantly. Spike concentrations may be toxic to the microorganisms in the filter bed, pass through the bed untreated, or pass through only partially treated. If such contaminant spikes are likely to occur, performing load equalization on the gas should be considered as a way to create a more steady-state inlet biofilter load. Activated carbon can be used prior to the biofilter to adsorb and desorb these spikes in a timely manner (Graham, 1996; Medina et al., 1995; Weber and Hartmans, 1995; Bentz, 1992). Such load equalization can also be effective in supplying contaminant to the reactor during plant process stoppages or shutdowns.

Biofilter Sizing

Once an accurate problem definition is available, biofilter sizing is begun. Biofilter sizing determines the type and volume of packing material required as well as the contact time needed to attain the desired contaminant removal efficiency. For any given application, the performance of a biofilter system (the target compound removal efficiency) is generally proportional to the volume of biofilter medium used. As the volume of media increases, the overall target compound removal efficiency also increases. Since the cost of a biofilter system is also proportional to the volume of media used, a balance between system cost and system performance must be established.

Unfortunately, no comprehensive and accurate biofilter sizing algorithm is publicly available. Individuals responsible for sizing biofiltration systems require an understanding of the biofiltration process, and mathematical modeling is a useful and sometimes necessary tool. In general, the vapor contact time required for treatment is the single most important design parameter.

Terminology. Several key terms used in biofilter system design are defined and explained in this section. The *vapor flow rate* to the biofilter is expressed as unit volume of gas per unit time (m^3/h or ft^3/min). *Contaminant concentrations* are expressed as mass per unit volume (g/m^3 or lb/ft^3). If contaminant concentrations are given in a volumetric ratio, i.e., parts per million by volume (ppmv), they should be converted to an equivalent mass-per-volume form. *Contaminant mass loading* is defined as the total mass of contaminant per unit time (g/h or lb/h).

The term *empty-bed vapor contact time,* or simply *contact time,* is most often used to calculate the volume of packing media required for a given application. The contact time (actually calculated in units of hours or minutes, respectively, but commonly converted and referred to in units of minutes or seconds) is usually defined as the nominal biofilter bed volume occupied by media, divided by the actual volumetric vapor flow rate through the bed. This definition for vapor contact time neglects the bed volume actually occupied by solid matrix material, hence its more proper definition as the *empty-bed vapor contact time.*

Vapor contact time is also sometimes defined in the literature as the *void volume of media* divided by the volumetric flow rate through the bed, creating some confusion when one is comparing results for one type of media to another. Reported contact times that use this void volume convention will be shorter than empty-bed vapor contact times, as defined above, for the same size system and gas flow rate. Caution should therefore be used when reviewing and reporting contact times in the literature to avoid misunderstandings.

Empty-bed vapor contact time can be thought of as the *relative time* that the vapor spends in contact with the media before exiting the system. Generally, as the vapor contact time increases, either by supplying a lower volumetric flow rate to the system or by increasing the volume of media, the pollutant removal percentage in the system increases. Typical vapor contact time requirements for industrial applications range from 25 s for the treatment of low-concentration VOCs and odors to 60 s for the control of high-concentration ethanol emissions (Leson et al., 1993). Table 8.8.1 lists contact times for a variety of applications and the reported biofilter performance at those contact times.

Specific terms are also used to describe the mass rate of contaminant entering into and being removed by a specific volume of packing media. The *contaminant loading rate* is used to describe the contaminant mass loading to the biofilter per unit volume of medium. It is the product of the inlet gas concentration and the gas flow rate divided by the bed volume. The contaminant loading rate is usually expressed as mass of contaminant per cubic volume of media per unit time [g/(m^3/h) or lb/(ft^3/h)]. Likewise, *elimination capacity* is used to describe the rate of contaminant removed per unit volume of biofilter media [g/(m^3/h) or lb/(ft^3/h)].

Instantaneous and *overall* elimination capacities for various biofilter media can be found in the literature. As originally used in the mathematical modeling of biofilter performance, the *instantaneous elimination capacity* is defined as the vapor velocity through a biofilter bed multiplied by the concentration gradient of the target contaminants at any point in the bed (Deshusses and Hamer, 1993; van Lith et al., 1990).

TABLE 8.8.1 Examples of Contact Times and Removal Efficiencies for Various Compounds

Contaminant	Contact time, s	Removal efficiency, %	Reference
Composting odors	90	90	Kuter et al., 1993
Composting odors	25	Not given	Kuter et al., 1993
BTX	90	>95	Tahraoui et al., 1994
BTEX	42	90	Severin et al., 1993
Foundry alcohols	24	>99	van Groenestijn and Hesselink, 1993
Foundry alcohols	30	80	van Groenestijn and Hesselink, 1993
Jet fuel (JP-4)	180	95	Peters et al., 1993
Iso-pentane	180	40	Togna and Singh, 1994
Hexane	60	99	Morgenroth et al., 1995
Hydrogen sulfide	35	>99	Allen and Yang, 1991
Methyl mercaptan	18	90	Allen and Phatak, 1993
Dimethyl disulfide	30	>90	Arpacioglu and Allen, 1996
Alpha-pinene	30	>99	Mohseni and Allen, 1996
THC*	26	82	Gilmore and Briggs, 1997

*Total hydrocarbons from a batch chemical manufacturing operation; primarily methanol, acetone, and heptane.

Assuming that the vapor velocity remains constant, the instantaneous elimination capacity changes with depth through the bed, and provides information about only one specific point within the media. In contrast, the *overall elimination capacity,* or EC, is defined as the overall, or total, mass of target contaminant removed from a selected airstream per unit time per unit volume of media [(g/(m^3 • h) or lb/(ft^3 • h)].

It follows that the elimination capacity can only be equal to or less than the contaminant loading rate. When the loading rate is equal to the elimination capacity, 100 percent of the contaminant is removed. When the loading rate is greater than the elimination capacity, less than 100 percent of the contaminant is removed. The overall elimination capacity is a useful parameter for biofilter design and sizing. However, the elimination capacity of a biofilter system does not imply any level of target compound removal efficiency. The *removal efficiency* is defined as the difference in influent and effluent contaminant concentrations, divided by the influent concentration. It can also be calculated as the ratio of the elimination capacity to the load. Removal efficiencies are typically expressed as a percent value rather than as a decimal number.

Figure 8.8.8 shows a typical graph of biofilter loading rate versus overall elimination capacity at two different contact times. Each vapor contact time results in a different

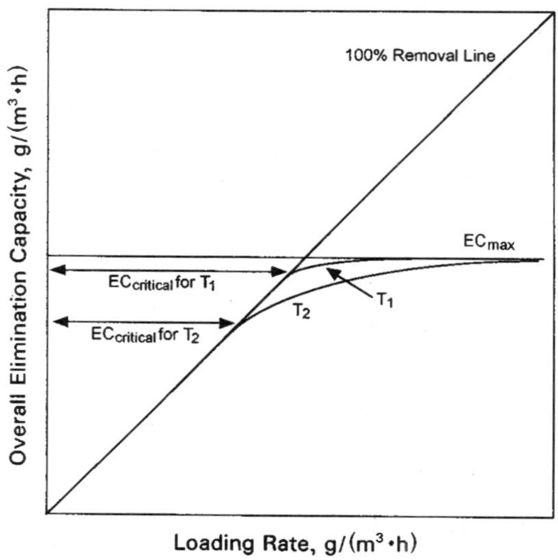

FIGURE 8.8.8 Graph showing the relationship between loading rate and elimination capacity for a target contaminant at two different empty-bed vapor contact times ($T_1 > T_2$). Also shown: (1) the 100 percent removal line, (2) maximum elimination capacity, and (3) critical elimination capacity.

curve on the graph, although the curves tend to overlap at low and high loading rate values. The 100 percent contaminant removal line, corresponding to equal values for the loading rate and elimination capacity, is also shown. The trends shown are typical for most applications. As illustrated, the overall elimination capacity is a direct function of both the concentration and time variables present in the contaminant loading rate; i.e., the elimination capacity increases as the contaminant concentration and/or vapor contact time increases.

Under low-loading conditions, the overall elimination capacity is typically equal to the loading rate, resulting in 100 percent contaminant removal. As the loading rate increases (either by increasing flow rate or by increasing inlet concentration), a point is reached beyond which the loading rate exceeds the overall elimination capacity of the bed, and less than 100 percent of the contaminant is removed. For some compounds and media, this deviation from the 100 percent removal line is sharp. For other compounds and media, the deviation from 100 percent removal is more gradual. The *critical elimination capacity,* or EC_{crit}, is the elimination capacity value at which the removal efficiency starts to deviate significantly from 100 percent. The critical elimination capacity value is not numerically determined in a precise manner. Rather, it is selected by visually interpreting the plotted graph of loading rate versus elimination capacity. Critical elimination capacity values remain a function of both contaminant concentration and vapor contact time.

If the loading rate continues to increase, the elimination capacity eventually reaches a maximum value called the *maximum elimination capacity,* or EC_{max}. Once the maximum elimination capacity of a given media is reached, any further increase in inlet contaminant concentration and/or contact time will not cause a further increase in overall elimination capacity. Therefore, the maximum elimination capacity value is independent of contaminant concentration and vapor contact time.

Biofilter Sizing Based on Literature Information. One approach to biofilter sizing relies on using elimination capacities and vapor contact times available from the technical literature. Keep in mind that elimination capacities are media- and contaminant-specific; their use may not successfully transfer to other projects with different contaminants, concentrations, site conditions, and cleanup levels. Typically the values reported in the literature are maximum elimination capacities, since these values are independent of concentration and contact time. Table 8.8.2 shows some representative maximum elimination capacities obtained from the literature. In general, the more soluble the contaminant, the higher the elimination capacity. Care must be taken when using literature values for design, however, since the conditions under which the values

TABLE 8.8.2 Examples of Elimination Capacity (EC) for Various Compounds

Contaminant	EC, $g/(m^3 \cdot h)$	Maximum EC, $g/(m^3 \cdot h)$	Reference
Acetone	120	—	Gilmore and Briggs, 1997
Acetone	—	40	Dharmavaram et al., 1993
Butyl acetate	—	35	Ottengraf, 1986
Dimethyl disulfide	—	7–10	Arpacioglu and Allen, 1996
Ethanol	120	—	Kiared et al., 1996
Ethanol	220	—	Hodge and Devinny, 1994
Ethyl acetate	—	30	Ottengraf, 1986
Heptane	70	—	Gilmore and Briggs, 1997
Hydrogen sulfide	—	130	Yang and Allen, 1994
Methanol	100	—	Gilmore and Briggs, 1997
Methanol	—	95	Baltzis and Shareefdeen, 1993
Alpha-pinene	30–35	—	Mohseni and Allen, 1996
Styrene	—	80	Ottengraf et al., 1986
Toluene	60	—	Severin et al., 1993
Toluene	—	25	Ottengraf, 1986
Toluene	70	—	Kiared et al., 1996
Triethylamine	—	140	Tang et al., 1996

were determined may not have resulted in high removal efficiencies, and many of the elimination capacity values reported in the literature are obtained from laboratory treatability studies using synthetic waste streams that may not accurately reflect actual field conditions.

Sizing biofilters from published information can be done in several ways. It is possible to simply copy the dimensions of a field system already in use. To do so, it is important to compare the compounds, concentrations, and removal efficiencies reported in the literature with those of the new application. For example, suppose the new system is required to treat a 500 ft^3/min (850 m^3/h) waste stream containing 50 ppmv (0.0014 g/m^3) of hydrogen sulfide at a removal efficiency of 95 percent. The literature gives an account of a biofilter installed at a wastewater facility treating a 1000 ft^3/min (1700 m^3/h) waste stream containing 75 ppmv (0.0021 g/m^3) of hydrogen sulfide with an average removal efficiency of 98 percent. Since the new project's critical parameters compare favorably with those of the active system, the new application has a good chance of success by simply duplicating the old system. If information on the composition of the active system's packing media and inoculum is also available, the risks associated with copying the system will be minimized.

It is also possible to design a system by using reported elimination capacities. While this approach is simple in practice, keep in mind that elimination capacities are contaminant-, inoculum-, and packing-media specific. It is also beneficial to know specifically how the tests were performed to arrive at the reported values. To use this approach, the mass flow rate is calculated by multiplying the contaminant concentration by the vapor flow rate. Next, the mass flow rate is divided by the maximum elimination capacity to obtain the volume of biofilter packing medium required. The packing volume is then divided by the desired volumetric flow rate to arrive at an empty-bed vapor contact time. The calculated vapor contact time should be compared with the actual contact time at which the maximum elimination capacity was determined. If the calculated contact time is greater than or equal to the contact time reported in the literature, it may be appropriate for use. However, if the calculated contact time is less than the contact time reported in the literature, the calculated contact time should be increased if a margin of safety is required.

A more conservative biofilter sizing approach uses the critical elimination capacities, if available, rather than maximum elimination capacities, in performing the calculations above. If the maximum elimination capacity is much larger than the critical elimination capacity, significantly shorter contact times will be calculated by using the maximum elimination capacity.

Biofilter Sizing Based on Mathematical Models. Mathematical models can also be used to estimate elimination capacities. Intrinsic properties of the biofilter media, such as the interfacial area, and of the microbiology, such as the biokinetic constants for the specific contaminants of interest, and biofilm thickness, are usually required when such mechanistic mathematical models are used. These parameters are not always known, and individuals responsible for the sizing of biofilters often have to either estimate or rely on parameters determined from similar biofilter systems. For these reasons the use of mathematical models can be complicated or unreliable. However, if the appropriate model is used and if model parameters are known, estimation of elimination capacities as functions of concentration and contact time can be achieved, so that design conditions resulting in high removal efficiencies can be determined. In any case, the design engineer using models to size a biofilter should add a safety factor to the bed volume to ensure that the biofilter is not undersized.

A number of mathematical models have been developed to describe biofilter operation (Andrews and Noah, 1995; Deshusses et al., 1995; Hodge and Devinny, 1995;

Shareefdeen et al., 1993; Ottengraf, 1986). A detailed mathematical analysis of these and other models is beyond the scope of this section, and the reader is advised to consult the specialized literature for more detailed information.

Ottengraf's biophysical model (Ottengraf, 1986) is probably the most widely used model for design purposes. It allows the prediction of pollutant gas-phase concentrations as a function of height in a biofilter bed. This model is based on a planar biofilm geometry (see Fig. 8.8.2) and relies on intrinsic parameters such as Henry's law coefficients for the target pollutants, packing media interfacial area, biofilm thickness, and biokinetic rate constant. The latter constant usually needs to be determined experimentally. At low contaminant concentrations, and for the treatment of sparingly soluble compounds, this model may not be appropriate. In addition, the use of this model depends on whether the biofiltration process is limited by the diffusion of the pollutant into the biofilm or by the biodegradation rate of the pollutant within the biofilm. Once these determinations or assumptions have been made, the use of the model is straightforward. A single equation is used to calculate concentration along the bed under diffusion-limited conditions, and a second equation is used similarly for reaction-limited conditions (Ottengraf, 1986). In any case, the design engineer will have to cope with the fact that key parameters may be unknown. For example, biokinetic data for a specific contaminant may be unknown, requiring the assumption that its biokinetic parameters are approximately equal to those of a structurally similar compound with known biokinetic data. Slight corrections of biodegradation rate constants using analogies and common sense may also be necessary.

Recently, a new and promising generation of models have been developed (Govind et al., 1997; Johnson and Deshusses, 1997; Choi et al., 1996). These models use quantitative structure activity relationships (QSARs); that is, they seek to predict the behavior of biofilters based on the structure and/or physicochemical properties of the pollutant undergoing treatment. Such an approach has already proved useful for wastewater treatment. So far, first-order degradation rate constants (Choi et al., 1996), maximum elimination capacities and critical elimination capacity (Johnson and Deshusses, 1997), and biokinetic constants (Govind et al., 1997) have been modeled using QSARs. It is likely that the use of models will become common practice for biofilter design once the accuracy of their predictive results is verified using data from full-scale systems.

Biofilter Sizing Based on Project-Specific Data. A technically superior approach to biofilter sizing is to use data generated from laboratory or field pilot tests. Field pilot tests using a slipstream of the actual waste gas as well as the packing media to be installed in the full-scale system provide the best data for system design. During these tests, elimination capacities and removal efficiencies can be determined by varying the gas flow rate and contaminant concentrations.

Field studies can be conducted to merely confirm a particular design (i.e., contact time and resultant removal efficiency) or to provide design matrix data relating flow rates and concentrations to removal efficiencies. A study to confirm a particular contact time and removal efficiency need only be performed at a single flow rate with the target contaminant concentration. This approach provides limited process optimization options in exchange for a relatively low cost. In contrast, design matrix data provide information relating several different flow rates and contaminant concentrations to corresponding removal efficiencies. At a minimum, at least two different flow rates and two different contaminant concentrations must be evaluated. This approach is more time-consuming and expensive, but provides the data needed to mathematically model the biofiltration process. The model can then be used to predict system behavior at flow rates and concentrations different from those actually tested.

Design matrix data generated from such studies can be used to obtain the intrinsic biodegradation order of the system. Determining the kinetic order allows the use of models to predict biofilter performance under different conditions, such as removal efficiencies and contact times, other than those tested. The order, either first or zero, can be determined by graphically plotting influent contaminant concentration (x axis) versus removal efficiency (y axis) for each of the contact times tested. For first-order biodegradation kinetics, the graph will yield a series of lines parallel to the x axis. For zero-order biodegradation kinetics, the graph will generate a series of curves.

The order of the biodegradation rate has significant implications for design. With zero-order kinetics, the rate of contaminant removal is independent of concentration. The removal efficiency, however, is a function of both contaminant concentration and vapor contact time. Under these conditions, the following relationships hold among contaminant concentration, contact time, and removal efficiency:

1. If the contact time is held constant, an increase in contaminant concentration will result in a decrease in removal efficiency, while a decrease in contaminant concentration will result in an increase in removal efficiency.
2. If the contaminant concentration is held constant, an increase in contact time will result in an increase in removal efficiency, while a decrease in contact time will result in a decrease in removal efficiency.
3. If the contaminant concentration increases and the contact time decreases, the removal efficiency will decrease.
4. If the contaminant concentration decreases and the contact time increases, the removal efficiency will increase.
5. If the contaminant concentration and contact time both increase or decrease together, the effects on removal efficiency are uncertain.

With first-order biodegradation kinetics, the removal efficiency is a function of contact time only. The implications of first-order biofilter design are simple:

1. If the contact time increases, the removal efficiency also increases.
2. If the contact time decreases, the removal efficiency also decreases.
3. If concentrations increase or decrease, the removal efficiency remains unchanged.

If the biokinetic constants for the target contaminants are not known, and an estimate of reaction order is needed for design, an approximation of reaction order can frequently be made using Henry's law constants. With this approximation method, soluble compounds (those with low Henry's law constants) will typically exhibit zero-order behavior, while sparingly soluble compounds (those with high Henry's law constants) will typically exhibit first-order behavior. While this approach may be an oversimplification (since the reaction order is strictly a function of liquid-phase contaminant concentration and not Henry's law constant), for many typical biofilter applications this approximation has been useful.

The use of zero- and first-order kinetics to model or predict biofilter behavior is rigorous only for the treatment of a single vapor-phase contaminant. For waste streams containing mixtures of contaminants, plots of the overall elimination capacity versus influent concentration will provide an aggregate kinetic order. For multiple contaminants with different physical properties, the aggregate kinetic order will be a combination of the kinetic orders for the individual contaminants. While the use of the aggregate kinetic order may not accurately predict the removal efficiency for each specific contaminant present, it will tend to reflect the overall behavior of the system.

Engineering the Physical Biofilter System

Biofiltration systems consist of the following major components: (1) ducting and mechanical air handling systems, (2) the pretreatment system, and (3) the biofilter reactor vessel with internal components. An air distribution system, packing media (with or without microbia inoculum), and some select process control systems (i.e., media moisture control) are typically present within the reactor, depending upon the design configuration selected. Based on past knowledge and designs, present-day biofilters are generally configured with similar major components along with some modifications to meet the specific needs of the particular waste airstream. This section discusses the major process issues and physical components of biofilter systems.

Process Issues. Three major process issues significantly control the success or failure of biofiltration systems: moisture control, temperature, and pressure drop. These issues must be addressed throughout the design of the biofilter system's ductwork, pretreatment system, and reactor vessel.

Moisture Control. Moisture control is important for a biofilter to function efficiently. The media must contain appropriate amounts of inorganic nutrients and water, and severe fluctuations in moisture content must be avoided. Media drying can lead to the formation of shrink cracks and gaps between the media and the walls of the biofilter vessel, resulting in short-circuiting of untreated gases. While the air supplied to a biofilter is usually completely humidified in the pretreatment system prior to coming in contact with the media, the oxidation of organic contaminants is an exothermic reaction, and for a completely adiabatic system, the temperature of the air exiting the system will be slightly higher than that of the air entering the media. Therefore, the passing air will tend to remove moisture from the media on a continuous basis, leading to bed drying and associated problems if not corrected. In summer months, biofilter media is more prone to moisture loss, while during cold winter months, water from the humidified air may condense out in the media. All these factors will contribute to severe fluctuations in media moisture content unless an effective moisture control strategy is implemented within the biofilter housing. The requirement for effective moisture control within the biofilter packing, especially for the treatment of high-concentration waste streams, is increasingly being realized and documented (van Lith et al., 1997; Pinnette et al., 1996). Many biofiltration systems now employ two complementary moisture control systems: Moisture is added during air pretreatment and also periodically directly to the packing media.

Temperature. Temperature must be considered for its effect on the microorganisms present as well as on bed moisture control and physical integrity. Microorganisms have a temperature threshold which should be closely maintained. Most biofilters operate in the mesophilic temperature range (25 to 35°C), because the majority of heterotrophic species which are metabolically active against "hazardous" chemicals dominate within this range. Microbial growth and biodegradation rates increase with increasing temperatures, up to a maximum specific for each species present. Therefore, influent gas temperature is a variable that directly affects microbial activity and therefore compound removal efficiencies. If the influent gas stream has a relatively low temperature, a larger biofilter (more packing) will be needed to provide a longer contact time to compensate for the lower level of biological activity. If this waste stream is heated, a smaller filter, and hence lower capital costs, will be required. Waste streams at moderate temperatures (15 to 20°C) may only exhibit a modest increase in performance for the additional costs of heating.

The temperature of the gas also directly influences the control of moisture in the bed. High inlet gas temperatures will affect the relative humidity of the gas entering the biofilter and may cause the filter bed to dry out. In addition, high gas temperatures contribute to faster breakdown and aging of the media. If the bed temperature is too low, incoming moisture in the gas stream may condense out in the packing, interfering with gas flow and mass transfer.

Pressure Drop. Pressure drop is responsible for the major continuous operating cost of biofilters, due to the energy costs associated with moving air through the system. Usually the pressure drop across the media is the largest contributor to the overall system pressure drop. Pressure drops occur throughout a biofilter system, including all ducting, piping bends, pretreatment equipment, and exiting ductwork and stacks.

The pressure drop through any solid support media increases as both the linear velocity of the flowing air and the height of the media increase. The operating moisture content of the media also affects the media pressure drop. As the moisture content increases and the void space becomes filled with water, the pressure drop through the media increases. Typically, the pressure drop also increases as the media decomposes and compacts over time. Since the fan and motor sizes are selected at the beginning of the project, it is important to accurately estimate the starting pressure drop across the entire biofilter system, and to plan for anticipated pressure increases over time. Initial pressure drops through new media can be less than 0.5 in (125 Pa) of water, increasing to 4 to 8 in (1000 to 2000 Pa) of water (depending on the packing and system design) before replacement is needed.

Hardware. Once the biofilter media has been properly selected and sized, attention focuses on the physical biofilter system. The complete physical biofilter system consists of influent ducting and mechanicals, air pretreatment system, biofilter structure and internal components, effluent ducting and mechanicals, process control equipment, and monitoring equipment. Each of these items is discussed in the sections that follow.

Ducting and Mechanical Air Handling Equipment. These items consist of the ducting needed to convey the gas to be treated to and from the biofilter reactor, as well as the mechanical fans and/or blowers needed to move the gas through the system. This equipment must be chemically compatible with the gases to be handled, and the mechanical components are sized based on the current and anticipated future overall system pressure-drop requirements. Fiberglass, stainless steel, and galvanized steel are common ducting materials. If the biofilter cannot be physically located close to the gas discharge point, extensive ducting runs may be required. Ducting may have to be insulated to stop condensation from forming during winter months, and may also need provisions for cleaning depending upon the characteristics of the waste gas stream. Depending upon the materials of construction and size of the air handling equipment needed, the budget for ducting and mechanical system components may be a significant portion of the cost of a complete biofiltration system.

Pretreatment System. The pretreatment system initially handling the waste gas stream is essential to the overall success of the biofiltration process. The pretreatment system is designed to help optimize the particulate load, temperature, and moisture level of the gas entering the biofilter. In order to achieve proper system design, it is critical that all characteristics of the waste gas be known as fully as possible, including the gas composition and concentration, flow rate, concentration of particulate matter, temperature, and relative humidity. Because a biofilter is a biologically active system, the importance of maintaining optimal growth conditions is paramount. Hence, failure to treat the waste gas so that these optimal conditions are maintained will lead to suboptimal biological removal and possibly system upsets.

PARTICULATES. Biofilter media inadvertently acts as a filter for particulate matter, but such an action severely reduces the life span of the filter material (Standefer and Willingham, 1996). The removal of particulate matter prior to its entering the biofilter will prevent large pressure drops from forming and channels from developing. For example, particulate grease can enter the biofilter and attach to the bed material. Although such grease may be biodegradable, the rate of biodegradation may not exceed the rate of accumulation, and over time the bed may prematurely clog. These clogging conditions will result in an increase in system pressure drop, straining the blower and increasing power costs. These clogging conditions can also create channels where air will pass untreated, gradually lowering the performance of the system until either the media is entirely clogged or removal becomes severely reduced.

Methods to remove particulates from the waste airstream prior to its entering the biofilter typically include water sprays, filtering systems, and packed towers. Many particulate removal processes can also be used to adjust the temperature and moisture levels of the influent gas simultaneously. Electrostatic precipitators or high-energy venturi scrubbers can also be used for particulate removal, but at significantly higher capital and operating costs.

TEMPERATURE. As discussed previously, very high inlet gas temperatures not only will affect the viability of the microorganisms within the media, but also will contribute to drying of the packing media. If required, cooling of the waste gas may be performed through adiabatic cooling during humidification of unsaturated air or by forced cooling processes. Water sprays and packed beds offer effective and cost-effective treatment under many conditions, and the use of a heat exchanger may be necessary to remove excess heat. If gas preheating is required before entering the biofilter, gas-fired in-line heaters, steam injection, or heated recirculating water in a packed tower may be used for temperature adjustment. Keep in mind that heating or cooling the waste gas stream, especially at high volumetric flow rates, may significantly increase system capital and operating expenses (Swanson and Loehr, 1997).

MOISTURE. Perhaps the most critical parameter of the waste gas that affects the success or failure of a biofilter system is the relative humidity of the gas stream to be treated, especially if short gas residence times are used. If unsaturated air passes through a biofilter bed, the air will have a tendency to absorb moisture from the filter material until an equilibrium is reached with the air humidity, air temperature, and bed temperature. If this occurs over extended periods and supplemental water irrigation to the filter bed is not supplied, a gradual drying out of the filter media will occur. A bed with insufficient moisture will not support microbial growth, causing poor treatment of the waste gases. As the filter bed dries out, the material may shrink and channeling of untreated gas may occur. The air pretreatment system provides the primary moisture control adjustment needed for the waste gas stream, and should be designed to produce an inlet airstream to the media near 100 percent relative humidity. By doing so, the rate of dry-out of the filter bed will be minimal, while further moisture control can be performed using direct irrigation of the filter media. Typical humidification devices include steam injection systems, spray chambers, venturi scrubbers, or packed-bed towers.

Biofilter Structure and Internals. These items include the influent gas air distribution system, packing media, secondary humidification system, monitoring and control systems, and physical structure supporting and enclosing the media. Biofilter structures come in either open or closed configurations. The open design places the media on a supporting structure, which includes an air distribution system, with the filter bed open to the atmosphere. Open-bed systems are generally used where space is not a constraint and natural irrigation is acceptable. Enclosed biofilters are designed in many different shapes, including cylindrical and rectangular, and allow for better operational control of critical parameters such as filter bed moisture content and temperature. This increased control generally enhances performance and allows enclosed beds to be designed smaller than those in open-bed systems.

Open and closed systems are available in modular, semicustom, and custom designs. Modular designs typically have all system design engineering and fabrication complete; the user merely purchases the correct number of modules needed for a given application. These off-the-shelf systems can be provided at low cost because expensive customized engineering and fabrication are eliminated. Semicustom systems include the use of prefabricated units as well as standard engineering designs that allow flexibility in enlarging the size of a system. For example, prefabricated, standard-size structures holding packing media can be used with customized in- or aboveground air distribution systems. Semicustom biofilter designs can also employ the use of standardized packing "bays." Each bay is sized to contain a specific volume of packing media. Once the volume of media required is known, the correct number of bays can be integrated into a single biofilter unit. In this way, engineering design costs for each bay are minimized; the project engineering costs are those associated with finishing the integrated design. Last, biofilters can also be custom-designed for a specific application. Customization may be required for chemical compatibility issues, site space limitations, or specialized process design requirements.

MATERIALS OF CONSTRUCTION. The open and closed systems may be simple single beds or multiple-story beds constructed from a variety of materials. Single-bed systems are often used below grade and may require a liner system to contain and collect any leachate generated. This liner may be constructed of a variety of materials, and materials such as high-density polyethylene will prevent water leaching while withstanding microbial deterioration. Sidewalls may be constructed of concrete, stainless steel, or wood. However, excavated pits may not need structured siding at all, with the filter material just placed inside the pit on top of the air distribution system (van Lith et al., 1997). Multiple-story systems may also be useful where space constraints are limiting, larger gas residence times are required, and the isolation of certain chambers of the reactor is needed for maintenance (Wolstenholme and Finger, 1994). For closed-bed, aboveground systems, suitable vessel construction materials include concrete, polyvinyl chloride (PVC), polyethylene plastic, fiberglass-reinforced epoxy resin, or stainless steel (304 or 316). For smaller applications, it may be possible to purchase premolded containers made of these materials. Such containers will be less expensive and require minimal labor to set up. Larger systems may also be constructed using predesigned modules. These predesigned modules may be added to or removed from a biofilter application as needed. Whether an open or closed, single-or multiple-level bed, the choice and amount of material for construction will require careful consideration by the design engineer so that the chemical and physical characteristics of the incoming waste stream do not interfere with the structural integrity of the vessel.

AIR HANDLING AND AIR DISTRIBUTION SYSTEMS. A number of questions related to airflow through a biofilter must be addressed during the design phase of the project. First, will air be pushed or pulled through the media bed? Open-bed designs have only one choice; the air must be forced by positive pressure up through the packing media. Closed-bed designs can operate with either positive or negative pressure to move contaminated vapors through the media. The use of positive pressure with a closed-bed system means that the contaminated vapors will come into direct contact with the blower, possibly raising compatibility issues. In addition, if the reactor shell develops leaks, untreated vapors can be forced from the unit to the outside air. In a negative pressure situation, the blower is physically located downstream of the biofilter bed, ensuring that only treated air comes in direct contact with the blower. In addition, any leaks in the biofilter housing will result in clean air being brought into the biofilter, rather than contaminated air being released.

If a closed-bed design is selected, a decision must then be made as to whether the gas will flow up or down through the bed. Each flow design has advantages and disadvantages, particularly with respect to moisture control within the packing. A downflow biofilter has both contaminated vapor and water supplied to the top of the filter bed where the majority of degradation is occurring.

Care should be taken to design an influent air plenum that maximizes even air distribution through the filter bed(s). For downflow systems the headspace volume, gas flow velocity, and chamber geometry need to be considered. Upflow systems use a variety of designs to evenly distribute air up into the filter bed. In simple open-bed systems, gravel or stone above slotted pipes can be used to disburse the gas evenly. Openings in the pipes can be differentially placed to correspond to different gas pressures within the ducting (Williams and Boyette, 1997). More sophisticated air distribution systems utilize specialized air distribution plates made out of concrete, plastic, or metal. While the materials of construction and costs differ, the goal of all the systems is to provide even distribution of air into the biofilter bed to help ensure even treatment of all gases.

The choice of whether to use a blower or fan to feed a biofilter system will be dependant on the flow rates needed to treat the waste gas effectively. Whether under pressure or vacuum, all air feeding systems are rated for certain flow rates at a maximum head. This head that the air will encounter as it moves through the system will need to be calculated by the design engineer. It will be a function of the air distribution inlet and outlet piping length and size (diameter), filter bed material and porosity, biomass buildup, and any other in-line piping additions (particulate dropout, prescreen filters, etc.). Predicted head-loss calculations performed before a blower or fan is purchased will ensure that it operates at the desired flow rate without being impeded. Underestimating head loss may cause inadequate sizing of fans or blowers. Difficulty in estimating head loss occurs because it will slowly increase as biomass increases, hence a margin of safety must be implemented in such calculations. In many cases in may be advantageous to use a variable-speed blower, in spite of its higher capital cost. As the pressure drop increases during system operation, the variable-speed blower can accommodate these changes exactly. A fan or blower at a fixed speed will waste energy during the early phases of biofilter operation when the pressure drop across the system is relatively low.

MEDIA. The biofilter media houses the microorganisms and serves as the microbial growth environment. The composition and qualities of the media influence microbial functionality as well as the physical properties of the system. Ideally, the properties of a selected packing formulation should be well known. Physical properties include pressure drop, effective surface area (void volume), adsorptive/absorptive characteristics, and water-holding capacity. Biological properties include microbial diversity, biomass density, organic and inorganic nutrient content, and pH-buffering capacity.

Each media formulation has distinct strengths and weaknesses, making it more suitable for some applications than others (Bohn, 1996; van Lith, 1996; VDI 3477 Richtlinien, 1991). It is highly unlikely that a single media formulation will have optimal qualities for use with all contaminants and under all treatment conditions. The exact media selected for use for a given application is best made through field pilot evaluation, though laboratory testing and/or past experience may also prove effective. Media formulations can also be selected based on information found in the technical literature.

Many different biofilter media formulations have been described in the literature and used in the field. These formulations usually contain peat, compost, or heather (separately or in mixtures) as the primary support material. Organic material, such as soils and compost, contains an indigenous microbial population and nutrient supply. Soils are inexpensive and plentiful and have a large indigenous microbial population (Bohn, 1996; Bohn and Bohn, 1988). Compost is inexpensive, provides a consortium of microorganisms, contains adequate nutrients, has good adsorptive characteristics, and has a structure which is conducive to good airflow. To help ensure even airflow through the packing at low pressure drops, bulking agents (such as wood chips, bark, perlite, polystyrene beads, and to a lesser extent sand and sawdust) can be added. Inorganic nutrients, if desired, can be added in the form of a dry fertilizer mixed into the bulk packing or as a liquid through the secondary humidification system. If acidic metabolic end products are expected, pH-buffering compounds can also be added to the media. Powdered lime or pellets of calcium carbonate will offer a finite buffering capacity to the bed. In some cases an organic food source, such as manure, is added. The manure also provides a microbial inoculum and inorganic nutrients. Last, it may be desirable to add adsorptive agents to the packing, especially if shock loads or high fluctuating contaminant concentrations are expected. Powdered, granular, and pelletized activated carbon have been used as suitable adsorbents, along with more expensive ceramic materials (Webster et al., 1996). Combinations of organic and inorganic materials may provide the benefits of both material types at lower capital and operational costs.

Many biofilter vendors have also developed proprietary media formulations for use within their systems. This media is usually more expensive than "homemade" media mixed from local ingredients. The main advantage to the use of proprietary media is that the vendor usually has extensive knowledge and experience using the media with different contaminants under different operating conditions. This media may offer more consistent performance, and its effectiveness may be more amenable to accurate modeling (van Lith, 1996).

In order to keep operating costs reasonable, biofilter media height is usually limited to 3 to 5 ft, although higher media heights can be used if the airflow through the media is low (corresponding to greater vapor contact times). The operating moisture content of the media also effects media pressure drop. As the moisture content increases and the void space becomes filled with water, the pressure drop through the media increases. The organic media pressure drop increases as the media is degraded and compacts over time, typically over 3 to 5 years. The initial pressure drop through 4 ft of biofilter media may be less than 0.5 in of water column (Bohn, 1996; VDI 3477 Richtlinien, 1991). Pressure drop increases over time as the media compacts and decomposes. The pressure drop through the remaining ducting is usually 3 to 4 in of water. For economic considerations, the pressure drop through a biofilter system, including the ducting and media, is typically limited to approximately 12 in of water, and therefore the media is replaced after the biofilter bed pressure drop exceeds the design specification of the blower. Depending upon the packing composition, waste gas to be treated, and operation of the system, media replacement typically occurs every 2 to 5 years for wood- and compost-based media (Bohn, 1996). Biofilters that are operated improperly or use poorly designed media may require bed replacement in significantly shorter times.

INOCULUM. Soil, peat, and compost used in biofilter media naturally contain healthy, metabolically diverse microbial populations. For many odor applications or for the treatment of low levels of readily biodegradable organics, the supplemental addition of microorganisms to the packing media is not necessary. The indigenous microorganisms will be able to effectively treat the waste stream, especially if the mass loading to the filter is slowly increased during the start-up period. Under these conditions, potentially low starting concentrations of the metabolically suitable microorganisms will continue to increase in number in response to the presence of the additional food (the chemical wastes present in the gas stream) until a new steady-state concentration is reached. This adaptation period may require from a few days to several months to complete.

However, if high concentrations of organics are expected, if the target contaminants are difficult to readily biodegrade, or if system start-up and maximum removal efficiency must be achieved in the shortest time possible, then inoculating the packing media with exogenous microbial cultures may be necessary. The major benefit of inoculation is the shortening of the adaptation period required before maximum filter efficiency is reached. In the case of difficult-to-degrade organics, inoculating with specialized cultures specifically selected for their ability to degrade the target organics may be required for project success.

Inocula suitable for packing media can be obtained from different sources. The most common inocula used is activated sludge from a wastewater treatment plant. Activated sludge contains a wide variety of metabolically capable microorganisms. Activated sludge can be added when the media is first formulated and mixed prior to its loading into the biofilter, or can be applied as the filter material is placed into the reactor. Commercial microbial cultures, in either a liquid or powder form, can also be purchased for use. These cultures are typically enrichment cultures selected for their improved ability to biodegrade specific categories of pollutants (such as petroleum hydrocarbons and phenols), and can be added to the media when it is initially prepared or can be sprayed onto the media as it is being loaded into the reactor. Last, specialized cultures, selected and supplied by biofilter vendors, universities, or custom biological supply companies, can be used. The cost of inoculation increases from activated sludge to commercial cultures to specialized cultures.

Regardless of whether inoculation is used, the microbial populations in the biofilter change over time. The biofilter is a continuous microbial enrichment process, constantly selecting and rewarding the growth of those microbial and fungal strains best able to grow under the given environmental conditions and with the organics and inorganics present in the waste gas stream.

MOISTURE CONTROL. The control of the packing media moisture content will dictate the success of any biofilter. If the filter bed moisture content is below optimum, microbial activity will decline and the filter bed may dry out, causing bed shrinkage and channeling of air. If the moisture content is excessive, high back pressures may occur, anaerobic zones can form, and nutrients may be quickly washed out. Hence, careful control of the filter bed moisture content is essential to good biofilter performance.

Different biofilter media can be most readily differentiated based on their water-holding capacities. Biofilter media that contains a high percentage of compost and small-diameter organic material tends to have a higher water-holding capacity than media which is composed of bark, wood chips, and large-diameter organic material. Water tends to drain more readily from media containing less absorptive and larger-diameter particles. Most biofilter media is maintained within the 40 to 65 percent moisture range by weight on a wet weight basis (VDI 3477 Richtlinien, 1991), though specific types of media may function best outside this range (Barshter et al., 1993).

Humidification of the air leaving the pretreatment system should provide near 100 percent saturation. However, the moisture content of the gas may change as it passes through the bed. The degree of humidity, temperature, and retention time for the waste gas entering a biofilter will cause either evaporation or condensation of water between the filter bed and the passing air. This loss or addition of water over extended periods of time will create suboptimal operating conditions in the biofilter reactor. A secondary moisture control system, which provides water directly to the packing media, is required if exact moisture control is critical.

For open- or closed-bed biofilters using vapor upflow through the media, additional water can be added through sprinklers or mister heads mounted above the packing media. The difficulty with this arrangement is that the highest concentration of influent contaminants, and hence the maximum biological heat generation and production of metabolic by-products, occurs in the bottom of the packing media. Excess water must be added to the top of the filter in order for it to percolate down to the region where it is needed most. Water can be added under manual control, automatically with timers, or automatically under computer control depending upon the degree of instrumentation and control associated with the biofilter.

Alternatively, water can be added by pressure-compensated irrigation hoses placed on top of or buried within the media. Irrigation hoses allow for differential watering at specific depths within the media and offer the ability to "wash" the bottom portions of the bed to remove metabolic by-products, particulates, or condensable materials (Togna et al., 1997). With this arrangement, limited additional water can be added to the top of the packing media, while supplying larger water quantities to the most biologically active zone. As with the sprinkler systems, water can be added under manual control, automatically with timers, or automatically under computer control depending upon the degree of instrumentation and control associated with the biofilter.

For closed-bed systems with downflow gas passage, sprinkler or mister heads are typically used above the packing media. In this manner, water can be added directly to the most biologically active zone, that is, the top of the bed. It is also possible to use water hoses buried within the bed. Water addition can be manually or automatically controlled.

Monitoring the moisture level in a biofilter bed can be a difficult process. The moisture content can be manually determined by collecting a sample of the packing and performing a dry-weight analysis, although this is impractical for many biofilter systems. Since direct readout moisture probes that can be remotely placed into the media have not been successfully developed, indirect methods of monitoring bed moisture levels have become established. It is possible, e.g., to monitor the volume of excess water leaching from a bed. The watering system can then be adjusted to add just enough water to replenish bed moisture but not so much as to form a leachate. This process can be performed manually, with cycling timers, or automatically via a programmable logic controller (PLC). The cycle time can be adjusted so that more water is added to the media during the summer months and less water during the winter months. When the timer method of moisture control is used, a media that drains well is typically used, and the cycle time is usually set so that a moisture content at the high end of the moisture range is used. Load cells can also be used to very precisely control the moisture content of biofilter media (van Lith et al., 1997). When this approach is used, the entire media, or sections of media, is continually weighed using load cells. Water is added to the media through spray nozzles above the media when the media weight drops below some preset level, either manually or via PLC. While the PLC increases the capital cost of a biofiltration system, more consistent operation is possible and operating costs in terms of time of operating labor may be significantly reduced.

pH CONTROL. The packing media pH can be measured directly by taking a sample of the material for analysis or indirectly by measuring the pH in the bed leachate water. As discussed previously, pH-buffering agents can be added as solids directly to the packing media as it is manufactured. However, this method will have a finite capacity, and once depleted, the media pH will change. Depending upon the type of packing and biofilter design employed, additional buffering agents can be added as a solid or liquid to the bed, or the bed material may have to be replaced.

INSTRUMENTATION, CONTROL, AND MONITORING. At least a certain amount of analytical equipment will be necessary to effectively monitor and evaluate the performance of a biofilter. However, the extent and sophistication of monitoring and control equipment used to analyze the biofilter performance will directly affect the capital cost of the system installed. Parameters which can be measured include (1) filter bed moisture content, pH, temperature, and pressure drop; (2) gas flow rate; and (3) influent and effluent relative humidity, temperature, and organic concentrations. Monitoring instruments can be placed in-line to evaluate many of these parameters on a continuous basis. Alternatively, the parameters may be periodically measured manually by an operator reading from gauges. For instance, a gas sampler with a flame ionization detector might continuously monitor air concentrations from the inlet and outlet of the reactor. Grab samples of the inlet and outlet air can also be taken manually and sent off to an analytical laboratory for assessment. The economics of how often certain parameters will need to be measured should be addressed by the design engineer before implementing any instrumentation system.

Parameters which will be measured with an in-line system may require the use of a PLC. Commonly PLCs are employed with advanced biofilter systems where maximum consistent system performance is required, and they offer an advantage in that data are automatically collected and saved on a timely basis. In addition, the PLC can use a feedback control mechanism program to document and resolve any peculiarities in operation. For example, the PLC may add water to the system if the filter bed moisture content is too low, or the PLC may adjust the temperature if the bed gets too cold. This type of computer logic system minimizes the need for human interaction and provides better control of system conditions.

DUCTING AND EXITING AIR DISTRIBUTION. In the United States, it is typical to duct the treated air from the biofilter up and out a suitable stack. This is especially true in odor control situations where not all the odor may have been removed. Air distribution modeling can be used to determine the location and height of a suitable discharge stack (Wang and Skipka, 1993).

START-UP, OPERATION, AND MONITORING

Biofilter vendors typically supply site-specific operation and maintenance (O&M) manuals covering the start-up, operation, and monitoring procedures to be used with their particular systems. Presented below are some general start-up, operation, and monitoring procedures common to most biofilter systems.

Start-up of Biofilter Systems

If media inoculation is desired to decrease the start-up and acclimation periods, the media must be inoculated prior to start-up. The inoculum source can be a general inoculum with broad substrate specificity (such as those supplied for use in wastewater treatment and field bioremediation projects), or it can be a specialized culture tailored for the particular application. There are no set rules for the amount of inoculum to use; the more inoculum that is used, the shorter the start-up period required. The inoculum can be mixed into the media during its preparation, or can be spread on top of the media prior to start-up if the media is open enough to allow for good water percolation. If the inoculum is spread on top of the media, water must be used to wash the microorganisms into the media. The use of activated sludge from wastewater treatment plants as a biofilter inoculum source is being discouraged due to the presence of potential pathogens as well as raw or partially treated sewage. If activated sludge is used as an inoculum source, disposal of spent media and discharge water may become problematic.

Biofilters are usually started up using an initial airflow rate significantly below the design flow rate so that the microbial populations present do not become shocked or inhibited by high initial contaminant loads. The flow rate is stepped up to the full design flow in two or three increments over a one- to two-week period (Barshter et al., 1993).

During the first few days of start-up, the biofilter media will typically adsorb/absorb the contaminants, resulting in a high observed "removal efficiency" (Barshter et al., 1993; Yang, Caratura, and Empfield, 1993). However, after the media becomes saturated with the contaminants (usually within 3 or 4 days), full or partial contaminant breakthrough may be observed until the microbial populations within the media increase to a sufficiently high level. Within a week following this initial breakthrough, effluent contaminant concentrations usually decrease due to microbial activity.

Biofilter Operations

The O&M manual should contain detailed instructions on the day-to-day operation of the biofilter system, including process control, maintenance, and monitoring. The amount of labor involved with operating a biofilter system depends to a great extent on the degree of instrumentation and control equipment installed. The simplest systems will have very limited automated process control, and gauges for airflow rate, air and bed temperatures, and pressure drop will need to be read and recorded manually. Secondary moisture control with these systems may also need to be manually started and stopped, although timers are frequently used as an alternative. Highly instrumented and automated systems using PLCs may automatically collect, evaluate, and adjust biofilter operation as necessary with minimal human intervention. Maintenance of mechanical system components as recommended will also help to ensure uninterrupted operation of the system.

While system operating labor can be minimized, it is important to realize that biofilters are biological systems, and as such overall performance will likely change slightly as the system continues to operate. This change may require that adjustments be made in terms of day-to-day system operation. For example, seasonal effects on moisture levels within the biofilter may require watering periods to be adjusted either manually or via timer. Biofilter systems that operate consistently usually have one or more individuals responsible for daily or frequent review of operational parameter data. Plots of such data can be used to discern trends in performance (either good or bad), usually providing advance warning of potential problems.

During start-up, and after the system is operating at its full-flow capacity, the biofilter should be monitored in accordance with the maintenance and monitoring requirements found in the O&M manual. Typically, these monitoring requirements will include (1) inlet temperature, (2) inlet relative humidity, (3) flow rate, (4) pressure drop across the media, (5) pressures at the inlet and/or outlet of the system, and (6) media moisture content. Most of the above parameters can be monitored using on-line instruments. The media moisture content can be monitored on-line using either load cells, to measure the media weight, or moisture probes (van Lith et al., 1997). If no on-line media moisture monitoring is available, then grab samples of media must be periodically taken for analysis. Additional monitoring requirements may include (1) media pH, (2) drainwater pH, (3) bed temperature, (4) particle size distribution of the media, and (5) media nutrient levels (Barshter et al., 1993). The monitoring frequency for each of the above parameters will be set by the supplier. Media pH, particle size distribution, and moisture content measurements (if using grab samples for moisture content) must be taken from various locations within the bed in order to get representative data. Monitoring of media pH levels is especially important when the biofilter is treating contaminants that can create acidic by-products, such as ammonia (due to nitrification), sulfur-containing compounds (due to the formation of sulfuric acid), chlorinated compounds (due to formation of hydrochloric acid), and high concentrations of ethanol (which can form organic acids under oxygen-limited conditions) (Devinny and Hodge, 1995).

The pressure drop across the filter bed(s) normally increases as the system operates until it approaches the upper design limit. Depending upon the biofilter design and type of media employed, it may be possible to remix and continue to use the media, since remixing typically restores porosity on at least a temporary basis. If mixing is not an option, the media must be replaced.

Depending upon the air distribution system used with the biofilter, it may be possible to shut off gas flow to only part of the system, allowing treatment to continue on at least a limited basis during bed replacement. Normally, the flow of contaminated vapors to the bed is stopped, but aeration is continued for a time to allow the microorganisms to finish degrading any adsorbed/absorbed contaminants. Depending upon local requirements, the media may need to be tested to verify that it is not hazardous and/or treated to reduce the viability of the microbial populations present.

Spent media can be disposed of in several ways. For example, some biofilter system vendors may reclaim the media for a fee. In most cases the material can be used for landscaping purposes on-site or disposed of off-site in a landfill as a nonhazardous material. Again depending upon local requirements, the use of certain inocula (such as sewage sludge) and/or the presence of certain bulking agents may preclude disposal in a standard landfill.

Modular biofilter systems may offer the advantage of easy media replacement by merely exchanging modules containing fresh media for modules containing the spent media. Where this is not possible, the spent media must be removed by laborers and replaced with fresh material. Depending upon the size and design of the biofilter system as well as the type of media used, media replacement can be a tedious and expensive process.

Monitoring Overall System Performance

If required, biofilter monitoring methods and frequency are usually set by state and local regulatory agencies and are specified in the permit for the system. These monitoring requirements vary from state to state and from application to application. Monitoring requirements may be as simple as maintaining records of biofilter operation such as media pH, media moisture content, inlet temperature, inlet humidity, and airflow rate to verify that the biofilter is operating within design requirements. Alternatively, monitoring requirements may include continuous or grab-sample vapor monitoring during a specified compliance testing period after start-up, along with the above recordkeeping after the compliance test is passed. Additional compliance testing may be required on an annual or biannual basis. Since effluent concentrations from biofilters do not react instantaneously to changes in inlet concentration, grab samples for compliance testing must be used with caution, especially for testing biofilters treating batch process operations. Influent and effluent vapor grab samples should be taken over time periods that will allow for the averaging of concentration fluctuations. Evacuated canisters for EPA T0-14 analysis can be set to sample for periods as long as 60 h. The major benefit of grab-sample analysis is that the removal of individual contaminants can be monitored within multicontaminant off-gas streams. Alternatively, continuous off-gas monitoring devices can be used for measuring total hydrocarbon removal. Typically, these devices employ a flame ionization detector (FID) or photoionization detector (PID) to measure total organic carbon (TOC) concentrations. However, for treatment of waste streams containing many different compounds, different FID and PID response factors and removal rates for different components can result in inaccurate treatment efficiency results. For these cases, grab samples should be used along with the on-line measurements. For enclosed systems, influent and effluent vapor samples can be collected from the influent feed ducting and effluent stack. For open systems, a collection hood must be used.

Grab samples must be taken for compliance odor measurements using odor panels. These samples are usually collected into gas sampling bags made of chemically inert and nonreactive materials (i.e., Tedlar or a similar material) (O'Brian, 1994). Once collected, the samples are usually diluted with dry nitrogen gas and then are transported to an odor panel for analysis within 24 h. A description of the use of an odor evaluation panel for olfactometry analysis can be found in Ziminski and Ferrara (1993). In general, olfactometry involves the use of a panel of people (usually seven to nine) to detect the presence or absence of odor within diluted gas samples. Odor-free air is used for diluting the samples. The odor concentration within the samples is expressed in odor units. An odor unit is defined as the number of dilutions required so that a sample would be detected by 50 percent of the population (Ziminski and Ferrara, 1993). Statistical analysis must be used in interpreting the odor panel results (van Doorn and van Harreveld, 1994). Due to the inherent variability in odor measurements, standardization of the odor measurement technique is constantly being sought (O'Brian, 1994), including the use of "standard noses" (van Doorn and van Harreveld, 1994), i.e., the use of individuals with defined odor-sensing capabilities.

BIOFILTRATION MARKETPLACE

The biofiltration marketplace in the United States has generally evolved into three major categories: treatment of odors, treatment of petroleum hydrocarbons, and treatment of VOCs, HAPs, and other organic compounds. While in many situations two or more of these categories may overlap, the validity of this generalization will become more apparent from examples of systems actually installed in the field.

Odor Control Systems

By far the greatest acceptance and use of biofiltration are for the treatment of odors, and the odor control marketplace can be subdivided into two major operationally defined categories: simple and complex odor problems. In general, *simple* odor problems involve (1) the treatment of readily biodegradable compounds, (2) waste streams that maintain a relatively consistent composition in terms of contaminants and concentrations, (3) low to moderate concentrations of contaminants, and (4) regulatory situations where low to moderate removal efficiencies are required. In contrast, *complex* odor problems typically involve (1) the treatment of compounds relatively resistant to rapid biodegradation, (2) complex waste streams that have highly variable composition in terms of the contaminants and concentrations present, (3) high influent contaminant concentrations, and (4) regulatory situations where very high and consistent removal efficiencies are required.

For example, the treatment of odors from wastewater treatment plants and composting facilities generally falls into the simple odor category because biofiltration technology has been extensively applied to these waste streams with favorable results for numerous years in both Europe and the United States. Major contaminants of concern include sulfides, mercaptans, carbonic acids, amines, and ammonia. Waste gas streams from these operations typically have very high airflow rates (up to several hundred thousand cubic feet per minute), putting a premium on the selection of a biofilter design with low capital and operating costs. Open-bed biofilters have traditionally been the reactor design of choice for these applications.

FIGURE 8.8.9 The photograph shows an open-bed biofilter treating composting odors. The city of Davenport (Iowa) operates a 28 dry ton/day aerated static pile biosolids and yard waste composting facility. The mixing and composting areas are totally enclosed in an insulated preengineered building. Exhaust gas from the entire building is mixed with compost pile process exhaust and is treated by the biofiltration system. The biofilter treats a total flow of 210,000 ft^3/min (357,143 m^3/h) and consists of eight independently controlled blowers and biofilter cells. Yard waste compost and wood chips are used as the media. Performance testing was conducted in October 1995 using odor control analysis according to ASTM method E679-91, and an average odor removal efficiency of 86 percent was achieved. No odor complaints have been received from neighbors since the facility began operation. (*Courtesy of E & A Environmental Consultants, Inc., Cary, North Carolina.*)

Figure 8.8.9 shows a full-scale in-ground open-bed biofilter system used to treat composting off-gases. The cost per cubic foot of air treated per minute with in-ground biofilters decreases as the total volume of air to be treated increases. Reported costs range from \$10 to \$40 per cubic foot per minute of air treated for airflows less than 15,000 ft^3/min, to \$3 to \$5 per cubic foot per minute of air treated at flow rates exceeding 100,000 ft^3/min (Williams and Boyette, 1997). Capital costs have been kept to a minimum by using a simple process design and local materials for construction and packing medium. Operating expenses are more difficult to accurately obtain because they are composed of labor; utilities (primarily electricity to power the blowers); monitoring, analytical expenses, recordkeeping and reporting (if applicable); and bed replacement and disposal charges. However, they have been estimated at \$2 to \$14 per cubic foot per minute of air treated (Williams and Boyette, 1997).

At many wastewater treatment plants it is common to see aboveground open-bed biofilters constructed of concrete sidewalls and slotted flooring for the treatment of high-volume waste streams. For the control of hydrogen sulfide at pump stations, headworks, and other locations with more limited gas flows, prefabricated modular biofilter units are becoming more common. Figure 8.8.10 shows a modular system treating odors from a pump lift station. Modular units are frequently constructed of corrosion-resistant materials such as plastic, and each unit typically handles 100 to 250 ft^3/min of vapor.

FIGURE 8.8.10 The photograph shows a 1200 ft^3/min (2041 m^3/h) biofilter system used to treat odors at a master pump station in Jacksonville, Florida. The system consists of two identical treatment trains, each handling one-half of the gas flow. Each train consists of one sparging-type humidifier and four Pentpack biofilters. The Pentpack biofilter consists of 5 ft of compost-based medium spilt into 1-ft layers placed in individual trays. The trays are one-piece molded high-density polyethylene, insulated to greater than R20, and approximately 6 ft in diameter. (*Courtesy of EG & G Biofiltration, Saugerties, New York.*)

Since the engineering design and fabrication costs associated with prefabricated units can be kept to a minimum, biofilter systems that treat several hundred cubic feet of air per minute can be purchased for $15,000 to $20,000.

For other simple odor control applications, soil bed biofilters can provide an attractive reactor design. Soil beds systems are typically used for low to moderate airflow volumes. Figure 8.8.11 shows a soil bed system treating gases from a wastewater treatment plant. Soil bed system installed costs have been estimated from $8 to $15 per cubic foot of air treated per minute (Bohn and Bohn, 1988). Periodic watering and tilling of the soil can improve system performance as needed. Many do-it-yourself simple odor control systems are constructed without the use of technical consultants and/or biofilter vendors.

To date, the treatment of complex odors (primarily for the food, flavor, and fragrance industries) has typically resulted in the use of more sophisticated, closed-bed biofilter designs. Manufacturing operations for these industries are frequently located in areas of high population density and regulatory interest. Batch production processes are commonly used, resulting in a waste gas stream highly variable with respect to both the vapor-phase compounds present and concentration. The biofilter may be exposed to hundreds of different compounds during the production year, and these systems may be required to treat not only odors but VOCs to specific levels as well. Closed-bed systems, fabricated from insulated metal panels and/or concrete, have appreciably higher capital costs than open-bed designs, but offer consistently high removal efficiencies in return.

FIGURE 8.8.11 The picture shows an odor application soil bed biofilter installed at an industrial manufacturing facility in Jamestown, New York. The system treats 6000 ft^3/min (10,204 m^3/h) of air containing glycols and some hydrocarbons released from a wastewater treatment facility. The 1996 installed cost was $56,000, or approximately $9.33 per cubic foot of air treated. (*Courtesy of Bohn Biofilter Corporation, Tucson, Arizona.*)

The closed configuration allows for better temperature, moisture, and air distribution control, which in return can provide near optimal growth conditions for the microorganisms present in the media. Figure 8.8.12 shows a closed-bed system fabricated from insulated metal panels; this system also includes a PLC system. Total installed costs for this type of biofilter design range from $20 to $50 per cubic foot of air treated per minute, depending upon the total volume of gas to be treated (Yavorsky, 1997). The price per cubic foot of air treated per minute decreases as the total gas flow increases.

Hydrocarbon Treatment Systems

Petroleum-contaminated vapor streams are most commonly encountered during field hazardous waste remediation projects, in particular, associated with off-gasses produced during soil vacuum extraction (SVE) and/or subsurface sparging of contaminated soils and groundwater. Gasoline, jet fuel, and diesel waste gas streams are generated at many small sites around the country (primarily gasoline service station remediation projects), as well as at storage terminals and military bases. Gasoline, jet fuel, and diesel fuel are in reality complex mixtures of hundreds of aliphatic, aromatic, and alicyclic compounds. Each of these compounds has different physical, chemical, and biological properties, and will behave differently in a biofilter system.

Many state and federal regulatory agencies subdivide petroleum waste streams into operationally defined aromatic (benzene, toluene, ethylbenzene, and xylenes) and nonaromatic fractions (remainder of compounds). With SVE projects, relatively low total gas flows (less than 200 ft^3/min) are common, and activated carbon as well as thermal processes are commonly used to treat the contaminated vapors. Using biofilters as an alternative treatment technology, the aromatic fraction of petroleum waste streams biode-

FIGURE 8.8.12 The picture shows a ClairTech Bioton biofilter treating 33,000 ft^3/min (56,122 m^3/h) process gases containing a variety of flavor and fragrance compounds. The unit was designed for 90 percent efficiency and contains approximately 8482 ft^3 (240 m^3) of medium. Note the eight roof hatches for use in loading and replacing the media. (*Courtesy of Ambient Engineering, Matawan, New Jersey.*)

grades much faster and more completely than the corresponding nonaromatic fraction (Togna et al., 1994). In situations where specific contaminant concentrations are at acceptable levels but petroleum odors are prevalent, biofilters can be used for odor control.

Several modular closed-bed biofilter designs, as well as many "home-grown" biofilter systems, have been installed to service this marketplace. Figure 8.8.13 shows a modular system treating SVE off-gases. Prefabricated modular systems typically are designed to treat 50 to 200 ft^3/min of vapor and usually cost between $15,000 and $20,000. These systems are designed primarily to treat the aromatic petroleum fraction, usually at contact times of 1 min or less. If both the aromatic and nonaromatic petroleum fractions require treatment, the vapor-phase contact times required increase from 1 min to several minutes (Togna et al., 1994).

The use of biofilters to treat either high-volume petroleum waste gas streams at refineries or sporadically high concentration waste streams from storage terminals has been investigated by many petroleum companies. To date, the use of biofilters in these situations has usually been found to be impractical for a variety of reasons.

VOC and HAP Control Systems

In a manner analogous to the treatment of odors, the treatment of VOCs, HAPs, and other organic compounds can be subdivided into simple and complex categories. In general, *simple* VOC and HAP problems involve (1) the treatment of readily biodegrad-

FIGURE 8.8.13 The photograph shows three modular biofilters, operating in parallel, treating petroleum-contaminated vapors from a soil vacuum extraction system. The packing media consists of finished municipal compost amended with pine bark nuggets and crushed limestone. The average airflow rate to each biofilter is approximately 100 ft^3/min (170 m^3/h), resulting in a vapor contact time of 1 min. Peak flow rates of 150 ft^3/min (255 m^3/h) can also be accommodated, but with lower removal efficiencies. The hydrocarbon loading rates to the biofilters vary with the groundwater elevation and the presence of free-phase gasoline. Removal of total VOCs generally ranges from 30 to 50 percent, while BTEX removal efficiencies range from 80 to greater than 95 percent. The air discharged from the biofilters at a stack height of 20 ft has had minimal to nondetectable hydrocarbon odors at ground level. (*Courtesy of Matrix Environmental Technologies Inc., Orchard Park, New York.*)

able compounds, (2) waste streams that maintain a relatively consistent composition in terms of contaminants and concentrations, (3) low to moderate concentrations of contaminants, and (4) regulatory situations where low to moderate removal efficiencies are required. In contrast, *complex* VOC and HAP situations typically involve (1) the treatment of compounds relatively resistant to rapid biodegradation, (2) complex waste streams that have highly variable composition in terms of the contaminants and concentrations present, (3) high influent contaminant concentrations, and (4) regulatory situations where very high and consistent removal efficiencies are required.

Both open- and closed-bed biofilter designs are commonly used for VOC and HAP applications. Depending upon the contaminants of concern and site-specific regulatory requirements, closed-bed systems (with and without automated sampling and analysis capabilities) may provide the only acceptable biofilter design. The installed cost for aboveground, open-bed designs can range from \$25 to \$40 per cubic foot of air treated per minute. Closed-bed systems, especially those with advanced process control and automated monitoring requirements, have installed capital costs in the range of \$30 to \$100 per cubic foot of air treated per minute. The per cubic foot per minute cost of air treated decreases as the total volume of gas increases.

An aboveground open-bed biofilter system treating a mixed chemical stream is shown in Fig. 8.8.14. Monitoring of this type of filter is accomplished by using a sampling hood to collect gas emanating from a known area of bed surface. Figure 8.8.15

FIGURE 8.8.14 The photograph shows an open-bed biofilter used to treat a mixed-organics stream from an industrial manufacturing process. This system has been in operation for 5 years. (*Courtesy of Solutia, Inc., Trenton, Michigan.*)

FIGURE 8.8.15 The photograph shows two closed-bed biofilters treating a combined 140,000 ft^3/min (238,095 m^3/h) of industrial process air from the Weyerhaeuser facility in Adel, Georgia. The overall system includes pretreatment humidification and particulate removal, and each biofilter unit (constructed from concrete) contains 44,000 ft^3 (2,490 m^3) of packing media. The major organic species treated are alpha- and beta-pinene. (*Courtesy of PPC Biofilter, Longview, Texas.*)

shows a closed-bed biofilter, constructed primarily from concrete, designed to treat a mixed-organic waste gas from a wood products manufacturing facility. The insulated metal-panel biofilter construction method has also been used to treat complex VOC and HAP streams (Gilmore and Briggs, 1997). Monitoring the performance of closed-bed systems is much easier to accomplish because representative samples can be obtained for analysis from the filter discharge point(s).

SUMMARY

When properly applied and operated, biofiltration systems offer an efficient and cost-effective method of treating odors, VOCs, HAPs, and other organic and inorganic compounds. Biofiltration relies on the beneficial metabolic activity of microorganisms to transform noxious and/or hazardous chemicals to harmless end products. This "green" technology continues to gain acceptance among the public, regulatory agencies, and the private sector. A variety of complementary biofilter designs and materials of construction can be optimized for each specific application.

While biofiltration technology is best understood and applied to the treatment of odors, much effort is currently underway to extend its usefulness to the treatment of complex organic waste streams generated by industrial manufacturing processes. Theoretical and practical knowledge regarding the treatment of VOCs and HAPs is increasing rapidly, as are efforts to predictably model the biofiltration process. By combining these improvements with past field implementation successes, biofitration technology will continue to provide the pollution control field with a valuable treatment technology.

REFERENCES

Allen, E. R., and S. Phatak, "Control of Organo-Sulfur Compound Emissions Using Biofiltration—Methyl Mercaptan," Paper 93-Wa-52B.03, 86th Annual Meeting and Exhibition of the Air & Waste Management Association, June 13–18, 1993.

Allen, E. R., and Y. Yang, "Biofiltration Control of Hydrogen Sulfide Emissions," Paper 91-103.10, 84th Annual Meeting and Exhibition of the Air & Waste Management Association, June 16–21, 1991.

Andrews, G. F., and K. S. Noah, "Design of Gas-Treatment Bioreactors," *Biotechnology Progress,* **11**(5): 498–509, 1995.

Arpacioglu, B. C., and E. R. Allen, "Control of Organo-Sulfur Compound Emissions Using Biofiltration: Dimethyl Disulfide," Paper 96-RP87C.02, 89th Annual Meeting and Exhibition of the Air & Waste Management Association, June 23–29, 1996.

Baltzis, B. C., and Z. Shareefdeen, "Modeling and Preliminary Design Criteria for Packed-Bed Biofilters," Paper 93-TP-52A.03, 86th Annual Meeting and Exhibition of the Air & Waste Management Association, June 13–18, 1993.

Barshter, D. W., S. W. Paff, and A. B. King, "Biofiltration—Room Temperature Incineration," Paper 93-WP-94.07, 86th Annual Meeting and Exhibition of the Air & Waste Management Association, June 13–18, 1993.

Bentz, R., "Waste Air Purification," U.S. Patent 5,082,472, Jan. 21, 1992.

Bohn, H. L., "Biofilter Medium," Paper 96-WP87A.01, 89th Annual Meeting and Exhibition of the Air & Waste Management Association, June 23–28, 1996.

Bohn, H., and R. Bohn, "Soil Beds Weed Out Air Pollutants," *Chemical Engineering,* **95**(6): 73–76, 1988.

Choi, D. S., T. S. Webster, A. N. Chankg, and J. S. Devinny, "Quantitative Structure-Activity Relationships for Biofiltration of Volatile Organic Compounds," in F. E. Reynolds, Jr., ed., *1996 Conference on Biofiltration* (*An Air Pollution Control Technology*), The Reynolds Group, Tustin, Calif., 1996, pp. 231–238.

de Beer, D., P. Stoodley, F. Roe, and Z. Lewandowski, "Effects of Biofilm Structures on Oxygen Distribution and Mass Transport," *Biotechnology and Bioengineering,* **43**: 1131–1138, 1994.

de Castro A., D. G. Allen, and R. R. Fulthorpe, "Characterization of the Microbial Population during Biofiltration and the Influence of the Inoculum Source," in F. E. Reynolds, Jr., ed., *Proceedings of the 1996 Conference on Biofiltration* (*An Air Pollution Control Technology*), The Reynolds Group, Tustin, Calif., 1996, pp. 164–172.

Deshusses, M. A., and G. Hamer, "The Removal of Volatile Ketones Mixtures from Air in Biofilters," *Bioprocess Engineering,* **9**: 141–146, 1993.

Deshusses, M. A., G. Hamer, and I. J. Dunn, "Behavior of Biofilters for Waste Air Biotreatment. 1. Dynamic Model Development," *Environmental Science & Technology,* **29:** 1048–1058, 1995.

Devinny, J. S., and D. S. Hodge, "Formation of Acidic and Toxic Intermediates in Overloaded Ethanol Biofilters," *Journal of the Air & Waste Management Association,* **45**: 125–131, 1995.

Dharmavaram, S., "Biofiltration—A Lean Emissions Abatement Technology," Paper 91-103.2, 84th Annual Meeting and Exhibition of the Air & Waste Management Association, June 16–21, 1991.

Dharmavaram, S., J. Casey, T. Timmermans, and C. van Lith, "Experimental Evaluation of a Biofiltration Unit for Removal of Acetone," Paper 93-WP-52C.01, 86th Annual Meeting and Exhibition of the Air & Waste Management Association, June 13–18, 1993.

Gilmore, G. L., and T. G. Briggs, "Experiences with a Compost Biofilter for VOC Control from Batch Chemical Manufacturing Operations," Paper 97-RA71C.02, 90th Annual Meeting and Exhibition of the Air & Waste Management Association, June 8–13, 1997.

Govind, R., Z. Wang, and D. F. Bishop, "Biofiltration Kinetics for Volatile Organic Compounds and the Development of a Structure-Biodegradability Relationship," Paper 97-RA71C.07, Annual Meeting and Exhibition of the Air & Waste Management Association, June 9–13, 1997.

Graham, J. R., "GAC Based Gas Phase Biofiltration," in F. E. Reynolds, Jr., ed., *Proceedings of the 1996 Conference on Biofiltration* (*An Air Pollution Control Technology*), The Reynolds Group, Tustin, Calif., 1996, pp. 85–93.

Hodge, D. S., and J. S. Devinny, "Biofilter Treatment of Ethanol Vapors," *Environmental Progress,* **13**: 167–173, 1994.

Hodge, D. S., and J. S. Devinny, "Modeling Removal of Air Contaminants by Biofiltration," *Journal of Environmental Engineering,* **121**(1): 21–32, 1995.

Hugler, W. C., J. G. Cantu-De la Garza, and M. Villa-Garcia, "Biofilm Analysis from an Odor-Removing Trickling Filter," Paper 96-RA87A.04, 89th Annual Meeting and Exhibition of the Air & Waste Management Association, June 23–29, 1996.

Institute of Clean Air Companies, Inc., "Biofiltration for Air Pollution Control," Institute of Clean Air Companies, Inc., Washington, 1993.

Johnson, C. T., and M. A. Deshusses, "Quantitative Structure-Activity Relationships for VOC Biodegradation in Biofilters," in Bruce C. Alleman and Andrea Leeson, eds., *In-Situ and On-Site Bioremediation Symposium,* vol. 5, Battelle Press, Columbus, Ohio, 1997, pp. 175–180.

Kiared, K., L. Bibeau, R. Bizezinski, G. Viel, and M. Heitz, "Biological Elimination of VOCs in Biofilters," *Environmental Progress,* **15**: 148–152, 1996.

Kuter, G. A., J. E. Harper, L. M. Naylor, and P. J. Gormsen, "Design, Construction and Operation of Biofilters for Controlling Odors at Composting Facilities," Paper 93-WP-52C.02, 86th Annual Meeting and Exhibition of the Air & Waste Management Association, June 13–18, 1993.

Leson, G., and A. Winer, "Biofiltration: An Innovative Air Pollution Control Technology for VOC Emissions," *Journal of the Air & Waste Management Association,* **41**: 1045–1054, 1991.

Leson, G., D. S. Hodge, F. Tabatabai, and A. M. Winer, "Biofilter Demonstration Projects for the Control of Ethanol Emissions," Paper 93-WP-52C.04, 86th Annual Meeting and Exhibition of the Air & Waste Management Association, June 13–18, 1993.

Mason, C. A., G. Hamer, and J. D. Bryers, "The Death and Lysis of Microorganisms in Environmental Processes," *FEMS Microbiological Reviews,* **39**: 373–401, 1986.

Medina, V. F., T. Webster, M. Ramaratnam, and J. S. Devinny, "Treatment of Gasoline Residuals by Granular Activated Carbon Based Biological Filtration," *Journal of Environmental Science and Health,* **A30**(2): 407–422, 1995.

Mohseni, M., and D. G. Allen, "Biofiltration of Alpha-Pinene Using Wood Waste and Activated Carbon Medium," in F. E. Reynolds, Jr., ed., *Proceedings of the 1996 Conference on Biofiltration (An Air Pollution Control Technology*), The Reynolds Group, Tustin, Calif., 1996, pp. 45–54.

Moller, S., A. R. Pedersen, L. K. Poulsen, E. Arvin, and S. Molin, "Activity and Three-Dimensional Distribution of Toluene-Degrading *Pseudomonas putida* in a Multi-Species Biofilm Assessed by Quantitative *in situ* Hybridization and Scanning Confocal Laser Microscopy," *Applied and Environmental Microbiology,* **12**: 4632–4640, 1996.

Morgenroth, E., E. D. Schroeder, D. P. Y. Chang, and K. M. Scow, "Nutrient Limitation in a Compost Biofilter Degrading Hexane," Paper 95-TP9C.05, 88th Annual Meeting and Exhibition of the Air & Waste Management Association, June 18–23, 1995.

O'Brian, M. A., "Standardization of Odor Measurement Techniques," Paper 94-TP52.07, 87th Annual Meeting and Exhibition of the Air & Waste Management Association, June 19–24, 1994.

Ottengraf, S. P. P., "Exhaust Gas Purification," in H. J. Rehm and G. Reed, eds., *Biotechnology 8,* VCH Verlagsgesellschaft Weinheim, Germany, 1986, pp. 426–452.

Ottengraf, S. P. P., and J. H. G. Konings, "Emission of Microorganisms from Biofilters," Paper 92-116.08, 85th Annual Meeting and Exhibition of the Air & Waste Management Association, June 21–26, 1992.

Ottengraf, S. P. P., J. J. P. Meesteis, A. H. C. van den Oever, and H. R. Rozema, "Biological Elimination of Volatile Xenobiotic Compounds in Biofilters," *Bioprocess Engineering,* **1**: 61–69, 1986.

Peters, D. A., G. T. Hickman, J. G. Stefanoff, and M. B. Garcia, Jr., "Laboratory Assessment of Biofiltration for Fuel-Derived VOC Emissions Control," Paper 93-WA-52B.06, 86th Annual Meeting and Exhibition of the Air & Waste Management Association, June 13–18, 1993.

Pinnette, J. R., C. A. Dwinal, M. D. Giggey, and G. E. Hendry, "Design Implications of the Biofilter Heat and Moisture Balance," in F. E. Reynolds, ed., *Proceedings of the 1996 Conference on Biofiltration (An Air Pollution Control Technology*), The Reynolds Group, Tustin, Calif., 1996, pp. 85–98.

Severin, B. F., J. Shi, and T. Hayes, "Destruction of Gas Industry VOCs in a Biofilter," Institute of Gas Technology Sixth International Symposium on Gas, Oil, and Environmental Technology, Nov. 29–Dec. 1, 1993.

Shareefdeen, Z., B. C. Baltzis, Y.-S. Oh, and R. Bartha, "Biofiltration of Methanol Vapor," *Biotechnology and Bioengineering,* **41**: 512–524, 1993.

Smet, E., H. Van Langehove, and W. Verstraete, "Long-Term Stability of a Biofilter Treating Dimethyl Sulfide," *Applied Microbiology and Biotechnology,* **46**: 191–196, 1996.

Standefer, S., and R. Willingham, "Experience with Pilot and Full-Scale Biofilter Operations," in E. F. Reynolds, J., ed., *Proceedings of the 1996 Conference on Biofiltration (An Air Pollution Control Technology*), The Reynolds Group, Tustin, Calif., 1996, pp. 77–84.

Swanson, W. J., and R. C. Loehr, "Biofiltration: Fundamentals, Design and Operations Principles, and Applications of Biological APC Technology," *Journal of Environmental Engineering,* **123**(6): 538–546, 1997.

Tahraoui, K., R. Samson, and D. Rho, "Biodegradation of BTX from Waste Gases in a Biofilter Reactor," Paper 94-TA260.07P, 87th Meeting and Exhibition of the Air & Waste Management Association, June 19–24, 1994.

Tang, H.-M., S.-J. Hwang, and S.-C. Hwang, "Waste Gas Treatment in Biofilters," *Journal of the Air & Waste Management Association,* **46**: 349–354, 1996.

Togna, A. P., and M. Singh, "A Comparative Study of Biofilter and Biotrickling Filter Performance for Isopentane Removal," Paper 94-RP115B.04, 87th Annual Meeting and Exhibition of the Air & Waste Management Association, June 19–24, 1994.

Togna, A. P., G. S. Skladany, and J.M. Caratura, "Treatment of BTEX and Petroleum Hydrocarbon Vapors Using a Field-Pilot Biofilter," in *49th Purdue Industrial Waste Conference Proceedings,* Lewis Publishers, Chelsea, Mich., 1994, pp. 437–448.

Togna, A. P., W. J. Fucich, R. E. Loudon, M. Del Vecchio, D. W. Barshter, and A. J. Nadeau, "Treatment of Odorous Toxic Air Pollutants from a Hardwood Panel Board Manufacturing Facility Using Biofiltration," Paper 97-WA71A.01, 90th Annual Meeting and Exhibition of the Air & Waste Management Association, June 8–13, 1997.

van Doorn, W., and A. Ph. van Harreveld, "Sampling Strategy and Statistical Aspects for Measuring the Odour Removal Efficiency of a Biofilter," in *VDI Berichte 1104—Biological Waste Gas Cleaning,* VDI-Verlag GmbH, Dusseldorf, Germany, 1994, pp. 415–427.

van Groenestijn, J. W., and P. G. M. Hesselink, "Biotechniques for Air Pollution Control," *Biodegradation,* **4**: 283–301, 1993.

van Lith, C., "Long Term Experiences with Biofilter Material Stabilized with Inert Additives," Paper 96-RA87B.06, 89th Annual Meeting and Exhibition of the Air & Waste Management Association, June 23–28, 1996.

van Lith, C., S. L. David, and R. Marsh, "Design Criteria for Biofilters," *TransIChemE,* **68B**: 127–132, 1990.

van Lith, C., G. Leson, and R. Michelsen, "Evaluating Design Options for Biofilters," *Journal of the Air & Waste Management Association,* **47**: 37–48, 1997.

VDI Richtlinien 3477, "Biological Waste Gas/Waste Air Purification, Biofilters," in *VDI-Handbuch Reinhaltung der Luft, Band 6,* Verein Deutscher Ingenieure, Dusseldorf, Germany, 1991. Available from Beuth Verlag GmbH, Berlin, Germany.

Wang, J., and K. J. Skipka, "Dispersion Modeling of Odorous Emissions," Paper 93-RA-114A.05, 86th Annual Meeting and Exhibition of the Air & Waste Management Association, June 13–18, 1993.

Weber, F. J., and S. Hartmans, "Use of Activated Carbon as a Buffer in Biofiltration of Waste Gases with Fluctuating Concentrations of Toluene," *Applied Microbiology and Biotechnology,* **43**: 365–369, 1995.

Webster, T. S., and J. S. Devinny, "Biofiltration Technology for Air Pollution Control," in R. A. Meyers, ed., *Encyclopedia of Environmental Analysis and Remediation,* Wiley, New York, 1998.

Webster, T. S., J. S. Devinny, E. M. Torres, and S. Basrai, "Biofiltration of Odors, Toxics and Volatile Organic Compounds from Publicly Owned Treatment Works," *Environmental Progress,* **15**(3): 141–147, 1996.

Webster, T. S., J. S. Devinny, E. M. Torres, and S. S. Basrai, "Microbial Ecosystems in Compost and Granular Activated Carbon Biofilters," *Biotechnology and Bioengineering,* **53**: 296–303, 1997.

Williams, T. O., and R. A. Boyette, "A New Generation of Biofilters for Controlling Odorous Emissions from a Wide Range of Municipal Sources," in *Proceedings of the Control of Odors and VOC Emissions Conference,* Water Environment Federation, Alexandria, Va., 1997, pp. 3-47 to 3-54.

Wolstenholme, P., and R. Finger, "Long-Term Odor and VOC Performance Tests on Biofilters," 67th Annual Conference and Exposition of the Water Environment Federation, Oct. 15–19, 1994.

Yang, Y., and E. R. Allen, "Biofiltration Control of Hydrogen Sulfide. 2. Kinetics, Biofilter Performance, and Maintenance," *Journal of the Air & Waste Management Association,* **44**: 1315–1321, 1994.

Yang, Y., J. M. Caratura, and D. Empfield, "Use of Biofiltration to Control Hydrogen Sulfide Emissions," in *Proceedings of the 16th International Symposium on Gas, Oil, and Environmental Biotechnology,* Nov. 29–Dec. 1, 1993, Institute of Gas Technology, Chicago, 1993.

Yavorsky, J., "Odor and VOC Control from Flavor Manufacturing through Advanced Biofiltration," Paper 97-FA159.04, 90th Annual Meeting and Exhibition of the Air & Waste Management Association, June 8–13, 1997.

Ziminski, R. W., and G. Ferrara, "The Control of Odors and VOCs from Flavor Manufacturing Facilities by a State-of-the-Art Commercial Biofilter," Paper 93-WP-52C.05, 86th Annual Meeting and Exhibition of the Air & Waste Management Association, June 13–18, 1993.

SECTION 8.8.2

BIOLOGICAL SYSTEMS: AERATION BASINS

Robert P. G. Bowker
Bowker & Associates, Inc.
Portland, Maine

INTRODUCTION 8.192
PROCESS MECHANISMS 8.193
  Absorption 8.193
  Adsorption 8.193
  Condensation 8.193
  Biological Oxidation 8.193
LIST OF FACILITIES 8.193
REPORTED PROBLEMS 8.197
  Corrosion 8.197
  Filamentous Organisms 8.197
  Limitations on Odorous Air Volumes 8.197
PERFORMANCE 8.199
ECONOMIC ANALYSIS 8.199
DESIGN CONSIDERATIONS 8.201
REFERENCES 8.202

INTRODUCTION

Many alternatives are available to reduce odor emissions from wastewater treatment plants. These typically consist of (1) adding chemicals to the wastewater or sludge to oxidize, precipitate, or neutralize odorous compounds and (2) containing the odorous air and treating it in one or more odor control devices. Traditionally, these devices have included wet scrubbers, activated carbon adsorbers, and thermal oxidizers. However, these conventional systems are often costly to operate due to high chemical consumption, frequent medium replacement, or high energy requirements. Within the past 5 years, there has been increased interest in biological odor control systems, including biofilters, biological scrubbers, and biotrickling filters.

Diffusion of odorous air into activated sludge basins has been practiced for over 30 years as a technique for controlling odors. Figure 8.8.16 shows a schematic diagram of that process. However, the use of the approach has been limited by concerns over blower corrosion and diffuser plugging, the lack of performance data on its effective-

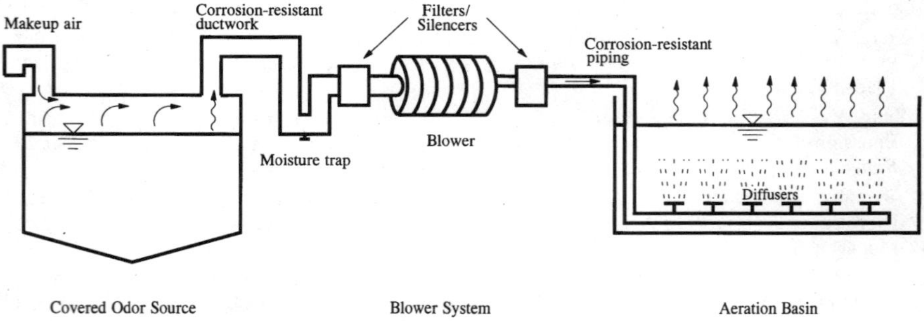

FIGURE 8.8.16 Schematic of activated sludge scrubbing system.

ness for controlling odors, or the observation that airflow rates from domed tanks and process buildings often exceed the demand for mixed-liquor aeration.

The purpose of this section is to provide a summary of information from selected facilities that employ activated sludge diffusion for odor control, present performance data, evaluate the capacity of existing activated sludge basins to handle major sources of odor, and analyze the relative costs of this technique as an alternative to conventional methods.

PROCESS MECHANISMS

Several mechanisms are likely responsible for the removal of odorous compounds:

Absorption

Odorous compounds are transferred from the gas phase into the bulk liquid. Water-soluble compounds such as hydrogen sulfide can readily be absorbed into the liquid. The mass of material transferred is a function of the surface area of the bubbles, the contact time, and the diffusivity coefficient.

Adsorption

Some high-molecular-weight compounds with lower solubility may be physically adsorbed onto the biological floc that is generally at concentration levels of 1000 to 2000 mg/L in the liquid.

Condensation

Relatively warm, odorous air that is transferred into a liquid of lower temperature will result in condensation of volatile organic compounds.

Biological Oxidation

Due to the high concentration of active aerobic microorganisms, biological oxidation is likely responsible for a significant conversion of odorous compounds that are initially absorbed by the liquid, adsorbed onto the floc, or condensed.

LIST OF FACILITIES

Table 8.8.3 summarizes pertinent information from selected facilities that currently use, or have used, activated sludge diffusion as a technique for treatment of odorous air. Many of these facilities listed were summarized in a paper by Ryckman-Siegwarth and Pincince (1992). This is only a partial listing. Many municipalities have implemented this approach while using their own resources, and such cases are difficult to identify.

TABLE 8.8.3 Summary of Selected Facilities Using Activated Sludge Diffusion for Odor Control

Treatment plant and location	Plant capacity, mgd	Source of odor controlled	Year installed	Odorous airflow, cfm	Comments
Bonny Brook WWTP Calgary, AB	132	Fermenters for biological nutrient removal process	1994	1,400	Corrosion of inlet filters occurred, causing damage to blower. New guide vanes are nickel-plated.
Kyrene Water Reclamation Facility Tempe, AZ	3	Aerated equalization tank	1993		Fiberglass reinforced plastic fan transfers foul air to centrifugal blowers for diffusion via jet aerators. High H_2S (>50 ppm). No corrosion problems reported.
Los Coyote WWTP Cerritos, CA	37	Primary clarifiers Influent wet well	1970	10,000+	Filters cleaned every 6 months. Blowers rebalanced and cleaned every year. No corrosion reported. Coarse bubble diffusers. Steel blowers w/coal tar epoxy. Stainless steel ducting.
Long Beach WWTP Long Beach, CA	25	Primary clarifier	1973	6,000+	No filters on compressor suction. Have to steam-clean suction and compressor once a year. Coarse bubble diffusers; no clogging. Concrete corrosion; 100% removal of H_2S.
Hyperion WWTP Los Angeles, CA	300	Headworks Primary clarifiers Dissolved air flotation thickeners Effluent pump station	1959	180,000	Have had to clean blowers twice since start-up. Fine bubble diffusers. No corrosion reported. Clean plus foul air passes through one system. Estimate odor removal efficiencies 96–99%, VOC removal 80%.
LA Glendale WWTP Los Angeles, CA	20	Bar screens Influent pump station Primaries	1976	20,000	Originally coarse bubble diffusers. From 1979–1991 transitioned to fine bubble. No corrosion or clogging reported. Foul air is only source of air to aeration tanks.
Terminal Island WWTP Los Angeles, CA	21	Headworks Primary clarifiers Centrifuge bldg. Sludge-blending tank	1977	30,000	Fine bubble diffusers; steel blower; polyvinyl chloride piping. No plugging. Corrosion only of concrete in covered tanks.
Tillman WWTP Los Angeles, CA	40	Primary clarifiers Base of screw pumps Bar screens	1984		Fine bubble diffusers, no clogging. No corrosion reported.

Pomona WWTP Pomona, CA	13	Primary clarifiers	1965	6,000+	Fine bubble diffusers. Change filters quarterly.
Ventura WWTP San Buena, CA	4	Primary clarifier launders Gravity thickener DAF thickener	1972	14,000	Fine bubble diffusers, cleaned every 2 years as regular maintenance. No corrosion reported. Blowers sandblasted every 2–3 years.
Whittier Narrows WWTP So. El Monte, CA	15	Primary clarifiers	1962	5,000+	Clean or replace filters on blower suction quarterly to annually. No corrosion reported. Fine bubble diffusers, no clogging.
San Jose Creek WWTP Whittier, CA	62	Primary clarifiers	1971	20,000+	Recently switched from coarse to fine bubble diffusers. New filter system on blower suction. Steel blower. Odors from seams of aluminum covers.
Reedy Creek WWTP Orlando, FL	9	Composting facility	1988	5,600	Odor control blowers, piping, and diffusers separate from main system aeration. Coarse bubble diffusers. No corrosion reported.
Vista Royale WWTP Vero Beach, FL	0.2	Lift station	1993		No corrosion apparent after 3 years. Positive displacement blower, coarse bubble diffusers.
Lexington WWTP Lexington, KY	18	Zimpro system Vacuum filter bldg.	1980	1,500	Odor control piping and diffusers separate from aeration. Coarse bubble diffusers. No filters; blowers cleaned every 3–6 months. No corrosion or plugging. Very effective system.
Lowell WWTP Lowell, MA	32	Gravity thickeners Sludge dewatering	1984	9,000	Switched from coarse to fine bubble diffusers in 1991. All stainless steel piping. No filters. Some particulate buildup at diffuser heads.
Springfield Regional WWTP Springfield, MA	67	In-vessel composting	NA	4,500	Interim measure before construction of thermal oxidizer. Stainless steel coarse bubble diffusers; impellers coated w/synthetic resin. Stainless steel and high density polyethylene piping; no corrosion or plugging.
Scarborough WWTP Scarborough, ME	1.8	Aerated sludge holding tank	1983	100	Foul air is separate system from clean air supply; coarse bubble diffusers, steel blower. Corrosion only on ductile iron intake pipes and steel inlet silencer.

TABLE 8.8.3 Summary of Selected Facilities Using Activated Sludge Diffusion for Odor Control (*Continued*)

Treatment plant and location	Plant capacity, mgd	Source of odor controlled	Year installed	Odorous airflow, cfm	Comments
Concord WWTP Concord, NH	10.2	Sludge holding tanks	1994	700	Interim solution until scrubbers built in 1996. High odor and H_2S levels to aeration. Fine bubble diffusers performed better than coarse. No corrosion after 2 years.
Durham WWTP Durham, NH	2.5	Sludge holding tanks	1993	1,300	Corrosion of steel inlet filter; replaced with stainless steel. Fine bubble diffusers; no plugging. Foul air is 100% of aeration air.
Roche Chemicals Belvedere, NJ	3	Primary clarifier	?	16,000	Industrial WWTP. Coarse and fine bubble diffusers; condensate traps; tarlike substance on blower impellers.
Musonetcong WWTP Stanhope, NJ	2.5	In-vessel composting	1992	5,000	Coarse bubble diffusers, 6 ft deep strictly for odor control (basins are mechanically aerated). Blowers have synthetic resin coating on impellers, heads, and sections.
Newberg, OR	4	In-vessel composting	1988	2,400	Dedicated system for diffusion of odors into oxidation ditch; coarse bubble diffusers; no corrosion.
Valley Forge WWTP Phoenixville, PA	8	Influent structure Primary clarifiers	1996	2,200	Dedicated fine bubble diffuser system for odor control. Aeration tanks mechanically aerated. Problems with "tar" accumulation in positive displacement blowers.
Walnut Creek WWTP Austin, TX	60	Headworks Primary clarifiers Flow equalization Sewer tunnel	1977	37,500	Fine bubble diffusers. Filters on blower intake. No corrosion reported. Fiberglass fans transfer air to blower plenum. "Greasy film" on blower lobes removed with steam.
Arlington Co. WPCP Alexandria, VA	30	Headworks	1995	20,000	Steel blowers, coarse bubble diffusers. No corrosion problems. Regular inspection program for blower alignment, oil change.

Source: Bowker (1997).

REPORTED PROBLEMS

Corrosion

Concern over corrosion of blowers is likely the single largest impediment to using activated sludge diffusion for odor control. However, although problems with corrosion of inlet filters and steel inlet ducts have been experienced, there is little documentation of corrosion of the internal components of the blower due to the presence of hydrogen sulfide or sulfuric acid. Several plants reported corrosion of steel inlet filters or silencers that in some cases were replaced with stainless steel. At Concord, New Hampshire, no corrosion was observed in a steel positive displacement blower after 2 years of handling foul air with high levels (up to 100 ppm) of H_2S. At the Mattabassett district in Cromwell, Connecticut, foul air from enclosed belt presses was ducted to the suction side of the fluidizing blower of a fluidized-bed sludge incinerator. Within 5 years, pitting of the aluminum lobes occurred, and the steel guide vanes were badly corroded. The blower was not equipped with a moisture trap. A Los Angeles County plant also experienced corrosion of steel guide vanes, which were replaced with stainless steel vanes (Bowker, 1997).

At least six facilities reported the accumulation of a "tarlike substance" or "greasy film" on the internal components of blowers. This is apparently water-soluble and readily removed by steam-cleaning. At the Valley Forge MTP, the tarlike substance caused the positive displacement blowers to shut down.

In Lowell, Massachusetts, foul air is drawn from a sludge dewatering and lime stabilization process into the blowers without filtration. Some accumulation of particulates was observed at the fine bubble diffuser heads. None of the plants surveyed reported diffuser clogging, although blowers are often equipped with inlet filters to protect fine bubble diffusers downstream.

Filamentous Organisms

Laboratory-scale "bioscrubber" reactors operated at the Metropolitan Wastewater Treatment Plant in St. Paul, Minnesota, did show an increase in a number of filaments when a mixed-liquor bioscrubber received high loadings of hydrogen sulfide (Johnson et al., 1995). None of the full-scale installations reported any cause-and-effect relationship between odorous air diffusion and proliferation of filamentous organisms.

Limitations on Odorous Air Volumes

One of the limitations of activated sludge diffusion is that the volume of odorous air to be treated may exceed the demand of the aeration tanks. Particularly in cases where domes are used for covering tanks, or where odorous air from a large room must be treated, it is likely that the required ventilation rates will be much greater than the aeration air requirements. However, the flow rate of odorous air can be dramatically reduced by using flat, gasketed covers or individual equipment (e.g., belt press) enclosures.

Table 8.8.4 provides estimates of odorous air volumes for major odor sources at a hypothetical 5 mgd conventional wastewater treatment plant. This facility is assumed to have two aerated grit chambers, two circular primary clarifiers, activated sludge basins with fine bubble diffusers, one gravity sludge thickener, two sludge holding tanks, and two 2-m belt presses. Based on a theoretical aeration demand of 3500 cfm, it is possible

TABLE 8.8.4 Estimated Odorous Air Volumes for Major Odor Sources at Hypothetical 5 mgd Plant

Source of odor	Design basis	Number and dimensions	Minimum air change rate, AC/h	Volume of air to be treated, cfm
Influent channels	—	2 @ 3 ft × 50 ft	6	100
Aerated grit chambers	3 min det. @ peak flow	2 @ 10 ft × 30 ft × 10 ft deep	6	200
Primary clarifiers	800 gal/(day • ft^2) @ avg. flow	2 @ 70-ft diameter; 30-in-wide launder	4	2000
Quiescent surface			6	600
Effluent launder				
Gravity thickener	—	1 @ 30-ft diameter	6	200
Sludge holding tanks	3 days storage of 4% Thickened waste activated sludge in 1 tank	2 @ 20 ft × 20 ft × 15 ft deep	6 (empty)	1200
Belt press enclosures	Two 2-m presses	2 @ 10 ft × 16 ft × 12 ft high	18	1200
		Total with full primary clarifier covers		5500
		Total with primary launder covers only		3500

Assumptions: Q_{avg} = 5 mgd
Q_{peak} = 15 mgd
Flat covers on all tanks; 3-ft freeboard except sludge storage tank
Aeration basin air requirements using fine bubble diffusers of 3500 to 4000 cfm (average)

that when using tight-fitting covers at relatively low air exchange rates (6 air changes per hour), a significant number of major odor sources can be treated by activated sludge diffusion. For 3500 cfm of potential airflow to the activated sludge basins, the following major odor sources could be enclosed and treated given the assumptions in Table 8.8.4 (Bowker, 1997):

- Influent channels (two)
- Aerated grit chambers (two)
- Primary effluent launders (two)
- Gravity thickener (one)
- Sludge holding tanks (two)
- Belt presses (two)

These represent sources of some of the most intense and objectionable odors at wastewater treatment plants. This is an illustration only. Plants receiving significantly less than their design flow may have odorous air volumes that exceed the aeration demand. Further, plant designs and tank sizing criteria will vary depending upon diurnal and seasonal flow patterns, wastewater characteristics, and other factors.

PERFORMANCE

Table 8.8.5 summarizes available performance data on the use of activated sludge diffusion for odor control. The data from Orlando, Florida and Springfield, Massachusetts, show a reduction in odor detectability down to background levels associated with activated sludge basins (Ostojic et al., 1992). Odor detectability (dilutions to threshold, or *D/T*) refers to the number of times an odorous air sample must be diluted with odor-free air until it is undetectable by one-half of an eight-member odor panel. These data indicate that the odor emissions from the aeration basins did not increase as a result of introducing foul air.

The Concord, New Hampshire, study is of interest because (1) the activated sludge basins received odorous air with very high odor detectability and H_2S levels and (2) the effectiveness of both coarse bubble and fine bubble diffusers was measured. With coarse bubble diffusers at 10-ft submergence, odor reduction was over 95 percent and H_2S reduction was 92 percent. However, the inlet levels were so high that odor and H_2S levels at the surface were elevated significantly above background. When fine bubble diffusers were tested, odor and H_2S reduction exceeded 99.5 percent, down to background concentrations measured when no odorous air was diffused into the basins.

ECONOMIC ANALYSIS

At the Valley Forge wastewater treatment plant (WWTP) in Phoenixville, Pennsylvania, the sewer authority chose activated sludge diffusion to treat odorous air from the influent chamber, primary clarifier feed wells and effluent launders, and a primary effluent splitter box. These areas had been identified as major odor sources during an odor emissions inventory conducted by Bowker & Associates, Inc. Table 8.8.6 summarizes the design criteria for the system. The new blower and diffuser system is dedicated for odor control, as the basins were designed with mechanical aerators to supply the oxygen necessary for biological oxidation.

TABLE 8.8.5 Performance of Activated Sludge Diffusion for Odor Control

Location	Odor source	Odor detectability, *D/T*			H_2S, ppm			Ref.
		In	Out	Back-ground	In	Out	Back-ground	
Concord, NH	Sludge holding tank							
	Coarse bubble	24,150	850	10	~100	7.8	0.2	1
	Fine bubble	39,000	18	14	~100	0.3	0.2	1
Los Angeles, CA	Headworks				6	<1	—	5
Orlando, FL	In-vessel composting reactor	2,600	74	18				4
			9	8				
		3,900	27	30				
			49	83				
Springfield, MA	In-vessel composting reactor	230	130	180				4
		430	230	240				
		3,500	110	130				

TABLE 8.8.6 Key Design Criteria for Odor Control System at Valley Forge WWTP

Parameter	Value	
Odor sources	Influent chamber Primary clarifier feed wells Primary clarifier effluent launders Primary effluent splitter box	
Airflow	2200 cfm	
Air change rate below covers	12 AC/h	
Inlet H_2S (estimated)	120 ppm (summer)	
Blowers	2–60 hp positive displacement	
Diffusers	394 tubular, flexible-membrane type at 14-ft depth	
Materials of construction	Covers	Fiber-reinforced plastic
	Ductwork	Polyvinyl chloride, 316 stainless steel
	Blower filters, silencers	316 stainless steel
	Blowers	Steel
	Discharge piping	316 stainless steel
Concrete protection	Min. 40-mil vinyl ester coating above waterline	

TABLE 8.8.7 Economic Analysis of Odor Treatment Alternatives for Valley Forge Sewer Authority

	Activated sludge diffusion	Two-stage wet scrubbing
Capital, $	230,000	360,000
Operation and maintenance, $/yr		
Power	72,000	7,000
Chemicals	0	30,000
Labor	1,000	7,000
	73,000	44,000
Total annual cost, $/yr (20 yr, 7%)	95,000	78,000

If a diffused aeration system already exists, odor treatment is virtually free since the only capital expenditures are for conveying the odorous air to the blowers, installing moisture traps, and possibly replacing some inlet components with corrosion-resistant materials. However, if a system is constructed strictly for odor control, as in the case of the Valley Forge Sewer Authority, both capital and operating costs must be considered. Table 8.8.7 shows a comparison of capital and operating costs for a blower and diffuser system versus a two-stage wet scrubber for the conditions at the Valley Forge plant. While the capital costs for the diffusion alternative are one-third lower than those for the scrubber option, annual costs for blower power are more than twice the estimated annual costs for scrubber chemicals. This analysis does not include any energy cost credit for possibly reducing the speed of mechanical aerators in those basins containing diffusers. On a total annual cost basis, the two-stage scrubber is estimated to be about 20 percent less costly than activated sludge diffusion. For airstreams containing lower odor and H_2S levels than those anticipated at the Valley Forge WWTP, a single-stage scrubber would likely be adequate, increasing the cost differential between the two approaches.

DESIGN CONSIDERATIONS

The following design considerations must be addressed for diffusion of odorous air into activated sludge aeration basins:

- Odorous air characteristics

  Concentration of constituents

  Solubility in water

  Biodegradability

- Odorous air volumes

  Compare desired odorous airflow rate with minimum aeration air demand

- Corrosion resistance

  Use materials resistant to hydrogen sulfide, sulfuric acid

  Inlet piping and ductwork, silencers, filters, moisture traps, guide vanes, etc.

- Moisture trap

  Install on inlet side of blower

- Aeration system design

  Oxygen demand

  Oxygen transfer efficiency

  Coarse versus fine bubble diffusers

  Diffuser type (ceramic dome, flexible membrane, slotted pipe)

  Diffuser depth, spacing

  Blower selection

  Blower control system

  Miscellaneous appurtenances (moisture trap, inlet filters, silencers, dampers, pressure gauges)

REFERENCES

Bowker, R. P. G., "Biological Odor Control by Diffusion into Activated Sludge Basins," in *Proceedings of WEF Conference on Odor and VOC Control,* Houston, TX, April 1997.

Ryckman-Siegwarth, J., and A. B. Pincince, "Use of Aeration Tanks to Control Emissions from Wastewater Treatment Plants," *Proceedings of 65th Annual Conference, Water Environment Federation,* vol. 10, Alexandria, VA, 1992.

Johnson, L. K., C. E. G. Waskow, P. A. Krizan, and R. C. Polta, "Suspended Growth Bioscrubber for Hydrogen Sulfide Control," *Proceedings of Specialty Conference on Odor/VOC Control,* Air & Waste Management Association, Pittsburgh, PA, 1995.

Ostojic, N., A. P. Les, and R. Forbes, "Activated Sludge Treatment for Odor Control," *Biocycle,* vol. 33, no. 4, April 1992, pp. 74–78.

Stillwell, S. A., D. E. Hans, and P. C. Katen, "Biological Scrubbing of Foul Air in Activated Sludge Treatment Reduces Odors and ROGs from Headworks and Primary Clarifiers," *Proceedings, Odor and Volatile Organic Compound Emission Control for Municipal and Industrial Wastewater Treatment Facilities,* Water Environment Federation, Alexandria, VA, 1994.

SECTION 8.9

HYBRID SYSTEMS

Avinash N. Patkar
ILA Consultants and
Jacobs Engineering
Cincinnati, Ohio

BACKGROUND 8.203
DESIGN BASIS 8.204
Airflow 8.204
Temperature 8.204
Gas-Phase Composition 8.204
GUIDELINES FOR STANDARD SYSTEMS 8.205
Absorption 8.205
Adsorption 8.206
Biofiltration 8.206
Condensation 8.206
Cryogenic Condensation 8.207
Flaring 8.207
Catalytic Oxidation 8.207
Thermal Oxidation 8.207
FIVE HYBRID SYSTEMS 8.209
Absorption and Recycling 8.209
Absorption in Water and Biotreatment 8.209
Combustion Air Supplement 8.209
Concentration and Condensation 8.209
Fuel Oil Absorption before Burning 8.209
GENERAL HYBRID RULES 8.210
BUDGETARY COSTS 8.210
FINAL SELECTION 8.211
CASE STUDIES 8.211
No. 1: Batch Dryer with Ethanol Emissions 8.211
No. 2: Acrylonitrile Manufacturing 8.212
No. 3: Organic Chemical Manufacturing 8.213
No. 4: Polymer Manufacturing 8.213
No. 5: Groundwater Stripping 8.214
CONCLUSIONS 8.214
REFERENCES 8.214

This chapter has so far discussed separate emission control technologies. It has been proven that hybrid emission control systems may be more cost-effective than single standard emission control systems in many industrial applications. Because of the upcoming stringent Environmental Protection Agency (EPA) ambient standards for ozone, it is essential to evaluate all possible control alternatives. We first present some pollution control strategies which are hybrids or combination of different ideas. We then present a method for screening and selecting the candidate hybrid systems for additional review. The overall evaluation approach ensures selection of the most cost-effective and optimal alternative for the specific application. Finally, five specific case studies illustrate where hybrid systems have been more cost-effective in meeting stringent air emission standards. In most cases, a cost comparison with standard systems for the same application is also given.

BACKGROUND

The three common EPA regulations applicable to the industry are the Hazardous Organic NESHAP, or HON rule for air toxics; the New Source Performance Standard (NSPS) for process vents; and maximum achievable control technology (MACT) standard for specific industries. Although emission control required under one of these regulations can be achieved by a standard control system, the cost-effectiveness value of a single control unit (expressed in dollars per ton of pollutant removed) may not be "reasonable" to the industry concerned. It is, therefore, beneficial to consider a hybrid system which combines some standard control technologies to be more cost-effective.

Although hybrid systems can offer improved cost-effectiveness, they require a higher degree of preliminary engineering and understanding of each component of the hybrid system. Extra effort may also be required to convince the technical managers within the company and the regulatory agency about the hybrid system's advantages. Without careful study, you may select a hybrid system which is less cost-effective and even more complex than a standard control system. It is, therefore, important to carefully select the cases in which hybrid air emission control systems are employed.

DESIGN BASIS

The process releasing the volatile organic compound (VOC) emissions should be studied by the environmental engineer with the process engineer and operating staff. The current regulations, and potential future regulations, will determine the control efficiency required. It is crucial for the team to also evaluate the following items as a part of the design basis.

Airflow

Traditionally, the worker exposure standards are met by increasing the building ventilation rate. Such dilution is expensive: It will incur an additional capital cost of \$50 to 200 per actual cubic foot per minute of the airflow to the control device. The annual operating cost is also dependent on the flow rate. The airflow should be minimized without raising the VOC concentration above 25 percent of the lower explosive limit (LEL) value, or above the threshold limit value (TLV). Ease of access and clear visibility requirements of the operating personnel should also be considered in the design of the capture system and the openings for the airflow.

Temperature

Since every 10°F drop in temperature would typically reduce the vapor pressure by a factor of 2, it is beneficial to reduce the operating temperature of paint spraying equipment. The viscosity of the paint can be held at the value equal to that at the higher temperature by adding few parts per million of a viscosity-reducing agent. Another successful approach to reduce the emissions has been to increase the level of the catalyst in the gel coat spray application so that the amount of styrene reacted increases. The same concept is valid for liquid-phase reactions using a VOC as a reactant.

Gas-Phase Composition

If the process is batch- or time-dependent, it is important to quantify the variations in the emissions with time. The peak value, with a safety factor of 2, can be used as the design basis. The VOC, oxygen, and moisture concentrations should be measured by source testing. If the oxygen level is below 5 percent, as is the case with nitrogen purging, the VOC level can be raised to LEL (or higher) without danger of explosion. The relative humidity should not exceed 50 percent when carbon adsorption is used, since it hinders VOC adsorption. The total moisture level is important if condensation is considered. The condenser load will depend largely upon the amount of water condensed, and only slightly on the amount of VOC condensed.

GUIDELINES FOR STANDARD SYSTEMS

The earlier sections of this chapter have discussed the standard VOC control systems in detail. A brief set of guidelines for the application of standard systems is given in Table 8.9.1 (modified from Patkar and Laznow, 1992). The airflow limit in Table 8.9.1 is per unit or module. If the total airflow is higher, two or more units will be needed in parallel. The VOC concentration range is one in which the particular technology is cost-effective. These guidelines are mainly applicable for VOCs and not for odors.

Absorption

If the VOC is water-soluble, valuable (price >$0.30 per pound) and does not form an azeotrope, water absorption with steam stripping should be considered. For inexpensive and water-soluble VOCs, water absorption and routing the absorber bleed stream to the

TABLE 8.9.1 Application Guidelines for Standard VOC Control Systems

System	Airflow, acfm per unit	Concentration, ppmv	Used for compounds	Not used for compounds
Absorption (water)	500–50,000	250–1,000	Alcohols, ketone	Aromatic hydrocarbons
Adsorption	Up to 30,000	250–1,000	Molecular weight 45 to 150	Boiling point >300°F, ketones
Biodegradation	500–30,000	20–600	Hydrocarbons	Halogenated
Condensation (Rankine cycle)	Up to 5,000	>2,000	Cost >$0.30/lb	Which need temp. <−40°F
Condensation (Brayton cycle)	Up to 5,000	>2,000	Cost >$0.30/lb	Highly volatile ones
Condensation (Cryogenic)	Up to 500	>200	Most VOCs	Mixtures which form azeotrope
Flare	500–1,000 Up to 5,000	>300 >2,000	Variable flow, mixed VOCs	<50 Btu/scf heating value
Oxidation—catalytic	1,000–50,000	20–400	Nonhalogenated VOCs	VOCs with S, N, or halogen
Oxidation—thermal	1,000–50,000	300–1,000	Halogenated	Nonhalogenated

on-site water treatment system could be cost-effective. The amount of water required and its treatment cost, especially for concentrations of 100 ppm or less when an open-loop absorber is necessary, are limiting factors (see the section "Five Hybrid Systems").

Adsorption

If the VOC has good adsorptivity (>0.10 lb/lb of adsorbent) and is sparingly soluble in water, adsorption followed by on-site steam regeneration is a good choice. An example is xylene which is used as a thinner in spray painting operation. Although activated carbon is the leading adsorbent, it has the following limitations: Ketones are known to form hot spots leading to fire hazard due to their high heat of adsorption. The adsorption capacity of carbon is adversely affected by relative humidity (RH) and drops off rapidly above an RH of 50 percent. Therefore, humidity control is required in addition to dust control to avoid bed plugging. Finally, high-boiling-point compounds do not desorb well from the carbon bed which is regenerated by using low-pressure steam (up to 300°F). The advantages of activated carbon include the wealth of adsorption data for many VOCs and its ability to adsorb many compounds from a VOC mixture. The alternatives to activated carbon are synthetic organic resins and zeolites. Both are hydrophobic so that the RH of the airstream has less impact on their adsorption capacity. The resins are usually used in a pressure-swing mode with the regeneration step conducted under vacuum. Zeolites, which are silica- or alumina-based, can be regenerated at higher temperatures (800°F) than in the carbon bed and can be used for high-boiling-point compounds. The two disadvantages with the synthetic resins or zeolites are that they are 4 to 5 times more expensive per pound than activated carbon and are designed to adsorb a specific VOC.

Biofiltration

This technology has been widely applied in controlling odors from municipal sludge handling operations. Since the 1980s, it has been developed further for VOC control, especially for low concentrations. The concept is to use a medium which will adsorb the VOCs, which are then degraded by the biomass. The biomass is fortified by nutrients which are supplied by a trickling flow of water. The process gas is saturated with water in a humidifier ahead of the bioreactor so that the biomass does not dry up. Some biomass may also adsorb VOCs. A primary limitation of biofiltration is the extent of biodegradability of the VOC. Usually, complex and heterogeneous molecules (containing sulfur and halides) are less biodegradable. In addition, the residence time required is higher (30 to 60 s) than that for absorption (5 to 6 s) or adsorption (10 to 12 s). The bioreactor, therefore, is much larger. Commercial experience using the technology is limited but growing rapidly.

Condensation

Condensation (using indirect Rankine cycle) is typically cost-effective for airflows up to 5000 acfm and VOC concentrations greater than 2000 ppm. The water present in the process air is condensed in a separate condenser, contributing most of the load on the refrigeration unit. A typical Rankine refrigeration unit can achieve a coolant temperature of −40°F. Depending on the VOC vapor pressure at −30°F and its inlet concentration, a condensation system can achieve 70 to 80 percent removal. The direct Brayton cycle can be used for airflows higher than 5000 acfm. The higher thermal efficiency of the cycle can partially compensate for its high capital cost.

Cryogenic Condensation

These systems use vaporization of cryogenic liquid nitrogen for VOC control. In one variation of the system, the nitrogen is vaporized and mixed with the process air directly in one vessel. In the indirect system, nitrogen vaporization and VOC condensation are carried out in separate heat exchangers with a water-glycol mixture used as the heat-transfer fluid. The cost of liquid nitrogen is high, but it can be considered "free" if it is being vaporized at the plant to create a nitrogen blanket.

Flaring

Flares are a special case of thermal oxidation. They are recommended for process flows up to 5000 acfm and VOC concentrations above 2000 ppm. The major limitation of a flare is the cost of the supplementary gas required to bring the heating value of the process exhaust to 300 Btu/scf. The advantage is that the monitoring and recordkeeping requirement is simple: make sure that the pilot flame is lit. Although condensation is also applicable above 2500 ppm, the flare is preferable when the process gas has low but variable flow and contains several VOCs which are not valuable. It is usually not cost-effective to recover each VOC from the liquid condensate by distillation.

Catalytic Oxidation

This technology is effective in destroying most VOCs. Usually, a VOC containing sulfur or halogens was considered unsuitable for catalytic oxidation due to the acids formed during combustion and their effect on the catalyst. This constraint is alleviated by some of the new catalysts which can withstand the inorganic acids. Catalytic oxidation is typically applied for low concentrations so as to limit the rise in temperature across the catalyst bed to 100°F.

Thermal Oxidation

Thermal oxidation is cost-effective when the VOC concentration is greater than 1000 ppm. At this level, the supplementary gas requirement is low. The heat recovery system used is either recuperative (indirect heat exchange between combustion air and exhaust gases) or regenerative (direct heat exchange between a solid medium and the hot off-gases or the combustion air). The recuperative units are less expensive but can recover only 70 percent of heat. The regenerative units can recover 90 to 95 percent of the heat. Thermal oxidizers are available with low NO_x burners and as a portable unit for low flow ($<$1000 acfm). These units are also called *enclosed flares.*

The basic application guidelines are depicted in Fig. 8.9.1. The boundaries between the regions are not to be interpreted as sharp. If an application falls near a boundary, it probably is a good candidate for either control system. This is especially the case in the very low concentration (up to 200 ppm) region. For example, adsorption and off-site regeneration is shown as the choice for low airflows, and biofiltration or catalytic oxidation is the choice for higher airflows. The VOC and its adsorption and biodegradation potential will influence the final choice. In the concentration range of 300 to 1000 ppm the choices for flows higher than 1000 acfm are absorption, adsorption, and thermal oxidation. Flare and cryogenic condensation are suggested for an airflow of less than 1000 acfm. A detailed analysis will narrow the choice further. In the high concentration range

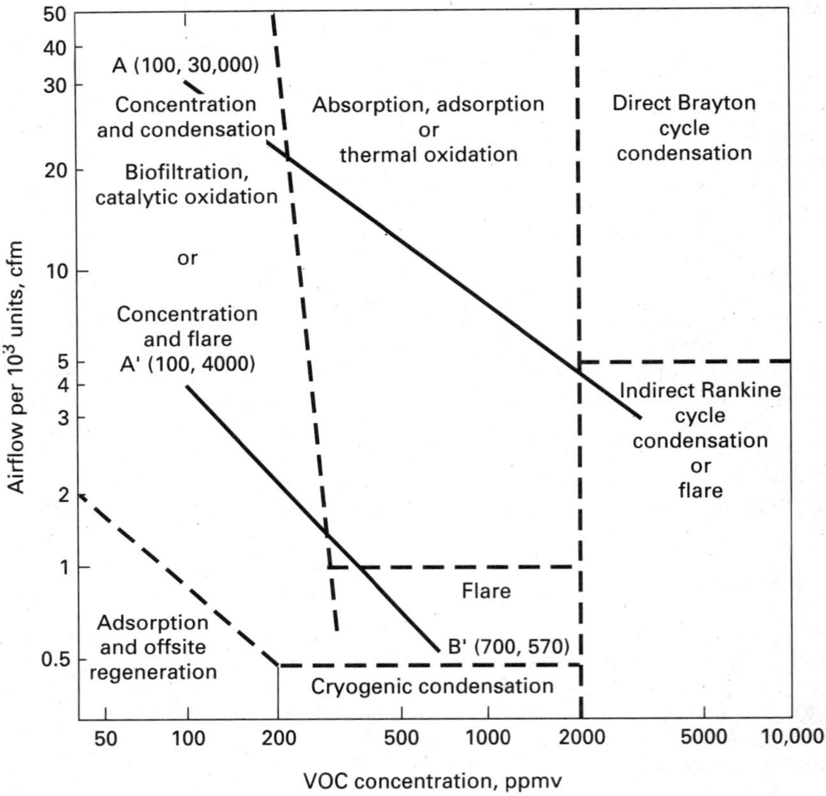

FIGURE 8.9.1 Regions of applicability for standard VOC control systems.

(above 2000 ppm), the figure gives three choices: direct Brayton cycle condensation (for airflow above 5000 acfm) and indirect Rankine cycle condensation or flare below 5000 acfm. This boundary is preliminary. The feasibility of reusing the VOC and its value will affect the cost-effectiveness of the condensation options as discussed above.

Concept of Hybird Systems

Consider a special case, in the upper left-hand region of the figure, where the airflow is 30,000 acfm and VOC concentration is 100 ppm (point *A*). Let us suppose that the VOC is not biodegradable and is a chlorinated organic worth $1.50 per pound. Concentration of the VOC by (zeolite) adsorption and hot inert gas regeneration can transfer the operating point to *B* (3000 ppm with 3000 acfm), where condensation is an attractive option. If the compound, however, is worth only $ 0.20 per pound or is mixed with other undesirable VOCs, a flare should be considered. The degree of concentration depends on the VOC. If the original point is at *A′* (concentration of 100 ppm and flow of 4000 acfm) and it can be concentrated only by a factor of 7 to point *B′*, again a flare should be considered. These two cases are examples of cost-effective hybrids: adsorption/regeneration and condensation, or adsorption/regeneration and flare.

FIVE HYBRID SYSTEMS

The following five hybrids should be evaluated for suitability and regulatory agency acceptance (Patkar and Reinhold, 1993; Sylvester, 1991). They can be more cost-effective than alternative hybrids presented.

Absorption and Recycling

It is very cost-effective to absorb the VOC in a solvent which will be used directly in the process. In a special case the solvent can be water. The level of VOC in the solvent which are acceptable to the process will determine the applicability of this hybrid.

Absorption in Water and Biotreatment

If the VOC is soluble in water, evaluate by using a water absorber which has a low liquid-to-gas ratio (e.g., mist spray tower versus packed tower). The water can be routed to the on-site activated sludge basin or to a small bioreactor nearby.

Combustion Air Supplement

The VOC-laden air may be used as combustion air in on-site furnaces and boilers. The primary considerations are the distance of the VOC source from the boiler, the amount of airflow, and the VOC concentration. Even if the VOC level is below 25 percent of LEL, flashback protection should be provided by using flame arresters. If the airflow is higher than the combustion air requirement, it can be reduced using adsorption followed by hot air regeneration. This technique can reduce the airflow by a factor of 10 to 15 (see points A' and B' in Fig. 8.9.1).

Concentration and Condensation

A variation on the adsorption and hot air regeneration is to use inert gas (nitrogen) for the regeneration. The concentrated VOC-laden gas can be routed to a condenser for VOC recovery. This is demonstrated in Fig. 8.9.1 by points *A* and *B*. Nitrogen use allows high VOC concentrations in the VOC-rich gas.

Fuel Oil Absorption before Burning

The VOC can be absorbed in the fuel oil before routing the oil to the furnace. The limitations of this idea are that (1) the amount of fuel oil available is based on the furnace rating and may not be sufficient for absorption and (2) the secondary fuel oil emissions in the exhaust process air from the absorber. If the VOC-laden fuel oil makes the boiler subject to the boilers and industrial furnace (BIF) regulations, this idea should not be used. The BIF regulations would require extensive and expensive testing of the secondary VOC emissions from the boiler.

GENERAL HYBRID RULES

Some application rules for selecting hybrid components are given in Table 8.9.2 (a modified version of one given by Govind and Bishop, 1996). The rules will help you create some more cost-effective hybrid systems. The table assumes that maximizing the cost-effectiveness is the only goal. For your specific application, select three standard systems, one of the hybrids in the previous section, and one hybrid system from these rules. The five systems are the preliminary candidates for your application. The list can be shortened to three based on cost-effectiveness. The section "Final Selection" identifies other important issues in final selection of the control system from the three finalist systems. Table 8.9.2 gives guidance for cases when all, some, or none of the VOCs are biodegradable. It also covers the case when oxidation or carbon adsorption is essential to meet the regulatory goals. Finally, it covers the case where one VOC is valuable and water-soluble. Not all combinations of VOC characteristics or applications are covered. The intent is to stimulate your thinking so as to synthesize an optimal hybrid for your case.

BUDGETARY COSTS

Of all the published sources for cost estimation of standard VOC control systems, two are recommended: the EPA manual (Vatavuk, 1990) and a recent summary (Bahner et al., 1996). Both these references give simple equations for the purchased equipment costs of standard VOC control systems. In addition, most vendors will provide budgetary costs based on the total airflow, VOC concentrations, oxygen and moisture levels, and degree of control required. These costs can be used, along with the site-specific installation factors, to estimate the total installed cost of the control system. The next

TABLE 8.9.2 Rules for Selection of Hybrid Components

If this applies	Then consider	Because
All VOCs are moderately biodegradable, one in high concentration.	Biofiltration, which needs RH of 100 percent, is the last unit (condensation first).	Condensation will render RH of 100 percent before biofiltration unit.
Some VOCs are not biodegradable, but are adsorbable.	Biofiltration, which needs RH of 100 percent, is the last unit (adsorption first).	Adsorption needs low RH. The gas can be humidified after it.
Not all VOCs are biodegradable, total <50 ppm.	Concentration and condensation ($>\$0.30$/lb)	Concentration and oxidation ($<\$0.30$/lb)
Oxidation has to be used to meet treatment goal. Total VOCs <100 ppm.	Concentration and oxidation (on-site heater or flare, oxidizer)	The cost of the hybrid will be less than cost of oxidation. NO_x and CO will be less.
One VOC is water-soluble and valuable, others are biodegradable.	Water absorption/stripping for the valuable VOC	Other VOCs can be treated by biodegradation with the saturated gas.
Carbon adsorption can meet treatment goal.	Using condensation to reduce the load on carbon	This combination may be more cost-effective.

step is to determine the direct and indirect operating and maintenance costs. The total annualized cost (TAC, \$/yr) is the sum of these costs and the capital recovery charge. The TAC can be converted to the cost-effectiveness value (CE, \$/ton of VOC removed), knowing the annual VOC removed (AVOC, tons/yr). These three are related by the simple equation

$$CE = \frac{TAC}{AVOC} \tag{8.9.1}$$

The control systems with the three lowest CE values can be analyzed further (see below).

FINAL SELECTION

The three final systems should be compared with respect to the following criteria (Patkar and Reinhold, 1992):

- The cost-effectiveness value
- The complexity of operation
- The space requirement for the entire system
- Acceptance by the regulatory agency
- The delivery period required by the vendors

Each criterion should be given a certain weight factor based on its importance to the user on a scale of 1 to 5, with 1 being the least important and 5 the most important. Then each team member should rate each control system on a scale of 1 to 10, with 1 for the least desirable and 10 for the most desirable. The scores should be averaged and multiplied by the weight factor. The ranking will be clear by the final total scores.

CASE STUDIES

Five case studies are cited here from literature and the author's experience. These are meant as examples which illustrate the usefulness of hybrid systems.

No. 1: Batch Dryer with Ethanol Emissions

A batch dryer with an exhaust of 7000 acfm was used for a pharmaceutical product with ethanol as the solvent. Ethanol can be controlled by water absorption (scrubbing), biofiltration, or oxidation. The choice of system depended upon the mode of operation, airflow, VOC concentration, fuel and wastewater treatment costs, and control efficiency required by the regulations.

A biofilter or an oxidizer would usually meet the 95 percent removal efficiency for ethanol. During the initial part of the drying cycle, the ethanol concentrations were very high (well over 10,000 ppm), so biofiltration was not considered. Oxidation was not considered since the peak was higher than 25 percent of the LEL of ethanol. For water scrubbing, a nonrecycling water system was required so as to achieve the high removal, with the associated higher biochemical oxygen demand (BOD) loading to the municipal sewer system. The company did not want to add to its existing outfall flow and BOD loading. It was also decided not to use water absorption and azeotropic distillation to recycle the ethanol due to process purity considerations. Since the ethanol concentration in the air varies by two orders of magnitude, it was decided that it would be unsafe to divert the air to the nearby boiler.

Combining a wet scrubber, an air stripper, and an oxidizer resulted in a hybrid system meeting the regulations, but at a much lower operating cost than an oxidizer alone. The water absorber/air stripper/incinerator hybrid system handled the variable concentration by storing the absorber effluent (of 45 gal/min) in a storage tank. The absorber typically runs only 8 h/day. The air stripper and the catalytic oxidizer were sized to handle a constant feed of 15 gal/min continuously. The solution is first heated by steam to increase the ethanol vapor pressure, and stripped by hot air saturated with water by direct injection of steam. This reduces the air volume sent to the catalytic oxidizer by a factor of 7, and also evens out the ethanol concentration. The capital cost of this hybrid system, as shown in Table 8.9.3, is equal to the capital cost to a catalytic oxidizer for the entire drier exhaust flow. But the hybrid has less than one-eighth the operating cost. Over 5 years, the hybrid system saves more than $80,000.

No. 2: Acrylonitrile Manufacturing

Some emission sources contain valuable chemicals that could be reused within the same or another process at the manufacturing facility. Use of condensation or carbon adsorption to recover the valuable chemical, however, may not meet the required 98+ percent removal efficiency. A carbon adsorber or condenser could be the first stage of a hybrid emission control system to recover as much of the valuable chemical as possible. This first-stage recovery device could be followed by an oxidizer or biofiltration unit to destroy the remaining air contaminants to meet a 98+ percent removal efficiency, and could be much more cost-effective for such an application. A hybrid carbon adsorber and catalytic oxidizer system for acrylonitrile control is described.

The acrylonitrile (ACN) monomer (1000 ppm) is removed from an airflow of 2000 acfm by carbon adsorption at 95 percent efficiency. The carbon bed is regenerated with steam, and ACN is condensed along with the regenerating steam. The ACN is recycled to the process after purification, and a credit from the reuse more than offsets the hybrid system operating cost. The value used for credit is $0.25 per pound. The remaining ACN is catalytically oxidized to reduce its emissions by another 95 percent. The overall efficiency of the hybrid is more than 99 percent, a level achieved only by thermal oxidation.

TABLE 8.9.3 Cost Comparisons: Case Study No. 1

Cost item	Absorber/air stripper/catalytic oxidizer	Catalytic oxidizer only (7000 acfm)
Capital cost*	$278,000	$275,000
Operating cost†	$ 2,350 per year	$ 19,100 per year
Total 5-year cost‡	$289,750	$370,600
Total savings	$ 80,750	Base case

*Installation included and assumed to be 25 percent of capital cost, 7000 acfm airflow.
†Based on $4.00/million Btu fuel cost, $0.05 per kWh electric cost, and 2000 h/yr.
‡Total is installed capital cost plus 5 years' operating cost. Interest on the money is not considered.
Source: Kenson, 1996.

TABLE 8.9.4 Cost Comparisons: Case Study No. 2

Cost item	Adsorber/oxidizer	Thermal oxidizer
Capital Cost*	$500,000	$375,000
Extra Cost	$125,000	Base case
Operating cost†	$26,400/year	$57,600/year
Reuse credit‡	$78,600/year	Base case
Net annual cost	($52,200/year)	$57,600/year
Annual savings	$109,800/year	Base case
Payback period	1.15 years	Base case

*Installation included and assumed to be 25 percent of capital cost, 2000 acfm airflow.
†Based on $4.00/million Btu fuel cost, $0.05 per kWh electric cost, and 8000 h/yr.
‡Based on 314,400 lb/yr monomer recovered at $0.25 per pound value.
Source: Kenson, 1996.

The hybrid system costs are compared with the thermal oxidizer costs in Table 8.9.4. Although the hybrid system has higher capital cost than the thermal oxidizer, the credit of acrylonitrile recovered by the hybrid pays back the extra capital cost in 1.1 years. This payback analysis is different from the 5-year cost comparison presented in the original reference (Kenson, 1996).

No. 3: Organic Chemical Manufacturing

In a batch operation, toluene and acetone were emitted as a mixture in an airstream. The airflow was 3500 acfm. The standard system selected was carbon adsorption. It was found that the acetone adsorption was much less than the value predicted by the vendor. Two modifications were considered: water absorption of acetone, distillation and recycle; or routing the adsorber tail gas to an on-site boiler. The cost analysis showed that the second option was much more cost-effective. The successful hybrid system was carbon adsorption and on-site boiler with a control efficiency of 99+ percent.

No. 4: Polymer Manufacturing

The emissions of aromatic hydrocarbons from a vacuum system at a polymer plant were being controlled by a condenser using chilled water at 40°F. Source testing data revealed that the airflow was 50 acfm; and the emissions were higher than the state permit limit, and higher than the NSPS for such plants. NSPSs require that if an add-on secondary condenser is used as the final control device, the exhaust gas temperature be below −20°C. Three systems were considered: a secondary condenser, routing the condenser off-gases to on-site gas-fired heaters, or to a flare.

The chiller-condenser was not selected because of the complexity of operations; the twin condensers need to intermittently thaw the ice formed on the tubes from the moisture in the gas. The capital cost of the flare was higher than the cost of ductwork and other modifications required to the heaters. The operating cost of the flare was also higher since it had to maintain a heating value of 300 Btu/scf. Still, the flare was selected because of quick acceptance by the agency and ease of operation.

No. 5: Groundwater Stripping

For a groundwater air stripping application, the following design basis (adapted from Govind et al., 1996) was considered:

- Total airflow: 10,000 acfm; relative humidity: 90 percent; temperature: 25°C
- Toluene and ethyl benzene: 400 ppm each; methylene chloride: 300 ppm
- Chlorobenzene: 200 ppm (not biodegradable)
- Treatment goal: total VOC <10 ppm

Since chlorobenzene is not biodegradable, and total VOC concentration of 1300 ppm is too high for catalytic oxidation, only thermal oxidation was considered as a standard system. The hybrid system selected was carbon adsorption and biofiltration. The total annualized cost of the thermal oxidizer was $330,000 per year, whereas that of the hybrid system was $143,000 per year with a net saving of $187,000 per year. The capital costs are not cited by the authors. They are estimated to be $600,000 and $800,000 for the two systems, respectively. The extra $200,000 for the hybrid system is paid back in 1 year.

CONCLUSIONS

In some cases, hybrid VOC control systems have been more cost-effective than the standard systems. The logical approach presented here can help you select some cost-effective hybrids. Other important issues in selecting a VOC control system include ease of operation; space, time, and utilities required; and agency approval. The evaluation method recommended gives a weight to these issues along with costs. Such an approach will let you select an optimal control system—standard or hybrid.

REFERENCES

Bahner, M., E. Kong, S. Turner, and N. Kaplan, An assessment of styrene emission control technologies, in *Proceedings of the Specialty Conference of Air and Waste Management Association,* Clearwater Beach, FL, March 1996.

Govind, R., and D. Bishop, Overview of biofiltration—Basic technology, economics and integration with other technologies, in *Proceedings of the Specialty Conference of Air and Waste Management Association,* Clearwater Beach, FL, March 1996.

Govind, R., X. Shao, and D. Bishop, Biofiltration for Treatment of VOCs in Air, in *Biodegradation Technology Development,* Subhas K. Sikdar and Robert L. Irvine (eds.), Technomic Publishing Co., Lancaster, PA, 1997.

Kenson, R. E., Hybrid and simple control systems for organic air emission sources subject to HON/MACT, CTG/RACT and NSPS regulations, in *Proceedings of the Specialty Conference of Air and Waste Management Association,* Clearwater Beach, FL, March 1996.

Patkar, A. N., and J. Laznow, *Hazardous Air Toxic Control Technologies, HAZMAT World,* Tower-Burner Publishing, Glen Ellyn, IL, April 1992.

Patkar, A. N., and J. M. Reinhold, Evaluation of control systems for VOC emissions, in *Proceedings of Specialty Conference of Air and Waste Management Association,* Boston, MA, October 1992.

Patkar, A. N., and J. M. Reinhold, Novel and hybrid control systems for air toxic emissions, in *Proceedings of 86th Annual Meeting of Air and Waste Management Association,* Cincinnati, OH, June 1993.

Sylvester, R. W., Choose the right technology for VOC control, in *Proceedings of 84th Annual Meeting of Air and Waste Management Association,* Vancouver, BC, June 1991.

Vatavuk, W. M., *VOC Control Cost Manual,* 4th ed., Environmental Protection Agency, Office of Air Quality Planning and Standards, Research Triangle Park, NC, 1990.

SECTION 8.10.1

PARTICULATES: PHYSICAL COLLECTION

Robert R. Banks
Technical Consultant
Pittsburgh, Pennsylvania

PARTICULATE REMOVAL 8.216
TYPES OF EQUIPMENT 8.216
Mechanical Collectors 8.216
Fabric Filters 8.217
Filter Bags 8.217

PARTICULATE REMOVAL

A gas stream or airstream that is to be treated to remove odors and/or volatile organic compounds (VOCs) may also contain particulate matter which may have to be removed, to prevent its emission as a pollutant or to avoid problems in the VOC or odor control equipment, such as plugging of solid beds or contaminating or building up in liquid scrubbing streams. In most cases the particulate removal equipment would be upstream of the VOC and odor removal equipment, but in some cases it would be advantageous to have it downstream. In some processes for removing acid gases, a baghouse in the downstream position acts as an additional reaction stage for removing the contaminant.

TYPES OF EQUIPMENT

The principal types of particulate collection equipment are *mechanical collectors, fabric filters, electrostatic precipitators,* and *wet scrubbers.* Wet scrubbers are not covered in this section because in those cases where they might be applicable they would be covered with the wet scrubber system for removing odors or VOCs. Electrostatic precipitators are covered in Sec. 8.10.2.

Mechanical Collectors

Mechanical collectors have the advantages of being simple in construction and compact and made of materials suitable for operation at extremely high temperatures. In addition, they do not cause variations in flow, such as occur when a compartment in a baghouse is taken off line for cleaning. Process variations in flow unfortunately affect particulate removal efficiency, particularly in the case of cyclones. The major shortcoming of mechanical collectors is their low efficiency to collect fine particles.

Although mechanical collectors include settling chambers and impingement collectors, for practical purposes the main type used is the cyclone. Settling chambers are large and are capable of removing only large, heavy particles. Impingement collectors are used to some extent to remove sparks and/or coarse particles from a gas stream before it enters a fabric filter so that they cannot damage the bags.

There is little published information on the design of settling chambers. They are occasionally found in older systems, sometimes in the form of a duct of very large cross section. Impingement collectors are either proprietary or designed in-house based on the experience of the designer.

Cyclones come in two main types: conventional cyclones comprising one unit with a tangential inlet (although these may be ganged in groups of two, four, or even more) and multitube cyclones in which large numbers of small-diameter cyclones are mounted in a housing so that they work in parallel. The *conventional cyclones* may be made of a wide variety of materials and are often built into equipment such as catalytic cracking vessels and fluidized-bed boilers. In the latter case the cyclones are refractory-lined or may be made with membrane walls that provide some of the heat-transfer surface for the boiler.

A great deal has been published on conventional cyclones, but, except in the case of small ones for experimental purposes, it is probably better to purchase proprietary equipment as there are small details of the design that can affect their performance.

Conventional cyclones are usually designed for a velocity of 3000 ft/min (15 m/s) in the inlet scroll and a pressure drop of 3 inWC (water column) (747 Pa) as base conditions. A second cone is sometimes put below the primary cone to provide a place for the dust to settle before it is removed through a rotary air lock or similar device. The clean air spins at high speed as it enters the vertical duct at the top of the cyclone, so it is common practice to have it exit through a scroll which converts the kinetic energy of the spin to static pressure. In-leakage degrades the efficiency of a cyclone, so it is important that it is carefully built and that any leaks that develop in the air lock through abrasion, or other cause, are repaired promptly. The tubes in the multitube cyclones are usually made of chilled cast iron and are installed in carbon steel housings so their maximum operating temperature is set by those materials. Multitube cyclones are all proprietary.

In the design of the multiple-tube cyclones, the inlet to the individual tubes is in the form of ramps, around the top of the tube, that direct the gas flow in a helical path around and down the inside of the tube, thus creating the spin that throws the particulate out of the gas stream. The supplier usually has tubes with different designs and angles of ramps to give a range of efficiencies and pressure drops. The tubes are usually rated, by the supplier, for a specific flow, and the required capacity is attained by using the appropriate number of tubes. It is very important that the flow be uniformly distributed among the tubes, or else gas from the dirty side may go backward through one or more of the lightly loaded tubes and seriously degrade the efficiency of the unit.

Fabric Filters

Fabric filters can give very high collection efficiencies, even for very fine particulate in the submicrometer size range. On multicompartment baghouses much of the maintenance can be carried out on an individual compartment without shutting down the system. Disadvantages are that the units are bulky for their capacity, that with usual filter media they are limited to maximum operating temperatures in the range of 400 to 500°F (200 to 260°C), and that the pressure drop is about 10 inWC (250 mmWC). Furthermore, the pressure drop goes through a cycle, except in the case of on-line cleaned pulse-jet collectors, when it increases as the dust cake builds up on the media, suddenly increases substantially when a compartment is taken off line for cleaning, and then is sharply reduced when the compartment is returned to service after cleaning. This pressure drop variation can be important; and in cases where the flow in the odor and VOC removal section of the system must be maintained at a steady value, a flow control system may be needed.

Fabric filters can be classified by whether they employ *low-intensity* or *high-intensity* cleaning. *Low-intensity cleaned or conventional baghouses* collect the dust on the inside of bags that are suspended from their tops and have their open bottom ends connected to a tube sheet or cell plate. At intervals, when the accumulation of dust has raised the pressure drop across the dust cake to a predetermined value, a compartment is isolated for cleaning, usually by closing the compartment outlet damper, to stop the gas flow through the bags. The dust is then cleaned from the bags, by shaking from the top or reversing the gas flow through the bags, and is allowed to drop into the hopper at the bottom of the unit, from which it is removed via a gastight seal.

Baghouses can also be designed as pressure or suction types depending on whether the dirty gas is blown or sucked through them. The suction type is more common, as leakage is inward so dust does not escape through any small openings and the fan handles clean gas. The pressure type is used in some large collectors, as the clean gas side does not have to be airtight, the fan handles dirty gas and, as a result, is usually of a design with lower efficiency.

Conventional cleaning systems use *shakers* or *reverse gas flow* for cleaning the bags. The common diameters of bags used in them are 5¼, 8, and 11½ in (13.3, 20.3, and 29.2 cm). Their strange diameters were selected to minimize waste when making the bags from standard-width cloth. The lengths vary, but the longest bags ordinarily used in each diameter are about 171, 264, and 408 in (434, 671, and 1040 cm). There are no standards for bag dimensions, and bags for two baghouses from the same manufacturer may not be interchangeable.

In a *shaker baghouse,* the bags are suspended from tubes or frames that run from the front to the back of the compartments and are supported so that they can oscillate about their axes. The oscillating motion is produced by a system comprising an eccentric, levers, and links driven by a motor. If the baghouse is intended for use with synthetic bags (bags made from organic fibers), the bags are suspended from each shaker tube in two rows parallel to the axis of the shaker tube, but spaced horizontally from it. This geometry results in the top of the bag moving vertically as well as horizontally when shaken and gives a snapping action on the bags that aids in removing the dust cake. Bags are installed in this type of collector at zero tension; i.e., there is no slack, but the bags are not pulled up tight. If the bags are slack, they wrinkle at the bottom and tend to fail at the wrinkles; in addition, the wrinkles may obstruct the opening at the bottom of the bag and interfere with its cleaning. If the bags are too tight, they may stall the shaker mechanism and/or deflect the cell plate excessively.

In shakers for use with fiberglass bags, only one row of bags is suspended per tube, and this row is on the axis. The purpose is to minimize the vertical motion so as to treat the more delicate bags gently. This design is also used for shaking 11½-in synthetic bags, as these bags may have residual dust cakes weighing in excess of 100 lb each after having been in service for a while, so that this load would require a more robust and expensive shaker mechanism if the other shaker geometry were used.

In a *reverse-flow baghouse,* clean gas, from the compartments that are still on-line or from the stack, is blown by an auxiliary fan into the clean side of the compartment being cleaned. This action fractures the dust cake, which falls downward through the bags and into the hopper. To prevent the bags from collapsing flat, and thus holding the dust in place, the bags have rings sewn in pockets on their outside at intervals along their length. A typical 32-ft bag might have seven rings. Different systems are used to suspend the bags, usually involving a chain and a spring connected to the rigid cap at the top of the bag to maintain tension in the bag.

Ultrasonic vibration, produced by air-powered sonic horns installed in the clean side of the compartments, is sometimes used to augment the cleaning, particularly in the case of reverse-flow baghouses. There are mixed opinions about the long-term efficacy of ultrasonic vibration, as it appears that when sonic horns are used, the residual dust cake may be lighter, but have a permeability comparable to that of the dust cake when horns are not used.

High-intensity cleaning or pulse-jet baghouses collect the dust on the outside of the bags that are suspended from a tube sheet to which their open-top ends are fastened. As the gas flow would collapse the bags, wire cages are inserted into them to hold them open. Cleaning is by a pulsed jet of compressed air directed into the open end of the bags. Pulse-jet collectors were originally always cleaned on-line; i.e., the pulse of air, directed into one or more rows of bags at a time, would stop the flow of gas through the bags and snap them radially away from the cages, thus shaking the dust off. Operated this way, the pressure fluctuations caused by cleaning are much lower than those with off-line cleaning. Over a period of years, the designs of pulse-jet collectors became less conservative so that it became necessary, in some cases, to clean them off-line. It is prudent, in buying a collector that is to be cleaned on-line, to be sure that it is also equipped with dampers so that it can be cleaned off-line. The dust is removed from the hoppers as with the conventional collectors.

Pulse-jet cleaning is very different from those methods described for conventional baghouses. Cleaning is effected by directing a short (about 100-ms) puff of air into the open end of the bag to snap the bag outward from the cage and thus shake the dust from the bag. Pulse-jet collectors are classified by the air pressure supplied for cleaning into three types: *high-pressure* at about 90 lb/in^2 (620 kPa), *medium-pressure* at about 35 lb/in^2 (240 kPa), and *low-pressure* at about 8 lb/in^2 (55 kPa).

The high-pressure, low-volume system is the most widely used in pulse jets, being applied for everything from bin vent collectors with a few short bags to large collectors with 450 or more bags per module, each as much as 20 ft (610 cm) or more in length. The bags usually range in diameter from approximately $4\frac{1}{2}$ to 6 in (11.4 to 15.2 cm). The pulsing system comprises an air header, which is often mounted across the front of the casing at about the level of the tube sheet and is equipped with pneumatic pulse (dump) valves, one for each row of bags. The discharge from the dump valves is piped through the wall of the collector and connected to the pulse tubes (sometimes called manifolds) by means of some type of coupling. Each pulse tube extends over a row of bags and has holes in its underside, each hole centered over the bag beneath it. A venturi is frequently installed in the top of each bag. The function of the venturi is to augment the flow of cleaning air by using the momentum of the high-velocity air jet to entrain gas from the clean air plenum and thus give improved cleaning. It is necessary that the venturis be held in place, by fastening them to either the cages or the tube sheet, or else they will be popped out of the bags by the pulse and cause trouble.

The medium-pressure, medium-volume system is set up similarly to the high-pressure system, but the air headers, valves, and blowpipes must be larger to accommodate the larger volumes of gas used. The air headers are often rectangular in section and have the dump valves incorporated into them. The pulse tubes are also larger, have larger holes, and are usually connected to the dump valve by a slip joint. Venturis are not usually used.

Low-pressure, high-volume cleaning systems are substantially different. Rather than being arranged in parallel rows, the bags are usually placed in concentric circles. Only one large dump valve and blowpipe are used per module. The valve is located on the axis of the circles of bags, with the very large tapered blowpipe rotating around it. The blowpipe has one hole, often rectangular, per ring of bags. The bags used are oval, with the long axis tangent to the rings. Each ring has more bags than the one inside it, so that the spacing around the circles is the same. Since the pulses occur at equal time intervals, the air jet from every hole does not go into a bag every time, but sometimes hits the tube sheet; this is considered in the design.

Filter Bags

In constructing filter bags, we should consider the media, construction, finishing, properties and testing, characteristics, design parameters, and use as chemical reactors.

Filter Media. Selection of the filter media to be used is based on a number of factors:

- The filtration efficiency must meet the requirements of the application.
- The material must be appropriate for the cleaning method being used.
- The material must withstand the environment in which it will be used.

The media considered here is confined to the "ordinary" types of fibers and constructions. Rigid ceramic or metallic filters and cloth made of ceramic fibers require a great deal of specialized knowledge for their application and are beyond the scope of this book.

The main fibers used in filter fabrics are *polyester, polypropylene, acrylic, aramid, sulfar, P84, PTFE,* and *fiberglass.* Most bag suppliers' literature contains a table showing the applicability of these and other fibers to enable the user to select the appropriate fiber for the application. Making the final selection solely on the basis of such tables has led to expensive problems. It is much safer to contact the technical departments of the producers of the fiber, the companies needling the felt or weaving and finishing the woven fabrics, and people using, for the same application, the cloth being considered. As an example, *aramid* fiber is recommended for use at temperatures up to 350 to 375°F (177 to 191°C), but the fiber is sensitive to oxygen and sulfur oxides under some conditions, so there can be problems in its use. The supplier has a very large database and can advise whether this fiber will have an acceptable life under the specific conditions in question.

Polyester is suitable for use up to 275°F (135°C) but may hydrolyze in the presence of water vapor and acids or alkalies from about 190°F (90°C). The exact conditions where problems may be encountered are not clearly defined, so it is best to check with people who have experience in the exact conditions being considered.

Polypropylene is suitable for use up to 190°F (88°C) and is extremely resistant to water and inorganic acids and bases except for strongly oxidizing acids; it may be attacked by hydrocarbons. Homopolymer *acrylic* is good for use up to about 260°F (125°C) and is resistant to some acids and fairly resistant to some alkalies. It is difficult to heat-set acrylic, so it may stretch or shrink in service, and it is subject to attack by some metallic chlorides, particularly zinc chloride.

Sulfar is said to be good for use up to 375°F (190°C) and to be resistant to sulfur oxides. *P84* is claimed to be good to 500°F (260°C); as there is little published operating experience with bags made from this fiber, it is best to get the information from the supplier. *PTFE* is probably the most chemically resistant fiber and is rated at 500°F (260°C); however, few suppliers have the capability of heat-setting it at the high temperature that it requires. In addition, since the dust tends to slide on the PTFE fibers, the penetration may be higher than with other fibers. A proprietary felt, Tefair, which is made from a blend of PTFE fiber and fiberglass, overcomes some of these shortcomings.

Fiberglass is suitable for use at 500°F (260°C). As fiberglass fabric is only available made from E (electrical) glass, it is not as chemically resistant as C (chemical) glass that is used for chemical equipment. It is always necessary to finish fiberglass cloth to protect the fibers from internal abrasion (flex abrasion) and to improve its chemical resistance. These coatings are proprietary and usually contain PTFE, silicones, and/or graphite. Woven cloth is not usually satisfactory for use in pulse jets, but special double-woven glass fabrics can be used. Fiberglass felt, which contains materials to bond the fibers together, is also available.

Great care must be taken in designing fabric filters to be used in VOC control systems so that there is no opportunity for the volatile organic compounds, e.g., tar vapors, to condense in the filter media or permanent damage to result. This condition can be particularly bad in the case of membrane media where tar vapors have passed through the membrane and condensed in the supporting fabric, thus ruining a set of very expensive bags.

Construction. The constructions commonly selected for cloths employed in fabric filters are *woven, seamless tubular knit, needle felt,* and *membrane.* Woven and knit bags are used in collectors with low-energy cleaning systems, i.e., shaker and reverse-flow cleaning, with or without sonic horns. Pulse-jet collectors, which are classified as high-energy cleaning, are usually dressed with bags made of felt although seamless pile knit bags (Beane Bags) are occasionally used. Membranes, usually made of expanded, microporous PTFE, are thin and in use must be supported by being bonded to felt or woven cloth, depending on the type of collector in which they will be used. The substrate cloth must be made of a fiber selected to withstand the environment in the collector.

Woven fabric is made by interlacing yarns that run lengthwise, called *warp yarns* or *ends,* with those that run crosswise, which are called *fill* (sometimes *weft*) *yarns* or *picks.* In describing the fabric, *the number of ends per inch by the number of picks per inch* is usually given, such as 44 × 24 (spoken 44 by 24). The *weaves* most frequently used in baghouses are 1 × 3 and 2 × 2 twills and satins. In a 1 × 3 twill, a fill yarn will go over 3 warp yarns and then under 1 and continue in that way; the next pick would go under 1 warp yarn, then over 3 and under 1 and continue. This progression results in diagonal ridges, called *twill lines,* running across the cloth. Because in the 1 × 3 twill and satin weaves (among others) the fill yarns go over and under different numbers of fill yarns, these fabrics have more fill yarns exposed on one face and warp yarns on the other. By using different yarns for warp and fill (usually the same fiber), e.g., multifilament (smooth) for the warp and spun (fuzzy) for the fill, a combination cloth is produced that is smooth on one side and fuzzy on the other. The fuzzy side is usually more efficient and is placed facing the dust, so that the dust does not penetrate the structure of the cloth where it may be trapped.

Knit bags are knit in the form of a tube and so have no seam. They are produced by only one or two companies and mostly in polyester and some in acrylic.

Felt used in fabric filters is properly called *needle felt,* as true felt can only be made from animal fibers, such as wool, since they have a scalelike surface that causes the fibers to lock together in the felting operation. Synthetic fibers have a smooth surface so that it is necessary to lock the fibers together mechanically in the needling operation. In producing needle felt, the fiber, in 2- to 3-in length, is formed into a very thin web by a machine called a *card.* Many layers of web, forming the batt, are deposited on a band of scrim, a very open-weave fabric. The band is then passed through the needling machine where thousands of barbed needles mounted in boards reciprocate vertically at high speed and penetrate the batt and the scrim, locking the whole together with tiny tufts of fiber. The roll is then turned over and the batt built up on the other side and then needled. With this method of fabrication the batt on the two sides of the scrim can be of different weights or composed of fibers of different deniers (thickness). The latter is called *duo-density felt.* It is possible to put a thin layer of a different fiber on the surface of the felt to impart special properties at lower cost. An example is to needle a thin layer of P-84 fiber onto a felt made of a lower-cost fiber to take advantage of the high filtration efficiency and cleanability claimed for the very expensive P-84.

Membrane media is made by stretching a film of PTFE, by a proprietary process, thus producing a microporous membrane. The membrane, which is very delicate, must be bonded to a substrate so that it can be made into bags and used in a collector. The pores are very fine so the dust does not penetrate the media and the dust cake is thus easier to remove. The substrate is usually selected from the normal filter media, on the basis of the operating conditions. The bags must be handled with great care during installation in the baghouse, and the baghouse must be designed to protect the bags from abrasion. While liquid droplets will not penetrate the membrane, vapors can penetrate the membrane and may then condense in the substrate.

Finishing. After the cloth is produced, usually it must go through some type of finishing operation. In the case of woven synthetic fabrics, this may consist of washing, to remove the sizing that protects the yarn during the weaving operation, followed by heat setting to stabilize the cloth for use at elevated temperatures. Glass cloth cannot withstand washing so it is heat-cleaned, followed by the application of a protective finish in one finishing line.

Felt is usually calendered to adjust the permeability, to give the cloth a smooth finish, and sometimes to heat-set it as well. The calender has two heated rolls running at different speeds to give an ironing effect. Felt is often singed to burn off the fuzz that has a tendency to hold the dust to the cloth.

Synthetic cloth is sometimes treated with silicone, PTFE, or similar material, to increase its oil and/or water repellency.

Properties and Testing. The properties that are usually specified and tested are *weight per square unit, air permeability, burst strength,* and sometimes *stretch* at temperature. Fiber is specified but not always tested, although it can be identified by employing a special dye that dyes different fibers different colors and by other tests. In addition, the count is given for woven cloth and for felt, the thickness.

In testing for *weight,* a sample of the cloth, of known dimensions, is cut and weighed, and the number of ounces per square yard (grams per square meter) is calculated.

Air permeability is measured using a Frazier permeometer. Other permeometers may be used, but they all give slightly different results, so it is important to know if an instrument other than the Frazier was employed. In the test the cloth is clamped in the permeometer, air is drawn through it at an increasing flow rate until a pressure drop of 0.5 in (20 mm) WC is attained, and the pressure drop through an orifice meter is noted. The permeability in cubic feet per minute per square foot at 0.5 inWC [L/(dm^2 • min)] is looked up in tables. It is customary to test five locations on the sample of cloth and report the individual values as well as the average.

Permeability is a more or less general quality control measurement rather than a direct indicator of the filtration efficiency of the cloth. Permeability measurements on used bags are of great importance in evaluating the condition of the bags and the efficacy of the cleaning system.

The *Mullen burst test* is a general measure of the strength of the cloth and is widely used, as the equipment is more compact and less expensive than a tensile tester and the specimens do not have to be specially prepared. The sample is clamped between two plates having a hole in each. One hole is fitted with a rubber diaphragm. For the test, glycerin is pumped under the diaphragm, forcing it against the cloth, which is deflected into the other hole until it bursts. A pressure gauge with a telltale indicates the highest pressure attained. This pressure, in pounds per square inch (kilopascals), is the gross Mullen burst and is the figure usually quoted. To get the net Mullen burst, the clamp holding the cloth is released and the pressure required to deflect the rubber diaphragm is noted. This figure is subtracted from the gross reading to get the net Mullen burst.

For the shrinkage or stretch test, a sample is heated in an oven under specified conditions, and the percentage of shrinkage or stretch is reported. In considering stretch and shrinkage figures, it is important to remember that a stretch of 1 percent in a 30-ft bag can leave the bag very slack.

Characteristics. The characteristic of fabric filters that seems to get the most attention is *pressure drop.* Many, if not most, baghouses are operating at a higher than anticipated pressure drop, generally because of unreasonable expectations, too little cloth area, higher than anticipated flow, and/or factors that were not known, or recognized as pertinent, when the system was designed. For baghouses on normal applications, it is considered good practice to design for an operating pressure drop across the cloth (actually across the tube sheet) of about 6 in (150 mm) WC. In some cases the expected pressure drop should be far higher than this, but such differences can only be anticipated on the basis of previous experience with the specific application, or one very similar. For the pressure drop across the entire baghouse (flange-to-flange pressure drop), a good design should normally have a pressure drop of 2 in (50 mm) for the inlet and outlet together, and it is best to allow at least 2 in (50 mm) more for contingencies, thus giving an overall pressure drop of at least 10 in (250 mm). Designing for a lower pressure drop will usually lead to excessively frequent cleaning, causing shortened bag life, and, if it is not possible to attain the projected pressure drop, inadequate fan capacity for the necessary airflow (gas flow).

Although the flow in the filter cake is in the viscous regime and the pressure drop varies linearly with the clean airflow rate, the pressure drop will vary as the square of the filter ratio as a heavier dust cake is being produced. This ratio is defined as the number of cubic feet per minute of gas to be filtered through one square foot of cloth and thus has the dimensions of feet per minute (meters per second). It is important to consider carefully the situation regarding the amount of dust and gas reaching the cloth area.

In collectors using off-line cleaning, the pressure drop varies during the cleaning cycle. As a compartment is taken off line, the pressure drop jumps up as the filter ratio in the other compartments is increased; when it goes back on line, the pressure drop will be lower than before. If this causes a problem with the rest of the VOC control system, consideration should be given to using an on-line pulse-jet cleaning system.

Penetration is defined as the fraction or percentage of the mass of the particulate entering the baghouse that leaves in the exit gas stream. *Percent efficiency* is defined as 100 minus the percent of penetration. Most of the penetration in a fabric filter takes place immediately after the cleaning cycle. The dust that forms the real filter has been removed from the fabric so conditions are best for the dust to penetrate the interstices in the cloth. Factors that permit longer cleaning cycles, such as lower filter ratios, will thus tend to increase the efficiency.

Design Parameters. After the type of baghouse is decided on, or perhaps as part of that decision, the first design parameter to be set is the filter ratio, also known as the *air-to-cloth ratio.* Since, for a given gas flow rate, the filter ratio sets the size of the collector, it has a direct impact on the cost. There is no way, other than experience, to decide on this ratio. As a *very* rough guide, a suitable ratio for a shake-and-deflate collector might be 10 to 20 percent higher than that for a reverse gas-cleaned collector on the same application. A suitable ratio for a pulse jet is usually considered to be about twice that for a conventional collector.

After the filter ratio is selected and bids are received from vendors, it is necessary to consider many points in evaluating these bids. The velocities at various locations in the collector can have a large effect on how satisfactorily it performs. In the inlet plenum, a velocity of 3500 ft/min (17.8 m/s) is generally considered good practice in balancing pressure drop, keeping the particulate in suspension, and having acceptable abrasive wear on the equipment by the particulate. For the outlet plenum, a velocity of 2500 ft/min (12.7 m/s) can be used as it gives a lower pressure drop and there is no particulate that needs to be kept in suspension. Similar velocities should be used in the dampers. In some cases the dampers are located such that while the apparent velocity in the dampers is acceptable, there is inadequate space around the damper, thus limiting the flow.

For the inlet dampers, a manually operated butterfly type is usually best, as its main function is to isolate a compartment to work on it and to throttle the flow to it after it has been rebagged. Inlet dampers are sometimes designed so that they leak slightly, thus providing ventilation in the open compartment by flow into the inlet manifold, which is below atmospheric pressure.

Outlet dampers are usually power-operated and controlled by the cleaning cycle timer. They should seal tightly so that there is no flow through the bags during cleaning. They may be the butterfly type on small collectors, but are usually poppets on larger ones.

All dampers should be designed so that they can be locked in the closed position, preferably by a mechanical block, for safety when a compartment is entered.

As the gas stream entering a compartment usually does so via the hopper that holds the dust cleaned from the bags, baffles must be provided, to prevent the incoming stream from resuspending the dust in the hopper and at the same time to slow the inlet stream and distribute it uniformly to the bags. To accomplish this, it is necessary to dissipate the kinetic energy in the stream; but many baffle designs only deflect the stream and will recirculate the dust.

The dust is removed from the hoppers via an air lock so that air cannot leak into the collector. Leakage at the bottom of the hopper is particularly harmful as it carries the fines out of the dust at the bottom of the hopper so that it redeposits on the bags. Air locks can be of the rotary type or double-flap type.

It is important that the bags be properly tensioned. If they sag, they crease and tend to fail at the creases as they flex during cleaning. In setting the tension on the bag the weight of the dirty bag must be considered. It does little good to have a bag at 90-lb (40.8-kg) tension at the top when the bag weighs 100 lb (45.4 kg), causing it to be wrinkled at the bottom. A similar situation exists with pulse-jet bags; they must not be too long for the cage, or else they will crumple at the bottom and fail prematurely. An excess length of ½ in (12 mm) is considered acceptable.

Use of a Fabric Filter as a Chemical Reactor. Fabric filters may be used as reactors to remove gaseous contaminants from a gas stream. They act as packed beds and give excellent contact because of the very fine reactant which is uniformly distributed over the surface of the cloth. In the spray drier system for removing acid gases from flue gas, most of the scrubbing takes place in the drier, but a significant additional amount is removed in the baghouse. In the aluminum industry the gas from the pots (electrolytic cells) is scrubbed by the alumina feed, which is then collected and given a final treatment in baghouses. The basic patents on using a fabric filter in this way have expired, but it might pay to check on some specific application.

SECTION 8.10.2

PARTICULATES: ELECTROSTATIC PRECIPITATORS

Chen-Lu Yang and Dave Meier
Beltran Associates, Inc.
Brooklyn, New York

INTRODUCTION 8.225
BRIEF THEORY OF ELECTROSTATIC PRECIPITATION 8.226
DESIGN PARAMETERS 8.228
TYPES OF ELECTROSTATIC PRECIPITATORS 8.228
Plate 8.229
Tubular 8.229
SYSTEM DESCRIPTION 8.229
Power Supply 8.231
Electrode System 8.231
Electrode Cleaning System 8.233
Housing 8.235
ELECTROSTATIC PRECIPITATORS IN INDUSTRY 8.235
Acid Mist Recovery 8.235
Cement Industry 8.235
Drying Processes 8.235
Iron and Steel 8.236
Municipal Incinerators 8.236
Nonferrous Metals 8.236
Power Plants 8.236
Paper and Pulp 8.236
REFERENCES 8.236

INTRODUCTION

Methods used in particulate collection from exhaust gas streams may be broadly categorized as mechanical and electrical. The basic difference between the two methods is reflected in the application of the electric power. Here, the electric energy is directly applied to the gas streams, and the separation forces are applied only on the particles. On the other hand, mechanical methods convert electric energy to mechanical energy and are applied to the entire gas stream. When particles suspended in a gas stream are exposed to gas ions in an electrostatic field, the particles will become charged and migrate under the action of the field. The functional mechanisms of electrical precipitation are as follows:

1. Gas ionization
2. Particle collection
   a Production of electrostatic field to cause charging and migration of particles
   b Gas retention to allow particle migration to a collection surface
   c Prevention of reentrainment of collected particles
   d Removal of collected particles from the equipment

Because of the direct use of electric separation forces, electrostatic precipitators (ESPs) are limited neither to large coarse particulate, as are gravitational systems and cyclones, nor by resistance to the motion of the gas, as are filters and scrubbers.

In comparison with other particulate control methods, electrostatic precipitators have the following strong points and shortcomings.

Strong Points

1. High efficiencies for submicrometer particles
2. Able to handle large gas volumes with low pressure drop
3. Applicable to sticky, corrosive fumes and mists
4. Can be operated at a wide range of gas temperatures
5. Low operating costs

Shortcomings

1. High capital costs
2. Will not control pollutants other than particulate
3. Lack of flexibility to respond to changes in operating conditions
4. Space-consuming
5. Might not work on particulates with extreme electrical resistivity

BRIEF THEORY OF ELECTROSTATIC PRECIPITATION

Efficiency can be defined as the percentage of particulates entering the precipitator that is collected. Efficiency is primarily related to the total surface area of the collecting electrodes per unit volume of gas and is directly proportional to the particle rate velocity. Consider a dusty airflow in a rectangular channel confined by two parallel plates, as shown in Fig. 8.10.1. The relationship can be derived with the following assumptions.

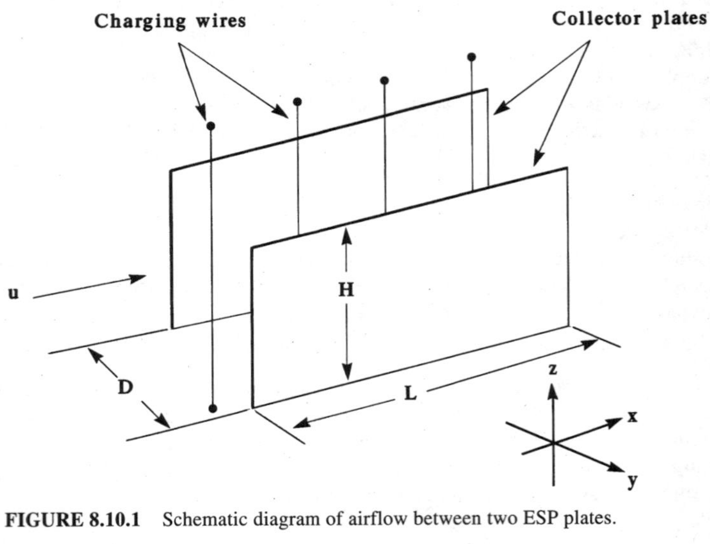

FIGURE 8.10.1 Schematic diagram of airflow between two ESP plates.

1. Gases and particles move in the x direction at a constant velocity u, with no longitudinal mixing.
2. The particles are uniformly distributed in the y and z directions at every x location.
3. The charging and collecting fields are constant and uniform, and the particles quickly attain terminal velocity w in the y direction.
4. Reentrainment of collected particles is negligible.

A material balance of particles flowing into and out of a small cross section of the channel shows that the difference between the mass of particles flowing into and out of the slice must equal the mass of particles collected at the plate. Thus,

$$uH \frac{D}{2} C_x - uH \frac{D}{2} C_{x+\Delta x} = \text{mass removed} \tag{8.10.1}$$

where u = gas velocity, m/min
H = plate height, m
D = channel width, m
C = particle concentration or loading, g/m^3

However, the mass collected is equal to the flux of particles in the y direction times the area normal to the flux. Thus, Eq. (8.10.1) becomes

$$uH \frac{D}{2} (C_x - C_{x+\Delta x}) = wC_{x+\Delta/2} H \, \Delta x \tag{8.10.2}$$

where w is the draft velocity in the y direction in meters per minute. Dividing through Eq. (8.10.2) by Δx and taking the limit as Δx approaches zero, we obtain

$$\frac{-uHD}{2} \frac{dC}{dx} = wHC \tag{8.10.3}$$

Equation (8.10.3) can be separated and integrated from 0 to L to give

$$\ln \frac{C_L}{C_0} = \frac{-2wHL}{uHD} \tag{8.10.4}$$

or

$$\ln \frac{C_L}{C_0} = \frac{-wA_p}{Q_c} \tag{8.10.5}$$

where A_p = area of one plate (both sides), m^2, and Q_c = volumetric gas flow in one channel, m^3/min.

For the whole ESP, we can use the total collection area and the total gas flow rate, and Eq. (8.10.5) can be written as

$$\frac{C_L}{C_0} = e^{-wA/Q} \tag{8.10.6}$$

Using the usual definition of collection efficiency, Eq. (8.10.6) becomes the Deutsch equation

$$\eta = 1 - e^{-wA/Q} \tag{8.10.7}$$

where η = fractional collection efficiency. The equation had previously been suggested by Anderson on the basis of experimental data (Cooper et al., 1986).

Note that any consistent set of units can be used for w, A, and Q.

DESIGN PARAMETERS

The complete design of an ESP entails sizing and determining the configuration of the plates, calculating the needed electrical energization, determining the structural needs, specifying the dust removal units, and designing the electrical control and performance monitoring systems.

Although there have been improvements in computer models, ESP design still relies on experience and pilot-scale studies. Some practical design parameters are listed in Table 8.10.1. Using the information given in Table 8.10.1 and a basic understanding of the configuration of an ESP, the preliminary geometry and size of an ESP can be specified (White, 1977).

TYPES OF ELECTROSTATIC PRECIPITATORS

Electrostatic precipitators are configured in several ways. Some of these configurations have been developed for special applications, and others have evolved for economic reasons. There are two classes of electrostatic precipitators: one-stage and two-stage. The one-stage precipitators are the most widely used in air pollution control. The basic elements are a source of unidirectional voltage, corona discharge electrodes, collecting

TABLE 8.10.1 Design Parameters

Parameter	Range of values	
Drift velocity w_e	1.0–10 m/min	
Channel width D	15–40 cm	
Specific collection area (SCA) (plate area/gas flow)	0.25–2.1 $m^2/(m^3/min)$	
Gas velocity u	1.2–2.5 m/s (70–150 m/min)	
Aspect ratio R (duct length/height)	0.5–1.5 (not less than 1.0 for $\eta > 99\%$)	
Corona power ratio P_c/Q (corona power/gas flow)	1.75–17.5 $W/(m^3/min)$	
Corona current ratio I_c/A (corona current/plate area)	50–750 $\mu A/m^2$	
Power density versus ash resistivity		
	Power density, W/m^2	Ash resistivity, $\Omega \cdot cm$
	43	10^4–10^7
	32	10^7–10^8
	27	10^9–10^{10}
	22	10^{11}
	16	10^{12}
	10.8	10^{13}
Plate area per electrical set A_s	460–7400 m^2	
Number of electrical sections N_s		
In the direction of gas flow	2–8	
Total	1–10 bus sections/(1000 m^3/min)	

electrodes, and a means of removing the collected particles from the precipitator. One-stage systems combine both ionizing and collecting in one unit. There are two main types of one-stage electrostatic precipitators.

Plate

In a plate-type ESP, gas flows between parallel plates of sheet metal and high-voltage electrodes. These electrodes are long wires weighted and hanging between the plates or are supported there by rigid frames. Within each flow path, gas flow must pass each wire in sequence as it flows through the unit. The plate-type ESP allows many flow lanes to operate in parallel, and each lane can be quite tall; therefore, this type of precipitator is well suited for handling large volumes of gas. Primarily, plate-type ESPs are used for dry dust removal (Oglesby et al., 1970).

Tubular

A tubular ESP is essentially a one-stage unit and is unique in having all the gas pass through the electrode region. The tubes may be formed as a circular, square, or hexagonal honeycomb with gas flowing upward or downward. Primarily, tubular-type ESPs are operated with wet walls and are most commonly applied where the particles are either wet or sticky. The water flow may be applied intermittently or continuously to wash the collected particles into a sump for disposal. The advantage of the wet precipitator is that it has no problems with rapping reentrainment or back corona. The disadvantages are the increased complexity of the washing system and the fact that the collected slurry must be handled more carefully (Beltran Associates, 1994).

In the previously described single-stage precipitators, the discharge and collecting electrodes are located side by side. In two-stage precipitators, the discharge and collecting electrodes follow each other; corona discharge takes place in the first stage between two electrodes having a nonuniform field. This is generally obtained by a fine-wire discharge electrode and a large-diameter receiving electrode. In this stage the potential difference must be above that required for corona discharge. The second stage involves a relatively uniform electrostatic field in which charged particles are caused to migrate to a collecting surface. This stage usually consists of either alternately charged parallel plates or concentric cylinders with relatively close clearance between their diameters. For indoor applications, the unit is operated with positive polarity to limit ozone generation. Advantages of this configuration include more time for particle charging, less propensity for back corona, and economical construction for small sizes. This type of precipitator is generally used for gas flow volumes of 50,000 acfm or less and is applied to submicrometer sources emitting oil mists, smokes, fumes, or other sticky particles, because there is little electric force to hold the collected particles. Cleaning may be by water wash of modules removed from the system up to automatic, in-place detergent spraying of the collector followed by air blow drying (Buonicore and Davis, 1987).

SYSTEM DESCRIPTION

Electrostatic precipitators consist of four main components (Beltran Associates, 1994):

1. Power supply unit, which provides high-voltage direct current
2. Electrode system

3. Electrode cleaning system
4. Housing that provides an enclosed precipitator zone

These components are illustrated with a wet tubular electrostatic precipitator. A cutaway view of a tubular ESP is shown in Fig. 8.10.2.

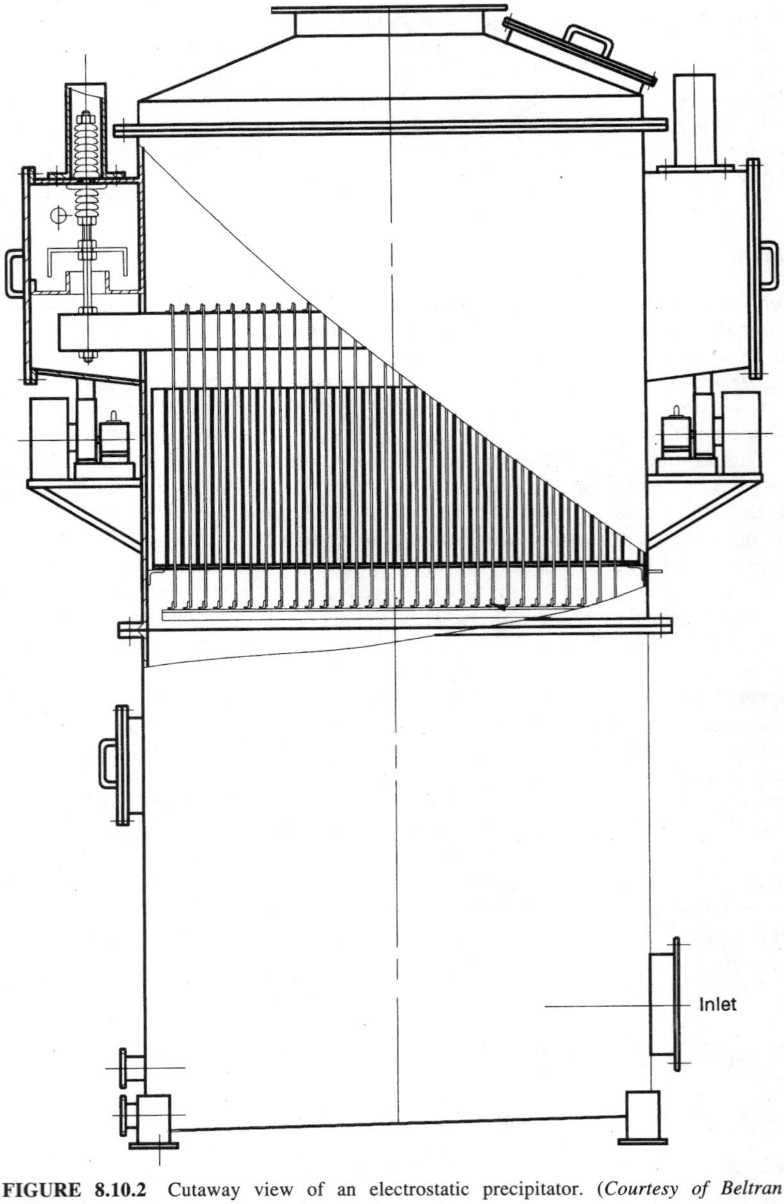

FIGURE 8.10.2 Cutaway view of an electrostatic precipitator. (*Courtesy of Beltran Associates, Inc., Brooklyn, New York.*)

Power Supply

Electrostatic precipitators are generally energized by rectified alternating current of commercial frequency. The voltage is stepped up to the required value by means of a transformer and then is rectified. Silicon rectifiers are the most widely used type, since they provide high efficiency and reliability. Automatic controls commonly are used to control voltage, current, spark rate, or some combination of these parameters. Modern precipitators use control circuits similar to those shown in Fig. 8.10.3. A high-voltage

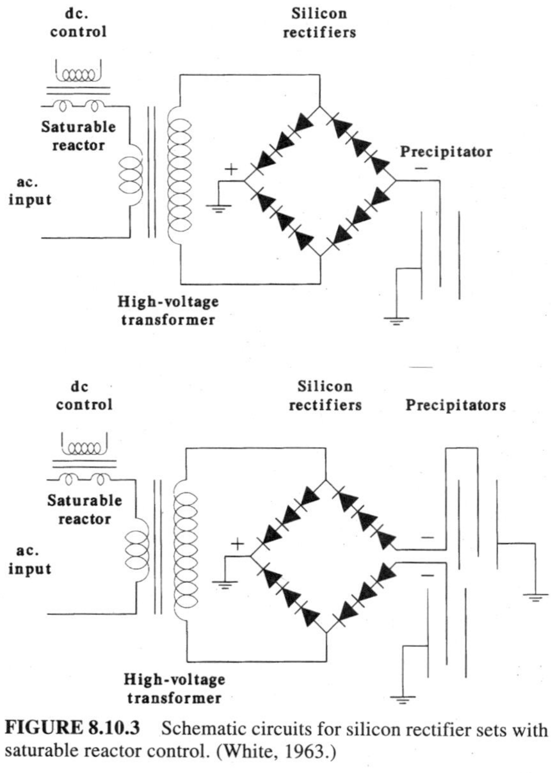

FIGURE 8.10.3 Schematic circuits for silicon rectifier sets with saturable reactor control. (White, 1963.)

silicon rectifier is used together with a saturable reactor and a means for limiting current and controlling voltage and/or spark rate. One popular method adjusts the voltage to give a specified sparking rate (typically 50 to 150 sparks per minute per bus section). Half-wave rectification is sometimes used because of its lower equipment requirement and power consumption. It also has the advantage of longer decay periods for sparks to extinguish between current pulses (White, 1963).

Electrode System

There are two types of electrodes. The discharge is a high-voltage ionizer. These are generally negatively energized. The collecting electrodes are grounded. The description and function of the major components of an electrode system follow (Beltran Associates, 1994).

Discharge Electrode. The discharge electrodes provide the corona, without which the precipitator cannot function. These may be round wire, square twisted rods, star rods, ribbons, barbed wire, and so forth. Practice differs between U.S. and European manufacturers in the method of supporting the discharge electrodes. Typical European designs provide a frame or tubular support for the electrodes, whereas most U.S. manufacturers suspend the electrodes from a support and maintain them in position by weights and guides at the bottom.

The shape and size of the discharge electrodes are governed by the corona current and mechanical requirements of the system. Where high concentrations of fine particles are encountered, space charge limits the current flow, especially in the inlet sections. In such cases, special electrodes that give higher currents may be used to achieve a high power density within the inlet sections. Variation in the current flow and electric field within limits is possible by controlling the type and size of discharge electrode.

Discharge Electrode Support. The main functions of the discharge electrode support are to provide the necessary high-voltage electrical insulation and to give mechanical support to the discharge electrode frame. There are several types of support systems currently used in precipitator design. One type is a support insulator in a separate chamber from the ESP housing, as illustrated in Fig. 8.10.4. A bus beam is mounted on the bottom of the insulators, and the discharge electrode assembly is suspended from the bus beam by hanger rods. Figure 8.10.5 is a cutaway of the upper housing which shows how the discharge electrodes and the supporting system are installed. The porcelain pin-type

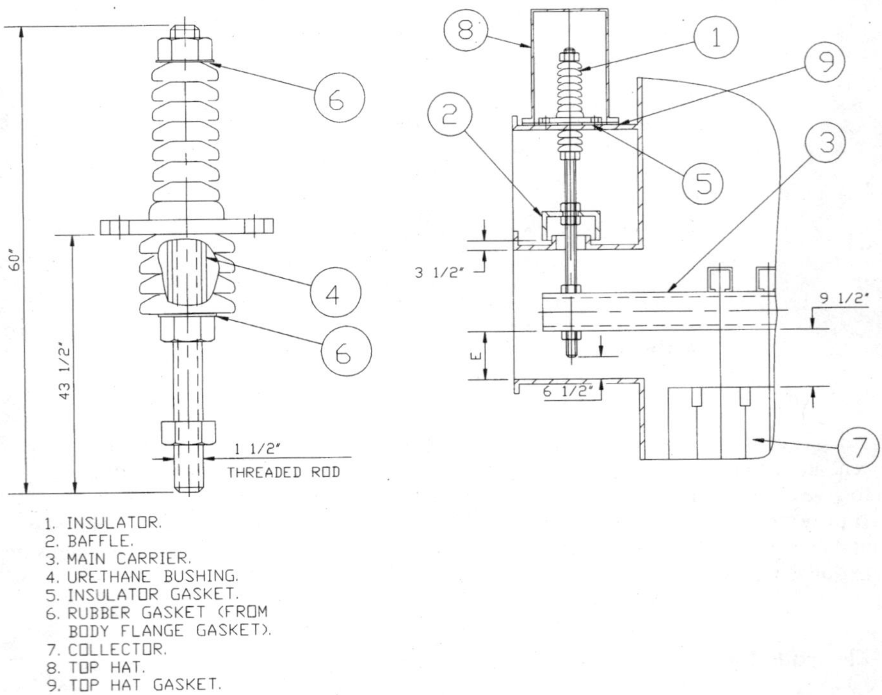

FIGURE 8.10.4 Insulator component and main carrier assembly. (*Courtesy of Beltran Associates, Inc., Brooklyn, New York.*)

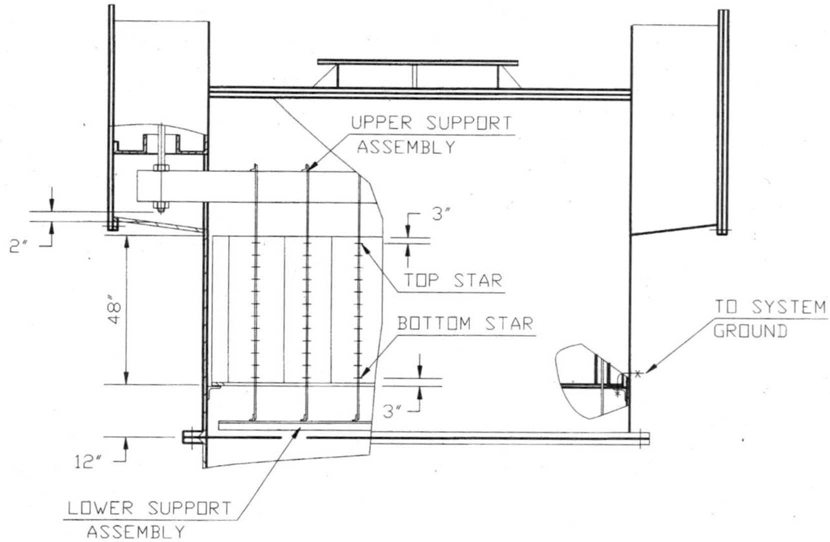

FIGURE 8.10.5 Discharge electrodes and supporting system. (*Courtesy of Beltran Associates, Inc., Brooklyn, New York.*)

insulators support the mechanical load of the internal framework and are located in separate chambers. These chambers are not airtight, and it is common practice to provide a purge air to keep the insulator chamber clean.

Collecting Electrodes. Various types of collecting electrode structures have been used in precipitator design. It is desirable that (1) the electrode act as a shield to provide protection of the collected particles and minimize reentrainment by scouring from the gas stream, (2) the electrode be free of points or sharp edges that may cause sparking and reduce the operating voltage, and (3) the design provide mechanical rigidity. Based on the design, the collecting electrodes can be categorized as plate collecting electrodes and tubular collecting electrodes. While many of these special shapes have important advantages, the use of smooth plates, with fins to strengthen them and produce quiescent zones, has become most common in recent years. Plate-type electrodes are usually preferred because they can handle a larger volume of gas than the tubular type without creating a big pressure drop in the system. Tubular-type precipitators lend themselves more readily to wet collection and are preferred for acid mist recovery and tar collection from coke oven gas. Figure 8.10.6 shows how a tubular collecting electrode module is installed in a wet electrostatic precipitator. The tubular-type precipitator is also commonly used in the steel industry to clean combustible gas from blast furnaces, to prevent fouling of the gas burners.

Electrode Cleaning System

Once the dust or fume has been precipitated on the collecting electrode or plate, it must be removed to a hopper or storage depository. To do this, rappers are commonly used. The plates are struck with sharp, hammerlike blows to dislodge the collected dust,

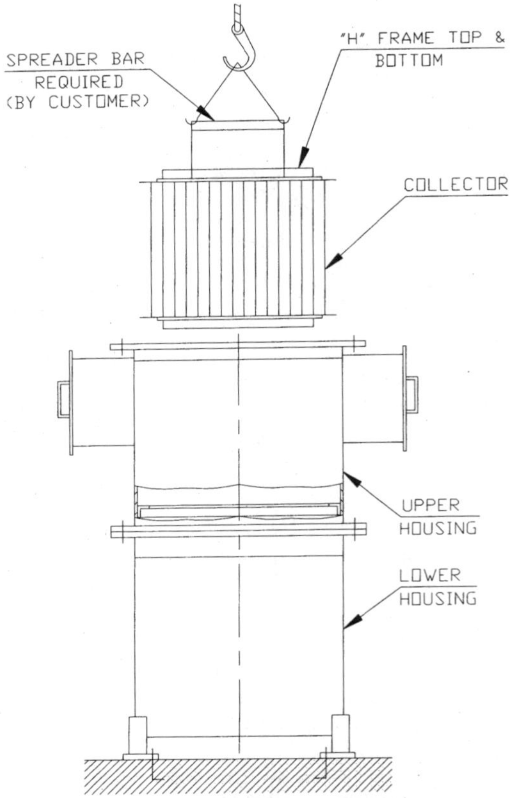

FIGURE 8.10.6 Collecting electrode module installation. (*Courtesy of Beltran Associates, Inc., Brooklyn, New York.*)

which then falls by gravity into the collecting hopper. Reentrainment of a portion of the dust at this point must be held to a minimum. Frequently, unsatisfactory collection efficiency is caused by improperly operated or adjusted rappers.

For fly ash precipitation, the dust buildup on the collecting electrodes should be allowed to reach about $\frac{1}{4}$ to $\frac{1}{2}$ in before it is rapped off. Discharge electrode rappers are also necessary in treating ashes predominantly composed of fine particles less than 10 μm in diameter (Research and Education Association, 1978).

Some electrostatic precipitators, such as tubular types, are used to precipitate mists from the gas stream. Most of these applications are found in chemical, petroleum, and paper plants. The mist droplets which collect on electrodes run together to form larger droplets, and these run down the electrode surface where they drain away. In some cases, water or another liquid is introduced at the inlet stream to improve mist removal. This is called *flue gas conditioning*. Tubular electrodes are more adaptable than plates for this application (Beltran Associates, 1994).

Housing

The housing serves as the containment vessel for the particle-laden gases. The structure is designed to funnel the air through the corona discharge zone and to support both collecting and discharge electrode assemblies.

For a tubular ESP, each housing consists of two sections: upper housing and lower housing. The upper housing provides support for the collecting electrode, discharge electrodes, and discharge. The upper section contains outlet and wash headers, and it serves as a transition to the outlet duct. The lower housing serves as a transition and air distribution section between the inlet and the collection area (Beltran Associates, 1994).

ELECTROSTATIC PRECIPITATORS IN INDUSTRY

Some of the recent installations of electrostatic precipitators in industry are briefly examined in this section (*PRECIP Newsletter,* 1995, 1996). This is not meant to be a complete list of electrostatic precipitator applications. However, it indicates the wide range of uses of electrostatic precipitators. Electrostatic precipitators are used for both air pollution control and product recovery. Still another use is found in the cleaning of a gas stream of impurities, where the gas is the desired product.

Acid Mist Recovery

One of the unique applications of a tubular ESP is acid mist recovery. Most of the acid precipitators are used to collect either sulfuric or phosphoric acid. The stream usually passes through a scrubber in order to saturate the gas and remove a portion of the acid. The gas is then sent into the ESP, where the acid is recovered as a fine mist. The collected material is added to the product stream. The treated gases are released either to the atmosphere or to a secondary scrubber for removal of any remaining SO_2 and SO_3.

When SO_3 encounters water vapor, H_2SO_4 is formed, which condenses to acid mist in cold units. This acid mist can corrode steel equipment and pipelines severely. The wet electrostatic precipitator (WESP) widely used in sulfuric acid production has always been made of lead, hard polyvinyl chloride (PVC), or fiber-reinforced plastic. A layer of sulfuric acid formed on the surface of the hard PVC tubes acts as collecting electrodes (*PRECIP Newsletter,* Jan. 1995).

Cement Industry

In cement plants, precipitators serve double duty, recovering process metals and reducing air pollution. Precipitators are used efficiently in both wet-process and dry-process cement kilns, shale and stone driers, raw grinding mills, and various other cement mill applications. Efficiency as high as 99.9 percent is common in the cement industry (*PRECIP Newsletter,* Dec. 1995 and Mar. 1997).

Drying Processes

Drying processes are employed in the production of such materials as cement, gypsum, bauxite, and various ores. The object is to drive off the free moisture by heating while the material is agitated slowly. A rotary kiln is commonly used. Generally, a large portion of the total feed is carried off as dust in the gas stream. This gas stream also includes products of combustion, excess air, and moisture. The collected material is added to the main raw material stream. The cleaned gas stream is discharged to the atmosphere (*PRECIP Newsletter,* Jan. 1995).

Iron and Steel

Iron and steel production, more than any other industry, involves complex, interdependent operations and constantly varying conditions. Control systems must be precisely matched to handle these. Furthermore, the systems must be designed to accommodate possible future alterations (Association of Iron and Steel Engineers, 1985).

Specific areas in this industry where precipitators are employed include

1. Basic oxygen furnaces—to remove dust and fumes with efficiencies around 99.9 percent
2. Open-hearth furnaces—in both hot and cold metal shops
3. Blast furnaces
4. Coke oven

Municipal Incinerators

Electrostatic precipitators have proved to be ideal in obtaining more efficient collection of particulates from incineration combustion gases. Efficiencies of 95 percent and better have been obtained in cleaning incinerator combustion gases, without causing any serious water pollution problems. Precipitator designs for steam-generating or refactory-lined incinerator furnaces handle capacities of 25 tons/day or greater (*PRECIP Newsletter,* Nov. 1995).

Nonferrous Metals

As in the iron and steel industry, nonferrous metal refining warrants proper pollution control techniques. Severe operating conditions are often encountered. Gas temperatures are usually high, and acids are often employed. Thus proper steps in corrosion prevention are necessary. Care must be taken not to operate precipitators and mechanical collection systems below the dew point.

Nonferrous applications primarily involve the treatment of off-gases from roasters, converters, blast furnaces, cupolas, and sinter machines. The main function is to reduce ores of copper, zinc, lead, tin, nickel, chromium, molybdenum, and precious metals (*PRECIP Newsletter,* Sept. 1996).

Power Plants

Precipitators have been designed and are being employed with practically every type of boiler and for all types of coal- and oil-fired boilers. Careful consideration must be given to operating parameters. Moisture content, sulfur content, ash content, and various other characteristics of coal and the ash that results from various combustions—all greatly influence precipitator design (*PRECIP Newsletter,* Mar. 1996 and Apr. 1997).

Paper and Pulp

Precipitators are employed in handling air pollution problems generated in kraft and soda mills. These plants pull hard- and softwoods, in addition to processing bagasse and bamboo.

Precipitator designs include wet or dry bottom, depending on the particular process in this industry. In the wet-bottom type, captured salt cake dislodged from the collecting electrodes falls into a flat bottom that is flushed with black liquor. A rotating agitator is used to keep solids in suspension. In the dry-bottom design, trough hoppers or flat bottoms are used. Collected matter is removed by a drag scrapper (*PRECIP Newsletter,* Apr. 1995 and Apr. 1997).

REFERENCES

Association of Iron and Steel Engineers, *The Making, Shaping and Treating of Steel,* Herbick & Held, Pittsburgh, Pa., 1985.

Beltran Associates, Inc., "Installation and Service Manual for Beltran Tubular Wet Electrostatic Precipitator," project no. PWT-1254R, Brooklyn, N.Y., 1994.

Buonicore, Anthony, and Wayne T. Davis, eds., *Air Pollution Engineering Manual,* Van Nostrand Reinhold, New York, 1987.

Cooper, C. David, and F. C. Alley, *Air Pollution Control: A Design Approach,* Waveland Press, Inc., Prospect Heights, Ill., 1986.

Oglesby, Sabert, et al., *A Manual of Electrostatic Precipitator Technology, Part I and II,* National Technical Information Service, U.S. Department of Commerce, Springfield, Va., 1970.

Perry, Robert, *Perry's Chemical Engineers' Handbook,* McGraw-Hill, New York, 1984.

PRECIP Newsletter, Nos. 227–256, Dec. 1994–May 1997.

Research and Education Association, *Modern Pollution Control Technology,* New York, 1978.

White, H. J., "Electrostatic Precipitation of Fly Ash—Parts I, II, III, and IV," *Journal of the Air Pollution Control Association,* vol. 27, nos. 1, 2, 3, 4, Jan.–Apr. 1977.

White, J. Harry, *Industrial Electrostatic Precipitation,* Pergamon Press, New York, 1963.

CHAPTER 9
EVALUATING EMISSION CONTROL SYSTEMS

Harold J. Rafson
Consultant
Highland Park, Illinois

Joseph R. Popeck
Greeley and Hansen
Chicago, Illinois

CHOOSING EMISSION CONTROL SYSTEMS 9.1
- **Understanding the Emission Problem** 9.2
- **Solving the Problem** 9.3

EVALUATING THE PERFORMANCE OF EXISTING EMISSION CONTROL SYSTEMS 9.9
- **Testing Performance to Design Criteria** 9.9
- **Testing Installation Factors** 9.10
- **Testing Equipment Factors** 9.10
- **Warranties** 9.11
- **Using a Consultant** 9.11

The purpose of this chapter is to assist the reader in evaluating emission control systems, which is discussed in two parts: choosing between emission control alternatives and evaluating how well that equipment works. This chapter is complementary to the discussion of prior chapters.

CHOOSING EMISSION CONTROL SYSTEMS

The major factors to be considered are as follows:

1. Understanding the emission problem
   - *a.* Sources of emissions
   - *b.* Quantities
   - *c.* Composition
   - *d.* Impact on neighbors
   - *e.* Conformance to regulations
   - *f.* Prioritizing sources
2. Solving the problem
   - *a.* Process changes—Pollution prevention
     - (1) Odors
     - (2) VOCs

b. Control technology—Selection factors

(1) Logical choices
(2) Effectiveness
(3) Life-cycle costs
(4) Experience in use
(5) Practicality of O&M
(6) Continuity of operation
(7) Life of equipment
(8) Ability to treat varying sources
(9) Flexibility
(10) Controls
(11) Hybrid systems

Understanding the Emission Problem

Although there are difficulties in identifying the problem, it is critical to identify, estimate, prioritize, and define the information and objectives for emission control.

Sources of Emissions

Existing Facilities. In most emission control issues, it is seldom the case that contaminants come from one point only, and usually there are many emission points in a process. It is desirable to begin by testing to determine the emissions at every stage in the process. Where possible, analytical methods can be used, or if that is too complex and costly, and it is applicable, odor panel test methods are available. All sources should be considered, including fugitive emissions. (Refer to Chap. 4.)

Proposed Facilities. With facilities which are not yet built or are being considered, you may need to estimate the emissions. This is discussed in Sec. 5.2, primarily for volatile organic compounds (VOCs). For odors, modeling results may be considered, but you must also rely on experience. The results of these emission estimation models should be reviewed and evaluated for reasonableness by an experienced specialist in the practice of emission control.

Quantities. Estimates of the amounts and concentrations of emissions should be made from either the testing at existing facilities or modeling of proposed facilities, as described above. Airflow rates should be based upon actual conditions in existing facilities, or on design values which comply with building code requirements for new or proposed facilities.

Composition. This may be relatively straightforward for VOC emissions and very complex for odor emissions. However, it is very important to identify the major compounds from analysis, modeling, or the literature.

Impact upon Neighbors. Detailed records should be kept of complaints: data, time, location of receptor, and nature of complaint. If possible, complaints should be verified by plant personnel. Refer to Sec. 4.2. Weather data, particularly wind velocity and directions, should be obtained. In many cases, it is desirable to perform dispersion modeling for existing or contemplated facilities. Dispersion modeling is discussed in Sec. 5.1. By reviewing this information and estimates, it is possible to gain a perspective of the impact on the community and of the concentration of the emissions.

Conformance to Regulations. A determination should be made of whether any emission regulation is being violated and the level to which emissions must be controlled.

Refer to Chap. 10, which reviews federal VOC regulations and comments on state regulations. Some state and local agencies have more severe requirements than the federal Clean Air Act. Odor emissions are regulated by state and local agencies which differ widely in the regulations and enforcement. It is emphasized that the individual responsible for the design of emission control must investigate the regulations thoroughly to make sure that all applicable rules are addressed properly and adequately.

Prioritizing Sources. This accumulated information can be used to prioritize sources to be treated. It is usually the case, and especially in odor control, that the one or two major odor control sources will be dealt with first in retrofits. In new plant designs, provisions should be made to treat all reasonable sources. Information obtained is prioritized, and a list of sources to be treated, and the sequence in which they will be treated, is established. The elapsed time over which this project will be accomplished varies from case to case.

Solving the Problem

Pollution Prevention. Decisions are determined by economics more than any other factor. In both odor control and VOC control, the costs of pollution prevention must be balanced with the benefits.

Odors. It is usually a simpler comparison for odor control. When improvements can be made by better housekeeping or temperature control, clearly those improvements should be made. When odor emissions decreases can be made by additions of chemicals or by addition of specialized bacteria to the process, these costs must be balanced against the costs of treating those emissions in the gas stream. When the sources are hard to collect and small in quantity, the benefit will frequently be toward treatment in the liquid stream (typically referred to as *source control*). When gases are easily collected, it is usually advantageous (and economical) to treat in the gas stream since chemicals added to the liquid stream are often consumed while treating nonodorous compounds. Sometimes, it is economical and logical to practice some level of source control in combination with off-gas collection and treatment. In cases where the concentration of emissions from the liquid stream exhibits wide swings from minimum to maximum values on an erratic basis, source control can be used to dampen the peaks and off-gas treatment can then be effective at consistently treating a more uniform inlet concentration.

VOCs. Concerning VOC emission reduction, this is usually a major process change, well discussed in Sec. 6.1.

Control Technology—Selection Factors

Logical Choices. In Chap. 2, Fundamentals, there is a discussion of the choice of control technologies based on consideration of fundamental properties of the contaminants. This is step 1 in many control technology evaluations since control must be based on the contaminant's natural characteristics. Therefore, a water-based scrubber will treat only contaminants which are water-soluble to some degree. An adsorbent will adsorb only certain compounds, and hold them only if they have low volatility. Any control system that does not consider these basics has an uphill battle to achieve removal rates.

Effectiveness of System in Removing Contaminants. If the system is not effective in removing the required amount of odor or VOCs, the major purpose for the equipment, then it is a failure. We recommend to all purchasers that the equipment be specified including a performance warranty. The performance warranty has to be clearly stated so that both buyer and seller agree on inlet and outlet levels, test methods, and reasonable operating conditions. More will be said about performance warranties, sampling, scope of work, and testing.

Effectiveness is often stated as percentage removal. We believe that is not the preferred way to state performance. As far as odor is concerned, satisfactory performance is achieved at a stated outlet odor level (*D/T*). For VOCs an acceptable outlet concentration is that which fulfills the requirements of the regulations for emission quantities. In both cases, a percent removal requirement does not achieve the desired goal. When the inlet concentration is low, a percent removal requirement requires unnecessarily and often unachievably low outlets, or outlets which are below reliable testing concentrations. When the inlet concentration is high, a percent removal requirement can mean that the odor or VOC emission is above nuisance odor levels or above regulatory requirements for VOCs.

Life-Cycle Cost

CALCULATIONS. Life-cycle costing considers all costs involved with an emission control system—initial capital cost, installation cost, ongoing operating costs of power, chemical, replacement or regeneration of adsorbents, the need for redundant equipment, a credit for recovery of usable materials, or heat, etc. Initial capital usually represents only a fraction of the total cost of an emission control system.

The following is a detailed discussion of life-cycle costing, which also accounts for the cost of money. By using life-cycle costing, many differing systems can be compared. Net present value is a measure which calculates cost as the sum of the present values of cash flows. Each cost can be valued as of a stated evaluations date. This method enables comparison of differing streams of costs.

It is not always easy to arrive at good estimates of operating costs before a system has demonstrated its costs in operation. This may be due to inestimable items—the actual amount of contaminants and the compounds of reaction that are formed; the usable operating life of carbon in a specific application; and for different technologies there may be significant unknowns. Nevertheless, reasonable and fair estimates must be made.

In the net present-value approach, each future capital and operating cost or credit, during the period of analysis, has an equivalent value in the base year. The sum of the present values of each cost or credit will be the net present value and will reflect the cost of capital involved.

After a base value is calculated, a sensitivity analysis, which evaluates the impact of variables on this calculation, allows for a better interpretation.

The net present value (NPV) of each cost or benefit is calculated as follows:

$$\text{System NPV} = \text{sum of nPV of each cost or credit}$$

$$= \sum_{1}^{x} \text{PV} \tag{9.1}$$

The present value of each cost or benefit is calculated as follows:

$$\text{PV} = \frac{\text{FV}}{(1 + i)^n} \tag{9.2}$$

where PV = present value; the amount of money which, if it were on hand today, would grow at interest to equal the future value

FV = future value; the cost or credit required at some future date

i = discount rate; rate of interest used to value cash flows, sometimes referred to as *cost of capital* since companies use the borrowing rate

n = time in years of cash flow being valued from valuation date

By neglecting the effect of inflation, maintenance and operating costs are often modeled at a level rate of expenditure. But inflation should be considered for each cash flow. The inflation factor can be reflected in an effective interest rate by combining discount rate and inflation rate into one factor. This allows one to simplify the calculation of maintenance costs at a level amount.

$$\text{Effective rate } r = \frac{1 + \text{discount rate}}{1 + \text{inflation rate}} - 1 \tag{9.3}$$

Equation (9.1) is modified to

$$\text{PV} = \frac{\text{FV}}{(1 + r)^n} \tag{9.4}$$

The value of the series of maintenance expenses may be calculated as follows:

$$\text{PV} = A\,\frac{1 - 1/(1 + r)^n}{r} \tag{9.5}$$

where A = the initial annual amount. The entire calculation begins from a base year, which we assume to be the year in which the comparison is made. Decisions must be made about

- *Discount rate.* This is usually the long-term actual cost of funds or capital, say, 10 percent.
- *Term of analysis.* This may be the useful life of the project, say, 20 years.
- *Inflation rate.* Let us assume 3 percent.

And data must be available on

- Initial equipment cost, including cost of installation.
- Ongoing operating costs, such as labor, utilities, maintenance, chemicals, or other consumables. Should there be a recapture value (such as the value of recovered solvent), a credit should be included.
- Replacements, which may occur periodically.
- Recovery value or disposal cost at end of period.

It is clear that by making certain assumptions, even if they are not precise, all equipment will still be compared equally. But every attempt should be made to ensure that the assumptions are as close to reality as possible; otherwise, the calculation may favor a system with higher initial costs or higher ongoing costs.

It is also clear how difficult it is to have all the information at the time of the comparison. For example, the ongoing operating costs, labor costs, replacement of parts, failure of equipment, and need for replacement before the assumed term will only be learned from use. The engineer is urged to be as realistic as possible. Information on operating costs, replacement parts, and life of equipment should be verified with users as well as from supplier estimates.

Because this calculation is affected by the varied assumptions, it is desirable to perform a sensitivity analysis, which is nothing more than a comparison of the calculation when an assumption is varied. For example, one can measure the effect of a different inflation rate, or cost of electricity.

	System A	System B
Initial cost (installed)	$500,000	$1,000,000
Annual operating cost	$150,000	$75,000
Term	20 years	20 years
Discount rate	10%	10%
Inflation rate	3%	3%
Recovery value	0	0
Present value	$2,144,082	$1,822,041

This example illustrates how it is sometimes "pennywise and pound foolish" to jump at the low-initial-cost alternative, and highlights the importance of using life-cycle costing when evaluating alternatives.

SCOPE OF WORK. In the above discussion of cost comparisons, it has been assumed that there is a similar responsibility for work, for all equipment suppliers being compared, This is usually not the case. Often one equipment supplier will provide equipment (e.g., a scrubber) as a standard catalog item sold off the shelf, and there is no responsibility for the performance of the equipment. A second supplier may sell the same type of equipment with performance warranties, field service, start-up and adjustment, training of operating and maintenance personnel, O&M manuals customized for the application, and sometimes testing services. Obviously there has to be a price difference between the two levels of supply. The purchaser has a choice. If there is experience "in house" with this kind of equipment for the process, and willingness to take all responsibility, then the purchaser may accept the lower price from a competent fabricator. But without such experience, a mistake can be made by rejecting the know-how of the equipment manufacturer. Of course, not all equipment manufacturers have comparable experience for varied industry applications, and it is suggested that you carefully review and interview references from a potential supplier's client list.

There are many other aspects of scope of work that should be considered to ensure equitable comparison: delivered pricing, provision of spare parts, types of motors used, type of materials of construction used, design choices to limit operating costs, etc. It is almost always possible to cheapen a system, and you do not want to be comparing a "price" supplier with a "quality" supplier without some recognition of the added values. The initial capital cost difference is often small for the many quality features; and this cost difference no doubt will be far less than the first major repair required. An old adage says, "You get what you pay for." Even if you think you are buying the most standard item, there are often quality and price differences between suppliers. There is no substitute for a well-researched, thorough, and concisely written specification which sets forth uniform requirements for all bidders to adhere to.

A Final Word. While cost calculations are certainly significant, they are not the whole story of any equipment selection. In any purchase there are three factors: price, quality, and service. The above-noted calculation treated the issue of price properly by evaluating long-term cost effects.

Quality is also a factor which affects cost by providing equipment with longer life and lower operating costs, but also by obtaining superior performance. The level of quality desired must be well defined and required equally of all competing vendors. Quality is the factor which is most controllable by the system purchaser.

Service is a significant factor, particularly with complex systems, where continuing support will be required from the manufacturer. Without good service from the manufacturer, a plant can experience extraordinary costs of engineering and maintenance time, parts replacement, as well as downtime. Service is the least controllable factor. Good service cannot be ensured simply by writing a specification which requires it. Good service is built by reputation and may be used in screening as a prequalification factor for consideration.

No cost factor has been assigned, in the above calculations, for equipment which is often out of service or fails performance. However, the costs associated with equipment which is often out of service or fails to perform can include the following:

- Loss of public confidence and goodwill
- Fines for noncompliance by regulatory agencies
- Monetary judgments as a result of legal action
- Excessive costs for labor and parts

Nevertheless, life-cycle costing, or present-value costing, is the best approach yet to make comparisons of different systems. When the results of life-cycle costing (LCC) of various alternative systems yield small differences (less than 10 percent LCC difference), the various systems may be considered equal. In such cases, it is particularly important to evaluate nonmonetary factors such as service, system simplicity, and aesthetics.

Experience in Use. Experience with equipment in use should be sought, but there also are some cautions about this information. If there is a disaster, you will usually find out about it. But if there is poor or middling performance, you may or may not find out about it without careful questioning. There are differences between equipment suppliers as well as technologies. There are differences between applications, not only in sources of odors, but also in topography and distance to neighbors. There are differences in plant operations; some plants have a schedule of conscientious preventive maintenance, while others only respond to crises. Nevertheless, much good information and guidance can be obtained, and if they are reviewed critically, good guidance can be provided by checking prior installation experience.

Practicality of Operation and Maintenance. An emission control system should operate for 20 to 30 years, with reasonable maintenance, and not just for the first few months after start-up. The difficulties and costs to maintain a system can be so great that a plant will not maintain the system adequately and will bypass it. One supplier's system may require more care than another's. It is strongly recommended that serious consideration be given to the ongoing costs of use, replacements, maintenance time, and operator's time. These costs and considerations should all be included in the life-cycle cost analysis and the purchasing decisions.

Another major factor is the ability of the operator to understand the operation of the system—he or she should not need extensive training to run a system well. If an operator does not know a system well and then becomes the trainer of a replacement, the system operation can go off track. But a system that is overly simple may be too limited. If the process has a wide range of emissions and concentrations, it must be expected that a system will have some complexities.

A great deal of the practicality of an emission control system is determined by the decision of the equipment supplier and the design engineer. For example, the materials of construction must be of a type that will withstand the corrosive chemicals used in the process. Sensitive electronics and motor control centers should be kept separate from areas which are wet and contain corrosive gases. Often equipment installations are located in freezing areas, and consideration should be given to freezing or increased viscosity of chemicals and protection of workers in inclement conditions. Parts which require frequent maintenance should be easily removable and replaceable, and should have conveniently located service platforms and ladders. Control instruments that require frequent and complex calibrations may prove to be impractical. The elevation of the chemical storage tank in relation to a point of use can often be troublesome because of lack of chemical, or air, and elevations which require excessive suction, or generate excessive pressure.

The engineer is advised to obtain reports from operating and maintenance personnel as they are most likely to be aware of the difficulties and often can comment constructively on new systems under consideration.

Continuity of Use. An emission control system must function continuously. It is an agreement with regulators to control emissions and with neighbors to control odors. Depending upon the location of the installation and the sensitivity of the surrounding neighbors, even shutdowns of an hour can lead to significant nuisance odor complaints. It must be determined just how critical the continuous operation of a control system is in every application. From this determination, allowable downtime either for malfunction or for planned maintenance can be established. This information will then come to play in determining the amount of redundant equipment and systems required to satisfy this requirement.

Life of Equipment and Components. In the evaluation of a system, it is critical to review the expected life of equipment for continued efficient operation.

Materials of construction are important. It is common that emissions which are to be treated are corrosive, and certainly the chemicals used to treat emissions are corrosive. The choice of materials of construction can change a system from continuous operating simplicity to a maintenance nightmare. All systems will require parts replacements (e.g., metering pumps) and should be designed to allow easy access for removal and replacement. Further, some systems (e.g., adsorbers) are based on regeneration and replacement of components, and this regeneration scenario must be realistic.

Ability to Treat Variable Sources. To design an emission control system, it is usually necessary to know the maximum and average emission rates. The control system has to be designed for and function adequately at both conditions. Some technologies are not "adjustable." To overcome this limitation, costly operating decisions may have to be made. This highlights the need to have specific design criteria. It is very helpful to the equipment designer to have as much meaningful inlet contaminant information on variations as possible—not just over 24 h, but on weekdays, weekends, during the hottest time of the year, during other events which may generate unusual levels of odor concentrations.

Flexibility of the System. Emissions from plant processes usually are very variable and may change. Equipment to control such emissions must be responsive to changes and easy to modify to accommodate major changes. Many systems have no flexibility and are difficult to modify. Therefore, if a system's emissions are vaguely defined, one should be very cautious and avoid choosing an inflexible emission control system.

Controls. If the performance of an emission control system is largely determined by manual control and the decisions of the operator, you can be fairly sure that the system will not work well some of the time. The more difficult it is to monitor performance, the more difficult it will be to adjust manually.

Systems which automatically monitor some parameter related to performance and *automatically* adjust conditions to ensure effective emission control are superior. Every effort should be made by equipment manufacturers to develop effective, practical control systems able to maintain good emission control at lower operating costs.

Hybrid Systems. Section 8.9 contains a discussion of hybrid systems in which optimal performance requires a combination of technologies. These multiple-stage systems are subject to the same reviews as discussed above.

EVALUATING THE PERFORMANCE OF EXISTING EMISSION CONTROL SYSTEMS

There is a close relationship between this evaluation and the prior discussion, except before there was the opportunity to avoid the problem system before purchase, while now there is a selected system, and it is necessary to estimate performance.

The following will serve as a checklist of items pertinent to evaluating a system, and hopefully this list is made prior to acceptance by the purchaser. The main emphasis is placed on testing of performance, but additional factors should not be overlooked.

1. Testing performance to design criteria
2. Testing installation factors
3. Testing equipment factors
4. Warranties

Testing Performance to Design Criteria

It is highly desirable that the equipment supplier warrant performance of equipment, not only mechanically, but also in its function of emission control. That warranty, agreed to before purchase, should clearly state how a test of performance will be done to ensure compliance with the criteria established.

Test Procedures. The following is a listing of some factors that should be considered in establishing a testing agreement and protocol.

1. It is very hard, during a test, to arrive at a desired, specified odor level. Therefore, frequently a simulation of an odor level is used, e.g., the controlled addition of a single contaminant compound. It should be understood by the engineer and user that a test using a single component does not ensure adequate performance in treating a complex mix of odors. It is quite easy to establish operating conditions that can successfully remove one compound. These conditions may be only moderately successful in treating the complex mixture. It is most important that performance be based on the real-world situation of the types of odors to be treated. Tests should be done at the process exhaust concentrations that usually occur.
2. Temperature.
3. Airflow rate. Tests must be done at the design flow rates and at approximately correct emission temperatures.
4. Acceptable level of outlet odors or VOCs. The acceptable level of outlet odors and VOCs should be stated as concentrations and not in terms of percentage of removal.
5. Analytical or testing procedure. The test procedure must be specified in detail, and a written test protocol agreed to, in advance of beginning any performance test. The analytical procedure should be done according to the test methods referred to in Secs. 4.1, 4.2, and 4.3.
6. Record all operating conditions.
7. Effluent liquid, chemical concentration, pH, and contaminant concentration (if applicable). For the record, all operating conditions must be recorded, and the system must be operated in a way that reflects the continuing operation at the plant.

8. Performance at equilibrium. All systems must be tested at equilibrium, if at all possible, since this gives a truer demonstration of expected continued performance.
9. Number of samples. No decision should be made based on one or two samples. The number of samples that will result in a reasonable measure of exhaust conditions must be evaluated in each case.
10. Calculation methods of acceptable performance. It is important to be specific about the calculation method. Is a system acceptable if the average treated emission is adequate, or must every sample fall below the specified performance? If there is a requirement that every sample fall below a specified level, then the designer must know this before the equipment is designed. Is performance satisfactory if it falls within the statistical variation of the results? How are nondetects included in the calculation? It is better to be clear, very early in the project, on all the details of the test and calculation procedure, to avoid disputes or misunderstandings concerning installed equipment.
11. Where automatic control is present, performance testing in the automatic mode should be performed at start-up. The test of an automatic operation should be done where the inlet is varied and where the equipment, without being touched by the operator, automatically adjusts to new conditions to maintain a specific outlet.
12. Action to be taken if equipment does not meet criteria. Performance warranties should state the responsibilities of the supplier, and the purchaser to properly define design criteria and that these comport with actual conditions of tests and then operate the system with supplier-recommended procedures.
13. Resolution procedure in case of dispute.

Testing Installation Factors

There are standards of good construction and installation practices and legal contracts that specify such performance. A typical agreement is the American Institute of Architects (AIA) contractual agreement.

These installations should specify requirements for such factors as strength of supports, tie-down strengths, vibration, level, drainage, leakage, safety, and any other features that impact upon the proper operation of the equipment. In many cases it is prudent to develop individual details which depict specific requirements to obtain a satisfactory installation.

There should be a thorough walk-through inspection of all completed installations, by the supplier as well as design and plant engineers.

Testing Equipment Factors

Following installation and before performance testing, all equipment should be checked for operation. All mechanical warranties are to be provided for components. All technical literature of components is to be provided. Whether a system is custom-designed or a standard package-type system, an O&M manual should be supplied as well as training of operation and maintenance personnel.

Warranties

Warranties for individual components and the complete system should be provided. Performance warranties have been previously discussed.

Using a Consultant

Emission control equipment, design, and testing are a specialized field. A plant engineer may wish to use the services of a consultant for all or some portion of this work. Nevertheless, the plant engineer will always know more about the plant process and the details of plant operation, and so she or he should be a participant in the emission control project. This relationship, along with that of the selected equipment supplier, should be one of cooperation to achieve a successful project completion. Qualified engineering consultants can be found among listings in either the trade or professional society publications, or in air pollution industry listings. Credentials should be checked with other clients. Odor control is a more limited specialization, and since there are fewer specialists, they usually cover many industry applications.

CHAPTER 10
REGULATIONS

Editor's Note: A decision had to be made about the content of this book. It would not be possible in this broad-ranging book topic to adequately cover all international air environment protection laws. We decided to limit the content to the U.S. and Canadian laws which regulate this field. However, to assist the reader who may be interested in relevant international environmental regulations in places other than these two countries, we suggest checking with Internet sites, as many governments provide assistance through information on the Internet to find out the regulations in a particular country.

SECTION 10.1

U.S. FEDERAL AND STATE LAWS

G. Vinson Hellwig
Manager, Air Quality Programs, S.E.
Dames & Moore
Atlanta, Georgia

FEDERAL LAWS AND REGULATIONS 10.2
Clean Air Act 10.2
Resource Conservation and Recovery Act 10.7
Toxic Substances Control Act 10.8
Comprehensive Environmental Response, Compensation, and Liability Act 10.8
Occupational Safety and Health Act 10.9
STATE LAWS AND REGULATIONS 10.9
Criteria Pollutants 10.9
Air Toxics 10.9
ODOR REGULATIONS 10.10
REFERENCES 10.10

FEDERAL LAWS AND REGULATIONS

There are numerous statutes and regulations that in effect regulate air emissions from facilities. Among the statutes that regulate volatile and toxic compounds are

- Clean Air Act (CAA)
- Resource Conservation and Recovery Act (RCRA)
- Toxic Substances Control Act (TSCA)
- Comprehensive Environmental Response, Compensation, and Liability Act of 1980 (CERCLA)
- Occupational Safety and Health Act (OSHA)

Clean Air Act

Although the federal regulations governing volatile organic compounds (VOCs) and air toxics do not deal directly with odor control, they indirectly control odors because these are sometimes the sources of detectable odors from stationary sources. The CAA regulates stationary sources through a series of regulatory requirements if they apply to a given facility. These are:

- Section 110 state implementation plan (SIP)—regulations adopted by the state and approved by the Environmental Protection Agency (EPA)
- Nonattainment areas—regulations requiring further reductions in emissions than the existing SIP
- Section 112 standards—National Emission Standards for Hazardous Air Pollutants (NESHAPs) and maximum achievable control technology (MACT) standards

- Section 111 standards—New Source Performance Standards (NSPSs)—emission standards for categories of sources built after a specified date

The Clean Air Act Amendments (CAAA) of 1990 greatly increased the regulatory framework for control of air emissions, and specifically the control of volatile organic compounds and hazardous air pollutants (HAPs). Title I of the CAAA dealt with nonattainment areas and principally is focused on the reduction of VOCs with the resulting reduction in ozone. After more than 20 years of regulation, approximately 100 urban areas in the United States still exceed the ozone National Ambient Air Quality Standards (NAAQS), referred to as *ozone nonattainment areas.* The CAAA require that state programs be revised to control more sources of ozone-causing substances in these nonattainment areas, including both mobile and stationary sources. In a number of respects, states may also be required to more stringently regulate sources than is currently done. In addition, EPA has proposed lowering the ozone NAAQS, and this will greatly increase the number of nonattainment areas and thereby require even greater reductions in emissions and more stationary sources affected by reduction requirements.

The CAAA impose compliance deadlines on ozone nonattainment areas. The CAAA require that each state designate its air quality control regions as being in either attainment or nonattainment. Each nonattainment area is further classified based on the area's *ozone design value* compared to EPA's standard. If the area does not meet these design values, the area is designated nonattainment and more controls must be imposed on stationary sources in that area through a SIP revision and require emitting sources to apply reasonably available control technology (RACT).

The amendments require that each state with one or more ozone nonattainment areas impose controls on sources of VOCs sufficient to allow those areas to comply with the ambient standard on an aggressive schedule. States must modify their SIPs as necessary to require interim and ultimate reductions of VOCs until the standard is attained. If a state fails to meet either the interim reduction targets or the ultimate attainment date, EPA is authorized to require more stringent offset requirements for new source construction, withhold highway funds, or even impose both of these sanctions. The amendments also contain additional specific requirements that must be incorporated into each state's SIP.

RACT is a requirement by EPA of the state or local agency to implement more stringent regulations in ozone nonattainment areas than in attainment areas for VOC emissions. This is done through a SIP revision by the state agency in which categories of source must meet more stringent emission limits than prior to the SIP change, and the SIP contains a schedule for compliance plans for the emitting sources. The RACT plan establishes the revised VOC emission limits, and in some cases the technology to achieve the emission reductions required.

Stationary sources in nonattainment areas are subject to RACT, and many of these were in source categories that were covered by control technique guidelines (CTGs). EPA has issued CTGs for years to guide states in identifying control options for VOC sources. The CTGs principally rely on add-on control equipment, but some CTGs also include process changes, spill control measures, good housekeeping, and reduction of fugitive emissions. EPA has issued CTGs for approximately 42 categories of sources. EPA has more recently issued alternative control technique (ACT) guidelines. The ACT development process is less rigorous and therefore less costly to develop. This leads the ACTs to be issued not only at reduced expense, but also more quickly than the CTGs were issued. There are approximately 14 ACTs presently available from EPA. Both CTGs and ACTs are documents that describe the process, control alternatives, and potential cost to control for the various options. Each of the CTGs or ACTs applies to a specific category of sources, and these categories are specified in the document.

Although they are guidance documents, they are viewed by most regulators as part of the rule, and exceptions or alternative control techniques must be justified to the regulatory agency and EPA.

Major VOC sources, as discussed previously, are defined by the classification of their ozone nonattainment area and are required to meet RACT. Also VOC sources located anywhere in the ozone transport region encompassing the eastern seaboard from northern Virginia to Maine are required to achieve RACT if they have the potential to emit more than 50 tons/yr VOCs or are covered by a CTG regardless of the attainment or nonattainment classification. States were required to amend their SIPs to require RACT for major VOC sources no later than May 31, 1995.

Where RACT is required for a source and it does not fall within a CTG or ACT, then the facility must prepare and submit a RACT plan for a "non-CTG" source. This means the control strategy to reduce the VOC emissions is a site-specific plan that addresses how the reductions will be made, what level of reductions will occur, how this will be documented, and what records will be kept to verify these reductions on an ongoing basis.

In addition, the CAAA require states to go beyond RACT by imposing additional controls where necessary to meet the interim goals and the applicable attainment deadlines. States with serious and severe nonattainment areas may not be able to meet these interim goals without imposing additional controls on numerous stationary sources.

Section 112 of the CAAA deals with the control of HAPs through regulations requiring technical controls that will result in emission reductions within industrial categories, MACTs. Congress listed 188 chemical compounds or groups of compounds for which EPA must develop MACT categories. One compound, caprolactam, was subsequently delisted by rulemaking. Of these 188, approximately 40 are groups of compounds yielding a list of almost 400 regulated chemical compounds. Most of the 400 regulated chemicals are also VOCs, and technology used to reduce these compounds will also reduce some VOCs and some odorous compounds. Prior to 1990, EPA had not implemented a workable federal program for identifying and controlling emissions of HAPs. When EPA attempted to list and develop controls for air toxics on a chemical-by-chemical basis, the process proved so cumbersome and time-consuming that only seven HAPs were regulated. As a result, many states decided to develop their own air toxics programs, based largely on allowable ambient levels of toxics. The CAAA essentially scrapped the existing federal framework and replaced it with a system calling for development of technology-based standards for HAPs, applicable to facilities within a specific industrial or commercial source category. Each categorical standard will serve as the basis for federally enforceable emission limits in the facility's operating permit. This regulatory system was patterned after that used to regulate direct discharges.

Under section 112 of the CAAA, the EPA has already promulgated MACT standards for more than 20 categories and proposed standards for others. The most prominent final MACT standard is the imposing Hazardous Organic NESHAP for the SOCMI category, commonly known as the HON. This was published in the *Federal Register* on April 22, 1994, and the entire list of MACT standard categories in the *Federal Register* on June 4, 1996 at page 28197.

The Hazardous Organic NESHAP (HON). The HON was promulgated based upon section 112 of the CAAA, and it was the first major rule to be issued for HAPs under the 1990 amendments. The HON is one of the most complex and pervasive rules ever issued by EPA. It covers more processes and pollutants than all previous EPA air toxics programs. For example, one major portion of the rule applies to sources that produce any of nearly 400 commercial organic chemical products and use any of 112 organic HAPs either in a product or as an intermediate or reactant. An additional 37 HAPs are regulated under another part of the HON.

The focus of this rule is the synthetic organic chemical manufacturing industry (SOCMI). For purposes of the MACT standard, a SOCMI manufacturing plant is viewed as an assortment of equipment, process vents, storage tanks, transfer racks, and wastewater streams, all of which emit hazardous air pollutants. The HON would require such a plant to monitor and repair leaks to eliminate fugitive emissions, and would require controls to reduce the toxics coming from discrete emission points by 95 to 98 percent or to minuscule concentrations.

These are unusually stringent requirements which may be indicative of measures to be imposed in subsequent MACT standards. The HON also brings with it a labyrinth of recordkeeping and reporting requirements destined to frustrate the best intentions of those required to use them.

One of the most conspicuous features of the HON is that it offers the source two different options for determining whether the facility is compliant with the MACT standard. Under the first option, the specified control technology would be applied to each regulated emission point and each would have to meet the standard. The second option, called *emissions averaging,* would allow the plant to control certain emission points through a system of emission credits and debits across its inventory of emission points. Thereby above-standard emissions from one source would be averaged with the more tightly controlled source over a given averaging period. Since the HON calls for a quarterly averaging period, emissions averaging would be helpful to chemical manufacturing facilities that find it necessary to make frequent operational changes. In such facilities, a moderate investment in emissions tracking might be rewarded by significant gains in operating flexibility without jeopardizing their compliance status. Without emissions averaging, compliance would likely be much more expensive.

HON MACT Controls

- Incineration of the toxics in process vents
- Modifying storage tanks or equipping them with vapor recovery systems
- Incineration of transfer vapors
- Leak monitoring and repair
- Treating wastewater streams to reduce air toxic emissions

The first major MACT standard that has extensive work practice standards is the wood furniture MACT, and EPA has publicly stated that this will be the model for future MACT standards of this type. The wood furniture MACT contains not only work practice standards, but also a training requirement for every employee involved in the surface coating of wood furniture as part of an implementation program. This training program is so extensive that it must train even experienced workers on the use of a spray gun in a manner that will minimize VOC emissions, and on the importance of the required recordkeeping. This work practice standard is similar to some OSHA programs, but goes further than any previous EPA standards. Compliance with this program must be in place in November 1997, with ongoing recordkeeping and annual training refresher courses.

Section 183(e) of the CAAA is being drawn into the MACT standards by EPA in the consideration of its rulemaking. Section 183(e) deals with the VOC emissions from consumer and commercial products. EPA has determined these product classes contribute to ozone nonattainment and will partially regulate them through a seemingly unrelated manner, MACT standards. Because many of the HAPs regulated through the MACT standards are also VOCs, and they may use commercial products such as inks and coatings, EPA is incorporating the emission reduction requirements by limiting the VOC content of some of these materials. This is being done by specifying VOC limits as well as HAP content limits in the inks and coatings covered by certain MACT categories.

NSPS regulations are required by section 111 of the CAA, and these are regulations dealing with facilities built after the date specified in the regulation. The NSPS regulations deal with both industries and industrial processes, and the applicability and emission limitations are written in the regulation. NSPS regulations are usually enforced by the state agencies once they have delegation from EPA; otherwise they are enforced and can be enforced by EPA at any time after a violation is found.

The federal air regulations dealing with these types of sources are NSPS, NESHAPs, and SIPs dealing with the VOCs. The SIPs are federally required, approved, and enforceable, but these are actually state regulations and are implemented on the state and local levels.

In order to simplify the process of determining the regulations that apply to your facility, see Fig. 10.1.1, and follow the steps. By answering the questions, you can deter-

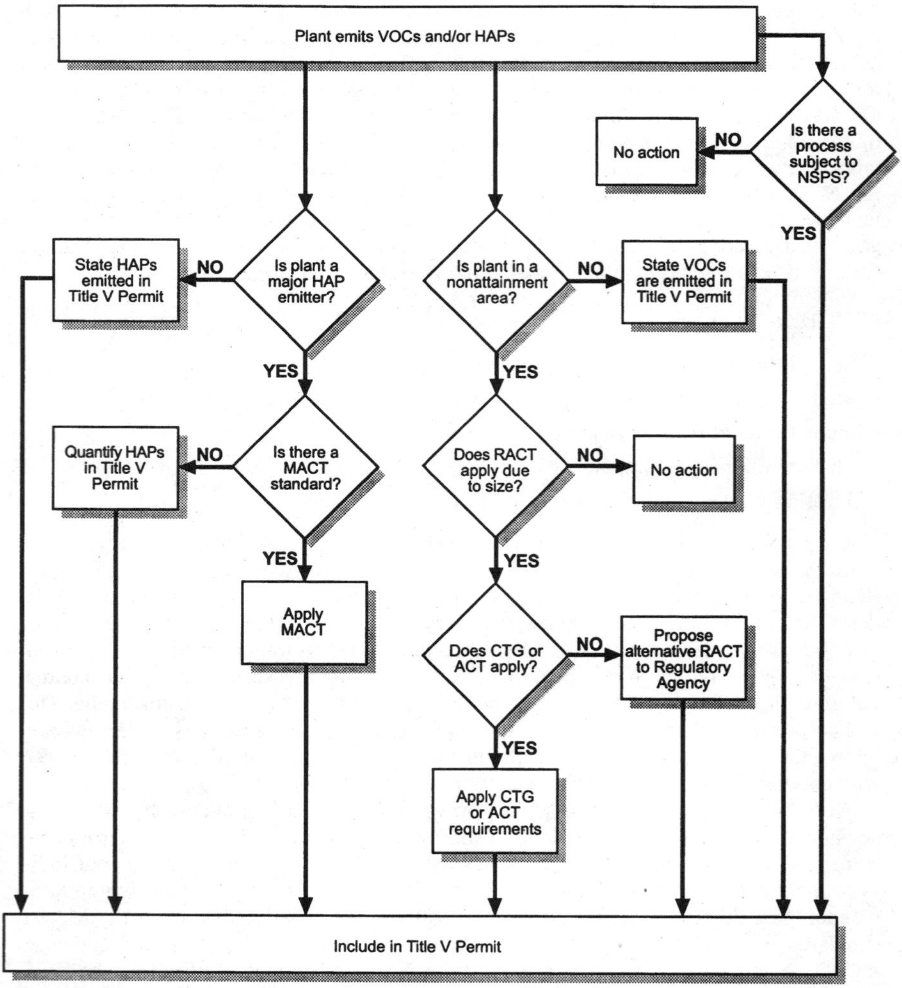

FIGURE 10.1.1 Regulatory applicability.

mine what regulations your facility must meet and which of these regulations must be addressed in a Title V application, if it applies to your facility.

Resource Conservation and Recovery Act

The Resource Conservation and Recovery Act was enacted as law in 1976 as an amendment to the Solid Waste Disposal Act (SWDA). The primary objective of RCRA is to protect human health and the environment, and the secondary objective is to conserve valuable material and energy resources while regulating the management of hazardous waste and encouraging recycling. RCRA provides for "cradle to grave" tracking of hazardous waste, from generator to transporter to treatment, storage, or disposal. The overlap of RCRA with CAA regulations relates to emissions arising from both the transportation and storage of hazardous waste and the disposal, e.g., combustion, of hazardous waste.

Transportation, Storage, and Disposal Facilities. There are requirements for transportation, storage, and disposal facilities (TSDFs) to limit VOCs and HAPs from their operations. Increasingly EPA is regulating the volatile emissions from these facilities through the RCRA and CAA rulemaking. The most widely encompassing of these regulations is subpart CC that regulates organic air emissions from tanks, surface impoundments, and containers. Under subpart CC certain requirements must be met:

- Controls for hazardous wastes containing >100 ppmw VOC meeting a 95 percent removal efficiency
- Sampling of each waste stream (allowing averaging periods up to 1 year)
- Routine reference method 21 testing
- Development of inspection and monitoring procedures
- Preparation of recordkeeping documents

EPA has also issued Part 63 regulations dealing with air toxic emissions from TSDFs and other facilities. The rule exemplifies the crossing of medium boundaries by regulating RCRA and some water discharge facilities. Although they apply to RCRA facilities, they are actually air toxic or MACT regulations issued subsequent to Title III of the CAAA. These regulations were designed to reduce VOCs and air toxic organic chemical emissions. These regulations apply to TSDFs, industrial wastewater treatment facilities, recycled used oil management facilities, and other facilities that provide waste management support services or recover and/or recycle spent materials. The rule requires specific air emission controls for affected tanks, containers, surface impoundments, process vents, and equipment leaks.

Promulgated under RCRA there are separate air rules in 40 CFR 264 limiting emissions at a TSDF from process vents, subpart AA; equipment leaks, subpart BB; and tanks, surface impoundments, and containers, subpart CC. Subpart H rules deal with the combustion of certain hazardous materials when used as fuels, and these are known as the *boiler and industrial furnace* (BIF) regulations. A facility that uses a listed hazardous waste as a fuel, even if it is a supplementary fuel or a primary fuel for heat or process steam generation, falls under these regulations. These regulations require a stringent air emissions source test program, and recently EPA has required that most permit renewals have modeling and a multipathway risk assessment performed as well. The permits issued under this rule set very strict air emission limits that include HAPs and VOCs. There are also extensive monitoring and recordkeeping requirements for all these permits.

Part B Facilities. Facilities that are handling or disposing of a RCRA waste must obtain what is referred to as a RCRA *Part B permit.* These permits, like the BIFs, have stringent HAPs and VOC limits, but the recordkeeping requirements are more extensive, especially for acceptance and handling of hazardous waste. There are overlaps in emission requirements between the RCRA Part B and Title V permits. Often the limits may cover pollutants normally not regulated by a Title V permit, or they may contain limitations different from the Title V permit. There is no regulatory requirement that they be the same, so long as the facility is in compliance with the regulatory requirements of each program and the facility is meeting the respective permit conditions.

Where the facility has an incinerator on site that is permitted as a RCRA Part B permit, there are specific air emission limits for toxic materials including organic compounds and metals. A trial burn must be conducted to demonstrate compliance with limits set forth in a construction permit. Often these limits are set by a risk assessment study to ensure no harm to humans and in some cases ecological risk as well. Existing air pollution regulations are considered in this phase of the permitting process to determine that the limits in the Part B permit will be at least as stringent as existing state and federal regulations.

Toxic Substances Control Act

The Toxic Substances Control Act (TSCA) was enacted on October 11, 1976. The purpose of TSCA was to ensure that adequate data would be made available to the EPA on adverse health and environmental effects of chemical substances and mixtures. The importance of the subject to this book is that this act in some cases limits the distribution, use, or disposal of chemicals that can have adverse health and environmental effects. TSCA requires the testing and reporting of tests of any new commercial chemical substances to EPA. Also TSCA regulates the manufacture, use, and disposal of polychlorinated biphenyls.

Comprehensive Environmental Response, Compensation, and Liability Act

Superfund Act Reauthorization Amendments. On October 17, 1986, the Superfund Amendments and Reauthorization Act (SARA) was signed into law. One part of SARA, Title III, is known as the Emergency Planning and Community Right-to-Know Act (EPCRA) of 1986. EPCRA requires states to establish a process for developing local emergency preparedness programs and to receive and disseminate information on hazardous chemicals present at facilities within local communities. Most of the information gathering and dissemination is performed by the federal government through the Toxics Release Inventory System (TRIS) database. The information in TRIS is acquired through the annual reporting requirements of EPCRA. This reporting is required in SARA Section 313 and applies to facilities in standard industrial classifications 20 through 39 with 10 or more full-time employees. This reporting requirement was amended by the Pollution Prevention Act of 1990, and recycling and pollution prevention measures were quantitatively required to be reported annually as part of the Section 313 reports.

This reporting requirement has led to solvent substitution and pollution prevention measures because the industrial facilities now must quantify all releases to the environment and this information is made public. This annual reporting of emissions to all environmental media has identified the emissions by chemical species and the media into which they are released. In all, there are more than 300 SARA reportable chemicals. See the current year's EPA guidance document for the complete list.

Occupational Safety and Health Act

The Occupational Safety and Health Act became law in July 1971, and it provided the basis for regulations protecting workers in the workplace. The premise under which OSHA usually works is to remove the hazard from the worker's area. Although this is not always possible, when it is possible, it sometimes means that an odorous or volatile compound is vented to the atmosphere, and this creates an environmental control situation regulated by other statutes. There are exposure levels established for listed compounds that are not to be exceeded for an 8-h period. OSHA regulations do not deal with the control of VOCs directly, but the regulations may either require control or cause them to be discharged in a manner that will make them subject to an environmental regulation and thus control.

STATE LAWS AND REGULATIONS

As discussed previously, most of the state air pollution laws are patterned after, and most were developed as a result of, the mandates in the Clean Air Act. Even though a number of states enacted clean air legislation before there was a federal statutory requirement for limiting air emissions, the vast majority of state air regulations are driven by the CAA. The state implementation plan is the vehicle used to adopt and approve, by EPA, state regulations and federally mandated state regulations.

Criteria Pollutants

The SIP is used to regulate criteria pollutants, SO_x, NO_x, particulate matter (PM), and PM_{10}, ozone (through regulation of VOCs and NO_x), CO, and lead for which there are National Ambient Air Quality Standards. The SIP must demonstrate how the NAAQS will be met, if an area is in nonattainment, and how the standards will be maintained. The states must enact emission-limiting regulations for these pollutants that will limit the emissions from stationary and in some cases mobile sources.

Air Toxics

The SIP is also used to regulate HAPs under Section 112 of the CAA. Prior to the 1990 CAAA, there were only nine HAPs regulated by the NESHAPs, and this led to a significant problem for state agencies. A number of states were pressured from the public to regulate HAPs. Without federal guidelines or a federal mandate to control HAPs, more than 20 states promulgated their own air toxic regulations, and most of these were health-based exposure standards. The methodology employed in regulating air toxics in most of these states follows a listing of the regulated compounds with some exposure limit based on the threshold limit value (TLV), lethal dose for 50 percent (LD_{50}) of the test animals, maximum allowable concentration (MAC), acceptable ambient level (AAL), or some other determined exposure limit. Usually the state agency sets the exposure limit to a fraction of one of these concentrations. The next step is to perform a screening model of the emission point of the air toxic, and this is a very conservative technique that will overpredict any potential exposure. States that require the air toxic modeling also allow for refined modeling, i.e., if the source fails the conservative screening step, the applicant may use models that have more refined mathematical representations of the emissions. Often a source may fail the screening step, but pass on refined modeling. The model output is compared with the allowable air toxic concentration to determine if it is within the allowable exposure range. The enforcement of these provisions is almost always through the construction or operating permit issued by the state or local regulatory agency. To gain access to the specific state regulations, see the references on the Internet at the end of this book, since most states maintain a Web site with their regulations and contacts for questions posted.

ODOR REGULATIONS

Odors are always regulated through the state and local agencies. These state and local regulations are based on nuisance law and are enforceable at the state and local levels. However, as a matter of policy, EPA has not approved any portions of a SIP containing odor regulations. There are no federal regulations dealing with odorous compounds per se.

REFERENCES

29 CFR 1910, Occupational Safety and Health Standards.

40 CFR 51, Air Programs.

40 CFR 60, Standards of Performance for New Stationary Sources.

40 CFR 61, National Emission Standards for Hazardous Air Pollutants.

40 CFR 63, National Emission Standards for Hazardous Air Pollutants for Source Categories.

40 CFR 68, Chemical Accident Prevention Programs.

40 CFR 70, State Operating Permit Programs.

40 CFR 264, Hazardous Waste Standards for Owners and Operators of Hazardous Waste Treatment, Storage, and Disposal Facilities.

40 CFR 372, Toxic Chemical Release Reporting Community Right-to-Know.

NESHAPS; revision of initial list of categories of sources and schedule for standards under Sections 112(c) and (e) of the CAAA. *Federal Register,* vol. 61, no. 108, June 4, 1996, pp. 28197–28208.

Superfund Amendments and Reauthorization Act, Pub. L. 99-499, Oct. 17, 1986, Title III, Emergency Planning and Community Right-to-Know.

The Clean Air Act, amended by the Clean Air Act Amendments, Pub. L. 101-549, Nov. 15, 1990, 104 Stat. 2399.

Toxic Substances Control Act, as amended, 7 U.S.C. Section 136, *et seq.*

U.S. Environmental Protection Agency, *Toxic Chemical Release Inventory Reporting Form R Instructions,* rev. 1995 version, EPA 745 K 96-001, March 1996.

SECTION 10.2

ODOR AND VOC REGULATIONS IN CANADA

Michael Rix
ZORIX Consultants Inc.
Mississauga, Ontario

ODOR REGULATIONS 10.11
- **Alberta** 10.12
- **British Columbia** 10.12
- **Manitoba** 10.12
- **New Brunswick** 10.13
- **Newfoundland** 10.13
- **Northwest Territories** 10.14
- **Nova Scotia** 10.14
- **Ontario** 10.14
- **Prince Edward Island** 10.15
- **Quebec** 10.15
- **Saskatchewan** 10.16
- **Yukon Territory** 10.16

VOC REGULATIONS 10.16
REFERENCES 10.16

ODOR REGULATIONS

Odor issues in Canada are regulated generally at the provincial level and in some cases locally. Federal regulations do not address odor. The provinces have a variety of approaches to odor regulation ranging from general clauses prohibiting air pollution to specific regulations limiting odor emission rates or off-property odor levels; and these may include both permitting requirements, and prohibitions. This range of approaches may be further categorized as follows:

- General clauses against adverse effects—clauses which prohibit causing one or more types of adverse effects, e.g., "interference with the normal conduct of business"
- Odor nuisance—a clause prohibiting obnoxious or offensive odors
- Industry-specific regulations—regulations prescribing operating condition or emission controls for specific industries traditionally associated with odor nuisances, e.g., rendering plants, pulp mills
- Pollutant-specific regulations—regulations setting ambient air standards for odorous chemicals, e.g., hydrogen sulfide
- Odor limits—regulations setting odor emission rate or ambient odor standards

Many provinces have more than one type of regulation which could be applied to nuisance odors.

Examining the regulations themselves does not always give a clear picture of how a particular jurisdiction deals with odor issues. Several provinces have essentially similar regulations, but different levels of sophistication in addressing odor issues. Some provinces are much more likely to include special conditions in operating permits for potentially odorous processes. The diversity in understanding, and sophistication in dealing with, nuisance odor issues may arise for several reasons—the size of the jurisdiction, past experiences with nuisance odors, staff available to address the issues, and the size and diversity of industries in the area.

Also note that changes are occurring in all jurisdictions in Canada which could impact the regulations themselves in some cases, and the administration of these regulations in virtually all cases. Several provinces have new regulations in the works, and most jurisdictions are coping with substantive changes in operating structures and staffing.

The following discussions summarize the regulations applying to odor pollution in each province and territory. Where possible, an attempt has been made to indicate how the regulations are administered and other policies or procedures which are followed.

Alberta

The Environmental Protection and Enhancement Act (EAPA) in Alberta includes prohibitions against the release of compounds which cause a "significant adverse effect." The main regulation concerning odor, however, is Section 105 of the EAPA, which authorizes the issuance of an environmental protection order "where a substance or thing is causing or has caused an offensive odor." The order could include requirements for investigating, monitoring, or controlling the offensive odor.

The permitting ("approvals") requirements in Alberta are based on specific industry categories, and they include a number of industries which may have odorous emissions. Alberta also has ambient air quality guidelines which include a guideline of 14 $\mu g/m^3$ hydrogen sulfide as a 1-h average concentration and 4 $\mu g/m^3$ as a 24-h average.

Agricultural operations are excluded from Section 105 of the EAPA, where they are carried out in accordance with generally accepted practices. However, agricultural regulations stipulate separation requirements for various types of livestock operations, with these separation distances varying with the number and type of animals.

British Columbia

Air pollution in British Columbia is regulated under the Waste Management Act. Odor is not addressed directly in the act. The act defines an *air contaminant* as a substance that, among other things, "interferes or is capable of interfering with the normal conduct of business" or "causes or is capable of causing material physical discomfort to a person." *Pollution* is defined as the presence of contaminants in the environment that "substantially alter or impair the usefulness of the environment." There is a general prohibition on the unlicensed or unapproved release of contaminants into the environment. There is provision for enforcement where a substance is being released, or may be released, in such a manner as would cause "pollution."

New regulations are under development including emission limits for pulp and paper plants.

Manitoba

The Environment Act of Manitoba includes odor in its definition of *pollutant,* where it may "interfere with or is likely to interfere with the comfort, well-being, livelihood or enjoyment of life by a person." There is no direct prohibition on the release of pollutants, but there are requirements for licensing or permitting industries or activities which may release emissions. The licensing provision has been used in Manitoba to address the potential for odor nuisance very specifically.

In 1978 Manitoba introduced guidelines for a number of air pollutants, including hydrogen sulfide and odors. These included an emission criterion of 150 odor units (dilutions to threshold) and ambient air criteria of 2 dilutions to threshold in a residential zone and 7 dilutions to threshold in an industrial zone, based on two tests conducted between 15 and 60 min apart. Measurements to determine compliance with these guidelines were made using a Scentometer, or a similar portable olfactometer developed by the Manitoba Department of Environment specifically for this purpose.

A new approach to managing odor problems was recently drafted in Manitoba (Manitoba Department of Environment, 1997). This document describes problems with the odor-level guidelines and outlines a number of principles which were used in establishing the proposed new Odor Nuisance Management Strategy. This approach is based on the following definition of odor nuisance:

> "Odor nuisance" means a continuous or repeated odor, smell or aroma, in an affected area, which is offensive, obnoxious, troublesome, annoying, unpleasant or disagreeable to a person:
>
> a. residing in the affected area;
> b. working in the affected area;
> c. present at a location in the affected area which is normally open to members of the public;
>
> if the odor or smell or aroma
>
> d. is the subject of at least 5 written complaints in a form satisfactory to the Director and from 5 different persons falling within clauses (a), (b) or (c), who do not live in the same household, received by the director within a 90 day period; or
> e. is the subject of at least one written complaint in a form satisfactory to the Director from a person falling within clauses (a), (b) or (c) and the Director is of the opinion that if the odor, smell or aroma had occurred in a more densely populated area there would have been at least 5 written complaints from 5 different persons and who do not live in the same household within a 90 day period.

This definition of odor nuisance is now being incorporated into licensing conditions for industries with a potential odor impact.

New Brunswick

The Clean Environment Act of New Brunswick includes odor in a definition of *contaminant,* with a nuisance clause including effects on the comfort of a person, or interference with the "normal conduct of business or the normal enjoyment of life or use or enjoyment of property." A similar clause is found in the Air Quality Regulation.

New Brunswick also has emission regulations for the pulp and paper industry which set limits for total reduced sulfur compounds.

A new Clean Air Act has been written, but there it contains no substantive changes in its approach to odor.

Newfoundland

The Environment Act in Newfoundland includes odor in a definition of *pollution.* Air Pollution Control Regulations were published in 1996. The heart of these regulations is pollutant-specific standards for a large number of compounds. These are mainly of two types: ambient air standards and point-of-impingement standards. The ambient air standards include some 24-h standards, as well as 1-h and 10-min levels. Point-of-impingement concentrations are based on a 1-h average. Many of these standards, particularly those with 1-h and 10-min time periods, are for odorous compounds.

Northwest Territories

The Northwest Territories defines a contaminant as a substance, including odor, which can interfere with normal enjoyment of life or property, and there is a prohibition against the discharge of contaminants into the environment, except under prescribed conditions.

Nova Scotia

Nova Scotia prohibits the release of substances into the environment in such a way as to cause an "adverse effect." The definition of *adverse effect* includes an adverse effect on the "reasonable enjoyment of life or property."

Ontario

Ontario's Environmental Protection Act includes a general prohibition on the release of contaminants into the environment that is likely to cause an "adverse effect." The definition of *adverse effect* includes the following effects which might be caused by an odor nuisance:

- Harm or material discomfort
- Loss of enjoyment of normal use of property
- Interference with the normal conduct of business

The General Air Pollution Regulation (Regulation 346) under the Environmental Protection Act has two sections which address odor indirectly. One section is similar to the more general prohibition in the EPA—a prohibition on the emission of any air contaminant "to such extent or degree as may cause discomfort, loss of enjoyment of normal use of property, or interfere with normal conduct of business." Another section prescribes maximum point-of-impingement concentrations for a variety of compounds. A number of these are based on the odor potential of these compounds. Dispersion models are included in the regulation for calculating maximum point-of-impingement concentrations from emission rate data. In addition to the 87 compounds with point-of-impingement standards in Regulation 346, there are more than 100 additional compounds with interim standards or guidelines. There is a 3-year plan for the development of additional standards.

Odor issues are routinely addressed in permits (Certificates of Approval) in Ontario. Requirements for odor emission tests are often included as conditions for industrial sources which are judged, by the Ministry of the Environment and Energy, to have the potential for an odor impact. Emission test results are used with regulatory dispersion models in Regulation 346 to estimate maximum point-of-impingement odor levels.

An odor sampling and evaluation procedure was drafted by the Ontario Ministry of the Environment and Energy (MOEE) in 1989 (Ontario Ministry of the Environment, 1989). The procedure focuses more on sampling than on evaluation and emphasizes the use of dilution sampling ("predilution") to ensure that loss of odor does not occur due to condensation. The MOEE requires the collection of a series of samples at different dilution ratios from each emission source. These are each evaluated for odor detection threshold, and an optimum dilution ratio is chosen for collection of a final set of samples, where the optimum predilution ratio is the dilution resulting in the highest net odor detection threshold value.

Prince Edward Island

The Environmental Protection Act in Prince Edward Island includes odor in its definition of a contaminant, if it interferes with or is likely to interfere with the comfort, well-being, livelihood, or enjoyment of life of a person. The Air Quality Regulation under this act prohibits the discharge of a contaminant into the air from any industrial source without a permit. Provision is included for revoking or changing a permit for several reasons, including the belief that the continued operation of the source is "an interference to normal business" or "obnoxious to the public."

Quebec

Quebec has odor emission standards in the Quality of the Atmosphere Regulation under the Environment Quality Act. The relevant part of the regulation, Division VI: Emission of Odors, is as follows:

> Standard. All odors discharged by a fried food plant or coffee roasting plant, by a smoke house with a capacity of greater than 250 kilograms of meat per week, by a brewery, a distillery, a rubber recycling plant and an asphalt saturation plant shall be ducted and treated so that the concentration of odor discharged to the atmosphere does not exceed 120 degrees of odor per cubic meter.
>
> Odors emitted by the processes and general ventilation of a dismembering plant must be ducted and treated by equipment for the treatment of gas. The concentration of the odors emitted by that equipment must be less than 100 degrees of odor per cubic meter.
>
> The operating areas for the processes and the stocking areas must be located within closed premises and must be maintained under negative pressure.

The method for determining odor levels is currently stipulated in the regulation as ASTM D1391-57 (1967) in the 1974 book of standards.

There are a number of changes being considered to this regulation. One is a change to an off-property limit of 1 odor unit, based on dispersion modeling. There is ongoing review of olfactometry, and the Ministère de l'Environnement et de la Faune has contributed to the development of an advanced butanol reference olfactometer (Technodor) being marketed by Technovir International Inc.

Quebec also has specific emissions regulations for reduced sulfur compounds from kraft paper mills. Agricultural odors have been a concern in certain regions, and guidelines for separation distances have been developed, and are under active review.

Within the greater Montreal area, air quality is regulated by the Communauté Urbaine de Montréal (CUM). Regulation 90 of the CUM includes several provisions affecting odorous emissions. The most direct of these is a limit of 1 odor unit beyond the property limit for all stacks located on that property. This is calculated based on odor detection threshold determinations on samples collected from the sources and a dispersion formula given in the regulation. There are also point-of-impingement and ambient air limits for several hundred compounds, including many odorous species. In addition, there are specific requirements for levels of emission control for a variety of types of industrial processes, including common odor nuisance sources such as rendering processes, paint spraying, and asphalt operations.

Odor emission measurements are conducted by the CUM using an olfactometer which it has developed. The olfactometer uses a forced-choice triangle protocol, with a flow rate of 15 L/min.

Saskatchewan

Saskatchewan, in the Clean Air Act, defines air pollution to include any air contaminant "in quantities that are or are likely to: ...be injurious to the comfort or well-being of the public;...or interfere with normal business; or that has an offensive or obnoxious odor, regardless of its concentration." The Clean Air Regulations include 1-h (15 $\mu g/m^3$) and 24-h (5 $\mu g/m^3$) ambient air quality standards for hydrogen sulfide.

Yukon Territory

The Yukon includes odor in a definition of *contaminant.*

VOC REGULATIONS

Regulations regarding VOC emissions in general or specific organic compounds are currently found in several jurisdictions in Canada. Coordinated efforts for reducing VOC emissions have been developed through the Canadian Council of Ministers of the Environment (CCME) NO_x/VOCs Management Plan (CCME, 1990). This discussion is a brief summary of some of these regulations and initiatives.

Some of the most specific VOC regulations at present are found in Quebec. The Quality of the Atmosphere Regulation includes requirements for VOC levels in paints used in automotive assembly paint shops. Several jurisdictions Ontario, Newfoundland, and the Montreal Urban Community have ambient air criteria for a number of volatile organic compounds.

Smog reduction plans are being developed in several regions of the country. These form part of a National Smog Strategy. A National Smog Management Plan addresses products, processes and transportation, and other initiatives which are best considered at a national level. Regional Smog Management Plans are being developed in Ontario, Quebec, the Lower Fraser Valley in British Columbia, and the south Atlantic region (Nova Scotia and New Brunswick). The most advanced smog reduction plan is in the Greater Vancouver Regional District (GVRD) in British Columbia. Ontario has produced a draft plan, which includes a summary of regional initiatives and changes to the original 1990 CCME management plan (Ontario Ministry of Environment and Energy, 1996). A number of studies have been conducted through the CCME, leading to the development of Codes of Practice for different industries, e.g., automotive assembly paint shops (CCME, 1995).

REFERENCES

Canadian Council of Ministers of the Environment (CCME). 1990. *Management Plan for Nitrogen Oxides (NO_x) and Volatile Organic Compounds (VOCs)*, Phase 1, Ottawa, Ontario, Canada.

Canadian Council of Ministers of the Environment (CCME). 1995. *New Source Performance Standards and Guidelines for the Reduction of Volatile Organic Compound Emissions from*

Canadian Automotive OEM Coating Facilities, Ottawa, Ontario, Canada.

Manitoba Department of Environment. 1997. *Odour Nuisance: Proposed Management Strategy Background Paper,* Winnipeg, Manitoba, Canada.

Ontario Ministry of the Environment. 1989. *Source Sampling for Odours (Version 2) Draft,* Toronto, Ontario, Canada.

Ontario Ministry of Environment and Energy. 1996. *Towards a Smog Plan for Ontario,* Queen's Printer for Ontario, Toronto, Ontario, Canada.

CHAPTER 11
AIR POLLUTION CONTROL MANAGEMENT

Joseph P. Botos
Cargill
Minneapolis, Minnesota

LINKING ENVIRONMENT, HEALTH, AND SAFETY REQUIREMENTS, PRODUCTION REQUIREMENTS, AND ENVIRONMENTAL PROTECTION 11.2
ENVIRONMENT, HEALTH, AND SAFETY POLICY 11.2
Environment, Health, and Safety Management Systems 11.3
General Requirements 11.4
Environmental Protection 11.7
Occupational Health 11.11
Operational Safety 11.12
ENVIRONMENT, HEALTH, AND SAFETY EDUCATION 11.14
REPORTING SYSTEMS 11.15
AUDITING SYSTEMS 11.15
WASTE MINIMIZATION 11.16
RECOGNITION FOR EXCELLENCE IN ENVIRONMENT, HEALTH, AND SAFETY 11.16
INTERNATIONAL STANDARDS 11.18
CONCLUSIONS 11.19

A major challenge for any organization is to transform environmental policy from words to actions. Implementing the requirements of an individual process and protecting the environment at the same time are a delicate job. The requirements of the business unit to improve yields while being a low-cost producer steer one toward economic solutions. Corporate environmental responsibility now directs the process toward "zero emissions." Thus, balancing these requirements can seemingly create a conflict.

Environmental management requires an aggressive approach that considers both day-to-day and long-term effects of an operation. It requires active participation of all employees and regular interaction with customers, the community, and regulators. It requires an understanding of regulations of the local, state or province, and federal levels of government. Above all, it takes management commitment to establish and implement policy, education programs, reporting systems, auditing systems, and technical resource capability. This is the environmental management system that will enable the organization to improve yields, be a low-cost producer, and do it in an environmentally sound manner.

Managing environmental affairs is a critical issue for any organization because environmental issues can affect the employees, customers, and community in which the organization operates. Environmental issues have a financial impact by requiring people's time and capital. Environmental issues determine where businesses are located, how and when businesses are expanded, and which businesses survive in the marketplace. Environmental issues affect the products and services produced. It takes a top-down, bottom-up commitment in the organization. Each organization must do those unique things that will create the opportunities for long-term success.

The following presentation outlines the environmental management process which has been developed over several years at Cargill. It is a system that works, in that it has shown substantial environmental improvements and a mobilization toward constructive improvements. It is also an example from which certain basic principles can be learned. In the following sections, the program is discussed, and in the "Conclusions" section, certain key principles for an effective program are highlighted.

LINKING ENVIRONMENT, HEALTH, AND SAFETY REQUIREMENTS, PRODUCTION REQUIREMENTS, AND ENVIRONMENTAL PROTECTION

Designing, building, and operating plants and processes that protect the environment and operate in a cost-effective manner are critical to an organization's success. Linking corporate environment, health, and safety requirements with production requirements and environmental protection is commonly called *sustainable development.* The corporation that can successfully link these requirements will achieve a competitive advantage in the marketplace. Sustaining a competitive advantage means survival.

How does an organization build this link? Each organization is unique. It must develop its own unique approach to managing environmental and health and safety issues. There are individual differences in company size, products and processes produced, the industry the organization labors in, the regulatory requirements, the local or global aspects, and the customer base.

At Cargill, we have a vision to raise living standards around the world by delivering increased value to producers and consumers. The Corporate Environment, Health, and Safety Department was developed with a mission to establish worldwide basic requirements, education programs, information reporting systems, auditing systems, and a waste minimization process and to provide an environment, health, and safety technical resource capability.

At the other end of the spectrum, we established a Cargill President's Award for *environmental excellence* and for *safety excellence.* This recognition system calls for setting goals, measuring progress, demonstrating management commitment, implementing sound business practices, and encouraging employee and community involvement and compliance.

Linking Cargill's vision and the Corporate Environment, Health, and Safety mission to action and recognizing environment, health, and safety excellence is the management system that Cargill believes will give us the competitive advantage we must have for survival.

While environmental issues and health and safety issues are different, they are not mutually exclusive. Environmental issues usually end up outside the plant gate. Health and safety issues are normally contained inside the plant gate. However, this is not as clear-cut as it appears since gaseous emissions affect the working environment as well as neighbors. Both issues require a similar management.

ENVIRONMENT, HEALTH, AND SAFETY POLICY

The first step in our journey was to establish an environment, health, and safety policy. The policy started the foundation on which we built our environmental management systems. "Cargill's policy is to conduct all business activities in a manner that protects the environment and the health and safety of our employees, contractors, customers, and the public."

- *Environment:* Comply with all applicable laws and regulations; promote waste reduction, resource and property conservation, and environmental protection; and train employees to be knowledgeable about all environmental matters relevant to their work.
- *Health and safety:* Design, construct, maintain, and operate facilities to protect human and physical resources; provide and require use of adequate protective equipment and measures; and insist that all work, however urgent, be done safely.
- *Food and product safety:* Provide safe food, feed, and materials for people and animals by using production, storage, and transportation facilities that ensure the safety of products; choose suppliers that meet our requirements for safe raw materials and ingredients; and provide employees the training and resources they need to meet their individual responsibilities.

This policy guides the development of basic requirements that cover environment, health, and safety management systems, general requirements, environmental protection, occupational health, operational safety, vehicle safety, and food and product safety.

Environment, Health, and Safety Management Systems

The basic requirement is that environment, health, and safety management systems be designed, implemented, and maintained to ensure that all business activities comply with government and internal requirements. Supplementary requirements include eight elements:

1. *Environment, health, and safety goals.* Establish appropriate environment, health, and safety goals and include those goals in the strategic plan of each business unit. A business unit may be described as any particular activity, plant, office, or laboratory. By integrating environment, health, and safety goals into the business unit strategic plan, the organization is assured that environment, health, and safety issues will receive attention at all levels of the organization.

2. *Accountability.* All employees of the organization must be advised of the environment, health, and safety requirements of the organization and be held accountable for their individual performance. Everyone in the organization must know what effect his or her work has on the environment for a business unit to succeed in meeting the environment, health, and safety goals.

3. *Operating standards.* Each business unit must establish an ongoing environment, health, and safety management process to enhance knowledge, improve technology, and comply with government and internal requirements.

4. *Education.* Each business unit must provide for employee education on environment, health, and safety issues to protect environmental, human, and physical resources.

5. *Design and management of operations.* Each business unit must design and manage operations to minimize adverse environmental and health effects and to provide a workplace free of recognized safety hazards. By taking into consideration the impact of environmental or safety issues, the organization can design in preventive action.

6. *Products.* Each business unit must include environment, health, and safety considerations in the development and production of new and existing products.

7. *Management and technical support.* Each business unit must provide knowledgeable management and technical support for environmental protection and health and safety efforts.

8. *Performance measurement.* Each business unit must understand and report its performance in environment, health, and safety activities through reviews, inspections, reports, and audits, as appropriate, designed to evaluate compliance with these requirements, measure progress, establish goals, and provide assurance that products and operations comply with applicable government and internal requirements and that deficiencies are identified and promptly corrected.

These elements are the foundation of an environment, health, and safety management system. An environment, health, and safety system by definition is leveraging environment, health, and safety management issues for competitive advantage. A cost-effective environment, health, and safety management system integrates environment, health, and safety considerations into all relevant business processes. There is accountability for environment, health, and safety performance throughout the organization. The result is continuous improvement in environment, health, and safety performance of the organization. It takes strong corporate commitment to allow this to happen.

General Requirements

General requirements include emergency action plans, responsibility for on-site contractors, capital project review, acquisition and divestiture review, product stewardship, plant inspection, housekeeping and sanitation activities, incident reporting, and auditing.

Emergency Action Plans. The purpose of a written emergency action plan is to cause each location to evaluate its individual situation and identify all potential emergency situations. These emergency situations include accidents, fires, explosions, process upsets, uncontrolled hazardous material reactions, unplanned releases of hazardous or nonhazardous materials, bomb threats, natural disasters, or other external factors including hazardous material releases from neighboring facilities.

The plan should include a written communication plan that enables the location to tell its own story through the media and to the local community or affected public promptly, accurately, and as thoroughly as possible. A capable location spokesperson and backup should be assigned. A written emergency notification procedure that lists names and phone numbers of local managers, the National Emergency Response Center, other local and state government authorities, and other appropriate agencies and personnel should be maintained. A background file of key information about the location should be in place as well.

The plan should contain floor plans, maps, or listed locations of emergency escape routes and means of egress. An effective alarm or signaling device should be included. The plan should include an inventory of the types and quantities of materials stored at the site. A system to account for people should be in place.

Education on the purpose of the emergency action plan and the duties of all plant, office, and laboratory personnel should be conducted, and regular drills should follow to ensure that the plan will work. Fire departments, rescue teams, and similar organizations should be invited to participate in these drills. When an emergency happens, people must automatically know what their duties are. It is too late to read the book when the event is going on.

On-site Contractors. Programs should exist and provide assurance that independent contractors perform their work in compliance with government regulations and internal requirements. To provide this assurance, a system should be in place to evaluate a contractor's past performance. While on site, the contractor's employees must know and understand location requirements. This is accomplished by setting up the criteria in advance, striking an agreement on the requirements during negotiations, and holding the contractor accountable while the work is performed.

Capital Project Review. A process to identify the environment, health, and safety impact of capital projects will provide assurance that the projects comply with applicable government and internal requirements and do not present unacceptable risks to the environment or health and safety of employees, customers, and the community. This is a system to review capital projects to identify potential environmental, health, and safety hazards and to identify the potential of operational environmental, health, or safety events; chronic adverse effects on employees, customers, the community, or the environment; and noncompliance with government and internal requirements. The reviews should be documented and include the title and description of the project, names and location of person(s) responsible for performing the review, and a report of findings and conclusions, including

- Assessment of potential environmental, health, or safety risks
- Assessment of potential adverse effect on employees, customers, the community, or the environment
- Assessment of compliance with government and internal requirements including permits needed, how permits will be applied for, and when they will be obtained
- Assessment of what wastes will be generated, what waste minimization techniques will be implemented, what waste disposal methods will be applied, and what impact the project is making on waste minimization for the entire facility
- Corrective action and control measures identified in the review should be included as conditions placed on the project

Recommendations, terms of endorsement, and any conditions placed on the project should be summarized and included with the request for funding of the capital project.

Acquisition and Divestiture Review. All acquisitions and divestitures of a business or real estate should be reviewed to identify environmental issues and evaluate any potential risks or liabilities. A written summary of the review should include

- Title and description of the project
- Name and location of person(s) responsible for performing the review
- Review of findings and conclusions, including

  Assessment of potential environmental risk

  Assessment of impact of past and present operations or conditions of the property on employees, customers, the community, or the environment

  Assessment of compliance with government and internal requirements
- Recommendations, terms of endorsement, and any conditions placed on project

The acquisition or divestiture review gives a baseline of information about property you are acquiring or selling. In most cases, these reviews should be conducted by environmental consultants who are experts in this process. This baseline of information allows appropriate decisions to be made relative to property. It is helpful to use a form or checklist to develop consistent information in this process. The checklist should be considered a guideline. The checklist should cover

- Site history, including ownership and uses of the property
- Permit status for air, water, waste of chemical use
- Information requests that cover environmental compliance history of the site
- Site visit results, including the environmental sensitivity of the location; proximity of neighbors relative to air issues; water situation, including sources of water used, methods of water disposal, and the proximity of surface water; any visible contaminated soil; waste disposal methods; chemical and spill prevention management and whether any asbestos or polychlorinated biphenyl (PCB) equipment is there, any pesticides used now or in the past, and whether any nuclear devices are used.

Product Stewardship. Environment, health, and safety considerations should be incorporated into the decision process used to develop new products and services, modify existing products, and/or acquire new business or product lines. Formal procedures should be established to review product safety risks. The review incorporates all stages of the product life cycle and considers foreseeable accidents and product misuse. Hazard evaluations should be conducted by knowledgeable persons who review available safety literature and, where necessary, conduct research to determine any significant product safety or environmental hazards. Products should be reevaluated whenever new potential hazards are suspected or identified. Product hazard evaluations must be conducted in accordance with applicable government requirements.

Plant Inspection. A program should be established outlining the practices and conditions to be checked during routine plant inspections. A hazard classification system should be established to prioritize any substandard conditions identified during the plant inspection so that needed corrective action is taken on a priority basis. The plant inspection should be conducted by qualified plant personnel (such as the manager and superintendent), who are adequately trained to identify the practices and conditions and prioritize the needed corrective action. A written report of the inspection, including the substandard conditions and practices which have been properly classified, should be submitted to the top management person at the plant for review, and needed corrective action scheduled, completed, and documented.

Housekeeping and Sanitation. A written program should be developed to establish the standard of housekeeping and orderliness. The written housekeeping program should be designed to establish the standards of housekeeping and orderliness, identify the areas of the plant or office to be cleaned, establish the frequency of cleanup for the identified areas, and assign responsibility for maintaining the established standards. Pest and rodent control activities that supplement housekeeping and sanitation efforts should be conducted only by licensed applicators.

Environmental Activity Reporting. Environmental activity should be reported to location management immediately and an incident investigation started within 8 h. Environmental activity includes spills and releases to the environment that are reportable to government agencies, complaints, and inspections. For any of these types of activities, an environmental activity report stating facts relevant to the activity should be completed within 24 h.

In the event of any type of spill or release to the environment, the local plant management should follow a procedure similar to this:

> Secure the source of the spill or release to prevent any further damage. Immediately document the facts relevant to the spill or release, including the location, quantity, time, and

exact chemical composition of the spill or release. If the incident is reportable to any local, state, or other government agency, the person in charge of a facility at the time of the spill or release or his or her designate must immediately contact those relevant regulatory agencies and report the incident.

When complaints are made about an operation, determine who is making the complaint and the nature of the complaint. Investigate the situation, and if the situation warrants, develop immediate corrective action to alleviate the problem.

Inspections by regulatory agencies will occur. It is usually beneficial to cooperate with government inspectors making reasonable routine inspections. Local management should be knowledgeable about laws and regulations pertaining to the operation or business. Require the inspector to present appropriate agency credentials and an official Notice of Inspection, and outline the purpose of the inspection. Exercise the right to protect confidential and proprietary information and the right to reasonably limit the scope of the inspection. Inspectors must follow local safety requirements and be accompanied at all times.

Auditing. An effective internal audit system will accomplish several purposes. It is a measurement of compliance with internal and external requirements. It is an educational tool both for the audit team and for the location being audited. It provides needed information to management, determines on-site conditions, evaluates regulatory compliance, identifies toxic or hazardous substances on site, evaluates emergency needs and capability, provides legal background, and reduces liability potential.

The scope of an environmental evaluation should cover site history, permit status, compliance status, and a site visit that includes an evaluation of the environmental sensitivity of the location; should cover any air, water, or waste issues that are present; and should evaluate prevention programs and hazardous material and waste management processes in place.

The basic cause approach should be used in the audit to identify problems, analyze exceptions for cause and effect, develop corrective action, and communicate the results.

Consultants can be used to do regulatory analysis and site evaluations and to conduct environmental media analysis and occupational health assessments or other risk assessments. An effective consultant will have specific industry capability, listen to your needs, and provide information that is needed and can be used.

Environmental Protection

At Cargill, we have divided our basic environmental protection requirements into the elements of waste management, air emissions, surface water protection, groundwater protection, polychlorinated biphenyls, asbestos, and liquid storage tanks and secondary containment. We believe the management of these elements will ensure that each business unit will operate in a manner that will not adversely affect the health and safety of our employees, the public, or the environment.

Waste Management. The intention is to both minimize waste and manage waste that is created. Wastes generated from normal activity must be identified and summarized. An inventory that includes the description of waste generated, quantity generated, sources of the waste, and treatment, storage, recycling, or disposal methods used should be developed. Waste minimization techniques that reduce the total waste should be identified. Techniques include reduction at the source, recycling, reuse, and similar practices. An assessment of waste minimization opportunities should be incorporated in the design of new processes or operations.

A system to manage the selection and use of practical treatment, storage, recycling, or disposal methods to minimize risks to human health and the environment should be developed. Procedures for waste transported off site should include the following:

- Written agreements are reviewed for legal and technical acceptability.
- Waste contractors and their equipment and facilities should be reviewed and selected on a sound business and technical basis.
- Waste contractors should be provided adequate technical information about the waste.
- Disposal of hazardous and special wastes should be reviewed by a knowledgeable party external to the location.

An annual report should be developed at each location that describes and quantifies the waste generated, waste minimization techniques used, annual waste reduction achieved, and future goals.

This forms the basic requirement of a waste management process. It allows the location to identify waste, focus on waste minimization targets, and provide necessary data on wastes that are properly managed.

Air Emissions. Air emissions from a location must be characterized and controlled, and where needed, a operating permit must be in place.

Air emission inventories should be developed and include at least the following:

- Nature of the substance emitted
- Source of the emission
- Quantities emitted under normal operating conditions, including start-up and shutdown
- Measurement protocol
- Toxicity level
- Summary of applicable legal requirements

An emission evaluation program should be established for all emissions of extremely hazardous substances. The evaluation should include

- An assessment of the impact of the emissions on the environment
- A description of the control technology used
- Identification and implementation of appropriate actions to reduce emissions

Each location should submit an annual summary of air permit status to product line management. A process should be established that allows for a periodic review of the status of all permits. This can be in the form of a list or operating logbook or some means to allow the operator to know what the requirements of the permit are and when the renewal is due.

Surface Water Protection. This element is focused on ensuring that operations do not have an adverse effect on surface water and that any discharge of process, sanitary, or storm water is permitted or authorized. An inventory should be developed that specifies the types, quantities, and discharge points of potential pollutants that are or could be discharged to surface waters from the operation. An assessment should be developed that describes the existing surface water volumes, flow, quality, present and future uses, and government classification for the potential affected surface waters. An evaluation should be completed that describes potential discharges, corrective action needed and implemented, and prevention control and control measures needed or in use. Engineering design of new and modified equipment and facilities should seek zero discharge. Each location should maintain a summary of water discharge permit status.

Groundwater Protection. This element is aimed at preventing adverse effects to groundwater from an operation. Develop an inventory identifying location activities including

- An assessment of current operations and existing groundwater quality, uses of groundwater, and other relevant information
- Appropriate remedial or control measures that are implemented to prevent adverse effects of operations on groundwater

Engineering design of new or modified operations and facilities should minimize both use of groundwater and releases to groundwater. Each location should maintain a summary of groundwater permit status.

Polychlorinated Biphenyls. PCBs pose a potential environmental risk that can be managed. Each business unit should conduct a thorough evaluation to determine if PCBs are present and, if so, in what form and concentration.

A documented survey should be completed, and all electric units with liquid dielectric coolants should be certified as to PCB concentration. All PCB units (over 500 ppm PCB) should be removed. In the event physical circumstances at the business unit prohibit removal, retrofit technology may be used. All PCB-contaminated units (50 to 500 ppm PCBs) that pose a potential problem of contamination to food, feed products, or the environment should be removed. All PCB-contaminated units not posing a potential problem of contaminating food, feed products, or the environment must be monitored per regulatory requirements. All PCB units and PCB-contaminated units must be labeled and registered with the local fire response personnel. New or replacement electric power units containing a liquid dielectric coolant must be certified "PCB-free" by the manufacturer.

Any leaks or spills from PCB units or PCB-contaminated units should be reported to the appropriate regulatory officials immediately, and cleanup begun within 24 h. Cleanup should be in accordance with appropriate regulatory spill policy.

Each location should maintain, at a minimum, the following records:

- Certification of all units
- Plot plan showing location of all liquid dielectric coolant units with any PCB units or PCB-contaminated units conspicuously marked
- Logs of all inspections
- Notations of any unusual circumstances
- Annual report of PCB use, storage, and disposal activity
- Annual records and annual documentation logs if required by regulations
- PCB waste disposal records and reports if required by regulations, including manifests, certificates, exception reports, discrepancy reports, and confirming telephone logs

Any PCB abatement, removal, or disposal should be under contract, and the following requirements incorporated into any such contract:

- Any contractor providing transportation, interim storage, or ultimate disposal of PCBs should list, in detail, all materials being handled or disposed.
- Each originating location should establish a minimum time from date of contract for transfer of materials.
- The contractor should provide supervision for handling of PCB materials and appropriately trained personnel for the draining, removal, and loading. The contractor should accept all liability and responsibility to complete the project.
- The contractor should accept legal title to the PCB material when the material leaves the plant site.
- The contractor should provide all necessary and applicable completed uniform hazardous waste manifests.
- The contractor should provide complete details regarding the identity of the transporter of the material, the site of storage and disposal, the method of disposal, the regulatory identification numbers of the carriers and disposal facilities, and any other information that may be requested.
- The contractor should provide a completed Certificate of Destruction/Disposal before payment for services is made.

Asbestos. Again, the purpose of this element is to ensure that asbestos is identified and, if it exists, its form and condition. Asbestos can be managed in place. If it is disposed of, it must be done properly. A documented survey of the location should be completed. Asbestos or asbestos-containing material (ACM) is often found in cement products, acoustical plaster, fireproofing textiles, wallboard, ceiling tiles, vinyl floor tiles, thermal insulation, and other materials.

Sampling and analysis:

- Samples for analysis must be taken carefully to ensure that airborne dust is not generated.
- Documentation of where samples were obtained should be maintained on file.
- Samples must be sent to licensed analytical laboratory for the preliminary analysis.

If sample material contains asbestos, do the following:

- Determine whether the asbestos material will be removed or will remain in place. Consideration should be given to the future likelihood of exposure and need for later removal.
- If it is determined not to remove the asbestos or ACM, it must be either in a nonfriable form or properly covered according to local regulations so that no asbestos fibers can become airborne. The surface covering the asbestos must be in good condition at all times in all areas. Periodic inspection of the covering is necessary to ensure that it has not been broken or has not deteriorated. If the surface covering has deteriorated, necessary repairs or replacement must be made. Records of the inspections and repairs should be maintained.
- If it is determined that the asbestos will be removed, use an experienced contractor qualified in asbestos removal and disposal, and obtain a signed contract before any work is commenced.
- Prior to releasing any contractor from an asbestos removal or encapsulation contract, the contractor should provide documentation that asbestos fiber concentration in the air does not exceed local regulations.

Liquid Storage Tanks and Secondary Containment. The primary purpose of this element is to ensure that liquid storage tanks are identified and that secondary containment is provided where the material in the tank has the potential to cause an adverse effect on people, animals, property, products, or the environment. A list of all aboveground, below-ground, temporary, and mobile storage tanks should be maintained. For each tank, identify the size, material of construction, type of construction, design specifications, original fabrication date, and subsequent relocation and remodel dates. All changes and modifications to storage tanks should be properly engineered and changes approved by structural experts. Every 5 years, evaluate the structural integrity of existing tanks to determine whether they are suitable for present and continued use.

The location of each tank holding hazardous or flammable liquids should be plotted in relation to property lines for buildings, public roads, major buildings, mainline railroad tracks, parks, schools, institutions, and sources of ignition (such as vaporizers, incinerators, driers) to determine if they meet current code requirements. Note prevailing winds on the plot.

A list of secondary containment systems for each aboveground tank should be maintained. This list should include the volumetric capacity, materials of construction (earth, concrete, concrete block, steel, plastic, wood) and the method of drainage. Secondary containment capacity should be at least 110 percent of the tank protected.

A list of items and conditions that affect product containment and safety should be maintained. Typical items include insulation, atmospheric vents, pressure and vacuum relief devices, rupture disks, tank lining, high and low liquid level-sensing gauges, excess-flow alarms, breakaway connections, pressure-sensing alarms, and incompatible products stored.

Where multiple tanks are protected by a common system, the secondary containment capacity should equal the capacity of the largest tank plus 10 percent. Take into consideration the space occupied by the tanks. Provisions for draining or pumping both rainwater and spilled products from within the secondary containment area should not affect the integrity of the secondary containment structure. Location of emergency valves should be outside the secondary containment. These valves should be designed so that they can be locked closed.

Occupational Health

Occupational health programs are intended to protect employees from adverse health effects that may occur from overexposure to physical, chemical, and biological agents in the workplace. An analysis to identify significant potential health hazards (acute and chronic) which may result from plant operations and processes should be completed. Sources of information used to identify hazards include known health hazards associated with physical, chemical, and biological agents in the workplace. In addition, a process hazard analysis and/or management of change review can be used. The results of industrial hygiene exposure assessments and/or monitoring and results of employee medical surveillance and/or epidemiology studies, plus worker health complaints and concerns, are all sources of information.

Written exposure assessment programs should be developed which provide guidelines on how employee exposures to potentially hazardous chemical, physical, and biological agents are evaluated. The program should address the responsibilities for conducting the exposure assessments, guidelines for conducting both qualitative assessments and exposure monitoring, and methods used for data interpretation and analysis. Then formal qualitative exposure assessments should be conducted within the workplace to identify jobs, processes, or procedures which have the potential for employee overexposure.

Where the potential for employee overexposure exists, representative industrial hygiene monitoring should be conducted to quantitatively determine employee exposures. Exposure assessments and monitoring should be conducted by a knowledgeable, qualified person, and assessments should be reviewed annually.

Employee exposure to chemical, physical, and biological agents should not exceed government limits, and where government limits do not exist, consideration should be given to supplier recommendations or exposure values established for occupational health organizations or governmental agencies.

Exposure assessment and monitoring records should be maintained and made accessible to employees in accordance with any government and internal requirements.

Health and safety design reviews should be conducted prior to construction of a new process or proposed modification of equipment, work practices, or materials to minimize potential employee exposure to chemical, physical, or biological agents.

Where potentially significant occupational health hazards are identified, immediate actions should be taken to reduce employee exposures within acceptable limits. Methods to accomplish this include substitution for less hazardous or toxic materials, engineering design modifications, administrative controls including work practice modifications, the use of personal protective equipment, and removal of workers at risk.

Procedures and practices should be in place to ensure hazard information is appropriately communicated and that employees are trained in safe work practices and relevant health and safety policies and requirements.

Information regarding the potential hazards of chemical, physical, and biological agents in the workplace can be communicated through a variety of methods, including the results of exposure assessments and monitoring relevant to their work assignments; education on product safety literature and information; the potential significant hazards of materials used in the workplace; protective controls or practices to be utilized to avoid adverse health effects; and actions to be taken in foreseeable emergency situations. Training should be repeated annually or when new potential hazards have been introduced or identified.

The safe use of respiratory protection is part of an overall program to protect employees from overexposure to chemical and biological agents. Exposure assessment information is utilized as the criteria for respiratory selection and is based on the hazards to which employees can be exposed. Periodic surveillance of work area conditions should be conducted to ensure proper respirator usage.

Where respiratory protection is utilized, written standard operating procedures should be developed and should address several issues: respirator program responsibility and authority, specific respirator selection criteria utilized for tasks requiring respirator use, procedures for employee training and instruction, respirator cleaning and disinfection, inspection and storage, proper use and fit testing requirements for all respirators, and restrictions on use of respirators when facial hair or other factors are present which may interfere with respirator face fit. Prior to an assignment to a task requiring the use of respiratory protection, medical assessments should be conducted to provide assurance that an employee is physically capable of wearing the respirator without suffering adverse health effects.

Operational Safety

The umbrella that we have chosen to place our operational safety program under is *process safety and risk management* (PSRM). A properly executed PSRM process will provide assurance that potential operational hazards are identified, that operational safety incidents are avoided, that operational safety requirements are considered in evaluating business decisions, and that unacceptable risks to persons, property, products, and the environment are avoided. The process should include all phases of plant operations from design to decommissioning.

Employee input and involvement in the analysis of process hazards and the development of accident prevention plans should be encouraged and utilized. Employees should have access to all PSRM information affecting the safe performance of their duties and tasks.

Information should be developed on materials and substances used in the process to allow for a thorough process hazard analysis and the completion of other loss prevention activity. This information would include toxicity and properties inherent to the materials or substances; operation specifications of the process and the consequences of their deviation; process and equipment technology, diagrams, and flow sheets; and design specifications, codes, and standards employed.

Documentation verifying that equipment is being operated, inspected, and maintained in a safe manner and complies with recognized and generally accepted good engineering practices needs to be in place.

Processes should be assessed to identify, evaluate, and control potential hazards. Risk assessments should be performed and documented to determine the need and priority order for process hazard analyses. The assessment should take into account the extent of the hazard, number of potentially affected individuals, age of the process, and operating history.

Process hazard analyses should be performed by a team sufficiently trained to utilize a methodology appropriate for the complexity of the process and potential risk. Systems should be in place to promptly address findings and recommendations, developing a written schedule for completing any actions to be taken, and communicating those actions to affected employees. Process hazard analyses should be revalidated at intervals appropriate for the hazard of the process or based upon other internal or government standards.

A *risk management plan* (RMP) should be prepared and implemented for processes utilizing hazardous materials to prevent or minimize the results of accidental releases and provide prompt response when releases occur. Factors to consider would include an assessment of potential effects of an accidental releasee, a program for preventing accidental releases, and a response program for actions to be taken following a release.

Written operating procedures that provide clear instructions for operating the process safely would include steps for all operating phases and conditions, including foreseeable emergencies; critical operating limits and consequences of their deviation; environment, health, and safety considerations of the materials and substances used in the process; critical safety devices or systems and their function; and confined-space entry, welding and burning, and lockout and tag-out permit systems.

As necessary, additional safe work practices should be developed to control activities by contractors, maintenance, laboratory, or other support employees around the process. Operating procedures and safe practices should be updated as needed and certified annually to ensure they are accurate and current. Contractors' safety programs and performance should be evaluated prior to the contractors being awarded work. Environment, health, and safety implications of a contractor's activities should be reviewed with the contract employer prior to the commencement of any work. While on site, contractor's safety performance should be evaluated and documented.

Education on applicable tasks and procedures should be provided to all employees to enable them to perform their jobs safely and effectively under a variety of operating conditions. Education should include a comprehensive verification and skill proficiency evaluation prior to task assignment, and it should be repeated at least annually.

Prestart-up safety reviews should be conducted for all new and modified process construction to ensure that construction and equipment are in accordance with design specifications, all operating procedures are in place, and any necessary employee training is complete.

For new construction or modified equipment, a process hazard analysis should be performed and recommendations resolved prior to start-up.

Testing, inspection, and maintenance procedures should be developed and implemented, which are consistent with applicable codes and good engineering practice, to ensure the mechanical integrity of critical process equipment is maintained. Testing and inspection documentation should be retained which identifies the equipment tested, the test or inspection method used, and test or inspection results. Maintenance personnel should be trained in the procedures applicable to their job tasks. A predictive or preventive maintenance program should be established.

Critical equipment deficiencies must be corrected before further use, and steps taken in a timely manner to ensure safe operation. Checks and inspections of new plants should be conducted to ensure the equipment is designed, installed, and maintained according to the specifications of the process applications.

A permit system for cutting and welding should be established consistent with recognized fire protection practices for cutting and welding and other hot work operations on or near processes involving high-hazard chemicals or substances.

Procedures should be developed and implemented to ensure a suitable review of any process modification which requires a change in process safety information. The complexity of the review should consider the technical basis for change, impact of change on health and safety, modification to operating procedures, item required for change, and authorization procedure required to initiate the change. Modifications to operating procedures, process safety information, and employee training procedures should be made prior to start-up of the process.

All process-related incidents or near accidents which did or could have resulted in an environment, health, and safety incident should be investigated immediately. Reports should ascertain information including a complete description of the incident, the immediate and basic causes of the incident, and the recommendations or actions needed to prevent recurrence. Systems should be in place to promptly address findings and recommendations, document actions to be taken, develop a written schedule for completing any actions to be taken, and communicate those actions.

An emergency action plan should be in place for the operation which clearly defines plant personnel duties and responsibilities to mitigate the consequences of any incident that has the potential to impact people or the environment. Employees should be trained to fulfill their roles and responsibilities during an emergency.

PSRM audits should be conducted at least every 3 years to assess PSRM implementation and effectiveness. More frequent audits will be based upon the risks, the process, and any government or internal standards. All audit findings should be resolved.

ENVIRONMENT, HEALTH, AND SAFETY EDUCATION

In the 1970s Cargill management determined that a basic seminar on environmental, health, and safety issues was needed to help employees understand the environment, health, and safety requirements and their responsibilities. Over the years this education effort has evolved into a Cargill Environmental Seminar and a Cargill Safety and Loss Control Management Seminar. More than 20,000 employees have attended these seminars, which are key parts of the learning process for employees.

The Environmental Seminar agenda consists of modules on Cargill environmental requirements and a review of the Clean Air Act, Clean Water Act, Resource Conservation and Recovery Act, Superfund, Hazardous Materials Transportation Act, and the Toxic Substances Control Act. The learning process is facilitated by a combination of case studies, films, lectures, and audience participation.

The two-day seminar is taught by personnel from Cargill environment, health, and safety; law; public affairs; and plant operations. This enables the discussions to focus on Cargill examples led by Cargill experience. Attendees at these seminars are shown the link between Cargill requirements and environmental regulations and a practical approach to compliance.

Our Safety and Loss Control Management Seminar focuses on modules covering causes, effect, and control of loss; incident investigation; critical tasks; safety behavior management; process safety; inspection systems; controlling property damage, measuring progress, and building an effective safety system.

These two seminars are integral to our continuous improvement process in environment, health, and safety.

REPORTING SYSTEMS

Preventing spills or releases that have the potential to adversely affect the environment or health of people or animals is very important. Spill-reporting plans and procedures should be written and implemented, and follow-up drills conducted. An internal reporting system should be in place and used as a management tool.

An Environmental Activity Report should be completed by all facilities and submitted to appropriate management within 24 h of a spill or release. Events which would require an Environmental Activity Report include

- All spills or releases. A log for recording information should be provided as an addition to the report. This log can be used to record the names of the people at any regulatory agency where the incident is reported, what information was given, what questions were asked, any report numbers the agency would give to the spill or release, and any other pertinent information.
- Complaints, from neighbors or from government agencies.
- Any type of environmental inspection.
- Notification of noncompliance or violation, administrative orders, or any other type of enforcement action, including any government request for information.

AUDITING SYSTEMS

An internal audit or environmental assessment is another management tool that should be used. The audit system can be used to review compliance with internal and external requirements. The audit protocol should be developed and should focus on identification of issues, developing a corrective action plan, implementing that plan, and following up to ensure that the corrective action worked. The audit protocol can cover several areas including administrative areas; waste management; hazardous waste management; air, water, groundwater, and surface water protection; spill prevention; chemical handling and hazardous substance reporting; pesticide use; underground storage tanks;

PCBs; asbestos; radiation devices; and the Toxic Substances Control Act. Locations should be audited once for the baseline data and then with a regular frequency depending on risk.

WASTE MINIMIZATION

Waste minimization or pollution prevention is an opportunity to improve productivity of people, plants, and processes; increase product yields; decrease waste disposal costs; and conserve energy and resources. Total quality and pollution prevention go hand in hand. The management of quality and the management of waste both deal with the same techniques. Pollution prevention can be used as a project for quality improvement teams or quality work groups.

Pollution prevention provides opportunities for employee involvement, employee empowerment, team building, improved public-community relations, improved customer relations, as well as reduction in costs of raw materials and disposal. With opportunities like these, planning is a must to realize the potential successes. Plans need to be site-specific and formalized so that all employees can understand the plan and be given opportunities for involvement in the process.

A successful pollution prevention plan begins with a statement of commitment and support by senior management, usually corporate as well as division and local management. It should include some operation- and site-specific definitions. The definitions become the critical terms used in the program. The plan outlines the goals that are set for the program as well as a measurement system that can be meaningful and useful to track progress of the process. The plan often sets the short- and long-term priorities and strategies that give specific direction and guidance to the process. Last, the plan outlines both the internal and external communication channels that can be used to stimulate interest, participation, and recognition.

Pollution prevention is an essential element of any short-term or long-term strategy for environmental management. You must have management interest and support to begin any pollution prevention efforts. Pollution prevention requires careful planning, creative problem solving, changes in attitude, capital investment, and a real commitment at all levels in the organization. You must have a dedicated team representing all functions at a facility working together to accomplish your pollution prevention goals.

Use the quality process. The management of quality and the management of pollution prevention both utilize the same techniques. You must define your goals, establish requirements, educate all employees, and measure progress.

RECOGNITION FOR EXCELLENCE IN ENVIRONMENT, HEALTH, AND SAFETY

An effective environment, health, and safety management system is a combination of procedures and beliefs. The procedures are the visible expressions of the objectives of the company to operate safely and in an environmentally sound manner. Procedures aim toward predictability. The beliefs are what people are thinking as they do their jobs. Beliefs result in day-to-day behaviors that individuals exhibit. If people do not believe in the objective of the organization, they will not perform in a consistent manner.

The environment, health, and safety recognition program that we have chosen is both a reward and an incentive. The incentive is to achieve the recognition. The reward is a simple plaque, along with inclusion in corporate videos, annual reports, and other communication tools as an example of excellence.

The direct recognition is nonfinancial. We aim the recognition at the product line level to encourage participation from the entire product line team. It stands to reason that to achieve recognition as an excellent performer, you must have all participants on your team involved. Participation yields predictability. Predictability yields customer and community and corporate respect which does have financial implications. The result of environmental excellence helps improve organizational effectiveness by motivating employees and increasing employee commitment.

This will result in opportunities for the organization.

Our recognition system is built on a system that includes six basic requirements:

1. *Environmental risk reduction goals.* The business unit should establish specific goals for reducing permitted discharges to water, air, and land; reducing waste of product, energy, and natural resources; and reducing unintentional releases to water, land, and air. Goals need to be measurable. Using past performance to set future goals is a recommended practice (that is, "20% reduction in reported releases on SARA chemicals from 1994 levels," rather than "reduce SARA reportable releases"). A waste minimization plan would list waste sources and quantities and set specific reduction goals.

2. *Clear and consistent measures of environmental performance.* The business unit should establish a system to measure progress against risk reduction goals. The information developed must be communicated in a clear, concise manner to management and to the Corporate Environment, Health, and Safety Department. Performance is regularly measured and reported. This can be done in a monthly recap or in an annual report. An annual report on waste minimization and pollution prevention is one example of how to meet the requirements of this section. Performance should be compared to goals and other benchmarks. Comparison with several past years to show positive trends is a plus.

3. *Environmental commitment and support.* The risk reduction goals and measurement systems should be clearly understood and supported at all levels of the organization. There should be regular communication from the top to facilitate achievement of the goals. Top management of the business unit demonstrates support of risk reduction goals by its actions. Evidence of this support should be provided. This can be in the form of meeting minutes, cover letters for capital requests, or other actions from top management used to communicate support.

4. *Environmental sound business practices.* The business unit should integrate environment, health, and safety into its strategic business plan and consider environmental factors in deciding what products to manufacture, what materials to use, and what industrial process to adopt. The business unit should have in place a system to determine the life-cycle effects of the resource used, products manufactured, and waste. A cost accounting system to determine the costs of regulatory compliance, liability costs, allocation of environmental costs to processes and production, and related environmental costs also should be established.

The requirement is to demonstrate that environment, health, and safety goals are included in the strategic business plan and to describe how the facility incorporates environment, health, and safety principles in the product development process and how risk reduction goals fit into the capital expenditure planning process. Inclusion of capital dollars spent on risk reduction over the past 3 to 5 years is one method to demonstrate this. A comparison of spending on risk reduction to total capital employed could also be included. A description of how the business unit accounts for all environmental costs, such as inclusion of a waste account for all expenses that are incurred for waste disposal, can also be used to meet the requirements of this section.

5. *Community and employee involvement.* The business unit should support and be actively involved in information transfer between the employees, customers, community, regulatory officials, and the media to solve environmental concerns. Any specific environmental project completed in alliance with community organizations should be included. List the organizations participating, number of employees participating, and the estimated hours of work involved or cost, if applicable. General community involvement efforts should be included. Examples are management involvement in business or other service organizations, in the local Chamber of Commerce, or in any local group where our company values can be shared.

The education process for employees should be explained.

6. *Compliance.* The business unit should demonstrate a commitment to compliance by promoting and measuring environmental compliance. The business unit should demonstrate committed resources and management processes to achieve and maintain compliance with environmental requirements. This includes management attention to compliance, compliance policies and procedures, auditing and measuring systems, training, recognition for compliance, and disciplinary procedures for noncompliance.

Efforts by the location to ensure compliance with internal and external requirements should be spelled out. This can include assignment of environmental responsibility as a principal job function to one or more individuals, percentage of management time spent on environmental issues, dollars spent on better measurement devices and other equipment, and meetings held to ensure permits and regulations are understood. Auditing and measurement systems for compliance with regulations and permits should also be included. Permit checklists can be used to show compliance.

INTERNATIONAL STANDARDS

There are currently several external environmental management system standards including British Standards 7750, the European Eco-Management and Audit Scheme (EMAS), and ISO 14001. These systems all have one thing in common: The purpose is to force industry to both understand and exhibit environmental responsibility. By their very nature, international standards develop consistency.

The International Organization for Standardization (ISO) has been in existence since 1947. The ISO mission is to promote worldwide standardization in order to facilitate an exchange of goods and services. The ISO standards are published for voluntary acceptance.

The process for standards development generally follows the path of assessing and recognizing the need for a standard, identifying the scope of the standard, generating international consensus, publishing the standards, and reviewing and updating the standards. This is a well-accepted model for any process. ISO has more then 200 technical committees that do the functional work of the organization. ISO 14001 is a product of one of these committees. As with all other ISO standards, ISO 14000 is being reviewed by all industry, and as we go forward, the standards will likely gain acceptance as a de facto environmental regulation. The basic premise of ISO 14001 is to require an organization to establish and maintain an environmental management system that contains some basic elements. There will be much written and discussed on this standard over the next several years. It goes without saying that any organization that has recognized the link between environment, health, and safety requirements, production requirements, and environmental requirements will have little trouble meeting the ISO requirements. It remains to be seen how involved and detailed this certification will become for large and middle-size global organizations.

CONCLUSIONS

In conclusion, application of several key principles will result in an effective environmental management program. First, there must be commitment from the top of the organization that managing environmental affairs is important, which is manifested in an environmental policy statement. Clear direction from the top will cascade through the organization. It can be very useful to coordinate this effort with health and safety requirements for the organization.

Second, an internal communication network to convey policy and best management practices to all employees across business unit and geographic boundaries is necessary. Employees need to know what is expected, and they are valuable resources to tap for new and better ideas.

Third, the organization needs to integrate environmental goals into the business strategy. Integrating environmental goals will enhance the long-term viability of the business.

Fourth, environmental regulations at the federal, state or province, and local levels must be followed. Compliance with environmental regulations is not optional; it is a basic requirement. This can be accomplished through an internal staff that has knowledge of the regulations and responsibility for designing oversight programs, or it can be accomplished by outsourcing through environmental consulting firms.

Fifth, capital improvements must be managed by comprehensive short-term and longer-term planning to ensure that the necessary tools are in place to meet the environmental and productivity goals and to comply with environmental regulations.

Sixth, an environmental education system to present an overview of environmental requirements, discuss individual responsibilities and introduce employees to available technical, legal, public affairs and operations resources is needed. A formal training session supplemented by ongoing communication accomplishes this.

Seventh, a reporting and auditing system will be helpful in developing a system to measure progress with the program. Measurement ensures that attention is focused on the environmental goals. Measurement should include positive events such as numbers of employees trained and capital improvements.

Eighth, waste minimization results in direct bottom-line improvement. A process that identifies waste sources, develops alternatives for reduction or elimination of waste, and implements action will show immediate payback.

Ninth, any successful business will be a good citizen in the community and industry in which it operates. Community outreach programs, working with customers on environmental improvement efforts, and talking to regulators before a problem occurs all have lasting positive effects on the business.

Tenth, recognition of excellence for environmental accomplishment will add value to the program. Individual employees, local business units, and entire product lines all benefit from their efforts being recognized.

Employing these ten principles will help establish an environmental program that will stand the test of time and enable a process of continuous improvement in managing environmental issues.

CHAPTER 12
STEPWISE EXAMPLES OF COMPLIANCE

Suresh Relwani
Melissa A. Hickey
RK & Associates, Inc.
Glen Ellyn, Illinois

EXAMPLE 1: OFFSET LITHOGRAPHIC PRINTER VOC COMPLIANCE 12.1
Background 12.1
Work Plan for Preparation of Permit Requirements 12.2
Conclusion 12.10

EXAMPLE 2: TREATING ODOROUS EXHAUSTS FROM A CHEMICAL MANUFACTURER 12.10
Odor Study 12.10
Prioritization 12.13
Selecting Control Technology 12.13

The purpose of this chapter is to follow real examples which illustrate the problems that this handbook aims to help solve. This first example relates to volatile organic compound (VOC) control and centers on the regulations. The second example relates to odor control and follows the procedures of evaluating the problems and devising the solutions.

EXAMPLE 1: OFFSET LITHOGRAPHIC PRINTER VOC COMPLIANCE

Background

LithoGraph is the fictitious name for a cold-set offset lithographic printer located in the Chicago suburban area that began operating two offset lithographic printing facilities in January 1991. Plants A and B began operations with three and four presses, respectively. Each plant was located in the Chicago nonattainment region for VOCs. Plants A and B were originally permitted by the Illinois Environmental Protection Agency (IEPA) to emit a maximum of 22 and 24 tons/yr of VOCs, respectively.

To increase the overall production capacity, LithoGraph decided, in January 1995, to consolidate all press operations to the plant A site. Plant B would be phased out gradually, over a 2-year period. Presses began moving from plant B to plant A in January 1995 and were completed by December 1996.

In September 1996, LithoGraph received a Compliance Inquiry Letter (CIL) from the IEPA concerning noncompliance with the state Environmental Protection Act and Air Pollution Control regulations. LithoGraph did not apply for, and failed to secure, construction and operations permits for the installation and operations of new lithographic presses at plant A. In addition, LithoGraph failed to demonstrate compliance with the offset lithographic printing reasonably available control technology (RACT) requirement, which became effective in Illinois on March 15, 1996. Upon receiving this CIL, LithoGraph contacted an environmental consultant to assist in air permitting and compliance.

Work Plan for Preparation of Permit Requirements

Determination of Requirements. After the facility walk-through, data gathering was completed, and the consultant outlined the regulations applicable to the LithoGraph facility: the Clean Air Act Amendments (CAAA) of 1990 and state air quality regulations.

CLEAN AIR ACT AMENDMENTS. As part of preparing the application, a regulatory audit of operations at plant A was conducted. Existing and future state and federal regulatory requirements applicable to emitting units were reviewed. This information, combined with the available air emission-related information for each unit, allowed the consultant to determine if the facility (plant A) was required to obtain a Title V permit; a federal enforceable state operating permit (FESOP); or an IEPA Division of Air Pollution Control operating permit.

To determine the local, state, and federal requirements applicable to the facility or air emissions unit, the regulations were analyzed with designated LithoGraph representatives, to determine which were applicable to the plant operations. This process involved comparing regulatory requirements with plant-specific information collected during the emissions inventory task, including facility and unit operations and regulatory history, age and location, existing permit conditions, and air pollutant(s) and their emission levels.

STATE AIR QUALITY REGULATIONS. Enhanced RACT requirements were triggered by the CAAA of 1990 since RACT was required for all major sources of VOCs in nonattainment regions by 1995. A major source of VOCs, in the Chicago nonattainment area, was defined as 25 tons/yr VOCs. In addition, enhanced RACT requirements were being imposed by the IEPA on smaller sources (<25 tons/yr VOCs) to attain ozone air quality standards. The enhanced RACT rule is based on the Control Techniques Guideline series: *Control of Volatile Organic Compound Emissions from Offset Lithographic Printing* (Draft), U.S. Environmental Protection Agency, Office of Air Quality, November 1993, and the Alternative Control Techniques Document, entitled *Offset Lithographic Printing,* EPA, Office of Air Quality Planning and Standards, September 1993.

In Illinois, the IEPA enhanced lithography RACT requirements became effective March 15, 1996, for all facilities in the Chicago land and East St. Louis nonattainment areas. RACT is defined as the lowest emission limit that a particular source is capable of meeting, by the application of a control technology that is reasonably available considering technological and economic feasibility.

The RACT requirements for cold-set offset lithographic printers in the Chicago nonattainment region are applicable to all lithographic facilities with actual VOC emissions greater than 100 lb/day before the application of controls. Actual daily emissions are to be determined each month based on the number of operating days per month.

Sources subject to the lithographic RACT requirements by March 15, 1996, had to certify to the IEPA that the source is subject to the RACT requirements. In the RACT certification letter, each source must identify each lithographic printing line, declare whether each printing line is in compliance, the compliance methods employed, and examples of record keeping and monitoring forms used. A new certification letter shall be submitted upon the start-up of a new printing line, or within 30 days of changing the method of demonstrating compliance.

INFORMATION REQUIRED. For sheet-fed offset lithographic printers, the following VOC content limitations apply:

- Fountain solution

  5% by volume VOCs as applied, if not refrigerated in the reservoir or fountain tray, *or*

  8.5% by volume VOCs as applied, if refrigerated to $<60°F$ in the reservoir or fountain trays
- Cleaning solutions (or wash solvents)

  VOC content as applied is limited to $<30\%$ by weight, or a VOC partial vapor pressure of <10 mmHg at 20°C.

  All cleaning materials must be stored in closed containers when not in use.

To demonstrate compliance with the fountain solution VOC content limitations, any of the following methods can be used to determine VOC content:

- Maintain records of each batch prepared.
- Analyze each batch using a refractometer, hydrometer, or conductivity meter.
- Determine the VOC content of solution based on the settings of automatic feed equipment.

To demonstrate compliance with the cleaning solution VOC content limitations, any of the following methods can be used:

- Maintain records of each batch of cleaning solution used.
- Determine the VOC content of the solution based on the setting of automatic mixing equipment.
- Use manufacturer-specified VOC composite partial vapor pressure and batch preparation data.

The following additional recordkeeping is required:

- VOC content of each batch of as-prepared fountain solution
- VOC content or vapor pressure of each as-applied cleaning solution
- Temperature records of refrigerated fountain solution reservoirs
- Documentation of settings, calibrations, and any changes to automatic feed equipment
- Scheduled inspections for compliance with cleaning materials work practices (storage of rags in closed containers)
- All records for at least 3 years

Step 2: Establish Regulatory Applicability. After a review, a list of applicable regulations for emissions units for the facility was developed. A summary of all pertinent information was also compiled during this step. This included the following items:

- Identification and description of the emissions units
- Relevant permitting information, such as:

  Source description, design operating conditions, and actual operating conditions

  Date of construction, reconstruction, or modification

Existing state and/or federal permit terms and conditions

Actual and potential-to-emit emissions for all pollutants of concern

- List of applicable regulations and the justification for applicability
- Description of federal and/or state enforcement procedures for each applicable regulation
- Identification of future projects and the associated regulatory impacts

IEPA Title V application forms and, to the maximum extent possible, existing operating permit applications provided emission inventory information and a basis for compiling the information required.

Step 3: Identify Compliance Requirements. Applicable regulations for emissions units were evaluated, and a list of the specific requirements for each emissions unit and the entire facility was then developed. This information includes all applicable emissions standards or limits, process monitoring requirements, and recordkeeping and reporting requirements.

Step 4: Identify Noncompliance Issues. The applicable regulatory requirements and emission limits defined in steps 2 and 3 were compared to the emissions inventory, source information, permit files, actual emissions statements, process monitoring and recordkeeping, and other information gathered during the emissions inventory. Potential noncompliance issues were then identified in Exception Reports, which briefly describe the emissions source, applicable regulations, compliance requirements, and actual data that indicate noncompliance. As part of the evaluation of the current compliance status, potential problem areas were identified in existing permit terms and conditions of the current operation and future consolidation project.

Facility Review and Assessment of Significant Sources

Facility Walk-through. The facility walk-through is the primary data-gathering phase of the project. At the kickoff meeting between the consultant and LithoGraph personnel, a preliminary list concerning the facility review and assessment of significant air emission sources was provided to LithoGraph. To perform the facility review, the consultant relied primarily on existing records but utilized, with authorization, state and federal agency records where necessary. It was necessary to identify and resolve where agency records might be in conflict with facility records. In addition, the consultant specified the documentation needed to obtain any missing data and to validate the data to be incorporated into the emission inventory.

It was requested that records be made available as soon as possible after the kickoff meeting. The information requested from LithoGraph included the following:

- Correspondence with the state regulatory agency
- Air pollution permits and permit applications
- Past annual emission inventory reports and annual inspection forms
- Source descriptions
- Actual raw material use information for the past 5 years (including ink, coating, varnish, cleaning solvents, process solvents, prepress solvents, and fountain solution additives)
- Material Safety Data Sheets (MSDSs) for all raw materials used
- Press information (press number, manufacturer, model number, serial number, stack information, number of color and/or coating stations)

- Dates of construction and installation modification
- Combustion unit rating and fuel use data for all combustion units
- Typical and maximum operating hours per month and year
- Facility layout map, roof plans, site surveys, and flow diagrams

It is preferable to have all the above information prior to the site visits. However, due to the aggressive schedule identified for this project, the site visits were scheduled as soon as possible after the kickoff meeting, even before some of the requested data were available. Information provided prior to the visits was verified for accuracy during the site visit.

Upon conclusion of the site visit, a preliminary list of potential emission sources was prepared. This list included the source code identification number, source location, brief description, air pollutants, physical characteristics, and operating hours. This list was submitted for review and discussion with LithoGraph. Based on this review, the information was revised as necessary.

In addition to collecting the above-mentioned data, the operations manager and press room operator gathered information pertaining to operating, monitoring, and recordkeeping practices:

- Preparation of fountain solution as applied

  What is the ratio of fountain solution additive to water?

  Is the solution prepared manually or by using automatic feed equipment?

  What are the recordkeeping practices for documenting the VOC content of the solution?

  Is the fountain solution chilled in the reservoirs, and if so, to what temperature? What monitoring is done to ensure that the temperature remains constant?

- Preparation of cleaning solution

  What is the ratio of cleaning solution to water?

  Is the solution prepared manually or by using automatic feed equipment?

  What are the recordkeeping practices for documenting the VOC content of the solution?

Emission Inventory Development. Templates were developed identifying the required information to determine actual and potential emissions from each emissions source identified in the facility walk-through. The templates included the data necessary to calculate emission rates of all regulated pollutants. The templates included data to define and quantify emissions from typical and maximum operating scenarios. The plant-specific information required for these templates was then completed by facility personnel, and the consultant calculated the air emissions.

Data contained in the templates were used to develop a detailed and comprehensive air emission inventory, including detailed calculations, addressing both point and fugitive emission sources for current operating scenarios. Where actual emission data did not exist or were not known, the consultant quantified both potential and actual emissions, using the emissions estimating techniques which were described in detail.

The completed emissions inventories, supplemented or corrected where required, were then formatted to facilitate future emissions reporting. Calculations and information used to estimate emissions were documented, including emission factors and/or assumptions used and the appropriate references.

Once the inventory was complete and emissions were quantified, each emission unit was identified as either a significant or an insignificant emissions source. This determination guided which source would be addressed in detail within the permit application and which sources would simply be listed.

Actual Emissions. For LithoGraph, there was little information pertaining to the actual facility emissions over the past 5 years. To determine the actual facility emissions, both source recordkeeping and supplier information for raw materials were used. Usually, suppliers can obtain purchasing records from the past 2 years. In the case of LithoGraph, the actual emissions for both plant A and plant B had to be determined to evaluate the applicability of the CAAA Title V permits and new source review.

Once the complete actual material use information was collected for the two plants, the actual facility emissions were estimated using the following equation:

$$\text{Actual VOC emissions} = \text{material used, lb} \times \text{VOC content} \times \text{emission factor}$$

where the following emission factors apply:

Material	Emission factor
Ink	0.05
Varnish	0.05
Coating	1
Fountain solution additive	1
Wash solvents	0.5

Note that for inks used in a cold-set printing line, 95 percent of the VOC in the ink is assumed to be retained in the paper. For wash solvents where manual press cleaning is performed using cleaning solutions with a VOC composite partial vapor pressure of less than 10 mmHg at 20°C, 50 percent of the VOCs can be considered retained in the rags. If the rags are kept in a closed container after use and the VOC is recovered or disposed of, then it is not emitted at the facility. At LithoGraph, cleaning operations are performed manually; rags are stored in closed containers and shipped off-site for cleaning. For the determination of the applicability of the RACT, no retention factor may be used for the cleaning solvents.

Table 12.1 shows the actual combined 1996 facility emissions from plant A and plant B. In addition to determining the applicability of a new source review, the actual facil-

TABLE 12.1 Plant A and B Combined Actual VOC Emissions

Material	Average use, tons/yr	VOC content, % by weight	Emission factor	Average emissions, tons/yr
Fountain solution additive	10	100	1	10
Cleaning solvents	47.6	100	0.5	23.8
Inks	200	20	0.05	2
Coatings	10	10	1	1
			Total	36.8

ity emissions for both plants were determined. The following shows the 5-year historical record of actual emissions for each facility:

Year	Air emissions, tons/yr	
	Plant A	Plant B
1992	7	28
1993	8	27
1994	9	26
1995	10	25
1996	17.6	19.2

The following calculation was performed to determine the applicability of the RACT requirements:

Material	Use, lb/mo	VOC content, % by weight	Emission factor		VOC, emissions, lb
Fountain solution additive	379.33	100	1		379.33
Cleaning solvents	2500	100	0.5		2500
Inks	400	20	0.5		4
Coatings	500	10	1		50
				Total	2933.33

Days operated per month: 24
VOC emissions (lb/day): 122.22

Potential to Emit. It is also necessary to determine the potential-to-emit (PTE) VOC emissions for the facilities to determine the applicability of the CAAA. To determine the PTE VOCs from LithoGraph, the consultant looked at the actual facility emissions from 1992 to 1996 and took the actual or typical emissions as 80 percent of the PTE. Typical annual emissions are based on the year of operation which best represents the facility's typical operating schedule. For LithoGraph, 1996 was chosen to represent the typical yearly emissions:

$$\text{PTE plant A: } 17.6 \text{ tons/yr} \times 100/80 = 22 \text{ tons/yr}$$

$$\text{PTE plant B: } 19.2 \text{ tons/yr} \times 100/80 = 24 \text{ tons/yr}$$

To determine the PTE for consolidation of the two plants, it was determined that the summary of the PTE for each plant individually would be accurate.

Plant	PTE emissions, tons/yr
A	22
B	24
	46

TABLE 12.2 Maximum Material Use and VOC Emissions

Material	Average use, tons/yr	VOC content, % by weight	Emission factor	Average emissions, tons/yr
Fountain Solution Additive	12.5	100	1	12.5
Cleaning Solvents	59	100	0.5	29.5
Inks	250	20	0.05	2.5
Coatings	15	10	1	1.5
				46

Once the PTE has been established, the corresponding maximum material use must be determined. Because offset lithographic printers may use many different inks and solvents, each having a different VOC content, the maximum VOC content for each category of material was used when writing the permit. Maximum material use corresponding to the PTE emissions was then determined using these maximum VOC contents for each material in Table 12.2.

Operating Scenario Evaluation. Title V of the CAA provides several ways to allow for operational flexibility. One option is to submit different operating scenarios as part of the operating permit application. Once approved, these operating scenarios will be incorporated into the operating permit, allowing operational flexibility without permit revision.

The consultant worked closely with LithoGraph personnel to define reasonable operating scenarios that would provide the necessary operating flexibility required to meet the facility's future 5-year strategic plan. They then applied the relevant emission factors to the facility's processing projections through emission inventory templates and spreadsheets, and they compared them to the applicable requirements to ascertain compliance. If any of the units were found to exceed their limits, the consultant worked with the facility staff to limit emissions to applicable requirements. For example, such scenarios could include process changes resulting from (1) variations in operational rates, (2) changes in operating hours, (3) variations in raw materials used, and (4) any other changes that would maximize the facility's flexibility with respect to operating permit compliance.

In concert with the LithoGraph staff, the consultants ensured that the proposed compliance alternatives were consistent with the overall 5-year strategic plan. Once the facility was satisfied with the proposed scenarios, it integrated the finding into operating permit applications.

Compliance Program Development. Once all calculations were completed and the PTE for the consolidated facility was established, a compliance program was developed. First, the consultants defined the work which needed to be completed to bring LithoGraph into compliance. This included

1. Application for a state operating permit for seven presses at plant A which included the demonstration of avoidance of new source review
2. Demonstration of compliance with enhanced RACT requirements
3. Submittal of the application for a CAA Title V permit

A meeting was held with the IEPA to discuss the consolidation project, a summary of the above work to be completed, and an estimate of the combined facility emissions.

Meeting with the IEPA staff helped to show that LithoGraph was now aware that it was out of compliance and that it was making every effort to comply with the IEPA and CAA regulations. In addition, this meeting provided LithoGraph with input from the IEPA concerning what type of information it expected in the permit applications.

Of particular concern to the IEPA was the demonstration of compliance with the enhanced lithographic RACT requirements and the avoidance of new source review. Since LithoGraph had not kept the required data for compliance with the state operating permit and RACT requirements, a recordkeeping system was developed. Of concern was the recordkeeping of the fountain solution as prepared and of cleaning solution as prepared, and monthly records of materials used and the estimation of monthly emissions.

To keep track of the fountain solution as prepared, it proved beneficial to have an automatic feed unit installed to regulate the amount of fountain solution additive mixed with water. The mixed solution is then distributed to a holding tank at each press. Daily inspections verify that the feed equipment is set at the appropriate rate and are recorded in a daily logbook.

To comply with the cleaning solution limitations, all cleaning solutions were evaluated for VOC content and VOC partial vapor pressure. At LithoGraph, cleaning solutions are used as supplied, so no records of mixing were required. A list of compliant cleaning solvents was made. Each time a new cleaning solvent will be introduced, it must first be evaluated for compliance with the RACT VOC limitations. Any changes in cleaning solvents used must be recorded in a logbook.

To keep track of monthly materials used and VOC emissions, tables are used by LithoGraph. The tables are in a spreadsheet format, which aids in the determination of emissions and compliance.

Permit Application Preparation

State Operating Permits. For LithoGraph, construction permit applications were submitted for those presses which had not yet been moved at the time of application submittal. An operating permit was applied for, to revise the present plant A operating permit to include the four additional presses from plant B. The IEPA was concerned that this move would not increase significantly the facility VOC emissions. For VOCs, a significant increase is an increase of 25 tons/yr emissions. To demonstrate this, LithoGraph began keeping material use records for each press. At the end of each month, the emissions from each press were calculated to verify that the emissions did not exceed 22 tons/yr VOCs for presses 1 to 3 and 24 tons/yr for presses 4 to 7.

A state operating permit was applied for, using the PTE and the maximum use numbers included in Table 12.2 and press and stack information. Included in the permit application was a narrative describing the consolidation project, complying materials used, detailed emissions calculations, process flow diagrams, a facility plot plan, and examples of records kept to demonstrate compliance with the RACT requirements. In addition, included were copies of the MSDSs for all complying inks, coatings, and solvents used.

Clean Air Act (Title V) Permit Application. Since the LithoGraph facility's PTE VOC emissions are greater than 25 tons/yr, a Title V permit application had to be submitted. Maximum material use, PTE VOC emissions, and recordkeeping activities were the same as for the state operating permit application. In addition, the Title V permit application included the information collected concerning insignificant activities and also hazardous air pollutant (HAP) emission information.

To determine the PTE HAPs for LithoGraph, MSDSs were used to determine the HAP content of each material used. The PTE HAP was estimated by multiplying each material's typical HAP content by the maximum material use reported in Table 12.2.

Based on this estimation method, LithoGraph was not considered a major source of HAPs.

Follow-up. Once the permit applications were submitted, it was important to follow up with the state regulatory agency to determine the status of the permit issuance process.

Conclusion

With these modifications, LithoGraph obtained permits and is operating in compliance with federal and state laws.

EXAMPLE 2: TREATING ODOROUS EXHAUSTS FROM A CHEMICAL MANUFACTURER

This example relates to treating odorous exhausts from a chemical manufacturer, who will not be identified. This project is still in process, but the steps taken are typical of many odor control problems. This midwest manufacturer has a 1-mi buffer zone between its fence line and the neighbors. Nevertheless, the neighbors have complained about odors and health problems, blaming the plant, over a period of 5 years. The EPA has encouraged the manufacturer to solve the complaint problems.

The manufacturer hired an outside consultant, and the work was divided into four stages.

1. Perform an odor study.
2. Prioritize sources and select which would be the first source to be treated.
3. Select the odor control technology to be used.
4. Make specific design recommendations.

The plant processes were much more complex than will be described in this example. About 30 different processes and the wastewater plant were identified as odor sources, and there were several options as to how each might be run, or combined with other sources, affecting mass estimates, equipment costing, etc. For simplicity, the number of processes and options to be considered will be greatly reduced.

Odor Study

Over a period of 2 weeks, samples were drawn from about 30 different odor sources and at the wastewater treatment plant, from grit chamber through secondary treatment, and were tested for odor, dilution to threshold, intensity, pervasiveness, and characterization. Other elements of the odor control project are listed in Fig 12.1. The results of the threshold or intensity tests are summarized in Fig. 12.2.

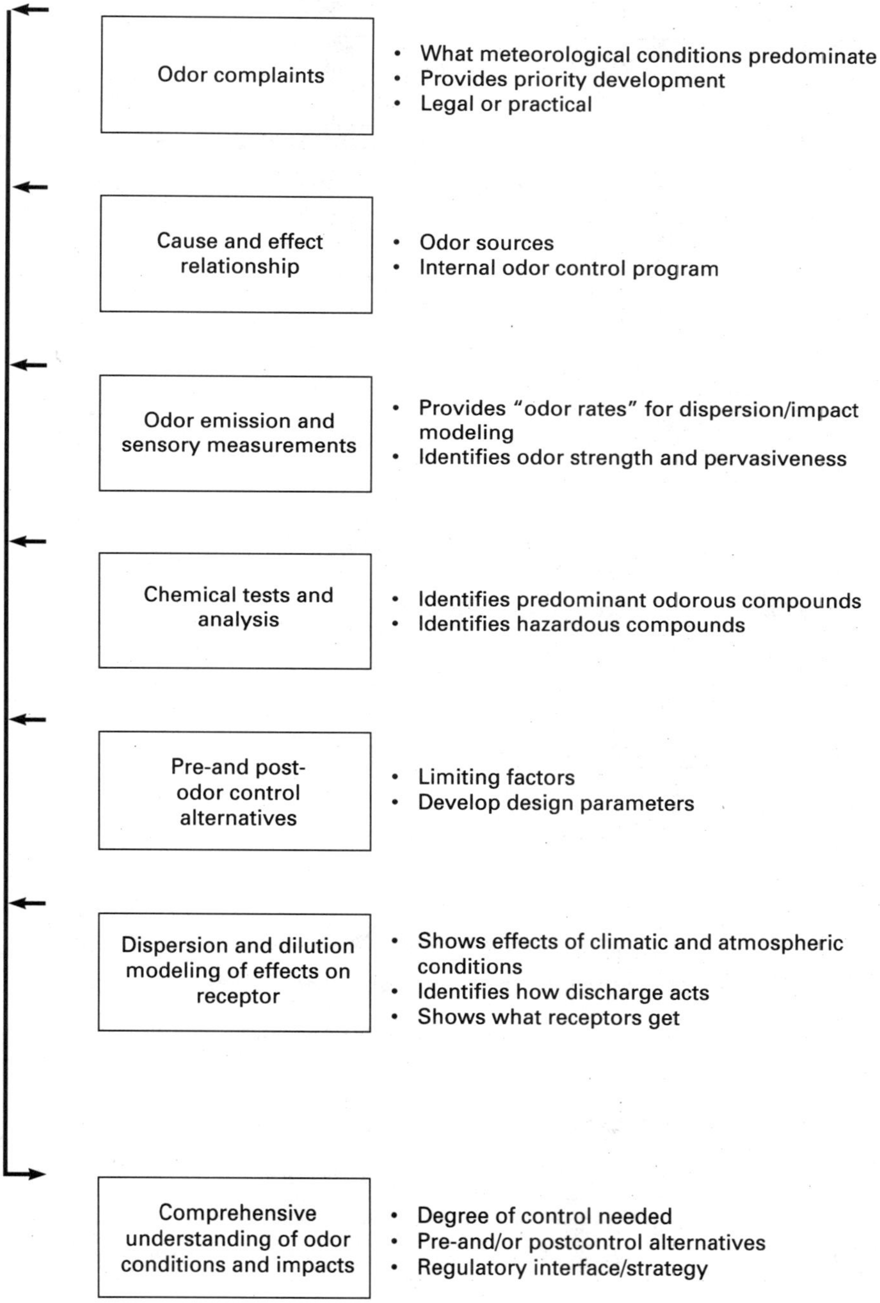

FIGURE 12.1 Overall odor control project.

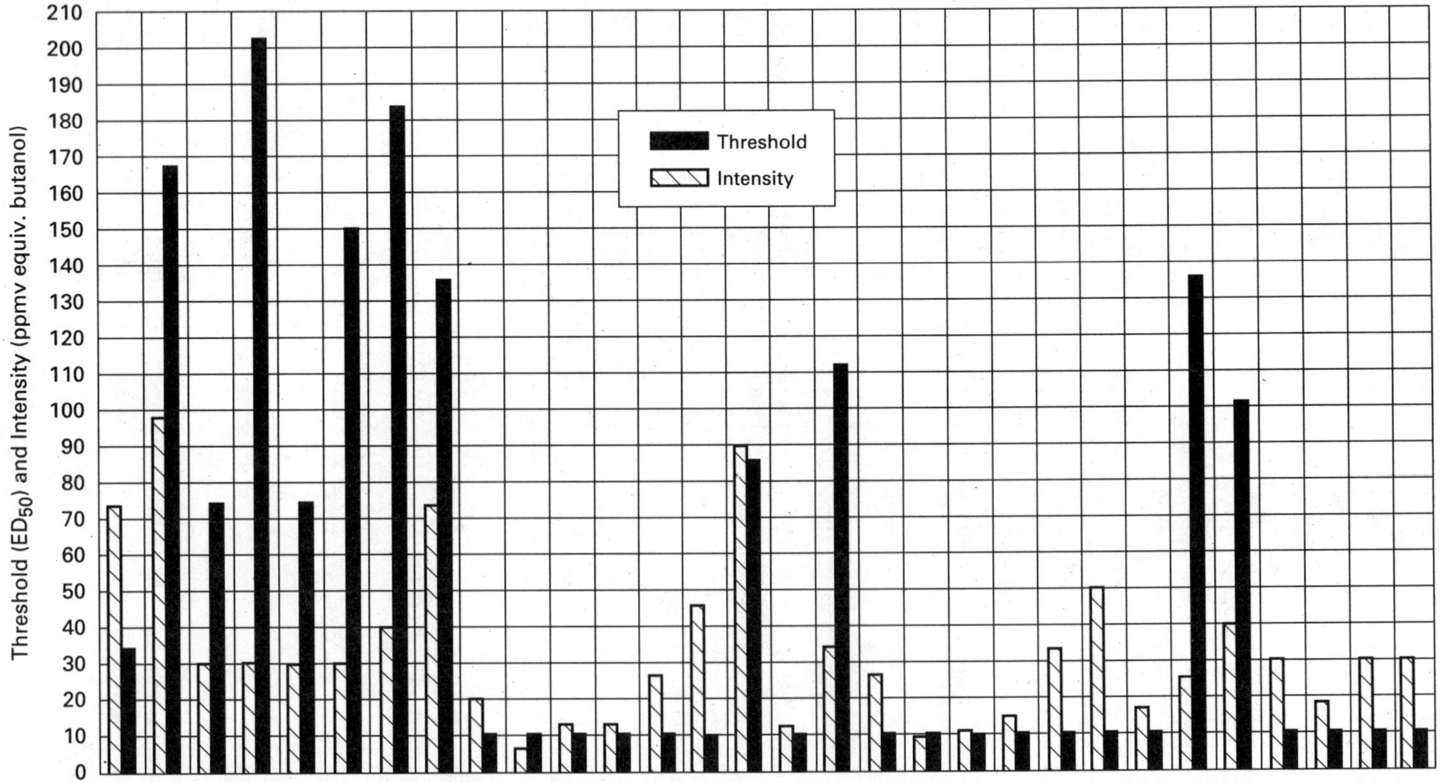

FIGURE 12.2 Odor characterizations results—threshold and intensity (August 20 to 28 sampling event).

Prioritization

If one uses, as a separation line, a threshold or intensity of 100, it is seen that the problem sources are reduced from 30 to 7. The pervasiveness values of all these seven odor sources were very similar, in the range of 0.6 to 0.65. With further consideration of mass flow rates and process conditions, one source was selected as first priority.

Selecting Control Technology

The overall objective is to reduce the odor emissions to less than an ED_{50} of 75. Alternative technologies were evaluated to arrive at an effective, economic decision. Consideration was given to the following:

Thermal Oxidation, with or without Energy Recovery. Various designs of recuperative, regenerative systems were considered including fixed-bed and catalytic. Because of acid exhausts, it was necessary to pretreat one of the streams prior to the thermal oxidizer.

Carbon Adsorption. Two of the streams feeding the adsorber required pretreatment, one to condense moisture and one to condense moisture plus an impregnated carbon adsorber to remove some compounds that would not effectively be removed by the activated carbon adsorber.

Scrubbers. No pretreatments were required, and two designs were considered, a packed column and a mist scrubber.

An economic comparison was made, and the clear advantage based on economic considerations favored wet scrubbing technology.

	Capital cost	Annual operating cost
Thermal oxidizer	$500,000	$200,000
Carbon adsorption	350,000	151,000
Wet scrubber	130,000	83,000

Specific Design Recommendations. Two designs of wet scrubbers were considered, packed columns, and mist scrubbers. Because the process allowed for the possibility of a process upset where large quantities of sticky particulates would be discharged into the scrubber, it was decided, without an economic comparison, that the mist scrubber design would be the most practical alternative.

At this point, before the project progressed to final equipment sizing and design, the client decided to pilot-test the proposed mist scrubbing technology. Mist scrubbing had never been previously applied to this process application, and the client wanted some additional assurance and demonstration of the successful performance of the equipment. Further, the client felt that, with a pilot on site, numerous other odor sources could be tested, and optimum operating conditions for effective odor removal would be determined. That is where the project stands, and pilot tests are projected for the spring of 1998. The consultant, who has more experience with this technology and the types of odors to be treated, is confident of success, but understands the client's need for reassurance before initiating such major expenditures. The final installation is not projected until the spring of 1999.

Additional work will be required to obtain permitting, although no obstacles are foreseen. Plant personnel will be working with the community to be sure that the neighbors are reassured that their interests are being considered, and the neighbors will be advised of the progress of the project.

This example represents the path of resolution that is typical for this kind of problem. It is expected that if the odor problem persists after this first source has been treated, other measures will be taken to eliminate odors from additional sources.

CHAPTER 13
SOCIAL AND BUSINESS ISSUES

SECTION 13.1

PUBLIC RELATIONS ASPECTS

Alan K. Leahigh
Executive Administration, Inc.
Arlington Heights, Illinois

PUBLIC OPINION 13.3
FACTORS OF COMMUNITY SUPPORT 13.4
GUIDELINES FOR ACTION 13.4
EMPLOYEES FIRST 13.5
COMPLETE DISCLOSURE 13.5
COALITION BUILDING 13.6
HANDLING THE CRISIS 13.6
PUBLIC RELATIONS COUNSEL 13.7
GOOD CORPORATE CITIZEN 13.8

There may have been a time when companies could do as they pleased with no regard to public opinion. That time, if it ever existed, is long gone. Companies, today, operate by consent of their customers, shareholders, and the local community.

This is especially true for air quality issues and the local community. A storehouse of corporate goodwill and accomplishments can be quickly emptied by a cavalier attitude toward a nuisance odor or a careless behavior about an airborne health hazard. These are issues with high impact on the local community. A well thought out and executed community relations program is essential to a successful pollution control strategy.

The rules for good relations with neighbors or any public are simple enough:

1. Act responsibly.
2. Communicate aggressively.

Acting responsibly means having manufacturing processes and procedures that avoid wastes and do not permit environmental hazards and nuisances. It requires constant monitoring of operations and embracing the latest methods for emission controls.

Aggressive communications means an ongoing dialogue with local political leaders, community organizations, the media, and local residents about your company's environmental policies and procedures.

All this requires no small amount of time, energy, and company resources. Not doing this risks even greater amounts of time, energy, and company resources.

3M

Often 3M is cited as a company with enlightened environmental policies and a commitment to the communities where it has plants. Since 1972, the company has pursued its *3P*s program, Pollution Prevention Pays, which eliminates waste from manufacturing processes—not only because it is the right thing to do for the environment, but because it has been shown to achieve savings that go to the bottom line.

A local television station in one of 3M's communities was preparing a report on VOC and toxic emissions from area manufacturing plants. A video crew set up a camera at the top of an overpass next to the 3M manufacturing plant and began taping the facility, its smoke stacks, and its activity. A security guard spotted the crew and chased them away.

You could write the script: "3M company officials bar Channel 5 cameras from taping emissions at their plant."

But when the security office immediately reported the incident to the plant manager, he called the station and invited the crew back. The crew was en route to home base when radioed to return to the 3M plant. In a taped interview, the plant manager was able to explain how millions had been spent in recent years to reduce emissions, and he told about staff training to prevent the likelihood of an environmental accident. He described the useful products the plant manufactured and the 800 jobs that it provided to the community.

A modest community relations victory? Sure. But the kind that compounds and strengthens a reputation for environmental responsibility. The plant manager could not have taken the action he did if he were not supported by a responsible environmental policy and an enlightened corporate public relations policy.

Asea Brown Boveri

A division of Asea Brown Boveri had been invited to build a solid municipal waste processing facility in a suburban community. It did not take much to stir concerns about VOC and heavy-metal emissions from the plant, noxious odors from rotting municipal garbage, increased traffic of garbage trucks making deliveries, and a host of other concerns about the facility.

The company unrolled a comprehensive public relations program to win the support of the community. An opinion survey was conducted to discover local perceptions about the proposed facility. In cooperation with the mayor, a citizens' advisory board was appointed so that community leaders could evaluate the facility plan and visit other communities where similar plants had been built.

Company officials personally called on community leaders and provided them with a portfolio of information so they could speak knowledgeably to their constituents about the plant.

An "environmental fair" was held at a community center so that anyone concerned about the facility could come to learn and ask questions about the plant's processes for recycling, composting, incineration, emissions control, and monitoring. A newsletter was sent to all residents in the community to keep them abreast of developments in the planning and permitting of the facility.

The result: When a public hearing was held, only four local residents showed up to speak against the facility. Without any major confrontation, the daily newspaper did not bother to cover the hearing. The city council voted unanimous approval for siting the waste processing facility.

Sybron Chemicals, Inc.

An accidental release of a noxious chemical from a Sybron Chemicals, Inc., plant caused the evacuation of more than 60 nearby residents. The company received negative publicity and widespread criticism from community leaders, the state's department of environmental protection, and a U.S. senator.

To compound the problem, a month later, a processing accident seriously burned three workers, causing concern among the plant's 250 other workers.

A door-to-door survey of 350 area residents led to the development of a community relations and safety plan, including an emergency telecommunications system. Called PINS (the Prompt Inquiry and Notification System), the computerized calling system gave neighbors immediate information about the chemical plant 24 hours a day. The company also established a Community Advisory Board.

The outcome: A follow-up survey of residents by an independent polling firm found that 91 percent felt more secure about the plant and confident that PINS would notify them about any problems. An even more dramatic 97 percent said that the company and the plant had improved communications with its neighbors.

PUBLIC OPINION

You have heard all the pejoratives for describing local opponent groups: the NIMBYs (Not In My Back Yard) and the folks who are LULU (Locally Undesirable Land Use) about residential property values, as well as their elected leaders who become NIMTO (Not In My Term of Office), not to mention the militant BANANAS (Build Absolutely Nothing Anywhere Near Anybody).

They all are acting on fears, justified or not, and all must be taken seriously.

Public opinion can be immensely complex. Some people spend their entire professional careers studying, monitoring, evaluating, and sometimes attempting to influence what the public knows, thinks, and does. But you do not need to be a public relations expert to be sensitive to public interests. The following five principles encompass nearly everything you need to know:

- *You are part of the public.* It is "the public *we*"—not "the public *they*." The public includes you, your family, your neighbors, and your property values that are at stake when there are risks of airborne health hazards or nuisance odors.
- *People fear what they do not understand.* You have the authoritative information, so it is your responsibility to inform and create awareness—about your company, the benefits you create for the community, your concern about the environment, the actions that you take to protect community health, and your plans for continual environmental improvements. The more effective and frequent your communication, the more your company will benefit.
- *People want control of their lives.* They may expose themselves to many health hazards and take risks with their personal and real property, but they will not tolerate risks that others impose on them. You can expect vigorous opposition if the actions of your company are perceived as a threat to their life (hazardous emissions), pursuit of happiness (odors that interfere with comfortable indoor living or stop outdoor activity), or pocketbook (decline in property values).

- *The public relations sequence is always, first, do good and, second, take credit for it.* Companies that try to work the formula in reverse place at risk their franchise to conduct business in the community. A good public relations plan is part of, but can never be a substitute for, a comprehensive pollution control program.
- *Companies that act responsibly are defended, not attacked, by their neighbors.* Residents come to the aid of companies that are perceived as adding value to the quality of life in the community.

If you understand and base your program on these principles, the rest of your public relations plan is mostly tactics.

FACTORS OF COMMUNITY SUPPORT

Many companies have achieved high levels of respect and support within their communities. They understand the requisites of creating goodwill and being a good corporate citizen:

- *Awareness of benefits.* They create awareness of the benefits of the company to the local community.
- *Continuing dialogue.* They go out of their way to create channels of two-way communications with the community and with their own employees.
- *Prompt response.* They are responsive to the interests and concerns of the community and move quickly on those issues.
- *Openness to new ideas.* They welcome and embrace new ideas and recommendations from the community and from their own employees.
- *Opportunity to participate.* They create ways for members of the community and their leaders to advise and participate in company decisions affecting the community.
- *Sincerity.* They do all the above not because it makes the company look good, but because members of management believe in and respect their community, their neighbors, and their workforce.

A company with this community relations attitude has a predictable response when it receives, for example, a complaint from neighbors about an odor. It collects information about the time, place, and description of the event. It sets up a "sniff patrol" with employees making a circuit of the property or the neighborhood on a regular basis. It involves community representatives in an evaluation of its odor controls. When new odor control equipment is installed, it lets neighbor representatives smell the exhaust. It invites the media to witness the event.

GUIDELINES FOR ACTION

The tactics of a successful public relations program will vary, depending on the community and the situation, but they are likely to be based on these principles:

1. *Be honest and straightforward in all communications.* This is rule number one. It takes only one dishonest, misleading, or deceptively incomplete statement to pillage a company's credibility. The rebuilding process can take years and may never be fully accomplished.

2. *Begin now.* It is much better to begin your community relations program before there is a major problem. It is easier to communicate with neighbors who are concerned than with those who are irate. Reasonable solutions and timetables are best achieved when there is no organized and inflamed local opposition.

3. *Establish a public image.* Companies that traditionally keep a low profile within their community are always vulnerable. Companies well known for their good deeds are given the benefit of the doubt when new issues arise.

4. *Anticipate issues and problems.* Many environmental problems can be anticipated and addressed before they become crises. Rather than depend on luck that the crisis will not occur, rely on your skilled early management of the problem.

5. *Offer solutions.* No one and no company is exempt from error or flaws. Mistakes can happen, and problems inadvertently may be overlooked. But everyone and every company can be expected to have a reasonable plan and timetable for solving its problems. For example, any discussion of an emission issue should include your efforts and timetable for attaining air quality goals.

6. *Build coalitions with community organizations.* Coalition building means becoming involved in the community and helping its institutions and people. Then, when it is your turn to have a problem, reciprocal help can be expected as you work your way toward a solution.

7. *Communicate aggressively.* Take credit for the good things you do. Employees, neighbors, and the public should be allowed to join with you in pride for your good deeds. Talk effectively and frequently to your community.

EMPLOYEES FIRST

Employees are part of your community and undoubtedly the most important part, because…

- They care the most about the company.
- They are most directly affected by any odors and airborne pollutants.
- They are in a position to help you minimize emissions by their operation of plant processes.
- They are your ambassadors to the rest of the community.

Many companies rely mostly on supervisors and foremen to handle internal communications. Some of those supervisors and foremen do a good job; others, merely adequate or less so. Traditional house organs often do only a superficial job.

A good employee relations program accomplishes more than telling and informing. It accomplishes understanding. It communicates management's thinking and encourages feedback. Employees need to know and believe in their company's mission and the company's strategies for protecting the air that they breathe and the environment.

COMPLETE DISCLOSURE

There is an old saying: "Feed the bears or they'll eat you." Establish your company as the primary and authoritative source of local information about air quality and environmental matters.

Holding back information so as not to alarm the public or employees is an enormous disservice to those you are claiming to protect. They will be much more alarmed when they find out you have withheld information. They will imagine the reasons why and will assume the worst.

If full disclosure is the policy of your company, there will be no need for members of the community or the local media to look for information elsewhere.

COALITION BUILDING

Your company benefits from being identified with the concerns of the community. This means involving your company in community activities—blood drives, United Way, food pantry, Halloween safety, athletic team sponsorships—and involving the community in your company issues, especially its environmental programs. People on the same team support one another.

If you are facing a problem that has implications for the community, form a citizens' advisory committee composed of opinion leaders representing disparate stakeholder groups. Show them the problem you are addressing, the steps that you have already taken, and the additional remedies that you are planning. Invite their input. Share the results.

Meet regularly with local government officials. Keep them up to date on your environmental activities so they are not blindsided by questions from constituents and so they can address any misinformation that they hear.

Invite representatives from academia to participate in environmental programs. Consumer opinion surveys always rank teachers and professors high in credibility, insight, and impartiality.

Hold open-house events and plant tours. Issue an annual environmental stewardship report. Establish a speakers' bureau and provide your company representatives with formal speaker training. Get to know reporters and be a dependable, willing news source.

In summary, welcome every opportunity to build bridges to the community. Inform. Educate. Participate.

HANDLING THE CRISIS

Now for the really bad news. Despite your best efforts, a crisis can still occur. You may have moved the odds in your favor, but you still need to be prepared for that single telephone call that causes the bottom to drop out. Everything you have worked for now depends on the decisions you make and the actions you take during the next several hours.

A poorly handled crisis can obliterate all cumulated public relations equity. If you do not believe this, think what first comes to mind when someone mentions Union Carbide or Exxon or Dow-Corning. Sometimes the present and future integrity of an entire company is defined by how it responds to a single public relations crisis.

An ad lib response to a crisis is a recipe for disaster. Instead, practice and have a checklist ready so you can quickly and comprehensively move through the necessary steps for a crisis communications response:

1. *Define the issue.* What will be of greatest concern to internal and external publics?

2. *Determine who is affected.* Those most affected should receive the first attention.

3. *Put people first.* Think and act in terms of the impact of the crisis on people: community, employees, neighbors, customers. People must always be placed ahead of product, profits, property, or politics.

4. *Tell the truth.* Accept the fact that the truth will become known. Deceptions, incomplete information, and half-truths invariably worsen a crisis and devastate your credibility when the truth later comes to the surface.

5. *Get the best information.* To act effectively, you need to know as much as possible about the crisis, and as quickly as possible. What happened? When did it happen? How did it happen? Who was involved? What is happening at present? What most likely will happen next?

6. *Create an inventory of what you do not know.* The inventory becomes your checklist of information you must have to complete your response to the crisis.

7. *Pull together your team.* Surround yourself with a preselected team of people who can help you gather and analyze information, develop communications strategy, and manage information dissemination to internal and external publics.

8. *Decide on the best remedy.* You may not be able to completely undo the damage, but there probably are steps you can take to help restore the situation and demonstrate your sincere concern about it.

9. *Designate a spokesperson.* In times of crisis, the best spokespersons are at the top and, best case, at the scene.

10. *Identify all audiences.* Some people need to be addressed before others. Who are the most affected? Give particular attention to internal audiences such as employees and their families.

11. *Release bad news completely, clearly, and quickly.* Tell the whole story up front. News that is released in dribbles prolongs the crisis and causes everyone to wonder what is next and when the series of bad news announcements will end.

12. *Listen as well as speak.* Determine the best ways to receive information from your publics. The inbound information will help you determine how well you are communicating and what else needs to be said or done.

A word of caution about legal advice and legal consultants during a crisis. Legal issues are very important, but your legal consultant has a primary responsibility: to protect your interests. It is up to you to make sure that those self-interests are balanced with the public interest. As one company chief executive officer stated it, "I may win in a court of law, but if I lose in the court of public opinion I've lost everything."

Be especially wary of any recommendations to withhold information or, worse yet, refuse comment. The withheld information invariably becomes known, amid the wreckage of the company's credibility and reputation.

There are exceptional lawyers who are sensitive to public opinion and the public interest. Treasure them as you would any other immensely valuable company asset.

PUBLIC RELATIONS COUNSEL

It is wise to have assistance from people with experience. If you have not had much interaction with the public, if you have never faced a public relations crisis, if your company is turning a new corner in its effort to be responsive to the community, you may want to seek the advice of a public relations professional.

The rationale is similar to the individual who is having a problem with her or his golf game. It is smart to see a pro. Make sure that the pro has experience with the kinds of issues that you are facing.

But do not expect the golf pro or the public relations pro to show you some secret moves that you otherwise would not know how to do. Rather, the pro goes back and reviews the fundamentals. The public relations consultant can be

- An early warner, identifying and interpreting the mood of the community
- A conscience, to be certain that you are doing the right things
- A communicator, with both internal and external publics
- A monitor, to measure whether your community relations plan is working and the company is living up to public expectations

The public relations professional may suggest a communications audit—a formal study of how you are perceived by the community and the ways in which the community receives information about your company. The audit should include recommendations on tactics for enhancing communications and relations with your most important publics.

GOOD CORPORATE CITIZEN

It all comes down to this: Companies today cannot afford to ignore their public relations. The risks are just too great. Whether the issue is plant emissions, other environmental issues, or any community relations concern, companies need to build coalitions with community organizations; cultivate relations with local opinion leaders, the media, and citizens; and communicate aggressively.

SECTION 13.2

COMPARATIVE RISK*

Ken Jones
Sue Thomas
Michele Witten
Green Mountain Institute for Environmental Democracy
Montpelier, Vermont

Sarah T. Fox
Western Center for Environmental Decision Making
Boulder, Colorado

WHY GOVERNMENTS REGULATE VOCS 13.9
HOW GOVERNMENTS REGULATE VOCS 13.10
IN COMES COMPARATIVE RISK 13.14
WHAT IS COMPARATIVE RISK? 13.14
RESULTS OF COMPARATIVE RISK PROJECTS 13.19
HOW COMPARATIVE RISK SEES VOCS 13.23
RISKS TO ECOSYSTEMS 13.23
ADDRESSING THE DRIVING FORCE 13.24
HOW COMPARATIVE RISK ADDRESSES ODORS 13.24
UNDERSTANDING RISK 13.25
APPENDIX: SEVERAL COMPARATIVE RISK PROJECTS 13.27
Washington Environment 2010 13.27
Colorado Environment 2000 13.27
Louisiana Environmental Action Plan 13.28
Seattle Environmental Priorities Project 13.29
Hawaii Environmental Risk-Ranking Project 13.30
California Comparative Risk Project 13.31
Maine Environmental Priorities Project 13.32
Ohio Comparative Risk Project 13.33
Northeast Ohio and Cleveland Regional Environmental Priorities Project 13.34
Priorities '95—Columbus, Ohio 13.35
Elizabeth River Project—Norfolk, Virginia 13.36
Houston Environmental Foresight Project 13.36
Minnesota Comparative Risk Project 13.37
Mississippi Comparative Risk Project 13.37
Other Projects 13.38
REFERENCES 13.39

WHY GOVERNMENTS REGULATE VOCS

Volatile organic compounds (VOCs) cause a range of impacts and potential impacts to the environment including acute human health problems, contribution to the generation

*The development of this section is supported by cooperative agreements between the U.S. Environmental Protection Agency and both the Green Mountain Institute for Environmental Democracy and the Western Center for Environmental Decison-making. In addition, parts of this section were developed with the support of the American Industrial Health Council and included in an unpublished paper titled "A Retrospective on Ten Years of Comparative Risk."

of ground-level ozone, long-term health problems associated with cancer, ecological damage, aesthetic problems and economic damages associated with lower property values, lost tourism revenues, and health care costs.

However, the simple identification of threats to the environment associated with problems such as VOCs is only the first step toward the corresponding regulatory actions necessary to control those threats. As with other environmental policies, VOC control activities result after the observations from the natural world get translated through a public discourse. This discussion initially brings pressure to elected or appointed officials whose responsibility it then becomes to make legislation. Environmental agencies must translate the statutory response to a collection of regulations, rules, and compliance and enforcement efforts.

HOW GOVERNMENTS REGULATE VOCS

Translating the observations regarding air pollution to specific regulatory activities is anything but straightforward. Today, the regulatory actions to address specific VOC emissions are based upon the specific human health impacts of carcinogenesis due to direct exposure or, in certain cases, the possibility of an acute health impact. We take just one example—benzene. The regulations affect the composition of petroleum products, the transportation and delivery of motor fuels, controls upon its use as a solvent, its presence in water for thousands of publicly supplied drinking water systems, and the release due to industrial processes. How those specific control levels were established is the subject of this discussion, in terms of both regulatory reform (for those outside of environmental agencies) and results-based environmental management internal to environmental agencies.

In general, each determination of regulatory limits uses one or more of the following decision models:

- Best-available control technology (BACT)
- Cost-benefit analysis
- Risk assessment and acceptable exposure limits

Best-available control technology (BACT) is often applied as the decision guidance for establishing limits on the release of chemical contaminants to the environment. BACT is based upon the premise that polluters should be expected to do all that they can do to control emissions, but should not be asked to do more than is technologically feasible. A problem with this approach identified by the regulated community is that over time, control technologies will advance, but so may the costs, and at some point an absurd investment may be required to achieve a minimal gain in environmental quality.

Partly in response to the weakness in BACT, cost-benefit analysis is based upon the economic logic that a regulatory action should not cost more than the benefits that result. According to this logic, whenever costs are greater than benefits, greater societal benefits than those accomplished through the regulation should take place via market forces. Unfortunately, in cases where there is a lack of fungibility of benefits, consistent use of cost-benefit analysis is difficult to implement.

The use of risk assessment is in some regards a middle course between cost-benefit and BACT. A risk assessment threshold, such as one excess death per million population, is used as a guideline toward which some regulations are steered. In this way, absurd levels of risk reduction, which may be required by BACT, are not required, and there is no need to rely on the market to find alternative risk reduction opportunities. Unfortunately, the determination of the threshold has no economic or scientific basis, and may, in fact, be quite arbitrary. Similarly, the estimation of risk from particular exposures is a grossly inexact science. (See Note 13.2.1.)

NOTE 13.2.1 The whys and hows of human health risk assessment.

Human health risk assessments serve as the sentinel information that is included in comparative risk projects. It is worthwhile to consider where these assessments find their roots.

One example of legislation that has required the use of risk assessments is Section 112 of the Clean Air Act which requires that EPA "allow emissions only to the extent that human health and the environment are not unduly harmed." The scientific community has been charged by government agencies to determine that level of exposure which will not harm human health or the environment.

The easy, and correct, answer from any competent toxicologist is that the only level of exposure that can be known with certainty to produce no negative health effect is zero. Unfortunately, this is an unacceptable policy answer in an economy that is built on the use of chemicals including those classified as volatile organic compounds (VOCs).

Therefore, the question translated from the statute by policy makers is altered to ask, What is the concentration of a particular chemical that will provide an "acceptable" risk? *Acceptable* requires further policy clarification outside the realm of the toxicologist or epidemiologist. Ultimately, the question is refined even further. For example, What is the concentration of a particular chemical that will provide a risk of less than 1 excess cancer death per year per million of an exposed population?

It is from that starting point that toxicologists return to their laboratories to investigate the toxicity and/or carcinogenicity of compounds. In the ideal, epidemiological studies would be carried out to compare the health effects of a large population in the presence and absence of a particular exposure. However, the need to determine risks at the 1 in a million range and the impossibility of any controlled human exposure typically banish the epidemiologist from the exercise. The toxicologist relies upon rodent testing of elevated concentrations of the chemical in question to determine the potential for carcinogenicity.

The use of these studies requires answering other policy questions, including, How do we translate laboratory rodent responses to humans? and How do we extrapolate high-dose studies to low-dose exposures? In the past, policy makers have interpreted statutory intent and asked scientists to make cautious assumptions with regard to evaluating data. These conservative assumptions are the fuel for much of the regulatory reform debate taking place in Congress in the 1990s. Extrapolating high-dose, significant-frequency observations to low-dose, low-frequency outcomes—combined with comparing the impacts on genetically pure rodents to a human population with a widely variable genetic composition—combined with exposure assumptions that may only be valid for a small population leads to a loss of confidence that these risk assessments are sufficient to create regulatory decisions.

Despite the lack of confidence such efforts provide for predicting the outcomes of chemical exposures, risk assessments are a source of information for many regulatory decisions. However, many regulatory decisions are not based upon the standard practice of risk assessment. Important for the discussion of the regulation of VOCs, ozone standards are not the result of accumulating conservative assumptions about exposures and dose-response relationships. There is a raft of information about the pulmonary system response to ozone levels that are actually realized in urban environments around the world. Current debates on revising the ozone standards are the result of further review of real exposure and health effects, not extrapolations from studies with artificially high exposures.

Proponents of change recognize the inefficiencies and inconsistencies in utilizing any of the decision models to address any particular environmental problem. Not only are considerations currently used in regulation very complicated and inexact and controversial, but they may not even be addressing the environmental damages the legislation initially intended. In addition, the basic science is extremely uncertain. In summary, it appears that regulation driven by legislation driven by reaction of people to visible problems does not yield action levels corresponding to risk levels. With the luxury of observing a collection of environmental regulatory decisions, new approaches to develop future regulations become apparent, and one of these is comparative risk. (See Note 13.2.2.)

NOTE 13.2.2 How benzene is regulated.

Hundreds of tons of benzene are released to the atmosphere each year from human activities. Section 112 of the Clean Air Act* instructed the EPA administrator in 1970 to identify and set standards for air pollutants that she or he judged hazardous to human health.

While the language in Section 112 was meant to be straightforward, it in fact raised significant questions within the newly formed environmental agency. All chemicals at high enough dose are hazardous to human health, but clearly, Congress did not intend EPA to set standards for each one.

Therefore, even though benzene had been recognized as a liver toxin for decades, the evidence pointing to benzene-induced health problems at lower concentrations has only been documented since the 1970s, and it was not until 1977 that EPA listed benzene as a hazardous air pollutant and started the mechanism to determine emissions standards.

When those standards were initially proposed for four different sources of benzene emissions, the standards were based on the ability to reduce emissions, a technology-based standard. In 1984, EPA proposed to eliminate one of the standards based on its finding that the health risk was low and available controls would not significantly reduce those risks. The Natural Resources Defense Council challenged the ruling, noting that Section 112 did not allow EPA to drop a standard just because EPA found no available technology to reduce risks beyond some undefined level. Partly, as the result of that challenge, the courts responded in another case and requested EPA to formalize its policy (the "vinyl chloride case"). EPA proposed four different approaches for making decisions under Section 112 of the Clean Air Act.†

- Approach 1: Case by case, consider all health risk measures, the uncertainties associated with the measures, and other health information.
- Approach 2: Total cancer incidence, with 1 case per year as the limit for acceptability
- Approach 3: Maximum individual risk (MIR) with a limit of 10^{-4} for acceptability
- Approach 4: MIR with a limit of 10^{-6}

In setting the 1989 benzene standards,‡ EPA used a policy approach based on approaches 1 and 3, but incorporating consideration of incidence from approach 2 and consideration of health protection for the general population of 10^{-6} from approach 4. Following this integrated approach, the standards for benzene were designed to ensure that greater than 99 percent of the population within 50 km of each source would be exposed to risks less than or equal to 1 in a million and incidence of cancer is "kept low."

The Notice publishing the 1990 benzene standards restates the above approach quite clearly, as follows.

> In protecting the public health with an ample margin of safety under section 112, EPA strives to provide maximum feasible protection against risks to health from hazardous air pollutants by (1) protecting the greatest number of persons possible to an

---

*42 U.S.C.S. §7412 (1997).

†*Natural Resources Defense Council, Inc. v. U.S. Environmental Protection Agency,* 824 F. 2d 1146 (1987).

‡National Emission Standards for Hazardous Air Pollutants; Benzene Emissions from Maleic Anhydride Plants, Ethylbenzene/Styrene Plants, Benzene Storage Vessels, Benzene Equipment Leaks, and Coke By-Product Recovery Plants, 54 Fed. Reg. 38,044 (1989).

individual lifetime risk level no higher than approximately 1 in 1 million and (2) limiting to no higher than approximately 1 in 10,000 the estimated risk that a person living near a plant would have if he or she were exposed to the maximum pollutant concentrations for 70 years. Implementation of these goals is by means of a two-step, standard-setting approach, with an analytical first step to determine an "acceptable risk" that considers all health information, including risk estimation uncertainty, and includes a presumptive limit on maximum individual lifetime risk (MIR) of approximately 1 in 10,000. A second step follows in which the actual standard is set at a level that provides "an ample margin of safety" in consideration of all health information, including the number of persons at risk levels higher than approximately 1 in 1 million, as well as other relevant factors including costs and economic impacts, technological feasibility, and other factors relevant to each particular decision.

...A principle that accompanies these numerical goals is that while EPA can establish them as fixed numbers, the state of the art of risk assessment does not enable numerical risk estimates to be made with comparable confidence. Therefore, judgment must be used in deciding how numerical risk estimates are considered with respect to these goals. Uncertainties arising from such factors as the lack of knowledge about the biology of cancer causation and gaps in data must be weighed along with other public health considerations. Many of the factors are not the same for different pollutants, or for different source categories.§

In fact, the standards for each source category are quite different. The 1989 *Federal Register* notice (54 Fed. Reg. 38044) relays detailed technical considerations involved in the standard setting for each category. McQuaid¶ (1991) offers a table summarizing the risk levels for all the benzene source categories, which range from 4×10^{-6} to 6×10^{-4}.

§National Emission Standards for Hazardous Air Pollutants; Benzene Emissions from Chemical Manufacturing Process Vents, Industrial Solvent Use, Benzene Waste Operations, Benzene Transfer Operations, and Gasoline Marketing Systems, 55 Fed. Reg. 8,292 (1990).

¶McQuaid, Janet L., *Risk Assessment of Hazardous Air Pollutants under the EPA's Final Benzene Rules and the Clean Air Act Amendments of 1990,* 70 Tex. L. Rev. 427 (1991).

Simply considered, comparing the risks between different human activities could serve as a replacement for BACT, cost-benefit analysis, and risk assessment thresholds. Those activities which pose greater risk should be regulated first or more strictly, and those posing smaller risks should not be addressed until the greater risks are under control or society requests greater overall activity to reduce risks.

The practice of comparative risk, as described in detail later in this section, shows that the difficulty in applying comparative risk findings exposes its shortcomings as a decision tool. Despite the weaknesses, proponents recognize the strength of comparative risk as an analytical tool which can be integrated into any decision-making framework including BACT, cost-benefit analysis, and risk assessment thresholds.

The key strength uncovered while carrying out comparative risk analysis lies in the multiple attributes of risk which reflect the public's observation that things are wrong in the environment. For example, the observation leading to a public clamor for reduction of air pollution includes many kinds of risk. We are not interested in just the reduction of cancer deaths from chemical exposures, but also in the chronic health effects, impacts on ecosystems, and impacts on our quality of life. Most other decision tools do not consider these multiple attributes.

Therefore, comparative risk finds a willing regulatory environment to address the weaknesses of BACT, cost-benefit analysis, and risk assessment and the need to consider the multiple attributes of risk.

IN COMES COMPARATIVE RISK

At the Environmental Protection Agency (EPA) in the mid-1980s, administrator Lee Thomas asked agency staff to determine if the agency was utilizing resources on those problems that pose the greatest risks. Because different regulatory actions are based upon the result of such different analytical approaches as described above, it is not surprising that the answer was an emphatic "no!" The analytical exercise resulted in the ranking of 23 environmental issues [e.g., drinking water, abandoned hazardous waste (Superfund) sites, etc.], published in a report titled, "Unfinished Business." The ranking showed that most of EPA's efforts were directed toward relatively low-risk problems, but there was little Thomas or his successor, William Reilly, could do to reverse the trends that had developed as the result of 25 years of federal legislation. Several interesting discussions are based on what that mismatch means for environmental policy, but those discussions fall outside the scope of this chapter, generally, *Worst Things First? The Debate over Risk-Based National Environmental Priorities* (Finkel and Golding, 1994); also, *Setting Environmental Priorities: The Debate about Risk* (U.S. EPA, 1991), and *Can Problems Shape Priorities? The Case of Risk-Based Environmental Planning* (Fiorino, 1990). The ranking exercise that the agency carried out has, over time, garnered great attention, perhaps more attention than the actual ranking.

The Science Advisory Board (SAB) to EPA reviewed the agency's original risk ranking, yielding a document titled *Reducing Risk* in 1990. In *Reducing Risk,* a panel of scientific experts supported both the methodology that the EPA used to rank risks and the need to consider risk as an important piece of the priority-setting exercise. The SAB is finishing a second comparative risk type of process called the Integrated Risk Project in 1998. Since EPA's *Unfinished Business,* the potential for risk ranking has sparked interest in EPA and among other environmental agency officials as a way to focus agency efforts on higher-risk environmental problems. Starting with four state pilot studies in 1988, EPA began a program to support others interested in doing comparative risk projects. Including these pilot projects, EPA has supported more than 24 state, 12 local, and a handful of tribal comparative risk projects as well as carried out comparative risk in each of the 10 EPA regional offices. Outside the United States, the EPA and U.S. Agency for International Development have sponsored comparative risk projects in central Europe, Africa, Latin America, and the Pacific Rim.

WHAT IS COMPARATIVE RISK?

Are the human health impacts from the emission of volatile organic compounds greater than the threat from heavy-metal contamination of fish?

Does ground-level ozone pose a greater risk than the possibility of global climate change due to increased concentrations of greenhouse gases?

These questions are but two of many that could be posed to compare the many environmental problems that are subject to regulatory control. In general, the comparative risk project is based on a more general question:

Is some environmental problem worse than some other?

Moving from this general question to more useful comparisons requires the development of a list of problems to analyze and a second list of criteria with which to evaluate the problems. Table 13.2.1 shows a two-dimensional matrix with both the list of problems and the list of criteria which form the analytical structure for the comparative risk process. Considering a series of problems (e.g., drinking water, criteria air pollutants) against criteria (cancer risks, biodiversity) leads to questions which can be answered to begin the process of comparing risks, such as

> Are the cancer risks from drinking water greater than the cancer risks from criteria air pollutants?

or

> Is biodiversity more threatened from radioactive waste or habitat destruction?

Refining the criteria further leads to a finer level of specificity in the questions that are posed:

> Is the number of people exposed to increased levels of ground-level ozone greater than the number of people exposed to the risks of biologically contaminated drinking water?

In addition, considering each problem against multiple criteria can help to present information about the many aspects of environmental problems. In the case of volatile organic compounds, such a structure will showcase information about the cancer and noncancer health effects in terms of susceptible or heavily exposed populations, impacts on ecosystems, impacts on recreational access, or economic damages.

Developing the list of criteria is not a particularly scientific process. The method used in most comparative risk projects is to expand upon the question, What environmental problems are of concern? by asking the second question: Why? From this exercise, people express their concern in terms of human health and future generations and aesthetics and fairness and impacts on ecosystems.

For example, an elevated level of ozone may have particular impacts on the growth rate of spruce trees, particularly at higher elevations where soil characteristics make the growth of these species sensitive to many environmental factors. However, simply measuring growth rates does not reflect the values that the productivity of spruce trees represent in forest ecosystems. A decreased growth rate decreases the viability of the tree, weakens the tree so that it is more susceptible to climatic and biological stresses, and changes its role as food and shelter to other species within that system. Understanding the potential of decreased growth rates to the full range of impacts within the ecosystem is an important part of ecological risk assessment that starts with the question, What values do the forest ecosystems represent?

Recent concepts developed at the Western Center for Environmental Decision-making suggest a nine-step process for considering ecological risk:

1. Define the ecosystems in the study.
2. Identify ecological goals for the ecosystems.
3. Designate assessment endpoints.
4. Compile baseline ecological information.
5. Identify stressors to the ecosystems.
6. Define risk criteria.
7. Designate measurement endpoints.
8. Analyze and evaluate the information.
9. Rank the issues.

TABLE 13.2.1 Analytic Matrix for Comparing Risks

	Human health risks			Ecological risks			Quality-of-life risks		
Environmental problems (examples)	Cancer	Noncancer	Special populations	Biodiversity	Endangered species	Reversible	Aesthetics	Economic	Fairness
Drinking water									
Criteria air pollution									
Radioactive waste									
Habitat destruction									
Hazardous waste									
Mercury									
Lead									
Pesticides									

The first step of a comparative ecological risk assessment is a discussion of how the project participants would like to define ecosystems. A working definition of ecosystems is necessary to help guide a state or community analysis. Technical experts soon recognize that the public perception of ecosystem may be different from their own. Understanding the differences in definitions of ecosystem is a necessary initial step in comparative ecological risk assessments.

The second step is the development of goals which help set the tone and structure of the whole analysis. Goals should be established from consideration of ecological concepts, not human use. Goals that are human-use-oriented, such as the preservation of the aesthetic beauty of a preserve, do not represent the structure and function of ecosystems and therefore become difficult to evaluate by any ecological criteria. An example of an ecologically oriented goal is "(to) promote continued growth in restoration and reconstruction of ecosystems" (Iowa Dept. of Natural Resources, 1997).

Assessment endpoints are an important part of projects and establish the rigor of the ecological assessment. Assessment endpoints are guidelines which direct the analysis of the function and structure of ecosystems. They should be developed at the landscape or ecosystem level. Goals are represented and communicated through the assessment endpoints. An example of assessment endpoints might be to maintain prairie biodiversity. In this case prairie biodiversity is one aspect of the goal to continue research and growth of ecosystem restoration and reconstruction.

Compiling baseline data is necessary to evaluate the current state of ecological systems. Trend data provide the basis for determining whether a particular system is close to or far from conditions that reflect its ecological integrity prior to human intervention. Baseline data will also lead to a better understanding of the natural variability of ecological structure and function.

The identification of stressors is a list of threats that may harm ecosystems. Included in this list are the physical threats of soil erosion, grazing, mining, and stream channelization; biological threats such as exotic species introductions and species extinctions; as well as the more traditional chemical threats.

Risk criteria are developed to aid the technical committee in the procedure of ranking. Criteria are also used to help with the data collection process. Examples of criteria are the uncertainty of the effect, recoverability, and the extent and uniqueness of the ecosystem.

Measurement endpoints or indicators define and are used to analyze assessment endpoints. Indicators will often be designated at the community or population level of the biological hierarchy. Data are readily available on single species; however, that does not always adequately represent the function of whole ecosystems. Therefore, focus on ecosystem evaluation by measurement endpoints is a unique component of comparative ecological risk assessments.

The analysis and evaluation step is the point at which all gathered measurement endpoints are synthesized and evaluated. Conclusions are interpreted through the threat posed by each stressor on each ecosystem.

The final step in comparative ecological risk assessment is the ranking process. Rankings can be done in many different ways. One example would be to rank the risk of a stressor on all analyzed ecosystems. Therefore, each ecosystem would be evaluated by the issues or risks separately. When ranking, ecological criteria and goals must be remembered and used to provide more structure to the ranking procedure.

Comparative risk has also introduced a formal mechanism for considering quality-of-life risks in addition to the human health and ecological risks. By considering the specific values of future generations, aesthetics, economic damages, and peace of mind, comparative risk goes to the core of environmental values and helps analysts better understand the motives that members of the public hold for environmental quality.

One important topic in this book is the issue of odors. Odors have been at the heart of several grass-roots efforts to regulate air emissions, but the analysis of odors does not provide the same kind of objective information as do health risk assessments or even the impacts on certain nonhuman species as a proxy for ecological risk assessment. Despite the lack of analytic rigor, the issue of odors, or visibility, or fairness must be considered in the development of any regulatory response to an environmental problem.

Quality-of-life analysis is the scientifically poorer cousin to human health risk analysis and ecosystem risk analysis. As a result, some people discount the use of quality-of-life analysis in considering the relative risk of different environmental issues. Convincing skeptical listeners of the importance of quality of life is critical for the success of developing a truly comprehensive description of environmental harm. Fortunately, those individuals involved in comparative risk projects quickly understand the need to consider quality-of-life analysis in their projects. Unfortunately, some individuals outside of comparative risk projects who may not understand the objectives for doing comparative risk criticize quality-of-life analysis as "unscientific."

For states using comparative risk, an important issue arose regarding the credibility and confidence behind the use of risk assessment as the source of information for the ranking. Some members of the public are skeptical about the use of risk assessment as the means to measure the relative harm posed by different environmental threats. Therefore, as the pilot states began their comparative risk projects, the project participants, with the support of EPA, developed mechanisms which brought the technical analysis of environmental risks to representatives of the public for their review, consideration, and even guidance. It is this role of public advisory groups in state comparative risk projects that distinguishes those efforts from the EPA exercises carried out at headquarters and in the ten regional offices.*

State and local projects typically have three types of committees that develop the analytical structure and resulting risk ranking. A steering committee oversees the project, setting project budgets, determining membership of other committees, and fostering the results within government and nongovernment decision-making structures. A public advisory committee ensures that the technical analysis, ranking process, and communication of ranking results reflect the values of the public. Technical committees gather the technical information that forms the basis for the risk-ranking exercise. Optimally, there is extensive communication between all three committees, particularly between the technical and public advisory committees so that the technical information both is understood by the public advisory committee members and reflects public values regarding the environment.

The steps of a comparative risk project include these:

- Develop a problem list.
- Develop analytical criteria.
- Prepare technical information.
- Incorporate public values.
- Rank risks.

*The role of public advisory groups also distinguishes state and local comparative risk projects from many of the academic discussions about comparative risk, including those proposed by Justice Stephen Breyer in his book *Breaking the Vicious Cycle.* In most of those academic discussions, the emphasis is on gathering together experts in the field of environmental assessment for their contributions regarding information about the relative risks caused by different problems. For Breyer and others, the role of nonexperts in the development of that analysis is either ignored or scorned for allowing too much subjective judgment into an otherwise objective process.

- Develop risk management strategies.
- Communicate project results.
- Implement new strategies.

The different interactions within and between the three committees in carrying out these steps provide a range of flavors for the different projects. For a short description of some of the EPA-supported state and local comparative risk projects, see the Appendix to this section.

Since 1987, 24 states and localities, supported in part by EPA, have completed risk-ranking exercises. In most of, if not all, these projects, priority setting was an initial, primary objective for project organizers and participants. However, as each of these projects evolved, additional objectives arose and, in most cases, eclipsed priority setting as the primary objective. Many of these additional objectives are strikingly common across projects and include these:

- Strengthen linkages between different environmental agencies.
- Develop linkages between government agencies and stakeholders.
- Provide a baseline of information about the environment for multiple (often undefined) uses.
- Provide a starting point from which to develop new strategies to address environmental problems.
- Bring more science into policy discussions.
- Change public misperceptions about relative risk.
- Build trust between environmental agencies, the regulated community, and the public.
- Improve environmental conditions.

The achievement of these objectives has yielded rich stories about the successes of comparative risk as a process. See Table 13.2.2. A few papers generated by the Green Mountain Institute for Environmental Democracy (GMI, formerly the Northeast Center for Comparative Risk) and the Western Center for Environmental Decision-making (formerly the Western Center for Comparative Risk) are available for interested readers [*Comparative Risk at the Local Level: Lessons from the Road* (Delhagen et al., 1996), *A Retrospective on Ten Years of Comparative Risk* (Jones, 1997) and *Building A Foundation for Change: Opportunities and Challenges in State Comparative Risk Projects* (Dea and Thomas, 1997)]. Important among the conclusions of these papers is that the comparative risk process must include a wide range of participation in order to accomplish its objectives.

RESULTS OF COMPARATIVE RISK PROJECTS

Because most state and local comparative risk projects look at a broad range of environmental problems (e.g., global climate change, solid waste management, hazardous air pollutants), the results tend to either be general or represent only a first step toward effecting specific environmental program change. For example, the state of Washington developed a series of legislative activities from its comparative risk project including a sweeping Clean Air Act. This legislation resulted from extensive deliberation and negotiations with comparative risk serving as a starting point for discussion. The project also

TABLE 13.2.2 Successes of State and Local Comparative Risk Projects

Project name	Resulting management activities
Washington Environment 2010	• Multiple pieces of legislation including Washington State Clean Air Act (1991) and the Growth Management Act (1990, 1991). • Budget changes: $16 million in requests to support recommendations; Department of Ecology internally reprogrammed nearly $7 million from lower- to higher-risk management priorities. • Contributed to evolution of Department of Ecology's planning system to a focus on strategic planning based upon measurable objectives and results and the development and use of environmental indicators.
Colorado Environment 2000	• Goals, strategies, and assessment information as input to planning and budgeting in the Department of Health and the Environment over past 5 years (including the development of a performance partnership Agreement with U.S. EPA). • Provided support for the development of pollution prevention programs.
Louisiana Environmental Action Plan (LEAP to 2000)	• Developed an agreement among representatives from very diverse and often antagonistic groups within the state on a vision for the state's environmental future and the relative severity of environmental risks, and a commitment to take steps to address those risks. • Many project participants involved in a subsequent project funded by the McKnight Foundation for evaluating, communicating, and addressing risks in the Mississippi River corridor.
Hawaii Environmental Risk Reduction Project	• New programs for indoor air and statewide testing for blood-lead levels in children. • Built communication and trust between department and community. • Initiated a discussion of multiagency issues which has evolved from an informal "brown-bag task force" to a cabinet-level committee to coordinate issues.
Utah's Environment: The 21st Century	• Partnership between Department of Environmental Quality and local health departments to identify and address areas of commonality between local and state priorities. • Informational input to departmental strategic planning process.
Maine Environmental Priorities Project	• Steering committee members participated in and influenced the Department of Environmental Protection's strategic plan and performance partnership agreement. • Contributed to the governor's environmental agenda and subsequently multiple agency plans. • Modeled and won support for a collaborative, proactive approach to environmental dialogues; transferred this approach to other settings (e.g., dioxin stakeholder group, Maine Forest Health Task Force).

Ohio Comparative Risk Project	• Created an Office of Environmental Education with $500,000 budget to integrate results and information from the project into environmental education activities in the state. • Involved the public in producing a State of the Environment report.
Kentucky Outlook 2000	• The integration of risk information in the development of futures scenarios for considering air pollution issues into the next century. • Creation of a new division of information services to improve the quality of data services. • Establishment of information centers and libraries within local conservation districts.
Comparative Environmental Risks in Mississippi	• Creation of Environmental Resources Center to improve public outreach.
Seattle, WA	• Adoption of the mayor's environmental action agenda with action by the city on over 120 initiatives • Budget review of city programs in light of the action agenda • Innovative public and private partnerships such as the Master Home Environmentalist Program
Columbus, OH	• Final report with 192 recommendations to reduce risk for the city • A new nonprofit organization working with the city to reduce risk • City commitment to develop an "environmental report card" using environmental indicators to track risk reduction
Elizabeth River Project (Norfolk, VA area)	• Consensus by 120-member watershed action team and 35-member leadership review board on an 18-point watershed action plan • $30,000 for bioremediation testing • $250,000 to enhance Virginia DEQ toxics monitoring capabilities • Wetland restoration project
Cleveland, OH	• Public input into the regional transportation planning organization's criteria for regional road building • Municipal cooperation to develop a common agenda addressing urban sprawl

provided a technical base informing the deliberations. In another example, the Elizabeth River Project in the tidewater area of Virginia completed a risk ranking which identified common ground for designing actions to address the environmental problems for the river. However, before detailed strategies were developed, consultants developed a more focused and comprehensive technical analysis of the problems.

It is useful to note that the information and linkages between project participants resulting from a comparative risk project can incrementally affect the allocation of resources. Therefore, instead of considering priorities as the result of a zero-based budgeting approach where all budget decisions are linked to a single exercise, these results suggest that priorities may result from the sum of many decisions. Comparative risk projects have informed some of these decisions.

Preventing Program Cuts. In Vermont, the state legislature considered taking action to cut funding for the state heritage program. This program supports efforts at monitoring the status of threatened and endangered species. Because of the importance of threatened and endangered species identified in the Vermont comparative risk project, the legislature decided against these cuts.

Directing New Resources to New Programs. Washington state expanded its efforts in reducing criteria air pollutants in the early 1990s. New legislation and increased staffing in the Department of Ecology resulted from the comparative risk project's strategy development.

Focusing Attention on a Few Problems. In northeastern Ohio, the participants in a comparative risk project have taken a leading role in restructuring local transportation and energy policies after the identification of these issues as driving forces behind many of the high-risk issues in the area.

Justification for Budgets. In Seattle, for the two years after the comparative risk project, each city agency submitted its budget request to the mayor and City Council together with a rationale for the request based on the findings of the city's comparative risk project.

Elevating Administration Initiatives. In Maine, the governor's environmental agenda was formed in parallel with the findings from the state's comparative risk project. When announcing his ideas, Governor King repeatedly cited the project results as rationale for new state initiatives.

A common use of the risk ranking is to initiate programs or enhance existing programs which address highly ranked risks. Participants in many projects have noted that the comparative risk process winnows down the list of potential management activities to a smaller number. While this is not necessarily a dollar-for-dollar shift away from any low-risk issue, it does serve the purpose of establishing some flavor of priority.

In addition to these incremental changes, the risk ranking and technical analysis supporting it become an additional part of the information base regarding a state's or locality's environmental condition. Changes in policy decisions as the result of changes in broad information bases are usually difficult to track. However, the past 10 years may offer some cases to consider. A clear example of the use of the change in an information base is the general increased understanding of the relative role of radon exposure to lung cancer compared with other air pollutants, where a significant portion of this discussion was developed within EPA's *Unfinished Business* (U.S. EPA, 1987) report of 1987. Other issues that are receiving more attention from a broader public are the relative importance of indoor air versus outdoor air pollutants and the importance of habitat destruction as a risk to ecological health.

HOW COMPARATIVE RISK SEES VOCS

In comparative risk projects, risk estimations for VOCs as hazardous air pollutants are unavoidably sketchy, with projects generally identifying a short list of five to ten stressors of concern based on their potential to cause health problems (derived from the literature) and the likelihood of exposure in the project area (derived from TRI or similar emissions data sets). Cancer risks are then estimated based on populations in proximity (e.g., the surrounding county) to major sources. Because the toxics of concern tend to be state- or locality-specific, few generalizations can be made about risk. Benzene and formaldehyde stand out as fairly universal VOCs of concern. In California, project members estimated an additional 100 cancer cases attributable to outdoor exposures to benzene, followed by 1,3-butadiene at 60 cases and formaldehyde at 14 cases.

Most projects presented the risks from VOCs in indoor versus outdoor air for comparison. Indoor environments contain and concentrate VOCs emitted from dry-cleaned fabrics, building materials, and tobacco smoke resulting in exposures to much higher levels than would be found outdoors. California's estimate of 14 additional cancers due to outdoor sources of formaldehyde was dwarfed by its indoor estimate of 124 excess cancers.

In contrast to the severe effects of direct exposure to hazardous VOCs, the health effects of ozone are generally not considered life-threatening. Nevertheless, ozone levels have driven the regulatory agenda for VOCs, presumably because of the sheer magnitude of the problem. Exposure to high concentrations for several hours induces respiratory inflammation and decreased lung function, and animal evidence suggests repeated exposures to high levels may result in permanent lung damage.

To arrive at a risk estimate for ozone, comparative risk projects typically total the populations living in areas classified as nonattainment based on ambient concentrations of ozone exceeding the National Ambient Air Quality Standard (NAAQS) of 0.12 ppm. In some cases, a threshold below the NAAQS was used to reflect a state standard (Maine) or project-derived risk-based threshold (Washington). In many cases, the exposed populations total in the millions: over 7 million in Texas and 3 million in the Houston project area alone. Projects did not assume attainment of the national standard to equal no risk. The fact that studies have noted health effects at levels as low as 0.07 ppm combined with the substantial number of areas very near the standard compelled many projects to judge ozone a relatively high risk whether there were exceedences or not.

Projects relate considerable uncertainty in the human health risk estimates for VOCs and ozone. Conservative assumptions, extrapolations, etc., may significantly overstate risk. On the other hand, long-term or synergistic effects are unaccounted for, as are higher-risk activities such as commuting on a congested highway.

RISKS TO ECOSYSTEMS

The effects of specific VOCs on organisms other than laboratory animals and humans appear to be relatively unstudied, thus project technical reports offer little help in estimating the ecological risks from VOCs in the ambient air. Some project reports offer the logical argument that wildlife would be similarly affected at concentrations of hazardous VOCs harmful to humans.

The impacts of ozone on ecological health have not been documented to the same extent as human health effects have, but existing evidence is among the strongest examples of chemical-induced environmental damage. California ozone studies in the San Bernardino and Sierra Nevada mountains have documented "significant amounts of tree injury" to Ponderosa and Jeffrey pines, suggesting that years of exposure to current levels do result in ecological damage. Observed variations in damage levels during 2 years of study in a Vermont wilderness area also corresponded with known variations in ozone concentrations.

Most projects lacked state-specific evidence and relied on the literature for assumptions about plant injury associated with increasing concentrations of ozone. Projects used this information to develop estimates based on local measurements (or extrapolations) of ozone levels and percentages of sensitive species.

ADDRESSING THE DRIVING FORCE

After 25 years of regulation, over 70 million people are living in nonattainment areas. Moreover, recent studies indicate that respiratory effects may be occurring at levels below the current standard. During this same time period, U.S. population increased by 28 percent and vehicle miles traveled (VMT) increased 116 percent. As documented in project reports, the challenges encountered in characterizing and controlling the VOC and ozone problem will intensify under a scenario of tightening standards along with increasing population and sprawl.

While accurate emissions inventories have been elusive, a major culprit distinctly emerges. Motor vehicles, according to EPA numbers cited by several projects, account for 35 percent of VOCs and 56 percent of air toxics. The California project estimates 100 to 1000 cancer cases per year can be attributed to mobile sources of air pollution, with VOCs such as benzene being major constituents of the problem. Michigan reported 48 percent of its VOC emissions attributable to mobile sources. Overall, reviewed projects suggest that motor vehicles account for one-third to one-half or more of the problem. Compounding the situation are trends like this one: The Ohio project documented annual VMT had increased from 72 billion in 1980 to 93 billion in 1990. Project reports agree that air quality improvements that have been achieved via emissions controls will likely be counteracted in the near future if these trends continue.

HOW COMPARATIVE RISK ADDRESSES ODORS

Odors, by themselves, do not cause the same direct risks to human health or ecosystems as do the chemicals which may lie at the source of the odor problem. For this reason, most comparative risk projects often do not explicitly consider the problem area of odors. This does not mean that project participants do not see the importance of odors in considering risk reduction. Many of the comments in comparative risk projects reflect the importance of risk, both as a very public signal about environmental problems and as a factor which must be addressed when reducing the risks from airborne environmental stressors.

In a small number of projects, odors have been considered explicitly in the quality-of-life analysis. As with other air pollution issues, the local nature of odors requires a discussion of the size of the population exposed. While this factor leads to less risk than problems that may cover a more general population, the responses of the public indicate the importance of the issue. In many cities and states, odors represent up to one-half of the complaints filed with air pollution control agencies. These complaints are difficult to evaluate against health impacts but point out both the difficulty and importance of evaluating quality-of-life risks.

The management of odors is often considered with its source problem. For example, the odors from a solid waste facility are managed as the result of a permitting process which also considers other health issues such as rodent control. Industrial pollutants which cause odors will also often have health impacts which allow them to be controlled in a manner that focuses on the health endpoints.

UNDERSTANDING RISK

Comparative risk projects have not been dramatically successful in changing broad perceptions of relative risk between environmental problems. However, participants consistently report their own personal knowledge of environmental problems is enhanced and, to the extent those individuals are involved in further policy decisions, their learning translates to policy change.

Most comparative risk projects do not address very specific regulatory policies. In most cases, as represented by comparative risk projects taking place at the state or local level, comparative risk is intended to inform a broad range of policy decisions across all the media affected by environmental damage. Furthermore, comparative risk projects intend to include broad representation from the technical community and the different stakeholders. However, this does not mean that the lessons from those exercises cannot inform more specific regulatory decisions.

The need for better decision tools, as described at the beginning of this section, probably finds its solution in considering multiple attributes of environmental problems.

Comparative risk projects across the country open the discussion of environmental impacts to the full range of potential endpoints and thus may provide the opportunity to structure flexible decisions in different jurisdictions. The possibility of flexibility in decision making is not without problems. A broader range of endpoints resulting from comparative risks application of a broad range of criteria could open up a regulatory can of worms in the development of new regulations.

But the analytical exercise of using multiple criteria is only a part of the comparative risk process. The delineation of the decision arena and the participation in the process of selecting the policy questions are even more important in beginning a discussion about comparative risk.

The true challenge in comparative risk projects is either to involve the decision makers in the risk ranking or to convince them to embrace the democratic principle that an informed and representative public should be allowed into the decision-making process. Meeting the former challenge has yielded many of the results occurring in projects, although these are largely outside of existing agency responsibilities. Meeting the latter challenge is difficult, but the possibilities are blossoming, as evidenced by debates on regulatory reform and the EPA's initiatives such as Project XL.

Comparative risk and regulatory reform activities are moving toward decisions such as represented in the following (still fictional) scenario.

> A petroleum tanker loading facility in a remote location emits large amounts of VOCs. There is no significant human exposure to the airborne chemicals due to its remote location

and no evidence of harm to local flora or fauna. However, because the concept of "risk" does not require the damage to be obvious or proved, there is a risk from the emissions from this facility recognized by ecologists.

Again, because of its remote location, in fact, near thousands of acres of forests, the contribution of VOCs from the facility has no measurable impact on local ozone concentrations.

This facility is subject to significant regulatory action because of its high volume of emissions. Existing regulatory mechanisms would require the facility to meet specific standards which were designed and applied for denser population locations, such as Houston or Newark.

Can an alternative regulatory decision be designed which will actually result in greater environmental protection and economic cost savings? The answer is probably yes, but getting to that answer raises the process issues that comparative risk project experience can inform.

In terms of the human health impacts, it is not sufficient to calculate the total assuming an even distribution across the population. In order to consider special populations, the alternative regulatory process (i.e., comparative risk) would include representatives from those potentially affected, including low-income and indigenous populations.

The impacts to ecosystems require special consideration as well. While damage to flora and fauna may be currently unknown, future research may uncover insidious effects. In order to ensure that the regulatory action improves the environment, the goal of the collaborative assessment process should be to identify problems that all agree result in significant risk. From this point, the regulatory strategies can be designed to address higher risks, higher risks as seen by all participants.

The results of a grand change in policy formation which looks at a landscape of environmental conditions will probably be a recognition of the role of flexibility in controlling environmental harm. In addition, flexibility in regulation will be necessary to effectively achieve further environmental protection while allowing for variations in local geographic or political conditions.

Flexibility is a central theme in the evolution of comparative risk as a management tool. The original EPA comparative risk studies were initiated because of concerns about the relative magnitude of risks as described by a limited set of human health endpoints. Therefore, the initial studies focused on numerical risks that could be attributed to specific program areas. As states and localities tried their hand with comparative risk, public involvement tended to expand both the type of problem areas (to include individual stressors such as lead and classes of activities such as agricultural runoff). In addition, the criteria for considering risks expanded to include many of the quality-of-life risks, such as recreational access and aesthetics.

By expanding the range of analysis, comparative risk projects appealed to a broader audience than those who are most directly responsible for implementing specific federal and state statutes. However, broadening the analysis also makes it harder to arrive at answers to focused policy questions. Therefore, the consequence of increased flexibility is a more difficult application of risk ranking to specific agency decisions. It is for this reason that comparative risk and other environmental information tools are forcing agencies to consider alternative decision-making systems. Collaborative assessments and partnerships in decision making are an unintended consequence of the practice of comparative risk.

The evolution of comparative risk and new decision structures is not complete, but the early results suggest that any increase in the use of technical information in decision making will only occur with a parallel increase in the development of flexible decision processes.

APPENDIX: SEVERAL COMPARATIVE RISK PROJECTS

Washington Environment 2010

The most commonly referenced state comparative risk project was a pilot study sponsored by the EPA in 1988. One of the first state comparative risk projects, it was also one of the first to actively engage stakeholders in the discussion of risk. The hallmark of the Washington project is a series of legislative initiatives that reflect the priorities identified in the project.

According to an analysis carried out by the Northeast Center for Comparative Risk in 1992, the success of the Washington project in initiating legislation and focusing department activities on high-risk issues was largely due to the commitment of the department's director and adherence to a strong project structure with buy-in from broad representation within and outside of Washington state government.

Although conceived as an interdepartmental effort to set environmental priorities, Washington Environment 2010 focused most of its attention on the pollution-related issues for which the Department of Ecology was responsible. An interagency steering committee, made up of the leaders of 19 state agencies and representatives from two federal agencies, was nominally in charge, although it appears that the Department of Ecology took the lead on designing and implementing the project. A technical advisory committee of 26 state employees, project staff, and several consultants identified and analyzed the risks of 23 threats to six key environmental resources—air, water, land, wetlands, fish and shellfish, and wildlife. A public advisory committee (PAC) of 34 citizens representing many of the key interest groups and the Legislature reviewed the technical information and ranked the problems approximately one year after the start of the project. An Action Strategies Analysis Committee composed of 28 agency planners and policy analysts studied the risk management options.

Following the ranking, the PAC and the state hosted a 2-day public environmental summit in November 1989. The nearly 600 participants reviewed the ranking and suggested possible risk reduction strategies. The committees then conducted a series of town meetings across the state to discuss the results and to give attendees the opportunity to rank the threats as well as suggest possible risk management solutions. In June 1990, the project published an action agenda based on the problem analyses, the rankings, an analysis of possible action strategies, and the public response to the rankings and action proposals.

Many project participants give credit to Environment 2010 for contributing to the passage of a number of new state laws the following year (1991), including revisions to the Washington Clean Air Act and the Growth Management Act. In addition, the Department of Ecology targeted its budget requests toward the higher-priority issues and internally reprogrammed over $6 million from lower to higher risk management priorities.

Colorado Environment 2000

The Colorado Environment (CE) 2000 project, conducted from 1988 to 1990 under the auspices of the Colorado Department of Health's Office of the Environment, was another of the early pilot comparative risk projects supported by EPA. Like other projects, CE 2000 aimed to identify and build consensus around the most pressing environmental issues facing Colorado. To do so, the project ranked environmental problems based on risk and developed a full set of recommendations for action. The risk ranking was used by Governor Roy Romer during his election campaign in 1990, and the recommended actions formed the basis for several activities within the state.

The project structure for CE 2000 included two distinct phases: the technical analysis and development of recommendations. The technical work groups, made up largely of state government experts, conducted their analyses of environmental problem areas according to the environmental medium affected (air, water, land, natural resources). In June 1989, the technical work groups ranked problems by medium using human health, ecological, and economic criteria. After an initial ranking, the project's work groups selected representatives to meet jointly and develop an integrated ranking. At the end of the analysis phase, the project issued its Environmental Status Report describing the risks from 31 issues.

To begin the second phase, the governor appointed a citizens advisory committee (CAC) to review the work of the technical committees, set goals for Colorado's environment, and propose actions to meet those goals. This diverse group, representing business, industry, environmental groups, and community organizations, adopted a consensus approach to encourage collaboration. The CAC members formed and refined a "wish list" for their vision of the Colorado environment, and developed 26 environmental goals distributed among 11 problem areas. Later, the project held a set of environmental summits joined by the governor to which members of the general public were invited. During these sessions, CAC members and public participants generated potential solutions to the state's problems. This list of recommendations was not strongly connected to the previous risk ranking, and the group did not attempt to suggest priorities—two decisions made to reduce the potential for conflict among members of the CAC.

In June 1990, upon release of the project's final report, Governor Romer asked state agencies to consider the results of CE 2000 when planning their policies. In subsequent years, the Department of Health has continued to refer to the products of the project and used the results in developing its annual work plans. One notable result took place in 1992 when EPA granted flexibility to the Department of Health for reducing the threat from high-risk issues. In this experimental pilot agreement, approximately 5 percent of EPA funding was to be targeted to such areas as environmental lead, indoor air pollution, and groundwater wellhead protection. The process encountered significant internal resistance, and was eventually dropped.

In 1993, the Colorado Board of Health convened a meeting of several commissions (air quality control, hazardous waste, water quality control) to review the state's progress on implementing the recommendations produced in the CE 2000 final report. To assist that discussion, the Department of Health prepared a report reviewing its own progress. Although this progress report was never released to the public, the report did identify important changes in the status of the environment that updated the original comparative risk technical assessment report and also listed actions toward the goals developed by the CAC that pertain to the Department of Health. One goal listed in the original Final Report, e.g., was to reduce the amount of industrial and commercial hazardous waste currently generated annually in Colorado by 30 to 35 percent by the year 2000. According to TRI data, pollution prevention efforts by the department and industry allowed Colorado to reach that goal.

Louisiana Environmental Action Plan

Louisiana has a history of conflict between environmental interests and the petrochemical industry, and the state has a reputation for relatively lax environmental protection.

Furthermore, until the beginning of the Louisiana Environmental Action Plan (LEAP) process, the state lacked any tradition of collaborative policy making. Beginning in April 1990, former Governor Buddy Roemer and his Secretary of Environmental Quality initiated a comparative risk process designed to address these weaknesses.

As the LEAP project began, project planners sought to reach consensus on environmental issues among the environmental and industry communities. Participants sought to find a balance between scientific and public perspectives and, in the process, hoped to determine environmental priorities based on risk. In the end, they intended to create a broad coalition that would support the final Louisiana Environmental Action Plan. The project reached a milestone in November 1991, when the members of the project's joint public advisory and steering committee (PASC) reached consensus on a final ranking of environmental issues. That result has been called the first "environmental treaty" for the state.

LEAP evolved through four distinct phases: a technical phase of risk evaluation, a public phase of risk evaluation, the public outreach phase, and the risk management and future-options phase. The technical analysis, covering 35 issues, was conducted by staff from Louisiana's Department of Environmental Quality (DEQ) and other state agencies through the middle of 1990. Simultaneously, the PASC reached consensus on its vision statement. In the spring of 1991, the technical work groups briefed the PASC on their findings and rankings. Subsequently, the PASC ratified these rankings and used the information to develop its own provisional priorities ranking. The PASC then conducted an environmental summit and a series of town meetings during which members of the public reviewed the provisional ranking, produced their own ranking, and generated potential risk management options. During its final ranking, the PASC included the public feedback and incorporated other considerations such as legal authority and technological affordability, although how these factors were used remains unclear.

Following the ranking process, the project consolidated its results and released its project report. During this period, the DEQ started to draft action strategies published in 1992. After a new administration halted the project, little change emerged in DEQ. However, in December 1992, newly elected Governor Edwards revived the process with an executive order. The DEQ then began considering risk management strategies for the top 10 ranked issues in 1993. To date, it is not clear how much the project actually affected the DEQ. After the project was finished, the project staff left the state, and little attention has been paid to advancing the project's recommendations. However, a number of initiatives outside DEQ, notably a project focusing on environmental quality in the Mississippi River corridor between New Orleans and Baton Rouge, evolved.

Seattle Environmental Priorities Project

The Seattle Environmental Priorities Project (SEPP), the first EPA-supported local comparative risk (CR) project, helped identify the possible outcomes for CR at the local level. In 1990, Seattle Mayor Norman Rice adopted comparative risk to help unite a fragmented set of city agencies with environmental responsibilities. Through the SEPP, the city identified key environmental concerns and developed an environmental action agenda for the mayor. The end product was a set of recommendations approved by project participants and the Seattle City Council. Through 1993 and 1994, the city regularly used the results of the project by requiring budget submissions to be supported by the environmental action agenda.

The SEPP approached the task of assessing and managing risks in a fashion similar to the process used in the Washington state project. The project's technical analysts, working in land, air, water, and cross-media work groups, considered risks to human health, ecosystems, and quality of life. At the end of their assessment phase, the technical committees presented rankings and information on risks to the policy advisory committee. The rankings included both a relative risk ranking and a subsequent priorities ranking, which the technical committee produced after a brief consideration of the need for further action and the city's ability to address the problem given its jurisdictional and legal constraints.

Upon completion of the technical analyses and rankings in late 1991, the PAC worked over the next eight months to develop and prioritize an environmental action agenda. The PAC's government agency and nongovernment representatives produced 156 agreed upon recommended action strategies. Over the next 2 years, the city took action on over 120 of the proposals, and continues to use the action agenda to a limited extent today.

The Seattle project was noteworthy not only because it demonstrated the ability for cities to use comparative risk internally, but also because of the process to engender broad community support leading to public-private partnerships such as the city's Master Home Environmentalist Program initiative, a joint effort between the city and community members.

Hawaii Environmental Risk-Ranking Project

The Hawaii Comparison of Environmental Risks was initiated as an exercise at the East-West Center and the University of Hawaii. While it was under way, the Hawaii Department of Health adopted the project. The Department of Health planned to use the risk analysis of various environmental problems for the purpose of identifying higher management priorities and increasing the role of science in environmental decision making. Over the course of the project, a third goal emerged—to engage communities in a dialogue about the risks they faced for the purpose of increasing their awareness of the technical information and the department's awareness of community concerns.

Although implemented by the Department of Health (DOH), a significant portion of the work was conducted by contractors. Scientists at the East-West Center and the University of Hawaii analyzed the risks to human health, the state's ecosystems, and quality of life. While the human health assessment was fairly standard, both the ecological and quality-of-life assessments took somewhat unique approaches. For the ecological analysis, analysts recorded the severity and frequency of impact of 12 stressors for each of more than 400 sites. The quality-of-life assessment focused upon the impacts of environmental changes on native Hawaiian culture.

Initial drafts of the analysis were completed by the fall of 1992 and formed the basis for the second phase of HERR—community outreach. Under the direction of a community action team made up of community leaders from across the state, a series of meetings were held in 14 different communities. Community participants were introduced to the project at an initial meeting and in follow-up meetings were invited to raise additional issues of concern and to rank the environmental problems.

Based on the results of the technical analyses, rankings, and input from the communities, HERR produced a number of outcomes. In response to higher-risk priorities, DOH established an indoor air program and instituted a statewide testing program for blood-lead levels in children. In addition, DOH has sought to disseminate a strong conclusion coming from the analyses—in general, the risks to ecosystems are much greater than impacts to human health from environmental problems. Finally, project leadership notes that HERR has opened channels of communication and built trust between the community and the department to a degree that has not existed in the past.

California Comparative Risk Project

One of the most frequently cited examples of state comparative risk projects, the California EPA initiated the California Comparative Risk Project (CCRP) in early 1992. Initially, the project's primary goal was to evaluate and rank the most important environmental issues for the purpose of setting management priorities and evaluating methods for risk mitigation. As the project progressed, a second and, in relation to other comparative risk projects, somewhat unique goal emerged—to evaluate comparative risk and to consider other alternative organizing frameworks for priority setting such as environmental justice, pollution prevention, and sustainability.

An interagency management cooperative composed of directors, deputy directors, and other designees of 30 state departments, boards, and commissions was set up to provide general oversight and agency coordination and support to the project. In addition, the cooperative and project staff prepared a series of nine case studies examining the role of various types of information in existing priority-setting processes within California state government. A 38-member statewide community advisory committee (SCAC) provided input on the project's goals and strategies, worked with the project leadership and technical committees to develop a problem list, and sought to coordinate and integrate the various pieces of information being generated by project participants. (Plans for a series of regional advisory committees across the state were dropped due to budget constraints.)

Six different technical committees with membership from government agencies as well as various nongovernment sectors within the state carried out the project's analytical work. Three risk-ranking committees—health, social welfare, and ecological health—developed the methodology, collected the data, and analyzed and ranked the risks. Three other committees provided other perspectives and information. The environmental justice committee critiqued the comparative risk process as it had been practiced previously and suggested alternative methods for setting priorities based upon environmental justice and pollution prevention objectives. The economic perspectives committee identified the economic factors that influence environmental decision making and developed a framework for analyzing potential environmental quality goals or management actions. The education committee assessed the state of environmental and public health education in the state and made recommendations for enhancing the role of education as a risk awareness and reduction tool. In addition, the education committee provided recommendations for public participation within the CCRP.

The information and recommendations compiled by the six committees were used by the SCAC and project staff to develop a final report, entitled "Toward the 21st Century: Planning for the Protection of California's Environment." The California EPA released the report for public comments in May 1994. The report was never finalized, nor is there evidence that the project results have been used in any significant manner. While the CCRP did not produce any tangible policy results, the project did contribute new approaches to conducting a comparative risk project. The technical analyses are among the most comprehensive yet conducted and utilized assessment methodologies that went beyond those of most other projects.

The two project objectives—to achieve a risk ranking for priority setting and to critique risk ranking as a means to priority setting—were never reconciled. However, reviewing the results of the technical committees identifies no conflicting conclusions. Unfortunately, a small, probably nonrepresentative sample of the project's results were reported in the media and generated a furor from which the Wilson administration retreated. The inability of the California EPA to maintain support for the project has not allowed the SCAC the opportunity to synthesize the results. Without that synthesis, there is no means to translate the analysis into discussions affecting policy decisions.

Maine Environmental Priorities Project

In 1990, a group of business and environmental leaders forming a group called ECO/ECO, looking for a way to build a more thoughtful, proactive, and collaborative approach to environmental management in the state, became interested in developing a comparative risk project. After 2 years of planning by ECO/ECO members, Governor McKernan initiated the Maine Environmental Priorities Project (MEPP) through executive order in May 1993.

MEPP has been governed by a steering committee (SC) of approximately 40 members representing key environmental, business, and government parties. Recognizing that MEPP would have a lasting impact only if all parties supported its decisions and recommendations, the SC decided from the beginning to conduct all its business by consensus. The result has been a long and at times slow process, but SC members held firm to this commitment throughout the project, and all decisions regarding project structure (e.g., committee membership), substance (e.g., risk ranking), and process have been made by consensus.

The MEPP established three technical working groups (TWGs) selected from a list of volunteers with expertise in environmental issues and representing a wide spectrum of interests. In consultation with the TWGs, and with public input, the SC developed the list of issues to be studied. While TWG members assessed the risks from these issues to human health, ecological health, and quality of life, SC members conducted two rounds of public involvement to gather input from the public regarding their concerns, information, and values. A third public involvement effort was conducted to review the analysis after completion of the TWG reports.

Using the technical information, its own expertise, and the information gathered from the public, the SC spent 3 days discussing and developing a consensus ranking of the risks. In addition, SC members agreed to a statement for each issue explaining the rationale for the committee's ranking of that issue. The consensus ranking document was issued in January 1996.

Following the ranking, the SC agreed to focus its efforts. The narrowing process, while not based upon explicit criteria, took into account the information regarding its impacts or risks (i.e., how bad the problem is) and feasibility of the MEPP to effect change in this area (i.e., feasibility of making an impact). The SC agreed to focus on four areas and convened two task forces to develop recommendations for further action. Two reflect specific issues—indoor air and patterns of development—and are now in the process of developing specific recommendations. A third, volunteer monitoring, reflects the SC's recognition of the relative lack of data necessary for conducting the comparative risk project (and environmental management more generally) and both the need and potential for utilizing citizens more effectively in developing this information. The fourth task force, MEPP Futures, is looking for ways to institutionalize the MEPP process into the state's environmental management structure.

While the results of the individual task forces are still ahead, the strong collaborative process has already yielded a number of unanticipated outcomes. The risk ranking served as the basis for Governor King's environmental agenda, communicated early in 1996. In addition, SC members have participated as reviewers of Maine DEP's strategic planning process and the performance partnership agreement with the U.S. EPA. As a third example, in response to a growing controversy over forestry management practices in the state culminating in a proposed ban on clearcutting, both environmental and forest industry members of the MEPP steering committee served as key players within the Maine Forest Health Task Force, a group that developed a set of forestry management rules acceptable to a broader set of parties within the state. According to project participants, the experience in the comparative risk project, particularly the reliance upon consensus as a decision-making process, has been key to the success of the project as well as other related collaborative efforts that have become increasingly commonplace over the past 3 years.

Ohio Comparative Risk Project

The Ohio Comparative Risk Project (OCRP) commenced in 1994 with support from U.S. EPA, the Ohio General Assembly, and Governor George Voinovich. Sponsored by Ohio EPA, the project sought to involve citizens in a positive rather than confrontational process. A collaborative process was selected in part to help rebuild credibility with environmental and business interests in the state. The governor's executive order that created the project specified outcomes such as a State of the Environment report and two strategy documents: one outlining potential risk reduction strategies for the legislature to consider and another aimed at citizens.

The project was structured using the typical technical assessment component in the first phase, followed by the development of risk management recommendations. The OCRP completed its risk ranking in 1995 and finalized its recommendations in December 1996. As a direct result of the recommendation for increased support for environmental education in the state, Governor Voinovich linked the preexisting Ohio Environmental Education Fund with the CR project to form the new Office of Environmental Education at Ohio EPA.

The Ohio project had some unique features that make the initiative stand out. First, the state solicited input on environmental risks from a large cross section of Ohioans during its far-reaching public involvement activities. Using multiple outreach techniques, the project connected with more than 6000 people. Second, Ohio EPA proffered requests for proposals from communities around the state to conduct its own comparative risk projects. These smaller community projects were fed into the state project to provide a more balanced, detailed perspective on regional concerns. Third, the project made a significant effort to understand environmental justice considerations by supporting an independent study of concerns from African Americans in Cleveland, Cincinnati, Dayton and Columbus.

The Ohio project's public advisory group (PAG) was the first to *explicitly* use public information to produce an integrated ranking. Drawing on information gleaned from the public outreach process as well as input from the work groups and judgment of the PAG members, the PAG developed a set of criteria for ranking, some of which were weighted more than others. Once the PAG completed the categorical rankings, it used the public input once again to determine which risks were most serious. Accordingly, the group gave greater weight to human health concerns, followed by quality-of-life impacts, and finally impacts to Ohio's ecosystems. Upon receipt of final comments from the public on the draft rankings, the PAG reevaluated the quality-of-life ranking and the weight accorded to ecosystem effects in the overall ranking.

During the risk management phase, the technical and public groups merged into a single, expanded PAG with four task forces formed to develop risk reduction strategies (land, air, water, and waste). This work phase benefited from group members' previous experiences during the assessment stage. Their work was more efficient, and much of the analysis for potential strategies was tied to the State of the Environment report to provide context for the issue. As such, task force members focused more attention on the higher-risk issues.

PAG members agreed upon a set of criteria including measurability, feasibility, and fairness to sift through the recommendations. At the end of the process, the group recognized several themes such as education, compliance assistance and pollution prevention. One of these themes—the need for increased environmental education—received recognition in October when the governor authorized the formation of the new Office of Environmental Education. With a budget of $500,000, the office will be able to continue focusing attention on the needs identified in the project.

The project has produced some unexpected outcomes. For example, the results of the project will be included in Ohio's 1998 performance partnership agreement with the U.S. EPA, thereby enhancing the project's ability to influence agency priorities. In addition, the project will play a role in the agency's new strategic planning process. Finally, the agency will soon develop environmental indicators—another process that will be informed by the project.

Northeast Ohio and Cleveland Regional Environmental Priorities Project

Following more than a year of planning and building support for the project, the Northeast Ohio and Cleveland Regional Environmental Priorities Project kicked off in April 1994. Involving the four-county region around Cleveland, the project sought to build consensus for environmental improvement by establishing a common vision for the region's future, identifying the most urgent environmental problems threatening that future, and developing solutions to those problems.

Unlike other CR projects, the Northeast Ohio project could not rely upon the support of a single agency or political unit. Given the regional nature of the project and the multiple communities and organizations necessarily involved, a defining characteristic of the project was the development and maintenance of support from community leaders throughout the region. This played out in a variety of ways. A planning committee composed of 12 community leaders guided the initial design of the project and selection of participants. In addition, in recognition of the need for a "neutral" host, the project was housed within and staffed by the Center for the Environment at Case Western Reserve University. Finally, the process itself focused upon identifying common ground where it already existed and in building it where it was absent.

The project consisted of three technical committees (TCs) and a public committee (PC). The PC consisted of 75 members selected by the planning committee to represent the diversity of community interests. The PC generated a vision statement, modified the problem list, reviewed the information generated by the TCs, discussed and ranked the environmental problems, and agreed upon regional environmental priorities. In addition, the PC made all major decisions affecting the project. The TCs, with 90 volunteers from various sectors of the community (e.g., business, environment, government, academia), analyzed the risks, and generated the technical information that formed the basis for subsequent PC ranking and priority discussions, including their own ranking of the 16 problem areas.

The PC completed a provisional ranking in March 1995. Following further discussions with various community groups, soliciting their input on the ranking and relative urgency of various environmental problems facing the region, the PC revisited and confirmed the ranking. Using the ranking as the starting point, PC members agreed to focus their subsequent efforts on two of the high-risk issues—urban sprawl and energy use—for which there seemed a lack of coordination or focus of management activities and which the PAC felt the project could positively influence. Since the fall of 1995, two action subcommittees, composed of members from the PAC and additional volunteers not previously engaged in the project, sought to develop both short-term activities and long-term partnerships for addressing the problems.

Some initial positive results include the project's contribution to the adoption of new criteria for building roads by the regional transportation commission which are more sensitive to environmental and urban sprawl impacts. In addition, both the urban sprawl and the energy action subcommittees, working with other key actors in the region, have helped convene ongoing multistakeholder processes, examining policy options for more coordinated efforts at addressing specific components of these problems (e.g., preserving open space on a regional basis, designing a revolving loan fund to encourage energy efficiency investments). This project is also developing environmental indicators to track progress in addressing the region's environmental problems.

Priorities `95—Columbus, Ohio

The Columbus Department of Health initiated Priorities `95 in the fall of 1993. At the outset, the DOH sought to identify and define the community's environmental risks, assess and prioritize those risks to suggest where to target resources, and develop a plan for targeting resources to maximize risk reduction. As the project developed, the goals evolved to improve environmental and public health protection efforts in the Columbus metropolitan area; to build trust between government interests and area citizens; to educate participants and the public about aspects and concepts of environmental risk; and to use project results to seek increased local autonomy in identifying problems and implementing solutions.

Although initiated and staffed by the DOH, Priorities `95 was largely a community process. A steering committee made up of representatives from different community groups and organizations oversaw the key steps of the process, including selection and definition of issues to be assessed, selection of assessment criteria, completion of the risk ranking, and selection of management recommendations. A technical advisory committee (TAC), again staffed by community volunteers from all different sectors, helped to develop the project's issue list and assessment criteria and was responsible for completing the risk assessments. A public advisory committee (PAC) assisted in community education and outreach activities associated with the project. In particular, the PAC conducted six public forums to solicit input on perceived risks. The steering committee used this information in developing the final problem list for the project.

Following the agreement on a problem list in early 1994, the TAC proceeded with its data collection and assessment process. TAC members completed and communicated their assessments and technical rankings to the steering committee in early 1995. Using both this information and the input from additional public forums in which participants reacted to the technical information, the steering committee reached agreement on a final risk ranking.

Just prior to the ranking, project leadership formed a strategic planning committee. Charged with developing a list of possible risk reduction strategies, the committee produced a document outlining 192 recommendations corresponding to 30 ranked environmental problems. In response, the city agreed to internally evaluate the feasibility of implementing each recommendation. As a result of this review process, the city identified those recommendations which it was already pursuing, a second set that it would attempt to implement, and a third group that was currently not feasible to pursue.

One of the unique outcomes of the Columbus project is the formation of a nonprofit organization by some of the project's participants to oversee the city's implementation of the Priorities `95 recommendations. Formed in early 1995, Priorities Partners, Inc. seeks to educate the public about the risks identified in the project, provide guidance to city officials in interpreting the broadly stated recommendations, and establish links between city and agency leadership and the Priorities `95 recommendations.

Elizabeth River Project—Norfolk, Virginia

When citizens in the Norfolk region of northern Virginia recognized the limited ability of government to respond to environmental concerns, they developed a grassroots, non-finger-pointing project using comparative risk to build a consensus plan for addressing the Elizabeth River's highest-risk problems.

The Elizabeth River estuary, once a highly productive marine resource, has become one of Chesapeake Bay's most highly polluted reaches. At one of the nation's busiest commercial, military, and industrial ports, the river plays a significant role in the region's economy. Although there have been many improvements to environmental quality in the Elizabeth River, numerous studies identified continued impacts from high levels of contamination.

After a group of citizens decided to mobilize the community, they adopted comparative risk to help them assess risks to the river and to generate momentum to improve its health. Over a 7-month period in 1994, citizen volunteers from academia, environmental groups, business and industry, the military, and government produced white papers examining the range of problems. In the fall of 1994, this group of citizens produced a consensus ranking of 10 broad risks facing the river. Using the consensus list of high risks, the project assembled a planning team to develop a set of recommendations for reducing risk. Over the next 6 months, the project's four strategy development work groups used criteria including effectiveness, affordability, and acceptability of the proposed action or strategy to winnow their list to a manageable level. In the end, the group produced a consensus watershed action plan that received the full endorsement of both project members and a wide range of influential regional leaders without whose support little action would be possible.

Now in its implementation phase, the project has attracted nearly $1 million in investments including funding for toxics monitoring, a wetland restoration project, support for environmental education, and funding for an Army Corps of Engineers study on toxic sedimentation remediation actions.

Houston Environmental Foresight Project

Upon request from the Houston Endowment, the Houston Advanced Research Center (HARC) developed a comparative risk project designed to seek consensus on an action agenda of important regional concerns. HARC, an entity known for its neutrality, convened a broad group of stakeholders to analyze and rank risks for the eight-county Houston region.

This project, known as the Houston Environmental Foresight Project, began in June 1994 at approximately the same time that the state of Texas initiated its project, but the two were not linked. The foresight project consisted of a science panel with three subpanels which reviewed the 19-issue list of environmental concerns. These work groups submitted their volunteer assessments that used a standard range of human health, ecosystems, and socioeconomic considerations to the foresight committee, a broad group of 32 community leaders including conservationists, concerned citizens, and industry and government representatives.

The foresight committee used the information provided by the science panel as a basis for its consensus risk ranking. The project reported its findings in a 1996 report in which the risks were identified as regional, community, global, or affecting natural systems.

Currently, the project has begun soliciting additional funding to produce the foresight plan, an intended product that will link the risk information with recommended risk reduction activities.

Minnesota Comparative Risk Project

The Minnesota Pollution Control Agency (PCA) embarked on a comparative risk project to help the agency set environmental priorities. Beginning in 1994, the PCA began a strategic planning process aimed at improving the overall effectiveness of the organization. Like many other states' environmental agencies, the PCA faces increasingly limited resources, growing demands for its services, and a host of emerging problems. The agency adopted the comparative risk process in 1995 to help shift to risk-based management.

The PCA comparative risk project has several interesting features that differentiate it from other state projects. For example, the PCA structured the process to be completed in less than the typical 2-year period. This decision increased the need for efficiency in conducting the technical assessment and public involvement components of the project.

To elicit information directly from citizens regarding their perceptions of environmental risk, the PCA's comparative risk steering committee decided to conduct a set of "citizen juries" to engage the public in an evaluation of environmental risk. As with other projects, the PCA process entailed technical analysis by members of the PCA staff covering 12 broad issue areas. Unlike most other projects, the analyses considered only issues over which the agency has authority to regulate and manage. Through the late summer and into the fall of 1996, PCA program staff assessed a range of risks covering human health, ecological, and socioeconomic concerns to prepare a set of concise issue descriptions. These abbreviated technical analyses provided the information base for the first citizen jury conducted under the auspices of the Jefferson Center for Democratic Processes.

In the fall of 1996, the PCA and Jefferson Center convened a panel of 20 "average" Minnesotans to explore the issues presented by the PCA staff. Using the technical information and their own values, the citizens developed a consensus risk ranking that will be used in future discussions as the agency works to set priorities.

The same technical information base along with the results of the first citizen jury will form the background for a second jury that will be composed of stakeholders from the business, local government, and environmental communities. Once the PCA receives the results from these rankings, agency personnel plan to conduct program evaluations to consider how to address high-ranking risks and whether the current allocation of staff and resources in specific program areas are still appropriate. Using this information and other information on mandates and existing responsibilities, PCA staff anticipate development of a risk management strategy with specific recommendations for reducing the identified risks.

The PCA intends to complete the comparative risk project in 1997, but anticipates revisiting the process in subsequent years.

Mississippi Comparative Risk Project

In response to legislation passed in early 1994, the Mississippi Department of Environmental Quality (DEQ) initiated a comparative risk project in August 1994 as part of the development of its strategic plan. Project leadership formed a public advisory committee comprised of representatives of 32 organizations, including environmental, business, local government, academia, community, and agricultural interests.

DEQ staff and PAC members identified 23 issue areas as well as specific stressors associated with each issue. Project leadership then formed three technical committees—human health, ecosystems, and quality of life—to assess the impacts of each of the issue areas and then rank them. Technical committee membership came primarily from DEQ staff, but PAC members were invited to participate on the committees and did in some cases. The three technical committees completed their assessment and ranking in July 1995.

The human health and ecosystem technical committees accomplished their assessment tasks in a fashion somewhat unique in two ways for state comparative risk projects. First, committee members developed and applied an algorithm to calculate and numerically score the risks from each issue (often selecting individual stressors that appeared to be more significant). Second, the data used and therefore the scores calculated were broken down to a county level. As a result, these committees were able to make comparisons between counties for single problems as well as to calculate which counties were being impacted disproportionately by multiple risks.

Following the discussion and confirmation of the ranking results by the PAC during the fall of 1995, PAC members and DEQ staff worked together to develop goals and strategies for each of the issue areas. During this process, they focused on innovative approaches that would simultaneously increase risk reduction, decrease the regulatory burden, and, if possible, improve the quality, quantity, and accessibility of data available to support environmental management in the state. A draft set of goals and strategies was completed in the summer of 1996. DEQ staff are currently in the process of taking the results of the comparative risk project and integrating them into the strategic plan.

Other Projects

Other States with Completed Comparative Risk Rankings in 1997

Vermont
Alabama
Florida
Kentucky
Texas
Arizona
Maryland
Utah
New Hampshire

Other Localities with Completed Comparative Risk Rankings

Guam
Athens County (OH)
Clinton County (OH)

States Currently Developing a Comparative Risk Ranking

New York
New Jersey
Iowa
Tennessee
Alaska

REFERENCES

Dea, J., and S. Thomas, 1997. *Building a Foundation for Change: Opportunities and Challenges in State Comparative Risk Projects,* U.S. EPA and Green Mountain Institute for Environmental Democracy, Montpelier, Vt., 11 pp.

Delhagen, E., J. Dea, and K. Kramer, 1996. *Comparative Risk at the Local Level: Lessons from the Road,* Western Center for Environmental Decision Making, Boulder, Colo., Dec., 55 pp.

Finkel, A.M., and D. Golding (eds.), 1994. *Worst Things First? The Debate Over Risk-Based National Environmental Priorities,* Resources for the Future.

Fiorino, D.J., 1990. "Can Problems Shape Priorities? The Case of Risk-Based Environmental Planning," *Public Administration Review,* vol. 50, pp. 82–90.

Iowa Dept. of Natural Resources, 1997. *Iowa Comparative Risk Project: Technical Assessment.*

Jones, K., 1997. *A Retrospective on Ten Years of Comparative Risk,* Green Mountain Institute for Environmental Democracy, Montpelier, Vt., Jan., 30 pp. plus appendix.

U.S. EPA, 1987. *Unfinished Business: A Comparative Assessment of Environmental Problems Overview Report,* OPPE, Feb., 118 pp.

U.S. EPA, 1991. "Setting Environmental Priorities: The Debate about Risk." *EPA Journal,* 17:2, 64 pp.

U.S. EPA Science Advisory Board, 1990. *Reducing Risk: Setting Priorities and Strategies for Environmental Protection,* SAB-EC-90-021, Sept., 32 pp.

SECTION 13.3

POLLUTION LIABILITY AND LIABILITY INSURANCE

Harlow G. Ballard
The Harlow G. Ballard Insurance Agency
Columbus, Ohio

INTRODUCTION 13.40
COST AND BENEFIT OF RISK MANAGEMENT 13.40
Role of Insurance in Risk Management 13.41
Insurance Costs as a Factor 13.41
Other Risk Management Tools 13.42
POLLUTION LIABILITY INSURANCE CONTRACTS 13.43
PAYMENT OF POLLUTION CLAIMS 13.43
CLAIMS MADE VERSUS OCCURRENCE POLICIES 13.44
SELF-INSURANCE 13.44
EXPLORING THE INSURANCE MARKET 13.45
PRICING OF POLLUTION COVERAGE 13.46
CONCLUSION 13.47

INTRODUCTION

Everyone would like to avoid the cost and confusion brought about by the necessity of purchasing insurance. At the time premiums are due, or when insurance companies are demanding compliance with safety recommendations, it is difficult to remember that insurance is the only ultimate assurance that we will be able to continue business operations in the face of accidental catastrophe. Fortunately, we live in a country with a dynamic and widely available competitive insurance market in which insurance for almost every need is available to responsible buyers, and in which pricing can be bid down by sagacious buyers.

Relative to the characteristics of the entire insurance market, the band of companies seeking buyers of pollution liability is relatively narrow, although growing. Likewise there are relatively few buyers of pollution liability—enough, however, to make the marketplace competitive, innovative, and reasonably flexible for sophisticated buyers to generally find the coverages they need at affordable costs. Buying strategies are important, as is a thorough search of the marketplace. There are many variations of coverage, and pricing is not always based as much on actuarial science as it is on "what the traffic will bear." Shrewd buyers will be aware of the many factors involved in loss control, available coverage forms, and insuring company pricing approaches before they enter the marketplace. This process begins by understanding what the risks of loss are and determining the proper balance of risk management techniques via loss control versus risk management through the purchase of insurance.

COST AND BENEFIT OF RISK MANAGEMENT

As a professional concerned with the technicalities of air pollution control systems, one cannot avoid two additional roles, those of a *risk* manager and of a cost-benefit analyst.

The *risk,* in this case, which must be managed is the risk of a release of pollutants which could result in lawsuits and consequent cost—prohibitive defense costs and possible catastrophic judgments—against a company. The technological choices considered are bound inextricably to the evaluation of pollution liability risks before, during, and after the installation of a control system; indeed risk evaluation followed by cost assessment of loss control measures must be central to one's decision-making model. Likewise, the choices of technology bear directly on the degree and type of risk management (RM) choices that must be made *after* your purchasing decision is made. Unfortunately there is not a totally efficient failsafe air pollution control system available; if there were, the RM job would be complete when the ideal purchase decision was made.

It is clear that strict accident control and maintenance procedures must be in place from the outset, but here also one contends with the implications of human and mechanical error and finds further need to discover more risk management techniques. One finally arrives at a point that insurance must fill in for other RM deficiencies. Indeed, insurance is a risk management tool.

Role of Insurance in Risk Management

What is the cost of risk management, and, more important, what are the tradeoff costs of relying on insurance instead of other risk management techniques? The answers to these two intertwined questions are as varied as the number of sizes and types of companies that exist which need pollution control. What is known, however, is the following: (1) There are more variables weighing on projecting the long-term cost of insurance than there are in pricing of equipment, safe installations, and personnel to safely operate and maintain equipment. The varying costs of insurance potentially volatile. One policyholder's costs can be drastically increased due to adverse claims experience, or a whole population of policyholders can suddenly feel the effects of insurance industry loss experience, unscientific pricing of policies (actuarial science, notwithstanding), and insurance industry competition, or lack of it, especially in the air pollution insurance market. (2) Also, insurance can become virtually unavailable as it did temporarily for medical malpractice during the "good Samaritan" scare, or during the landslide of the asbestos liability abatement claims 20 years ago.

Insurance Costs as a Factor

Given the availability of insurance and stability in pricing, there is greater certainty of complete protection relative to other risk management techniques if the correct insurance plan is in place and the proper amount of protection is purchased. The question of "affordable insurance costs" is the great unknown, addressed in factors 1 and 2 above. What is affordable this year may not be next year. Thus, the need becomes clear to have nearly unlimited expertise in risk management through loss control, as well as in its silent partner, *insurance.*

While costs of insurance can be volatile, an understanding of its uses, types, purchasing techniques, and ancillary services of insurance carriers can, and must, lead to its most efficient use, at which point the *cost of loss control plus the cost of insurance equals the lowest amount of dollars spent on risk management.* That is, there are potential tradeoff costs such that premiums can be reduced when certain loss control techniques are used. Thus, it must be determined if the combined cost of loss control plus insurance premiums goes up or down when such loss control techniques are employed. Naturally, there may be external considerations to the overall cost-benefit model, as a company may be willing to bear an overall higher-cost alternative because there may be community or employee relations factors that are overriding.

In planning for air pollution control systems, the blithe assumption should never be made that insurance will simply fill in the risk management gaps. True, it can fill them in, but given inadequate loss control, costs of insurance will eventually skyrocket or insurance will become unavailable; or costs of insurance can change drastically due to industry reasons unrelated to any given policyholder. Generally, insurance should be purchased only to fill those gaps where other RM measures cannot be accomplished or are inadequate.

Other Risk Management Tools

The "other" risk management measures abound, ranging from the type of system(s) a company purchases, to the manner and placement of those installations, the caliber of the people hired to monitor and maintain those systems, the contractors (along with the warranties they offer) selected to do the installation and future maintenance, the use of backup systems and warning systems, planning to be able to meet increased risks that may arise with increased production or changes in production techniques, and to many other choices depending on the specific manufacturing type of company with which one is concerned. All these have a determinable cost, some of which then must be weighed against the cost of certain options which exist in the purchase of insurance and some of which will be necessary in addition to the costs of insurance. That is, not all risk management measures can, or should for economic reasons, be traded off for insurance; but some can and should be traded.

One risk management technique, which is almost cost-free to the buyer and which could result in major insurance premium savings, is to carefully scrutinize the insurance and warranty programs of the seller and/or installer of a new pollution control system. Here is a checklist of basic factors to look for in the insurance program of a control system manufacturer, as well as the contractor who installs the system:

- Liability limit should be no less than $1,000,000 with a $5,000,000 umbrella policy on top of the basic policy. Higher limits, up to $35,000,000, are available to pollution control system suppliers and installation contractors, but the costs of such high limits must be passed on to the system purchaser.
- Their insurance carrier should be rated by A.M. Best Company at no less than "A."
- Their general liability policy should be on an "occurrence" basis. (See further reference to occurrence policies versus claims-made policies in this section.)
- They must carry products and completed operations coverage, which provides coverage for pollution caused by their product or their completed work. This requirement could be more important than the buyer's own pollution coverage!
- They must provide a certificate of insurance, with all the above detail, which warrants that the holder will be notified with no less than 30 days' notice if the insurance should be canceled.

In addition to these insurance factors, the warranty provided by the system manufacturer should be thoroughly scrutinized by an attorney who understands the buyer's risk management goals and budget. The insurance program and warranty programs of the system manufacturer can go far to offset future risks and insurance costs of the buyer. The insurer of the buyer should be informed of the insurance program of a system provider as their subrogation possibilities could be a major factor in offering premium discounts.

It is the goal of the remainder of this section to understand the major purchasing options and purchasing strategies for insurance for the buyer of pollution control systems, such that economical tradeoffs can more easily be made with other risk management techniques.

POLLUTION LIABILITY INSURANCE CONTRACTS

Coverage within general liability insurance policies for liability arising out of pollution of land, water, or air went through a major transformation from 1966 through 1987. Prior to 1966, pollution damaging "accidents" were fully covered by general liability policies, and in 1966 the language of policies was actually broadened to include *occurrences,* which meant that gradually occurring incidents, over time, were covered the same as sudden and accidental incidents. However, in 1987 insurers accepted new wording in the general liability policy which has been defined as an *absolute pollution exclusion.*

Efforts were made by a few insurers to write a policy which would specifically insure against pollution hazards, but that was short-lived and now the approach is to remove or modify the pollution exclusion of the standard general liability policy. In this way it is now practical for a manufacturer to purchase pollution liability as part of its general liability policy provided its existing carrier is willing to remove the exclusion. Most carriers are not willing to modify their policies, which means the manufacturer has to find a special carrier who will both provide general liability at a reasonable cost and, for an additional cost, remove or modify the absolute pollution exclusion.

For smaller companies, where the air pollution risk is fairly low and well controlled, there are still a reasonable number of insurance companies which will remove the absolute pollution exclusion for a very reasonable cost. A company should always first approach its present carrier about removing the exclusion before making the assumption that a special policy has to be purchased. It may be that its insurance agent, or another agent, does have a standard carrier which will provide general liability *and* will remove the exclusion.

PAYMENT OF POLLUTION CLAIMS

In any litigated claim there are first defense costs and then, possibly, judgment costs. Pollution claims are almost always defense cost-intensive. For this reason it is important to be sure that the policy chosen allocates defense costs for payment *outside* the policy limit. The entire policy limit should be reserved for judgments against the polluter because the cost of limits, which are sure to be great enough to pay high and unpredictable defense costs, will be prohibitive. Where it is possible to arrange for an existing *standard* carrier to simply remove the pollution exclusion, it is probable that the policy limits do allow for defense costs to be paid outside the policy limits. However, where it is necessary to find a special carrier, as it will be in most cases, the insurance buyer must be careful to avoid insurers which make the defense cost part of the policy limit. This distinction should be specified by the buyer at the outset and should be documented in the insurance proposal.

CLAIMS MADE VERSUS OCCURRENCE POLICIES

Where at all possible, pollution insurance buyers should specify that they want an *occurrence* policy; this is almost as important as specifying that defense limits be paid outside the policy limit. There are two types of general liability policies: *occurrence* and *claims made*. The insurer of a claims-made policy is obligated only to honor claims that are made during the policy period within which the carrier has continuously provided insurance. Conversely, the carrier of an occurrence policy is obliged to pay claims, even if the policy has long since expired if the occurrence which gave rise to the claim happened while the carrier carried the insurance.

If a manufacturer has had an occurrence policy and then agrees to accept a claims-made policy, say, in order to add pollution liability, then the cost of the insurance is likely to be low in the first year but could increase annually to an exorbitant premium eventually. The reason for this is that in the first year the claims-made carrier has no reason to be concerned that occurrences in prior years, which could give rise to claims, will have to be paid because the previous occurrence carrier will be responsible; however, with each passing year, the risk increases that a claim will be submitted to the claims-made carrier from a previous year in which it was the carrier. Thus, costs over time tend to be much more stabilized under occurrence policies.

The second major reason to avoid claims-made contracts is because it becomes very expensive to ever switch back to an occurrence policy. That is, once the claims-made policy has expired, that carrier has no obligation to pay any claims that were not made during the policy period, unlike occurrence carriers which theoretically have an eternal obligation. Thus, if a switch is made, the new occurrence carrier also must be requested to provide retroactive coverage for "sleeping claims," known as *tail coverage,* and this can become very expensive.

Companies are now becoming fairly free in offering occurrence-based pollution liability policies since competition is heating up the marketplace. However, many carriers will still offer claims-made contracts, hoping that low pricing in the first year will attract buyers.

SELF-INSURANCE

The use of *deductibles* is the most common type of self-insurance. Deductibles are most common in property insurance, and almost all people are well acquainted with the fact that they are responsible for a small amount of any loss. However, pollution liability almost always has a relatively high deductible, the lowest even for a very small light-risk polluter being no less than $5000 to $10,000. Large companies may have deductibles as high as $500,000. Of course, the higher the deductible, the lower the premium.

It is clear that if a company is self-insuring, via a deductible, the higher the deductible, the greater is the importance of internal risk management techniques. Thus, the insurance buyer must request a proposal which offers a full range of premiums for a full spectrum of deductibles. The savings between each deductible category must be weighed against the cost of additional internal risk management practices and strategies.

The selection of a certain type of pollution control system, e.g., has a direct effect on the amount of deductible which should be selected, assuming that the choice of system is based on the reduction of risk of having a pollution accident or occurrence, or the avoidance of EPA fines (the amounts of which are more predictable than the cost of litigation of a real pollution liability suit). Of course, this decision is one that can only be wisely made through statistical probability methods, but where potential costs of the self-insured risk can vary by tens of thousands of dollars, the investment in statistical research simply must be accepted as part of overall risk management costs.

One would expect an insurer to be agreeable to increasing the amount of the deductible in all cases; unfortunately, there is a "Catch 22." While it is true that insurers prefer to avoid smaller claims, they also must be assured that their policyholders will be able to afford to pay the deductible. Therefore, if it appears that the manufacturer may not be able to afford the payment of a high deductible, the insurance company might set a deductible ceiling. The normal procedure is for the company to pay the entire claim and then to collect the deductible amount from the insured. Thus, for this and other reasons, any insurer will thoroughly examine the financial condition of any company which buys a large policy.

EXPLORING THE INSURANCE MARKET

The logical first step in reviewing an insurance program is to ask one's present agent to review the policies and to report on gaps, overlaps, and any possible need to add new coverages. The buyer should insist on this being a formal process with all information presented in writing. However, it may be possible that the present agent does not have the expertise to purchase and obtain the best pricing on pollution liability.

Unless the buyer of pollution liability insurance is insured with a standard carrier who is willing to remove the pollution exclusion of the present general liability policy, it will be necessary to explore special markets. There are about 15 serious and high-rated specialized pollution liability carriers operating in the U.S. market. The three largest are American International Group (through the Commerce and Industry Insurance Company), Reliance Insurance Company, and Zurich Insurance Company. The London market (Lloyd's) also captures a sizable share of the U.S. market.

The most important initial decision is the selection of an agent. A good agent may select a specialty broker who will have access to the insurance companies that can do the best job; or if one of the large national insurance agencies is selected such as Johnson & Higgens or Willis-Caroon, the agent will have direct contracts with these companies. A buyer may get better service and pricing from a moderately sized local agent, assuming he or she is well acquainted with the issues and knows which brokers or large national agents to contact. Frequently, a good local agent can combine technical expertise and understanding of the marketplace with the special attributes of clients and be in a better position to bargain and get the best prices. Also, local independent agents are more frequently in a position to spend the necessary time working on behalf of the local client than would a large national agent who may be diverted with numerous clients larger than the one at hand.

Pricing of pollution liability has been gradually falling, and most carriers have absolute minimum premiums of $5000 to $50,000 for the pollution coverage on top of whatever premium is developed for the general liability. Of course, the premiums for larger companies, with many plants, can run into the hundreds of thousands depending on the size of the deductibles.

An alternative to specialty insurers is to join a risk retention group (RRG). Such groups exist for most specialty insuring needs, but in the area of pollution liability they are quite esoteric and closed to new members unless the new members closely fit within the characteristics of the other members in terms of size, financial solvency, and risk level. Usually, premium savings are significant, and often coverage is superior to what may be available from insurance companies. RRGs are essentially insurance buyer groups which provide a relatively low amount of coverage to their members for small claims, but which have direct reinsurance contracts with major carriers to handle the large claims. Pollution liability groups (PLGs) have very high membership charges, high enough to discourage members who have made this investment from leaving the group. They operate with a high degree of confidentiality, as they do not want a barrage of companies wanting to join them; and, they have extremely high standards of loss control as the group must be protected from loss. They are essentially insurance pooling groups. Insurance rating organizations do rate RRGs. A company should exercise extreme caution before joining an RRG, but it can be a viable way to hold down insurance costs where all members of the group are of a single mind to control losses and limit membership to the most select candidates.

Most national brokers are acquainted with the available risk retention groups and may refer a company if the "fit" appears to exist. Most local agents will not be familiar with these groups, but they can make direct inquiry of their national brokers to determine if membership for their client is a possibility.

The selection process of any insurer should always include the investigating of its loss control services. First, a good loss control program not only assists a manufacturer in avoiding claims, which could mean the outlay of a large deductible, and helps to avoid EPA fines, but also keeps down the cost of future insurance premiums if guidelines are followed. With some carriers, the loss control procedures can be a nuisance, with inspectors making irrelevant or petty recommendations. Therefore, prospective policyholders would do well to go beyond reading the insurance company's literature about loss control procedures and, perhaps, insist on meeting a loss control manager, or to take time to interview the loss control inspector when the preliminary inspection is made.

PRICING OF POLLUTION COVERAGE

It has already been pointed out that pricing is always subject to predetermined minimum premiums. These minimums cannot be avoided, but many factors influence the amount of premium charged above the minimum. Unofficially, one of the greatest determinants of the cost of any ancillary coverage, such as pollution liability, is the amount of premium generated by the basic general liability account. On one hand, if the basic general liability coverage, exclusive of the pollution exposure, is large, has a good loss history, and is considered low-risk, the carrier will be greatly influenced in its pricing of the pollution liability coverage. On the other hand, if the pollution liability is large in relation to the remainder of the account, the pricing is likely to be high.

The type of manufacturing process and the products manufactured, along with the annual sales and payroll of the manufacturer, are the primary official determinants of the basic pricing, but many loss control factors influence final pricing. The type of pollution control systems is a major consideration, and the policy premium may be highly discounted for state-of-the-art systems. In fact, companies which have updated their systems, or plan to, and which already have pollution liability coverage in place should request that discounts be allowed to their present premiums in view of the upgraded systems.

Premium discounts should be available based on the type of warranty that is given with a pollution control system, and the insurer should be provided with the supplier's and the installer's certificates of insurance along with copies of any warranties.

Insurers should be made aware of the credentials of company personnel who manage and maintain the pollution control systems, of special warning devices, and of all matters relating to any special technology. Insurance companies tend to be suspicious and to generalize regarding their policyholders; it is important to be persistent and to let it be known that other carriers are being considered in order to get the special attention needed to obtain the best pricing. Companies have a great deal of latitude in pricing policies, and at this time the market is competitive and buyers should make this work to their advantage. However, a rash of claims could change this, and as the politics of the environment ebbs and flows, the litigious nature of our population changes.

The financial fate of pollution liability carriers, as with all insurers, is inextricably tied together via the use of common reinsurers. All reinsurers will feel the effects of increased losses, and the costs will be passed on to all the players in the direct insurance market, which in turn will be felt by all buyers of pollution liability insurance.

The role of deductibles has been explained, and in requesting premium quotations, the cost of a full range of deductibles should be specified. After the basic pricing and all discounts have been applied, the use of deductibles will be the major variant in pricing. It must be remembered, as was pointed out in the beginning of this section, that deductible savings is a tradeoff for the cost of other loss control measures. Insuring companies tend to encourage deductibles where policyholders can afford them, because deductibles give the policyholders an incentive to develop their own internal loss control procedures as well as to follow the company's loss control recommendations.

CONCLUSION

It is clear that buyers' expertise is essential, though difficult to achieve, in the purchase of exactly the right insurance program. Without close scrutiny, a buyer cannot even be sure that the agent has sufficient expertise in such specialized areas as pollution liability. It may be advisable to hire a consultant to review current policies and to put together proper submissions for a new insurance program.

SECTION 13.4

ENVIRONMENT, ECOLOGY, AND ETHICS*

Jim Kenney
Council for a Parliament of the World's Religion
Chicago, Illinois

WHY ASK ETHICAL QUESTIONS? 13.49
THE RIGHT AND THE GOOD 13.49
PARADIGM SHIFT 13.51
BACKGROUNDS: EMERGENCE OF AN ENVIRONMENTAL AND ECOLOGICAL ETHIC 13.52
PARADIGM SHIFT REVISITED 13.55
NEW PARADIGM THINKING: ENVIRONMENT, ECOLOGY, AND ECONOMICS 13.56
THE CHANGING RIGHT: RELIGION AND ECOLOGY 13.58
CONCLUSION: RIGHT LIVELIHOOD 13.59
REFERENCES 13.61

This section addresses the ethical dimension of the modern concern with the human relationship to the environment. We will touch on the essential structure of ethics, its modern transformation, and the relevance of that structure and transformation for the current struggle to find an ethical compass as the human-environment relationship grows more complex and more daunting. We will also touch on the influence of religion and spirituality on human attitudes toward the earth. While much of this discussion will focus on developments in the United States, it has important implications for the developing ethical conversation beyond our shores.

What is ethics or morality?† What makes a given course of action ethical or moral? How does an individual—committed to a given ethical or religious tradition—understand the implications of that tradition in the context of a particular ethical decision? And finally, how do we know what is right and/or good? Arguably, questions of this sort have animated the human quest for understanding since its first stirrings. Their original province was the religious dimension, and most of the great religious traditions have attempted to provide principles, norms, and authoritative guidelines to structure understanding and behavior.

When our attention shifts to modern environmental and ecological concerns, however, the teachings of the traditional religions often seem inadequate as guides to the complex earth-centered issues which confront the human community at the threshold of the twenty-first century. We will discuss changing religious attitudes toward these issues later in this section.

---

*It would perhaps be useful at the outset to clarify a distinction between the terms *environment* and *ecology*. In popular discourse these days, particularly with regard to ethical questions, *environment* or *environmental* is most often used to refer to situations where despoliation of a particular region, habitat, or sector is at issue (e.g., environmental pollution, toxic waste, fouling of the waterways). The use of the term *ecology* or *ecological*" suggests a focus on the entire earth system—and generally implies a concern with larger ethical and moral issues, including human life-style choices, overarching socioeconomic assumptions, and the responsibility of every person 1. It would perhaps be useful at the outset to clarify a distinction between the terms, for the stewardship of the planet.

†While the terms *ethics* and *morality* are generally used interchangeably, it is clear that the former often implies a societal and temporal context for judging behavior, while the latter somehow implies a reference to universal, timeless, or religious principles. In this section, however, we will not make a distinction between these terms.

WHY ASK ETHICAL QUESTIONS?

Clearly, we take up the study of ethics in order to find answers to the critical personal questions What should I do? and Why should I do it? As human beings, we are aware that we are called upon to make choices and that we have options with regard to our behavior. Often, those choices have profound ethical and moral implications. The Pulitzer Prize–winning biologist René Dubos put it this way, "Man makes himself through enlightened choices which enhance his humanness" (Dubos, 1968, p. xii). Some (e.g., behavioral psychologists such as B. F. Skinner) would argue that choice and free will are illusory and that human behavior is determined by past circumstance. Nevertheless, it is a basic principle of ethics that we enjoy at least some freedom of choosing those actions which we carry out.

Ethics argues in addition that we are *responsible* for our behavior. To be responsible in a given situation is, above all, to be *able* to *respond.* Since we are capable of reflection and deliberate choice, we are ethically presumed to be *able to respond* appropriately to life's challenges. We are the originators of our actions and can therefore be judged accordingly.

We are, moreover, aware of others. Most ethical decisions involve others, other persons, other beings, or—by extension—the environment in which we live and choose. Ethics asks about the effects a given action will have on others and demands that such effects be weighed in the decision-making process.

THE RIGHT AND THE GOOD

In the west, the sixteenth-century advent of the Scientific Revolution set in motion a very gradual shift away from the moral norms provided by the dominant religion, Christianity, and toward a new structure grounded in philosophy and its offspring, science. By the close of the eighteenth century, the European Enlightenment had given rise to a radically new approach to the determination of the proper course of human action in general and in particular circumstances. To understand this critical development, we should turn our attention for a moment to one of the most fundamental dichotomies in modern ethical theory: that which separates the *right* from the *good.*

In normal conversation, we tend not to distinguish what is *right* from what is *good.* In the study of ethics, however, the distinction is critical. To argue from the *right* is to invoke some "transcendent" principle. That is, a proposed course of behavior is right or wrong to the degree that it accords with a given (and usually unchallenged) standard of duty and obligation. The standard may be discovered in scripture (e.g., "the Bible says..."), in some system of philosophy (e.g., "as Kant demonstrates..."), in a political document (e.g., "the Constitution demands..."), or in revered local tradition (e.g., "we have always done it this way"). Clearly, the approach from the *right* directs our attention to the past, to some agreed-upon set of principles.

On the other hand, to argue from the *good* is to direct attention to the future consequences of an action. That action is good or bad depending on the impact it is likely to have on the individual, his or her immediate environment, or the larger social or even planetary context. The ethical approach which inquires into the "goodness" of a course of action directs our attention to the future, to the likely consequences of our actions. This ethical mode is often termed *utilitarian,* since its usual method is to attempt to predict an action's outcome and thereby to gauge its probable usefulness or appropriateness and, ultimately, its ethical acceptability.

Throughout most of human history, ethical and moral questions were resolved through recourse to the *right,* to known principles, to the accepted norms of religion, society, and tradition. But the cultural revolutions which produced the Renaissance, the scientific revolution, and the Enlightenment produced a radically new way of thinking about human behavior and its ethical and moral character. The new argument went like this: To the degree that we can make reasonable predictions with regard to the outcomes of proposed courses of action, we ought to be able to make reasoned judgments about their appropriateness. The argument from the *good* came gradually to the forefront.

Beginning with the eighteenth-century European Enlightenment, a new struggle emerged in western life and thought: to identify, to describe, and to guarantee essential human rights. In part, the struggle drew on notions of the *right,* but it was most deeply rooted in the developing concept of the *good.* In the United States, we can trace its development from the first stirrings of the American Revolution, with its focus on the civil and political rights of human beings, through the early nineteenth-century emergence of efforts to secure the rights of women and blacks, to the late nineteenth and early twentieth-century campaigns for the rights of workers, to the late twentieth-century fight for the rights of the poor (Burns, 1991). In each phase of the rights struggle, the attempt was made to link utilitarian concepts of the good society with widely accepted older principles, in other words, to ground modern ideas of the *good* in cherished notions of the *right.* The Declaration of Independence, e.g., affirmed "that all men are created equal and are endowed by their Creator with certain inalienable rights, and that among these are life, liberty, and the pursuit of happiness." The vision which gave life to these words, however, grew out of the Enlightenment conviction that such an affirmation would tend to yield a more equitable, beneficent, and productive social order.

As the struggle to identify and secure human "rights" proceeded, the conversation came more and more to be dominated by the effort to develop a comprehensive understanding of the *good.* Always, however, the effort persisted to tie the *good* to the *right,* to demonstrate that the newly understood *good* was consistent with the time-honored understanding of the *right.* American rights activists in the nineteenth and twentieth centuries were often at pains to invoke scriptural as well as Constitutional principles in support of their consequence-based arguments. By the late twentieth century, U.S. civil society faced an even more complex issue, as the notion of the rights of nature itself began to take shape.

Much of the modern ethical confusion in the face of environmental and ecological issues is a direct result of the fact that our decision making tends to be based on norms drawn from each of the two most basic ethical and moral modes. At times, of course, the dictates of the two approaches are in harmony. Often, however, they seem to be at odds.

Perhaps the most instructive example of the potential conflict between *right-based* and *good-based* approaches is to be found in the modern western debate on abortion. For the most part, advocates of the "pro-life" position argue from the *right,* holding up what they see as time-honored religious principles and supporting them with citations from scripture. On the other hand, the "pro-choice" argument has most often been based on attempts to describe the arguably negative consequences of limiting access to abortion.

To be sure, each side does from time to time cross over into the ethical territory of the other. Abortion foes speak of the long-term societal consequences of abortion on demand—among them a possible diminishment of the value placed on human life in general. Abortion rights proponents invoke what they see as the unchallengeable principle of a woman's right to govern her own reproductive life. But the debate remains largely focused on the seemingly unresolvable opposition of assertions of the *right* and claims about the *good.*

Reflection on the abortion controversy also provides an essential insight into a critical difference between the two ethical and moral approaches. Those who argue from the *right,* from traditionally accepted principles, can claim to their advantage that their position is known to be "true" or "right." They are at a disadvantage, however, when conventions begin to change. Those who argue from the *good,* from scientific or other projections of likely outcomes, find that their position is only as strong as popular acceptance of those projections. On the other hand, in an age dominated by science, the argument from the *good* has extraordinary appeal.

Finally, note that very few of us make ethical and moral decisions in a consistently *right*-based or *good*-based fashion. Instead, we often find ourselves, prompted by our developing understanding of the *good,* struggling to come to a new understanding of the *right.* This modern dilemma is nowhere clearer than in the swirl of facts, assertions, and complex issues which surrounds the environment-ecology discussion. Much of the modern argument for heightened ethical concern with the human impact on the natural world is based on new understandings of the likely consequences of disregard for the delicacy of the ecological system of the earth (the appeal to the *good*). On the other hand, resistance to environmental causes often finds its ethical ground in traditional notions of the human as the master of all he or she surveys (the appeal to the *right*).

PARADIGM SHIFT

It has been some 30 years since the philosopher Thomas Kuhn, in his landmark work *The Structure of Scientific Revolutions,* described the progress of science as marked by periods dominated by more or less fixed sets of assumptions but occasionally punctuated by noncumulative breaks with tradition or with "the norm." Kuhn used the term *paradigm shift* to denote the relatively abrupt departure from prevailing theoretical models, predispositions, experimental protocols, and preeminent hypotheses and the entry into a radically new scientific view of reality.

A paradigm is a pattern or structure of belief and understanding which consists of a network of associated ideas, concepts, ways of speaking, assumptions, beliefs, attitudes, values, songs, and stories which come together to influence the way one "sees" the world. Just as rose-colored glasses filter the light which reaches the wearer, paradigms (as sets of notions about what is good, true, possible, etc.) influence our experience of the world and our reaction to it. Paradigms are of vital importance for human experience and understanding. They condition our patterns of response to the world of everyday experience, to each other, and to the unseen universe. A culture necessarily depends upon a net of assumptions so interwoven that to challenge one is to threaten the entire structure. As a result, a real paradigm shift eventually touches and transforms every phase of the life of the individual or the society in which it takes place. It can be argued that ethical and moral decision making is one of the areas of human life and experience most profoundly affected by the change of the prevailing paradigm.

That ours is an age shaped by such a paradigm shift can scarcely be doubted. In virtually every field of expertise, inquiry, or endeavor, fundamental, paradigmatic changes are underway. The last major paradigm shift became apparent over 400 years ago with the coming of the scientific revolution. It was shaped by the elaboration of the new model of the universe as rational, lawlike, and mechanical. That model displaced the medieval notion of a universe governed solely by divine whimsy. The current shift, prompted in large part by the twentieth-century scientific revolution—and the relativity and quantum theories—proceeds from a radical insight into the essential interdependence of all events and processes in the universe. The mood swing has clearly begun, and the current paradigm shift may prove to be the most significant development of the modern era.

It is hard to imagine a more ironic cultural development: that the twentieth century would leave as its legacy the unlikely combination of devastating ecological disregard and the transformative discovery of the intimately interdependent character of all that exists. Before we take up the implications of the modern paradigm shift for the ethical approach to environment and ecology, however, we will turn to a review of the historical evolution of our attitudes toward the natural world.

BACKGROUNDS: EMERGENCE OF AN ENVIRONMENTAL AND ECOLOGICAL ETHIC

The sixteenth-century Copernican paradigm shift, which set the scientific revolution in motion, began with a significant diminishment of human significance in the scheme of the universe. Copernicus's most upsetting claim (later elaborated and supported by the work of Kepler and Galileo) was that the earth was not the fixed center of the universe, but rather a *planet* (from the Greek word for "wanderer") in orbit around the sun. It was a stunning blow to a Christian world which had long accepted the Aristotelian notion that the earth was the center of the scheme of creation. Humankind, it was believed, was "the crown of creation," the ultimate focus of the divine creative purpose. If earth were merely another planet and not the center of all, what was to be said of human beings? But the anxiety was to be relatively short-lived.

Over the course of the next century, the paradigm shift advanced dramatically. By the time of the death of Isaac Newton in 1727, a new intellectual and moral order had emerged, grounded in the elevation of the human to an unprecedented status as the unchallenged master of the physical world. Sir Francis Bacon summed up the new vision with the words, "Knowledge is power." Humankind now held the keys to knowledge and had begun to exercise what seemed likely to prove unlimited power.

The scientific revolution soon yielded the extraordinary transformation which came to be known as the Enlightenment or the dawn of the Age of Reason. This eighteenth-century movement brought the paradigm shift to bear on virtually every dimension of human philosophical, social, political, and moral life. It proclaimed the reality of natural law and universal order, championed a scientific approach to political and social issues, and celebrated the inevitability of human progress. It brought a promise of power and prosperity as the birthright of humans and the natural consequence of their intellectual attainments.

Just as the thinkers of the Enlightenment (e.g., Voltaire, Adam Smith, Rousseau, Hume, and Kant) took up the question of human rights, so, too, did they develop an elaborate vision of human mastery over nature. That human domination over the earth was an ethical *good* was scarcely questioned. Arguing from assertions of the consequences of such a course, the new philosophers and political economists described the limitless benefits of such mastery. Moreover, the movement found affirmation in appeals to the *right,* e.g., in enthusiastic citations of the biblical assertion of the principle of human "dominion" over nature. ("Be fruitful and multiply, and fill the earth and subdue it; and have dominion over the fish of the sea and over the birds of the air and over every living thing that moves upon the earth." Genesis 1:28.)

This line of thought would continue into the nineteenth century, growing in influence as the industrial revolution radically multiplied the technologies of control over nature. The nineteenth century, however, also produced another cultural dynamic, sometimes called *romanticism.* This movement, in large part a reaction against what many saw as the cold rationalism of the Enlightenment, celebrated the "spirit of the divine" or the "hand of God" as an active force not only in the creation of the universe but also in its continuing unfolding. A new concern for the natural world began to emerge.

By the standards of ecological activists in the late twentieth century, the romanticists' understanding of the human relationship to the earth left much to be desired. It remained thoroughly anthropocentric, focusing exclusively on nature as a source of human delight and spiritual uplift, rather than as the fragile arena of all life. Much of the effort which the movement produced lacked any real grounding in environmental science, which was for the most part a significantly later development. Nevertheless, especially in the United States, the romantic movement gave rise to an unprecedented degree of attention to and affection for nature.

The second half of the nineteenth century saw the burgeoning of the "nature essay" as a major U.S. literary genre. In 1849, the U.S. Department of the Interior was established. In 1851, Henry David Thoreau declared in an address to the Concord (Massachusetts) Lyceum that "in Wildness is the preservation of the World." Three years later, he would publish *Walden, or, Life in the Woods.* The period also witnessed a remarkable procession of concerned naturalists who called attention to the wonders of the U.S. environment and demanded a new ethical and moral understanding of a sacred human trust. They included John Wesley Powell, John Burroughs, William Cullen Bryant, John Muir, and Theodore Roosevelt. By the end of the century, Congress had passed legislation designating Yellowstone, Yosemite, and Sequoyah as "public parks."

The new environmental movement seemed at this point to promise a steady increase in understanding and conservation of the beauty and resources of nature. It was not to be.

World War I put an end to the romanticists' idyll with nature and severely damaged the progressive political movement which had supported many of the conservation initiatives advanced in the decades bracketing the turn of the century. Like the romantics, the progressives were sanguine about human nature, affirming its basic goodness, and were confident that the cultivation of a truly ethical, social, and political order was eminently possible. In a very real sense, widespread access to education was seen as the most essential ingredient of reform. At the heart of the progressive movement lay an ancient conviction—first voiced in the West by Socrates—that knowledge is the root of virtue. If people come to recognize and to understand the critical issues which face society and the world, progressives maintained, they will respond as the situation demands.

The progressive movement breathed the air of the Enlightenment. Characterized by a profound ethical and moral concern and commitment, it based the greater number of its positions on arguments from consequences, i.e., on assertions of the *goodness* of given courses of action. To be sure, progressives appealed from time to time to the *right,* to transcendent principles of the sort which had found a place in U.S. political rhetoric since the time of the Declaration of Independence (e.g., "All men are created equal"). In essence, however, progressives found their animating center in their understanding of and confidence in progress itself. World War I, however, which had left in its wake unprecedented human, economic, social, and environmental devastation, called the notion of human goodness and the concept of progress into grave question. By the early 1920s, the progressive movement and its ethic of the *good* had receded into insignificance.

The era of the depression, ironically enough, began with a further erosion of the belief in progress and ended—in the New Deal—with a restirring of progressive idealism and the ethics of the *good.* American consciousness of the fragility of nature was profoundly deepened by the dust bowl experience of the 1930s as skies were darkened across United States by swirling clouds of ancient topsoil. In addition, the work of the Civilian Conservation Corps brought a largely unanticipated heightening of environmental awareness by opening up hiking trails and campgrounds in some of the country's most scenic areas. Still, however, a comprehensive environmental and ecological ethic lay decades in the future.

The impact of World War II on the development of an earth ethic is discernible only in long retrospect. In the years immediately following the war, in the United States and the larger world, ethical attention and reflection focused much more on the human relationship to other humans than on that of humankind to nature. The Universal Declaration of Human Rights, promulgated in 1948 by the United Nations, became the most important social ethics document of its time. Not until the Rio Declaration on Environment and Development, published by the United Nations Conference on Environment and Development (UNCED) at the 1992 Earth Summit, was a comparable environmental and ecological ethical compact created.

It is important, however, to note that much of the modern conversation about human ethical responsibility for the stewardship of the planet earth draws deeply on World War II's demonstration of the awful power of unchecked human ingenuity—a demonstration which culminated in the devastation of two Japanese cities. The atomic mushroom cloud remains today one of the most powerful icons of the human threat to human and other life, and to the earth itself.

The turbulent 1960s saw the rebirth of the environmental movement. This extraordinarily complex period is most often remembered only for its more colorful aspects—flower children, dramatic opposition to the war in southeast Asia, and psychedelic excess. The 1960s, however, had another face, the understanding of which bears directly on the present discussion. As a generation of young people began to question—to be sure, often with considerable naiveté—the basic principles which had so long shaped U.S. culture, business, government, and education, the ethical common ground of the society began subtley to shift. Often misconstrued as simple irresponsibility, the dominant ethos of 1960s youth culture actually manifested one of the most dramatic of a series of modern shifts—beginning with the European Enlightenment—from the *right* to the *good* as the center of ethical decision making. This nascent culture did not lack values; rather, it sought urgently (and often unsuccessfully) to identify and to embody *new* values, even as it rejected some of the most cherished principles of the larger society.

While cultural historians tend to dwell on the antiwar dimension of the 1960s' dynamic, its deepening sensitivity to environmental and ecological issues may prove in the long run its most lasting and precious legacy. The communal experiments of the period have often been dismissed as primarily sexual in character. A great many, however, grew out of a burgeoning interest in the development of a new ethic of relationship to nature—living lightly and simply on the land. At the same time, awareness of a whole range of critical environmental issues increased. In 1962, Rachel Carson's landmark study, *Silent Spring,* first focused widespread attention on the disastrous impact of chemical pollution—fertilizers, herbicides, and pesticides as well as toxic effluents of every sort—on the world around us. By the end of the decade, topics including the population explosion, the "greenhouse" theory, and the steady poisoning of land, air, and water were as widely discussed among the youth as the Vietnam war itself.

In the early 1970s, the publication of *Limits to Growth,* the report of the Club of Rome—one of the original international think tanks devoted to the detailed study of the effects of human technology and population growth on the life systems of the earth—marked the first major comprehensive statement of the complex ecological peril facing the planetary community. While many of the direr forecasts contained in the report have not come to pass, the substance of its warning call has been validated and vindicated in countless ways in the past two decades. As Thomas Berry has pointed out, "Few books have had such a startling impact in awakening America to the absurdity of exponential rates of growth" (Berry, 1988, p. 68).

Gradually, the outlines of an ethic were beginning to take shape as scientific and popular understanding of the earth's ecosystem deepened. In the same period, another event dramatically demonstrated the potential of the new environmental consciousness. In April 1970, the first Earth Day astonished skeptics and true believers alike as hundreds of thousands of people in the United States participated in local, regional, and national gatherings, sharing information and celebrating a common commitment to the development of new ways of living and a new ethic of responsibility for the planet. Since 1970, Earth Day has waxed and waned in terms of organizational efficiency, numbers of participants, press coverage, and the like. It has, however, gradually evolved into an institution of considerable significance in the broadening of popular awareness of the essential features of an environmental and ecological ethic.

Over the last two decades, a series of developments at the international level has given greater form and depth to the ethical conversation. Organizations such as the United Nations Environment Programme (UNEP), the International Union for the Conservation of Nature and Natural Resources (IUCN), the World Wildlife Fund (WWF), World Watch, the Millennium Institute, and many others have made great strides in articulating the dimensions of the "global problematique" and setting forth essential ethical guidelines for what has come to be called *sustainable living* or *sustainable development.* In 1991, *Caring for the Earth: A Strategy for Sustainable Living* was published by a partnership including IUCN, WWF, and UNEP, as well as many other sponsoring and collaborating organizations. It offered the following simple ethical principle as a foundation statement for all subsequent reflection on the human relationship to nature:

> Be conservative in actions that could affect the environment, study the effects of such actions carefully, and learn from your mistakes quickly. (IUCN, 1996)

In June 1992, the Earth Summit, the United Nations Conference on Environment and Development (UNCED), convened in Rio de Janeiro, Brazil. As noted above, the Rio Declaration on Environment and Development was the first international compact to address the rights and ethical responsibilities of nation-states with respect to environmental and ecological issues.

As we move toward the threshold of the next century and the next millennium, we find ourselves in a paradoxical situation. Never before have environmental awareness and ecological understanding been keener; and yet never before has the complex of challenges to the biosphere been more alarming. The litany has become all too familiar: deterioration of the ozone layer (slowed but continuing), ever more daunting evidence supporting the threat of global warming, destruction of the equatorial rain forest, loss of species' habitat, and the increasingly apparent reality of the pollution of land, air, and water. Challenge, however, often keeps company with opportunity.

PARADIGM SHIFT REVISITED

As we have noted, it is often suggested of late that ours is another age of *paradigm shift*—the first such major transition since the scientific revolution—a culturewide transformation of our most fundamental values and assumptions about the universe, the planet, and the human condition. Although the environmental, social, economic, and spiritual problems which confront the planetary community seem at times insurmountable, there are reasons for hope. On every side and in virtually every discipline—in physics, medicine, social science, education, economics, business, and religion, to name but a few—one can encounter new ways of thinking, new models, and new modes of understanding, all animated by a shared appreciation of the radical interdependence of all life, the rediscovery of the ancient understanding that every thing, person, lifeform, place, species, habitat, and event influences and is influenced by every other.

A cultural transformation of the first order may well be stirring. It is, to a degree, self-empowering (with the all but irresistible energy of the proverbial "idea whose time has come"). Paradoxically, however, it demands the enthusiastic advocacy of concerned and informed members of the world community if it is to be fully *realized.* Visions of alternative futures must first be widely shared before they can be achieved.

We live in a time of transition from mechanistic and reductionistic models of experience (concerned with the analysis of moving parts) to models which may be characterized as holistic, or even *organic* (concerned with the description of the behavior of entire systems). The central fact of the modern paradigm shift is the rediscovery of the interdependence of all existence. Its characteristic theme is the convergence of what had seemed to an earlier age to be impossibly disparate and unrelated dimensions of reality and modes of human experience. Perhaps the old Zen maxim says it best: "Touch a blade of grass and shake the universe to its foundations."

It is clear enough that a rekindling of the ancient understanding of the wholeness of existence and the intimate interdependence of all life would be the most precious gift that earth could receive from her children. Our personal and communal commitment to that revisioning—with all its global and local implications—is the great and the only hope.

A good friend once asked me, "How many people do you think lived through the Renaissance?" I have never forgotten his question. He was not asking about population figures, but about the essence of the age. How many people knew then that they were living in an age of revolution? How many woke each morning eager to find out "how the Renaissance is coming along"? How many simply went about their daily tasks, "blissfully" unaware of the new world taking shape around them? Who responded to the opportunity and the challenge? Who exulted?

The situation that confronts us today is similar in many ways. We live in an era of revolutionary transformation, and yet we are often so preoccupied with simply "coping" that we fail to notice dramatic changes in the fundamental rhythms of life and thought. Gradually, however, we may be beginning to awaken. The steadily increasing response to the challenge facing the home planet expresses a growing consciousness of the vital reality of interdependence. Ironically enough, the despair we sometimes feel may be a sign of that growth. If we are dismayed at the mess we seem to have made of things, it may be because we are beginning to understand. Perhaps the step from lamentation to commitment will prove shorter than we might have imagined.

NEW PARADIGM THINKING: ENVIRONMENT, ECOLOGY, AND ECONOMICS

In the environmental and ecological context, the change in understanding seems to be proceeding from the emergence of a wide variety of new models, in virtually every discipline, which share a common emphasis on the interdependent, holistic character of the most basic dimensions, processes, and events of the universe. In view of the global threat to the environment, these new models provide the clearest indication that the old paradigm has dangerously outlived its usefulness. A number of new ethically significant developments seem worthy of note:

- Increasing understanding of the biosphere as the intricate and fragile web of all life
- Growing awareness of the very real limits to unconsidered growth in an interdependent system

- The gradual development of economic approaches in which environmental stewardship is reflected as a value and environmental pollution and degradation are clearly perceived as a significant cost
- The promotion of ecology—in the broadest possible sense—as a coherent, consistent, and comprehensive political platform, rather than simply a plank
- Changes in corporate culture, including increasing efforts to bring the environmental and ethical discussion to bear on all aspects of business and public policy
- The beginnings of a rediscovery within the world's major religious traditions of a spirituality of wholeness, interdependence, evolution, and transformation
- Heightened sensitivity to the metaphor of earth as Gaia, of the planet as living organism

Nowhere is new-paradigm thinking more vital or more relevant to ethical concerns than in the economic arena. Economics is the dominant discipline in modern society. Ours is an era dramatically characterized by the enshrinement of economic institutions at the very pinnacle of the social and cultural order. The bank in a sense occupies the promontory where once the temple stood. (It is no accident that the architecture of the former so often recalls that of the latter.) In light of this fact, one of the most significant indications that a new paradigm may be taking shape in modern U.S. and European society is the emergence of a challenge to the most fundamental assumption of prevailing economic theory and policy: that all economic growth is good and that more economic growth is better. The challenge takes form in the posing of a handful of simple but vital questions: Growth of what? Growth benefiting whom? Growth with what consequences?

An important feature of prevailing economic theory is its claim to be purely pragmatic and value-free, lying somehow beyond political, psychological, sociological, ecological, or ethical and moral realities. This notion is rejected by an increasing number of thinkers who insist that it grows out of the misguided focus on money and its movement as the primary objects of economic concern. Economics, they argue, is not just about money. Its "raw material" is resources—natural resources, human resources, and finally capital resources. In this view, the real task of the human science of economics is to increase human welfare, in all its social, environmental, emotional, and physical dimensions. Thus, economics demands a profoundly interdisciplinary approach if it is to avoid the pitfalls of purely mechanistic and reductionistic theorizing about abstract patterns which do not even attempt to take into account the full systems complexity of planetary reality.

One clear implication of some of the new approaches to economic theory is that individuals and communities must begin to move toward policies grounded in a holistic, systems view of the reality of the planetary "household." The Greek word *oikos* ("an inhabited house, a home") is the root of such words as *economics, ecology,* and *ecumenism,* and reflects a more ancient mode of understanding which seems not only relevant but also empowering in the modern situation. The call for a new economics, "as if people and the earth really mattered," is actually a summons to an ethical, moral, as well as political renewal.

The "new economists" reject the old model's uncritical reliance on abstract patterns (e.g., supply and demand) which do not even attempt to take into account the full complexity of planetary reality. And at the heart of the modern discussion lies the old assumption that "growth is always good." As Paul Ekins argues:

> It is only by:
>
> - Showing that the growth has taken place through the production of goods and services that are inherently valuable and beneficial,

- Demonstrating that these goods and services have been distributed widely throughout society,
- And proving that these benefits outweigh any detrimental effects of the growth process on other parts of society, that one can arrive at any sort of assessment as to whether a particular instance of economic growth is in fact a good or bad thing. Such an evaluation will vary, depending on personal attitudes and on the political perspective, and will still say nothing about economic growth in general. In time, perhaps, certain patterns of economic growth might come to be recognized as benign and life-enhancing, while others are perceived as wasteful, polluting, or inequitable, as the case may be. Conventional economic thinking makes little or no attempt to make this assessment, nor has it developed the conceptual or political tools for such a task. For the New Economics, their development and implementation is of the most fundamental importance. (Ekins, 1986, p. 8-9.)

To the extent that economic thinking fails to take seriously the radical interdependence of the life systems which make up the earth's biosphere and the finite nature of planetary resources, it is impossible even to begin to make the kinds of ethical choices which are demanded by the modern social and environmental crisis. When we define "progress" in terms of the increasing availability and consumption of products, when we think of global development as the industrialization of the Third World, when we accept the premise that the clearcutting of the tropical rain forest belt is economically advantageous, and when we deal with concepts of wealth and work in a strictly materialistic fashion, we sacrifice more than clean air and water. When ethical concerns are set aside or inadequately addressed, a desperate threshold is crossed. The news, however, is not all bad.

We are witnessing today a deepening of participation by individuals, groups, and communities in the struggle for planetary renewal and for social, scientific, political, economic, and ethical remodeling. Participation of this sort has genuine power not only because it brings about a concentration of efforts on the problem at hand but also because it offers to humankind the sort of shared project which we so desperately require.

THE CHANGING RIGHT: RELIGION AND ECOLOGY

As we have seen, the growth of ethical concern with the human relationship to the earth relates directly to the longer-term shift from the ethical dominance of notions of the *right* to the preeminence of ideas of the *good.* The modern environmental movement has, until recently, tended to distance itself from traditional ethical and moral principles, viewing them as hopelessly anthropocentric and, consequently, irrelevant to the realities of the ecological crisis. Many environmentalists have been particularly hostile to the world's religions. In a now-famous 1967 article in *Science,* Prof. Lynn White (1967) asserted that the environmental crisis was grounded in "the victory of Christianity over paganism." White went on to argue that, in the book of Genesis, the human is "not simply part of nature: he is made in God's image." Moreover, God "planned all this explicitly for man's benefit and rule.…By destroying pagan animism, Christianity made it possible to exploit nature in a mood of indifference to the feelings of natural objects." (White, 1967, pp. 1203–1207.)

To be sure, a cursory reading of the world's great religious scriptures provides precious little guidance for life in an age of ecological peril. A brief reflection on the relationship of the "right" and "left" of the modern U.S. and European political spectrum would seem to support such a conclusion. It is useful in this connection to note that the politics of the right tend to be grounded in an ethics of the *right,* while those of the left tend to espouse an ethics of the *good.* That is to say, modern conservativism appeals to traditional values, principles, and ideals, while modern liberalism argues most often from analyses of the future consequences of various courses of personal or public action. It is not accidental that the modern environmental and ecological movement has drawn far greater support from the left than from the right. Unfortunately, this clash of ethical models has given rise to a destructive divisiveness in the face of an increasingly urgent global threat.

The perceived long silence of religion on the human-earth relationship has exacerbated the ethical conundrum. Once again, however, the current paradigm shift offers reason for guarded hope. For the last several years, a new religious address to the question of human moral responsibility for the state of the earth has been emerging. In 1990, at the first International Conference on Caring for Creation, the Reverend James Parks Morton, of New York's Cathedral of St. John the Divine, a long-time leader in the environmental movement, reminded those present that only 10 years earlier there were very few "religious types" who demonstrated any level of real concern for the earth. He went on to say that we were witnessing the first stirrings of a "new collaboration between religion and environmentalism." At the same conference, Judge Russell Train, chairman of the World Wildlife Fund and Conservation Foundation, said:

> At long last religion seems to be awakening to the environment....And we really have no other option in this regard. If we truly care for the human condition, then we must necessarily care for the rest of creation on which humanity's well-being and even existence so clearly depend....After all, ecology is nothing more than creation at work.*

In 1993, in Chicago, the Parliament of the World's Religions brought together over 8000 representatives of the world's religious traditions for nine days of encounter, dialogue, and reflection. One of the principal themes of the gathering was the address by religion to the critical issues which face the human community at the threshold of the twenty-first century. Again and again, participants called for a thoroughgoing reinterpretation of the basic teachings of the religions with regard to issues of peace, social justice, and ecological responsibility. Numerous presentations demonstrated that religious wisdom, properly understood, speaks powerfully to the modern crisis. Beginning in 1999, the Parliament of the World's Religions will reconvene every 5 years at a new location. The 1999 parliament will be held in Cape Town, South Africa. One of the central themes of the gathering will be the relevance of traditional religious teaching to the elaboration of an ethical and moral approach to environment and ecology.

What is happening, in other words, is a convergence of understanding of the *right* with understanding of the *good.* It should, perhaps, come as no surprise that a deepening concern with desired outcomes would bring about a reevaluation of guiding principles. It is, in any case, one of the most promising features of the current struggle to arrive at an ethical consensus on the human-earth relationship.

CONCLUSION: RIGHT LIVELIHOOD

These days I find myself thinking more and more about a man I never met. The late E. F. Schumacher was, in the words of Theodore Roszak, an "anarchist," whose "deliberate intention was to subvert `economic science' by calling its every assumption into question, right down to its psychological and metaphysical assumptions" (Schumacher, 1973,

*The conference was organized by the North American Conference on Religion and Ecology (NACRE). Citations are from unpublished addresses.

pp. 2–3). He was also an extraordinary thinker and one of the most articulate spokespersons of "new paradigm" thinking in the field of economics and environmental ethics. What set Schumacher apart was his remarkable sensitivity to questions of product, progress, scale, and growth in the modern world. The "logic of production" is neither the logic of life nor that of society. It is a small and subservient part of both. The destructive forces unleashed by it cannot be brought under control, unless the "logic of production" itself is brought under control—so that destructive forces cease to be unleashed. It is little use trying to suppress terrorism if the production of deadly devices continues to be deemed a legitimate employment of human creative powers. Nor can the fight against pollution be successful if the patterns of production and consumption continue to be of a scale, a complexity, and a degree of violence which, as is becoming more and more apparent, do not fit into the laws of the universe, to which humans are just as much subject as the rest of creation. Equally, the chance of mitigating the rate of resource depletion or of bringing harmony into the relationships between those in possession of wealth and power and those without is nonexistent as long as there is no idea anywhere of enough being good and more than enough being evil (Schumacher, 1973, p. 295).

It might prove useful, in the context of the present discussion, to reflect on a very different cultural approach to ethical considerations. In his powerful book *Small Is Beautiful: Economics as if People Mattered,* Schumacher drew on one of the fundamental teachings of Buddha—that individuals and communities must deepen their commitment to *right livelihood.* In the Buddhist understanding, right livelihood entails simplicity, fairness, noninjury of other living beings, and a deeply grounded sense of the interdependence of all existence. (It is interesting from a modern perspective to note that Buddhists have traditionally regarded at least five occupations as specifically prohibited: trading in arms, in living beings, in intoxicating substances, in poisons, and in flesh.) It may not be immediately obvious, but such a spiritual-philosophical-ethical outlook implies an economic model just as surely as the materialist outlook gives rise to modern theories of the "movement of monies, goods, and services." As Schumacher put it:

> It is clear, therefore, that Buddhist economics must be very different from the economics of modern materialism, since the Buddhist sees the essence of civilization not in a multiplication of wants but in the purification of human character.... A Buddhist economist would consider (the modern) approach irrational: since consumption is merely a means to human well-being, the aim should be to obtain the maximum of well-being with the minimum of consumption. (Schumacher, 1973, pp. 55, 57)

Right livelihood means "living lightly," in ever-increasing awareness of the symbiotic delicacy of the earth community. It demands what Schumacher called "appropriate technology," employing strategies which take into account existing cultural patterns, levels of development, resources, and the like. It calls for what Willis Harman terms a commitment to "the learning society," moving beyond economic models based on production to models based on the deepening of human understanding and the enrichment of the planetary community as an organic whole (Harman, 1988). Above all, it means living in a way that promotes rather than hinders the growth of all beings. It is the commitment to the intensification of life and the enrichment of consciousness. No one can doubt that the choices which face the earth family are daunting, but as Schumacher reminded us:

> It is not a question of choosing between "modern growth" and "traditional stagnation." It is a question of finding the right path of development, the Middle Way between materialist heedlessness and traditionalist immobility, in short, of finding "Right Livelihood." (Schumacher, 1973, p. 62)

REFERENCES

Berry, Thomas. 1988. *The Dream of the Earth* (San Francisco: Sierra Club Books).

Burns, James Mcgregor, and Stewart Burns. 1991. *A People's Charter: The Pursuit of Rights in America* (New York: Alfred A. Knopf). Perhaps this is the best account of the long U.S. debate over "rights."

Dubos, René. 1968. *So Human an Animal* (New York: Charles Scribner's Sons).

Ekins, Paul (ed.). 1986. *The Living Economy* (London: Routledge & Kegan Paul).

Harman, Willis. 1988. *Global Mind Change* (Indianapolis: Knowledge Systems Inc.).

IUCN. 1996. *Caring for the Earth: A Strategy for Sustainable Living,* Gland, Switzerland.

Schumacher, E. F. 1973. *Small Is Beautiful: Economics as if People Mattered* (New York: Harper and Row).

White, Lynn. 1967. "The Historical Roots of Our Ecological Crisis," *Science,* Mar. 10.

ACRONYMS

8-HRTWA	Eight-hour time-weighted average
AAL	Acceptable ambient level
ACGIH	American Conference of Governmental Industrial Hygienists
ACS	American Chemical Society
ACT	Alternative control technology
AL	Action level
AMCA	Air Moving and Conditioning Association, Inc.
AOX	Adsorbable organic haides
ASHRAE	American Society of Heating, Refrigerating and Air Conditioning Engineers
ASME	American Society of Mechanical Engineers
AVOC	Annual VOC, tons/yr
A&WMA	Air & Waste Management Association
BACT	Best-available control technology
BIF	Boiler and industrial furnace
BOD	Biochemical oxygen demand
BOD_5	Biological oxygen demand (5 days), mg/L
B.S.	Bachelor of Science
BSIE	Bachelor of Science in industrial engineering
CAA	Clean Air Act
CAAA	Clean Air Act Amendments
CaO	Quicklime
CCME	Canadian Council of Ministries of the Environment
CCS	Common chemical sense
CE	Cost-effectiveness
CEM	Continuous emissions monitor
CERCLA	Comprehensive Environmental Response, Compensation, and Liability Act

Ch.E.	Chemical engineering
CIL	Compliance inquiry letter
CNS	Central nervous system
CO	Carbon monoxide
COD	Chemical oxygen demand
CTG	Control technology guidelines
CUM	Communauté Urbaine de Montreal
D/T	Dilution to threshhold
DAS	Data acquisition system
DO	Dissolved oxygen
DQO	Data quality objectives
EAPA	Environmental Protection and Enhancement Act (Alberta, Canada)
EC	Overall elimination capacity by biofilters
ECD	Electron capture detector
EMAS	Ecomanagement audit scheme
EPA	Environmental Protection Agency
EPCRA	Emergency Planning and Community Right-to-Know Act (1986)
ESP	Electrostatic precipitator
FESOP	Federal enforcible state operating permit
FID	Flame ionization detection
FPD	Flame photometric detector
FTIR	Fourier transfer infrared analyzer
GC	Gas chromatography
GC-MS	Gas chromatography-mass spectometry
GIS	Geographic information system
GMI	Green Mountain Institute (for Environmental Democracy)
GVRD	Greater Vancouver Regional District (British Columbia, Canada)
H_2S	Hydrogen sulfide
HAP	Hazardous air pollutant
HAPs	Hazardous Air Pollutants, as defined by the Clean Air Act
HON	Hazardous organic NESHAP
HPLC	High-performance liquid chromatography
HVLP	High-velocity, low-pressure
IEPA	Illinois Environmental Protection Agency
IIT	Illinois Institute of Technology
IITRI	Illinois Institute of Technology Research Institute
IPA	Isopropyl alcohol
IR	Infrared
ISC	Industrial source complex
ISO	International Standards Organization

IUCN	International Union for the Conservation of Nature and Natural Resources
LCC	Life-cycle costing
LCL	Lower control limit
LD_{50}	Lethal dose for 50 percent of test animals
LEL	Lower explosive limit
LEV	Local exhaust ventilation
MAC	Maximum allowable concentration
MACT	Maximum achievable control technology
MBA	Masters in business administration
MBK	Methyl butyl ketone
MEK	Methyl ethyl ketone
MIBK	Methyl isobutyl ketone
MIR	Maximum industrial risk
MOEE	Ministry of Environment and Energy (Canada)
MRI	Magnetic resonance imaging
MSDS	Material Safety Data Sheet
MSIE	Masters of Science in industrial engineering
NA_2CO_3	Sodium carbonate
NAAQS	National Ambient Air Quality Standard
NACRE	North American Conference on Religion and Ecology
NCCR	Northeast Center for Comparative Risk
NESHAP	National Emission Standard for Hazardous Air Pollutants
NFPA	National Fire Protection Association
NIOSH	National Institute of Occupational Safety and Health
NJIT	New Jersey Institute of Technology
NJTAP	New Jersey Technical Assistance Program
NO_2	Nitrogen dioxide
NO_x	Nitrous oxides
NPDES	National Pollutant Discharge Elimination System
NSPS	New source performance standards
ORP	Oxidation reduction potential
OSHA	Occupational Safety and Health Administration (or Act)
OU	Odor unit
OVA	Organic vapor analyzer
P2	Pollution prevention
PBS	Parallel-bed scrubbers
PCB	Polychlorinated biphenyl
PE	Professional engineer
PEL	Permissible exposure limit

PEM	Parametric emissions monitor
PERC	Perchloroethylene
Ph.D.	Doctor of philosophy
PIA	Printing Industries of America
P&ID	Process and instrumentation diagram
PID	Photo ionization detection
PLC	Programmable logic controller
PNS	Peripheral nervous system
PUF	Polyurethane foam
QSAR	Quantitative structural activity relationship
RACT	Reasonably available control technology
RCRA	Resource Conservation and Recovery Act
RTDM	Rough terrain diffusion model
RM	Risk management
RMP	Risk management plan
RRG	Risk retention (management) group
RTO	Regenerative thermal oxidizer
SAB	Service advisory board
SARA	Superfund Amendments and Reauthorization Act (1986)
SBR	Sequencing batch reactor
SIC	Standard industrial classification
Sigma Xi	The Scientific Research Society
SIP	State implementation act
SMACNA	Sheet Metal and Air Conditioning Contractors National Association, Inc.
SO_2	Sulfur dioxide
SOCMI	Synthetic organic chemical manufacturing industry
SWDA	Solid Waste Disposal Act (Administration)
TAC	Total annualized cost
TCA	Trichloroethylene
TCD	Thermal conductivity detector
THM	Trihalo methane
TLV	Threshold limit value
TLV-C	Threshold limit value—ceiling
TLV-STEL	Threshold limit value—short-term exposure limit
TLV-TWA	Threshhold limit value—time-weighted average
TOC	Total organic carbon (compound)
TPY	Tons per year
TQM	Total quality management
TRI	Toxics release inventory

TRIS	Toxic release inventory system
TSCA	Toxic Substance Contol Act
TSDF	Transportation, storage, and disposal facilities
TSS	Total suspended solids
TURI	Toxics Use Reduction Institute
UCL	Upper control limit
UNCED	United Nations Conference on Environment and Development
UNEP	United Nations Environment Program
USEPA	United States Environmental Protection Agency
UTL	Upper tolerance limit
UV	Ultraviolet
UVM	University of Vermont
VOC	Volatile organic compound
VOST	Volatile organic sampling train
WEF	Water Environment Federation
WWF	World Wildlife Fund
WWTP	Wastewater Treatment Plant

CONTRIBUTOR BIOGRAPHIES

Anthony Earl is a partner with the Quarles & Brady law firm in Milwaukee, Wisconsin. He was the Governor of Wisconsin from 1983 to 1986; Secretary, Wisconsin Department of National Resources from 1976 to 1980; Secretary, Wisconsin Department of Administration in 1975; and Member, Wisconsin State Legislature from 1969 to 1974. Mr Earl was also City Attorney, Wausau, Wisconsin from 1966 to 1969 and Assistant District Attorney, Marathon County, Wisconsin in 1965. He was Co-founder and Chairman of Board of Directors for the Center for Clean Air Policy; Member of Boards of Directors of Great Lakes Protection Fund, Common Cause, Resources for the Future, and the State Capital Law Firm Group; and on the Advisory Board of the St. Lawrence Seaway Development Commission. (CHAP. 1)

Thomas J. Card is nationally recognized for the assessment and control of air toxics, volatile organic compounds (VOCs), and odors from industrial and municipal sources. He has written three books on VOC assessment and control and participated in over 150 air emission control projects over the past 20 years. Mr. Card is currently the chairman of the Air Quality Technical Committee for the American Society of Civil Engineers and the co-chair of the VOC technical session for the Water Environment Federation. (CHAP. 2)

Linda Forst, M.D., M.P.H. is an Assistant Professor in the Department of Emergency Medicine, in the College of Medicine and a Research Assistant Professor in the Division of Environment and Occupational Health Sciences, in the School of Public Health, at the University of Illinois at Chicago. Her research interests include environmental toxicology and occupational diseases of railroad and construction workers. She is Board Certified in Internal Medicine and Occupational Medicine. She received her M.S. in Genetics from the University of Arizona, her M.D. from the Michigan State University, and her M.P.H. from the University of Illinois at Chicago. (CHAP. 3)

Lorraine M. Conroy, SID, CIH, is an Associate Professor of Environmental and Occupational Health Sciences at the University of Illinois at Chicago. Her research interests include occupational exposure assessment, and design and evaluation of engineering controls, particularly ventilation, for chemical and biological hazards. She received her B.S. in Chemical Engineering from the University of Massachusetts in Amherst and her M.S. and Sc.D. in Environmental Science and Physiology, concentrating in Industrial Hygiene and Air Pollution, from Harvard University, School of Public Health. (CHAP. 3)

Frank H. Jarke is Manager, Analytical and Quality Assurance, Safety Director, and Head of Odor Sciences at Mostardi-Platt Associates, Elmhurst, Illinois. He has wide experience in the environmental field since 1969. He has been associated with the Illinois Institute of Technology Research Institute in scientific and project management positions, including heading IITRI's Odor Science Center. Mr. Jarke was with Waste Management Corp. from 1981 to 1994, establishing facilities and programs for shipment and analysis of hazardous waste samples, ground water monitoring, and quality assurance. He worked with the U.S. EPA in the development of Good Automated Laboratory Practices. Mr. Jarke has a B.A. from Southern Illinois University and a M.S. degree from IIT. He has 35 publications and one patent, and is active in ASHRAE, ACS, and Sigma Xi. (SEC. 4.1)

Samuel S. Cha is a principal consulting chemist and Manager of Odor Assessment and Control Programs at TRC Environmental Corporation. He has more than 25 years of experience in the environmental field. He has broad experience in manufacturing process, source, and ambient measurements, environmental chemistry, and odor control. He has served as principal investigator on numerous programs involving air, odor, water, and waste pollution control. While most of these projects were performed in North America, several of them were conducted in Southeast Asia, South America, and the Caribbean Islands. (SEC. 4.2)

Richard Trzupek has been active in the field of air quality and is currently the Managing Principal of Air Solutions, Inc. an air quality consulting firm in Oak Park, Illinois. His activities relate to measurement, research, permitting, and compliance issues of air quality. He has been active in numerous societies, such as A&WMA, Graphic Arts Technical Foundations, Environmental Conservation Board of the Graphic Communications Industry, and ACS. He has presented numerous papers. He is a graduate of Loyola University and has a degree in chemistry. (SEC. 4.3)

Dennis R. Weier received his Ph.D. in Statistics from the University of Missouri in 1978. He subsequently served on the faculty of the Department of Statistics at the University of South Carolina for 5 years where his research interests included nonparametric methods and reliability and life-testing applications. He also served as the department's Statistics Consulting Laboratory Director. He then was an internal statistical consultant for 13 years at the Rocky Flats Environmental Technology Site in Golden, Colorado, where he worked primarily in areas of environmental and waste management and nuclear materials safeguards. He continues to work on environmental and data quality objectives applications as well as nuclear materials safeguards. He is participating in the upgrade of safeguards in Russia. (SEC. 4.4)

Kalliat T. Valsaraj, Ph.D., M.Sc., is an Associate Professor in the Department of Chemical Engineering at Louisiana State University. He received his M.Sc. from the Indian Institute of Technology, Madras, India and his Ph.D. from Vanderbilt University, Nashville, Tennessee. His research is focused on the fate and transport of contaminants in the natural environment and development of processes for dilute solutions separation from industrial wastewaters. He has authored or co-authored over 80 publications, 9 book chapters and 1 book on environmental engineering. He is a member of the American Society for Environmental Toxicology and Chemistry. (SEC. 5.1)

Richard L. Corsi, Ph.D., is a faculty member in the Environmental and Water Resources Engineering program in the Department of Civil Engineering, The University of Texas at Austin. He teaches courses entitled "Environmental Pollution Engineering," "Air Pollution Engineering," "Air Pollution Sources," and "Atmospheric Dispersion Modeling." He also serves as lead faculty member in a professional short-course entitled "VOC Emissions from Industrial and Municipal Wastewater: Estimation Methods." Prof. Corsi's research is focused on the characterization and passive control of sources of hazardous air pollution. Sources which are currently being investigated include diesel-contaminated soil, gasoline stations (refueling and underground storage tanks), municipal sewers, industrial process drains, industrial machines). Prof. Corsi and his research team have published over 100 technical papers and reports related to their research, and have delivered over 75 professional presentations. (SEC. 5.2)

David A. Olson received a Bachelor of Science in Mechanical Engineering from The University of Kansas in 1994. He then entered graduate school at The University of Texas at Austin, earning a

Master of Science in Environmental Engineering in 1996. His master's thesis was titled "Gas Exchange Rates between Industrial Process Drains and the Ambient Atmosphere." After graduation, he worked for one year as a research engineer at the university, managing five projects related to estimation of fugitive emissions of volatile organic compounds (VOCs). He is co-author of the naUTilus model, a model which predicts VOC emissions from entire industrial sewer networks. Research conducted by David Olson while at UT-Austin has led to nearly 20 publications in the air pollution field. He currently works for Waid and Associates in Austin, Texas. (SEC. 5.2)

Marcus J. Healey, Ph.D., M.B.A., is Director of the New Jersey Technical Assistance Program (NJTAP) for Industrial Pollution Prevention at the New Jersey Institute of Technology (NJIT), which provides a public service outreach program. Dr. Healey has a diverse background. He received his M.B.A. in Finance from Rider University; his M.S. in Chemical and Biochemical Engineering from Rutgers University; his Ph.D. and M.S. in Environmental Science from Cook College, Rutgers University, and his B.A. in Biology from the University of Pennsylvania. His primary interests are in the environmental management and business interface. Dr. Healey has authored over 50 publications and reports and has received over $2.8 million in funding from various governmental agencies and commissions. (SEC. 6.1)

Douglas Kretkowski graduated from the New Jersey Institute of Technology (NJIT) with a B.S. in Chemical Engineering. His undergraduate experiences have led to several journal publications as well as a strong background in instrumentation and application software. Currently, Mr. Kretkowski is a pollution prevention analyst with the New Jersey Technical Assistance Program (NJTAP) performing site assessments for small and midsized manufacturing facilities throughout the state. In additional, Mr. Kretkowski is a member of the NJIT HazMat Response Team and is NJTAP's Webmaster. (SEC. 6.1)

Jeffrey Lewis graduated from Princeton University with a B.S. in Chemical Engineering. Mr. Lewis, was a pollution prevention analyst with New Jersey Technical Assistance Program for two years before moving on to investment banking. He co-authored several publications including "Hazardous Waste Reduction in New Jersey: How Much Has P2 Contributed?" (SEC. 6.1)

Mark Walsh graduated from New Jersey Institute of Technology with an M.S. in Environmental Policy. Mr. Walsh's master's thesis integrated publicly available New Jersey hazardous waste databases with Geographic Information Systems (GIS). He is currently enrolled in Penn State University as an environmental law student. (SEC. 6.1)

Gayle VanDurme obtained bachelor's and master's degrees in civil engineering at the University of Missouri–Columbia. In his 25-year career at Black & Veatch, he acquired an extensive background in wastewater process design and developed specialized expertise in the field of odor control. From his office in Kansas City, he now spends full time directing odor and corrosion control activities on projects throughout the United States and abroad. He has performed extensive research in both pilot and full-scale studies in wastewater collection systems and treatment facilities. He has written numerous trade journal articles and presented seminars and papers at national environmental conferences. He is an active member of the Water Environment Federation and currently serves on the Air Quality Impacts Committee. (SEC. 6.2)

Philip L. Wolstenholme, P.E., is a professional mechanical engineer with over 20 years of design and project management experience. His main areas of expertise in wastewater design are in odor control, biosolids systems, and materials handling. He has been project manager on the design and construction engineering of many odor and bestialities projects and published papers on odor control, ventilation and air movement, and production of fertilizer products from sludge. His experience in odor control encompasses numerous studies and designs including chemical, carbon, and biological odor treatment systems. He has developed practical expertise and published several papers on bulk-media biofiltration as a cost-effective method of odor control. (CHAP. 7)

Edward A. Bonanni is currently a senior Technical Sales Engineer at Eaglebrook, Inc. Edgemoor, Delaware. In 1974, he graduated from Saint Joseph's University, Philadelphia, with a B.S. in Chemistry. In 1979 he graduated from Widener University, Chester, Pennsylvania with an M.B.A. He has been working on odor control in the wastewater treatment industry since 1986. (SEC. 8.1)

Timothy W. Planker is Vice-President and General Manager for Hinsilblon Laboratories, Cape Coral, Florida. He has wide experience in marketing of air pollution control and odor control equipment and chemicals. He formerly was Manager of Odor Control Systems of the Duall Division of Met Pro Corporation, and a Regional Engineering Manager for Wheelabrator-Fry. He holds a B.S.I.E. and M.S.I.E. from Rutgers University. (SEC. 8.2)

Richard E. Waldrop is President of Edwards Engineering Corp. Mr Waldrop was educated at East Carolina University, where he received a Bachelor of Science in Industrial Technology (1977). His 20 years' experience with Edwards Engineering encompasses all aspects of the design and manufacturing process. He is a member of ASME (American Society of Mechanical Engineers), ASHRAE (American Society of Heating, Refrigerating and Air-conditioning Engineers), and the Hydronics Institute. His knowledge in the use of vapor recovery using condensation has offered him the opportunity to lecture internationally and author numerous papers. (SEC. 8.3)

Gene H. Anguil, Bachelor of Science, Mechanical Engineering and Master of Science, Applied Mathematics, has been a Fulbright Scholar at the University of Rome, Italy. He was a recipient of a National Defense Fellowship at the Virginia Polytechnic Institute. He was formerly Vice-President of Engineering with the Kelley Company, Inc. where he was responsible for the development of the Pyrolytic Incinerator, a solid waste thermal oxidizer with heat recovery. Prior to that, Mr. Anguil was affiliated with General Motors and with the Singer Corporation as an Aerospace Engineer where he worked on the Apollo project. Mr. Anguil is president of Anguil Environmental Systems, Inc., an engineering and manufacturing company in the air pollution control industry, that specializes in controlling VOCs and toxic chemicals. His company, started in 1979, has hundreds of environmentally approved installations throughout the United States, Mexico, Canada, Europe, and Asia. (SEC. 8.4)

Charles S. Parmele is a senior technical specialist with more than 30 years of chemical engineering experience, including more than 27 years of experience in environmentally related technology development and application activities. He holds a B.S., Chemical Engineering, from Northwestern University in 1967. He has worked and/or consulted in several different industries that use volatile organic chemicals. He has evaluated and applied vapor-phase adsorption for controlling emissions since 1975. These projects have included comparison of vapor-phase adsorption with other control devices; design, construction and operation of an automated pilot plant; survey of full-scale applications; development of design bases for new systems and troubleshooting full-scale systems to identify methods to improve performance. (SEC. 8.5.1)

T. Kovalcson is a registered professional engineer with nine years of chemical engineering experience in environmentally related technical evaluation and design activities. She holds a B.S. in Chemical Engineering from the University of Kansas in 1989. She has provided consulting for federal and commercial clients regarding disposition of volatile organic chemicals. She has evaluated and applied her knowledge of vapor-phase carbon adsorption for controlling emissions since 1990. Projects have included comparison of vapor-phase adsorption with other control devices; multiple designs of full-scale groundwater treatment systems that included vapor-phase carbon as secondary treatment for emissions; and evaluating the status of currently operating systems for ways to increase system performance. (SEC. 8.5.1)

Robert N. Rafson, P.E., has achieved recognition in several fields, including innovative soil remediation technologies, modern insulation technology, and gaseous emission controls. He serves as C.E.O. of three technology based companies: Rafson Engineering, Inc. (Environmnental Consulting), Greenfield Partners (Brownfield Remediation), and Service Insulation (Modern

Industrial Process Insulation). Over the past 17 years, he has been involved in engineering, design, and accomplishment, of over 100 projects in environmental improvements and over 800 projects of insulation applications. (SEC. 8.5.2)

Charles McGinley, principal of McGinley Associates and the technical director of the odor evaluation laboratory, St. Croix Sensory, Inc., is a chemical engineer with over 25 years of practical experience in the field of environmental engineering. Mr. McGinley is a graduate of the University of Minnesota and holds an M.B.A. from the University of St. Thomas. St. Paul, Minnesota. Since 1980, McGinley Associates has provided engineering services to industries, municipalities, and engineering firms throughout the United States and to projects in Australia, Canada, Denmark, Germany, Great Britain, Japan, Korea, Saudi Arabia, and Central and South America. McGinley Associates specializes in environmental health engineering, indoor air quality, air toxics, and odors. (SEC. 8.6.1)

Harold J. Rafson received his B.Ch.E. and M.Ch.E. from Polytechnic Institute of Brooklyn. He is now an independant consultant and is nationally recognized in the field of control of odors in toxic and volatile organic compounds. This is the second book he has authored. He has published and presented over 150 technical papers and speeches over the past 40 years. Mr. Rafson holds numerous U.S. and international patents concerning gas emission control technology. He has been an active member of Odor and Emission Control Committees of both the Water Environment Federation (WEF) and the Air & Water Management Association (AWMA). Mr Rafson developed mist scrubbing technology over a period of 20 years, as President of Quad Environmental Technologies. This method has been brought from a concept to a traditionally accepted technology by his company. He was directly involved in the design, sale, fabrication, installation, and start-up of over 300 gaseous emission control projects. His technical expertise in air emissions includes testing methods and all control technologies. Prior to his work in air pollution control Mr. Rafson was a food process engineer. (SEC. 8.6.2, *also* CHAP. 9 *and* EDITOR)

Robert P. G. Bowker, P.E., has over 20 years' experience in the planning, design, and operation of wastewater facilities, with a strong background in odor and corrosion control. He is the principal author of the 1985 EPA *Design Manual on Odor and Corrosion Control* and a contributing author of the 1995 WEF *Manual of Practice on Odor Control.* He has completed over 80 odor control projects in the United States and Canada. He has a B.S. in Civil Engineering from the University of New Hampshire and an M.S. in Environmental Engineering from Cornell University. He is active in the WEF and AWMA. Mr. Bowker is President of Bowker & Associates, an environmental consulting firm in Portland, Maine. (SECS. 8.6.3, 8.8.2)

Richard L. Davie obtained his degree in marine engineering in 1956. Subsequent graduate studies included advanced work in nuclear and mechanical engineering. Early engineering assignments included design of nuclear power plant systems and nuclear fuel. Later, as a contributing engineer and program manager, Mr. Davie was instrumental in initial development of fuel cell power systems for aerospace and deep submergence marine vehicle propulsion. Since 1970, he has concentrated on development and marketing of various processes and products for wastewater treatment, including fixed film biological systems. In 1988, he started the Environmental Systems Division of Purafil, Inc. As General Manager, he has been responsible for development and marketing of gas phase (dry chemical) scrubber products for control of odors, corrosion, and toxic gases in water/wastewater treatment facilities. Mr. Davie has published papers on odor control and emergency gas scrubbers through the Water Environment Federation (WEF) of which he is a member. He was also a member of the joint task force which developed the WEF manual of Practice No.22, *Odor Control in Wastewater Treatment Plants,* published in 1995. (SEC. 8.7)

George J. Skladany is Director of the Remediation Technology Facility for Flour Daniel, GTI, Inc. (Trenton, New Jersey) and is also a partner in Applied Air Systems Inc, (Linden, New Jersey), an air pollution engineering and construction company. Mr. Skladany previously commercialized biofilter and biotrickling filter technology for Envirogen, Inc. and developed the first commercial short-courses on biofiltration offered in the United States. (SEC. 8.8.1)

Marc A. Deshusses, Assistant Professor of Chemical Engineering, is leading the biological waste air treatment research group at the University of California, Riverside, where his interests include the modeling of biofilter and biotrickling filter processes. Dr . Deshusses is the author of several scientific publications on biological techniques for air pollution control, and is co-author of a book, *Biofiltration for Air Pollution Control.* (SEC. 8.8.1)

Joseph S. Devinny, Ph.D., is Professor of Civil and Environmental Engineering and Associate Dean for Academic Affairs in the School of Engineering at the University of Southern California. He is co-author of *Subsurface Migration of Hazardous Waste* and *Biofiltration for Air Pollution Control,* and cohost of the biennial USC-TRG conferences on biofiltration of air. (SEC. 8.8.1)

A. Paul Togna is Manager of the Air & Water Systems Development Group for Envirogen, Inc., (Lawrenceville, New Jersey, an environmental system and services company). Dr. Togna is responsible for the design, process optimization, and testing of Envirogen's air and waste biotreatment systems, and has been involved in the biofiltration area for six years. (SEC. 8.8.1)

Todd S. Webster, a group leader for Envirogen, Inc. (Lawrenceville, New Jersey, an environmental system and services company) for California-based biological remediation projects, has conducted both laboratory and field-pilot investigations of biofilter and biotrickling filter systems. The author of numerous articles on these forms of vapor-phase biological treatment, Dr. Webster is currently co-authoring the book, *Biofiltration for Air Pollution Control.* (SEC. 8.8.1)

Robert P. G. Bowker (SEC. 8.8.2; *see above* SEC. 8.6.3)

Avinash N. Patkar, Ph.D., obtained his B.S. (Ch.E.) at Bombay University and his M.S. and Ph.D., both in chemical engineering, at the University of Cincinnati. He is a registered chemical engineer in Ohio. He has worked in the field of air pollution control for over 20 years, and specializes in selection of cost-effective hybrid systems for VOC control. As the lead process engineer for several industrial projects, he has designed and installed VOC collection and control systems which were the first applications of their type in the industry. (SEC. 8.9)

Robert R. Banks, a technical consultant in Pittsburgh, Pennsylvania, specializes in the design, maintenance, and troubleshooting of air pollution control systems and equipment with special emphasis on fabric filters. He has presented papers on these subjects at a number of technical meetings. (SEC. 8.10.1)

Chen-Lu Yang is a research engineer at Beltran Associates. His current research includes decomposition of organic solvents using nonthermal plasmas, polystage chemical scrubber for the control of NO_x emissions from sewage sludge incinerators, and applications of tubular electrostatic precipitators in the contact process for sulfuric acid mist recovery. He is responsible for developing and designing plasma reactors, chemical scrubbers and tubular electrostatic precipitators for industrial activities and waste treatment facilities.

From 1994 to 1995 Dr. Yang was a postdoctoral Research Fellow in the Department of Chemical and Environmental Engineering at the University of Arizona. He conducted research on regeneration of flue gas desulfurization (FGD) waste liquors for production of ammonium and potassium sulfate mixed fertilizer and recovery of sodium and aluminum nitrates from Hanford radioactive waste liquor by crystallization.

From 1988 to 1994 Dr. Yang was at the New Jersey Institute of Technology. He was Research Associate of the Hazardous Substance Management Research Center responsible for setting up a rotary basket catalytic reduction unit for soot and NO_x control and directing graduate students in the investigation of kinetics of HCl reaction with CaO by FTIR. Earlier he was Research Assistant in the Department of Chemical Engineering, Chemistry and Environmental Science where he developed a scrubbing method for quantitative removal of NO_x from flue gases and obtained scale-up parameters.

Dr. Yang received his B.E. in Chemical Engineering from Feng-Chia University, Taiwan in 1984. He received his M.S. in Environmental Science and his Ph.D. in Chemical Engineering from New Jersey Institute of Technology in 1990 and 1994. He has been author or coauthor for journal articles and conference presentations on nonthermal plasma and electrostatic precipitators and on absorption and chemical scrubbers. (SEC. 8.10.2)

David C. Meier is currently working at Beltran Associates as a sales manager specializing in air pollution control equipment for municipal and heavy industry markets. Mr. Meier has also worked at Natec Resources specializing in dry sorbent technology; he was Regional Sales Manager at Fuel Tech, Inc. He sold the first successful utility water-in-oil emulsion program in New York State. He rose to Utility Sales Manager for Sybron Chemicals, Inc. marketing pollution control equipment utilizing mutant bacteria and ion-exchange systems for waste treatment plants. Mr. Meier was a District Representative for Calgon Corporation. He began his professional career as high school teacher in chemistry and earth science. Mr. Meier received his B.S. in Chemistry from The State University of New York, Oneonta, in 1965 and he was a graduate candidate for an M.B.A. at Temple University in 1965. (SEC. 8.10.2)

Joseph R. Popeck, P.E., has focused his career primarily on the study, design, and construction of wastewater treatment and collection facilities. Since the late 1970s, he has been involved with studies, designs, construction, start-up, testing, and operator training for over 20 odor control projects. Mr. Popeck is a partner with Greeley & Hansen, Engineers in Chicago. He also chairs Greeley & Hansen's nationwide air quality team. Mr. Popeck obtained his civil engineering degree from The Pennsylvania State University in 1971. (CHAP. 9)

Harold J. Rafson (CHAP. 9; *see above,* SEC. 8.6.2; *also* EDITOR)

G. Vinson Hellwig is manager of Air Quality Services for the Southeast for Dames & Moore in Atlanta, Georgia. He was formerly Senior Vice President and Air Director with TRC Environmental Corporation in Windsor, Connecticut. His responsibilities included project management of permitting and air engineering projects. Mr. Hellwig also served as Senior Project Manager on a number of air projects including MACT compliance issues with corporation and trade associations, emission measurements, permitting SARA Title III inventories, and control equipment engineering. He has served as Program Manager on several major Title V programs including GM, 3M, Allied Signal, and Procter & Gamble. He also successfully managed a hazardous waste disposal practice investigation of 43 plants throughout the United States and Canada for a major U.S. chemical company. Previously, Mr. Hellwig was a senior environmental engineer in the North Carolina office of a major environmental firm. He also worked for the U.S. EPA in the area of air compliance. Mr. Hellwig received his B.S. from Shorter College and did M.S. studies at Auburn University. (SEC. 10.1)

Michael R. E. Rix is an environmental consultant specializing in odor management. He graduated from the University of Toronto in 1978 with a B.S. in Biology. For many years he headed the odor and gas analysis capability program at a large research and consulting firm. In 1997, he and two colleagues formed ZORIX Consultants, an environmental management consulting company specializing in odor and VOC control, located in Mississauga, Ontario. Mr. Rix has 20 years of broad environmental experience covering many aspects of industrial odor and air emissions including odor assessment and control, continuous emission monitoring systems (CEMS), instrumental analysis, trace contaminant sampling, gas chromatography, dispersion modeling, and community odor surveys. He is routinely involved in regulatory liaison on behalf of industrial and municipal clients with respect to test plans, permits, and compliance issues. Mr. Rix has extensive experience in odor measurement which has included the design of sampling systems, olfactometers, and odor evaluation laboratories and the design and implementation of sampling and evaluation strategies both at the emission source and in the ambient air. Major projects have included comprehensive programs at automotive assembly plants, furniture manufacturing, refineries, coil coating plants, food processing plants, beverage can manufacturing plants, printing plants, and wastewater treatment plants. Mr. Rix is a member of the Air and Waste Management Association and the AWMA Odor Committee, has made presentations at many conferences, and has published several articles on odor. (SEC. 10.2)

Joseph P. Botos was appointed Vice President, Administrative Division, of Cargill in January 1989, and in June 1992 became manager of Cargill's Environment, Health and Safety Department. He is responsible for Cargill's worldwide activities in environment, health, and safety management. He is a director of Willknight, Inc. and an assistant vice president of Cargill International, S.A. Mr. Botos currently serves as co-chairman of the Cargill Cares Water Matters initiative, chairman of the Cargill Drug and Alcohol Screening Committee, and a member of the Cargill Workers' Compensation Committee. He is a member of the Conference Board's Chief Environment, Health and Safety Officers' Council I, and serves on the executive committee as Program Committee Chairman. He represents Cargill with the National Association of Manufacturers on environmental quality issues. Mr. Botos received a B.A. in Business from the University of Iowa in Iowa City in 1965 and attended the University of Minnesota Executive Program in 1988. (CHAP. 11)

Suresh R. Relwani has broad experience over the past 20 years in the field of air quality. He obtained his B.S. in Chemical Engineering from the University of Baroda, India (1978), and his M.S. in Environmental Engineering from the University of Cincinnati (1980). He is a principal of RK & Associates, Consultants. Formerly, he managed the Air Quality Group for Harding Lawson Associates. His recent work has centered on air emission source permitting, regulatory analysis, odor control guidelines, and modeling of atmospheric pollutants. Previously he managed the Air Quality Group for the Illinois Institute of Technology Research Institute (IITRI). Here his work covered services to a wide variety of industries. He is author, or co-author, of approximately 35 papers and presentations. (CHAP. 12)

Melissa A. Hickey is an environmental engineer with RK & Associates (Glenn Ellyn, Illinois) with experience in projects concerning emission inventories, compliance assessment, Title V operating permit applications, and abatement plans for several wastewater treatment plants. She has worked for Lithographic and Flexographic Printers and prepared several annual emission inventories and Title V permit applications. Ms. Hickey also has done work for airline maintenance facilities, coatings and resin manufacturers and a plastic manufacturer. She has also served as the resident project engineer during two of the construction phases of the vertical expansion of a RCRA hazardous waste landfill for a steel manufacturer and on other soil and groundwater remediation projects at a Superfund site, and a plastics manufacturer. She received a B.S. in Civil Engineering from the University of Illinois in Champaign-Urbana. (CHAP. 12)

Alan Leahigh is former executive vice president of Public Communications Inc., a Chicago-based public relations firm. He is the winner of national awards for programs in community relations, crisis communications, and environmental affairs. He is currently executive vice-president of Executive Administration, Inc., a national medical association management company in Arlington Heights, Illinois. (SEC. 13.1)

Ken Jones, Ph.D., is the Executive Director of the Green Mountain Institute for Environmental Democracy (GMI) in Montpelier, Vermont, and co-founder of the Northeast Center for Comparative Risk (NCCR), GMI's predecessor. Mr. Jones earned a Ph.D. in biochemistry from the University of Michigan, and a master's in public policy from Kennedy School of Government. NCCR/GMI is supported in part by the U.S. Environmental Protection Agency to assist cities and states in comparative risk projects. Dr. Jones was a policy analyst for the Vermont Office of Policy Research from 1988 to 1991 and a staff scientist for the Institute for Sustainable Communities. He is a member of the Vermont comparative risk project's health work group and an electric utility consultant for Green Mountain Power, of Vermont and New Hampshire. (SEC. 13.2)

Sarah Tollison Fox is an ecologist at the Western Center for Environmental Decision-Making. Her research focused on the impacts of human population increases on biodiversity and climate within regional and global ecosystems. She also worked on the development of a global change curriculum for high schools in the State of New York. Ms. Fox graduated from Trinity College, Washington, D.C. and the State University of New York, Albany. (SEC. 13.2)

Sue Thomas has been a Research Associate at the Green Mountain Institute for Environmental Democracy and the organization's predecessor, the Northeast Center for Comparative Risk, since 1995. Before coming to GMI, Ms. Thomas spent four years at the Ohio Environmental Protection Agency most recently as the coordinator for the Maumee River Remedial Action Plan, and formerly as an environmental planner for the state's Water Pollution Control Loan Fund. Ms. Thomas has a master's degree in natural resources policy from Ohio State University. (SEC. 13.2)

Michele Witten is a Research Associate at the Green Mountain Institute for Environmental Democracy. Ms. Witten came to GMI from the Environmental Partnerships in Communities project at the University of Vermont (UVM). She earned a master's degree in natural resources planning from UVM. Previously, Ms. Witten worked as a legal assistant at the firm of Crowell & Moring, Washington, D.C., after graduation from Bowdoin College. (SEC. 13.2)

Harlow G. Ballard, M.P.A., worked from 1962 to 1966 as a property/casualty commercial lines underwriter for Trinity Universal and The Home Insurance companies. From 1967 to 1975 he was Sales Manager for Ohio for Transamerica Insurance Group. From 1975 to 1980 he attended Ohio State University, obtaining a B.A. (1976) and an M.P.A. (1979). In 1980 he joined Town & Village Insurance Service, founding and managing their national commercial lines department. In 1992 Mr. Ballard started his own business, The Harlow Ballard Insurance Agency, which specializes in commercial lines for contractors and environmental insurance. (SEC. 13.3)

Jim Kenney, Ph.D., is Co-founder and Executive Director of Common Ground, an adult cultural-educational organization in Deerfield, Illnois (1988 to the present). He was vice-chairman of Council for a Parliament of the World's Religions, Chicago (1988 to 1993) and program chair for the organization's 1993 parliament in Chicago. He is a trustee for the International Interreligious Initiative Council for the Parliament, International Committee for the Peace Council, and the International Interfaith Center (Oxford, United Kingdom). He has held the following positions: Director of 1966 Conference of Society for Christian-Buddhist studies; Consultant for International Orientation Resources, Northbrook, Illinois; former chair for Department of Religions, Barat College, Lake Forest Illinois; and former faculty member for Departments of Religion, Lake Forest College and Mundelein College, Chicago, Illinois. He is an active lecturer on a wide range of cultural, religious, and philosophic subjects. Dr. Kenney has been Consulting Editor for *Conscious Choice* magazine and he was co-editor of *Fireball and the Lotus: Emerging Spirituality from Ancient Roots.* His doctoral thesis at Northwestern University was *A Comparative Study of Buddhism, Christianity, and Modern Physics.* (SEC. 13.4)

GLOSSARY

air scrubber An air pollution control device to remove contaminants from the air stream. (Chap. 1)

batch dyeing Dyeing process in which a given amount of textile material, usually 100 to 1000 kg, is loaded into a dyeing machine and brought to equilibrium (or near equilibrium) with a solution containing dye. (Sec. 6.1)

biochemical oxygen demand (BOD) A standardized means of estimating the degree of contamination of water supplies, especially those which receive contamination from sewage and industrial wastes. It is expressed as the quantity of dissolved oxygen (in milligrams per liter) required during stabilization of the decomposable organic matter by aerobic portions of the sample with water saturated with oxygen and measured with water saturated with oxygen and measuring the dissolved oxygen in the mixture both immediately and after a period of incubation, usually 5 days. (Sec. 6.1)

biofilter Uses microorganisms to metabolize volatile organic or inorganic contaminants passing through a packing media. Bioreactors with no free liquid phase are called biofilters (Sec. 8.8.1)

biogas Methane-rich gas produced from sludge digestion. (Sec. 6.2)

biotrickling filter Consists of pollutant degrading microorganisms immobilized on a packing medium, and employs continuous water recirculation over the media. (Sec. 8.8.1)

bioscrubber Initially absorbs vapor phase contaminants into a recirculating liquid, and the dissolved contaminants then undergo a subsequent biological treatment step. Bioscrubbers are also called biowashers. (Sec. 8.8.1)

caustic Any strongly alkaline material which has a corrosive or irritating effect on living tissue. (Sec. 6.1)

conservation Preserving and renewing, when possible, natural resources. (Chap. 1)

continuous dyeing Processes in which textile materials are fed continuously into a dye "range" at speeds of roughly 50 to 250 meters per minute. Continuous dyeing processes usually consist of dye application, dye fixation with chemicals or heat, and washing. (Sec. 6.1)

diffusion coefficients A measure of the rate of the movement of matter from a more concentrated to a less concentrated state. (Chap. 2)

effluent The aqueous mixture from secondary biological waste treatment facilities that enters a stream, river, or body of water, or such material from a textile operation or pretreatment system entering a sewer for treatment at a publicly owned treatment works (POTW). (Sec. 6.1)

exhaust Dye migration from a dyebath into a fiber or other textile material. (Sec. 6.1)

heavy metals Metallic elements with high atomic weights, e.g., mercury, chromium, cadmium, arsenic, and lead; can damage living things at low concentrations and tend to accumulate in the food chain. (Chap. 1)

hematopoietic Blood formation. (Chap. 3)

hybrid A pollution control system which may contain two or more elements from differing control technologies. (Sec. 8.9)

iron salts Generic term referring to ferrous and ferric chloride, and ferrous and ferric sulfate. (Sec. 6.2)

kappa number (K) The accepted measure of the lignan content of wood. The kappa number is of critical importance, since the process of pulping is the removal of lignan from wood. (Sec. 6.1)

kraft process The most common process used by the pulp and paper industry. (Sec. 6.1)

mineralization Complete bioconversion of a target compound to carbon dioxide, water, and possibly inorganic salts or acids. (Sec. 8.8.1)

oxidation/reduction potential The electrical potential required to transfer electrons from one compound or element (the oxidant) to another compound (the reductant). (Chap. 2)

particulate matter Fine solid particles such as dust, smoke, fumes, found in air or emissions. (Chap. 1)

partition coefficient Measure of the distribution of a compound between phases. (Chap. 2)

penetration theory This theory assumes that mass transfer coefficients are equal to the square root of the diffusion coefficients times the renewal rate. Also called surface renewal theory. (Chap. 2)

pollution prevention Changes in production technologies, raw materials or products, that result in the reduction of the demand for hazardous substances per unit of product manufactured and the creation of hazardous products or nonproduct outputs; or changes in the use of raw materials, products, or production technologies that result in the reduction of the input use of hazardous substances and the creation of hazardous by-products or destructive results; or on-site facility changes in production processes, products, or the use of substitute raw materials that result in the reduction of the amount of hazardous waste generated and disposed of on the land or hazardous substances discharged into the air or water per unit of product manufactured prior to treatment, and that reduce or eliminate, without shifting, the risks that the use of hazardous substances at an industrial facility pose to employees, consumers, and the environment and human health. Pollution prevention includes, but is not limited to, raw material substitution, product reformulation, production process redesign or modification, in-process recycling, and improved operation and maintenance of production process equipment. (Chap. 1, Sec. 6.1)

sanitary landfill Disposal sites for nonhazardous solid wastes. (Chap. 1)

SIC code The Standard Industrial Classification (SIC) is a standard used for all establishment-based federal economic statistics. The SIC codes facilitate comparisons between facilty and industry data. (Sec. 6.1)

spent material Any material that has been used and as a result of contamination can no longer serve the purpose for which it was produced without processing. (Sec. 6.1)

strip To remove color from a dyed fabric. This is usually a repair procedure and generally degrades the cloth, especially if cotton. (Sec. 6.1)

substrate Any materials upon which ink is impressed, such as paper, plastic, wood, or metal. (Sec. 6.1)

suspended solids Small particles of solid materials that are suspended in liquids. (Chap. 1)

toxic substances A chemical, or mixture, that may present an unreasonable risk of injury to health or the environment. (Chap. 1)

toxicology The branch of pharmacology dealing with poisons, their effects, detection, antidotes, etc. (Chap. 3)

tradeable emission credits The creation of surplus emission reductions at certain stacks, vents or similar emissions sources and the use of this surplus to meet or redefine pollution requirements applicable to other emissions sources. This allows one source to increase emissions when another source reduces them, maintaining an overall constant emission level. Facilities that reduce emissions substantially may "bank" their "credits" or sell them to other facilities or industries. (Chap. 1)

TRI facilities Manufacturing facilities that have 10 or more full-time employees and are above established chemical throughput thresholds. Manufacturing facilities are defined as facilities in Standard Industrial Classification primary codes 20 to 39. Facilities must submit estimates for all chemicals that are on the EPA's defined list and are above throughput thresholds. (Sec. 6.1)

vapor pressure The pressure exerted by vapor which is in equilibrium with the liquid. (Chap. 2)

volatility The ease with which a liquid or solid passes into its vapor state. (Sec. 6.1)

waste ink The ink removed from the ink fountain at the end of a run or following its contamination. (Sec. 6.1)

waste minimization A reduction in the total quantity and/or toxicity of the hazardous waste generated. A waste can be reduced or eliminated at its source, or used, reused, or reclaimed once it is generated. (Sec. 6.1)

wastewater treatment plant A facility by which pollutants are removed from wastewater. (Chap. 1)

INDEX

Absorber, using chelated iron solution, **8.**135
(*See also* Chemical scrubbing, other designs; Mist scrubbing; Packed column)
Adsorption, adsorbate **8.**73 to **8.**74
polarity, **8.**73
volatility, **8.**73
practical limit, **8.**74
Adsorption, adsorbent, **8.**73
disposal of spent media, **8.**73
residual for regeneration, **8.**73
Adsorption, carbon, **8.**66 to **8.**99, **2.**12
process and operations, **8.**66 to **8.**88
acceptable working capacities of, **8.**78
airflow, **8.**67
batch operation, **8.**67
breakthrough, **8.**73
capacity, reduced by water, **8.**74
concentration profile, **8.**69
condenser, amounts, **8.**67
condenser, water fouling, **8.**67
cost factors, **8.**87 to **8.**88
carbon replacement, **8.**88
valve selection, **8.**74
flammable vapors, **8.**74
flow diagram, **8.**67
full-scale performance, **8.**75
heat dissipation, **8.**74
heat of, **8.**74
less volatile materials, of, **8.**76
maximum practical concentration, **8.**74
operation, **8.**72
operation-concentration profile, **8.**72
practical limits of, **8.**76
refrigerated condenser application, **8.**67
scrubbers, using nonvolatile organics, **8.**67
scrubbing with subcooled solvent, **8.**67
temperature, rise, **8.**74
transfer zone, **8.**73
ventilation, air heat economy, **8.**66
Adsorption, fog droplets, **8.**92 to **8.**99
atmospheric chemistry, **8.**96
enhancement of removal, **8.**96
enrichment factor, **8.**96
mechanism, **8.**92 to **8.**93
absolute pressure, **8.**93
added surface active agents, **8.**93
scrubbing, absorption, **8.**98
(*See also* Mist scrubbing)
Adsorption, regeneration, **8.**67 to **8.**79
carbon, using steam, **8.**76
conventional, **8.**76
decreased partial pressure, **8.**97
desorption, **8.**73, **8.**76
condensation, **8.**76
incineration, **8.**76
desorption efficiency curve, **8.**78
hot gas, **8.**76
importance of cooling and drying, **8.**79
increased temperature, **8.**76
incineration, **8.**76
limited to low inlets, **8.**76
optimizing steam use, **8.**76
rate of steam usage, **8.**76
residual "heel," **8.**76
steam, **8.**76
steam requirements, **8.**76
superheated stream, **8.**78
thermal, **8.**76
vacuum and hot gas, **8.**76
Adsorption applications, **8.**66 to **8.**67, **8.**89 to **8.**90
from air strippers, **8.**66
from exhausts, **8.**66
high removal efficiencies, **8.**89
breakthrough prevention, **8.**89
improving desorption, **8.**90
upgrading system, **8.**90
recovery of solvents, **8.**67

Adsorption applications (*Cont.*):
 from soil vapor extraction, **8.**66
Adsorption equipment, **8.**78 to **8.**86
 alternative design, corrosion, **8.**80, **8.**83
  regeneration energy, **8.**80
  with thermal regeneration, **8.**80
  vacuum regeneration, **8.**80
 design, recycling exhaust, **8.**83
  piping requirements, **8.**78
 design data, **8.**78 to **8.**86
  capacity limits, **8.**84 to **8.**86
  design experience, **8.**85
  isotherm, **8.**85
  isotherm test, **8.**85
  safety consideration, **8.**86
  scale up, **8.**84
  working capacity, **8.**84
 full-scale design, **8.**83 to **8.**84
  amount of carbon, **8.**83
  cross-sectional area, **8.**83
  duty definition, **8.**83
  large flow, **8.**84
  minimum cycle, **8.**83
  pressure drop, **8.**83
  velocity limits, **8.**83
Adsorption media, **8.**67, **8.**71 to **8.**72
 alumina, **8.**72
 carbon performance, **8.**72
 granular carbon, **8.**67
 molecular sieves, **8.**72
 silica gel, **8.**72
Adsorption theory, **8.**69 to **8.**72
 choosing an adsorbent, **8.**70
 Freundlisch equation, **8.**69
 isotherm, **8.**70
 nonpolar compounds of, **8.**72
 polar compounds of, **8.**72
 Polyani's potential theory of, **8.**69
Aeration basins, **8.**192 to **8.**201
 activated sludge, **8.**201
 design considerations, **8.**201
 economic analysis, **8.**199
 list of facilities, **8.**193
 problems, **8.**197
  corrosion, **8.**197
  filamentous organisms, **8.**197
  limited airflow, **8.**197
 process mechanics, **8.**193
 references, **8.**199
 system schematic, **8.**192
Air pollution control management, **11.**1 to **11.**19

Air pollution control management (*Cont.*):
 international standards, **11.**18
 management and technical support, **11.**4 to **11.**11, **11.**16
  acquisition and divestiture review, **11.**5
  air emissions, **11.**8
  asbestos, **11.**10
  auditing, **11.**7, **11.**15
  capital project review,, **11.**5
  emergency action plans, **11.**4
  environmental activity repository, **11.**6
  groundwater, **11.**9
  housekeeping and sanitation, **11.**6
  liquid storage tanks, **11.**11
  on-site contractors, **11.**4
  PCBs (polychlorinated biphenyls), **11.**9
  performance requirements, **11.**4
  plant inspection, **11.**6
  product stewardship, **11.**6
  secondary contaminant, **11.**11
  strong corporate contact, **11.**4
  surface water, **11.**8
  waste management, **11.**7
  waste minimization, **11.**16
 policy, **11.**3 to **11.**4, **11.**14
  accountability, **11.**3
  contractors, **11.**3 to **11.**4
  customers, **11.**3
  design and management of operations, **11.**3
  education, **11.**3, **11.**14
  employees, **11.**3
  market systems, **11.**3
  morality, **11.**3
  operating standards, **11.**3
  products, **11.**3
  public, **11.**3
 process, safety and risk management, **11.**12, **11.**15
  occupational health, **11.**12, **11.**15
  operational safety, **11.**12
  reporting systems, **11.**15
 recognition of excellence, **11.**16 to **11.**17
  commitment and support, **11.**17
  community and employee involvement, **11.**18
  compliance, **11.**18
  environmental performance, **11.**17
  risk reduction goals, **11.**17
  sound business practices, **11.**17
 ten principles of environmental management, **11.**19

Air pollution prevention, **6.**1 to **6.**3
  benefits, **6.**3
Analysis, field techniques, **4.**10 to **4.**11
  absorption tubes, **4.**10
  monitors, **4.**10
  organic vapor analyzers, **4.**10
  portable GCs and GC-MS (gas chromatography–mass spectrometry), **4.**11
  specific analytical methods, **4.**11 to **4.**15
  titration, **4.**9
  treated media, **4.**6 to **4.**7
Analytical methods, **4.**1 to **4.**53,
  GC (gas chromatography), **4.**7 to **4.**8
  GC sniffing, **4.**9
  high performance liquid chromatography, **4.**9
  mass spectrometry, **4.**8 to **4.**9
  porous polymers, **4.**4 to **4.**5
    disadvantages, **4.**
    elution of media, **4.**5
  selection of media, **4.**4 to **4.**5
Antoine equation, **2.**3
Applications of other designs of chemical scrubbers, **8.**137

Bioaccumulation of toxins in the body, **3.**3
Biofilters, **8.**150 to **8.**185
  applications, **8.**151
    compounds, **8.**152
    list of facilities, **8.**193
  biological systems, **8.**151 to **8.**154
    acidic metabolites, **8.**153
    aerobic oxidations, **8.**154
    level of concentration, **8.**152
    mineralization, **8.**152
    mixed populations, **8.**154
    nutrient loading, **8.**154
    overview, **8.**151
    process, **8.**153
  configurations, **8.**155 to **8.**159
    biofilters, **8.**155
    biotrickling filters, **8.**155
    limited optimization of microbial growth, **8.**159
    multiple open bed biofilter, **8.**159
    open single bed, **8.**157
    open soil bed, **8.**156
    process control, **8.**156
  design criteria definition, **8.**162
  diffusion, **8.**154
  engineering, **8.**162 to **8.**77

Biofilters (*Cont.*):
    materials of construction, **8.**152, **8.**162, **8.**173
    structure and internals, **8.**172 to **8.**177
      air distribution, **8.**173
      air pretreatment, **8.**162
      controls, **8.**177
      design, **8.**162
      inoculum, **8.**175
      instrumentation, **8.**177
      media, **8.**162, **8.**174
      monitoring, **8.**170, **8.**176
      pH control, **8.**177
      sizing, **8.**162
  equipment, **8.**171 to **8.**172
    air handling, **8.**171
    ducting, **8.**171
    moisture, **8.**170, **8.**172, **8.**176
    particulates, **8.**171
    pretreatment, **8.**171
    temperature, **8.**170, **8.**172
  kinetics, **8.**155
  marketplace, **8.**181 to **8.**185
    evolving status, **8.**153
    hydrocarbon treatment system, **8.**184
    odor control systems, **8.**181
    VOC (volatile organic compound) and HAP (hazardous air pollutant) control systems, **8.**185
  operations, **8.**179
    efficiency, **8.**154
    monitoring, **8.**180
    start-up, **8.**178
  reactor designs, **8.**152
    closed bed, **8.**152
  sizing, **8.**163 to **8.**168
    data for design, **8.**168
    elimination capacity, **8.**166
    empty bed vapor contact time, **8.**163
    mathematical models for design, **8.**167
  typical system drawing, **8.**151
Biological oxidation, **2.**9
  biorates, **2.**9
  hydrolysis, **2.**9
  monod, **2.**9
Biotransformation of toxins in the body, **3.**3

Canadian regulations and laws (*see* Regulations and laws, Canada)
CCS (common chemical sense), **4.**16
Chemical additions to liquid, **8.**1 to **8.**16

Chemical additions to liquid (*Cont.*):
- advantages and disadvantages, **8.**15 to **8.**16
- biochemical enhancement, **8.**4, **8.**13 to **8.**15
  - nitrates, **8.**13 to **8.**15
  - nitrites, **8.**13 to **8.**15
- H_2S control, **8.**12 to **8.**13
  - ferrous salts, **8.**11 to **8.**12
- pH elevation, **8.**4
  - sodium hydroxide, **8.**6 to **8.**7, **8.**12 to **8.**13
- oxidation, **8.**3 to **8.**10
  - air, **8.**5
    - (*See also* Aeration basins)
  - chlorine, **8.**6
  - hydrogen peroxide, **8.**8 to **8.**10
  - oxygen, **8.**5
  - potassium permanganate, **8.**7 to **8.**8
  - reduction potential, **8.**4
- oxygen, precipitation and adsoption, **8.**3 to **8.**4, **8.**10 to **8.**11
  - ferris salts, **8.**10 to **8.**11

Chemical interactions, **2.**2 to **2.**3
- liquids, **2.**3
- gases, **2.**2

Chemical methods, **4.**1 to **4.**15
- analysis, **4.**1 to **4.**2
  - field methods, **4.**1
  - laboratory methods, **4.**2
- sampling, **4.**1

Chemical properties, **2.**2 to **2.**3
- ionization, **2.**2
- polar compounds, **2.**2
- predictions, **2.**3
- sources of data, **2.**3

Chemical scrubbing, other designs, **8.**131 to **8.**132
- ejector Venturi scrubbers, **8.**131
- performance, **8.**137
- plate scrubbers, **8.**132
- spray towers, **8.**131

Comparative risk, **13.**9 to **13.**38
- analytical matrix for comparing risks, **13.**16
- California Comparative Risk Project, **13.**31
- Colorado Environment 2000, **13.**27
- Columbus, Ohio Priorities '95, **13.**35
- committees, **13.**18
- comparative risk, description, **13.**14
- ecological risk, **13.**15
- Elizabeth River Project, Norwalk, Va., **13.**36
- Hawaii Environmental Risk-Ranking Project, **13.**30
- Houston Environmental Foresight Project, **13.**36

Comparative risk (*Cont.*):
- how benzene is regulated, **13.**12
- how comparative risk addresses odors, **13.**24
- how comparative risk sees VOCs, **13.**23
- how and why governments regulate VOCs, **13.**9 to **13.**10
- human health risk assessment, **13.**11
- Louisiana Environmental Action Plan, **13.**28
- Maine Environmental Priorities Project, **13.**32
- Minnesota Comparative Risk Project, **13.**37
- Mississippi Comparative Risk Project, **13.**37
- Northeast Ohio and Cleveland Regional Environmental Priorities Project, **13.**34
- other projects, **13.**38
- Ohio Comparative Risk Project, **13.**33
- quality of life, **13.**18
- risks to ecosystem, **13.**23
- results of comparative risk projects, **13.**19 to **13.**22
- Seattle Environmental Priorities Project, **13.**29
- several comparative risk projects, **13.**27
- Washington Environment 2010, **13.**27

Compliance, examples of, **12.**1 to **12.**13
- offset lithograph printer, **12.**1
  - emission limits, **12.**1
- facility review, **12.**4 to **12.**8
  - actual emissions, **12.**6
  - compliance program development, **12.**8
  - emissions inventory, **12.**5
  - operating scenario evaluation, **12.**8
  - potential to emit, **12.**7
- noncompliance issues, **12.**4
- odorous exhausts, **12.**10
  - odor study, **12.**10
  - prioritizing, **12.**13
  - selection of control technology, **12.**13
  - specific design recommendations, **12.**13
- permit applications, **12.**9
  - state operating permits, **12.**9
  - title V permits, **12.**9
- regulatory applicability, **12.**3
- work plan, requirements, **12.**2 to **12.**4
  - compliance, **12.**4
  - definition, **12.**2
  - information, **12.**3

Computer models, air dispersion, **5.**8
- complex terrain (complex I), **5.**8

Computer models, air dispersion (*Cont.*):
  industrial source complex (ISC), **5.**8
  rough terrain diffusion (RTD), **5.**8
Condensation:
  at higher temperatures, **8.**25
  at refrigerated temperatures, **8.**26 to **8.**30
    defrosting, **8.**29
    low-temperature coils, **8.**29
    precoolers, **8.**29
    process vent systems, **8.**27
    ranges of temperature, **8.**28 to **8.**29
    safety, **8.**30
      advantages, **8.**30
      life expectancy, **8.**30
    storage transfer systems, **8.**26
      efficiency, **8.**27
      safety, **8.**27
Conservation, **1.**1
  public ownership of lands, **1.**1
  (*See also* Ethics)
Continuous emission monitoring (CEM), **4.**50 to **4.**53
Control charts:
  false alarm rate, **4.**67
  limits, **4.**66 to **4.**67
    action, **4.**67
    lower, **4.**66
    upper, **4.**66
    warning, **4.**67
  quality control, **4.**67
Cryogenic techniques, **4.**6
  disadvantages, **4.**6
  (*See also* Condensation)

Data acquisition system, **4.**40 to **4.**42
  range of functionality, **4.**41 to **4.**42
Data problems, **4.**74 to **4.**75
  nondetects, **4.**75
  nonnormal population, **4.**75
  outlying values, **4.**75
Data quality objectives (DQO), **4.**55
  process steps, **4.**55
  web site, **4.**55
Direct measurement techniques, **4.**34 to **4.**42
Dispersion, **5.**1 to **5.**10
  adiabatic lapse rate, **5.**2
  environmental lapse rate, **5.**2
  Pasquill stability criteria, **5.**2
  wind velocity, **5.**2
Dispersion models, additional, **5.**23
(*See also* Computer models, air dispersion)

Dry chemical scrubbers, **8.**138 to **8.**148
  ammonia removal, **8.**148
  applicability, **8.**141
  applications, **8.**146
  deep bed:
    advantages, **8.**141
    face velocity, **8.**146
    gas concentration, **8.**146
    parameters, **8.**145
    scrubber, **8.**145
  disposal of media, **8.**148
  economics, **8.**148
    initial cost, **8.**148
    operating cost, **8.**148
  hydrogen sulfide (H_2S) reactions, **8.**142
  media, **8.**139 to **8.**141
    advantages and disadvantages, **8.**141
    chemicals, **8.**139
    impregnated, **8.**139
    multiple media, **8.**141, **8.**143
    physical properties, **8.**141
    sizing, **8.**139
    substitute materials, **8.**139
  modular units, **8.**145
  parallel units, **8.**145
  polishing, **8.**147
  pretreatment, **8.**144
  potassium permanganate, **8.**143
  small units, **8.**145
  specific gas applications, **8.**142
  wastewater odor, **8.**146
  VOC control, **8.**146

Electrostatic precipitators (ESP), **8.**225 to **8.**235
  advantages and disadvantages, **8.**226
  applications of, **8.**235
  design parameters, **8.**228
    one-stage, **8.**228
    two-stage, **8.**228
    typical range, **8.**228
  gas ionization, **8.**225
  housing, **8.**235
  particle collection, **8.**225
    electrostatic field, **8.**225
    particle migration, **8.**225
    reentrainment, **8.**225
    removal of collected particles, **8.**225
  plate, **8.**229
  system description, **8.**229, **8.**231 to **8.**233
    electrode cleaning, **8.**223

Electrostatic precipitators (ESP), system description (*Cont.*):
electrode system, **8**.231
power supply, **8**.231
theory, **8**.226
tubular, **8**.229
Emission, estimation of spills, **5**.7
low-volatile liquid, **5**.7
heavier than air gas, **5**.7
highly volatile liquid, **5**.7
Emission models, principles **5**.11 to **5**.14
rate of absorption, **5**.12
rate of biodegradation, **5**.13
rate of stripping, **5**.12
rate of volatilization, **5**.12
Environmental movement, **1**.1 to **1**.6, **13**.48 to **13**.59
adversarial, **1**.3 to **1**.4
citizen activism, **1**.2
compliance philosophy, **1**.4
Silent Spring, Rachel Carson, **1**.2
sustainable development, **1**.5
(*See also* Ethics)
EPA (Environmental Protection Agency), **1**.1 to **1**.2
establishment, **1**.2
test methods, **4**.50 to **4**.53
Ethics, **13**.48 to **13**.59
concepts, **13**.51, **13**.55
environment, ecology, economics, **13**.56
environmental ethic, **13**.52
questions, **13**.49
religion and ecology, **13**.58
right livelihood, **13**.59
the right and the good, **13**.49
Evaluating emission control systems, **9**.1 to **9**.11
consultant services, **9**.1
control technology factors, **9**.3 to **9**.5
continuity, **9**.8
controls, **9**.8
effectiveness, **9**.3
equipment, **9**.8
experience, **9**.7
flexibility, **9**.8
hybrid systems, **9**.8
life cycle cost, **9**.4
logical choices, **9**.3
selection, **9**.2
performance, **9**.9 to **9**.11
equipment factors, **9**.10
installation factors, **9**.10
tests, **9**.9
Evaluating emission control systems, performance (*Cont.*):
warranties, **9**.11
problem, **9**.2 to **9**.3
definition, **9**.2
solution, **9**.3

Flame ionization detectors (FID), **4**.34 to **4**.39
analyzer operation, **4**.35 to **4**.36
construction, **4**.36 to **4**.37
interference, **4**.38 to **4**.39
Flow rate measurement, **4**.24
estimation, **4**.24
Pitot tube, **4**.24
vane anemometer, **4**.24
Fog scrubbing (*see* Mist scrubbing)
Fogging technologies, **8**.22 to **8**.24
atomizing fans, **8**.23
compressed air, **8**.23
diffused air, **8**.23
high-pressure liquid, **8**.23
small foggers, **8**.23
vaporization, **8**.24
(*See also* Mist scrubbing)
Foul air collection, **7**.1 to **7**.14
building pressurization, **7**.2
covers, channel and tank, **7**.4
construction, **7**.5 to **7**.8
costs, **7**.8
testing, **7**.5
ductwork, **7**.9 to **7**.12
enclosures, **7**.9
fans, **7**.12
axial, **7**.14
centrifugal, **7**.13
hoods, **7**.9

Gas chromatography (GC), **4**.34 to **4**.40
detectors for, **4**.40
electron counter detectors (ECD), **4**.40
flame ionization (PID), **4**.40
flame photometric (FPD), **4**.40
Gas phase filtration (*see* Dry chemical scrubbers)
Global environmental issues, **1**.5
developing countries, **1**.5
(*See also* Environmental movement; Ethics)
Graphical displays, **4**.58 to **4**.60
box and whisker plot, **4**.58 to **4**.59
interquartile range, **4**.58 to **4**.59

Graphical displays, box and whisker plot (*Cont.*):
multiples, **4**.58 to **4**.59
histogram, **4**.60
multiple, **4**.60
looking at the data, **4**.58
probability plot, **4**.60
software packages for, **4**.60

Health assessment, **3**.4
control, **3**.4
evaluation, **3**.4
recognition, **3**.4
Health effects, **3**.13 to **3**.17
central nervous system, **3**.13 to **3**.14
hematopoiesis system, **3**.17
kidney, **3**.15
liver, **3**.15
peripheral nervous system, **3**.14
respiratory system, **3**.14 to **3**.15
skin and mucous membranes, **3**.17
Health hazards, **3**.4
biological, **3**.4
chemical, **3**.4 to **3**.5
control measures, **3**.5
physical, **3**.5
survey, **3**.5
Health of organs, targets for VOCs, **3**.16
Henry's law coefficient, **2**.5 to **2**.6, **8**.96 to **8**.97, **8**.155, **8**.168, **8**.169
nonvolatiles, **2**.6
prediction of, **2**.6
semivolatiles, **2**.6
Hybrid systems, **8**.203 to **8**.214
budgeting costs, **8**.210
case studies, **8**.211 to **8**.214
acrylonitrile manufacture, **8**.212
batch dryer with ethanol emission, **8**.211
groundwater stripping, **8**.214
organic chemical manufacture, **8**.213
polymer manufacturer, **8**.213
design basis, **8**.204
airflow, **8**.204
gas phase composition, **8**.204
temperature, **8**.204
final selection factors, **8**.211
guidelines for standard systems, **8**.205 to **8**.207
absorption, **8**.205
adsorption, **8**.206
biofiltration, **8**.206

Hybrid systems, guidelines for standard systems (*Cont.*):
catalytic oxidation, **8**.207
condensation, **8**.206
cryogenic condensation, **8**.207
flaring, **8**.207
thermal oxidation, **8**.207
ranges of applicability, **8**.208
rules for selection of hybrids, **8**.210
types of hybrid systems, **8**.209
absorption and recycling, **8**.209
absorption in water and biotreatment, **8**.209
combustion air supplement, **8**.209
concentration and condensation, **8**.209
fuel oil absorption and burning, **8**.209
Hydrogen sulfide, **8**.2
equilibrium, **8**.2
Hypothesis tests, statistical, **4**.68 to **4**.74
alternative hypothesis, **4**.68
applications, **4**.71
framework for testing hypothesis, **4**.68 to **4**.69
level of significance, **4**.71
null hypothesis, **4**.68
p value, **4**.71
paired *t* test, **4**.73
power curve, **4**.70
power of test, **4**.70
relationship to confidence intervals, **4**.73 to **4**.74
software for hypothesis tests, **4**.72
test for normality, **4**.73
two-sample *t* test, **4**.71
type I error (α), **4**.69
type II error (β), **4**.69

Indoor air pollution, **3**.24
Industrial health controls, **3**.10 to **3**.13
administrative controls, **3**.13
dilution, **3**.13
isolation, **3**.12
local exhaust ventilation, **3**.12
personal protections, **3**.13
substitution, **3**.13, **3**.12
Industrial health monitoring, **3**.6
area sampling, **3**.6
personal sampling, **3**.6
Industrial pollution prevention, **6**.1
Infrared analyzers (IR), **4**.34, **4**.39 to **4**.40
Fourier transfer infrared analyzers, **4**.39 to **4**.40
interference, **4**.39
method of selection, **4**.39 to **4**.40

Insurance (*see* Pollution liability insurance)
International standards organization (ISO), **6**.2
  ISO 9000, **6**.2
  ISO 14000, **6**.2
Ionization, **2**.7 to **2**.8
  constants, equations, **2**.7
    effect of, **2**.7 to **2**.8
    H_2S, **2**.7
    listing, **2**.7

Market incentives, **1**.4 to **1**.5
  tradeable emission credits, **1**.5
Masking odor, **8**.18 to **8**.22, **8**.24
Mechanical collection (*see* Physical collection)
Meteorology, **5**.1 to **5**.10
Mist scrubbing, **8**.111 to **8**.128
  application of fog scrubbers, **8**.127 to **8**.128
    concentration of airflow, **8**.127
    complexity, **8**.127
    listing, **8**.127
    types of contaminants: complexity, concentration variability, volume, **8**.128
  applications, **8**.112
    automatic controls, **8**.112
    complex mixtures, **8**.112
    compounds: insoluble, reactable, soluble, variable, **8**.112
    high concentration/low volume, **8**.112
    low concentration/high volume, **8**.112
    multiple stages, **8**.112
    (*See also* Adsorption, fog droplets)
  biological treatment comparison, **8**.127
  characterizing gas stream, **8**.116
  comparisons to alternate technologies, **8**.123 to **8**.4, **8**.126
    continuous operation, **8**.124
    disposal of wastes, **8**.124
    ease of maintenance, **8**.124
    efficiency of removal, **8**.123
    flexibility for change, **8**.124
    life cycle cost, **8**.123
    thermal oxidizers, **8**.127
    variable source treatment, **8**.124
  design, **8**.112 to **8**.113
    concentration to be treated, **8**.113
    contact time, **8**.113
    multiple stage, multiple chemicals, **8**.113
    removal efficiency, **8**.113
    single stage, multiple chemicals, **8**.112
    single stage, single chemical, **8**.112
    system configuration, **8**.113
Mist scrubbing (*Cont.*):
  design elements, **8**.111 to **8**.116
    atomization, **8**.113
    characterizing gas stream, **8**.116
    chemical metering, **8**.116
    fog, **8**.114
    fog character, **8**.114
    fog-mist technology, **8**.114
      collision theory, **8**.114
      contact time, **8**.115
      diffusion theory, **8**.114
      droplet size, **8**.113
      mass transfer equation, **8**.112
      surface comparison, **8**.114
    mixing of droplets and air, **8**.115
    power usage, **8**.115
    sonic atomizer, **8**.114
  detailed description, **8**.117 to **8**.118
    atomization system, **8**.117 to **8**.118
    chemical controls, **8**.119
    chemical metering, **8**.119
    chemical storage, **8**.118
    contact chambers, **8**.118
    fans, **8**.120
    flow diagram, **8**.117
    pump sizing, **8**.119
    water softener, **8**.120, **8**.123
  distinguishing factors, **8**.112
    contact surface, **8**.112
    controllable, **8**.112
    once-through, **8**.112
    reactivity, **8**.112
    settling rate, **8**.112
  hybrid systems, **8**.129
    with biofilters, **8**.129
    with carbon adsorption, **8**.129
    with conventional scrubbers, **8**.129
    with dilution, **8**.129
    with thermal oxidation, **8**.129
  packed columns, comparison to fog scrubber, **8**.124 to **8**.127
    capital cost, **8**.126
    chemistry, **8**.125
    design, **8**.125
    maintenance, **8**.126
    recirculation, **8**.126
    space requirements, **8**.126
    wastage, **8**.126
  performance results, **8**.128
  thermal oxidizers, compared to fog scrubbing, **8**.127
Modelling emissions, **5**.1 to **5**.10, **5**.11 to **5**.25

Modelling emissions (*Cont.*):
- area sources, **5.**3
  - box model, **5.**3
- line sources, **5.**7 to **5.**8
- stacks and point sources, **5.**7 to **5.**8
  - atmospheric dispersions, **5.**4
  - ground level concentration, **5.**5
  - receptor concentration, **5.**5
  - stack height, effective, **5.**5

Monitoring emissions, **4.**33 to **4.**53
- continuous, **4.**33
  - direct measurement, **4.**34
- criteria for monitor system selection, **4.**33, **4.**34, **4.**49
- parametric, **4.**33
  - indirect measurement, **4.**34
- stack test, **4.**33

Neutralization of odors, **8.**18 to **8.**22, **8.**24
- essential oils, **8.**20
- operating costs, **8.**20
- trace elements, **8.**21

Non–point source pollution, **1.**4

Occupational Safety and Health Administration (OSHA), **3.**8 to **3.**10
- hazard communication standard, **3.**10
- permissible exposure limits, **3.**8
- substance specific standards, **3.**10

Octanol-water partitioning, **2.**12

Odor, **2.**1 to **2.**36
- category scales, **4.**20
- character, **4.**20, **4.**30 to **4.**31
  - descriptors, **4.**20
- chemistry and characteristics, **2.**1 to **2.**36
- concentrations, expressions for, **4.**18
  - odor unit (OU), **4.**18
  - odor units per cubic feet, **4.**18
  - odor quotient, **4.**18
- detection threshold, **4.**16 to **4.**18, **4.**28 to **4.**29
  - dynamic olfactometer, **4.**28 to **4.**29
- emission rate, **4.**18
- hedonic tone, **4.**21
  - acceptability, **4.**21
- human response to, **4.**21 to **4.**22
  - desensitization, **4.**21
  - health effects, **4.**22
  - hypoactivity, **4.**21
  - other factors, **4.**21 to **4.**22

Odor, human response to (*Cont.*):
  - response to odor mixtures, **4.**21
  - sensitization, **4.**21
  - tolerance, **4.**21
- intensity, **4.**18, **4.**29 to **4.**31
  - butanol olfactometer, **4.**30
  - dose response, **4.**19
  - relation of intensity and concentration, **4.**18
  - static scale, **4.**30
- magnitude estimation, **4.**20
- measurement, sensory, **4.**26 to **4.**31
  - criteria, **4.**27
  - panel selection, **4.**27
    - guidelines, **4.**28
    - intensity test, **4.**27
    - screening, **4.**27
    - triangle test, **4,**27
  - parameters affecting, **4.**31
- perception, **4.**16, **8.**18
- pervasiveness, **4.**19
- reference scales, **4.**19 to **4.**20
  - 1-butanol, **4.**19
- surveys, community, **4.**22 to **4.**25
- threshold, **4.**16 to **4.**18, **4.**28 to **4.**29
  - detection, **4.**16 to **4.**18
  - dilution to threshold (D/T), **4.**17
  - recognition, **4.**16 to **4.**18
  - sources of data, **4.**18

Packed column, **8.**102 to **8.**108
- air-water system, **8.**103
- carbon dioxide, air-water system, **8.**103
- and catalyst reactor, **8.**135
- chemical scrubbing, **8.**102
- designs, **8.**106 to **8.**108
  - crossflow, **8.**106
  - example, **8.**108
  - multiple stage, **8.**108
  - single stage, **8.**108
  - vertical tower, **8.**106
- and electrolytic hypochlorite generation, **8.**132
- inlet concentration, **8.**104
- liquid chemical reactions, **8.**104
- mass transfer, **8.**102 to **8.**108
  - coefficients, **8.**104 to **8.**105
  - surface renewal model, **8.**102
  - two-film model, **8.**102
- media, **8.**104 to **8.**106
  - basic requirements, **8.**104

Packed column, media (*Cont.*):
height of transfer unit, **8.**106
packing factor, **8.**106
size, **8.**106
surface area, **8.**106
solubility, **8.**104
surface area, **8.**104
temperature, **8.**104
turbulence, **8.**104
Painting (*see* Surface coating)
Paper and pulp industry, pollution prevention, **6.**22 to **6.**26
gaseous emissions, **6.**22
in pulp bleaching, **6.**26
in pulp drying and paper making, **6.**26
Parametric emission monitor (PEM) and control systems, **4.**42 to **4.**53
carbon absorbers, **4.**46 to **4.**48
catalytic oxidizers, **4.**44 to **4.**46
thermal oxidizers, **4.**42 to **4.**44
wet scrubbers, **4.**48
Partitioning of gas-liquid, **2.**5
Physical collection, **8.**216 to **8.**224
cyclones, **8.**217
conventional design and operation, **8.**217
multiple, design and operation, **8.**217
design parameters, **8.**223
fabric filter, **8.**217, **8.**224
advantages and disadvantages, **8.**218
as chemical reactor, **8.**224
filter media, **8.**220 to **8.**223
construction, **8.**221
finishing, **8.**222
penetration, **8.**223
pressure drop, **8.**223
properties and testing, **8.**222
high intensity, **8.**219
high pressure, **8.**219
low intensity, **8.**219
medium pressure, **8.**219
pulse jet baghouse, **8.**219
impingement collectors, **8.**216
low intensity, **8.**218
cleaning, **8.**218
conventional baghouse, as, **8.**218
reverse gas flow, **8.**218
shakers, **8.**218
settling chambers, **8.**218
Pollution liability insurance, **13.**40 to **13.**46
insurance market, **13.**45 to **13.**46
exploring, **13.**45
pricing, **13.**45 to **13.**46
Pollution liability insurance, insurance market (*Cont.*):
risk retention group, **13.**45
specialty insurers, **13.**45
insurance program, **13.**40 to **13.**45
basic factors, **13.**40
claims made, **13.**44
contracts, **13.**43
payment of claims, **13.**43
self-insurance, **13.**44
warranty, **13.**42
risk management, **13.**41 to **13.**42
cost and benefit, **13.**40
insurance costs, **13.**41
other, tools, **13.**41
role of insurance, **13.**41
Pollution prevention, **1.**4, **6.**1 to **6.**55
Printing and publishing, pollution prevention, **6.**26 to **6.**28
environmental impacts, **6.**27
processes, **6.**26
recommendation, **6.**28
source reduction, **6.**28
waste description, **6.**27
waste minimization options, **6.**28
Public relations, **13.**1 to **13.**8
coalition building, **13.**6
community support, **13.**4
complete disclosure, **13.**5
employees first, **13.**5
five principles of public relations, **13.**3
good corporate citizenship, **13.**8
good relations, **13.**2
guidelines for action, **13.**4
handling the crisis, **13.**6
public relations council, **13.**7
rules for good relations, **13.**1

Quality management, **6.**3

Regulations and laws, Canada, **10.**11 to **10.**16
odor regulations, **10.**11 to **10.**16
Alberta, **10.**12
British Columbia, **10.**12
Manitoba, **10.**12
New Brunswick, **10.**13
Newfoundland, **10.**13
Northwest Territories, **10.**14
Nova Scotia, **10.**14
Ontario, **10.**14

Regulations and laws, Canada, (*Cont.*):
- Prince Edward Island, **10.**15
- Quebec, **10.**15
- Saskatchewan, **10.**16
- Yukon Territory, **10.**16
- VOC regulations, **10.**16

Regulations, and laws, U.S. federal and state, **10.**2 to **10.**10
- AAL (acceptable ambient level), **10.9**
- ACT (alternative control technology), **10.3**
- air toxics, **10.9**
- CAA (Clean Air Act), **10.2**
- CERCLA (Comprehensive Environmental Response, Compensation and Liability Act), **10.**2
- criteria pollutants, **10.9**
- CTG (control technology guidelines), **10.3**
- EPCRA [Emergency Planning and Community Right-to-Know Act (1986)], **10.**8
- historical development, **1.**2 to **1.**4
  - Clean Air Act, **1.**2
  - amendments 1990, **1.**4
  - cleaning air emissions, **1.**3
  - Clean Water Act, **1.**2
  - Command and Control, Economic impact, **1.**3, problems, **1.**4, successes, **1.**3
  - Council on Environmental Quality, **1.**2
  - EPA (Environmental Protection Act), **1.**1 to **1.**2
- odor regulations, **10.**10
- part B permits, **10.7**
  - President Richard Nixon, **1.**2
  - RCRA (Resource Conservation and Recovery Act), **1.**2
- regulatory applicability, **10.6**
- transportation, storage and disposal for, **10.7**
- wastewater treatment plants, **1.**3
- HON (hazardous organic NESHAP), **10.**4 to **10.**5
- MACT (maximum achievable control technology), **10.**2, **10.**4 to **10.5**
- NESHAP (National Emission Standard for Hazardous Air Pollutants), **10.**2
- NSPS (New Source Performance Standards), **10.**3
- OSHA [Occupational Safety and Health Administration (or Act)], **10.**2, **10.9**
- RACT (reasonably available control technology), **10.**3
- RCRA (Resource Conservation and Recovery Act), **10.**2

Regulations, and laws, U.S. federal and state (*Cont.*):
- SARA [Superfund Amendments and Reauthorization Act (1986)], **10.**8
- SIP (state implementation plans), **10.**2
- SOCMI (synthetic organic chemical manufacturing industry), **10.**4
- TLV (threshold limit value), **10.9**
- TRIS (Toxic Release Inventory System), **10.**8
- TSCA (Toxic Substance Control Act), **10.**2, **10.**8

Risk assessment and prioritization, **3.5**
- exposure rating, **3.**6
- hazard, **3.6**
- toxicity, **3.6**
- (*See also* Comparative risk)

Sampling, impingers, **4.**6
- disadvantages, **4.6**

Sampling, statistical, **4.**60 to **4.**62
- for emission compliance, **4.**62
- judgmental sampling, **4.**62
- probability sampling, **4.**61
  - adaptive, **4.**61
  - composite, **4.**61
  - cluster, **4.**61
  - grid, **4.**61
  - optimal, **4.**61
  - sequential, **4.**61
  - simple random, **4.**61
  - stratified, **4.**61
  - systematic, **4.**61

Sampling methods and techniques, **4.**2 to **4.**7
- capture techniques, application, **4.**2
- flexible bags, inert, **4.**2 to **4.**3
  - applications, **4.**3
  - limitations, **4.**3
  - pressure filling, **4.**3
  - vacuum filling, **4.**3
- glass bombs, **4.**4
  - advantages and disadvantages, **4.**4
- steel canisters, **4.**3 to **4.**4
  - advantages and disadvantages, **4.**3 to **4.**4

Sampling of odors, **4.**24 to **4.**26
- bag sampling, **4.**24
- flux chamber, **4.**24
- preconditioning bags, **4.**24
- sample dilution, **4.**24
- sample storage time, **4.**24

Secondary metals, pollution prevention, **6.**29
  material impacts, **6.**29
  options, **6.**29
  pollution outputs, **6.**29
  processes, **6.**29
Sensory test methods, **4.**16 to **4.**32
Sewage wastewater systems (*see* Wastewater conveyance systems)
Solubility, **2.**5 to **2.**6
  effect of, **2.**6
  in water, **2.**5
Solvent cleaning (*see* Solvent degreasing)
Solvent degreasing, **6.**12 to **6.**16
  aqueous cleaning, **6.**13
    benefits, **6.**16
  organic solvents used, **6.**12
  recycling solvents, **6.**15
    benefits, **6.**15
  ultrasonic cleaning, **6.**14
    benefits, **6.**14
  vapor degreasing, **6.**12, **6.**15
    benefits, **6.**16
    modification, **6.**15
Sorption, **2.**12
Statistics, **4.**54 to **4.**81
  comparison to a standard, **4.**65
  confidence interval for a proportion, **4.**67 to **4.**68
    binomial distribution, **4.**67 to **4.**68
    nonparametric, **4.**67 to **4.**68
  data types, **4.**55 to **4.**56
    internal, **4.**55 to **4.**56
    nominal (categorical), **4.**55 to **4.**56
    ordinal, **4.**55 to **4.**56
    ratio, **4.**55 to **4.**56
    [*See also* Data quality objectives (DQO)]
  emission applications for, **4.**80
    emission compliance, **4.**81
    odor testing, **4.**81 to **4.**81
  estimates, internal, **4.**63
    confidence intervals, **4.**63 to **4.**65
  experiment design, **4.**79
    factorial, **4.**79
    fractional, **4.**79
  measures of location and dispersion, **4.**56 to **4.**58
    log normal and log distribution, **4.**58
    mean (average), **4.**56 to **4.**57
    median, **4.**57
    normal distribution, **4.**57
    percentiles, **4.**58
    proportion, **4.**58

Statistics, measures of location and dispersion (*Cont.*):
    range, **4.**57
    standard deviation, **4.**57
    variance, **4.**57
  nonparametric methods, **4.**75 to **4.**77
    software for, **4.**77
    Wilcoxon rank-sum test, **4.**76
  regression and correlation, **4.**77
    least squares regression, **4.**77
    root mean square error, **4.**79
    software program, **4.**78
  sampling distributions, **4.**62 to **4.**63
    central limit theorem, **4.**63
    log inverse transformation, **4.**63
    Student's *t* distribution, **4.**63
  statistical applications, **4.**54
  statistical decision making, **4.**54
  statistical software, **4.**80
  tools for applications, **4.**54
  tolerance limits, **4.**65 to **4.**66
    confidence limits, **4.**65
    nonparametric, **4.**65 to **4.**66
    (*See also* Control charts; Hypothesis tests, statistical; Sampling, statistical)
Strategies for emission control, **2.**1 to **2.**2
  chemical separation, **2.**2
  physical separation, **2.**1
Surface coating, **6.**4 to **6.**10
  application improvements, **6.**10
    spray equipment, **6.**10
    transfer efficiency, **6.**10
  pollution problems, **6.**5
  powder coating, **6.**6
    benefits, **6.**7
  solvents used in typical VOC, **6.**4
  substitution, **6.**5
  waterborne coatings, **6.**8
    benefits, **6.**9

Textile industry pollution prevention, **6.**16 to **6.**22
  air emissions, **6.**19
  chemicals used, **6.**19
  environmental impacts, **6.**18
  opportunities, **6.**18
    available options, **6.**21
    material change strategies, **6.**22
    other techniques, **6.**22
    waste reuse, **6.**21

Textile industry pollution prevention (*Cont.*):
processes, **6.**16
waste reduction, **6.**20
wastewater emissions, **6.**19
Thermal oxidation, **8.**31 to **8.**65
burners, **8.**65
caloric value of organic compounds, **8.**48
catalytic recuperative, **8.**35, **8.**49 to **8.**52
catalysts, **8.**51
compound destruction, **8.**49 to **8.**50
P&ID (process and instrumentation diagram), **8.**52
typical process, **8.**52
catalytic regenerative, **8.**61 to **8.**62
operating temperature comparisons, **8.**62
ceramic filters, **8.**65
characterizing the process stream, **8.**33
contaminants, **8.**36
cost comparisons, **8.**36
destruction efficiency, **8.**37
equipment cost factors, **8.**33
flares, **8.**65
LEL (lower explosive limit), **8.**47
applicability limits, **8.**47
of organic compounds, **8.**47
lower flammability limit, **8.**47
multiple process lines, **8.**47
operating costs, **8.**36
organic loading, **8.**36
oxidation, heat generation, **8.**47
process flow reduction, **8.**35
recuperative, **8.**53 to **8.**57
forced draft:
illustration, **8.**56
P&ID (process and instrumentation diagram), **8.**57
oxidizer equipment fabrication, **8.**55
three Ts of combustion, **8.**53
effect on destruction efficiency, **8.**55
time, **8.**53
temperature, **8.**53
turbulence, **8.**53
regenerative, **8.**35, **8.**57 to **8.**61
bake out, **8.**61
destruction efficiency, **8.**59
diverter valves, **8.**61
fabrication, **8.**61
figures, **8.**58 to **8.**59
P&ID (process and instrumentation diagram), **8.**60
supplementary fuel, **8.**61
technologies, **8.**47
Thermal oxidation (*Cont.*):
VOCs major industrial sources, **8.**32
major VOC emissions, **8.**32
process emission volume ranges, **8.**32
Thermal oxidizer, preceded by concentrator, **8.**35, **8.**62 to **8.**64
destruction efficiency, **8.**63
figure, **8.**63
P&ID (process and instrumentation diagram), **8.**64
Threshold limit values, by ACGIH (American Conference of Governmental Industrial Hygienists), **3.**10
Topography, **5.**1 to **5.**10
Toxic effects, **3.**3
at distal sites, **3.**3
at point of entry, **3.**3
Toxic emissions, **1.**4
Toxicity, **3.**3 to **3.**4
acute, **3.**3
chemical of VOCs, **3.**17 to **3.**24
aldehydes, **3.**25
aliphatics, **3.**17 to **3.**23
aromatics, **3.**17 to **3.**22
chlorinated hydrocarbons, **3.**24
ketones, **3.**23 to **3.**24
summary, **3.**17 to **3.**22
chronic, **3.**3 to **3.**4
Toxicology, **3.**3
absorption in the body, **3.**3
dermal (skin), **3.**3
ingestion, **3.**3
inhalation, **3.**3
routes of entry, **3.**3
Toxins:
distribution in the body, **3.**3
dose response, **3.**4
threshold, **3.**4

Vapor pressure, **2.**3 to **2.**4
Antoine equation, **2.**3
effect of, **2.**4
prediction of vapor pressure, **2.**4
VOC emissions, from wastewater, **5.**11
VOCs (volatile organic compounds), **2.**1, **3.**1
absorption in the body, **3.**1 to **3.**2
chemistry and characteristics, **2.**1, **2.**36
sources of exposure, **3.**1
Volatilization/adsorption, **2.**10
definition, **2.**10
mass transfer theories, **2.**10 to **2.**11

Volatilization/adsorption, mass transfer theories (*Cont.*):
- diffusion coefficients, **2.**11
- mass transfer coefficients, **2.**11
- penetration, **2.**10
- surface renewal, **2.**10
- two film, **2.**10

Waste minimization strategies, **6.**11

Wastewater collection system emission prediction, **5.**14 to **5.**18
- chloroform mass balance, **5.**18
- Coral+, **5.**16
- dissolved oxygen contamination at saturation, **5.**15
- NaUtilus, **5.**17 to **5.**18
- oxygen deficit ratio, **5.**15
- Toxchem+, **5.**17
- WATER 8, **5.**16 to **5.**17

Wastewater conveyance systems, pollution prevention, **6.**36 to **6.**54
- chemical oxidation, **6.**39
  - chlorine , **6.**40
  - hydrogen peroxide, **6.**40
  - potassium permanganate, **6.**40
  - sodium hypochlorite, **6.**40
- crown treatment, **6.**42
- designs to reduce emissions, **6.**38
- emission sources, **6.**37

Wastewater conveyance systems, pollution prevention (*Cont.*):
- iron pretreatment, **6.**42
  - ferric chloride, **6.**42
  - ferrous chloride, **6.**42

Wastewater conveyance systems, pollution prevention (*Cont.*):
- mechanical oxidation, **6.**39
- metabolic modification, **6.**41
- nitrite addition, **6.**40
- pH elevation, **6.**42
- pipe cleaning, **6.**41
- summary, **6.**44

Wastewater treatment facilities, pollution prevention, **6.**43 to **6.**54
- advanced treatment, **6.**48
- disinfection, **6.**48
- land application, **6.**53
- preliminary treatment, **6.**45
- primary treatment, **6.**45
- secondary treatment, **6.**45
- solids dewatering, **6.**51
- solids stabilization, **6.**51
- solids storage, **6.**52

Wastewater treatment system emission models, **5.**18 to **5.**22
- Baste, **5.**20
- comparison to field data, **5.**21 to **5.**22
- Toxchem+, **5.**21
- WATER 8, **5.**21

ABOUT THE EDITOR

Harold J. Rafson is a nationally recognized authority on the engineering control of odors, toxics, and volatile organic compounds. He has extensive experience as an innovator and CEO of a company that manufactured air pollution control equipment. Mr. Rafson holds numerous U.S. and international patents on gas emission control technology and is the author of more than 150 technical papers and presentations in the field.